Cellular and Molecular Biology of the Renin-Angiotensin System

Cellular and Molecular Biology of the Renin-Angiotensin System

Edited by

Mohan K. Raizada
M. Ian Phillips
Colin Sumners

Department of Physiology
College of Medicine
University of Florida
Gainesville, Florida

CRC Press
Boca Raton Ann Arbor London Tokyo

Library of Congress Cataloging-in-Publication Data

Cellular and molecular biology of the renin-angiotensin system
edited by Mohan K. Raizada, M. Ian Phillips, Colin Sumners.
p. cm.
Includes bibliographical references and index.
ISBN 0-8493-4622-3
1. Angiotensin. 2. Renin. I. Raizada, Mohan K. II. Phillips, M. Ian. III. Sumners, Colin.
QP572.A54C45 1993
616. 1'32 — dc20 92-26890
CIP

Direct all inquiries to CRC Press, Inc., 2000 Corporate Blvd., N.W., Boca Raton, Florida 33431.

International Standard Book Number 0-8493-4622-3

Library of Congress Card Number 92-26890

Printed in the United States of America 1 2 3 4 5 6 7 8 9 0

Printed on acid-free paper

PREFACE

EARLY STUDIES ON HYPERTENSION

Research in the field of hypertension was initiated in 1834 by Richard Bright, a noted physician who observed that hypertrophy of the heart often accompanies kidney disease. Following the disclosure of Bright's observation, techniques for measuring blood pressure were designed, allowing scientists to link cardiac hypertrophy to renal hypertension. In 1898, the discovery of renin by Tigerstadt and Bergmann, and the demonstration that this kidney extract caused vasoconstriction, spotlighted the kidney, in addition to the adrenal medulla, as a hypothetical origin of hypertension.

The next phase of hypertension research began in 1939, when Page, Helmer, and Kohlstaedt maintained that renin itself was not a direct vasoconstrictor, but an enzyme that formed another substance, which they called angiotonin. At the same time the Page team was performing these renin studies, the Braun-Menendez group in Buenos Aires was isolating an identical substance, a small dialyzable material from the renal vein, which they named hypertensin. Page and Braun-Menendez later agreed to call the substance angiotensin. Almost two decades passed before angiotensin was purified and its structure elucidated. In 1956, during the purification stage, Skeggs and his group of investigators discovered angiotensin converting enzyme (ACE), and in the 1950s two peptides, angiotensin I and angiotensin II, were synthesized. When these peptides later became available in large quantities, their widespread supply allowed many teams of investigators to develop a higher level of understanding of the "renin-angiotensin system." Part of this understanding included the concept that angiotensin is not only a vasoconstrictor, but also a major participant in the processes of sodium homeostasis, catecholamine release, and thirst enhancement. Throughout this stage of investigation, Irvine Page continuously cautioned against the attitude of his clinical/scientific contemporaries, who maintained that hypertension was not a disease but a perfusion of the body controlled by several factors in dynamic equilibrium, each affecting the others.

Despite the vast amount of information about the physiology of the renin-angiotensin system acquired during the 1950s and 1960s, the majority of clinicians and scientists maintained that renin was involved only in "renal hypertension," which represents only one aspect of the total picture of the human hypertensive system. This single-aspect concept was challenged by the development of angiotensin antagonists and ACE inhibitors. Because these inhibitors were shown to lower blood pressure in patients with hypertension of unknown etiology, many who doubted that the renin system played a role in forms of essential hypertension became convinced of the involvement of angiotensin, even when circulating renin levels were not elevated. Evidence supporting the more encompassing concept of angiotensin-dependent hypertension was the discovery of renin-angiotensin system components in many tissues of the body, and the demonstration that by blocking one or more of the components of the system, blood pressure became lowered.

EXPLOSION OF KNOWLEDGE

This monograph presents the major achievements in the field of renin-angiotensin investigation during the last decade. So far, various components of the renin-angiotensin system have been sequenced and the genes cloned. Because the angiotensin receptor(s) has so far been particularly resistant to purification, it has not been directly sequenced. The gene(s) for the AT_1 receptor, however, have been sequenced and cloned, an accomplishment that allows an understanding of the structure and function of the elusive receptors.

The synthesis of potent nonpeptide angiotensin II antagonists has decisively established the existence of various types of angiotensin receptors and has stimulated the search for the genes of these receptor types and subtypes. Detailed studies have been performed on the signal transduction mechanisms for the AT_1 receptor, but despite the sophistication and intricacy of this work, the function of the AT_2 receptor is not yet clearly understood.

Since the experiments of the 1960s, which demonstrated that angiotensin II binds to the cellular plasma membrane, angiotensin binding has been assumed to cause a receptor conformational change that results in the activation of a second-messenger system. This hormonal system, as well as various ion channels that, in turn, are activated or inactivated by mechanisms within the hormone system, is contained in the cellular plasma membrane. This plethora of information, now becoming available through the use of methodologies carried out at the molecular and cellular levels, is beginning to explain the role of angiotensin and its metabolites throughout the organ systems. The consolidation of this current information presented in this monograph offers an exciting treatise to the scientists involved in blood-pressure control mechanisms.

F. Merlin Bumpus, Ph.D.
Chairman Emeritus, Research Institute
Department of Heart and Hypertension Research
Cleveland Clinic Foundation
Cleveland, Ohio

ACKNOWLEDGMENT

The editors extend sincere thanks to Mrs. Gayle Butters for invaluable assistance in putting together this volume.

CONTRIBUTORS

Kenneth J. Abel
Department of Internal Medicine
University of Michigan
Ann Arbor, Michigan

Greti Aguilera
NIH, Rockville Pike
Bethesda, Maryland

François Alhenc-Gelas
Inserm U-36
College de France
Paris, France

Jürgen Bachmann
German Institute of High Blood Pressure Research
University of Heidelberg
Heidelberg, Germany

Sebastian Bachmann
Department of Anatomy I
University of Heidelberg
Heidelberg, Germany

Michael Bader
German Institute of High Blood Pressure Research
University of Heidelberg
Heidelberg, Germany

Tamas Balla
Endocrinology and Reproduction Research
NIH, Rockville Pike
Bethesda, Maryland

Smriti Bardhan
Department of Biochemistry
Vanderbilt University
School of Medicine
Nashville, Tennessee

John D. Baxter
Metabolic Research Unit
University of California
San Francisco, California

Kathleen H. Berecek
Department of Physiology and Biophysics
University of Alabama
Birmingham, Alabama

Manfred Böhm
German Institute of High Blood Pressure Research
University of Heidelberg
Heidelberg, Germany

F. Merlin Bumpus
Cleveland Clinic Foundation
Cleveland, Ohio

Kevin J. Catt
Endocrinology and Reproduction Research
NIH, Rockville Pike
Bethesda, Maryland

Shigeyuki Chaki
Department of Biochemistry
Vanderbilt University
School of Medicine
Nashville, Tennessee

Andrew T. Chiu
Cardiovascular Diseases Research
DuPont Merck Pharmaceutical Company
Wilmington, Delaware

William N. Chu
Department of Medicine
Stanford University Medical School
Palo Alto, California

Eric Clauser
Department of Hypertension
Inserm U-36
College de France
Paris, France

Pierre Corvol
Department of Hypertension
Inserm U-36
College de France
Paris, France

Christian F. Deschepper
Clinical Research Institute of Montreal
Montreal, Quebec, Canada

Behrus Djavidani
German Institute of High Blood Pressure Research
University of Heidelberg
Heidelberg, Germany

Victor J. Dzau
Falk Cardiovascular Research Center
Stanford University
School of Medicine
Stanford, California

John R. Fabian
Department of Molecular and Cellular Biology
Roswell Park Cancer Institute
Buffalo, New York

Penelope Feuillan
NICDS/NIH, Rockville Pike
Bethesda, Maryland

Hiroaki Furuta
Department of Biochemistry
Vanderbilt University
School of Medicine
Nashville, Tennessee

Detlev Ganten
Center for Molecular Medicine
Berlin-Buch, Germany

Raul Garcia
Clinical Research Institute of Montreal
Montreal, Quebec, Canada

Kenneth W. Gross
Department of Molecular and Cellular Biology
Roswell Park Cancer Institute
Buffalo, New York

Deng-Fu-Guo
Department of Biochemistry
Vanderbilt University
School of Medicine
Nashville, Tennessee

Norman K. Hollenberg
Department of Medicine
Brigham and Women's Hospital
Boston, Massachusetts

Ly Q. Hong-Brown
Department of Physiology
University of California
San Francisco, California

Christine Hubert
Inserm U-36
College de France
Paris, France

Tadashi Inagami
Department of Biochemistry
Vanderbilt University
Nashville, Tennessee

Naoharu Iwai
First Department of Internal Medicine
Shiga University of Medical Sciences
Shiga-ke, Japan

Xavier Jeunemaitre
Howard Hughes Medical Institute
Euccles Institute of Human Genetics
University of Utah
Salt Lake City, Utah

M. Cecilia Johnson
IDIMI
University of Chile
Santiago, Chile

Craig A. Jones
Department of Molecular and Cellular Biology
Roswell Park Cancer Institute
Buffalo, New York

Sham S. Kakar
Department of Physiology and Biophysics
University of Alabama
Birmingham, Alabama

Judith E. Kalinyak
Department of Endocrinology and Hypertension
University of California
San Francisco General Hospital
San Francisco, California

Birgitta Kimura
Department of Physiology
College of Medicine
University of Florida
Gainesville, Florida

Steven J. King
University of Alabama at Birmingham
Medical Center
Birmingham, Alabama

Edyta M. Konrad
Clinical Research Institute of Montreal
Montreal, Quebec, Canada

Min Ae Lee
German Institute of High Blood
Pressure Research
University of Heidelberg
Heidelberg, Germany

Kam H. Leung
Cardiovascular Diseases Research
DuPont Merck Pharmaceutical
Company
Wilmington, Delaware

Richard P. Lifton
Howard Hughes Medical Institute
Euccles Institute of Human Genetics
University of Utah
Salt Lake City, Utah

Kevin R. Lynch
Department of Pharmacology
University of Virginia
Charlottesville, Virginia

Chantal Mercure
Clinical Research Institute of Montreal
Montreal, Quebec, Canada

Kathleen Michels
NIMH, Rockville Pike
Bethesda, Maryland

Monica Millan
Division of Endocrinology
OHSU
Portland, Oregon

Jimmy D. Neill
Department of Physiology and
Biophysics
University of Alabama
Birmingham, Alabama

Damian P. O'Connell
Department of Internal Medicine
Health Sciences Center
Charlottesville, Virginia

Jörg Peters
German Institute of High Blood
Pressure Research
University of Heidelberg
Heidelberg, Germany

M. Ian Phillips
Department of Physiology
College of Medicine
University of Florida
Gainesville, Florida

Richard E. Pratt
Falk Cardiovascular Research Center
Stanford University
Stanford, California

Mohan K. Raizada
Department of Physiology
College of Medicine
University of Florida
Gainesville, Florida

Timothy L. Reudelhuber
Clinical Research Institute of Montreal
Montreal, Quebec, Canada

Bartosz Rydzewski
Department of Physiology
University of Florida
College of Medicine
Gainesville, Florida

Juan M. Saavedra
NIMH/NIH
Rockville Pike
Bethesda, Maryland

Kathryn Sandberg
Endocrinology and Reproduction
Research
NIH, Rockville Pike
Bethesda, Maryland

Maike Sander
German Institute of High Blood Pressure Research
University of Heidelberg
Heidelberg, Germany

Katsutoshi Sasaki
Kyowa Hakko Kogyo Company, Ltd.
Tokyo Research Labs
Tokyo, Japan

Ernesto L. Schiffrin
Clinical Research Institute of Montreal
Montreal, Quebec, Canada

Nabil G. Seidah
Biochemical Neuroendocrinology
Clinical Research Institute of Montreal
Montreal, Quebec, Canada

Alicia Seltzer
NIMH, Rockville Pike
Bethesda, Maryland

Curt D. Sigmund
Internal Medicine
University of Iowa
Iowa City, Iowa

Ronald D. Smith
DuPont Merck Pharmaceutical Company
Experimental Station E400/3466
Wilmington, Delaware

Florent Soubrier
Department of Hypertension
Inserm U-36
College de France
Paris, France

Elisabeth A. Speakman
Department of Physiology
College of Medicine
University of Florida
Gainesville, Florida

Christer Strömberg
NIMH, Rockville Pike
Bethesda, Maryland

Colin Sumners
Department of Physiology
College of Medicine
University of Florida
Gainesville, Florida

Louise L. Théroux
Clinical Research Institute of Montreal
Montreal, Quebec, Canada

Pieter B.M.W.M. Timmermans
Cardiovascular Diseases Research
DuPont Merck Pharmaceutical Company
Wilmington, Delaware

Keisuke Tsutsumi
NIMH/NIH, Rockville Pike
Bethesda, Maryland

Mohan Viswanathan
NIMH, NIH, Rockville Pike
Bethesda, Maryland

Jürgen Wagner
German Institute of High Blood Pressure Research
University of Heidelberg
Heidelberg, Germany

Lei Wei
Inserm U-36
College de France
Paris, France

Johannes Wilbertz
German Institute of High Blood Pressure Research
University of Heidelberg
Heidelberg, Germany

Gordon H. Williams
Harvard Medical School
Brigham and Women's Hospital
Boston, Massachusetts

Magdalena Wozniak
Department of Pharmacology and Therapeutics
University of Florida
Gainesville, Florida

Jian N. Wu
Department of Physiology and Biophysics
Birmingham, Alabama

Yoshiaki Yamano
Laboratory of Metabolic Biochemistry
Faculty of Agriculture
Tottori University
Tottori, Japan

Karin Zeh
German Institute of High Blood Pressure Research
University of Heidelberg
Heidelberg, Germany

Yi Zhao
Department of Physiology
University of Southern California
Pasadena, California

Frank Zimmermann
German Institute of High Blood Pressure Research
University of Heidelberg
Heidelberg, Germany

Stefan Zorad
NIMH/NIH, Rockville Pike
Bethesda, Maryland

TABLE OF CONTENTS

SECTION I: RENIN

SECTION II: ANGIOTENSINOGEN

SECTION III: ANGIOTENSIN CONVERTING ENZYME

SECTION IV: ANGIOTENSIN RECEPTORS

Section I

Renin

Chapter 1

SUBCELLULAR SORTING AND PROCESSING OF PRORENIN

T.L. Reudelhuber, C. Mercure, L.L. Théroux, W.N. Chu, J.D. Baxter, and N.G. Seidah

TABLE OF CONTENTS

0-8493-4622-3/93/$0.00 + $.50

I. INTRODUCTION

Nearly one hundred years ago, Tiegerstedt and Bergman reported that extracts of kidney induced an increase in blood pressure when injected into rabbits.[1] This activity, which they named renin, was subsequently found to be a proteolytic enzyme which increased the blood pressure by generating the decapeptide angiotensin I, that is subsequently cleaved by angiotensin converting enzyme (ACE) to form the octapeptide angiotensin II (Figure 1). Angiotensin II (AII) modulates blood pressure directly by inducing vasoconstriction and it stimulates the release of aldosterone, which in turn causes sodium retention and potassium loss.[2] In addition, it has recently become apparent that AII has growth factor-promoting activity[3-5] which may be critical in the contributions of the renin-angiotensin system (RAS) to the pathological changes associated with heart failure, kidney failure, and chronic hypertension.[6-10] For these reasons, the RAS has become an important target for treating a number of cardiovascular diseases. While ACE inhibitors have been widely used for treating these pathological states,[6,8-10] new and potent inhibitors of both renin[11-15] and antagonists of AII receptors (see Chapter 9) may soon join these compounds in the treatment of a broad variety of diseases.

Within the circulation, renin is the rate-limiting component of the RAS. The release of renin into the circulation results from a number of intracellular steps, including organelle targeting and site-specific processing of the precursor to renin, prorenin. Both of these processes may eventually offer new targets for pharmacological interventions for controlling RAS activity. In this chapter we outline current knowledge regarding the intracellular processes that govern the ultimate release of active renin, in the context of recent information about the mechanism of secretion of other mammalian proteins.

II. BRIEF OVERVIEW OF MAMMALIAN SECRETORY PROCESSES

All mammalian messenger RNAs are translated into proteins on cytoplasmic ribosomes. Those proteins ultimately destined for insertion into the lumen of the endoplasmic reticulum (ER), lysosomes, Golgi apparatus or endosomes, and proteins which are secreted or are integrated into nuclear, lysosomal, or plasma membranes are synthesized on and integrated into the rough endoplasmic reticulum (RER). For these proteins, this first level of "sorting" occurs via the interaction between an amino-terminal hydrophobic peptide (the signal peptide) on the initial translation product with a "signal recognition particle" (SRP), which in turn binds to a high-affinity receptor on the membranes of the RER.[16] As translation proceeds, the nascent peptide chains are inserted into the lumen of the ER where the first chemical modifications to the protein occur: the signal peptide is cleaved off, disulfide bond formation is initiated between cysteine residues, and oligosaccharide

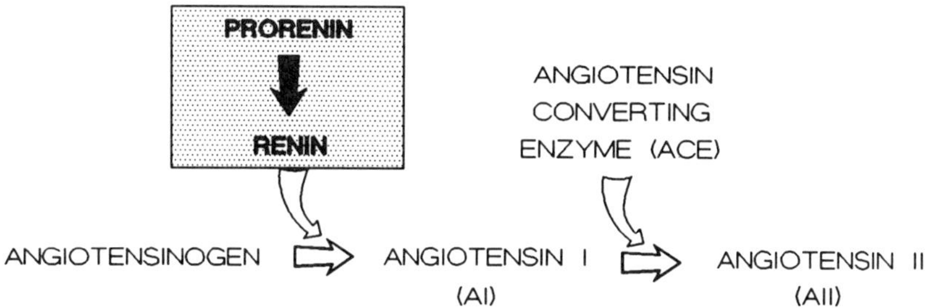

FIGURE 1. Schematic representation of the renin-angiotensin system. The shaded box depicts the area of focus of this chapter.

addition begins on asparagine residues contained in the sequence Asn-X-Ser/Thr (N-linked glycosylation). In addition, the growing peptide chain begins to fold into a native conformation with the help of "molecular chaperones," such as BiP/GRP78.[17] Such proteins are critical for the assembly of oligomeric protein complexes and may recognize specific amino acid sequences in incorrectly folded polypeptides to target them for destruction in the ER.[18,19]

Vesicles budding off from the ER transport the native protein to the first compartment of the Golgi lamellae, the so-called cis-Golgi (Figure 2). Proteins destined to remain in the ER are recaptured at this point by interaction of a specific amino acid sequence (Lys-Asp-Glu-Leu or KDEL) with the KDEL receptor.[20] Both the transport of proteins from the ER through the Golgi stacks and the recycling of proteins back to the ER occurs via vesicular transport in a process that requires hydrolysis of GTP.[20,21] Proteins destined for further transport through the medial and trans-lamellae of the Golgi apparatus may undergo additional chemical modifications including further processing of N-linked carbohydrate residues, sulfation, α-amidation, O-glycosylation, phosphorylation, lipid attachment, and proteolytic processing.[22-24] All of these modifications occur through recognition of specific peptide sequences or through specific positioning of amino acids in the target protein. The trans-most cisternae of the Golgi apparatus (commonly referred to as the trans-Golgi network or TGN) is the site of another major sorting decision (Figure 2): proteins are segregated depending on whether they are destined for inclusion in lysosomes, direct secretion or integration at the plasma membrane (constitutive secretion), or storage in dense core secretory granules for secretion in response to stimuli (regulated secretion).[25-28]

Targeting of proteins to the lumen of lysosomes requires two intracellular recognition events.[28,29] The first is the phosphorylation of mannose residues on N-linked oligosaccharides. In cathepsin D, a lysosomal aspartyl protease, a lysine residue and a noncontiguous 27-amino acid peptide sequence contained in a surface loop of the protein are required for recognition by the phosphotransferase.[30,31] The second recognition event is between the mannose-phosphorylated protein and a mannose-6-phosphate receptor located in

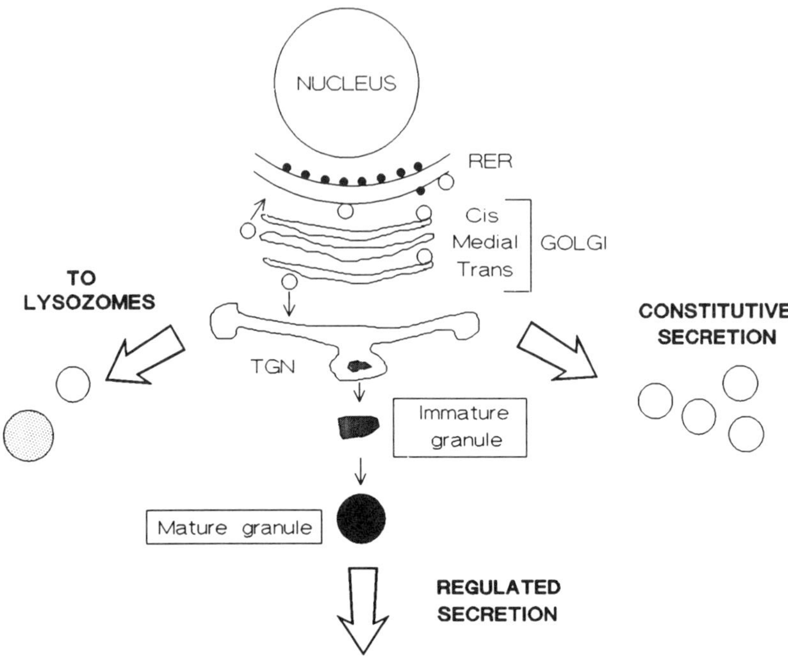

FIGURE 2. Schematic representation of the mammalian secretory pathway. Open arrows represent the major sorting pathways from the trans-Golgi network (TGN).

the TGN.[28] Recent evidence suggests that peptide signals may also play a role in targeting of mammalian lysosomal proteins to the proper intracellular compartment,[28] as they do in the lysosomal equivalent (the vacuole) of yeast and plants.[32]

Proteins are secreted from the cell by two major pathways. The first involves direct secretion from the TGN. Because there is no evidence of intracellular storage of proteins secreted by this route, it has been called the constitutive secretory pathway (Figure 2). Data suggest that this pathway constitutes the bulk flow or ''default'' pathway of protein secretion[33,34] and it is the dominant secretory pathway in most eukaryotic cell types. Proteins move to the plasma membrane through the constitutive pathway in microtubule-anchored low-density vesicles[35] with a relatively rapid transit time on the order of minutes to hours.[36]

Some specialized endocrine and neuroendocrine cells may also export proteins by the regulated secretory pathway[20,34,37] (Figure 2). In this pathway, certain proteins are sorted at the TGN to dense core secretory granules, where they are stored until the cell receives a signal which triggers their release. Little is known about the precise molecular signals that target proteins to dense core secretory granules. However, using the model system of mouse pituitary AtT-20 cells, Moore and co-workers obtained the first evidence that cells from different tissues use a common mechanism to sort proteins to secretory granules.[38] AtT-20 cells sort endogenous proopiomelanocortin (POMC) to secretory granules, process POMC to adrenocorticotrophic hormone (ACTH) and other peptides and release them in response to cAMP. When transfected with an expression vector for human proinsulin, these cells sort and process the encoded prohormone correctly and release human insulin in response to a secretagogue.[38] AtT-20 cells have subsequently been shown to correctly sort proteins destined for secretory granules from various species and tissues of origin and to exclude from granules proteins which are known to be secreted in a constitutive manner (Table 1). Further, fusion of a secretory granule-targeted protein with a protein secreted by the constitutive pathway results in secretory granule targeting of the fusion protein.[39,40]

Taken together, these results suggest that the signals for targeting of proteins to dense core secretory granules are (1) contained within the targeted protein; (2) universal (i.e., recognized across species and tissue barriers); and (3) dominant (i.e., not a bulk flow mechanism). Surprisingly, in spite of this apparent universal nature of the sorting machinery, there exists no extended sequence homology between proteins sorted to granules. There is, in fact, no

TABLE 1
Secretion of Exogenous and Endogenous Proteins by Mouse Pituitary AtT-20 Cells

Protein Expressed	Source		Regulated Secretion	Ref.
	Tissue	Species		
Proneuropeptide Y	Intestine	Human	+	131
Proinsulin	Pancreas	Human	+	38
Proenkephalin	Brain	Human	+	132
Prothyrotropin releasing hormone (TRH)	Hypothalamus	Human	+	133
Trypsinogen	Pancreas	Rat	+	134
Growth hormone	Pituitary	Human	+	36
Prosomatostatin	Hypothalamus	Rat	+	40
Procholecystokinin (CCK)	Small intestine	Pig	+	135
Provasopressin/neurophysin	Hypothalamus	Rat	+	136
Proatrial natriuretic factor (ANF)	Heart	Rat	+	137
Prorenin	Kidney submaxillary gland	Rat, mouse, human	+	105–108
Laminin	Ubiquitous	Mouse	–	134
Vesicular stomatitis virus G protein	—	Viral	–	36
Immunoglobulin (kappa light chain)	Lymphoid cells	Mouse	–	138
Angiotensinogen	Liver	Rat	–	139

consensus regarding the actual mechanism of sorting. One hypothesis is that proteins destined for secretory granules interact with a specific secretory granule receptor in much the same way as proteins are sorted to lysosomes. While evidence for the existence of such a "sortase" in dog pancreatic microsomes has been reported,[41] there has been no further confirmation of its role as a sorting receptor.

Another hypothesis advanced to explain the segregation of granule targeted proteins in the TGN is selective aggregation.[34,42] Many of the proteins which end up in dense core secretory granules show a tendency to aggregate in the presence of calcium and a slightly acidic environment, conditions thought to be present in immature secretory granules.[42-44] This proposed mechanism is also consistent with microscopic studies demonstrating aggregates in the trans-Golgi in the process of being encapsulated by membrane.[42] Regardless of the actual mechanism of sorting, the segregated protein is first packaged into a small, relatively electron-lucid "immature" granule which is often characterized by a clathrin "patch."[43] The maturation process for secretory granules is poorly understood but may involve fusion of immature granules and progressive condensation of the granule contents by formation of dense para-crystalline aggregates.[43] Mature secretory granules reside at or near the plasma membrane at "docking" sites[34] and fuse with the plasma membrane in response to an extracellular signal. The transit time of proteins in the regulated pathway is longer than that for the constitutive pathway, being in the range of hours to days, and depends on the timing of a stimulus for exocytosis.[36,45]

Sorting can also occur to two additional secretory pathways in specialized mammalian cells. The first is a recently demonstrated "basal" pathway in which presumably immature secretory granules fuse with the plasma membrane, resulting in a seemingly constitutive pattern of release of proteins which would normally be destined for dense core secretory granules.[46,47] The second is the secretion or membrane anchoring of proteins selectively by either the apical or basolateral membranes of polarized cells. This sorting mechanism, which has been most extensively studied in the targeting of polymeric immunoglobulin receptor to the basolateral surface of epithelial cells, requires a short peptide sequence adjacent to the membrane-spanning domain of the receptor.[48] The finding that thyroglobulin, a regulated secretory protein, is also segregated to the apical surface of thyroid epithelial cells has led to the suggestion that some relationship between the targeting signals for the two secretory pathways may exist.[49]

Cell-type specific proteolytic processing of proteins also plays an important role in the mammalian secretory pathway. Numerous hormones, bioactive peptides, and proteases are synthesized as precursors and are processed in the secretory pathway to the biologically active forms that are subsequently secreted.[50-53] The intracellular site of processing for these precursors may vary significantly, however. For example, proalbumin is a constitutively secreted protein and is processed to albumin in the Golgi lamellae of hepatocytes.[54]

In contrast, insulin appears to be activated only after its sequestration into secretory granules and is stored in these granules in the active form.[55] Atrial natriuretic factor (ANF), on the other hand, seems to be packaged in atrial cardiocyte secretory granules as the prohormone, and is activated only upon release of the granule.[56,57] These processing enzymes can be exquisitely selective in the sites they cleave in individual substrates. For example, while proalbumin and proinsulin are cleaved at sites that contain a pair of basic amino acids,[50,58] pro-ANF is cleaved at a site containing only a single basic amino acid.[57] Indeed, not all potential processing sites within a given substrate are cleaved. POMC is processed differentially in the anterior and intermediate lobes of the pituitary, in spite of the fact that all of the processing events occur at pairs of basic amino acids.[52,59] It is easy to see why prediction of endoproteolytic cleavage sites in complex mammalian proteins based on sequence comparison has not been very successful to date. Thus, processing specificity may be determined by a combination of cellular distribution of different processing enzymes, the particular architecture of the substrate, and effects on the chemistry of the reaction by the intracellular compartment or environment.

While there have been many attempts to purify and characterize processing enzymes from mammalian tissues and cell lines, a major advance was made in this field with the cloning of the gene which encodes the processing enzyme responsible for cleaving yeast pro-(alpha) mating factor.[60] Surprisingly, this enzyme, called Kex2, was also found to correctly cleave both proalbumin and POMC when expressed in mammalian cells.[61,62] Using sequence information from the yeast Kex2 gene, several related sequences have been identified in mammalian cells, including furin, PC1 (also called PC3), PC2, and PACE4.[53,63] Whereas furin and PACE4 are expressed ubiquitously in tissues and cell lines, PC1/PC3 and PC2 have to date only been detected in neural and endocrine cells.[53,64]

III. SECRETION OF PRORENIN AND RENIN IN WHOLE ANIMALS AND TISSUES

The genes encoding mouse, rat, and human renins have been cloned and characterized.[65-69] The genomic organization and deduced amino acid sequence of human renin confirm its close relatedness to other aspartyl proteinases. Human and rat genomes contain only one renin gene[67-70] and while all inbred strains of mice carry the Ren-1 structural gene (expressed at high levels in the kidney), some strains contain a second closely linked gene (Ren-2) which is expressed at high levels in the submaxillary gland (SMG; see Chapter 2). In all cases, these genes encode protein precursors of 400 to 406 amino acids.[70-72] A signal peptide encoded at the amino-terminus directs the nascent polypeptides to the ER. Upon insertion into the ER, the signal peptide is removed (Figure 5) to generate prorenin and posttranslational modification begins. There are two consensus sequences for N-glycosylation in human renin and three such sites in rat and mouse (Ren-1) renal renin.[70,72,73] These sites may be modified to different extents, resulting in multiple species of renin and prorenin that can be separated by either isoelectric focusing or lectin affinity chromatography.[74,75] In human renin, mannose residues on the carbohydrate side chains may also be partially phosphorylated,[76] creating a classical lysosomal targeting signal. The importance of the glycosylation of prorenin and renin is most obvious in clearance of the protein from the circulation,[77,78] although some evidence has accumulated that glycosylation may also affect intracellular transit time,[79] efficiency of intracellular sorting,[80] and stability[81] (protease sensitivity?) of prorenin. There is no direct evidence that prorenin undergoes post-translation modifications other than N-glycosylation. However, rat prorenin can be fractionated into multiple isoelectric species even after enzymatic deglycosylation.[82] In contrast, recombinant human prorenin in which the glycosylation sites have been eliminated by protein engineering migrates as a single isoelectric species.[143] It is currently uncertain whether this difference is explained by incomplete removal of carbohydrate side chains by the glycosidase, as has been demonstrated for human renin,[83] or whether rat and human prorenins undergo different post-translational modifications.

Evidence to date suggests that glycosylated prorenin is analogous to the circulating "big" or "inactive" renin and can be present at 3 to 5 times the level of "active" renin in the circulation of humans.[84] While circulating active renin is derived almost exclusively from the kidney, numerous nonrenal tissues in humans secrete prorenin.[84,85] In rats, the case seems a bit different: although several extrarenal tissues contain prorenin mRNA and/or protein[86] (see Chapter 4), nephrectomy results in the disappearance of both renin and prorenin from the circulation.[87] Thus, the kidney is clearly capable of releasing both prorenin and renin. How is this accomplished and what determines the relative proportions of the two proteins secreted? Much of what we understand about this process has derived from ultrastructural studies on the cells that secrete renin from the kidney.

The juxtaglomerular apparatus of the kidney is the primary site of synthesis for circulating renin. Juxtaglomerular (JG) cells are modified smooth muscle cells which make up approximately 0.1% of the cellular mass of the adult kidney.[88] While JG cells resemble other neuroendocrine cells in their rich cytoplasmic content of dense core secretory granules (Figure 3), they are distinguished from these other cell types by two rather striking characteristics. First, rhomboid, para-crystalline structures can be seen budding off from the TGN and in membrane-bound structures within the cytoplasm (Figure 4A). Second, the secretory granules of JG cells are atypical and display many similarities to lysosomes. At the ultrastructural level, the electron-dense matrix of these granules is sometimes seen to contain multiple vesicular inclusions and membrane fragments (Figure 4C) and some micrographs suggest that these granules are capable of micropinocytosis and autophagy of other cellular organelles.[89,90] In addition, JG cell granules are immunoreactive to antibodies

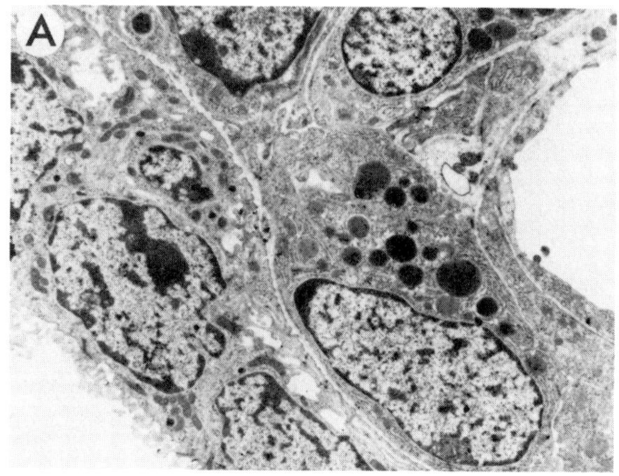

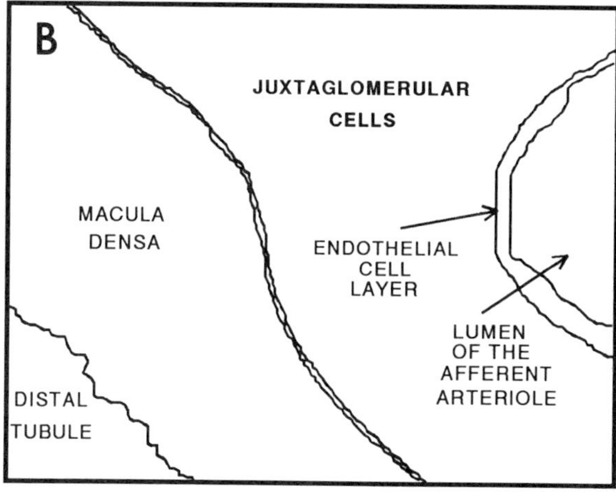

FIGURE 3. Ultrastructure of the juxtaglomerular (JG) apparatus in the 5-day-old mouse kidney. Tissue was fixed in 2% glutaraldehyde and post-fixed in 2% osmium tetroxide. Thin sections were stained in uranyl acetate and lead citrate. (A) Electron micrograph of the JG apparatus. Note the abundant rough endoplasmic reticulum and dense core secretory granules in the JG cells. Original magnification × 2500. (B) Schematic representation of the micrograph depicted in panel A, showing the location of the various cell types in the JG apparatus.

against a number of lysosomal enzymes, including acid phosphatase, β-glucuronidase, arylsulfatase, *N*-acetyl-β-glucosaminidase, and cathepsins B, D, H, and L.[90,91] These results, combined with the apparent lack of any recognizable, classical lysosomal structures, have led some investigators to suggest that JG cells do not contain the type of secretory granules seen in many endocrine cell types, but rather have adapted lysosomes for the processing and secretion of prorenin.[90]

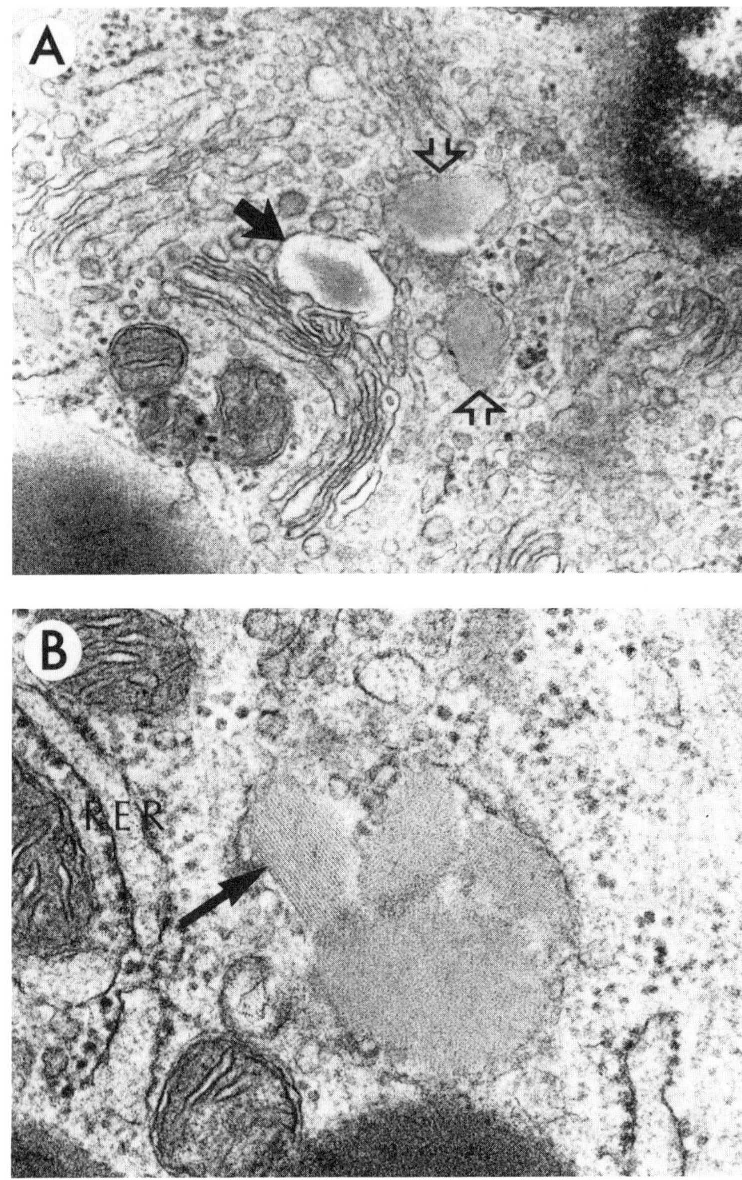

FIGURE 4. Detailed ultrastructure of the JG cells of 5-day-old mice. (A) Emergence of a para-crystalline protogranule (solid arrow) from the TGN. Note the presence of two membrane-encapsulated protogranules in the cytoplasm (open arrows). Original magnification × 46,600. (B) Detail of an immature secretory granule. Note the nonaligned crystalline structures of the recently fused protogranules which will eventually become lost as the granules mature. RER, rough endoplasmic reticulum. Original magnification × 73,200.

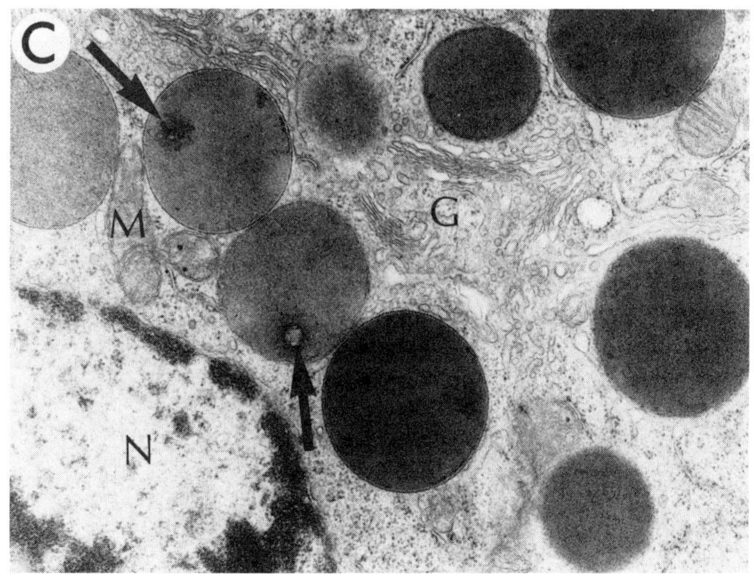

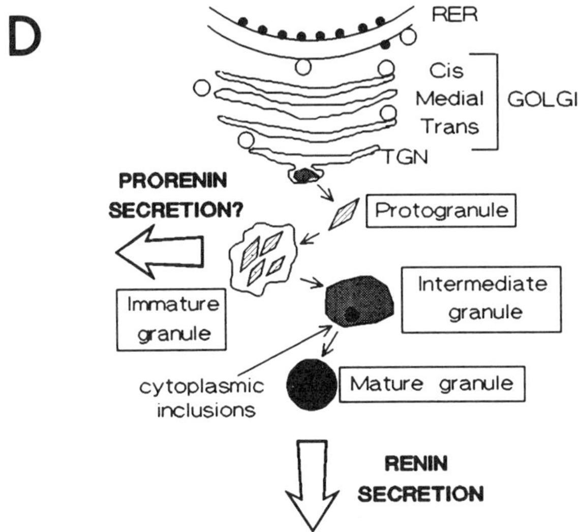

FIGURE 4 (*continued*). (C) Diversity of granule densities in JG cells. Arrows point to cytoplasmic inclusions in intermediate granules. G, Golgi apparatus; M, mitochondrion; N, nucleus. Original magnification × 17,200. (D) Schematic representation of the proposed routes of prorenin and renin secretion in JG cells. See text for details.

Using antibodies to various portions of the human prorenin prosegment, Taugner and colleagues have demonstrated that the para-crystalline structures budding off from the TGN of JG cells contain prorenin.[92] These structures (Figure 4A), which have been called protogranules, subsequently fuse together to form a membrane-bound structure with a relatively amorphous, low-density content, which has been termed the ''immature'' or ''juvenile'' secretory granule (Figure 4B). In some micrographs these immature granules can be seen to fuse directly with the plasma membrane,[90] which would presumably result in the release of prorenin and would correspond, therefore, to the ''basal'' pathway of protein secretion described above. Immature granules which are not released from the cell become progressively more electron dense and have been referred to as ''intermediate'' or ''mature'' granules, depending on their apparent degree of condensation (Figure 4C).

Evidence suggests that conversion of prorenin to active renin begins in the immature secretory granule. Antibodies specific for the prosegment of human prorenin stain predominantly protogranules and immature granules and show little or no staining of intermediate and mature granules.[92] In contrast, an antibody which reacts with both prorenin and renin stains all of these granular structures.[92] The role of the secretory granule in processing prorenin is further supported by biochemical studies: granule fractions purified from kidney homogenates contain predominantly active renin.[93] In addition, renin is secreted more slowly than prorenin from human renal cortical slices and with kinetics that are consistent with its storage in secretory granules.[94] Finally, pharmacologic stimuli which cause a release of secretory granules result in an acute and preferential release of active renin into the circulation.[95]

Amino-terminal sequencing of renin isolated from human kidney lysates suggests that activation occurs by the proteolytic removal of a 43-amino acid prosegment from the amino-terminus of prorenin.[96] This processing site follows a pair of basic amino acids in human renin (Figure 5). Processing of prorenin in the mouse SMG occurs at the analogous position,[97] but processing of rat renal renin seems to occur after a threonine residue which is located 7 amino acids toward the carboxy-terminus relative to the analogous site in human renin[82,98] (Figure 5). The processing site for mouse renal renin is currently unknown. Rat renal and mouse SMG renins subsequently undergo an additional internal processing event, converting ''one-chain'' active renin to a ''two-chain'' molecule[82,97,98] in which the two halves are held together by a disulfide bridge (Figure 5). Notably, while renal renins appear to be processed within secretory granules, mouse SMG renin may be processed to ''one-chain'' renin within the Golgi and only a portion of the protein is further processed in granules to yield the ''two-chain'' protein.[99] Thus, while renins from mice, rats, and humans share many similarities in protein structure and function, differences exist in the way these proteins are modified within the secretory apparatus.

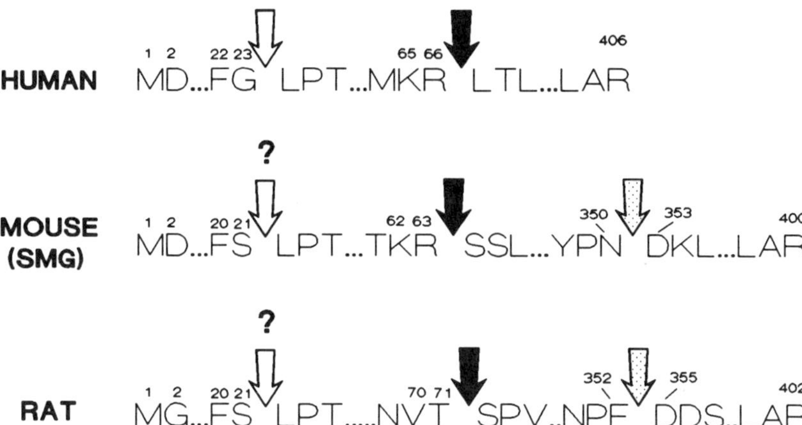

FIGURE 5. Proposed cleavage sites for human and rat renal renins and mouse submaxillary gland renin (SMG). Numbering is from amino acid 1 of preprorenin.[71,82,97,99,115] Open arrows, signal peptide cleavage site; closed arrows, cleavage of the prosegment; stippled arrows, cleavage to generate "two-chain" renin. Question marks indicate that assignment has only been made by homology to human prorenin.

IV. MODELS OF PRORENIN SORTING AND ACTIVATION

Clearly, a critical determinant in the exclusive ability of renal JG cells to secrete active renin is the proteolytic cleavage of prorenin to generate renin. However, since ultrastructural and biochemical studies suggest that this processing is granule specific, the sorting of prorenin to dense core secretory granules is also a crucial step in the secretion of active renin by the kidney. Investigations of the molecular mechanisms of prorenin processing and sorting in the kidney would be facilitated if large quantities of renal JG cells were available for study. Several approaches have been used to isolate and characterize JG cells, including density-gradient enrichment of primary cell preparations[100] and culture of cells from human renin-secreting tumors.[101] To date, these attempts have largely been frustrated by two major problems: the relative paucity of JG cells in the kidney (less than 0.1% of the cell mass) and the tendency of tumor-derived JG cells to dedifferentiate in culture.[102] In an alternative approach, Sigmund et al.[103] have used renin gene fragments to target expression of a viral oncogene to renin-producing cells of transgenic mice. Initial reports suggest that cells derived from a renal tumor in such mice express active renin and contain secretory granules.[104] However, more extensive characterization will be required to determine whether these transformed cells will retain sufficient terminal differentiation to be useful for studying all of the intracellular steps in renin biosynthesis.

In the absence of suitable quantities of JG cells, efforts have largely been directed at characterizing the biochemical properties of various prorenin-processing enzymes and at using model cell systems which correctly sort and process prorenin to obtain a better understanding of these two important intracellular processes.

A. SORTING DETERMINANTS IN PRORENIN

One possible explanation for the seemingly exclusive ability of JG cells to secrete active renin could be that only this particular cell type recognizes sorting and processing signals contained on prorenin. This hypothesis has been directly tested by transfection of a number of cell types with expression vectors encoding human, mouse, and rat prorenins. Transfection of Chinese hamster ovary (CHO) cells, which contain only a constitutive secretory pathway, with an expression vector encoding human preprorenin leads to secretion of prorenin[105,106] (Figure 6). As expected, this prorenin accumulates in a linear fashion in the transfected culture supernatants and secretion is not stimulated acutely by secretagogues. In contrast, transfection of AtT-20 cells, which contain secretory granules and process endogenous POMC, leads to secretion of both prorenin and renin. In addition, while prorenin accumulates in culture supernatants constitutively, treatment of the transfected cells with a secre-

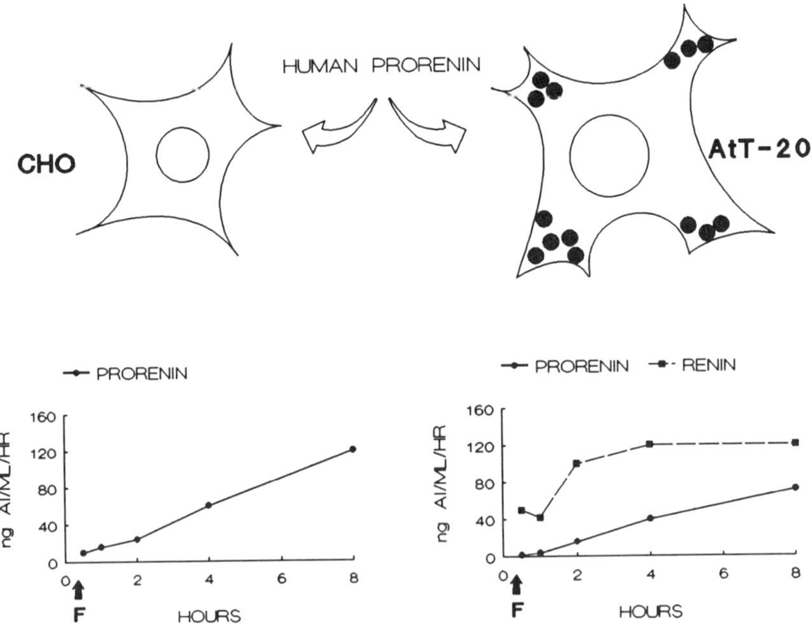

FIGURE 6. Pattern of secretion of prorenin and renin from transfected AtT-20 cells. F, point of addition of secretagogue to the cultures. Data from Fritz et al.[105]

tagogue causes a selective and acute release of active renin[105,106] (Figure 6). This result is characteristic of proteins secreted by the regulated secretory pathway[45] and implies that a portion of the prorenin is sorted to dense core secretory granules where it is processed and stored for later release. Mouse SMG[107] and renal prorenins,[108] as well as rat prorenin,[108] are also sorted to the regulated secretory pathway in transfected AtT-20 cells. Human prorenin is also sorted to the regulated secretory pathway in PC12 rat pheochromocytoma cells[109] and rat somatomammotrophic GH_4 cells,[110] although in these two cases, prorenin is not activated in granules. These results imply that sorting and activation of prorenin can be seen in cells other than JG cells and that prorenin must contain primary or higher-order structural information which directs the cell to carry out these processes.

What is the nature of the secretory granule sorting signal on prorenin? Although the physical segregation of proteins destined for secretory granules occurs at the TGN,[25] commitment to this pathway could theoretically take place as early as insertion of the nascent protein into the ER. However, replacement of the native signal peptide of human prorenin with a signal peptide from a constitutively secreted immunoglobulin M (IgM) did not impair the sorting of human prorenin to the regulated pathway in AtT-20 cells.[111] Likewise, eliminating the glycosylation sites on human prorenin did not prevent its targeting to secretory granules but, instead, increased the percentage of prorenin activated (sorted?) in AtT-20 cells.[111] The dispensability of the carbohydrate residues for granule sorting is also evidenced by the fact that mouse SMG prorenin (which is naturally nonglycosylated) is sorted to granules in AtT-20 cells.[107] The processing of prorenin to renin also is not required; prorenin molecules containing mutations in the paired basic amino acids at the native processing site cannot be activated in AtT-20 cells, but are nevertheless sorted to the regulated pathway and released as prorenin in response to secretagogues.[112] In addition, regulated secretion of prorenin is seen in transfected PC12 and GH_4 cells which contain granules but apparently lack of processing enzyme capable of activating prorenin.[109,110] Finally, the prosegment can also be deleted from human prorenin and the resulting "prerenin" is secreted in a regulated manner in both PC12 and AtT-20 cells.[109,111,113] By deduction, a sorting signal would appear to be located within the protein domain corresponding to active renin.

To further characterize this sorting sequence, we have constructed fusion proteins between a portion of an immunoglobulin constant region (which is constitutively secreted) and fragments of human prorenin and have examined the targeting of these fusion proteins to the regulated secretory pathway in AtT-20 cells (Figure 7A). The results indicate that human prorenin contains a peptide at the extreme amino-terminus of its prosegment that can direct the fusion protein to the regulated secretory pathway (Figure 7B). How do these results fit with the finding that the prosegment can be deleted without preventing correct sorting? The simplest explanation is that prorenin contains more than one domain involved in secretory granule targeting (Figure 7C).

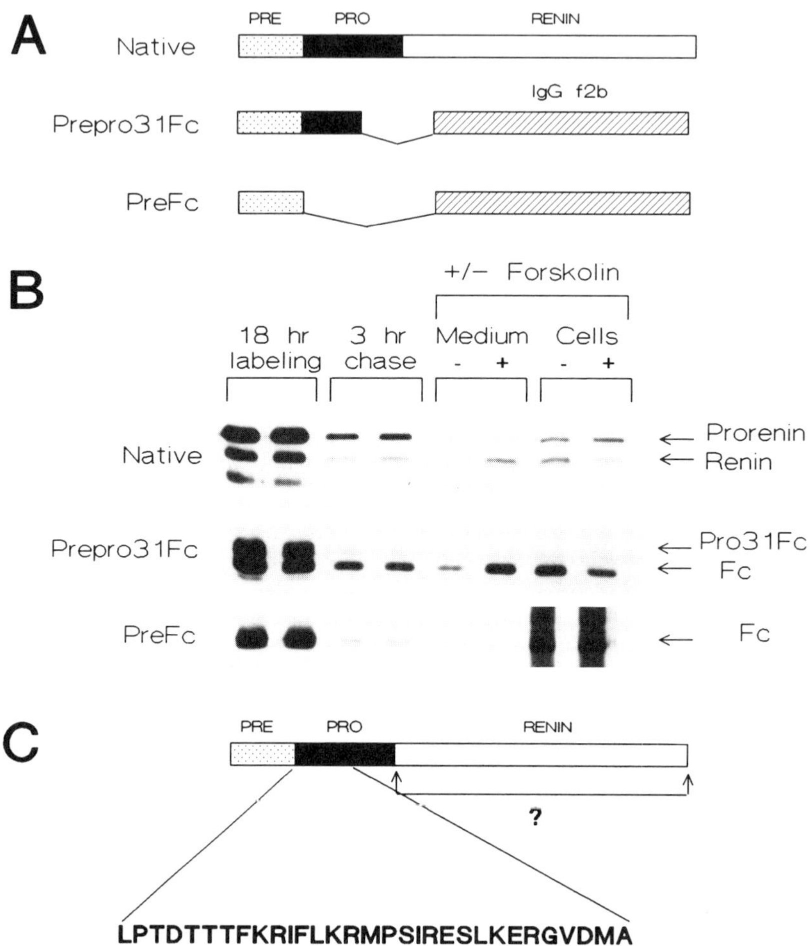

FIGURE 7. Location of a secretory granule-sorting peptide in the prosegment of human prorenin. (A) Construction of the fusion proteins. (B) Plasmids containing either native prepro-renin or the fusion proteins were transfected into AtT-20 cells and the expressed proteins were assayed by pulse-chase studies for unstimulated (−) and forskolin (10 μ*M*)-induced release (+) of the products, as described in Chu et al.[111] (C) Location of secretory granule-sorting peptides in human prorenin.

As previously mentioned, there is no obvious linear homology between the putative sorting peptide in the human prorenin prosegment and peptide sequences contained within other proteins destined for the secretory granule such as POMC, growth hormone, insulin, and ANF.[140] This result suggests that potential homologies in secretory granule-sorting peptides may either be in secondary or tertiary structure rather than the linear amino acid sequence. Alternatively, multiple sorting receptors may exist or a given sorting receptor may have the capacity to interact with a wide spectrum of binding sequences, thereby restricting homologies to subsets of secretory granule-sorted proteins. Discrimination between these possibilities will likely have to await the further characterization of minimal sorting peptides in both prorenin and other proteins secreted by the regulated pathway.

Is the use of an endocrine cell model a reasonable approach to identifying sorting determinants to the lysosome-like granules contained in JG cells? In this regard, it is interesting to note that up to 80% of human prorenin expressed in *Xenopus* oocytes (which do not contain dense core secretory granules) is directed to lysosomes.[76] Elimination of N-linked glycosylation sites in human prorenin by site-directed mutagenesis appears to eliminate lysosomal uptake of prorenin in injected oocytes and leads to the secretion of the prorenin,[114] but does not inhibit targeting of human prorenin to dense core secretory granules in AtT-20 cells.[80] Secretory granule targeting of the nonglycosylated prorenin in the mouse SMG must also occur in the absence of carbohydrate signals, suggesting that the lysosomal and granular targeting signals on prorenin are distinct and separable. Nevertheless, while only 5 to 6% of prorenin expressed in mammalian cells acquires phosphomannosyl residues,[76] it is an intriguing possibility that renal prorenins make use of both granular and lysosomal signaling mechanisms to ensure their efficient sorting to the lysosome-like secretory granules in JG cells. Clearly, it will be imperative to test for the function of any potential prorenin-sorting peptides in the JG cells of intact animals.

B. PROCESSING OF PRORENIN

Amino-terminal sequencing of renin isolated from human kidney lysates suggests that the cleavage of prorenin is highly specific and occurs at a pair of basic amino acids (Lys-Arg), resulting in removal of a 43-amino acid prosegment from the amino-terminus of prorenin.[115] There are 6 additional pairs of basic amino acids in human prorenin,[73] including Lys-Lys, Arg-Arg, and Arg-Lys within the body of renin, two Lys-Arg pairs in the prosegment, and the Lys-Lys-Arg triplet in the body of renin, none of which appear to be cleaved in the kidney. What determines this cleavage site selectivity?

Many proteases are capable of activating human prorenin *in vitro,* including trypsin, plasmin, tissue and plasma kallikreins, and cathepsin B.[84,116,117] While some of these enzymes cleave prorenin with the correct specificity,[118] most are likely to be physiologically irrelevant due to their tissue distribution. An exception to this rule is cathepsin B, which was recently purified as a prorenin-processing activity from human kidney lysates.[117,119] While cathepsin B is a lysosomal enzyme which is expressed in a broad variety of cell types,[120] it appears to co-localize with renin not only in the lysosome-like granules of JG cells,[90,91] but also in the more classical secretory granules of prorenin-containing human pituitary lactotrophs.[121] *In vitro,* cathepsin B cleaves human prorenin with the correct specificity and with a K_s in the nanomolar range.[117]

For these reasons, cathepsin B has been proposed as the renal prorenin-processing enzyme. In a similar effort, an enzyme capable of processing mouse SMG prorenin has been isolated from submaxillary glands.[122] This enzyme, which has been called PRECE, was subsequently revealed to be identical to the mGK-13 gene product[123] (also known as the epidermal growth factor-binding protein type B), a member of the kallikrein gene family. While PRECE can activate mouse SMG prorenin to generate ''one-chain'' renin, it is unable to carry out the second cleavage to yield the ''two-chain'' form (Figure 5).[122] In addition, kidney glandular kallikrein cannot activate mouse SMG prorenin and PRECE is unable to activate mouse renal or human prorenins.[122] Recently, a second enzyme capable of converting mouse SMG prorenin to ''one-chain'' renin has also been isolated from mouse submaxillary gland.[124] Thus, while it has been possible to purify and characterize candidate prorenin-processing enzymes by classical biochemical techniques, the occasional promiscuity displayed by processing enzymes *in vitro* and the tendency for cellular colocalization to be misleading regarding function[125,126] has complicated the unequivocal identification of prorenin-processing enzymes. For this reason, it is imperative that these studies be complemented with genetic or other experiments that specifically block the actions of the putative protease *in vivo* before a specific role in prorenin activation can be confirmed.

AtT-20 cells transfected with a human preprorenin expression vector also cleave prorenin at the same site as that reported for renin purified from human kidney lysates.[105] One hypothesis to explain the cleavage site selectivity displayed by enzymes *in vitro,* AtT-20 cells, and the kidney is that primary and/or higher-order structural determinants on prorenin render the native processing site uniquely sensitive to proteolytic cleavage. This hypothesis was directly tested by introducing single amino acid mutations in human preprorenin surrounding the natural cleavage site and expressing the resultant recombinant proteins to proteolytic activation either by trypsin or by the endogenous processing enzyme in AtT-20 cells.[127] The results suggest that amino acids in addition to the pair of basic amino acids surrounding the cleavage site affect the ability of both trypsin and the AtT-20-processing enzyme to cleave prorenin (Figure 8). Notably, while a proline at position -4 is essential for processing of human prorenin in AtT-20 cells and is correlated with predicted formation of a β-turn at this position, other site-directed mutations suggest that this structural feature in addition to a pair of basic amino acids is not sufficient to lead to proteolytic activation of prorenin.[127] In contrast to the case with human prorenin, neither mouse renal prorenin[108] nor rat prorenin[108,128] are processed at the analogous positions in transfected AtT-20 cells. In both cases, mutagenesis of the natural substrates has demonstrated that this is also due to the particular arrangement of amino acids immediately adjacent to the native processing site (Figure 9).

Is the AtT-20 cell-processing enzyme identical to the enzyme which activates prorenin in mammalian JG cells? The answer cannot be unequivocally positive, since AtT-20 cells are unable to cleave mouse or rat renal

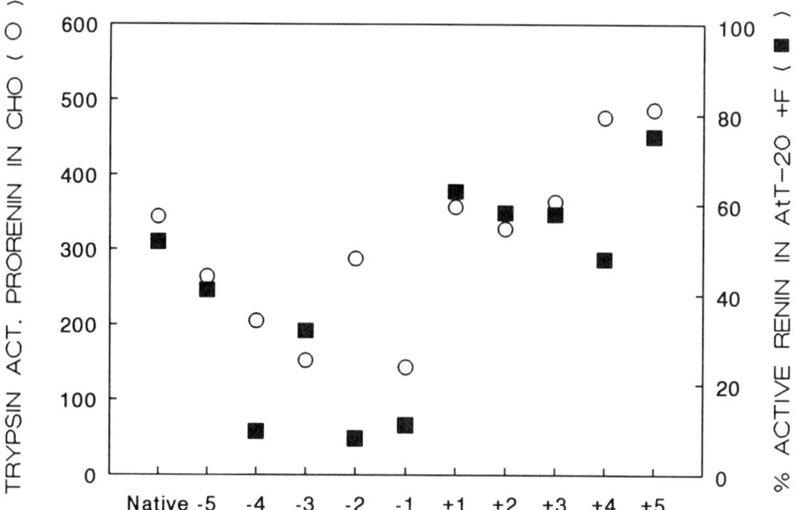

FIGURE 8. Effect of mutations on activation of human prorenin. Abcissa denotes the position (relative to amino acid 1 of human renin) at which the native amino acid has been changed to alanine. Left ordinate, trypsin-activatable prorenin (ng AI/ml/h) in culture supernatants of transfected CHO cells (○). Right ordinate, active renin secreted from transfected AtT-20 cells in the presence of 10 μM forskolin (■). Data from Chu et al.[127]

	SEQUENCE	ACTIVATION
Mouse Ren2	**...V F T K R↓S S L T...**	++
	...V [R] T K R S S L T...	+++
	...V F T K R [P] S L T...	−
Mouse Ren1	**...V F T K R P S L T...**	−
	...V F T K R [S] S L T...	+++
Rat	**...E F I K K S S F T...**	−
	...E F I K [R] S S F T...	+++
Human	**...Q P M K R↓L T L G...**	++
	...Q P M [R] R L T L G...	+
	...Q P M K [K] L T L G...	−
	...Q P M [R] [K] L T L G...	−
	...Q P M [A] R L T L G...	−
	...Q P M K [A] L T L G...	−
	...Q [A] M K R L T L G...	−
	...Q [F] M K R L T L G...	−
	...Q [G] M K R L T L G...	−

FIGURE 9. Summary of the effect of processing site mutations on the generation of active renin in transfected AtT-20 cells.[108,127] The native sequences of the analogous region of mouse, rat, and human prorenins are represented in bold type. Mutations induced are represented by shaded boxes. Where the native processing site is known, it is denoted by an arrow.

prorenins.[108,128] Nevertheless, it should be remembered that the processing site of rat renal renin is not in the same position as that of human renin[82,98] and the amino-terminus of mouse renal renin has not yet been determined. Therefore, it is possible that species-specific differences exist in the processing enzymes which activate prorenins. Could the AtT-20 enzyme be related to the human renal prorenin-processing enzyme? The first step in answering this question is to identify the human prorenin-processing enzyme in AtT-20 cells. Transfection of secretory granule-containing GH_4 cells with an expression vector encoding human prorenin results in the secretion of prorenin into culture supernatants, confirming the lack of a prorenin-processing enzyme in these cells[110] (Figure 10). When prorenin is cotransfected into GH_4 cells with an expression vector encoding the mammalian subtilisin-like endoprotease PC1, the cells are rendered capable of selectively processing prorenin at the identical site as that reported for the kidney. Interestingly, this processing event does not occur in cells devoid of secretory granules (CHO and BSC-40) and is not due to any apparent differences in processing of PC1 in these cell types.[128] When tested in a similar assay, neither furin,[110] PC2[128] nor human cathepsin B[141] are able to process human prorenin. Mutations which inhibit the processing of human prorenin in AtT-20 cells also inhibit processing of human

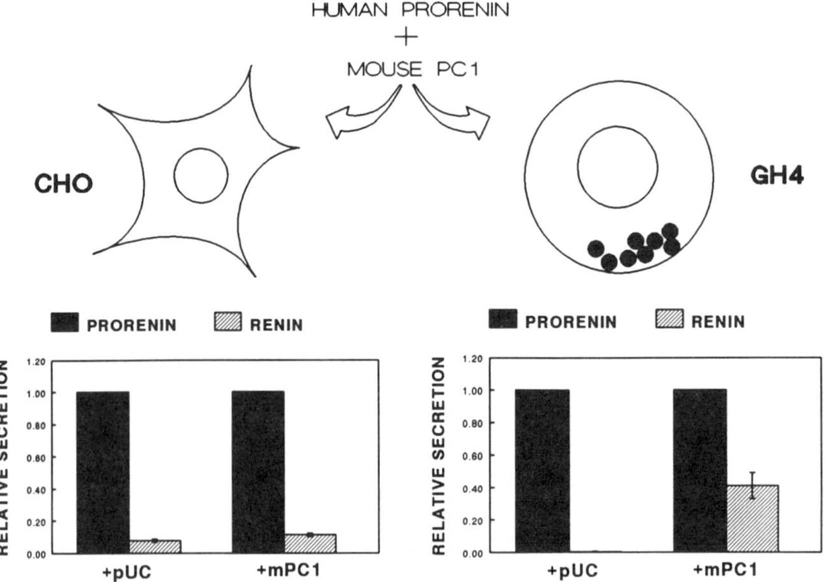

FIGURE 10. Cell type-specific processing of human prorenin by the endoprotease PC1.[128] CHO cells (lacking granules) and GH_4 cells (containing granules) were cotransfected with expression vectors encoding human prorenin and either carrier plasmid DNA (+pUC) or an expression vector encoding mouse PC1 (+mPC1). Cell culture supernatants were tested for prorenin and renin content 48 h after transfection.

prorenin by PC1.[142] This finding, combined with the fact that PC1 is expressed at high levels in AtT-20 cells,[64] makes it likely that the human prorenin-processing activity in AtT-20 cells is PC1. Whether or not the human homolog of PC1/PC3 colocalizes with prorenin in human JG cells remains to be determined.

V. FUTURE DIRECTIONS

What have we learned from studies to date concerning the sorting and processing of prorenins? First, the regulated secretory pathway in JG cells exhibits unique features, particularly with regard to its content of para-crystalline structures containing prorenin and the striking similarity of its secretory granules to lysosomes. In spite of these unique characteristics, prorenins contain information which direct unrelated neuroendocrine cells to sort them to the dense core secretory granules. Several questions remain however: are the sorting signals identified in model cell systems functional in JG cells? Is the secretory granule sorting event mediated by interaction of prorenin with a ''sortase'' receptor or by aggregation of prorenin in the TGN? Are there unique features of the sorting process for prorenin in the kidney which would make it amenable to pharmacologic intervention in an effort to control the production of circulating renin?

The second lesson we have learned from these studies is that processing of prorenin can be species-, organelle-, and substrate-specific. In the case of rat renal renin, processing occurs at a different amino acid position than that for human renin. In addition, mouse SMG prorenin is processed in the Golgi by an enzyme which will not cleave renal prorenin. Human prorenin can be activated *in vitro* by a widely distributed lysosomal protease (cathepsin B), but active renin is only secreted in the circulation from the kidney. In spite of these complexities, *in vitro* processing studies and model cell culture systems are giving us information which is useful in the ultimate characterization of the processing of prorenin in the kidney. *In vitro* assays are allowing the purification and characterization of candidate processing enzymes. In addition, prorenin molecules containing site-directed mutations provide a powerful tool to distinguish between processing activities, as different proteases appear to require different amino acids in the immediate area of the processing site. Finally, transfection studies in cultured cells will determine whether candidate processing enzymes carry out their putative function in the appropriate cellular compartment. In combination, these approaches will provide extremely useful information in designing experiments to directly test the activity of potential processing enzymes in JG cells.

Nevertheless, the lack of established cultures of fully differentiated JG cells in which prorenin sorting and processing determinants can be directly tested remains a major impediment in this field. For this reason, novel approaches to the development of JG cell cultures, such as by targeted oncogenesis in transgenic animals, should remain a high priority. In the meantime, targeted expression of native and mutated prorenin molecules to the JG cells of transgenic mice, which have the demonstrated ability to activate and secrete both rat and human prorenin,[129,130] should provide an alternative means of further characterizing the intracellular processes which determine the secretion of active renin.

ACKNOWLEDGMENTS

Because of the breadth of the subject matter covered in this review, it was not possible to cite all pertinent references. Rather, we have attempted to cite recent reviews and illustrative examples to guide the reader in a more in-depth analysis; we apologize to any of our colleagues whose contributions have not been cited for this reason. The authors wish to thank Vivianne Jodoin for typing the manuscript and Drs. Gaetan Thibault and Djamel Ramla for critical reading of the manuscript.

This work was supported by National Institutes of Health grant HL35706 and grants MA-11179, PG-2 and MA-11170 from the Medical Research Council of Canada. T.L.R. is the recipient of the Merck Frosst Canada, Inc. Chair in Clinical and Molecular Pharmacology.

REFERENCES

1. **Tiegerstedt, R. and Bergman, P.G.,** Niere und Kreislauf, *Skand. Arch. Physiol.*, 8, 223, 1898.
2. **Baxter, J.D., Perloff, D., Hsueh, W., and Biglieri, E.G.,** The endocrinology of hypertension, in *Endocrinology and Metabolism*, Felig, P.F., Baxter, J.D., Broadus, A.H., and Frohman, L.A., Eds., McGraw Hill, New York, 1987, 693.
3. **Dzau, V.J., Gibbons, G.H., and Pratt, R.E.,** Molecular mechanisms of vascular renin-angiotensin system in myointimal hyperplasia, *Hypertension*, 18, II100, 1991.
4. **Schelling, P., Fischer, H., and Ganten, D.,** Angiotensin and cell growth: a link to cardiovascular hypertrophy?, *J. Hypertension*, 9, 3, 1991.
5. **Campbell, J.H., Tachas, G., Black, M.J., Cockerill, G., and Campbell, G.R.,** Molecular biology of vascular hypertrophy, *Basic Res. Cardiol.* 86 (Suppl. 1), 3, 1991.
6. **Rodicio, J.L., Praga, M., Alcazar, J.M., Oliet, A., Gutierrez-Millet, V., and Ruilope, L.M.,** Effects of angiotensin converting enzyme inhibitors on the progression of renal failure and proteinuria in humans, *J. Hypertension Suppl.*, 7, S43, 1989.
7. **Clozel, J.P., Kuhn, H., and Hefti, F.,** Effects of chronic ACE inhibition on cardiac hypertrophy and coronary vascular reserve in spontaneously hypertensive rats with developed hypertension, *J. Hypertension*, 7, 267, 1989.
8. **Zannad, F.,** The emerging role of ACE inhibitors in the treatment of cardiovascular disease, J. *Cardiovasc. Pharmacol.*, 15 (Suppl. 2), S1, 1990.
9. **Schölkens, B.A. and Linz, W.,** Cardioprotective effects of ACE inhibitors: experimental proof and clinical perspectives, *Clin. Physiol. Biochem.* 8 (Suppl. 1), 33, 1990.
10. **Chapman, A.B., Gabow, P.A., and Schrier, R.W.,** Reversible renal failure associated with angiotensin-converting enzyme inhibitors in polycystic kidney disease, *Ann. Intern. Med.*, 115, 769, 1991.

11. **Haber, E.,** Why renin inhibitors?, *J. Hypertension Suppl.* 7, S81, 1989.
12. **Corvol, P., Chauveau, D., Jeunemaître, X., and Ménard, J.,** Human renin inhibitor peptides, *Hypertension,* 16, 1, 1990.
13. **Cody, R.J.,** Renin system inhibition: beginning the fourth epoch, *Circulation,* 85, 362, 1992.
14. **Kleinert, H.D., Baker, W.R., and Stein, H.H.,** Renin inhibitors, *Adv. Pharmacol.,* 22, 207, 1991.
15. **Samani, N.J.,** Molecular biology of the renin-angiotensin system: implications for hypertension and beyond, *J. Cardiovasc. Pharmacol.,* 18 (Suppl. 2), S1, 1991.
16. **Hann, B.C. and Walter, P.,** The signal recognition particle in S. cerevisiae, *Cell,* 67, 131, 1991.
17. **Gething, M.-J.,** Molecular chaperones: individuals or groupies?, *Curr. Opinion Cell Biol.,* 3, 610, 1991.
18. **Klausner, R.D. and Sitia, R.,** Protein degradation in the endoplasmic reticulum, *Cell,* 62, 611, 1990.
19. **Bonifacino, J.S. and Lippincott-Schwartz, J.,** Degradation of proteins within the endoplasmic reticulum, *Curr. Opinion Cell Biol.,* 3, 592, 1991.
20. **Rothman, J.E. and Orci, L.,** Molecular dissection of the secretory pathway, *Nature,* 355, 409, 1992.
21. **Goud, B. and McCaffrey, M.,** Small GTP-binding proteins and their role in transport, *Curr. Opinion Cell Biol.,* 3, 626, 1991.
22. **Rothman, J.E. and Orci, L.,** Movement of proteins through the Golgi stack: a molecular dissection of vesicular transport, *FASEB J.,* 4, 1460, 1990.
23. **Kemp, B.E. and Pearson, R.B.,** Protein kinase recognition sequence motifs, *TIBS,* 15, 342, 1990.
24. **Wold, F.,** In vivo chemical modification of proteins, *Annu. Rev. Biochem.,* 50, 783, 1981.
25. **Orci, L., Ravazzola, M., Amherdt, M., Perrelet, A., Powell, S.K., Quinn, D.L., and Moore, H.P.,** The trans-most cisternae of the Golgi complex: a compartment for sorting of secretory and plasma membrane proteins, *Cell,* 51, 1039, 1987.
26. **Tooze, S.A. and Huttner, W.B.,** Cell-free protein sorting to the regulated and constitutive secretory pathways, *Cell,* 60, 837, 1990.
27. **Pfeffer, S.R. and Rothman, J.E.,** Biosynthetic protein transport and sorting by the endoplasmic reticulum and Golgi, *Annu. Rev. Biochem.,* 56, 829, 1987.
28. **Figura, K. von,** Molecular recognition and targeting of lysosomal proteins, *Curr. Opinion Cell Biol.,* 3, 642, 1991.
29. **Kornfeld, S.,** Lysosomal enzyme targeting, *Biochem. Soc. Trans.,* 18, 367, 1990.
30. **Baranski, T.J., Faust, P.L., and Kornfeld, S.,** Generation of a lysosomal enzyme targeting signal in the secretory protein pepsinogen, *Cell,* 63, 281, 1990.
31. **Baranski, T.J., Koelsch, G., Hartsuck, J.A., and Kornfeld, S.,** Mapping and molecular modeling of a recognition domain for lysosomal enzyme targeting, *J. Biol. Chem.,* 266, 23365, 1991.
32. **Chrispeels, M.J. and Raikhel, N.V.,** Short peptide domains target proteins to plant vacuoles, *Cell,* 68, 613, 1992.
33. **Wieland, F.T., Gleason, M.L., Serafini, T.A., and Rothman, J.E.,** The rate of bulk flow from the endoplasmic reticulum to the cell surface, *Cell,* 50, 289, 1987.
34. **Kelly, R.B.,** Secretory granule and synaptic vesicle formation, *Curr. Opinion Cell Biol.,* 3, 654, 1991.
35. **Kelly, R.B.,** Microtubules, membrane traffic, and cell organization, *Cell,* 61, 5, 1990.
36. **Moore, H.P. and Kelly, R.B.,** Secretory protein targeting in a pituitary cell line: differential transport of foreign secretory proteins to distinct secretory pathways, *J. Cell Biol.,* 101 (5 Pt 1), 1773, 1985.
37. **Miller, S.G. and Moore, H.-P.,** Regulated secretion, *Curr. Opinion Cell Biol.,* 2, 642, 1990.

38. **Moore, H.P., Walker, M.D., Lee, F., and Kelly, R.B.,** Expressing a human proinsulin cDNA in a mouse ACTH-secreting cell. Intracellular storage, proteolytic processing, and secretion on stimulation, *Cell,* 35, 531, 1983.
39. **Moore, H.H. and Kelly, R.B.,** Re-routing of a secretory protein by fusion with human growth hormone sequences, *Nature,* 321, 443, 1986.
40. **Stoller, T.J. and Shields, D.,** The propeptide of preprosomatostatin mediates intracellular transport and secretion of alpha-globin from mammalian cells, *J. Cell Biol.,* 1647, 1655, 1989.
41. **Chung, K.N., Walter, P., Aponte, G.W., and Moore, H.P.,** Molecular sorting in the secretory pathway, *Science,* 243, 192, 1989.
42. **Reaves, B.J. and Dannies, P.S.,** Is a sorting signal necessary to package proteins into secretory granules? *Mol. Cell. Endocrinol.,* 79, C141, 1991.
43. **Tooze, S.A.,** Biogenesis of secretory granules: implications arising from the immature secretory granule in the regulated pathway of secretion, *FEBS Lett.,* 285, 220, 1991.
44. **Orci, L., Ravazzola, M., and Anderson, R.G.,** The condensing vacuole of exocrine cells is more acidic than the mature secretory vesicle, *Nature,* 326, 77, 1987.
45. **Burgess, T.L. and Kelly, R.B.,** Constitutive and regulated secretion of proteins, *Annu. Rev. Cell Biol.,* 3, 243, 1987.
46. **Matsuuchi, L. and Kelly, R.B.,** Constitutive and basal secretion from the endocrine cell line, AtT-20, *J. Cell Biol.,* 112, 843, 1991.
47. **von Zastrow, M. and Castle, J.D.,** Protein sorting among two distinct export pathways occurs from the content of maturing exocrine storage granules, *J. Cell Biol.,* 105, 2675, 1987.
48. **Casanova, J.E., Apodaca, G., and Mostov, K.E.,** An autonomous signal for basolateral sorting in the cytoplasmic domain of the polymeric immunoglobulin receptor, *Cell,* 66, 65, 1991.
49. **Arvan, P. and Lee, J.,** Regulated and constitutive protein targeting can be distinguished by secretory polarity in thyroid epithelial cells, *J. Cell Biol.,* 112, 365, 1991.
50. **Harris, R.B.,** Processing of pro-hormone precursor proteins, *Arch. Biochem. Biophys.,* 275, 315, 1989.
51. **Darby, N.J. and Smyth, D.G.,** Endopeptidases and prohormone processing, *Biosci. Rep.,* 10, 1, 1990.
52. **Andrews, P.C., Brayton, K.A., and Dixon, J.E.,** Post-translational proteolytic processing of precursors to regulatory peptides, in *Regulatory Peptides,* Polak, J.M., Ed., Birkhäuser Verlag, Basel, 1989, 192.
53. **Barr, P.J.,** Mammalian subtilisins: the long-sought dibasic processing endoproteases, *Cell,* 66, 1, 1991.
54. **Judah, J.D. and Quinn, P.S.,** Calcium ion-dependent vesicle fusion in the conversion of proalbumin to albumin, *Nature,* 271, 384, 1978.
55. **Orci, L., Ravazzola, M., Storch, M.J., Anderson, R.G., Vassalli, J.D., and Perrelet, A.,** Proteolytic maturation of insulin is a post-Golgi event which occurs in acidifying clathrin-coated secretory vesicles, *Cell,* 49, 865, 1987.
56. **Thibault, G., Garcia, R., Gutkowska, J., Bilodeau, J., Lazure, C., Seidah, N.G., Chrétien, M., Genest, J., and Cantin, M.,** The propeptide Asn1-Tyr126 is the storage form of rat atrial natriuretic factor, *Biochem. J.,* 241, 265, 1987.
57. **Thibault, G., Lazure, C., Schiffrin, E.L., Gutkowska, J., Chartier, L., Garcia, R., Seidah, N.G., Chrétien, M., Genest, J., and Cantin, M.,** Identification of a biologically active circulating form of rat atrial natriuretic factor, *Biochem. Biophys. Res. Commun.,* 130, 981, 1985.
58. **Russell, J.H. and Geller, D.M.,** The structure of rat proalbumin, *J. Biol. Chem.,* 250, 3409, 1975.
59. **Loh, Y.P.,** Kinetic studies on the processing of human β-lipotropin by bovine pituitary intermediate lobe pro-opiomelanocortin-converting enzyme, *J. Biol. Chem.,* 261, 11949, 1986.

60. **Julius, D., Brake, A., Blair, L., Kunisawa, R., and Thorner, J.,** Isolation of the putative structural gene for the lysine-arginine-cleaving endopeptidase required for processing of yeast prepro-alpha-factor, *Cell,* 37, 1075, 1984.
61. **Thomas, G., Thorne, B.A., Thomas, L., Allen, R.G., Hruby, D.E., Fuller, R., and Thorner, J.,** Yeast KEX2 endopeptidase correctly cleaves a neuroendocrine prohormone in mammalian cells, *Science,* 241, 226, 1988.
62. **Bathurst, I.C., Brennan, S.O., Carrell, R.W., Cousens, L.S., Brake, A.J., and Barr, P.J.,** Yeast KEX2 protease has the properties of a human proalbumin converting enzyme, *Science,* 235, 348, 1987.
63. **Kiefer, M.C., Tucker, J.E., Joh, R., Landsberg, K.E., Saltman, D., and Barr, P.J.,** Identification of a second human subtilisin-like protease gene in the fes/fps region of chromosome 15, *DNA Cell Biol.,* 10, 757, 1991.
64. **Seidah, N.G., Gaspar, L., Mion, P., Marcinkiewicz, M., Mbikay, M., and Chrétien, M.,** cDNA sequence of two distinct pituitary proteins homologous to Kex2 and furin gene products: tissue-specific mRNAs encoding candidates for pro-hormone processing endoproteases, *DNA Cell Biol.,* 9, 415, 1990.
65. **Holm, I., Ollo, R., Panthier, J.-J., and Rougeon, F.,** Evolution of aspartyl proteases by gene duplication: the mouse renin gene is organized in two homologous clusters of four exons, *EMBO J.,* 3, 557, 1984.
66. **Fukamizu, A., Nishi, K., Cho, T., Saitoh, M., Nakayama, K., Ohkubo, H., Nakanishi, S., and Murakami, K.,** Structure of the rat renin gene, *J. Mol. Biol.,* 201, 443, 1988.
67. **Hobart, P.M., Fogliano, M., O'Connor, B.A., Schaefer, I.M., and Chirgwin, J.M.,** Human renin gene: structure and sequence analysis, *Proc. Natl. Acad. Sci. U.S.A.,* 81, 5026, 1984.
68. **Miyazaki, H., Fukamizu, A., Hirose, S., Hayashi, T., Hori, H., Ohkubo, H., Nakanishi, S., and Murakami, K.,** Structure of the human renin gene, *Proc. Natl. Acad. Sci. U.S.A.,* 81, 5999, 1984.
69. **Hardman, J.A., Hort, Y.J., Catanzaro, D.F., Tellam, J.T., Baxter, J.D., Morris, B.J., and Shine, J.,** Primary structure of the human renin gene, *DNA,* 3, 457, 1984.
70. **Burnham, C.E., Hawelu-Johnson, C.L., Frank, B.M., and Lynch, K.R.,** Molecular cloning of rat renin cDNA and its gene, *Proc. Natl. Acad. Sci. U.S.A.,* 84, 5605, 1987.
71. **Murakami, K., Hirose, S., Miyazaki, H., Imai, T., Hori, H., Hayashi, T., Kageyama, R., Ohkubo, H., and Nakanishi, S.,** Complementary DNA sequences of renin. State-of-the-art review, *Hypertension,* 6, I95, 1984.
72. **Panthier, J.J. and Rougeon, F.,** Kidney and submaxillary gland renins are encoded by two non-allelic genes in Swiss mice, *EMBO J.,* 2, 675, 1983.
73. **Imai, T., Miyazaki, H., Hirose, S., Hori, H., Hayashi, T., Kageyama, R., Ohkubo, H., Nakanishi, S., and Murakami, K.,** Cloning and sequence analysis of cDNA for human renin precursor, *Proc. Natl. Acad. Sci. U.S.A.,* 80, 7405, 1983.
74. **Katz, S.A., Malvin, R.L., Lee, J., Kim, S.H., Murray, R.D., Opsahl, J.A., and Abraham, P.A.,** Analysis of active renin heterogeneity, *Proc. Soc. Exp. Biol. Med.,* 197, 387, 1991.
75. **Hosoi, M., Kim, S., and Yamamoto, K.,** Evidence for heterogeneity of glycosylation of human renin obtained by using lectins, *Clin. Sci.,* 81, 393, 1991.
76. **Faust, P.L., Chirgwin, J.M., and Kornfeld, S.,** Renin, a secretory glycoprotein, acquires phosphomannosyl residues, *J. Cell Biol.,* 105, 1947, 1987.
77. **Shier, D.N. and Malvin, R.L.,** Differential secretion and removal of multiple renin forms, *Am. J. Physiol.,* 249, R79, 1985.
78. **Kim, S., Hiruma, M., Ikemoto, F., and Yamamoto, K.,** Importance of glycosylation for hepatic clearance of renal renin, *Am. J. Physiol.,* 255, E642, 1988.
79. **Paul, M., Nakamura, N., Pratt, R.E., and Dzau, V.J.,** Glycosylation influences intracellular transit time and secretion rate of human prorenin in transfected cells, *J. Hypertension Suppl.,* 6, S487, 1988.

80. **Baxter, J.D., James, M.N., Chu, W.N., Duncan, K., Haidar, M.A., Carilli, C.T., and Reudelhuber, T.L.,** The molecular biology of human renin and its gene, *Yale J. Biol. Med.,* 62, 493, 1989.
81. **Hori, H., Yoshino, T., Ishizuka, Y., Yamauchi, T., and Murakami, K.,** Role of N-linked oligosaccharides attached to human renin expressed in COS cells, *FEBS Lett.,* 232, 391, 1988.
82. **Kim, S., Hosoi, M., Kikuchi, N., and Yamamoto, K.,** Amino-terminal amino acid sequence and heterogeneity in glycosylation of rat renal renin, *J. Biol. Chem.,* 266, 7044, 1991.
83. **Sielecki, A.R., Hayakawa, K., Fujinaga, M., Murphy, M.E., Fraser, M., Muir, A.K., Carilli, C.T., Lewicki, J.A., Baxter, J.D., and James, M.N.,** Structure of recombinant human renin, a target for cardiovascular-active drugs, at 2.5 Å resolution, *Science,* 243, 1346, 1989.
84. **Hsueh, W.A. and Baxter, J.D.,** Human prorenin, *Hypertension,* 17, 469, 1991.
85. **Hosoi, M., Kim, S., Tabata, T., Nishitani, H., Nishizawa, Y., Morii, H., Murakami, K., and Yamamoto, K.,** Evidence for the presence of differently glycosylated forms of prorenin in the plasma of anephric man, *J. Clin. Endocrinol. Metab.,* 74, 680, 1992.
86. **Suzuki, F., Ludwig, G., Hellmann, W., Paul, M., Lindpainter, K., Murakami, K., and Ganten, D.,** Renin gene expression in rat tissues: a new quantitative assay method for rat renin mRNA using synthetic cRNA, *Clin. Exp. Hypertension, A,* 10, 345, 1988.
87. **Kim, S., Hosoi, M., Nakajima, K., and Yamamoto, K.,** Immunological evidence that kidney is primary source of circulating inactive prorenin in rats, *Am. J. Physiol.,* 260, E526, 1991.
88. **Taugner, R., Bührle, Ch. Ph., and Nobiling, R.,** Ultrastructural changes associated with renin secretion from the juxtaglomerular apparatus of mice, *Cell Tissue Res.,* 237, 459, 1984.
89. **Taugner, R., Metz, R., and Rosivall, L.,** Macroautophagic phenomena in renin granules, *Cell Tissue Res.,* 251, 229, 1988.
90. **Taugner, R. and Hackenthal, E.,** On the character of the secretory granules in juxtaglomerular epithelioid cells, *Int. Rev. Cytol.,* 110, 93, 1988.
91. **Matsuba, H., Watanabe, T., Watanabe, M., Ishii, Y., Waguri, S., Kominami, E., and Uchiyama, Y.,** Immunocytochemical localization of prorenin, renin, and cathepsins B, H, and L in juxtaglomerular cells of rat kidney, *J. Histochem. Cytochem.,* 37, 1689, 1989.
92. **Taugner, R., Kim, S.J., Murakami, K., and Waldherr, R.,** The fate of prorenin during granulopoiesis in epithelioid cells. Immunocytochemical experiments with antisera against renin and different portions of the renin prosegment, *Histochemistry,* 86, 249, 1987.
93. **Kawamura, M., McKenzie, J.C., Hoffman, L.H., Tanaka, I., Parmentier, M., and Inagami, T.,** The storage form of renin in renin granules from rat kidney cortex, *Hypertension,* 8, 706, 1986.
94. **Pratt, R.E., Carleton, J.E., Richie, J.P., Heusser, C., and Dzau, V.J.,** Human renin biosynthesis and secretion in normal and ischemic kidneys, *Proc. Natl. Acad. Sci. U.S.A.,* 84, 7837, 1987.
95. **Toffelmire, E.B., Slater, K., Corvol, P., Menard, J., and Schambelan, M.,** Response of plasma prorenin and active renin to chronic and acute alterations of renin secretion in normal humans. Studies using a direct immunoradiometric assay, *J. Clin. Invest.,* 83, 679, 1989.
96. **Do, Y.S., Shinagawa, T., Tam, H., Inagami, T., and Hsueh, W.A.,** Characterization of pure human renal renin, *J. Biol. Chem.,* 262, 1037, 1987.
97. **Misono, K.S., Chang, J.-J., and Inagami, T.,** Amino acid sequence of mouse submaxillary gland renin, *Proc. Natl. Acad. Sci. U.S.A.,* 79, 4858, 1982.
98. **Campbell, D.J., Valentijn, A.J., and Condron, R.,** Purification and amino-terminal sequence of rat kidney renin: evidence for a two-chain structure, *J. Hypertension,* 9, 29, 1991.

99. **Pratt, R.E., Ouellette, A.J., and Dzau, V.J.,** Biosynthesis of renin: multiplicity of active and intermediate forms, *Proc. Natl. Acad. Sci. U.S.A.*, 80, 6809, 1983.
100. **Johns, D.W., Carey, R.M., Gomez, R.A., Lynch, K., Inagami, T., Saye, J., Geary, K., Farnsworth, D.E., and Peach, M.J.,** Isolation of renin-rich rat kidney cells, *Hypertension*, 10, 488, 1987.
101. **Pinet, F., Corvol, M.T., Dench, F., Bourguignon, J., Feunteun, J., Ménard, J., and Corvol, P.,** Isolation of renin-producing human cells by transfection with three simian virus 40 mutants, *Proc. Natl. Acad. Sci. U.S.A.*, 82, 8503, 1985.
102. **Pinet, F., Mizrahi, J., Laboulandine, I., Ménard, J., and Corvol, P.,** Regulation of prorenin secretion in cultured human transfected juxtaglomerular cells, *J. Clin. Invest.*, 80, 724, 1987.
103. **Sigmund, C.D., Jones, C.A., Fabian, J.R., Mullins, J.J., and Gross, K.W.,** Tissue and cell specific expression of a renin promoter-reporter gene construct in transgenic mice, *Biochem. Biophys. Res. Commun.*, 170, 344, 1990.
104. **Sigmund, C.D., Okuyama, K., Ingelfinger, J., Jones, C.A., Mullins, J.J., Kane, C., Kim, U., Wu, C.Z., Kenny, L., Rustum, Y. et al.,** Isolation and characterization of renin-expressing cell lines from transgenic mice containing a renin-promoter viral oncogene fusion construct, *J. Biol. Chem.*, 265, 19916, 1990.
105. **Fritz, L.C., Haidar, M.A., Arfsten, A.E., Schilling, J.W., Carilli, C., Shine, J., Baxter, J.D., and Reudelhuber, T.L.,** Human renin is correctly processed and targeted to the regulated secretory pathway in mouse pituitary AtT-20 cells, *J. Biol. Chem.*, 262, 12409, 1987.
106. **Pratt, R.E., Flynn, J.A., Hobart, P.M., Paul, M., and Dzau, V.J.,** Different secretory pathways of renin from mouse cells transfected with the human renin gene, *J. Biol. Chem.*, 263, 3137, 1988.
107. **Landenheim, R.G., Seidah, N., Lutfalla, G., and Rougeon, F.,** Stable and transient expression of mouse submaxillary gland renin cDNA in AtT20 cells: proteolytic processing and secretory pathways, *FEBS Lett.*, 245, 70, 1989.
108. **Nagahama, M., Nakayama, K., and Murakami, K.,** Sequence requirements for prohormone processing in mouse pituitary AtT-20 cells: analysis of prorenins as model substrates, *Eur. J. Biochem.*, 197, 135, 1991.
109. **Chidgey, M.A. and Harrison, T.M.,** Renin is sorted to the regulated secretory pathway in transfected PC12 cells by a mechanism which does not require expression of the propeptide, *Eur. J. Biochem.*, 190, 139, 1990.
110. **Hatsuzawa, K., Hosaka, M., Nakagawa, T., Nagase, M., Shoda, A., Murakami, K., and Nakayama, K.,** Structure and expression of mouse furin, a yeast Kex2-related protease. Lack of processing of coexpressed prorenin in GH4C1 cells, *J. Biol. Chem.*, 265, 22075, 1990.
111. **Chu, W.N., Baxter, J.D., and Reudelhuber, T.L.,** A targeting sequence for dense core secretory granules resides in the protein moiety of human prorenin, *Mol. Endocrinol.*, 4, 1905, 1990.
112. **Nakayama, K., Nagahama, M., Kim, W.S., Hatsuzawa, K., Hashiba, K., and Murakami, K.,** Prorenin is sorted into the regulated secretory pathway independent of its processing to renin in mouse pituitary AtT-20 cells, *FEBS Lett.*, 257, 89, 1989.
113. **Nagahama, M., Nakayama, K., and Murakami, K.,** Effects of propeptide deletion on human renin secretion from mouse pituitary AtT-20 cells, *FEBS Lett.*, 264, 67, 1990.
114. **Nakayama, K., Hatsuzawa, K., Kim, W.S., Hashiba, K., Yoshino, T., Hori, H., and Murakami, K.,** The influence of glycosylation on the fate of renin expressed in Xenopus oocytes, *Eur. J. Biochem.*, 191, 281, 1990.
115. **Do, Y.S., Shinegawa, T., Tam, S., Inagami, T., and Hsueh, W.A.,** Characterization of pure human renal renin: evidence for a subunit structure, *J. Biol. Chem.*, 262, 1037, 1987.
116. **Frohlich, E.D., Iwata, T., and Sasaki, O.,** Clinical and physiologic significance of local tissue renin-angiotensin systems, *Am. J. Med.*, 87 (Suppl. 6B), 6B-19S, 1990.

117. **Wang, P.H., Do, Y.S., Macaulay, L., Shinagawa, T., Anderson, P.W., Baxter, J.D., and Hsueh, W.A.,** Identification of renal cathepsin B as a human prorenin-processing enzyme, *J. Biol. Chem.*, 266, 12633, 1991.
118. **Heinrikson, R.L., Hui, J., Zürcher-Neely, H., and Poorman, R.A.,** A structural model to explain the partial catalytic activity of human prorenin, *Am. J. Hypertension*, 2, 367, 1989.
119. **Shinagawa, T., Do, Y.S., Baxter, J.D., Carilli, C., Schilling, J., and Hsueh, W.A.,** Identification of an enzyme in human kidney that correctly processes prorenin, *Proc. Natl. Acad. Sci. U.S.A.*, 87, 1927, 1990.
120. **San Segundo, B., Chan, S.J., and Steiner, D.F.,** Differences in cathepsin B mRNA levels in rat tissues suggest specialized functions, *FEBS Lett.*, 201, 251, 1986.
121. **Saint-André, J.P., Rohmer, V., Pinet, F., Rousselet, M.C., Bigorgne, J.C., and Corvol, P.,** Renin and cathepsin B in human pituitary lactotroph cells. An ultrastructural study, *Histochemistry*, 91, 291, 1989.
122. **Nakayama, K., Kim, W.-S., Nakagawa, T., Nagahama, M., and Murakami, K.,** Substrate specificity of prorenin converting enzyme of mouse submandibular gland, *J. Biol. Chem.*, 265, 21027, 1990.
123. **Kim, W.S., Nakayama, K., Nakagawa, T., Kawamura, Y., Haraguchi, K., and Murakami, K.,** Mouse submandibular gland prorenin-converting enzyme is a member of the glandular kallikrein family, *J. Biol. Chem.*, 266, 19283, 1991.
124. **Kim, W.S., Nakayama, K., and Murakami, K.,** The presence of two types of prorenin converting enzymes in the mouse submandibular gland, *FEBS Lett.*, 293, 142, 1991.
125. **Quinn, P.S. and Judah, J.D.,** Calcium-dependent Golgi-vesicle fusion and cathepsin B in the conversion of proalbumin into albumin in rat liver, *Biochem. J.*, 172, 301, 1978.
126. **Docherty, K., Hutton, J.C., and Steiner, D.F.,** Cathepsin B-related proteases in the insulin secretory granule, *J. Biol. Chem.*, 259, 6041, 1984.
127. **Chu, W.N., Mercure, C., Baxter, J.D., and Reudelhuber, T.L.,** Molecular determinants of human prorenin processing, *Hypertension*, 20, 782, 1992.
128. **Benjannet, S., Reudelhuber, T., Mercure, C., Rondeau, N., Chrétien, M., and Seidah, N.G.,** Pro-protein conversion is determined by a multiplicity of factors including convertase processing, substrate specificity and intracellular environment, *J. Biol. Chem.*, 267, 11417, 1992.
129. **Ohkubo, H., Kawakami, H., Kakehi, Y., Takumi, T., Arai, H., Yokota, Y., Iwai, M., Tanabe, Y., Masu, M., Hata, J. et al.,** Generation of transgenic mice with elevated blood pressure by introduction of the rat renin and angiotensinogen genes, *Proc. Natl. Acad. Sci. U.S.A.*, 87, 5153, 1990.
130. **Fukamizu, A., Hatae, T., Kon, Y., Sugimura, M., Hasegawa, T., Yokoyama, M., Nomura, T., Katsuki, M., and Murakami, K.,** Human renin in transgenic mouse kidney is localized to juxtaglomerular cells, *Biochem. J.*, 278, 601, 1991.
131. **Dickerson, I.M., Dixon, J.E., and Mains, R.E.,** Transfected human neuropeptide Y cDNA expression in mouse pituitary cells, *J. Biol. Chem.*, 262, 13646, 1987.
132. **Comb, M., Liston, D., Martin, M., Rosen, H., and Herbert, E.,** Expression of the human proenkephalin gene in mouse pituitary cells: accurate and efficient mRNA production and proteolytic processing, *EMBO J.*, 4(12), 3115, 1985.
133. **Sevarino, K.A., Goodman, R.H., Spiess, J., Jackson, I.M., and Wu, P.,** Thyrotropin-releasing hormone (TRH) precursor processing. Characterization of mature TRH and non-TRH peptides synthesized by transfected mammalian cells, *J. Biol. Chem.*, 264, 21529, 1989.
134. **Burgess, T.L., Craik, C.S., Matsuuchi, L., and Kelly, R.B.,** In vitro mutagenesis of trypsinogen: role of the amino terminus in intracellular protein targeting to secretory granules, *J. Cell Biol.*, 105, 659, 1987.
135. **Lapps, W., Eng, J., Stern, A.S., and Gubler, U.,** Expression of porcine cholecystokinin cDNA in a murine neuroendocrine cell line. Proteolytic processing, sulfation, and regulated secretion of cholecystokinin peptides, *J. Biol. Chem.*, 263, 13456, 1988.

136. **Cwikel, B.J. and Habener, J.F.,** Provasopressin-neurophysin II processing is cell-specific in heterologous cell lines expressing a metallothionein-vasopressin fusion gene, *J. Biol. Chem.*, 262, 14235, 1987.
137. **Shields, P.P., Sprenkle, A.B., Taylor, E.W., and Glembotski, C.C.,** Rat pro-atrial natriuretic factor expression and post-translational processing in mouse corticotropic pituitary tumor cells, *J. Biol. Chem.*, 265, 10905, 1990.
138. **Matsuuchi, L., Buckley, K.M., Lowe, A.W., and Kelly, R.B.,** Targeting of secretory vesicles to cytoplasmic domains in AtT-20 and PC-12 cells, *J. Cell Biol.*, 106, 239, 1988.
139. **Deschepper, C.F. and Reudelhuber, T.L.,** Rat angiotensinogen is secreted only constitutively when transfected into AtT-20 cells, *Hypertension*, 16, 147, 1990.
140. **Reudelhuber, T.L.,** unpublished results.
141. **Mercure, C. and Reudelhuber, T.L.,** unpublished observations.
142. **Mercure, C., Reudelhuber, T.L., and Seidah, N.,** unpublished observations.
143. **Su, Y.X. and Reudelhuber, T.L.,** unpublished observations.

Chapter 2

THE REGULATION OF RENAL AND EXTRARENAL RENIN GENE EXPRESSION IN THE MOUSE

C.A. Jones, J.R. Fabian, K.J. Abel, C.D. Sigmund, and K.W. Gross

TABLE OF CONTENTS

0-8493-4622-3/93/$0.00 + $.50

I. INTRODUCTION

From the earliest observations indicating the existence of a renal pressor substance,[1] we have come to appreciate that renin is an aspartyl protease which participates in the regulation of systemic blood pressure and electrolyte balance through its fundamental role in the renin-angiotensin system (RAS).[2] Classical systemic renin is produced and secreted by modified intrarenal arterial smooth muscle cells (juxtaglomerular [JG] cells) in response to appropriate physiological and neurological signals.[3-6] Clearly, a detailed understanding of the mechanisms governing how these signals are transduced within the JG cell to regulate the expression and elaboration of renin would assist our comprehension of the biology of the RAS and its role in arterial pressure regulation. Unfortunately, while considerable insight into general features of regulation has been gained from functional studies of the cell *in situ,* direct analysis of fundamental processes, such as gene transcription, has been hampered by the relative paucity of renin-expressing cells in the kidney and the lack of suitable cell culture models.

In addition to the unresolved issues of renal renin expression, it has also become apparent that renin or renin-like activities or immunoreactive renin can be detected at a number of extrarenal sites.[7-14] While the variety of these sites and the precise role subserved by renin expression at these extrarenal sites is not yet clear, the provocative observation that such expression can be found in association with other components of the classic RAS has prompted speculation on, and interest in, the existence of extrarenal tissue-renin-angiotensin systems.[15-19]

Indeed, it was the fortuitous discovery of unusually high levels of a renin-like activity at one extrarenal site, the submandibular gland (SMG) of mice,[20,21] that has proved to play a paramount role in providing molecular access to the renin genes.[22-28] Inbred strains can be divided into two classes: those that produce high levels of SMG renin, and those that produce significantly lower levels.[21] The two expression phenotypes serve to define two alleles, *Rnr*^{s} and *Rnr*^{b}, at this locus.[29] *Rnr*^{s} (after the type strain SWR) can exhibit SMG renin levels that correspond to as much as 2% of SMG protein.[20,21] Moreover, these abundant levels of renin protein are paralleled by high levels of mRNA.[30,31]

It was this copious abundance of a renin mRNA that facilitated development of the first cDNA clones and ultimately recovery of the corresponding genomic sequences encoding the mouse renin gene(s).[22,23,25,32,33] The availability of these cloned recombinant probes from mouse has permitted in turn isolation of the homologous sequences from other mammals and provided the tools necessary to conclusively demonstrate for the first time primary expression of renin mRNA in a spectrum of other tissues.[13,30,33-45]

Cloning of the mouse renin (*Ren*) sequences was also instrumental in definitively establishing that gene duplication provided the molecular basis for the high renin salivary phenotype and ultimately, the demonstration that the structural genes for renin, lie coincident with the *Rnr* locus on chromosome 1.[22,46,47] We now know that all mice have the *Ren-1* locus which encodes the classical circulating enzyme and has homologs in other species as well. The locus in mouse has two alleles: *Ren-1*^{c} after type strain C57BL/6 which is found in strains with a single renin gene and a low salivary renin phenotype (*Rnr*^{b}), and *Ren-1*^{d} after type strain DBA/2 which is found in strains harboring the renin gene duplication and exhibiting the high salivary renin phenotype (*Rnr*^{s}).[48] The duplicated locus is termed *Ren-2* and is found only in strains with the high salivary renin phenotype.

The species-specific duplication of the renin genes in mouse has been the bane as well as the boon of renin research in this organism. The multiplicity of renin genes in this system has complicated analysis of expression and made it incumbent upon investigators to develop methods to distinguish and quantitate gene-specific expression patterns. Moreover, the existence of the duplication locus has always raised the specter that somehow the mouse was a different and unacceptable model from which to generalize on renin expression and regulation. It may well be, however, that understanding the basis for the differences which superficially appear to distinguish the mouse from other organisms, for example the rat, may in fact provide a key to deriving a deeper understanding of renin's biological role(s) in general. In any event, currently much of our detailed knowledge of renin expression and regulation stems from work performed in mice, and it is evident that the mouse continues to offer an experimentally manipulable system with which to address fundamental issues of renin gene regulation. This reflects its tractability as a classical genetic system as well as the continuing development of a host of sophisticated modern molecular genetic tools, such as transgenic technology and targeted recombination strategies. These combined approaches permit rigorous evaluation of the role and regulation of single or multiple components in complex physiological systems, such as the RAS.

This chapter will review: (1) cellular sites in adult mice where renin mRNA accumulates; (2) spatial, developmental, and hormonal aspects of the differential expression of murine renin genes at renal and extrarenal sites; (3) current knowledge of murine renin gene and transcript structure; (4) what is presently understood about renin gene structure/expression correlations; (5) the use of transfected cells and transgenic mice to identify cis-acting regions of DNA involved in regulating renin expression; and (6) the development of cognate cell lines for examining regulation of renin expression.

II. RENIN GENE EXPRESSION

Since renin is present in the circulation and periarterial fluid, it is important to distinguish whether the presence of renin at a given site is due to uptake from extracellular fluid or serum vs. that actually synthesized at the site.[49,50] For instance, in murine kidneys, renin was detected by immunocytochemistry in cells of the afferent arteriole (JG cells, myoepithelioid cells) and in cells of the proximal tubule.[51-53] However, *in situ* hybridization assays using a renin cDNA probe located renin mRNA in the former cell types but not in the latter, thereby eliminating proximal tubule cells as a site of primary renin synthesis.[54-57]

A variety of direct approaches employing cDNA probes have been used to screen for the primary expression of renin mRNA in other tissues where renin had reportedly been localized. These investigations were able to confirm or expunge tissues as sites of renin transcription. Numerous reports have now clearly demonstrated the primary expression of renin mRNA in: kidney, adrenal gland, submandibular gland, testes, ovary, and coagulating gland using classical Northern blot assays.[30,33,36,37,42,43,58,59] These results demonstrate that the levels of renin mRNA in these tissues are relatively abundant. In addition, however, substantial controversy exists over whether renin mRNA is present in heart and brain as previously reported.[38,39,42,60] Other sites of renin expression have been detected using the highly sensitive but not quantitative polymerase chain reaction (PCR) assay.[45,61,62] These include brain, heart, hypothalamus, spleen, thymus, lung, prostate, and liver. The significance of these results remains unclear and therefore, for the purposes of this review, we will discuss only those sites where renin mRNA is relatively abundant.

A. DIFFERENTIAL RENIN EXPRESSION IN THE ADULT MOUSE

A priori, with the existence of multiple loci in the mouse, it became necessary to clarify the individual contributions of each gene to the expression pattern in each tissue. The *Ren-1*^{c}, *Ren-1*^{d}, and *Ren-2* genes encode highly similar yet unique renin transcripts. By taking advantage of minute sequence differences between them it proved possible to develop methods to discern the tissue specificity of each. These studies have revealed highly complex gene-specific patterns of differential expression. Details of these methods have been previously described.[30,43] The following is a summary of differential renin gene expression in renal and extrarenal tissues.

Kidney — Renin expression in the kidney has been shown to be roughly equivalent among *Ren-1*^{c}, *Ren-1*^{d}, and *Ren-2*.[30] Under normal physiological conditions, the expression is limited to a population of modified smooth muscle cells of the afferent arteriole proximal to the glomerulus (myoepithelioid or JG cells).[4,52,63] The population of cells expressing renin mRNA, as well as the renin mRNA levels in JG cells, can be modulated.[56,58,64-71] Both are induced by conditions of physiological stress such as sodium depletion, pathophysiologic stress such as uretal and renal artery obstruction, as well as by pharmacologic intervention with angiotensin-converting enzyme inhibitors. *In situ* hybridization assays localizing renin transcripts to specific cells have shown that vascular smooth muscle cells in the afferent arterioles and interlobular arteries can be recruited into a renin-expressing phenotype. These vascular smooth muscle cells have been termed intermediate cells because their ultrastructural appearance contains elements of both vascular smooth muscle cells and fully transformed JG cells.[4,5]

Adrenal gland — In adult adrenal glands, *Ren-1*^{d} and *Ren-2* expression is equivalent and higher in females than in males.[35,43] Expression occurs in the X-zone and *zona fasciculata* but undergoes shifts of cell specificity during the estrus cycle (see below). *Ren-1*^{c} is not detectable in adult adrenal glands of either sex at this site by Northern blot or *in situ* hybridization assay.[31,35]

Salivary glands — Submandibular gland *Ren-2* expression is 100-fold higher than *Ren-1*^{c} expression in male SMG as detected by primer extension analysis.[30] Miller et al.[42] reported *Ren-1*^{d} was detectable but was at very low levels by RNase protection assay. In this tissue, renin is expressed by the granular convoluted tubule cell (GCT, a glandular epithelial cell which makes up 20% of the cellular population) and becomes detectable by *in situ* hybridization at puberty. In the female SMG, renin transcripts accumulate to an approximately 5-fold lower level than is seen in males.[22,37,72] The sublingual gland also expresses renin mRNA which has been located to the striated ductal cell by *in situ* hybridization. These cells were similar to GCT cells in that they appeared to have comparable levels of transcripts on a per cell basis. Male sublingual glands of mice with *Ren-1*^{d} and *Ren-2* genes exhibited high levels of renin mRNA, while in females, transcripts were detectable by treatment with androgen. *In situ* hybridization did not detect any cells with renin transcripts in the parotid gland.[31]

Gonads — Renin expression in Leydig cells of the testes is roughly equivalent for *Ren-1*^{c} and *Ren-2* with *Ren-1*^{d} slightly in excess.[36,43] Renin expression in the ovary is equivalent between *Ren-1*^{d} and *Ren-2*.[45] Low levels of *Ren-1*^{c} were also detected by Sigmund and Gross.[111] The renin-expressing cell type in the mouse ovary has not been identified.

Coagulating gland — Renin gene expression is high for *Ren-1*^{c} but undetectable for *Ren-1*^{d} and *Ren-2*.[43,59] Renin expression in this tissue is limited to the glandular epithelial cells.

B. DEVELOPMENTAL SHIFTS IN RENIN EXPRESSION

In addition to the observed adult patterns of murine renin expression, developmentally regulated expression occurs during organogenesis in the metanephric kidney of mice and rats,[51,53-55,73] the mouse adrenal gland,[14,54] and also in subcutaneous tissues of mice and rats.[74] In the murine fetal kidney, expression of *Ren-1*^{c}, *Ren-1*^{d}, and *Ren-2,* as detected by *in situ* hybridization, is first observed 14.5 days post coitum (pc).[31,54] By 15.5 days pc, renin expression is clearly visible in cells surrounding the lumens of early-forming intrarenal arteries. This expression shifts with the newer portions of the elongating arteries, while the more mature portions of the vascular tree lose the ability to express renin. Expression becomes progressively restricted so that by one week of age, the expression sites become similar to those seen in adult kidney.

Fetal adrenal gland expression is characterized by accumulation of high levels of renin transcripts in both males and females. This stands in contrast to the adult adrenal gland where expression in females of *Ren-1*^{d} and *Ren-2* genes is higher than in males. Interestingly, *Ren-1*^{c}, which is not detectable in adult adrenal gland, is expressed at comparable levels to *Ren-1*^{d} and *Ren-2* at this time. Again, expression is first visible by *in situ* hybridization at 14.5 days pc.[31,54] Expression is located throughout the entire gland except for the outermost cell layers, as judged by accumulation of silver grains over the tissue. By 16.5 days, expression appears less intense, becomes limited to the cortical region and is clearly absent in the medulla. *Ren-1*^{c} expression disappears by birth, reflecting a developmental downregulation of steady-state levels of renin transcripts while detectable renin expression persists in strains with both *Ren-1*^{d} and *Ren-2*.[35]

Renin expression can also be detected by *in situ* hybridization in the testes during fetal development.[54] Accumulation of silver grains in this fetal tissue is lower relative to the kidneys and adrenal glands. The testes, along with the kidney and the adrenal gland derive from the same limited region of the intermediate mesoderm.[75,76] This leads one to speculate that perhaps some event predisposes cells derived from this embryonic tissue with the potential to express renin, provided the cells then follow defined paths of differentiation.

Renin expression in murine subcutaneous sites also appears to be developmentally regulated.[74,77] The expression of an SV40 large T antigen reporter gene under the control of renin regulatory elements in transgenic mouse fetuses first suggested the presence of renin at this extrarenal site. The reporter gene was expressed at all fetal sites known to express renin and in a mesenchymal cell type amid the muscle layers directly beneath the developing dermis. Interestingly, the fetal expression pattern of the reporter gene is remarkably similar to the pattern of angiotensin II receptors reported by Zemel et al.,[78] suggesting a possible developmental role for the RAS. The levels of expression were below the limit of sensitivity of *in situ* hybridization. However, Northern blot analysis of decapitated and eviscerated fetal carcasses revealed the presence of renin mRNA consistent with localization to extra-visceral tissue. Sigmund et al.[74] were able to show that transcripts from either the *Ren-1*^{c} or the *Ren-1*^{d} allele accumulated to higher levels than transcripts derived from the *Ren-2* gene.

C. HORMONAL INFLUENCES ON RENIN EXPRESSION

Renin expression in mouse SMG has been shown to be androgen and thyroxine responsive.[20,29,46,79] Wilson et al.[29] investigated renin activity in SMG of female mice with the *Ren-1*^{d}/*Ren-2* genotype. Onset of activity was observed around puberty (3 to 4 weeks of age), reaching maximum basal levels around 7 weeks of age. The basal activity in females treated with dihydrotestosterone was found to be increased 4- to 5-fold compared with untreated females, which is comparable to the levels in the male gland. Likewise, *Ren-1*^{c} exhibits androgen inducibility. Nuclear runoff transcription assays show androgen responsiveness to be a result of increased transcriptional activity, as opposed to merely increased message stability.[72,80,81] Administration of the thyroid hormone thyroxine has the same effect as dihydrotestosterone, resulting in an approximately 5-fold increase in the accumulation of renin mRNA.[72,80,81]

Another interesting fluctuation in renin expression which appears to be under hormonal influence has been observed in the adrenal gland of some closely related inbred strains carrying *Ren-1*[d] and *Ren-2*.[35] Female mice of this genotype exhibit shifts in renin expression between the X-zone and the *zona fasciculata* of the cortex as the animal cycles through estrus. The adrenal gland at various stages of the estrus cycle was examined by *in situ* hybridization and revealed that in proestrus, renin transcripts are evident in both the X-zone and the *zona fasciculata*. During the next stage, estrus, transcripts are found only in the *zona fasciculata*. At metestrus, expression is evident in the X-zone but not the *zona fasciculata*. In diestrus, expression at both locations is evident, with noticeably higher accumulation of transcripts in the *zona fasciculata*. Differential primer extension revealed that the level of *Ren-1*[d] and *Ren-2* remained equivalent at each stage of the estrus cycle. The molecular mechanisms regulating this change in cell specificity remain unclear.

D. INTERSPECIES CONSERVATION OF EXTRARENAL RENIN EXPRESSION

The important role of renal renin expression is widely conserved in vertebrate animals (see Nishimura[82] and Wilson[83] for reviews). The relevance of extrarenal renin expression is not as clear but conservation of expression at a given site suggests these sites may also have an important function. It seems probable that common sites of extrarenal expression found across mammalian species serve an important function in order to have persisted over an evolutionary time scale. Table 1 lists extrarenal tissues where renin mRNA has reportedly been detected by Northern or *in situ* analysis in mice, rats, and humans.

Renin expression in some extrarenal sites is apparently unique to the mouse, raising questions as to its functional relevance. Examples of this are the SMG and the coagulating gland.[84,85] In addition, there are differences in cell specificity among species. For instance, renin expression in the rat adrenal gland is restricted to the zona glomerulosa.[8] On the other hand, renin expression in the mouse adrenal gland is confined to the inner cortical zones, X-zone, and *zona fasciculata*.

III. PHYSICAL STRUCTURE OF THE MOUSE RENIN GENES

A. GENE STRUCTURE

Extensive genetic linkage information has permitted the formation of a relational map of mouse chromosome 1. Using the linkage information, the

TABLE 1
Comparison of Mouse, Rat, and Human Extrarenal Sites of Renin mRNA Synthesis

	Ren-1^{c}	*Ren-1*^{d}	*Ren-2*	Rat	Human
Submandibular gland	A	Negative[a]	A	Negative	No data
Adrenal gland	F	F, A	F, A	A	A
Coagulating gland	A	Negative	Negative	Negative	No data
Testes	F, A	F[b], A	F[b], A	A	No data
Ovary	A	A	A	A	No data
Subcutaneous tissue	F	F	F	F	No data
Chorion	No data	No data	No data	No data	Positive

Note: F, Present in fetal tissue; A, present in adult tissue.

[a] Detectable by RNase protection assay.
[b] Relative contributions of *Ren-1*^{d} and *Ren-2* not determined.

position of the renin structural genes has been established in relation to other chromosomal markers (Figure 1A).[86] Comparison of the three mouse renin genes revealed they have the same intron-exon arrangement and that this region spans a distance of roughly 10 kb (Figure 1B).[87] Sequence comparison of the respective cDNAs revealed that the coding regions of *Ren-1*^{c} and *Ren-1*^{d} are 99% identical, while *Ren-1*^{d} and *Ren-2* are 97% identical.[88] Interestingly, the *Ren-2* gene does not encode any of the potential N-linked glycosylation sites encoded in the *Ren-1* gene (and found in the renin-1 polypeptide).[26,87,89] The lack of glycosylation could explain the thermolability of renin seen in the SMG of mice with the duplicated gene. The mouse, rat, and human renin genes share significant structural organization and sequence similarity with each other. The renin gene coding regions in mouse and rat are approximately 88% identical, while between mouse and human renin genes they are approximately 78% identical.[28] These genes are members of the aspartyl protease family.[87]

The 5′ flanking regions of the mouse renin genes exhibit significant homology.[90-93] The renin genes are homologous for 150 bases upstream from the transcription start site (designated as +1), preserving the TATA boxes (−23 to −29) and a region of alternating purine pyrimidine bases (−30 to −45). Renin promoters do not contain a CAAT box but do have an AT-rich region conserved in mice and humans at approximately −60.

B. DNA INSERTIONS ASSOCIATED WITH MOUSE RENIN LOCI

The availability of rat and human genomic renin sequences permitted other interspecies comparisons of renin genes. These studies revealed that 5′ flanking regions of the mouse renin genes exhibit significant segmental homology with each other and regions of rat and human renin.[90-92,94,95] The segmental nature is due to a number of genetic events including not only the duplication of the gene, but also the presence of numerous insertional elements in the 5′ and 3′ flanking regions of the genes (Figure 1C).[90-93,95-99] Several of these elements are well-characterized repetitive sequences, namely B1, B2, and a partial intracisternal A particle (IAP). Others are anonymous insertions which are arbitrarily referred to as M1, M2, M3, and M4 (M, mouse). At least M1 is known to be repetitive in the mouse genome. The presence or absence of these insertions has helped distinguish between the mouse renin genes and contribute to the segmental homology of the 5′ flanking DNA through their breakup of the primordial flanking regions (Table 2).

The combined approaches of sequence analysis, southern blotting, and pulse field gel electrophoresis (PFGE) studies have permitted comparison of the murine renin genes. It has been found that *Mus hortulanus,* a wild derived variant with two renin loci, lacks M1, M2, M4, and the IAP found in DBA/2J while possessing the M3 and B2 insertions.[94,95] The lack of the insertions in *M. hortulanus* also suggests the insertional events probably occurred after the duplication event in the inbred strains. These genomic inserts have been proposed to have potential influences on renin gene expression. However, *M. hortulanus,* which lacks many of the inserts, showed the same patterns of tissue-specific expression as is seen in DBA/2J. One exception where a lack of insertions in *M. hortulanus* correlates with altered expression as compared with DBA/2J, is in the adult adrenal gland, where no mRNA from either renin gene is detected.

C. STRUCTURE AT THE DUPLICATION LOCI

The unique pattern of tissue-specific expression in mice carrying the *Ren-1*^{d} and *Ren-2* loci and in those with the solitary *Ren-1*^{c} locus has led to efforts to characterize the physical structure of the duplicate locus. Genetic analysis indicates that the two loci are tightly linked.[22,98,99] PFGE helped to determine that *Ren-2* lies upstream of *Ren-1*^{d}, that the respective coding regions are separated by approximately 21 kb and that the two genes are transcribed in the same direction (Figure 1D).[98,99] Abel and Gross,[99] using PFGE and sequence information, determined the precise site of the recombination event that resulted in the gene duplication. Duplication apparently occurred through nonhomologous recombination. The analysis by PFGE also identified clusters of rare cutting restriction enzyme sites (or HTF islands) in the vicinity of the renin gene. HTF islands have been shown to be associated with 5′ regions of many vertebrate genes. *Ren-1*^{d} and *Ren-2* genes have an HTF island at homologous positions in their 3′ flank which apparently has been duplicated along with the *Ren-2* locus. Additional HTF islands are located 21 kb upstream and 65 kb downstream of the locus.

D. TRANSCRIPT STRUCTURE

In the kidney, the size of the mature mouse renin mRNA is approximately 1450 bases.[33] Additional higher molecular weight species of renin mRNA have been detected in SMG and coagulating gland by northern blot analysis.

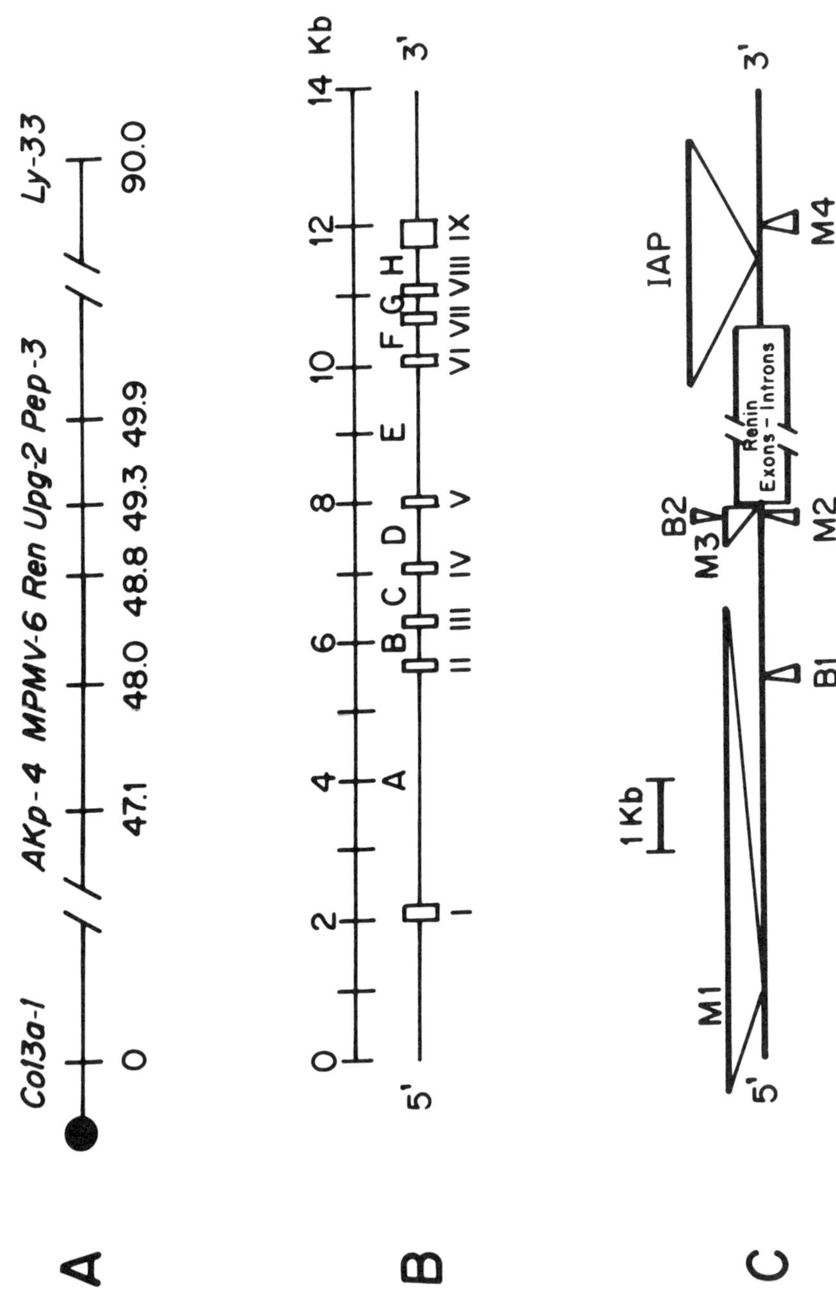
A
Col3a-1
AKp-4
MPMV-6
Ren
Upg-2
Pep-3
Ly-33
0
47.1
48.0
48.8
49.3
49.9
90.0
B
0
2
4
6
8
10
12
14 Kb
5'
3'
A
B
C
D
E
F
G
H
I
II
III
IV
V
VI
VII
VIII
IX
C
1 Kb
M1
B1
M3
B2
M2
Renin
Exons - Introns
IAP
M4
5'
3'

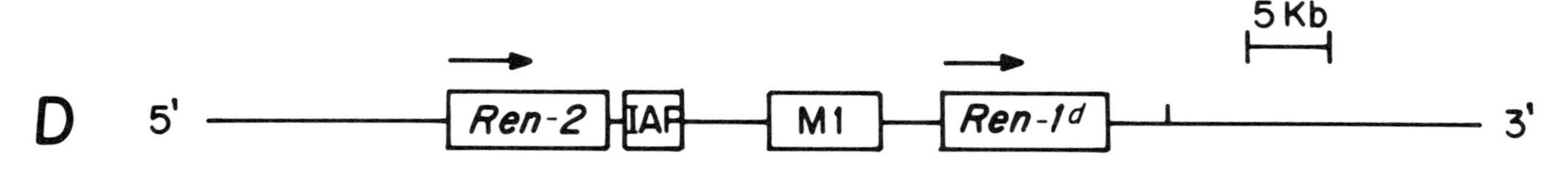

FIGURE 1. (A) Composite linkage map of mouse chromosome 1 illustrating the *Ren* locus and the placement of the closest known flanking loci in cM relative to the centromere. The two loci on the centromeric side of the renin locus *(Ren)* are alkaline phosphatase 4 *(AKp-4)* and modified polytropic murine leukemia virus-6 (*MPMV-6*). The two loci distal to the *Ren* locus are urinary pepsinogen 2 (*Upg-2*) and peptidase 3 (*Pep-3*). The collagen-3-α-1 (*Col3a-1* locus is most proximal to the centromere while lymphocyte antigen-33 (*Ly-33*) is most distal. (Adapted from Seldin et al.[86]) (B) Structure of the transcribed region of murine renin genes. This diagram illustrates the approximate arrangement and size of the exons (roman numerals) and introns (capital letters) of the mouse renin genes. (C) Location of murine insertional elements. This representation illustrates the approximate size and location of the insertional elements relative to the murine renin coding sequences. Refer to Table 2 for gene-specific insertions. (D) The renin locus containing the duplicate gene. This schematic illustrates the placement of the duplicated gene relative to the progenitor gene. Arrows indicate direction of transcription.

TABLE 2
Approximate Size and Location of the Insertional Elements Associated with the Mouse Renin Locus

Insertion	Size	Location	Ref.
M1	7.0 kb	−3.1 kb of *Ren-1*^{d}	98
M2	143 bp	−110 bp of *Ren-2**	92,97,98
M3	500 bp	−80 bp of *Ren-1*^{c}, *Ren-1*^{d}, *Ren-2*	97
M4	300 bp	+1.5 kb of *Ren-1*^{d}	98
B1	180 bp	−1.5 kb of *Ren-1*^{c}, *Ren-1*^{d}, *Ren-2*	96
B2	200 bp	Within the M3 element of *Ren-2*	90
IAP	3.5 kb	+1.0 kb of *Ren-2**	92

Note: A (−) indicates upstream distance from exon I if intervening insertions are not present. A (+) indicates downstream distance from exon IX. * Indicates the element is not associated with the *Ren-2* gene of *Mus hortulanus*.

S1 nuclease protection and primer extension assays have demonstrated the utilization of additional upstream transcriptional start sites.[90] These encode an open reading frame which potentially adds 23 amino acids to the N-terminus of the translated products. It remains unclear if this open reading frame is ever utilized.

IV. STRUCTURE/EXPRESSION CORRELATES

As has been shown, the mouse renin genes exhibit an array of complex expression patterns. Recent reports have speculated that the structural variations noted above may be responsible for several of the gene-specific expression patterns.[93,97,98,100] An opportunity to correlate specific structural features with gene expression patterns is afforded by comparing naturally occurring genetic variants (such as *Mus hortulanus*) with inbred strains (such as DBA/2).[95] Also, genetic crosses as well as transgenic analysis have demonstrated that the gene-specific expression differences are mediated by closely linked sequences in cis.[35,43,46] However, the available information does not satisfactorily limit which regions control the variable expression patterns of the renin genes. Therefore, in order to define the specific identity of the regulatory DNA sequences controlling these variations of expression, investigators have employed direct tests of recombinant DNA constructs in expression assay systems.

A. EXPRESSION ASSAY SYSTEMS

To identify regulatory DNA sequences, it is necessary to systematically examine the effects of discrete regions with an assay system which can directly measure the effects of linked DNA sequences on expression from a particular promoter. These fall into two categories, transfection into established cell lines, and more recently the ability to insert genes via transgenesis.[88,101-103] Each system has its own advantages and disadvantages.

The transgenic approach provides the opportunity to examine expression of transgenes temporally in all tissues, with ensuing physiological feedback regulation. However, this method is time consuming, expensive, and labor intensive. Furthermore, the integration of the transgene into a chromosome is a random event and therefore the site of insertion cannot be controlled. The chromosomal environment around the integration site can have significant influences on transgene expression; a position effect. Therefore, it becomes necessary to examine multiple independent founder lines for each transgenic construct to determine whether the pattern of transgene expression is being controlled by elements of the transgene or by endogenous flanking elements.

The transfection approach allows the rapid testing of many different DNA constructs, as will be illustrated below. This should facilitate systematic examination of DNA sequences derived from large regions known to regulate expression. Ideally, the assay cell line should elaborate the trans-acting factors which promote the transcription of the endogenous renin gene (a cognate cell). Until recently, there have been no suitable established cell lines available for fulfilling this criterion. Previous attempts at establishing cells which express renin *in vitro* have been unsuccessful because the resulting cells often lose the ability to express renin.

B. TRANSGENIC ASSAYS FOR IDENTIFICATION OF CIS-ACTING ELEMENTS

To date, several groups have undertaken informative studies utilizing transgenic animals containing various renin genes and constructs. Initial experiments centered on reconstructing two renin gene type mice from a single transgene on a *Ren-1*^{c} genetic background. Tronik et al.[104] used a *Ren-2* transgene with 2.5 kb of upstream flanking sequence, the exon-intron region, and 3 kb of downstream flanking sequence. The *Ren-2* transgene was expressed in a quantitative tissue-specific manner; and *Ren-2* expression in the SMG was inducible by androgen. Mullins et al.[35,105] performed similar studies using a *Ren-2* transgene with a more extensive upstream and downstream flanking sequence (approximately 5 kb of 5′ flank, the exon-intron region, and approximately 10 kb of 3′ flank). They found qualitatively similar results to those reported with the less extensive *Ren-2* transgene. This group was also able to extend these observations to the adrenal gland, where they showed that the estrus cycle-specific effects on *Ren-2* expression could be partially reconstituted in the transgenic mice. In a similar set of studies Miller et al.[42] examined the expression of a *Ren-1*^{d} transgene (spanning approximately 19 kb with approximately 5 kb of 5′ flank, the exon-intron region, and approximately 4 kb of 3′ flank) in a *Ren-1*^{c} genetic background. They showed that the expression differences between *Ren-1*^{c} and *Ren-1*^{d} were encoded in cis. All these studies are in agreement with genetic studies and support the notion that the DNA sequences conferring tissue specificity and hormonal regulation lie close to the body of the structural genes. This was evidenced by the conservation of gene-specific patterns of transgene expression in the different genetic backgrounds.

While the above experiments were gratifying, they failed to resolve the specific localization of important sequence determinants controlling tissue-specific expression. In order to further refine the location of these elements, different segments of the *Ren-2* 5′ flanking region were fused to an easily detectable reporter gene and used to construct transgenic mice. Tissue- and cell-specific expression of the reporter gene were analyzed to determine if the renin sequence could confer a tissue-specific expression profile. One approach to accomplish this has been to produce transgenic mice with constructs made by fusing renin 5′ flanking sequences to the coding region of SV40 large T antigen.[57,106] T antigen provides a reporter function because the transcripts are unique to the mouse and T antigen protein can be detected by conventional immunohistochemistry. The fact that T antigen is sequestered within the nucleus aids in the detection of cells which have it. An additional feature of T antigen is that it can predispose the cells in which it is expressed to transformation and thereby selectively amplify a small population of cells into a more easily detectable mass.

Sola et al.[106] fused the T antigen coding sequence to 2.5 kb of *Ren-2* sequence including the transcription initiating site and extending 5′ (−2500 to +7). They found that this transgene was not restricted to the correct spectrum of renin-expressing tissues but was in fact expressed in a number of inappropriate tissues and not in kidney or SMG. Using the same approach, Sigmund et al.[57] fused approximately 4.6 kb of *Ren-2* 5′ flank (−4600 to +6) to T antigen coding sequence and found that appropriate tissue- and cell-specific expression patterns were observed. Comparison of the results obtained with these two constructs suggests that important cis-acting elements, conferring cell- and tissue-specific expression are contained in the region between −2.5 and −4.6 kb of *Ren-2*. Furthermore, comparison of expression patterns of the *Ren-2*/T antigen fusion transgene of Sola et al.[106] and the *Ren-2* genomic transgene of Tronik et al.[104] reveals differences in the specificity of expression. While the 2.5 kb of 5′ flanking region was sufficient to confer appropriate tissue-specific expression in the genomic transgene,[104] it was insufficient for the fusion transgene.[106] This suggests that redundant regulatory elements exist within the *Ren-2* exon-intron and/or 3′ flanking regions and that their presence confers the proper tissue-specific expression pattern of the *Ren-2* gene in transgenes that have only 2.5 kb of *Ren-2* 5′ flanking sequences. The existence of redundant regulatory elements needs to be confirmed experimentally.

C. TRANSFECTION ASSAYS FOR IDENTIFICATION OF CIS-ACTING ELEMENTS

A second approach towards identifying cis-acting regulatory elements is through transfection in tissue culture. Routinely, this is performed by fusing various segments of a gene's promoter region upstream of a reporter gene sequence. Transfection assays of expression controlled by renin regulatory sequences have made use of chloramphenicol acetyl transferase (CAT) as a reporter gene. Previously these constructs have been introduced into a variety of noncognate cells (i.e., cells incapable of expressing their endogenous renin genes). However, expression data from established noncognate cell lines has limited applicability for deducing the role of cis-acting elements *in situ* and therefore the results obtained should be interpreted with caution.

Ekker and co-workers[107] found that promoters consisting of approximately 0.5 kb of *Ren-1*^{d} (−449 to +30) or 2.5 kb of *Ren-2* (−2500 to +7) were inactive in five different noncognate cell lines unless an SV40 enhancer (that has been shown to be active in these cells) was also present in the fusion construct. Therefore, the promoters could correctly initiate transcription only when directed by a functioning enhancer. They also reported that selected renin 5′ flanking sequences (−2500 to +7 of *Ren-2*) did not enhance the function of the SV40 or thymidine kinase promoters in the noncognate cells that were tested. Taken together, these experiments suggested that the failure of the enhancerless renin promoter-CAT constructs to express was due to a lack of renin-specific positive trans-acting factor(s) in the noncognate cells. Similarly, renin promoters exhibited equivalent transcriptional activity regardless of whether they were using 2500 bp of *Ren-2* or 500 or 180 bp (−150 to +30) of *Ren-1*^{d}. Additionally, no increase in promoter activity was noted when progressive deletions of *Ren-1*^{d} were made toward the transcriptional start site. This led them to conclude that an element with repressor-like function was not present in these regions.

In contrast, Nakamura et al.[100] reported that *Ren-1*^{d} sequences (−707 to −367), when fused to Tk-CAT in the sense orientation, repressed CAT activity greater than 60% when compared with the Tk-CAT construct alone. These authors suggested the presence of a negative regulatory element (NRE) in this *Ren-1*^{d} 5′ flanking region. The modulation of expression by these upstream mouse renin sequences was further investigated by Barrett et al.[108] They found two putative NRE sequences in both *Ren-1*^{d} and *Ren-2* by consensus comparison to sequences of chicken lysozyme silencer elements. In *Ren-1*^{d} these were located at −619 to −597 and at −557 to −544. The *Ren-1*^{d} (−707 to −367) or *Ren-2* (−1055 to −571) fragments, with and without the NRE sequences, were transfected into the noncognate JEG-3 cells. Constructs without the putative upstream NRE returned CAT activity to the levels of the Tk-CAT construct while deletion of the putative downstream NRE did not. Therefore, they concluded that only the upstream NRE was functional in *Ren-1*^{d}.

Interestingly, in *Ren-2* this NRE sequence is positioned next to the M2 insertion discussed above. Barrett et al.[108] suggested that disruption of the NRE sequence by the M2 element may explain the basis for the high levels of *Ren-2* in SMG. However, this supposition must be applied cautiously to regulation of expression *in vivo*. Recall *Ren-1*^{d} does not contain the M2 element nor is it expressed in the SMG. Also, as described by Abel et al.,[95] the M2 insert is absent in the *M. hortulanus Ren-2* allele. Yet, this *Ren-2* allele is expressed at a high level in the SMG despite the absence of the M2 element, leaving the apparently intact NRE.

Finally, as an additional caveat, all the transfection studies described above were in noncognate cells. It should be pointed out that these experiments are done with relatively small portions of renin upstream sequences which may not contain all of the elements needed to reflect full control of regulation *in vivo*. Recall that *Ren-2* 5′ flanking sequences between −4.6 and −2.5 were required for correct expression in transgenic mice. The renin upstream regions may interact differently with the heterologous promoters used in these experiments than with the homologous renin promoter.

V. RENIN-EXPRESSING CELL LINES

A. COGNATE CELLS DERIVED BY TRANSGENE-MEDIATED TUMORIGENESIS

Previous transfection studies have been performed in cell lines incapable of expressing their endogenous renin gene. This has severely limited our ability to assess the contribution of DNA sequence elements in the control of renin expression and to examine DNA-protein interactions at the renin locus. Recently, Sigmund et al.[109] have isolated a cell line (As4.1) from a mouse transgenic for a renin promoter-SV40 T antigen fusion construct. The cells not only expressed the transgene but their endogenous *Ren-1*^{c} gene as well.

The construct used to produce the transgenic mice, from which the cells were isolated, contained 4.6 kb of *Ren-2* 5′ flanking sequence fused to the coding region of SV40 large T antigen. As discussed above, this construct exhibited an appropriate tissue- and cell-specific expression profile in transgenic mice. Independent founder lines with this construct exhibited a varied spectrum of tumors in kidney, adrenal gland, testes, coagulating gland, and subcutaneous tissue. The cells were isolated from an individual with a unilateral kidney tumor. The precise cellular origin of this line could not be determined because the involved kidney was no longer identifiable. However, the tumor-derived cell line has features consistent with it being of JG cell origin.[101,109] These features include: (1) high levels of renin mRNA accumulation; (2) the presence of dense granules, as determined by electron microscopy, which are of similar size and morphology as those found in JG cells; (3) intracellular active renin; (4) immunoreactive renin in rounded structures (as determined by immunofluorescence) of a size and distribution consistent with the dense granules; (5) secretion of prorenin and active renin, as determined by activity assay; and (6) modulation of renin mRNA levels by culture conditions (decreasing serum in the culture medium leads to increasing accumulation of renin mRNA). These characteristics suggest this cell line will provide an excellent system with which to study the regulatory role of renin 5′ flanking DNA sequences, as well as the cellular processing mechanisms of renin protein.

B. TRANSCRIPTIONAL REGULATION IN As4.1 CELLS

A prerequisite for a gene to be transcribed is that it be free to interact with the transcriptional machinery of the cell. At the level of chromatin, this means that sequences to be transcribed, or that promote or enhance transcription, cannot be sequestered within nucleosomal complexes. Genes capable of being actively transcribed in a given cell type are often associated with sites of DNase I hypersensitivity. Presumably, these sites reflect an open chromatin structure where trans-acting factors interact with cis-acting elements. The As4.1 cells provided an opportunity to compare and contrast chromatin structure around the renin locus of renin-expressing cells with non-renin-expressing cells (Ltk-cells), which also possess a *Ren-1*ᶜ gene, but do not express it.[110] Three DNase I hypersensitive sites were found specific for the As4.1 cell in the 5′ flank of the gene at −3.4, −2.8, and −2.3 kb. Moreover, homologously positioned DNase hypersensitive sites were evident in the 4.6 kb of *Ren-2* 5′ flanking sequence in the renin/T antigen fusion transgene.

Since the As4.1 cells can transcribe their endogenous renin gene, these cells can be presumed to produce the complement of transcription factors required for expression of the gene and therefore provide a useful system in which to study transcriptional regulation. Transcriptional analysis has been performed with constructs consisting of a promoter, various renin 5′ flanking sequences and the CAT coding sequences.[110] For example, control constructs containing renin promoter and SV40 enhancer, which is active in a wide variety of cell types, were examined. These were transiently transfected into the As4.1 cells and into L cells. The results are shown in Figure 2. As expected, a high level of CAT activity was seen in both cell types when transfected with a construct containing the SV40 promoter-enhancer (row I). Only a background level of activity was seen in cells transfected with constructs lacking a functional promoter region (row II). The minimal renin promoter (−117 to +6) was active in both cell types as evidenced by moderate levels of CAT activity. The minimal renin promoter in the reverse orientation exhibited only background activity (rows IV and VI). High levels of CAT activity were seen in As4.1 cells but not L cells transfected with a CAT construct containing 4.1 kb of *Ren-1*ᶜ gene 5′ flanking sequence (row V). This strongly suggests the presence of a cell-specific enhancer element upstream of *Ren-1*ᶜ.

VI. CONCLUSIONS

Results from nuclease hypersensitivity analyses and transfection assays using the As4.1 cognate cell line are internally consistent with previous findings in transgenic mice. This is reassuring and provides compelling support

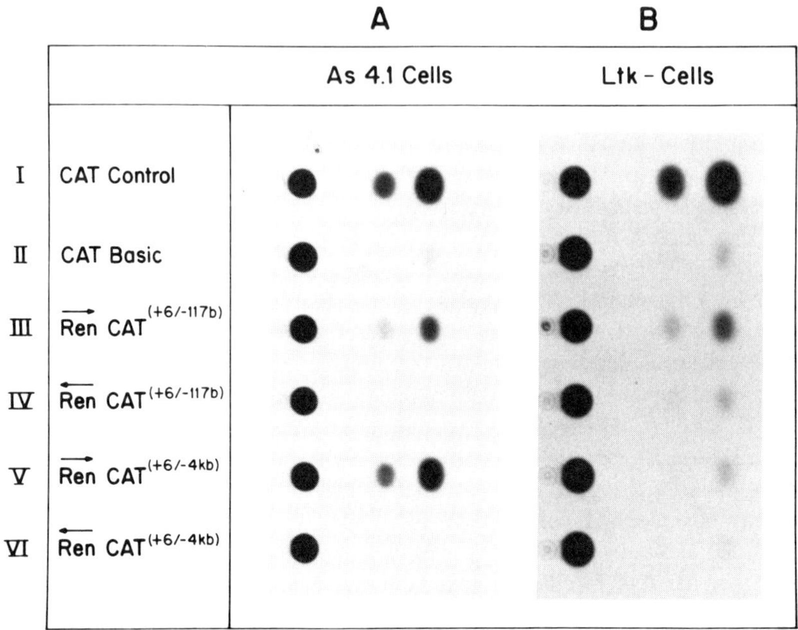

FIGURE 2. Cell-specific activation of Ren-CAT expression vectors. CAT coding sequences with or without SV40 promoter/enhancer elements and CAT coding sequences fused to the indicated renin 5′ flanking sequences were transfected into As4.1 or LtK cells. CAT activity was assayed 48 to 60 h after transfection.

for the notion that sequences located distal to the promoter are in fact required for tissue- and cell-specific expression of the mouse renin genes. It would thus appear that the As4.1 cell line provides a suitable model with which to determine the cis-acting elements and trans-acting factors that affect expression of the mouse renin genes at a key *in vivo* site of common expression. This should markedly facilitate delineation of these important mechanistic features and, when joined judiciously with testing of selected constructs in transgenic assays, establish a firm base for further studies. Indeed it will be of interest to ascertain whether these combined assays can expedite localization of the corresponding elements of renin genes from other species. The generalized evolutionary conservation of such important regulatory mechanisms and some very preliminary direct assessments suggest this will be the case.

These studies also highlight and reaffirm the utility of transgene-mediated oncogenesis as a means of developing cell lines to facilitate studies of renin gene regulation. Since mice harboring the renin/T antigen fusion gene also develop tumors in a spectrum of additional extrarenal sites of renin gene expression, it should be feasible to establish a library of renin-expressing cell lines representative of different cell specificities. Such cells should be invaluable tools for identifying the cis-acting elements mediating tissue-specific expression at these sites and dissecting the basis for the gene-specific expression observed at these extrarenal sites.

Finally, studies thus far indicate that the As4.1 cell line exhibits several salient and critical features appropriate to and expected of a JG cell model. It remains to be ascertained whether it will reliably model or afford insight into additional features such as the relevant receptor-mediated signal transduction pathways or processing and secretion pathways for the renin polypeptide.

Clarification of these issues will entail continual intercomparison between the cell as isolated and cells *in situ* in the organism. Nevertheless, the range and power of the current technology augers favorably that significant insights and advances will be forthcoming.

ACKNOWLEDGMENTS

We wish to thank Mary K. Ellsworth, Chuan Zhen Wu, Colleen Kane, and Frank Pacholec for excellent technical assistance and past members and colleagues of the Gross laboratory for their contributions. Aspects of the presented work were supported by NIH grant numbers HL-35792, GM-30248, NIH Fellowship HL-07963 (C.D.S.), and NIH Biomedical Research Grant SO7 RR-05648-23. We also wish to acknowledge Marcia Held, Cheryl Mrowczynski, and Mary Ketcham for expert secretarial assistance and J. Pablo Abonia and Dr. Anil Ratty for valuable contributions in preparation of this manuscript.

REFERENCES

1. **Tigersted, R. and Bergman, P.D.,** Niere und Kreislauf, *Skand. Arch. Physiol.*, 8, 223, 1898.
2. **Johnston, C.I.,** Biochemistry and pharmacology of the renin-angiotensin system, *Drugs,* 39(1), 21, 1990.
3. **Goormaghtigh, N.,** Existence of an endocrine gland in the media of the renal arterioles, *Proc. Soc. Exp. Biol. Med.*, 42, 688, 1939.
4. **Hackenthal, E., Paul, M., Ganten, D., and Taugner, R.,** Morphology, physiology, and molecular biology of renin secretion, *Physiol. Rev.*, 70, 1067, 1990.
5. **Taugner, R. and Hackenthal, E.,** *The Juxtaglomerular Apparatus,* Springer-Verlag, Berlin, 1989.
6. **Barajas, L. and Salido, E.,** Editorial: Juxtaglomerular apparatus and the renin-angiotensin system, *Lab. Invest.*, 54(4), 361, 1986.
7. **Taylor, G.M., Cook, H.T., Hanson, C., Peart, W.S., Zondek, T., and Zondek, L.H.,** Renin in human fetal lung — a biochemical and immunohistochemical study, *J. Hypertens.*, 6, 845, 1988.
8. **Deschepper, C.F., Mellon, S.H., Cumin, F., Baxter, J.D., and Ganong, W.F.,** Analysis by immunocytochemistry and in situ hybridization of renin and its mRNA in kidney, testis, adrenal and pituitary of the rat, *Proc. Natl. Acad. Sci. U.S.A.*, 83, 7552, 1986.

9. **Cohen, S., Taylor, J.M., Murakami, K., Michelakis, A.M., and Inagami, T.,** Isolation and characterization of renin-like enzymes from mouse submaxillary glands, *Biochemistry,* 11, 4286, 1972.
10. **Naruse, K., Murakoshi, M., Osamura, Y., Naruse, M., Toma, H., Watanabe, K., Demura, H., Inagami, T., and Shizume, K.,** Immunological evidence for renin in human endocrine tissues, *J. Clin. Endocrinol. Metab.,* 61, 172, 1985.
11. **Naruse, K., Takii, Y., and Inagami, T.,** Immunohistochemical localization of renin in lutenizing hormone-producing cells of rat pituitary, *Proc. Natl. Acad. Sci. U.S.A.,* 78, 7579, 1981.
12. **Howard, R.B., Purcell, A.G., Bumpus, F.M., and Hussain, A.,** Rat ovarian renin: characterization and changes during the estrous cycle, *Endocrinology,* 123, 2331, 1988.
13. **Kim, S.J., Shinjo, M., Fukamizu, A., Miyasaki, H., Usuki, S., and Murakami, K.,** Identification of renin and renin messenger RNA sequence in rat ovary and uterus, *Biochem. Biophys. Res. Commun.,* 142, 169, 1987.
14. **Kon, Y., Hashimoto, Y., Kitagawa, H., Sugimura, M., and Murakami, K.,** Renin immunohistochemistry in the adrenal gland of the mouse fetus and neonate, *Anat. Rec.,* 227, 124, 1990.
15. **Jin, M., Wilhelm, M.J., Lang, R.E., Unger, T., Lindpaintner, K., and Ganten, D.,** Endogenous tissue renin-angiotensin systems; from molecular biology to therapy, *Am. J. Med.,* 84(3A), 28, 1988.
16. **Sealey, J.E., Glorioso, N., Itskovitz, J., and Laragh, J.H.,** Prorenin as a reproductive hormone. New form of the renin system, *Am. J. Med.,* 81, 1041, 1986.
17. **Dzau, V.J.,** Circulating versus local renin-angiotensin system in cardiovascular homeostasis, *Circulation,* 77 (Suppl. I), I4, 1988.
18. **Lindpaintner, K., Wilhelm, M.J., Jin, M., Unger, T., Lang, R.E., Scholkens, B.A., and Ganten, D.,** Tissue renin-angiotensin systems: focus on the heart, *J. Hypertens.,* 5, 33, 1987.
19. **Dzau, V.J., Rosenthal, J., and Swales, J.D.,** Vascular renin — a consensus view, *J. Hypertens. Suppl.,* 5, S77, 1987.
20. **Wilson, C.H., Erdos, E.G., Dunn, J.F., and Wilson, J.D.,** Genetic control of renin activity in the submaxillary gland of the mouse, *Proc. Natl. Acad. Sci. U.S.A.,* 74, 1185, 1977.
21. **Bing, J., Poulsen, K., Hackenthal, E., Rix, E., and Taugner, R.,** Renin in the submaxillary gland: a review, *J. Histochem. Cytochem.,* 28, 874, 1980.
22. **Piccini, N., Knopf, J.L., and Gross, K.W.,** A DNA polymorphism, consistent with gene duplication, correlates with high renin levels in the mouse submaxillary gland, *Cell,* 30, 205, 1982.
23. **Rougeon, F., Chambraud, B., Foote, S., Panthier, J.J., Nageotte, R., and Corvol, P.,** Molecular cloning of a mouse submaxillary gland renin cDNA fragment, *Proc. Natl. Acad. Sci. U.S.A.,* 78, 6367, 1981.
24. **Mullins, J.J., Burt, D.W., Windass, J.D., McTurk, P., George, H., and Brammar, W.J.,** Molecular cloning of two distinct renin genes from the DBA/2 mouse, *EMBO J.,* 1, 1461, 1982.
25. **Imai, T., Miyazaki, H., Hirose, S., Hori, H., Hayashi, T., Kageyawa, R., Ohkubo, H., Nakanishi, S., and Murakami, K.,** Cloning and sequence analysis of cDNA for human renin precursor, *Proc. Natl. Acad. Sci. U.S.A.,* 80, 7405, 1983.
26. **Panthier, J.J. and Rougeon, F.,** Kidney and submaxillary gland renins are encoded by two non-allelic genes in Swiss mice, *EMBO J.,* 2, 675, 1983.
27. **Burnham, C.E., Hawelu-Johnson, C.L., Frank, B.M., and Lynch, K.R.,** Molecular cloning of rat renin cDNA and its gene, *Proc. Natl. Acad. Sci. U.S.A.,* 84, 5605, 1987.
28. **Hardman, J., Hort, Y., Catanzaro, D.F., Tellam, J., Baxter, J.D., Morris, B.J., and Shine, J.,** Primary structure of the human renin gene, *DNA,* 3, 457, 1984.

29. **Wilson, C.M., Cherry, M., Taylor, B.A., and Wilson, J.D.,** Genetic and endocrine control of renin activity in the submaxillary gland of the mouse, *Biochem. Genet.*, 19, 5, 1981
30. **Field, L.J. and Gross, K.W.,** Ren-1 and Ren-2 loci are expressed in mouse kidney, *Proc. Natl. Acad. Sci. U.S.A.*, 82, 6196, 1985.
31. **McGowan, R.A.,** An Analysis of the Temporal and Spatial Expression of Renin mRNA in the Mouse (thesis), Roswell Park Division of SUNY Buffalo, Buffalo, NY, 1987.
32. **Matsuda, T., Imai, T., Fukushi, T., Sudoh, M., Hirose, S., and Murakami, K.,** Molecular cloning of DNA complementary to mouse submandibular gland renin mRNA, *Biomed. Res.*, 3, 541, 1982.
33. **Field, L.J., McGowan, R.A., Dickinson, D.P., and Gross, K.W.,** Tissue and gene-specificity of mouse renin expression, *Hypertension,* 6, 597, 1984.
34. **Naruse, M., Susson, C.R., Naruse, K., Jackson, R.V., and Inagami, T.,** Renin exists in human adrenal tissue, *J. Clin. Endocrinol. Metab.*, 57, 462, 1983.
35. **Mullins, J.J., Sigmund, C.D., Kane-Haas, C., McGowan, R.A., and Gross, K.W.,** Expression of the murine Ren-2 gene in the adrenal gland of transgenic mice, *EMBO J.*, 8, 4065, 1989.
36. **Pandey, K.N., Maki, M., and Inagami, T.,** Detection of renin mRNA in mouse testis by hybridization with renin cDNA probe, *Biochem. Biophys. Res. Commun.*, 125, 662, 1984.
37. **Catanzaro, D.F., Mesterovic, N., and Morris, B.J.,** Studies of the regulation of mouse renin genes by measurement of renin messenger ribonucleic acid, *Endocrinology,* 117, 872, 1985.
38. **Dzau, V.J. and Re, R.N.,** Evidence for the existence of renin in the heart, *Circulation,* 75 (Suppl. I), I134, 1987.
39. **Dzau, V.J., Ellison, K.E., Brody, T., Ingelfinger, J., and Pratt, R.E.,** A comparative study of the distributions of renin and angiotensinogen messenger ribonucleic acids in rat and mouse tissues, *Endocrinology,* 120, 2334, 1987.
40. **Dzau, V.J., Brody, K.E., Ellison, R.E., Pratt, E., and Ingelfinger, J.R.,** Tissue specific regulation of renin expression in the mouse, *Hypertension,* 9 (Suppl. 3), 36, 1987.
41. **Lightman, A., Deschepper, C.F., Mellon, S.H., Ganong, W.G., and Naftolin, F.,** In situ hybridization identifies renin mRNA in the rat corpus luteum, *Gynecol. Endocrinol.*, 1, 227, 1987.
42. **Miller, C.C.J., Carter, A.T., Brooks, J.I., Badge, R.H.L., and Brammar, W.J.,** Differential extra-renal expression of the mouse renin genes, *Nucleic Acids Res.*, 17, 3117, 1989.
43. **Fabian, J., Field, L.J., McGowan, R.A., Mullins, J.J., Sigmund, C.D., and Gross, K.W.,** Allele specific expression of the murine Ren-1 genes, *J. Biol. Chem.*, 64, 17589, 1989.
44. **Brecher, A.S., Shier, D.N., Dene, H., Wang, S.H., Rapp, J.P., Franco-Saenz, R., and Mulrow, P.J.,** Regulation of adrenal renin messenger ribonucleic acid by dietary sodium chloride, *Endocrinology,* 124, 2907, 1989.
45. **Ekker, M., Tronik, D., and Rougeon, F.,** Extra-renal transcription of the renin genes in multiple tissues of mice and rats, *Proc. Natl. Acad. Sci. U.S.A.*, 86, 5155, 1989.
46. **Wilson, C.M., Erdos, E.G., Wilson, J.D., and Taylor, B.A.,** Location on chromosome 1 of Rnr, a gene that regulates renin in the submaxillary gland of the mouse, *Proc. Natl. Acad. Sci. U.S.A.*, 75, 5623, 1978.
47. **Chirgwin, J.M., Schaefer, I.M., Diaz, J.A., and Lalley, P.A.,** Mouse kidney renin gene is on chromosome one, *Somatic Cell Mol. Genet.*, 10, 633, 1984.
48. **Dzau, V., Baxter, J., Cantin, M., deBold, A., Ganten, D., Gross, K., Husain, A., Inagami, T., Menard, J., Poole, S., Robertson, J.I.S., Tang, J., and Yamamoto, K.,** A report of the Joint Nomenclature and Standardization Committee of the International Society of Hypertension, the American Heart Association and the World Health Organization, *Hypertension,* 10, 461, 1987.

49. **Swales, J.D. and Heagerty, A.M.,** Vascular renin-angiotensin system: the unanswered questions, *Hypertension,* 5 (Suppl. 2), S1, 1987.
50. **Kriz, W.,** A periarterial pathway for intrarenal distribution of renin, *Kidney Int.,* 31 (Suppl. 20), S-51, 1987.
51. **Minuth, M., Hackenthal, E., Poulsen, K., Rix, E., and Taugner, R.,** Renin immunocytochemistry of the differentiating juxtaglomerular apparatus, *Anat. Embryol.,* 162, 173, 1981.
52. **Taugner, R., Hackenthal, E., Inagami, T., Nobiling, R., and Poulsen, K.,** Vascular and tubular renin in the kidneys of mice, *Histochemistry,* 75, 473, 1982.
53. **Richoux, J.P., Amsaguine, S., Grignon, G., Bouhnik, J., Menard, J., and Corvol, P.,** Earliest renin containing cell differentiation during ontogenesis of the rat, *Histochemistry,* 88, 41, 1987.
54. **Jones, C.A., Sigmund, C.D., McGowan, R., Kane-Haas, C., and Gross, K.W.,** Temporal and spatial expression of the murine renin genes during fetal development, *Mol. Endocrinol.,* 4, 375, 1990.
55. **Gomez, R.A., Lynch, K.R., Sturgill, B.C., Elwood, J.P., Chevalier, R.L., Carey, R.M., and Peach, M.J.,** Distribution of renin mRNA and its protein in the developing kidney, *Am. J. Physiol.,* 257, F850, 1989.
56. **Sigmund, C.D., Jones, C.A., Jacob, H., Ingelfinger, J., Kim, U., Gamble, D., Dzau, V.J., and Gross, K.W.,** Pathophysiology of vascular smooth muscle in renin promoter-T antigen transgenic mice, *Am. J. Physiol.,* 260, F249, 1991.
57. **Sigmund, C.D., Jones, C.A., Fabian, J.R., Mullins, J.J., and Gross, K.W.,** Tissue and cell-specific expression of a renin promoter-T antigen reporter gene construct in transgenic mice, *Biochem. Biophys. Res. Commun.,* 170, 344, 1990.
58. **Miller, C.C.J., Samani, N.J., Carter, A.T., Brooks, J.I., and Brammar, W.J.,** Modulation of mouse renin gene expression by dietary sodium chloride intake in one-gene, two-gene and transgenic animals, *J. Hypertens.,* 7, 861, 1989.
59. **Fabian, J.F., Kane, C.M., Abel, K.J., and Gross, K.W.,** Expression of the mouse Ren-1 gene in the coagulating gland; localization and regulation, in preparation.
60. **Dzau, V.J., Ingelfinger, J., Pratt, R.E., and Ellison, K.E.,** Identification of renin and angiotensinogen messenger RNA sequences in mouse and rat brains, *Hypertension,* 8, 544, 1986.
61. **Lou, Y.K., Smith, D.L., Robison, B.G., and Morris, B.J.,** Renin gene expression in various tissues determined by single-step polymerase chain reaction, *Clin. Exp. Pharmacol. Physiol.,* 18, 357, 1991.
62. **Okura, T., Kitami, Y., Iwata, T., and Hiwada, K.,** Quantitative measurement of extrarenal renin mRNA by polymerase chain reaction, *Biochem. Biophys. Res. Commun.,* 179, 25, 1991.
63. **Taugner, R., Hackenthal, E., Nobiling, R., Harlacher, M., and Reb, G.,** The distribution of renin in the different segments of the renal arterial tree, *Histochemistry,* 73, 75, 1981.
64. **Lindop, G.B.M. and Lever, A.F.,** Anatomy of the renin-angiotensin system in the normal and pathologic kidney, *Histochemistry,* 10, 335, 1986.
65. **Gomez, R.A., Chevalier, R.L., Carey, R.M., and Peach, M.J.,** Molecular biology of the renal renin-angiotensin system, *Kidney Int.,* 38 (Suppl. 30), S18, 1990.
66. **Gomez, R.A., Lynch, K.R., Chevalier, R.L., Everett, A.D., Johns, D.W., Wilfong, N., Peach, M.J., and Carey, R.M.,** Renin and angiotensinogen gene expression and intrarenal renin distribution during ACE inhibition, *Am. J. Physiol.,* 254, F900, 1988.
67. **Gomez, R.A., Chevalier, R.L., Everett, A.D., Elwood, J.P., Peach, M.J., Lynch, K.R., and Carey, R.M.,** Recruitment of renin gene-expressing cells in adult rat kidneys, *Am. J. Physiol.,* 259, F660, 1990.
68. **El Dahr, S., Gomez, R.A., Gray, M.S., Peach, M.J., Carey, R.M., and Chevalier, R.L.,** In situ localization of renin and its mRNA in neonatal uretal obstruction, *Am. J. Physiol.,* 258, F854, 1990.

69. **El Dahr, S., Gomez, R.A., Khare, G., Peach, M.J., Carey, R.M., and Chevalier, R.L.,** Expression of renin and its mRNA in the adult kidney with chronic uretal obstruction, *Am. J. Kidney Dis.*, 15(6), 575, 1990.
70. **Samani, N.J., Godfrey, N.P., Major, J.A., Brammar, W.J., and Swales, J.D.,** Kidney renin mRNA levels in the early and chronic phases of two-kidney, one clip hypertension in the rat, *J. Hypertens.*, 7, 105, 1989.
71. **Moffet, R.B., McGowan, R.A., and Gross, K.W.,** Modulation of kidney renin messenger RNA levels during experimentally induced hypertension, *Hypertension,* 8, 874, 1986.
72. **Morris, B.J.,** Stimulation by thyroid hormone of renin mRNA in mouse submandibular gland, *Am. J. Physiol.*, 251, E290, 1986.
73. **Gomez, R.A., Chevalier, R.L., Sturgill, B.C., Johns, D.W., Peach, M.J., and Carey, R.M.,** Maturation of the intrarenal renin distribution in Wistar-Kyoto rats, *J. Hypertens.*, 4 (Suppl. 5), 31, 1986.
74. **Sigmund, C.D., Jones, C.A., Kim, U., Mullins, J.J., and Gross, K.W.,** Expression of the murine renin genes in subcutaneous connective tissue, *Proc. Natl. Acad. Sci. U.S.A.*, 87, 7993, 1990.
75. **Sadler, T.W., Ed.,** *Langmans Medical Embryology,* 5th ed., Williams and Wilkins, Baltimore, 1985, 247.
76. **Rugh, R.,** *The Mouse; Its Reproduction and Development,* Burgess Publishing, Minneapolis, 1968, 276.
77. **Sigmund, C.D. and Gross, K.W.,** Differential expression of the murine and rat renin genes in peripheral subcutaneous tissue, *Biochem. Biophys. Res. Commun.*, 173, 218, 1990.
78. **Zemel, S., Millan, M.A., and Aguilera, G.,** Distribution of angiotensin II receptors and renin in the mouse fetus, *Endocrinology,* 124, 1774, 1989.
79. **Wilson, C.M. and Taylor, B.A.,** Genetic regulation of thermostability of mouse submaxillary gland renin, *J. Biol. Chem.*, 257, 217, 1982.
80. **Wilson, C.M., Myhre, M.J., Reynolds, R.C., and Wilson, J.D.,** Regulation of mouse submaxillary gland renin by thyroxine, *Endocrinology,* 110, 982, 1982.
81. **Tronik, D. and Rougeon, F.,** Thyroxine and testosterone transcriptionally regulate renin gene expression in the submaxillary gland of normal and transgenic mice carrying extra copies of the Ren2 gene, *FEBS Lett.*, 234, 336, 1988.
82. **Nishimura, H.,** Physiological evolution of the renin-angiotensin system, *Jpn. Heart J.*, 19, 806, 1978.
83. **Wilson, C.M.,** Renin-angiotensin system in nonmammalian vertebrates, *Endocr. Rev.*, 5(1), 45, 1984.
84. **Morris, B.J.,** New possibilities for intracellular renin and active renin now that the structure of the human renin gene has been elucidated, *Clin. Sci.*, 71, 345, 1986.
85. **Morris, B.J., de Zwart, R.T., and Young, J.A.,** Renin in mouse but not rat submandibular glands, *Experientia,* 36, 1333, 1980.
86. **Seldin, M.F., Roderick, T.H., and Paigen, B.,** Mouse chromosome one, *Mammalian Genome,* 1, S1, 1991.
87. **Holm, I., Ollo, R., Panthier, J.J., and Rougeon, F.,** Evolution of aspartyl proteases by gene duplication: the mouse renin gene is organized in two homologous clusters of four exons, *EMBO J.*, 3, 557, 1984.
88. **Sigmund, C.D. and Gross, K.W.,** Structure, expression and regulation of murine renin genes, *Hypertension,* 18, 446, 1991.
89. **Kim, W.S., Murakami, K., and Nakayama, K.,** Nucleotide sequence of a cDNA coding for mouse Ren1 preprorenin, *Nucleic Acids Res.*, 17, 9480, 1989.
90. **Field, L.J., Philbrick, W.M., Howles, P.N., Dickinson, D.P., McGowan, R.A., and Gross, K.W.,** Expression of tissue-specific Ren-1 and Ren-2 genes of mice: comparative analysis of 5′-proximal flanking regions, *Mol. Cell. Biol.*, 4, 2321, 1984.

91. **Panthier, J.J., Dreyfus, M., Roux, D.T.W., and Rougeon, F.,** Mouse kidney and submaxillary gland renin genes differ in their 5′ putative regulatory sequences, *Proc. Natl. Acad. Sci. U.S.A.*, 81, 5489, 1984.
92. **Burt, D.W., Reith, A.D., and Brammar, W.J.,** A retroviral provirus closely associated with the Ren-2 gene of DBA/2 mice, *Nucleic Acids Res.*, 12, 8579, 1984.
93. **Burt, D.W., Mullins, L.J., George, H., Smith, G., Brooks, J., Piolo, D., and Brammar, W.J.,** The nucleotide sequence of a mouse renin-encoding gene, Ren-1d, and its upstream region, *Gene,* 84, 91, 1989.
94. **Dickinson, D.P., Gross, K.W., Piccini, N., and Wilson, C.M.,** Evolution and variation of renin genes in mice, *Genetics,* 108, 651, 1984.
95. **Abel, K.J., Howles, P.N., and Gross, K.W.,** DNA insertions distinguish the duplicated renin genes of DBA/2 and *Mus hortulanus* mice, *Mammalian Genome,* 2, 32, 1991.
96. **Soubrier, F., Panthier, J.J., Houot, A.M., Rougeon, F., and Corvol, P.,** Segmental homology between the promoter region of the human renin gene and the mouse Ren-1 and Ren-2 promoter regions, *Gene,* 41, 85, 1986.
97. **Tronik, D., Ekker, M., and Rougeon, F.,** Structural analysis of 5′-flanking regions of rat, mouse and human renin genes reveals the presence of a transposable-like element in the two mouse genes, *Gene,* 69, 71, 1988.
98. **Abel, K.J. and Gross, K.W.,** Physical characterization of genetic rearrangements at the mouse renin loci, *Genetics,* 124, 937, 1990.
99. **Abel, K.J. and Gross, K.W.,** Close physical linkage of the murine Ren-1 and Ren-2 loci, *Nucleic Acids. Res.*, 16, 2111, 1988.
100. **Nakamura, N., Burt, D.W., Paul, M., and Dzau, V.J.,** Negative control elements and cAMP responsive sequences in the tissue-specific expression of mouse renin genes, *Proc. Natl. Acad. Sci. U.S.A.*, 86, 56, 1989.
101. **Sigmund, C.D., Jones, C.A., Fabian, J., Wu, C., Kane, C.M., Ellsworth, M.K., Pacholec, F.D., and Gross, K.W.,** Transgenic mice and the development of animal models and resources for hypertension research. WHO/IPSEN Foundation, in *Genetic Approaches for the Prevention and Control of Coronary Heart Disease and Hypertension,* Springer-Verlag, Berlin, 1991.
102. **Mockrin, S.C., Dzau, V.J., Gross, K.W., and Horan, M.J.,** Transgenic animals: new approaches to hypertension research, *Hypertension,* 13, 394, 1991.
103. **Field, L.J.,** Cardiovascular research in transgenic animals, *Trends Cardiovasc. Med.*, 1, 141, 1991.
104. **Tronik, D., Dreyfus, M., Babinet, C., and Rougeon, F.,** Regulated expression of the Ren-2 gene in transgenic mice derived from parental strains carrying only the Ren-1 gene, *EMBO J.*, 6, 983, 1987.
105. **Mullins, J.J., Sigmund, C.D., Kane-Haas, C., Wu, C., Pacholec, F., Zeng, Q., and Gross, K.W.,** Studies on the regulation of renin genes using transgenic mice, *Clin. Exp. Hypertens.*, 10, 1157, 1988.
106. **Sola, C., Tronik, D., Dreyfus, M., Babinet, C., and Rougeon, F.,** Renin-promoter SV40 large T-antigen transgenes induce tumors irrespective of normal cellular expression of renin genes, *Oncogene Res.*, 5, 149, 1989.
107. **Ekker, M., Sola, C., and Rougeon, F.,** The activity of the mouse renin promoter in cells that do not normally produce renin is dependent upon the presence of a functional enhancer, *FEBS Lett.*, 255, 241, 1989.
108. **Barrett, G., Horiuchi, M., Paul, M., Pratt, R.E., Nakamura, N., and Dzau, V.J.,** Identification of a negative regulatory element involved in tissue-specific expression of mouse renin genes, *Proc. Natl. Acad. Sci. U.S.A.*, 89, 885, 1992.
109. **Sigmund, C.D., Okuyama, K., Ingelfinger, J., Jones, C.A., Mullins, J.J., Kim, U., Kane-Haas, C., Wu, C., Kenney, L., Rustum, Y., Dzau, V., and Gross, K.W.,** Isolation and characterization of renin expressing cell lines from transgenic mice containing a renin promoter viral oncogene fusion construct, *J. Biol. Chem.*, 265, 19916, 1990.

110. **Fabian, J.R., Sigmund, C.D., Kane, C.M., Wu, C., and Gross, K.W.,** Cell-specific activation of the mouse renin promoter, in preparation.
111. **Sigmund, C.D. and Gross, K.W.,** unpublished observation.

Chapter 3

RENIN GENE EXPRESSION AND HYPERTENSION IN TRANSGENIC ANIMALS

M. Bader, M.A. Lee, Y. Zhao, M. Böhm, J. Bachmann, M. Sander, B. Djavidani, S. Bachmann, F. Zimmermann, J. Wilbertz, K. Zeh, J. Wagner, J. Peters, and D. Ganten

TABLE OF CONTENTS

0-8493-4622-3/93/$0.00 + $.50

I. INTRODUCTION

More than 10 years ago, the technique was established for the production of transgenic animals by the introduction of foreign genes into the genome of mammals.[1-4] Most of the first experiments were done in mice (reviewed in References 5 to 13), but in the meantime the methodology has been extended to other vertebrate species like rats,[14-16] rabbits,[17-20] sheep,[17,21-25] goats,[26] cattle,[23-25,27] pigs,[17,20,24,25,28-30] fish,[31,32] and birds.[33] Transgenic domestic animals have been developed in order to increase the quality and the yield of their economically exploited products.[19,25,28-30,32-34] In addition, it is possible to express transgenes in the mammary gland of these animals, leading to the secretion of pharmaceutically interesting proteins in considerable amounts in the milk.[22-24,26,27,34] In order to produce transgenic animals other than mice, the peculiarities of the physiology of the other species and the distinct physical properties of their oocytes had to be taken into account.[35]

In principle, the technique for mice and rats involves several steps[35-38] (Figure 1), starting with the collection of fertilized oocytes. In order to increase the yield, young female animals are superovulated by the administration of follicle stimulating hormone (FSH) and two days later by the injection of human chorionic gonadotropin (hCG) several hours before mating. On the morning after, the fertilized oocytes are collected by opening the ampulla of the oviduct. Then several hundred copies of the foreign DNA in about 2 pl are injected into the male pronuclei using a microinjection facility. Directly afterwards or after an overnight incubation which allows the elimination of damaged zygotes, the surviving one- or two-cell embryos are transferred into the oviduct of a female animal (foster mother) made pseudopregnant by mating it with a vasectomized male.

The offspring is analyzed for the integration of the transgene by isolating the genomic DNA of somatic cells, e.g., from the tails of the animals and detecting the introduced gene by Southern blotting[39,40] or polymerase chain reaction[40,41] using transgene-specific probes or oligonucleotide primers, respectively. On an average, 10 to 40% of the littermates will be transgenic.

Normally, several copies of the transgene are integrated as concatameres at random into one site on a host chromosome. During the integration event, the foreign as well as the endogenous DNA at the target site can be rearranged. A high percentage of the transgenic founder animals, but not all, will pass the transgene on to the next generation, and a transgenic line can be established and bred to homozygosity. Homozygous animals may show an altered phenotype because the insertion of the transgene occasionally produces recessive mutations when it occurs in a functional gene.[6,42-46] This has to be taken into account when specific effects of the additional gene are studied. Only if at least two independent transgenic lines exhibit the same phenotype is it likely that this is not elicited by an insertional mutation.

In addition, the site of transgene integration sometimes changes the regulation of its expression dramatically. Chromosomal position effects may be responsible for an altered cell-type specificity, overactivity, or absence of transcription of the inserted gene.[5,9,47] Despite these pitfalls, transgenic methodology has produced a lot of insights into the regulation of genes and their relevance for the physiology of an organism which far exceeds the possibilities of *in vitro* assays. Most frequently this approach has been employed to study the tissue specificity and regulation of genes (reviewed in References 5, 9, and 10). Regulatory elements can be defined by including or excluding them from constructs used for the production of transgenic animals and analyzing the expression of the investigated gene. In this way, regulatory DNA regions in genes for the renin-angiotensin system (RAS) have been described.[48-54] The reports concerning the renin genes form the subject of the following sections. Most recently, the presence of a testis-specific promoter in the angiotensin-converting enzyme (ACE) gene was confirmed using hybrid constructs of the ACE gene with the β-galactosidase gene of *Escherichia coli* as a reporter gene,[40] which in transgenic mice led to β-galactosidase activity in spermatocytes.[53] A comparable approach was used in our laboratory to map brain-specific regulatory sequences in the rat angiotensinogen gene.[52] Clouston et al.[51] produced transgenic mice bearing constructs containing different portions of the 5′-flanking region of the mouse angiotensinogen gene linked to the rest of the gene, which was partially deleted to yield a minigene. With this construct, they could define DNA regions responsible for tissue-specific and hormonal regulation of the angiotensinogen gene.

An increasing number of animal models for human diseases has recently been developed by the transgenic technology which can now be used to elucidate further the etiology of these disorders and to establish an efficient therapy for them (Table 1). These include Alzheimer's disease, neoplasias, acquired immunodeficiency syndrome (AIDS), autoimmune diseases, and hypertension, all of which are abundant and accompanied by a high mortality rate. As they depend on the physiology of a whole organism and are often specific for the human system, where for obvious reasons they can hardly be studied, useful model systems for them were desirable but not yet available.

This applies especially to hypertension. Conventionally bred animal models for this disorder already exist, like the spontaneously hypertensive rat (SHR). However, they exhibit a major disadvantage: the genetic background for their elevated blood pressure cannot be defined easily as it is multigenetic, and the contribution of one particular gene is difficult if not impossible to determine. With newly developed methods, however, a region on the 10-chromosome has recently been defined, interestingly containing the ACE gene, which shows genetic linkage with the hypertensive phenotype.[99-102]

Transgenic animals allow the analysis of effects of a single additional gene on the physiology of an organism. In order to elucidate the etiology of hypertension, it is now possible to test genes which are candidates for being causative factors by this approach. Our laboratory focused on the RAS and successfully developed two hypertensive animal models, TGR(mREN2)27[14,103] and TGM(rAOGEN).[98] TGR(mREN2)27 was produced by introduction of the

1. Infusion of FSH by osmotic mini-pump

↓ 48 h

2. Injection of hCG and mating

↓ 12 h

3. Collection of fertilized oocytes

Infundibulum

Fat

Oviduct of donor animal

Ampulla with fertilized oocytes

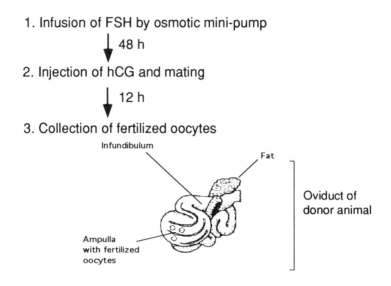

4. Microinjection of DNA into oocyte

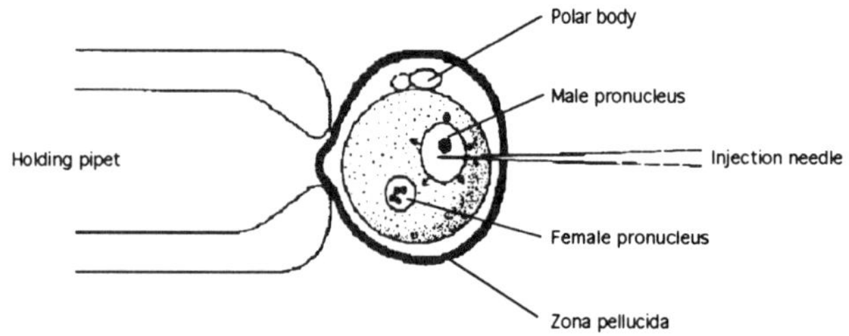

FIGURE 1. Technique for the production of transgenic rodents by microinjection of DNA into the male pronucleus of fertilized oocytes. The method is described in the text.

mouse *Ren-2* gene into the germline of rats. It will be described in more detail in the next section. TGM(rAOGEN) represents a transgenic mouse bearing the rat angiotensinogen gene and expressing it in the liver, kidney, and brain. This expression leads to the development of high blood pressure. As in TGR(mREN2)27 (see below), tissue RAS seem to be responsible for the elevation of the mean arterial pressure, as their plasma angiotensinogen levels are not higher than those of the not hypertensive transgenic mice described by Ohkubo et al.[97] containing the same gene under the control of the metallothionein promoter. This promoter leads to ectopic expression of the transgene, especially in the brain. TGM(rAOGEN) exhibit rat angiotensinogen mRNA in brain areas where it is normally found in a rat and which are claimed to be involved in cardiovascular regulation.[98] Therefore, we assume that the overexpression of the gene in these regions is responsible for the hypertensive

5. Transfer of injected oocytes into oviduct of foster mother

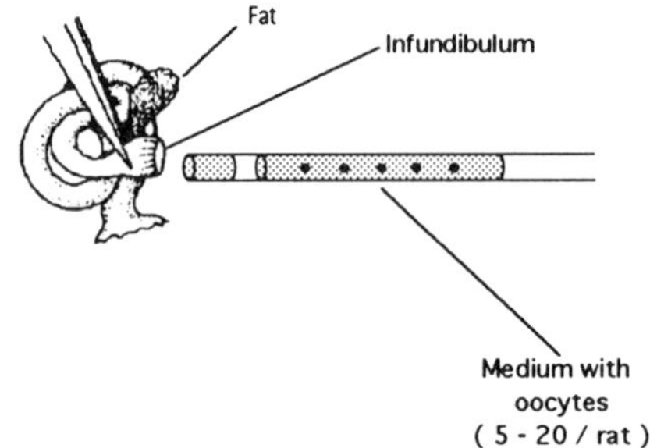

6. DNA-analysis of offspring

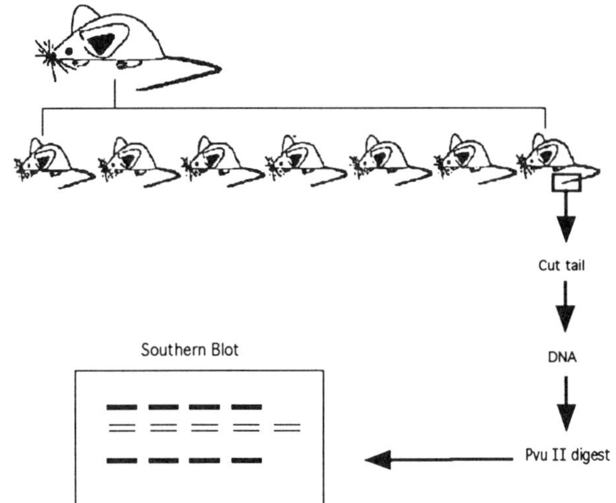

FIGURE 1 (*continued*).

phenotype of TGM(rAOGEN). The transgenic mice reported by Ohkubo et al.[97] can also be made hypertensive by mating them to transgenic mice harboring the rat renin gene (see below).

The following sections will focus on transgenic animals harboring renin genes of mice, rat, and humans which were produced to study the regulation of these genes and the effects of their overexpression in an animal.

II. HYPERTENSIVE, TRANSGENIC RATS WITH THE MOUSE *Ren-2* GENE [TGR(mREN2)27]

For cardiovascular research, the mouse, otherwise the most commonly used species for transgenic experiments, is less appropriate, mainly because of its physical size.[12,35] In this field, the rat has been extensively studied as

TABLE 1
Transgenic Animals as Models for Human Diseases

Disease	Species	Transgene	Promoter	Ref.
Alzheimer's	Mouse	Gene for β-amyloid precursor protein 751	Rat neural-specific enolase gene	55
	Mouse	Gene for 4-kDa β-amyloid protein	Homologous	56
	Mouse	Gene for 100 c-terminal amino acids of β-amyloid precursor protein	JC-virus early region	57
	Mouse	Gene for C-terminal fragment of β-amyloid precursor gene	Human *Thy*-1 gene	58
Gerstmann-Sträussler-Schinke syndrome	Mouse	Mutated mouse prion protein	Homologous	59
Chondrodysplasia	Mouse	Human type II collagen gene with central deletion	Homologous	60
Osteogenesis imperfecta	Mouse	Insertional mutant in the α1(I) collagen gene		42
	Mouse	Point-mutated gene for mouse pro α(1) collagen chain type I	Homologous	61
	Mouse	Minigene for human pro α(1) collagen chain type I with central deletion	Homologous	62
Multiple sclerosis	Mouse	Gene for murine class I H-$2K^b$ MHC-protein	Myelin basic protein gene	63,64
Lupus erythematosus	Mouse	Human BCL2-gene	Immunoglobulin heavy chain enhancer	65
AIDS	Mouse	HIV provirus DNA	Both LTRs	66
	Mouse	HIV-*tat*gene	3′ LTR of HIV provirus	67,68, reviewed in 69
	Mouse	Defective HIV provirus DNA	Both LTRs	70
Diabetes type I	Mouse	Several models		Reviewed in 71
α_1P_i-deficiency	Mouse	Mutated human α_1-antitrypsin gene	Homologous	72
Hepatitis B	Mouse	Hepatitis B surface antigen gene	Homologous	73,74

Arthritis	Rat	Genes for human HLA-B27 and β_2-microglobulin	Homologous	15
	Mouse	Human tumor necrosis factor gene, 3′-modified	Homologous	75
	Mouse	Total HTLV-I virus DNA	Both LTRs	76
	Mouse	Gene for granulocyte macrophage stimulating factor	Moloney virus LTR	77,78
Down's syndrome/brain edema and infarction	Mouse	Human CuZn superoxide dismutase gene	Homologous	79–81
Sickle cell anemia	Mouse	Human genes for α- and β^S-globin	β-Globin locus control region	82,83
	Mouse	Human genes for α- and β^{SAD}-globin	β-Globin locus control region	84
	Mouse	Human genes for α- and $\beta^{S\text{-}Antilles}$-globin	β-Globin locus control region	85
Dwarf syndrome	Mouse	Insertional mutant		46
Neoplasias	Mouse	Several models		Reviewed in 6,7, 86–92
Atherosclerosis	Mouse	Human apolipoprotein AI gene	Homologous	93,94
Nephrotic syndrome/renal failure	Mouse	Insertional mutation by Mpv17 virus		43
Endomyocardial fibrosis	Mouse	v-*fps* Oncogene	Human β-globin gene	95,96
Hypertension	Rat	Mouse *Ren-2* gene	Homologous	14
	Mouse	Rat renin and angiotensinogen genes	Mouse metallothionein I gene	97
	Mouse	Rat angiotensinogen gene	Homologous	98

an animal model for human cardiovascular diseases. Therefore, a methodology was developed for the production of transgenic rats in order to study the effects of an additional renin gene on the physiology of an animal. To achieve this goal, the techniques for superovulation, microinjection, and embryo handling had to be modified. Fertilized eggs were isolated from 4-week-old females derived from a cross between Sprague-Dawley (SD) female and Wistar-Kyoto male rats after superovulation and mating.[14,104] About 100 copies of an XhoI fragment containing the *Ren-2* gene of DBA/2 mice including 5.3 kb of the 5′-flanking region, 9 exons, 8 introns, and 9.5 kb of the 3′-flanking region[105] were injected into the male pronuclei of these zygotes. This construct was chosen as it directed tissue-specific renin gene expression in a transgenic mouse (see below).[105,106] Then, 37 eggs were reimplanted, of which 8 developed to term, and 5 turned out to be transgenic. Three of them transmitted the *Ren-2* gene to their offspring, thereby founding the lines TGR(mREN2)25, 26, and 27 (Table 2). The normotensive female 26 turned out to be mosaic for a transgene insertion site. All other founder animals exhibited fulminant hypertension (Table 2). In addition, this phenotype cosegregated with the transgene in all following generations of lines 26 and 27. The offspring of line 25, however, showed lower blood pressure values, the reasons for which are currently under investigation. In the meantime, lines 25 and 27 have been bred to homozygosity without revealing conspicuous abnormalities except a reduced weight gain in line 27.[99,107] This was impossible for TGR(mREN2)26 as all males of this line are sterile, most probably because of an insertional mutation at the transgene integration site which seems to reside on the X-chromosome as the phenotype is already apparent in the heterozygous state. To investigate the molecular mechanisms underlying this phenomenon, mapping of the chromosomal insertion sites of the *Ren-2* gene in the transgenic rats is being undertaken.

TABLE 2
Development of Transgenic Rat Strains by the Integration of the Mouse *Ren-2* Gene [TGR(mREN2)]

Animal	Sex	Transgenicity	Germline transmission	BP	BP of transgenic offspring
24	Female	No	—	Low	—
25	Female	Yes	Yes	High	Medium
26	Female	Yes (mosaic)	Yes (only females)	Low	High
27	Male	Yes	Yes	High	High
28	Male	Yes	No	High	—
29	Male	Yes	No	High	—
30	Male	No	—	Low	—
31	Male	No	—	Low	—

In particular, TGR(mREN2)27 has been studied in more detail. Heterozygous male animals of this line develop hypertension shortly after weaning at 5 weeks of age, and plateau values for blood pressure (about 240 mmHg) are reached at 10 weeks of age.[14,107] Homozygous TGR(mREN2)27 exhibit even higher values (up to 300 mmHg)[108] (see also Figure 3) and would die early of the consequences, if they were not treated with ACE inhibitors.[99,108]

Female TGR(mREN2)27 are far less hypertensive, with blood pressure values about 60 mmHg lower than males, reflecting the human situation, in which hypertension occurs less frequently in women.[107,109,110] Androgens are probably the main cause for this sexual dimorphism since dihydrotestosterone treatment of females increased the blood pressure to male values, and orchidectomy of males had the opposite effect.[110] To reveal the hypertensinogenic mechanisms active in TGR(mREN2)27, the tissue-specific expression of the transgene and its effects on their physiology were studied.

A. GENE EXPRESSION IN TISSUES

As detected by Northern blots[39,40] and RNase-protection assay,[40] the *Ren-2* transgene in TGR(mREN2)27 is expressed most abundantly in the adrenal gland, followed by thymus, intestine, fat, female and male genital tracts, kidney, brain, pituitary, thyroid gland, and eye[108,111,112] (Table 3). No specific transcripts could be detected in the liver and submandibular gland (SMG). This pattern seems at first sight unusual, but it reflects closely the natural expression pattern of the *Ren-2* gene in DBA/2 mice from whence it is derived[130] (Table 3). The main exception represents the SMG, which in the mouse is the organ with the highest expression.[48] The lack of transcription is best explained by the absence of trans-acting factors in the rat SMG which are present in the organ of the mouse and interact with cis-acting elements on the *Ren-2* gene.[125] The relatively low level of *Ren-2* mRNA in the kidney is most probably due to feedback inhibition of renin gene expression by the high blood pressure. Considerable amounts of *Ren-2* transcripts in the adrenal gland have also been described for DBA/2 mice, but predominantly in females.[105,117] Most recently, the intestine was shown to exhibit significant levels of renin gene expression in humans and mice, e.g., *Ren-2* in the DBA/2 jejunum.[112,120] Furthermore, *Ren-2* mRNA has been detected in the male and female reproductive tracts of mice.[48,105,131] The absence of transgene expression in the liver also reflects the situation found in a nontransgenic mouse[116] (Table 3).

The kidney- and adrenal-specific renin gene expression in TGR(mREN2)27 was the subject of more detailed investigations using *in situ* hybridization.[40,132] In the kidney, the mRNA for both renin genes, the endogenous and *Ren-2,* was hardly detectable, confirming the results obtained by Northern blot and RNase protection assay and showing that the rat renin gene is also downregulated (Figure 2a, b).[132] In the adrenal gland, the expression pattern resembles that of female DBA/2 mice[105] as the renin mRNA was found in two separate layers, the innermost region of the zona glomerulosa together with the outer

TABLE 3
Tissue-Specific Expression of the Renin Genes in Several Nontransgenic and Transgenic Mammals

Species/strain	Human	Rat	Mouse	Mouse	Mouse	TGR (mREN2)	Rn2	Tg(Xho)	Tag (T1-8)	*Ren1*Tag/ *Ren2*Tag	Ren-1*	84-4	hRN	TGR(hREN)
(trans)Gene	–	–	*Ren-1*^{c}	*Ren-1*^{d}	*Ren-2*	*Ren-2*	*Ren-2*	*Ren-2*	T-antigen	T-antigen	*Ren-1**	Rat renin	Human renin	Human renin
Promoter	–	–	–	–	–	*Ren-2*	*Ren-2*	*Ren-2*	*Ren-2*	*Ren-1*^{d}/ *Ren-2*	*Ren-1*^{d}	MT-I	Human renin	Human renin
Kidney	+ +	+ + +	+ + + +	+ + +	+ + +	+	+	+ +	+	+ +	+ + +	+	+ + +	+ + +
SMG	nd	–	+ +	+	+ + + +[1]	–	+ +[1]	+ + +	+ + +	+	+	nd	–	nd
Adrenal	+	+	–	+ +	+ +	+ + + +	nd	+ + +[2]	+ +	nd	nd	nd	nd	+
Liver	nd	+	–	+	–	–	+	–	–	+	+	+ +	–	–
Heart	+	+	+	+	+	+	–	–	–	nd	–	+	+	–
Brain	nd	+	+	+	+	+	–	+[3]	–	nd	+	+	+	+
Lung	nd	+	–	+	–	–	–	–	–	nd	nd	nd	+	+ + +
Intestine	+	+	+	–	+ +	+ + +	nd	nd	nd	+ +	nd	+	+	+ + +
Spleen	nd	+	–	+	–	–	–	–	–	+	nd	–	+	+ +
Muscle	nd	nd	–	–	–	–	–	–	–	nd	nd	–	nd	nd
Ovary	+	+	nd	+ +	+ +	+ +	+ +	nd	nd	nd	nd	nd	nd	nd
Testis	nd	+	+ +	+ +	+ +	+ +	+ +	nd	+ +	+ + +	+	+	+	nd
Sex access tissue	nd	–	+ +	–	–	1+	nd	+	nd	nd	+	nd	nd	nd
Thymus	nd	nd	–	+	+	+ + +	nd	nd	nd	+ +	–	+ +	+ +	+ +

Note: The number of + indicates semiquantitative estimations of the relative amounts of one particular mRNA in a tissue of an animal strain. –: at or under the limit of detection; nd: not determined; [1]: males > females; [2]: males < females; [3]: longer transcript.

Data from References 16, 48–50, 97, 105, 106, 111–131.

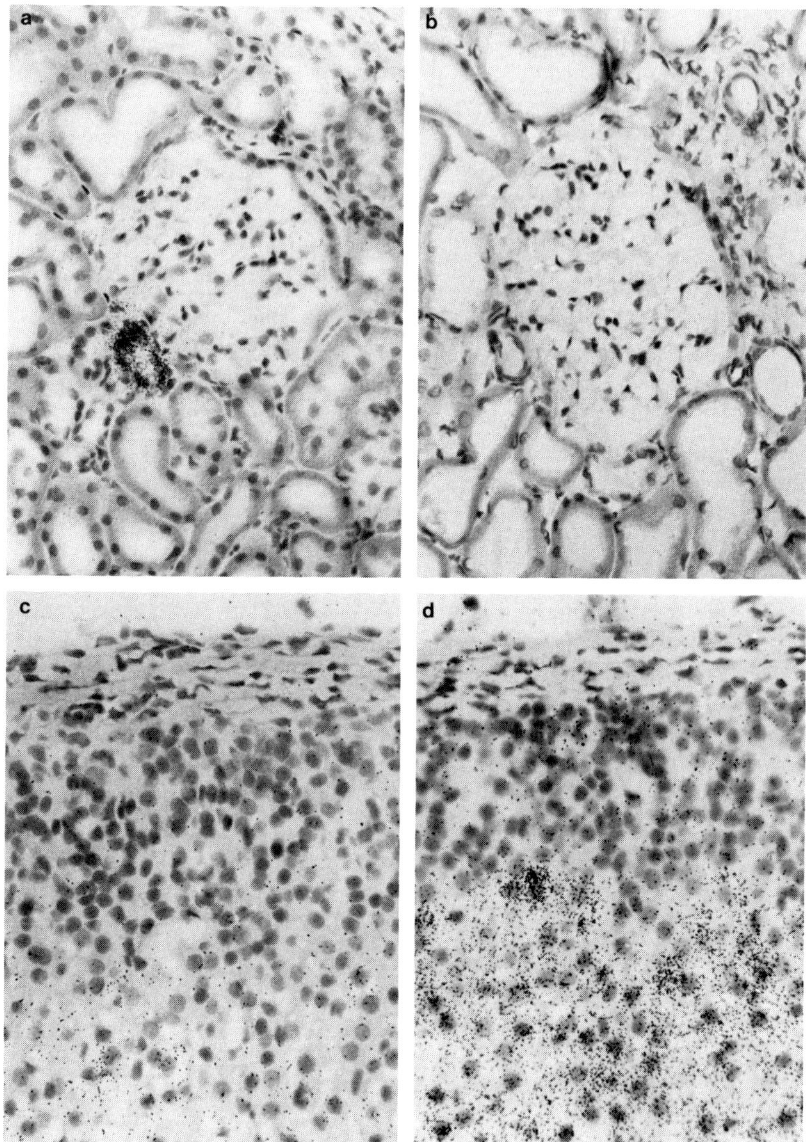

FIGURE 2. *In situ* hybridization of a glomerulus in the kidney (a and b) and the outer zones of the adrenal gland (c and d) of SD rats (a and c) and of TGR(mREN2)27 (b and d) using a ^{35}S-labeled *Ren-2* cRNA probe detecting both the *Ren-2* and the rat renin mRNA. Silver grains represent the presence of renin mRNA in the juxtaglomerular cells of the SD kidney but not in TGR(mREN2)27 and in the outer zona fasciculata and to a lesser extend in the zona glomerulosa of TGR(mREN2)27. Renin mRNA is hardly detectable in all zones of the SD adrenal gland. Original magnification: (a and b) 300×; (c and d) 350×.

zona fasciculata (Figure 2d) and patches of cells in the zona reticularis bordering the medulla.[132] In a nontransgenic rat adrenal gland, the renin gene expression was barely detectable (Figure 2c), indicating that the signal in TGR(mREN2)27 is mainly derived from *Ren-2* transcripts.

The tissue renin protein concentrations correspond to the mRNA measurements, being low in the kidney and enhanced in the adrenal gland, indicating that the transgene mRNA is accurately translated.[14] These striking peculiarities of transgene expression have consequences for the physiology of TGR(mREN2)27.

B. RAS IN TGR(mREN2)27

Keeping in mind the decreased renin gene expression in the kidney, it is not surprising that the active renin concentration in the plasma of TGR(mREN2)27 is lower than in control animals.[14,107] The same was true for nearly all other parameters of the RAS, like angiotensin (Ang) I, AngII, and angiotensinogen, but not for prorenin. This precursor is greatly enriched in the plasma, reaching about 20-fold higher levels than in SD rats. Most probably, it is mainly derived from the adrenal glands as its concentration is lowered to about 20% by bilateral adrenalectomy.[132] In addition, it is possible to show that isolated adrenal cells from TGR(mREN2)27 secrete considerable amounts of active renin and prorenin.[133,134] The physiological relevance of the high plasma prorenin levels, especially in the development of hypertension, is still a subject of research. The fact that the blood pressure was reduced to nearly normotensive values after adrenalectomy[132] does not prove a causal relationship of transgene expression and prorenin secretion in the adrenal gland with the high blood pressure, as it can also be explained by the total elimination of corticosteroid production. In order to clarify the role of this organ in the hypertensinogenic process further, the steroid status of TGR(mREN2)27 was analyzed.

C. STEROID HORMONES IN TGR(mREN2)27

AngII is well known as a stimulator of aldosterone synthesis in the adrenal gland.[135-137] With the concept of locally acting tissue RAS,[138-143] e.g., also in the adrenal gland,[144-151] the enhanced renin gene expression and activity in this organ of the transgenic rats may lead to an increased level of AngII and in turn to an elevation of steroid, especially aldosterone, secretion. While studies with adult hypertensive animals do not corroborate this hypothesis, exhibiting only insignificantly increased urinary steroid concentrations,[152] the excretion of deoxycorticosterone, corticosterone, and aldosterone is significantly elevated in young TGR(mREN2)27 during the development of hypertension compared with SD rats.[153] When the animals were treated with adrenocorticotropic hormone (ACTH), their urinary corticosterone excretion increased more pronouncedly than in SD rats.[152] In addition, the already very abundant plasma prorenin was further enhanced more than 10-fold by this treatment.[153] These data indicate that there is an intimate interplay between

ACTH, the intra-adrenal RAS, and steroid production by the adrenocortical cells in TGR(mREN2)27. These interactions are currently being studied in more detail *in vitro* with adrenal cells isolated from transgenic rat adrenals.[134] They may well be relevant for the development of hypertension, but not, as one would expect, via the elevated mineralocorticoid levels, as the mineralocorticoid receptor antagonist spironolactone did not lower the blood pressure in young TGR(mREN2)27.[108] However, the stimulated adrenal RAS is most probably involved in the pathogenesis of hypertension, since high-dose dexamethasone treatment inhibiting ACTH-secretion from the pituitary and, thereby, adrenal steroid production, blocked the blood pressure elevation in young TGR(mREN2)27 significantly.[153,154] In contrast, the same dose of dexamethasone had a marked hypertensinogenic effect in adult SD rats. In addition and in agreement with the ACTH experiment described above, prorenin in the plasma of TGR(mREN2)27 was markedly downregulated by this treatment.[153,154]

Taken together, these data argue in favor of a causal effect of a stimulated adrenal RAS in the development of hypertension, but via an up to now unknown mechanism including prorenin or steroids lacking mineralo- and glucocorticoid action.

D. MORPHOLOGY IN TGR(mREN2)27

The extreme systemic hypertension of TGR(mREN2)27 leads to morphological alterations in the kidney, heart, and vessels in adult animals (6 to 8 months of age).[132] In the kidney, glomerular damage is evident by the development of severe glomerulosclerosis, leading to death from end-stage renal failure, at least in homozygous animals with extreme hypertension. The observed pathologies closely resemble the pattern of glomerulosclerosis in other experimental hypertension models and human disease states.[155,156] Thus, TGR(mREN2)27 represents a valid model for the secondary effects of sustained hypertension on the kidney morphology.

The same holds true for the cardiovascular alterations. The heart and, particularly, the left ventricle, are hypertrophied, and focal perivascular fibrosis of the coronary arteries can be observed.[107,132] These and other muscular-type arteries of the body, including the aorta, exhibit an increased thickness of the tunica media based on the hypertrophy of vascular myocytes and a massive neogenesis of the connective tissue matrix.[107] This phenotype is much less pronounced in female transgenic rats, probably because of their less severe hypertension. For human hypertensive patients and genetically hypertensive rats, like SHR, it is well known that high blood pressure is regularly accompanied by hypertrophy of the heart and vessel walls.[157-161] Therefore, it was not surprising to ascertain the same symptoms also in TGR(mREN2)27. The mechanisms leading to cardiovascular hypertrophy are not yet fully understood. There is evidence for a direct growth-promoting effect of AngII on vascular smooth muscle cells and cardiomyocytes.[162,163] As the vessels exhibited detectable *Ren-2* gene expression in

TGR(mREN2)27,[111] an activated local tissue RAS in the vascular wall[143,164,165] could be responsible for the hypertrophic effects. In support of this, it was observed that isolated perfused hind limbs of the transgenic rats produce significant amounts of AngII.[166]

In conclusion, stimulated tissue RAS in the adrenal gland and vasculature seem to play a role in the development of hypertension and its accompanying symptoms in TGR(mREN2)27. In order to evaluate this hypothesis further, we treated the animals with drugs interfering with the RAS.

E. PHARMACOLOGICAL INTERVENTIONS IN TGR(mREN2)27

As hypertension in TGR(mREN2)27 is based on the expression of an additional renin gene in several tissues, ACE inhibitors should have a marked antihypertensive effect. Indeed, captopril (10 mg/kg body weight per day) lowered the blood pressure of heterozygous[14] (Figure 3) and homozygous[108] animals significantly after 4 weeks of treatment. Lisinopril turned out to be more active in decreasing the blood pressure markedly even at a dose of 0.5 mg/kg per day and normalized it with 2 and 10 mg/kg per day[167] (Figure 3). This decrease in mean arterial pressure was accompanied by an increase in the active plasma renin and renin gene expression in the kidney affecting both the *Ren-2* transgene and the rat renin gene.[167] This, again, shows that the depressed renin gene expression in the kidney may be caused by a negative feedback of the high blood pressure.

ACE inhibitors also influence the bradykinin system. In order to show convincingly that their inhibitory effect on AngII synthesis is influential, the

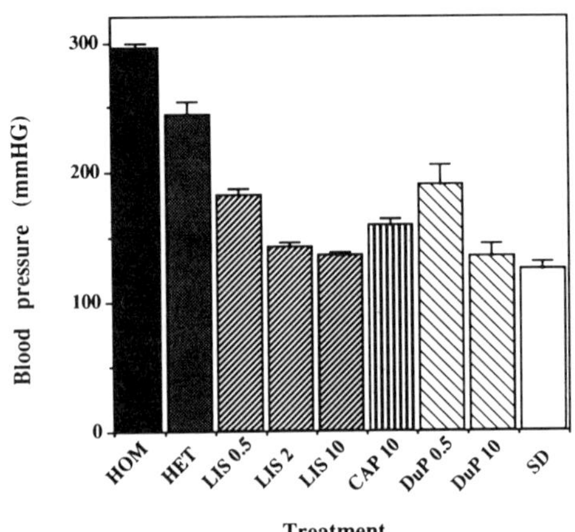

FIGURE 3. Blood pressure of untreated homozygous (HOM), heterozygous (HET) TGR(mREN2)27, and SD control rats and heterozygous transgenic rats after four weeks of treatment with captopril (CAP; 10 mg/kg per day), lisinopril (LIS; 0.5, 2, and 10 mg/kg per day), or losartan (DuP; 0.5 and 10 mg/kg per day) for 4 weeks.

animals were treated with the AngII receptor type I antagonist DuP 753 (losartan).[108] Just 0.5 mg/kg per day had a slight blood pressure lowering effect and a dosage of 10 mg/kg per day normalized the blood pressure after 4 weeks of treatment (Figure 3). Again, the plasma renin activity increased several fold, revealing the stimulating effect of the treatment on the kidney renin gene expression. The plasma prorenin concentration was unchanged, indicating that tissue RAS, especially in the adrenal gland, might be regulated differently.

In conclusion, these treatments show that AngII plays a crucial role in the hypertensinogenic process in TGR(mREN2)27. Additionally, all available data exclude the plasma as the relevant source of AngII, but support a local production in the adrenal gland and possibly other tissues as causative factors for the elevation of the blood pressure and the accompanying phenomena. Therefore, TGR(mREN2)27 is a valid model to study the specific regulation of these tissue RAS and their role in blood pressure homeostasis, which seems to be more important also for human hypertension than previously thought, as it might explain the long-term therapeutic effects of ACE inhibition in hypertensive patients mostly exhibiting a low plasma RAS.[168] In addition, these transgenic rats allow the investigation of sexual dimorphism in blood pressure regulation also observed in humans. Finally and more basically, TGR(mREN2)27 can be employed to study the tissue-specific regulation of renin genes and thereby belongs among the other transgenic animal model systems which are the subject of the following sections.

III. TRANSGENIC MICE WITH THE *Ren-2* GENE

Two laboratories have produced mice bearing the *Ren-2* gene as a transgene on a pure *Ren-1*^{c} background.[45,105,106,118,169] Tronik et al.[118] injected a construct containing the entire *Ren-2* gene of Swiss mice including 2.5 kb of the 5′-flanking and 3 kb of the 3′-flanking sequences[170] into oocytes derived from a cross between C57Bl/6 and CBA or BALB/c mice, all of which contain exclusively the *Ren-1*^{c} gene. They obtained three founder animals, one of which had two insertion sites, and thus they could establish four lines, Ren2-1, Ren2-6(a), Ren2-6(b), and Ren2-9.

Ren2-1, Ren2-6(a), and Ren2-6(b) were investigated for tissue-specific *Ren-2* gene expression by Northern blot, dot blot,[40] and primer extension analysis.[171] All lines exhibited *Ren-2* mRNA in the kidney and SMG, but not in the liver, heart, muscle, brain, lung, and spleen (Table 3). Basal levels of expression in the SMG of male animals were comparable to that in the Swiss mice from whence the transgene was derived, while in the SMG of females it was found to be considerably lower. In Swiss mice there is already a marked gender difference in their SMG *Ren-2* gene expression which can be equalized by androgen or thyroxine (T4) treatment of females. This stimulating effect of testosterone and T4 was even more pronounced in the transgenic female mice.[169] The authors conclude that the lower basal expression of the *Ren-2* gene in female transgenic mice is due either to cis-acting DNA elements absent from their construct or to pecularities of the strains used for transgene insertion.

Mullins et al.[105,106] produced transgenic mice with the DBA/2 *Ren-2* gene using the same XhoI fragment which was inserted into the rat germline, resulting in TGR(mREN2) (see above). They obtained 19 transgenic pups, of which 14 were successfully bred and founded distinct lines. One of the lines (TgX15) exhibited in the homozygous state an interesting neurological disorder caused by an insertional mutation in a gene important for motor function.[45]

Line TgX2 was analyzed for tissue-specific *Ren-2* gene expression, which revealed the mRNA in the SMG, kidney, adrenal gland, testis, sex-accessory tissues, and brain but not in the liver, heart, lung, spleen, and skeletal muscle.[105] Expression in the SMG and adrenal gland was studied further by *in situ* hybridization. In the SMG, *Ren-2* transcripts were restricted to the granular convoluted tubule (GCT) cells as in other mouse strains with two renin genes.[172] In the adrenal gland of female transgenic mice, the expression of the *Ren-2* gene was much higher than in males and exhibited a complex cycling pattern between the different zones directed by the estrus of the animals. This phenomenon, as well as the described tissue distribution of the *Ren-2* gene expression, corresponds to the situation in the DBA/2 mice from whence the gene was derived. According to published data, these two studies indicate that the correct tissue specificity of *Ren-2* gene expression is directed by DNA sequences residing between 2.5 kb 5′-flanking and 3 kb 3′-flanking regions. For the adrenal gland, this region might be more extensive, as Tronik et al.[118] did not analyze this organ.

Unfortunately, neither group mentioned the effect of the transgene on the plasma and tissue RAS or blood pressure, so a comparison with TGR(mREN2)27 is not possible. The main purpose of these studies, however, was to analyze the tissue-specific expression of the *Ren-2* gene, and, therefore, both groups continued, as described in the next section, by producing transgenic mice containing parts of their constructs linked to the SV40 large T-antigen gene as a reporter gene.[40]

IV. TRANSGENIC MICE WITH *Ren-2*/SV40 T-ANTIGEN FUSION CONSTRUCTS

In order to study in more detail the DNA elements on the *Ren-2* gene responsible for tissue-specific regulation, fusion constructs consisting of the *Ren-2* 5′-flanking region linked to the large T-antigen gene of simian virus 40 were used to produce transgenic mice. Correct tissue-specific expression of this reporter gene[40] which is easily detectable by immunocytochemistry as the protein is localized in the nuclei of expressing cells, would mean that the DNA sequences of the *Ren-2* gene contain all the elements necessary. In addition, SV40 T-antigen can transform the cells in which it is expressed.[86,88] From the resulting tumors, cell lines can be established which should produce renin and allow further investigations of its gene expression. Especially for the renin genes, it has been a major drawback for a long time not to have a suitable cell line available for *in vitro* expression studies.

Sigmund et al.[50,173] injected a fusion construct consisting of nucleotides −4600 to +6 of the DBA/2 *Ren-2* gene into oocytes of a hybrid mouse (BCF_2) containing only the *Ren-1*^{c} gene. Eight lines were established (T1 – 8).[174] The transgene was expressed in the kidney, SMG, testis, ovary, liver, brain, and skeletal muscle, but not in the spleen, heart, and lung (Table 3). The ontogenetic regulation of the *Ren-2* gene in the kidney and adrenal gland could be reproduced in the fetuses and embryos of the transgenic mice.[105,175] In addition, the T-antigen gene was expressed in the correct cell types, i.e., the juxtaglomerular (JG) cells of the kidney,[176-179] the GCT cells of the SMG,[172] and the interstitial Leydig cells of the testis.[180,181] Thus, the authors conclude that 4.6 kb of the 5′-flanking sequence of the *Ren-2* gene are enough to confer tissue specificity to a reporter gene in transgenic mice. Five of the eight lines developed tumors, as expected, in the kidney, SMG, and testis.[173,174] In addition, tumors were detected in the adrenal glands and subcutaneous connective tissue of three of the lines (T3, T4, and T7). The latter finding led to the detection of *Ren-1* and *Ren-2* gene expression in the skin of fetal and newborn nontransgenic mice and of the rat renin gene in early rat fetuses (15.5 days post-coitum) but not at more advanced stages in both species.[174] The physiological relevance of this expression or of a potential local RAS in the skin remains to be established, but as the authors mention, it might explain the derivation of rare renin-secreting tumors eliciting hypertension in this tissue of patients.[182]

The tumors in the kidney affected predominantly the myocytes in the vascular wall of nearly all arteries, often narrowing the vessel lumen and leading to the degeneration of tubules.[183]

Surprisingly, renin gene expression, renal renin concentration, and plasma renin activity were markedly suppressed to nearly undetectable levels.[183,184] This phenotype could not be influenced by ACE inhibition, which normally stimulates renin gene expression significantly. In addition, the transgenic animals exhibited an increased plasma volume and a marked azotemia although they were normotensive.[184] These phenomena are most probably based on the renovascular pathology observed, but their etiology is not yet clarified.

Meanwhile, the authors were able to isolate and propagate cell lines at least from kidney[185] which stably contain high amounts of renin mRNA, store active renin, and secrete prorenin. These cells may represent excellent tools to study the secretory pathways of renin and the cis-acting elements and trans-acting factors involved in the tissue-specific and hormonal regulation of its gene.

A comparable transgenic approach was used by Sola et al.[49] They fused the SV40 T-antigen to the 2.5 kb 5′-flanking region of the *Ren-2* gene or

0.45 kb of the *Ren-1* gene. In both cases, the resulting transgenic mice suffered from tumors derived exclusively from tissues which normally do not secrete renin and consisting of non-renin-secreting cells. They appeared in the choroid plexus, the pelvis epithelium of the kidney, the face, and the abdomen in the *Ren1* Tag mice and only in the abdomen of the *Ren2* Tag mice. In addition, the animals exhibited enlargement of the spleen, the liver, and the thymus and peripheral neuropathies. As stated by the authors, these symptoms had also been detected in other transgenic mouse strains harboring the SV40 T-antigen gene under the control of different promoters.[186,187] Apart from the tumor tissue, T-antigen expression was detected in the kidney, SMG, intestine, spleen, liver, heart, and lung of both the *Ren-1* and *Ren-2* T-antigen lines. The level of this mRNA in the SMG was neither different between males and females nor influenced by androgen treatment (see above). Thus, the expression of the transgene is not directed by the renin promoters used in the fusion constructs. As Sigmund et al.[50] found the correct tissue-specific expression of a comparable construct but with a 2 kb longer 5′-flanking region of the *Ren-2* gene, they speculate that important regulatory elements might reside in this region. However, they had to postulate that those elements are redundant and can be functionally replaced by sequences downstream of the start site as a construct containing only a short 5′-flanking region but the entire *Ren-2* gene and 3 kb of the 3′-flanking region directed the correct tissue-specific expression in transgenic mice[118] (see above). Strain differences in the *Ren-2* gene sequences or the mice used for the production of the transgenic lines may also account for the discrepant results. In addition, it is possible that dominant regulatory elements of prokaryotic or SV40 origin directing a different tissue specificity of gene expression are present on the constructs of Sola et al. In other transgenic animal systems, such phenomena have been described.[5,186-188]

V. TRANSGENIC MICE WITH THE *Ren-1* GENE

In order to study the molecular mechanisms underlying the differential expression of the *Ren-1* genes *Ren-1*^{c} and *Ren-1*^{d} (Table 3), Miller et al.[48] produced transgenic mice with the entire but mutated *Ren-1*^{d} gene (*Ren-1**), including 5 kb of upstream and 4 kb of downstream sequences on a pure *Ren-1*^{c} background. The mutation consisted of a 30-bp insertion into exon 2, creating a unique XhoI site which allowed the distinction of the *Ren-1** gene from the endogenous *Ren-1*^{c} gene. They obtained three founder animals and analyzed the expression of the transgene in the offspring. *Ren-1** mRNA was detected in the kidney, SMG, testis, liver, and brain in amounts which better reflected the pattern of the *Ren-1*^{d} gene than that of the *Ren-1*^{c} gene (Table 3). Thus, the authors suggest that different cis-acting regulatory elements responsible for tissue-specific expression are present in the two *Ren-1* genes. The expression of both genes in the kidney was stimulated by a low-sodium diet,[189] showing that the *Ren-1** transgene also contained all cis-acting elements responsible for this regulation.

In addition, the same group established transgenic mice bearing the chloramphenicol transferase (CAT) gene as a reporter gene under the control of the 5 kb 5′-flanking region of the *Ren-1*^{d} gene.[48] As they were not able to detect CAT activity in any organ, they conclude that sequences downstream of the start site are necessary for the correct expression of the renin gene. This would agree with the findings of Sola et al.,[49] who used the *Ren-2* promoter fused to the SV40 T-antigen gene and found only ectopic expression (see above). However, the problem with every fusion construct containing renin promoter sequences is that they cannot efficiently be tested *in vitro* because of the persisting lack of a suitable cell line. For example, the absence of an intron may well be responsible for a very low level of expression of a reporter gene, despite the presence of all elements important for tissue specificity.[190] As Miller et al. do not describe their CAT constructs in detail, this hypothesis cannot be evaluated.

VI. TRANSGENIC MICE WITH THE RAT RENIN GENE

Ohkubo et al.[97] applied the opposite approach in producing transgenic mice using the mouse metallothionein (MT-I) promoter fused to the whole rat renin gene from nucleotide −15 to 1.5 kb in the 3′-flanking region. They obtained 36 founder animals of the C57Bl/6 strain, 29 of which were shown to express renin in liver pieces collected by partial hepatectomy after the administration of $ZnSO_4$, which stimulates the MT-I promoter.[122] This site of expression was verified by immunocytochemistry using a renin-specific antibody. It demonstrates renin in the parenchymal and Kupffer cells of the liver. Additionally, the rat renin mRNA was detected in other tissues known to express genes governed by the MT-I promoter,[122] i.e., testis, brain, heart, kidney, muscle, intestine, and spleen (Table 3).

In the same way, the authors produced transgenic mice harboring and expressing the rat angiotensinogen gene under the control of the MT-I promoter.[97] Both transgenic strains did not develop hypertension, not even in the homozygous state. For the renin transgenic mice, this can be explained by the fact that rat renin is reported not to cleave mouse angiotensinogen.[191] Murine renin, however, does react with rat angiotensinogen,[191] as exemplified by TGR(mREN2)27 (see above), excluding the same explanation for the MT-I angiotensinogen transgenic mice, which contained considerable amounts of the transgenic protein in their plasma. Here, the authors postulate that a feedback downregulation of the mouse renin gene might occur, producing a new equilibrium in the AngII generation without affecting the blood pressure. Another possible explanation is that the plasma RAS in the mouse may not be relevant for blood pressure homeostasis (see below).

When both strains are crossed by *in vitro* fertilization techniques, the resulting double-transgenic animals become hypertensive, with blood pressure values of about 130 mmHg, which can be enhanced to 140 mmHg by $ZnSO_4$ treatment.[97] As blood pressure can be normalized within one day by the ACE inhibitor captopril, it is very likely that the interaction between the two transgene products elicits hypertension. Interestingly, the double transgenic mice showed much higher levels of angiotensinogen mRNA in the liver than pure MT-I angiotensinogen transgenic animals, caused by a stimulated expression exclusively of the rat gene. This finding may be explained by the activation of prorenin secreted in the liver, either locally or in the circulation, which leads to an enhanced AngII generation in this organ reported to stabilize the rat angiotensinogen mRNA[192] but obviously not the murine transcript.

The question has still to be resolved whether the increased plasma RAS is responsible for the increment in blood pressure or a local RAS in a tissue in which both transgenes are coexpressed owing to the identical promoter. This problem could be addressed by the measurement of angiotensin peptides in the plasma of the double-transgenic animals. The very fast effect of ACE inhibition argues in favor of the plasma RAS theory. In contrast, our laboratory has generated transgenic mice with the rat angiotensinogen gene under the control of its own promoter which become hypertensive when the transgene is expressed with the correct tissue specificity irrespective of the activity of the plasma RAS[98] (see also the Introduction).

Thus, both hypertensive transgenic mouse models can be employed to study the different roles of plasma and local RAS in blood pressure homeostasis.

VII. TRANSGENIC MICE WITH THE HUMAN RENIN GENE

The rodent renin genes mentioned in the preceding sections can easily be studied in nontransgenic animals, and much was known about their regulation and role in cardiovascular regulation before the transgenic methodology had emerged. The following sections will deal with the human renin gene, which is much more difficult to study in its natural environment for obvious ethical and practical reasons. Therefore, the transgenic approach was especially tempting for this gene. In addition, it enables the testing of human-specific renin inhibitors in rodents. Because of the species specificity of the enzymatic reaction of renin, such newly developed drugs had previously to be studied in primates.

The first group to apply this approach was Fukamizu et al.[121,193,194] They generated transgenic mice (C57Bl/6) bearing the human renin gene with 10 exons, 9 introns, 3 kb of the 5′-flanking and 1.2 kb of the 3′-flanking regions. Three founder animals transmitted the gene to their offspring. The most actively expressing line (hRN8-12) was used for further investigation. The human renin mRNA was detected in the kidney, brain, lung, thymus, spleen, heart, pancreas, stomach, and, as reported elsewhere,[120] intestine but not in the liver and SMG (Table 3). The transgene transcript is correctly spliced,[193] and the resulting protein is confined to the JG cells as detected by a human-specific anti-renin antibody[195] and its amount exceeds 7.5 times the renin protein in the human kidney.[121]

As human renin does not cleave mouse angiotensinogen,[191] the animals exhibited no phenotypical alterations upon the expression of the transgene, e.g., no hypertension. It would be worthwhile to mate them with the reported transgenic mice carrying the human angiotensinogen gene[196] to establish a possibly hypertensive mouse line based on the human RAS according to the mice with the rat RAS described in the preceding section.

VIII. TRANSGENIC RATS WITH THE HUMAN RENIN GENE TGR(hREN)

The entire human renin gene containing 3 kb of the 5′-flanking sequence, 10 exons, 9 introns, and 1.2 kb of the 3′-flanking sequence, used to generate the transgenic mice described in the preceding section, was microinjected into fertilized oocytes of SD rats.[16] From 87 rat eggs implanted, 15 progeny were obtained, of which 5 were found to carry the renin transgene. Two of these founder animals, TGR(hREN)1936 and TGR(hREN)1988, transmitted the transgene to their offspring, and the line TGR(hREN)1936 has been bred to homozygosity.

Human renin and prorenin, measured using a human-specific anti-renin antibody, was detected in the plasma of both TGR(hREN) lines. The enzymatic activity of the human renin in rat plasma as measured with a human substrate amounted to 18 ng AngI/mlxh in TGR(hREN)1988; this compares with about 1.5 ng AngI/mlxh in human plasma. Rat prorenin, rat renin, AngI, AngII, and angiotensinogen levels were unchanged in TGR(hREN) animals as compared with SD rats, indicating that the human renin did not react with rat angiotensinogen to generate angiotensins.

To demonstrate the response of human renin to physiologic stimuli, TGR(hREN)1936 animals were sodium depleted, raising their human active renin concentration from 4.8 pg/ml to 58.3 pg/ml. The rat renin was also stimulated, but there was no difference between animals with or without the transgene. These data clearly indicate that the transgene is transcribed and translated and the protein activated and secreted into the plasma in both lines. The specificity of human renin prevents reaction with the endogenous rat angiotensinogen, leaving the plasma AngI and AngII levels unchanged. Transgene expression was studied by a human renin-specific RNase-protection assay. A high expression of human renin was found in the kidney, restricted to the JG cells followed by lung, thymus, and gastrointestinal tract in both transgenic lines. The mean systolic blood pressure in both TGR(hREN) lines was normal (133 ± 3 mmHg). Homozygous TGR(hREN)1936 exhibited slightly elevated human active renin concentrations but no elevated blood pressure levels compared with the heterozygous animals.

TGR(hREN)1936 were mated with transgenic rats harboring and expressing the human angiotensinogen gene in large amounts [TGR(hAOGEN)].[16] The double-transgenic rats [TGR(hRENxhAOGEN)] did not spontaneously develop hypertension. This may be due to the fact that both renin genes under the control of their natural promoters may respond to a higher AngII formation in a similar fashion that is downregulation,[197] resulting in a normalized blood pressure. Acutely, the injection of partially purified human renin into TGR(hAOGEN) elicited a marked hypertensinogenic effect which could be blocked by the administration of the human renin-specific inhibitor RO 425892 (Figure 4).[16] This demonstrates that the species-specific activity of the human RAS can be maintained in rodents.

A second explanation for the unchanged blood pressure in the crossed animals may be that the comparably low amount of human plasma renin expressed in TGR(hREN)1936 prevents the development of arterial hypertension in TGR(hRENxhAOGEN). The higher expressing line TGR(hREN)1988 could not be used for crossing because of fertility problems which perhaps are transgene derived. Currently, experiments are under way challenging the human RAS in TGR(hRENxhAOGEN), i.e., by alteration of sodium homeostasis in order to elicit blood pressure effects. In addition, new TGR(hREN) lines with higher levels of expression of the transgene are being produced, which may create hypertensive rats based on the human RAS when

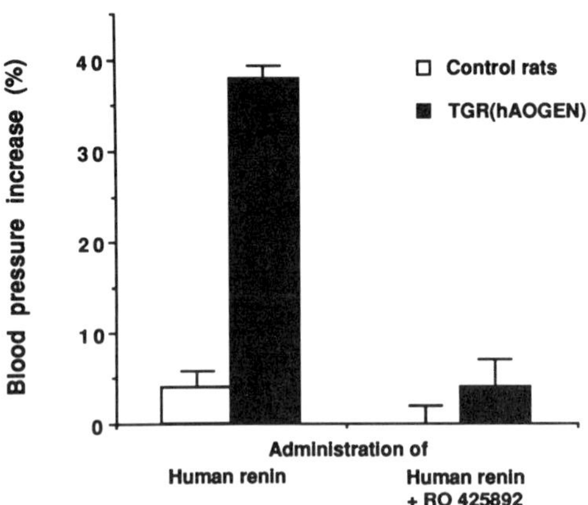

FIGURE 4. *In vivo* species specificity of the human renin-angiotensinogen reaction and of the human renin inhibitor RO 425892 in TGR(hAOGEN) rats. The percent increase of systolic steady-state blood pressure was determined in individual, conscious, unrestrained TGR(hAOGEN) rats (solid bars) or transgene-negative littermates (white bars) after infusion of human renin (5 μg ANG I/mlxh as i.v. bolus over 5 min). The human-specific inhibitor, RO 425892 (1000 μg/kg body weight i.v.) was infused 30 min after the human renin and normalized the blood pressure to pretreatment values.

mated with TGR(hAOGEN). Like the mice described in the preceding section, the TGR(hREN) and TGR(hAOGEN) and particularly the crossed animals are valid models for the testing of renin inhibitors and the investigation of the regulation of human RAS genes *in vivo*.

IX. CONCLUSIONS

What progress concerning the understanding of the molecular biology and physiology of renin was possible after the development of the transgenic methodology? First, we learned much about the tissue specificity of renin gene expression. Before the transgenic technique was available, such studies were hampered by the lack of a cell line expressing renin. One of the consequences of the developing of transgenic animals is that such a cell line is now available. Additionally, the DNA regions responsible for tissue-specific expression could be defined for the different renin genes. Thus, the 4.6 kb 5′-flanking region of the *Ren-2* gene or the 2.5 kb 5′-flanking region plus the entire rest of the gene direct correct tissue specificity; for *Ren-1*^{d} from 5 kb upstream to 4 kb downstream region is sufficient. For the *Ren-1*^{c} and rat renin gene, no new data are available, but the respective regions for the human renin gene could be localized between 3 kb of upstream and 1.2 kb of downstream sequences. The same regions of the human renin gene and the *Ren-1*^{d} gene were shown to contain the elements conferring salt regulation. For the human renin gene, such studies were made possible for the first time. The regulatory sequences for thyroxine and testosterone were mapped between −2500 and 3 kb of the 3′-flanking region of the *Ren-2* gene. With the help of the cell lines derived from the *Ren-2* T-antigen mice, these regions can be defined more precisely.

Second, additional sites of renin gene expression could be detected first in transgenic animals and subsequently also in normal mammals. Subcutaneous tissue expresses renin in fetal mice and rats, intestinal renin gene expression could be demonstrated in mice, rats, and humans. The relevance for these sites is still unclear.

Third, new animal models for hypertension and its associated symptoms are now available, i.e., TGR(mREN2)27, TGM(rAOGEN), the transgenic mice with the rat renin and angiotensinogen genes, and possibly the rats and the mice carrying both human genes. In most of the transgenic animals in which it was studied, the plasma RAS did not control the blood pressure regulation, but local tissue RAS in the adrenal glands, brain, or other tissues. This is in agreement with the long-term effects of ACE inhibition on hypertensive patients and indicates that the role of these tissue RAS may have been underestimated previously.

It should, however, not be forgotten that the RAS is not the only system involved in the pathogenesis of hypertension, especially in men. Therefore, other transgenic animals will be produced bearing genes for additional peptide systems like endothelin, in order to evaluate their relevance for blood pressure homeostasis. Mice harboring the gene for the atrial natriuretic factor have already been generated exhibiting hypotension and supporting the possible use of this peptide as a therapeutic drug.[198]

Transgenic animals, as described in this paper, have almost exclusively been produced in order to overexpress an additional gene in an animal (gain of function). However, the same technology can be employed to reduce the expression of endogenous genes (loss of function). Transgenic animals with antisense or ribozyme constructs[199-202] targeted against genes, e.g., of the RAS, may lead to a decrease in their expression and enable new insights into their function. In particular, this approach is suitable to reduce the gene expression in defined organs by governing the transcription of the antisense construct by tissue-specific promoters.

However, it is probably not possible to eliminate totally the expression and function of a gene by this method. This can be achieved using a recently developed technique based on totipotent stem cells isolated from early embryos, embryonic stem cells (ES cells), which can be genetically manipulated in culture and subsequently reintroduced into embryos to produce an animal.[203-205] Via homologous recombination, genes can be replaced by mutated versions[206,207] or reporter genes[208] generating transgenic animals lacking, and thereby revealing, the function of a gene. However, this method is up to now only available for mice, which is not the ideal model species for cardiovascular research (see above). In several laboratories, attempts are under way to extend the ES cell technique to other species, especially to the rat.

Despite the exciting progress made possible by the transgenic technology it should not be disregarded that hypertension is a polygenetic disease and each of the existing and forthcoming animal model systems will only give us partial answers. However, they are valuable tools in the investigation of the pathogenesis of this disease and will enable us to develop novel therapeutic concepts.

REFERENCES

1. **Gordon, J.W., Scangos, G.A., Plotkin, D.J., Barbosa, J.A., and Ruddle, F.H.,** Genetic transformation of mouse embryos by microinjection of purified DNA, *Proc. Natl. Acad. Sci. U.S.A.,* 77, 7380, 1980.
2. **Wagner, E.F., Stewart, T.A., and Mintz, B.,** The human β-globin gene and a functional viral thymidine kinase gene in developing mice, *Proc. Natl. Acad. Sci. U.S.A.,* 78, 5016, 1981.
3. **Costantini, F. and Lacy, E.,** Introduction of a rabbit β-globin gene into the mouse germ line, *Nature,* 294, 92, 1981.
4. **Harbers, K., Jähner, D., and Jaenisch, R.,** Microinjection of cloned retroviral genomes into mouse zygotes: integration and expression in the animal, *Nature,* 293, 540, 1981.
5. **Palmiter, R.D. and Brinster, R.L.,** Germ-line transformation of mice, *Annu. Rev. Genet.,* 20, 465, 1986.

6. **Connelly, C.S., Fahl, W.E., and Iannaccone, P.M.,** The role of transgenic animals in the analysis of various biological aspects of normal and pathologic states, *Exp. Cell Res.*, 183, 257, 1989.
7. **Hanahan, D.,** Transgenic mice as probes into complex systems, *Science,* 246, 1265, 1989.
8. **van Brunt, J.,** Transgenics primed for research, *Bio/Technology,* 8, 725, 1990.
9. **Rusconi, S.,** Transgenic regulation in laboratory animals, *Experientia,* 47, 866, 1991.
10. **Merlino, G.T.,** Transgenic animals in biomedical research, *FASEB J.*, 5, 2996, 1991.
11. **Rosenfeld, M.G., Crenshaw, E.B., III, Lira, S.A., Swanson, L., Borelli, E., Heyman, R., and Evans, R.M.,** Transgenic mice: applications to the study of the nervous system, *Annu. Rev. Neurosci.*, 11, 353, 1988.
12. **Field, L.J.,** Cardiovascular research in transgenic animals, *Trends Cardiovasc. Med.*, 1, 141, 1991.
13. **Mockrin, S.C., Dzau, V.J., Gross, K.W., and Horan, M.J.,** Transgenic animals: new approaches to hypertension research, *Hypertension,* 17, 394, 1991.
14. **Mullins, J.J., Peters, J., and Ganten, D.,** Fulminant hypertension in transgenic rats harbouring the mouse Ren-2 gene, *Nature,* 344, 541, 1990.
15. **Hammer, R.E., Maika, S.D., Richardson, J.A., Tang, J., and Taurog, J.D.,** Spontaneous inflammatory disease in transgenic rats expressing HLA-B27 and human β2m: an animal model of HLA-B27-associated human disorders, *Cell,* 63, 1099, 1990.
16. **Ganten, D., Wagner, J., Zeh, K., Bader, M., Michel, J.-B., Paul, M., Zimmermann, F., Ruf, P., Hilgenfeldt, U., Ganten, U., Kaling, M., Bachmann, S., Fukamizu, A., Mullins, J.J., and Murakami, K.,** Species specificity of renin kinetics in transgenic rats harboring the human renin and angiotensinogen genes, *Proc. Natl. Acad. Sci. U.S.A.*, 89, 7806, 1992.
17. **Hammer, R.E., Pursel, V.G., Reynolds, R.C., Rexroad, C.E., Jr., Wall, R.J., Bolt, D.J., Ebert, K.M., Palmiter, R.D., and Brinster, R.L.,** Production of transgenic rabbits, sheep and pigs by microinjection, *Nature,* 315, 680, 1985.
18. **Knight, K.L., Spieker-Polet, H., Kazdin, D.S., and Oi, V.T.,** Transgenic rabbits with lymphocytic leukemia induced by the c-myc oncogene fused with the immunoglobulin heavy chain enhancer, *Proc. Natl. Acad. Sci. U.S.A.*, 85, 3130, 1988.
19. **Müller, M., and Brem, G.,** Disease resistance in farm animals, *Experientia,* 47, 923, 1991.
20. **Weidle, U.H., Lenz, H., and Brem, G.,** Genes encoding a mouse monoclonal antibody are expressed in transgenic mice, rabbits and pigs, *Gene,* 98, 185, 1991.
21. **Cherfas, J.,** Molecular biology lies down with the lamb, *Science,* 249, 124, 1990.
22. **Wright, G., Carver, A., Cottom, D., Reeves, D., Scott, A., Simons, P., Wilmut, I., Garner, I., and Colman, A.,** High level expression of active human alpha-1-antitrypsin in the milk of transgenic sheep, *Bio/Technology,* 9, 830, 1991.
23. **Moffat, A.-S.,** Transgenic animals may be down on the pharm, *Science,* 254, 35, 1991.
24. **Wilmut, I., Archibald, A.L., McClenaghan, M., Simons, J.P., Whitelaw, C.B.A., and Clark, A.J.,** Production of pharmaceutical proteins in milk, *Experientia,* 47, 905, 1991.
25. **Ward, K.A., and Nancarrow, C.D.,** The genetic engineering of production traits in domestic animals, *Experientia,* 47, 913, 1991.
26. **Ebert, K.M., Selgrath, J.P., DiTullio, P., Denman, J., Smith, T.E., Memon, M.A., Schindler, J.E., Monastersky, G.M., Vitale, J.A., and Gordon, K.,** Transgenic production of a variant of human tissue-type plasminogen activator in goat milk: generation of transgenic goats and analysis of expression, *Bio/Technology,* 9, 835, 1991.
27. **Krimpenfort, P., Rademakers, A., Eyestone, W., van der Schans, A., van den Broek, S., Kooiman, P., Kootwijk, E., Platenburg, G., Pieper, F., Strijker, R., and de Boer, H.,** Generation of transgenic dairy cattle using 'in vitro' embryo production, *Bio/Technology,* 9, 844, 1991.

28. **Ebert, K.M., Low, M.J., Overstrom, E.W., Buonomo, F.C., Baile, C.A., Roberts, T.M., Lee, A., Mandel, G., and Goodman, R.H.,** A Moloney MLV-rat somatotropin fusion gene produces biologically active somatotropin in a transgenic pig, *Mol. Endocrinol.,* 2, 277, 1988.
29. **Miller, K.F., Bolt, D.J., Pursel, V.G., Hammer, R.E., Pinkert, C.A., Palmiter, R.D., and Brinster, R.L.,** Expression of human or bovine growth hormone gene with a mouse metallothionein-1 promoter in transgenic swine alters the secretion of porcine growth hormone and insuline-like growth factor-I, *J. Endocrinol.,* 120, 481, 1989.
30. **Pursel, V.G., Pinkert, C.A., Miller, K.F., Bolt, D.J., Campbell, R.G., Palmiter, R.D., Brinster, R.L., and Hammer, R.E.,** Genetic engineering of livestock, *Science,* 244, 1281, 1989.
31. **Culp, P., Nüsslein-Volhard, C., and Hopkins, N.,** High-frequency germ-line transmission of plasmid DNA sequences injected into fertilized zebrafish eggs, *Proc. Natl. Acad. Sci. U.S.A.,* 88, 7953, 1991.
32. **Houdebine, L.M., and Chourrout, D.,** Transgenesis in fish, *Experientia,* 47, 891, 1991.
33. **Shuman, R.M.,** Production of transgenic birds, *Experientia,* 47, 897, 1991.
34. **Westphal, H.,** Transgenic mammals and biotechnology, *FASEB J.,* 3, 117, 1989.
35. **Mullins, J.J. and Ganten, D.,** Transgenic animals: new approaches to hypertension research, *J. Hypertens.,* 8 (Suppl. 7), S35, 1990.
36. **Gordon, J.W. and Ruddle, F.H.,** Gene transfer into mouse embryos: production of transgenic mice by pronuclear injection, *Meth. Enzymol.,* 101, 411, 1983.
37. **Brinster, R.L., Chen, H.Y., Trumbauer, M.E., Yagle, M.K., and Palmiter, R.D.,** Factors affecting the efficiency of introducing foreign genes into mice by microinjecting eggs, *Proc. Natl. Acad. Sci. U.S.A.,* 82, 4438, 1985.
38. **Hogan, B., Costantini, F., and Lacy, E.,** *Manipulating the Mouse Embryo,* Cold Spring Harbor Laboratory, Cold Spring Harbor, New York, 1986.
39. **Sambrook, J., Fritsch, E.F., and Maniatis, T.,** *Molecular Cloning: A Laboratory Manual,* 2nd ed., Cold Spring Harbour Laboratory, Cold Spring Harbour, New York, 1989.
40. **Bader, M., Kaling, M., Metzger, R., Peters, J., Wagner, J., and Ganten, D.,** Basic methodology in the molecular characterization of genes, *J. Hypertens.,* 10, 9, 1992.
41. **Gilliland, G., Perrin, S., and Bunn, H.F.,** Competitive PCR for quantification of mRNA, in *PCR Protocols: A Guide to Methods and Applications,* Innis, M.A., Gelfand, D.H., Sninsky, J.J., and White, T.J., Eds., Academic Press, San Diego, 1990, 60.
42. **Schnieke, A., Harbers, K., and Jaenisch, R.,** Embryonic lethal mutation in mice induced by retroviral insertion into the α1(I) collagen gene, *Nature,* 304, 315, 1983.
43. **Weiher, H., Noda, T., Gray, D.A., Sharpe, A.H., and Jaenisch, R.,** Transgenic mouse model of kidney disease: insertional inactivation of ubiquitously expressed gene leads to nephrotic syndrome, *Cell,* 62, 425, 1990.
44. **Costantini, F., Radice, G., Lee, J.L., Chada, K., Perry, W., and Son, H.J.,** Insertional mutagenesis in transgenic mice, *Prog. Nucleic Acids Res. Mol. Biol.,* 36, 159, 1989.
45. **Ratty, A.K., Fitzgerald, L., Titeler, M., Glick, S.D., Mullins, J.J., and Gross, K.W.,** Circling behaviour exhibited by a transgenic insertional mutant, *Mol. Brain Res.,* 8, 355, 1990.
46. **Xiang, X., Benson, K.F., and Chada, K.,** Mini-mouse: disruption of the pygmy locus in a transgenic insertional mutant, *Science,* 247, 967, 1990.
47. **Bonifer, C., Vidal, M., Grosveld, F., and Sippel, A.E.,** Tissue specific and position independent expression of the complete gene domain for chicken lysozyme in transgenic mice, *EMBO J.,* 9, 2843, 1990.
48. **Miller, C.C.J., Carter, A.T., Brooks, J.I., Lovell-Badge, R.H., and Brammar, W.J.,** Differential extra-renal expression of the mouse renin genes, *Nucleic Acids Res.,* 17, 3117, 1989.

49. **Sola, C., Tronik, D., Dreyfus, M., Babinet, C., and Rougeon, F.,** Renin-promoter SV40 large T-antigen transgenes induce tumors irrespective of normal cellular expression of renin genes, *Oncogene Res.*, 5, 149, 1989.
50. **Sigmund, C.D., Jones, C.A., Fabian, J.R., Mullins, J.J., and Gross, K.W.,** Tissue and cell specific expression of a renin promoter-reporter gene construct in transgenic mice, *Biochem. Biophys. Res. Commun.*, 170, 344, 1990.
51. **Clouston, W.M., Lyons, I.G., and Richards, R.I.,** Tissue-specific and hormonal regulation of angiotensinogen minigenes in transgenic mice, *EMBO J.*, 8, 3337, 1989.
52. **Kaling, M., Bunnemann, B., Mullins, J., Wernicke, G., and Ganten, D.,** Requirements for expression of the angiotensinogen gene in tissue culture and transgenic animals, *J. Hypertens.*, 9 (Suppl. 6), S455, 1991.
53. **Langford, K.G., Shai, S.-Y., Howard, T.E., Kovac, M.J., Overbeek, P.A., and Bernstein, K.E.,** Transgenic mice demonstrate a testis-specific promoter for angiotensin-converting enzyme, *J. Biol. Chem.*, 266, 15559, 1991.
54. **Morris, B.J.,** Molecular genetics links renin to hypertension, *Mol. Cell. Endocrinol.*, 75, C13, 1991.
55. **Quon, D., Wang, Y., Catalano, R., Scardina, J.M., Murakami, K., and Cordell, B.,** Formation of β-amyloid protein deposits in brains of transgenic mice, *Nature*, 352, 239, 1991.
56. **Wirak, O., Bayney, R., Ramabhadran, T.V., Fracasso, R.P., Hart, T., Hauer, P.E., Hsiau, P., Pekar, S.K., Scangos, G.A., Trapp, D., and Unterbeck, A.J.,** Deposits of amyloid β protein in the central nervous system of transgenic mice, *Science*, 253, 323, 1991.
57. **Sandhu, F.A., Salim, M., and Zain, S.B.,** Expression of the human β-amyloid protein of Alzheimer's disease specifically in the brain of transgenic mice, *J. Biol. Chem.*, 266, 21331, 1991.
58. **Kawabata, S., Higgins, G.A., and Gordon, J.W.,** Amyloid plaques, neurofibrillary tangles and neuronal loss in brains of transgenic mice overexpressing a C-terminal fragment of human amyloid precursor protein, *Nature*, 354, 476, 1991.
59. **Vandenberg, P., Khillan, J.S., Prockop, D.J., Helminen, H., Kontusaari, S., and Ala-Kokko, L.,** Expression of a partially deleted gene of human type II procollagen (COL2A1) in transgenic mice produces a chondrodisplasia, *Proc. Natl. Acad. Sci. U.S.A.*, 88, 7640, 1991.
60. **Hsiao, K.K., Scott, M., Foster, D., Groth, D.F., DeArmond, S.J., and Prusiner, S.B.,** Spontaneous neurodegeneration in transgenic mice with mutant prion protein, *Science*, 250, 1587, 1990.
61. **Stacey, A., Bateman, J., Mascara, T., Cole, W., and Jaenisch, R.,** Perinatal lethal osteogenesis imperfecta in transgenic mice bearing an engineered mutant of pro-α1(I) collagen gene, *Nature*, 332, 131, 1988.
62. **Khillan, J.S., Olsen, A.S., Kontusaari, S., Sokolow, B., and Prockop, D.J.,** Transgenic mice that express a mini-gene version of the human gene for type I procollagen (COL1A1) develop a phenotype resembling a lethal form of osteogenesis imperfecta, *J. Biol. Chem.*, 266, 23373, 1991.
63. **Turnley, A.M., Morahan, G., Okano, H., Bernard, O., Mikoshiba, K., Allison, J., Bartlett, P.F., and Miller, J.F.A.P.,** Dysmyelination in transgenic mice resulting from expression of class I histocompatibility molecules in oligodendrocytes, *Nature*, 353, 566, 1991.
64. **Yoshioka, T., Feigenbaum, L., and Jay, G.,** Transgenic mouse model for central nervous system demyelination, *Mol. Cell. Biol.*, 11, 5479, 1991.
65. **Strasser, A., Whittingham, S., Vaux, D.L., Bath, M.L., Adams, J.M., Cory, S., and Harris, A.W.,** Enforced BCL2 expression in B-lymphoid cells prolongs antibody responses and elicits autoimmune disease, *Proc. Natl. Acad. Sci. U.S.A.*, 88, 8661, 1991.

66. **Leonard, J.M., Abramczuk, J.W., Pezen, D.S., Rutledge, R., Belcher, J.H., Hakim, F., Shearer, S., Lamperth, L., Travis, W., Fredrickson, T., Notkins, A.L., and Martin, M.A.,** Development of disease and virus recovery in transgenic mice containing HIV proviral DNA, *Science*, 242, 1665, 1988.
67. **Vogel, J., Hinrichs, S.H., Reynolds, R.K., Luciw, P.A., and Jay, G.,** The HIV tat gene induces dermal lesions resembling Kaposi's sarcoma in transgenic mice, *Nature*, 335, 606, 1988.
68. **Vogel, J., Cepeda, M., Tschachler, E., Napolitano, L.A., and Jay, G.,** UV activation of human immunodeficiency virus gene expression in transgenic mice, *J. Virol.*, 66, 1, 1992.
69. **Ruprecht, R.M., Bernard, L.D., Chou, T.-C., Gama Sosa, M.A., Fazely, F., Koch, J., Sharma, P.L., and Mullaney, S.,** Murine models for evaluating antiretroviral therapy, *Cancer Res.*, 50 (Suppl.), 5618s, 1990.
70. **Dickie, P., Felser, J., Eckhaus, M., Bryant, J., Silver, J., Marinos, N., and Notkins, A.L.,** HIV-associated nephropathy in transgenic mice expressing HIV-1 genes, *Virology*, 185, 109, 1991.
71. **Lipes, M.A. and Eisenbarth, G.S.,** Transgenic mouse models of type I diabetes, *Diabetes*, 39, 879, 1990.
72. **Dycaico, M.J., Grant, S.G.N., Felts, K., Nichols, W.S., Geller, S.A., Hager, J.H., Pollard, A.J., Kohler, S.W., Short, H.P., Jirik, F.R., Hanahan, D., and Sorge, J.A.,** Neonatal hepatitis induced by α 1-antitrypsin: a transgenic mouse model, *Science*, 242, 1409, 1988.
73. **Chisari, F.V., Pinkert, C.A., Milich, D.R., Filippi, P., McLachlan, A., Palmiter, R.D., and Brinster, R.L.,** A transgenic mouse model of the chronic hepatitis B surface antigen carrier state, *Science*, 230, 1157, 1985.
74. **Babinet, C., Farza, H., Morello, D., Hadchouel, M., and Pourcel, C.,** Specific expression of hepatitis B surface antigen (HBsAg) in transgenic mice, *Science*, 230, 1160, 1985.
75. **Keffer, J., Probert, L., Cazlaris, H., Georgopoulos, S., Kaslaris, E., Kioussis, D., and Kollias, G.,** Transgenic mice expressing human tumour necrosis factor: a predictive genetic model of arthritis, *EMBO J.*, 10, 4025, 1991.
76. **Iwakura, Y., Tosu, M., Yoshida, E., Takiguchi, M., Sato, K., Kitajima, I., Nishioka, K., Yamamoto, K., Takeda, T., Hatanaka, M., Yamamoto, H., and Sekiguchi, T.,** Induction of inflammatory arthropathy resembling rheumatoid arthritis in mice transgenic for HTLV-I, *Science*, 253, 1026, 1991.
77. **Lang, R.A., Metcalf, D., Cuthbertson, R.A., Lyons, I., Stanley, E., Kelso, A., Kannourakis, G., Williamson, D.J., Klintworth, G.K., Gonda, T.J., and Dunn, A.R.,** Transgenic mice expressing a hemopoietic growth factor gene (GM-CSF) develop accumulations of macrophages, blindness and a fatal syndrome of tissue damage, *Cell*, 51, 675, 1987.
78. **Metcalf, D.,** Transgenic mice as models of hemopoiesis, *Cancer*, 67, 2695, 1991.
79. **Minc-Golomb, D., Knobler, H., and Groner, Y.,** Gene dosage of CuZnSOD and Down's syndrome: diminished prostaglandin synthesis in human trisomy 21, transfected cells and transgenic mice, *EMBO J.*, 10, 2119, 1991.
80. **Kinouchi, H., Epstein, C.J., Mizui, T., Carlson, E., Chen, S.F., and Chan, P.H.,** Attenuation of focal cerebral ischemic injury in transgenic mice overexpressing CuZn superoxide dismutase, *Proc. Natl. Acad. Sci. U.S.A.*, 88, 11158, 1991.
81. **Ceballos-Picot, I., Nicole, A., Briand, P., Grimber, G., Delacourte, A., Defossez, A., Javoy-Agid, F., Lafon, M., Blouin, J.L., and Sinet, P.M.,** Neuronal-specific expression of human copper-zinc superoxide dismutase gene in transgenic mice: animal model of gene dosage effects in Down's syndrome, *Brain Res.*, 552, 198, 1991.
82. **Ryan, T.M., Townes, T.M., Reilly, M.P., Asakura, T., Palmiter, R.D., Brinster, R.L., and Behringer, R.R.,** Human sickle hemoglobin in transgenic mice, *Science*, 247, 566, 1990.

83. **Greaves, D.R., Fraser, P., Vidal, M.A., Hedges, M.J., Ropers, D., Luzzatto, L., and Grosveld, F.,** A transgenic mouse model of sickle cell disorder, *Nature,* 343, 183, 1990.
84. **Trudel, M., Saadane, N., Garel, M.-C., Bardakdjian-Michau, J., Blouquit, Y., Guerquin-Kern, J.-L., Rouyer-Fessard, P., Vidaud, D., Pachnis, A., Romeo, P.-H., Beuzard, Y., and Costantini, F.,** Towards a transgenic mouse model of sickle cell disease: hemoglobin SAD, *EMBO J.,* 10, 3157, 1991.
85. **Rubin, E.M., Witkowska, H.E., Spangler, E., Curtin, P., Lubin, B.H., Mohandas, N., and Clift, S.M.,** Hypoxia-induced in vivo sickling of transgenic mouse red cells, *J. Clin. Invest.,* 87, 639, 1991.
86. **Compere, S.J., Baldacci, P., and Jaenisch, R.,** Oncogenes in transgenic mice, *Biochim. Biophys. Acta,* 948, 129, 1988.
87. **Pattengale, P.K., Stewart, T.A., Leder, A., Sinn, E., Muller, W., Tepler, I., Schmidt, E., and Leder, P.,** Animal models of human disease, *Am. J. Pathol.,* 135, 39, 1989.
88. **Cory, S. and Adams, J.M.,** Transgenic mice and oncogenesis, *Annu. Rev. Immunol.,* 6, 25, 1988.
89. **Adams, J.A. and Cory, S.,** Transgenic models of tumor development, *Science,* 254, 1161, 1991.
90. **Berns, A.,** Tumorigenesis in transgenic mice: identification and characterization of synergizing oncogenes, *J. Cell. Biochem.,* 47, 130, 1991.
91. **Pastan, I., Willingham, M.C., and Gottesman, M.,** Molecular manipulations of the multidrug transporter: a new role for transgenic mice, *FASEB J.,* 5, 2523, 1991.
92. **Muller, W.J.,** Expression of activated oncogenes in the murine mammary gland: transgenic models for human breast cancer, *Cancer Metast. Rev.,* 10, 217, 1991.
93. **Rubin, E.M., Ishida, B.Y., Clift, S.M., and Krauss, R.M.,** Expression of human apolipoprotein A-I in transgenic mice results in reduced plasma levels of murine apolipoprotein A-I and the appearance of two new high density lipoprotein size subclasses, *Proc. Natl. Acad. Sci. U.S.A.,* 88, 434, 1991.
94. **Rubin, E.M., Krauss, R.M., Spangler, E.A., Verstuyft, J.G., and Clift, S.M.,** Inhibition of early atherogenesis in transgenic mice by human apolipoprotein Al, *Nature,* 353, 265, 1991.
95. **Chow, L.H., Yee, S.-P., Pawson, T., and McManus, B.M.,** Progressive cardiac fibrosis and myocyte injury in v-fps transgenic mice, *Lab. Invest.,* 64, 457, 1991.
96. **Anversa, P. and Capasso, J.M.,** Cardiac hypertrophy and ventricular remodeling, *Lab. Invest.,* 64, 441, 1991.
97. **Ohkubo, H., Kawakami, H., Kakehi, Y., Takumi, T., Arai, H., Yokota, Y., Iwai, M., Tanabe, Y., Masu, M., Hata, J., Iwao, H., Okamoto, H., Yokoyama, M., Nomura, T., Katsuki, M., and Nakanishi, S.,** Generation of transgenic mice with elevated blood pressure by introduction of the rat renin and angiotensinogen genes, *Proc. Natl. Acad. Sci. U.S.A.,* 87, 5153, 1990.
98. **Kimura, S., Mullins, J.J., Bunnemann, B., Metzger, R., Hilgenfeldt, U., Zimmermann, F., Jacob, H., Fuxe, K., Ganten, D., and Kaling, M.,** High blood pressure in transgenic mice carrying the rat angiotensinogen gene, *EMBO J.,* 11, 821, 1992.
99. **Ganten, D., Lindpaintner, K., Ganten, U., Peters, J., Zimmermann, F., Bader, M., and Mullins, J.,** Transgenic rats: new animal models in hypertension research, *Hypertension,* 17, 843, 1991.
100. **Hilbert, P., Lindpaintner, K., Beckmann, J.S., Serikawa, T., Soubrier, F., Dubay, C., Cartwright, P., DeGouyon, B., Julier, C., Takahasi, S., Vincent, M., Ganten, D., Georges, M., and Lathrop, G.M.,** Chromosomal mapping of two genetic loci associated with blood-pressure regulation in hereditary hypertensive rats, *Nature,* 353, 521, 1991.
101. **Jacob, H.J., Lindpainter, K., Lincoln, S.E., Kusumi, K., Bunker, R.K., Mao, Y.-P., Ganten, D., Dzau, V.J., and Lander, E.S.,** Genetic mapping of a gene causing hypertension in the stroke-prone spontaneously hypertensive rat, *Cell,* 67, 213, 1991.

102. **Kreutz, R., Higuchi, M., and Ganten, D.,** Molecular genetics of hypertension, *Clin. Exp. Hypertens. (A),* A14, 15, 1992.
103. **Mullins, J. and Ganten, D.,** Establishment of transgenic rats with fulminant hypertension, *Hypertension,* 14, 350, 1989.
104. **Armstrong, D.T. and Opavsky, M.A.,** Superovulation of immature rats by continuous infusion of FSH, *Biol. Reprod.,* 39, 511, 1988.
105. **Mullins, J.J., Sigmund, C.D., Kane-Haas, C., McGowan, R.A., and Gross, K.W.,** Expression of the DBA/2J Ren-2 gene in the adrenal gland of transgenic mice, *EMBO J.,* 8, 4065, 1989.
106. **Mullins, J.J., Sigmund, C.D., Kane-Haas, C., Wu, C., Pacholec, F., Zeng, Q., and Gross, K.W.,** Studies on the regulation of renin genes using transgenic mice, *Clin. Exp. Hypertens. (A),* 10, 1157, 1988.
107. **Lee, M., Zhao, Y., Peters, J., Ganten, D., Zimmermann, F., Ganten, U., Bachmann, S., Bader, M., and Mullins, J.J.,** Preparation and analysis of transgenic rats expressing the mouse Ren-2 gene, *J. Vasc. Med. Biol.,* 3, 50, 1991.
108. **Bader, M., Zhao, Y., Sander, M., Lee, M., Bachmann, J., Böhm, M., Djavidani, B., Peters, J., Mullins, J.J., and Ganten, D.,** Role of tissue renin in the pathophysiology of hypertension in TGR(mREN2)27, *Hypertension,* 19, 681, 1992.
109. **Bachmann, J., Feldmer, M., Ganten, U., Stock, G., and Ganten, D.,** Sexual dimorphism of blood pressure: possible role of the renin-angiotensin system, *J. Steroid Biochem.,* 40, 511, 1991.
110. **Bachmann, J. et al.,** unpublished data, 1992.
111. **Peters, J., Bader, M., Ganten, D., and Mullins, J.,** Tissue distribution of Ren-2 expression in transgenic rats, in *Genetic Approaches to Coronary Heart Disease and Hypertension,* Berg, K., Bulyzhenkov, V., Christen, Y., and Corvol, P., Eds., Springer-Verlag, Berlin, 1991, 74.
112. **Zhao, Y. et al.,** unpublished data, 1992.
113. **Field, L.J., McGowan, R., Dickinson, D.P., and Gross, K.W.,** Tissue and gene specificity of mouse renin expression, *Hypertension,* 6, 597, 1984.
114. **Samani, N.J., Swales, J.D., and Brammar, W.J.,** Expression of the renin gene in extra-renal tissues of the rat, *Biochem. J.,* 253, 907, 1988.
115. **Dzau, V.J., Ingelfinger, J.R., and Pratt, R.E.,** Regulation of tissue renin and angiotensin gene expressions, *J. Cardiovasc. Pharmacol.,* 8 (Suppl. 10), 11, 1986.
116. **Ekker, M., Tronik, D., and Rougeon, F.,** Extra-renal transcription of the renin genes in multiple tissues of mice and rats, *Proc. Natl. Acad. Sci. U.S.A.,* 86, 5155, 1989.
117. **Fabian, J.R., Field, L.J., McGowan, R.A., Mullins, J.J., Sigmund, C.D., and Gross, K.W.,** Allele-specific expression of the murine ren-1 genes, *J. Biol. Chem.,* 264, 17589, 1989.
118. **Tronik, D., Dreyfus, M., Babinet, C., and Rougeon, F.,** Regulated expression of the Ren-2 gene in transgenic mice derived from parental strains carrying only the Ren-1 gene, *EMBO J.,* 6, 983, 1987.
119. **Do, Y.S., Sherrod, A., Lobo, R.A., Paulson, R.J., Shinagawa, T., Chen, S., Kjos, S., and Hsueh, W.A.,** Human ovarian theca cells are a source of renin, *Proc. Natl. Acad. Sci. U.S.A.,* 85, 1957, 1988.
120. **Seo, M.-S., Fukamizu, A., Saito, T., and Murakami, K.,** Identification of a previously unrecognized production site of human renin, *Biochim. Biophys. Acta,* 1129, 87, 1991.
121. **Fukamizu, A., Seo, M.S., Hatae, T., Yokoyama, M., Nomura, T., Katsuki, M., and Murakami, K.,** Tissue-specific expression of the human renin gene in transgenic mice, *Biochem. Biophys. Res. Commun.,* 165, 826, 1989.
122. **Durnam, D.M. and Palmiter, R.D.,** Transcriptional regulation of the mouse metallothionein-I gene by heavy metals, *J. Biol. Chem.,* 256, 5712, 1981.
123. **Ganten, D., Kaling, M., Bader, M., Paul, M., and Peters, J.,** Molecular biology, pathophysiology and pharmacology of the renin-angiotensin system, *Naunyn-Schmiedeberg's Arch. Pharmacol.,* 344 (Suppl.), R24, S30, 1991.

124. **Dzau, V.J., Ingelfinger, J., Pratt, R.E., and Ellison, K.E.,** Identification of renin and angiotensinogen messenger RNA sequences in mouse and rat brains, *Hypertension,* 8, 544, 1986.
125. **Tronik, D., Ekker, M., and Rougeon, F.,** Structural analysis of 5′-flanking regions of rat, mouse and human renin genes reveals the presence of a transposable-like element in the two mouse genes, *Gene,* 69, 71, 1988.
126. **Dzau, V.J., Brody, T., Ellison, K.E., Pratt, R.E., and Ingelfinger, J.R.,** Tissue-specific regulation of renin expression in the mouse, *Hypertension,* 9 (Suppl. III), 36, 1987.
127. **Shionoiri, H., Hirawa, N., Ueda, S.-I., Himeno, H., Gotoh, E., Noguchi, K., Fukamizu, A., Seo, M.S., and Murakami, K.,** Renin gene expression in the adrenal and kidney of patients with primary aldosteronism, *J. Clin. Endocrinol. Metab.,* 74, 103, 1992.
128. **Paul, M., Wagner, J., Liu, Y., and Niedermaier, N.,** Application of the polymerase chain reaction for the mRNA measurement of the components of the renin angiotensin system in the heart, *Naunyn-Schmiedeberg's Arch. Pharmacol.,* 343 (Suppl.), R70, A277, 1991.
129. **Wagner, J., Paul, M., Ganten, D., and Ritz, E.,** Gene expression and quantification of components of the renin-angiotensin-system from human renal biopsies by the polymerase chain reaction, *J. Am. Soc. Nephrol.,* 2, 421, 1991.
130. **Sigmund, C.D. and Gross, K.W.,** Structure, expression, and regulation of the murine renin genes, *Hypertension,* 18, 446, 1991.
131. **Pandey, K.N., Maki, M., and Inagami, T.,** Detection of renin mRNA in mouse testis by hybridization with renin cDNA probe, *Biochem. Biophys. Res. Commun.,* 125, 662, 1984.
132. **Bachmann, S., Peters, J., Engler, E., Ganten, D., and Mullins, J.,** Transgenic rats carrying the mouse renin gene — morphological characterization of a low renin hypertension model, *Kidney Int.,* 41, 24, 1992.
133. **Bader, M., Peters, J., Ganten, D., and Mullins, J.J.,** Adrenal gland specific expression of the mouse Ren-2 gene in transgenic rats, Abstracts of the Meeting on Mouse Molecular Genetics, Aug. 29 – Sept. 2, Cold Spring Harbor Laboratories, Cold Spring Harbor, New York, 1990, 101.
134. **Peters, J., Münter, K., Bader, M., Hackenthal, E., Mullins, J. J., and Ganten, D.,** Increased adrenal renin in transgenic hypertensive rats, TGR(mREN2)27, and its regulation by cAMP, angiotensin II and calcium, *J. Clin. Invest.,* in press.
135. **Luft, F.C., Wilcox, C.S., Unger, Th., Kühn, R., Demmert, G., Rohmeiss, P., Ganten, D., and Sterzel, R.B.,** Angiotensin-induced hypertension in the rat: sympathetic nerve activity and prostaglandins, *Hypertension,* 14, 396, 1989.
136. **Haning, R., Tait, S.A.S., and Tait, J.F.,** In vitro effects of ACTH, angiotensins, serotonin and potassium on steroid output and conversion of corticosterone to aldosterone by isolated adrenal cells, *Endocrinology,* 87, 1147, 1970.
137. **Aguilera, G. and Catt, K.J.,** Loci of action of regulators of aldosterone biosynthesis in isolated glomerulosa cells, *Endocrinology,* 104, 1046, 1979.
138. **Campbell, D.J.,** Circulating and tissue angiotensin systems, *J. Clin. Invest.,* 79, 1, 1987.
139. **Jin, M., Wilhelm, M.J., Lang, R.E., Unger, Th., Lindpaintner, K., and Ganten, D.,** Endogenous tissue renin-angiotensin systems. From molecular biology to therapy, *Am. J. Med.,* 84 (Suppl. 3a), 28, 1988.
140. **Ganten, D., Peters, J., and Lindpaintner, K.,** Pathophysiology and molecular biology of tissue renin, in *Current Advances in ACE Inhibition,* MacGregor, G.A. and Sever, P.S., Eds., Churchill Livingstone, Edinburgh, 1989, 25.
141. **Frohlich, E.D., Iwata, T., and Sasaki, O.,** Clinical and physiological significance of local tissue renin-angiotensin systems, *Am. J. Med.,* 87 (Suppl. 6B), 19S, 1989.

142. **Paul, M., Bachmann, J., and Ganten, D.,** The tissue renin angiotensin systems in cardiovascular research, *Trends Cardiovasc. Med.*, in press.
143. **Dzau, V.J.,** Significance of the vascular renin-angiotensin pathway, *Hypertension,* 8, 553, 1986.
144. **Ryan, J.W.,** Renin-like enzyme in the adrenal gland, *Science,* 158, 1589, 1967.
145. **Ganten, D., Hutchinson, J.S., Schelling, P., Ganten, U., and Fischer, H.,** The isorenin angiotensin systems in extrarenal tissue, *Clin. Exp. Pharmacol. Physiol.*, 3, 103, 1976.
146. **Doi, Y., Atarashi, K., Franco-Saenz, R., and Mulrow, P.,** Adrenal renin: a possible regulator of aldosterone production, *Clin. Exp. Hypertens. (A),* 5, 1119, 1983.
147. **Nakamaru, M., Misono, K.S., Naruse, M., Workman, R.J., and Inagami, T.,** A role for the adrenal renin-angiotensin system in the regulation of potassium-stimulated aldosterone production, *Endocrinology,* 117, 1772, 1985.
148. **Husain, A., DeSilva, P., Speth, R.C., and Bumpus, F.M.,** Regulation of angiotensin II in rat adrenal gland, *Circ. Res.*, 60, 640, 1987.
149. **Horiba, N., Nomura, K., and Shizume, K.,** Exogenous and locally synthesized angiotensin II and glomerulosa cell functions, *Hypertension,* 15, 190, 1990.
150. **Sasamura, H., Suzuki, H., Kato, R., and Saruta, T.,** Effects of angiotensin II, ACTH, and KCl on the adrenal renin-angiotensin system in the rat, *Acta Endocrinol. (Copenhagen),* 122, 369, 1990.
151. **Oda, H., Lotshaw, D.P., Franco-Saenz, R., and Mulrow, P.J.,** Local generation of angiotensin II as a mechanism of aldosterone secretion in rat adrenal capsules, *Proc. Soc. Exp. Biol. Med.*, 196, 175, 1991.
152. **Sander, M., Bader, M., Djavidani, B., Maser-Gluth, C., Vecsei, P., Mullins, J., Ganten, D., and Peters, J.,** The role of the adrenal gland in the hypertensive transgenic rats TGR(mREN2)27, *Endocrinology,* 131, 807, 1992.
153. **Sander, M., Djavidani, B., Bader, M., and Peters, J.,** Role of the adrenal gland in the hypertensive transgenic rats TGR(mREN2)27, *Naunyn-Schmiedeberg's Arch. Pharmacol.* 345 (Suppl.), R91, 1992.
154. **Djavidani, B. et al.,** unpublished data, 1992.
155. **Raij, L., Azar, S., and Keane, W.,** Role of hypertension in progressive glomerular injury, *Hypertension,* 7, 398, 1985.
156. **Anderson, S. and Brenner, B.M.,** Role of intraglomerular hypertension in the initiation and progression of renal disease, in *The Kidney in Hypertension. Perspectives in Hypertension,* Vol. 1, Kaplan, N.M., Brenner, B.M., and Laragh, J.H., Eds., Raven Press, New York, 1987, 67.
157. **Olivetti, G., Melissari, M., Marchetti, G., and Anversa, P.,** Quantitative structural changes of the rat thoracic aorta in early spontaneous hypertension, *Circ. Res.*, 51, 19, 1982.
158. **Lee, R.M.K.W. and Smeda, J.S.,** Primary versus secondary changes of the blood vessels in hypertension, *Can. J. Physiol. Pharmacol.*, 63, 392, 1985.
159. **Smeda, J.S., Lee, R.M.K.W., and Forrest, J.B.,** Structural and reactivity alterations of the renal vasculature of spontaneously hypertensive rats to and during established hypertension, *Circ. Res.*, 63, 518, 1988.
160. **Clozel, J.-P., Kuhn, H., and Hefti, F.,** Decreases of vascular hypertrophy in four different types of arteries in spontaneously hypertensive rats, *Am. J. Med.*, 87 (Suppl. 6B), 92S, 1989.
161. **Krieger, J.E. and Dzau, V.J.,** Molecular biology of hypertension, *Hypertension,* 18 (Suppl I.), 3, 1991.
162. **Lyall, F., Morton, J.J., Lever, A.F., and Cragoe, E.J.,** Angiotensin II activates Na-H exchange and stimulates growth in cultured vascular smooth muscle cells, *J. Hypertens.*, 6 (Suppl. 4), S438, 1988.
163. **Schelling, P., Fischer, H., and Ganten, D.,** Angiotensin and cell growth: a link to cardiovascular hypertrophy?, *J. Hypertens.*, 9, 3, 1991.

164. **Swales, J.D. and Heagerty, A.M.,** Vascular renin-angiotensin system: the unanswered questions, *J. Hypertens.,* 5 (Suppl. 2), S1, 1987.
165. **Kvist, S., Mulvany, M.J., and Aalkjær, C.,** Studies of the renin-angiotensin system in the wall of rat femoral resistance vessels, *Eur. J. Pharmacol.,* 198, 77, 1991.
166. **Hilgers, K.F., Peters, J., Veelken, R., Sommer, M., Rupprecht, G., Ganten, D., Luft, F.C., and Mann, J.F.E.,** Increased vascular angiotensin formation in female rats harboring the mouse Ren-2 gene, *Hypertension,* 19, 687, 1992.
167. **Lee, M., Boehm, M., Bader, M., Bachmann, S., Zimmermann, F., Mullins, J., and Ganten, D.,** High sensitivity to converting enzyme inhibition in hypertensive transgenic rats (TGR) carrying the renin gene of the mouse (ren-2), *J. Hypertens.,* 9 (Suppl. 6), S458, 1991.
168. **Ganten, D., Lee, M., Sander, M., Peters, J., Bader, M., and Paul, M.,** The renin gene in hypertension: from molecular analysis to hypertensive transgenic rats, *Eur. J. Physiol.,* 419 (Suppl. 1), R124, 1991.
169. **Tronik, D. and Rougeon, F.,** Thyroxine and testosterone transcriptionally regulate renin gene expression in the submaxillary gland of normal and transgenic mice carrying extra copies of the Ren2 gene, *FEBS Lett.,* 234, 336, 1988.
170. **Panthier, J.J., Dreyfus, M., Tronik-Le Roux, D., and Rougeon, F.,** Mouse kidney and submaxillary gland renin genes differ in their 5′ putative regulatory sequences, *Proc. Natl. Acad. Sci. U.S.A.,* 81, 5489, 1984.
171. **Field, L.J. and Gross, K.W.,** Ren-1 and Ren-2 loci are expressed in mouse kidney, *Proc. Natl. Acad. Sci. U.S.A.,* 82, 6196, 1985.
172. **Bing, J., Poulsen, K., Hackenthal, E., Rix, E., and Taugner, R.,** Renin in the submaxillary gland, *J. Histochem. Cytochem.,* 28, 874, 1980.
173. **Sigmund, C.D., Mullins, J.J., Kim, U., and Gross, K.W.,** Transgenic mice containing renin SV40 T antigen gene fusions develop tumors expressing renin mRNA, *Hypertension,* 12, 339, 1988.
174. **Sigmund, C.D., Jones, C.A., Mullins, J.J., Kim, U., and Gross, K.W.,** Expression of murine renin genes in subcutaneous connective tissue, *Proc. Natl. Acad. Sci. U.S.A.,* 87, 7993, 1990.
175. **Jones, C.A., Sigmund, C.D., McGowan, R.A., Kane-Haas, C.M., and Gross, K.W.,** Expression of murine renin genes during fetal development, *Mol. Endocrinol.,* 4, 375, 1990.
176. **Camilleri, J.-P., Phat, V.N., Bariety, J., Corvol, P., and Menard, J.,** Use of a specific antiserum for renin detection in human kidney, *J. Histochem. Cytochem.,* 28, 1343, 1980.
177. **Levens, N.R., Peach, M.J., and Carey, R.M.,** Role of the intrarenal renin-angiotensin system in the control of renal function, *Circ. Res.,* 48, 157, 1981.
178. **Celio, M.R. and Inagami, T.,** Renin in the human kidney, *Histochemistry,* 72, 1, 1981.
179. **Hackenthal, E., Paul, M., Ganten, D., and Taugner, R.,** Morphology, physiology, and molecular biology of renin secretion, *Physiol. Rev.,* 70, 1067, 1990.
180. **Parmentier, M., Inagami, T., Pochet, R., and Desclin, J.C.,** Pituitary-dependent renin-like immunoreactivity in the rat testis, *Endocrinology,* 112, 1318, 1983.
181. **Naruse, K., Murakoshi, M., Osamura, R.Y., Naruse, M., Toma, H., Watanabe, K., Demura, H., Inagami, T., and Shizume, K.,** Immunohistological evidence for renin in human endocrine tissues, *J. Clin. Endocrinol. Metab.,* 61, 172, 1985.
182. **Fernandez, L.A., Olsen, T.G., Barwick, K.W., Sanders, M., Kaliszewski, C., and Inagami, T.,** Renin in angiolymphoid hyperplasia with eosinophilia, *Arch. Pathol. Lab. Med.,* 110, 1131, 1986.
183. **Sigmund, C.D., Jones, C.A., Jacob, H.J., Ingelfinger, J., Kim, U., Gamble, D., Dzau, V.J., and Gross, K.W.,** Pathophysiology of vascular smooth muscle in renin promoter-T-antigen transgenic mice, *Am. J. Physiol.,* 260, F249, 1991.
184. **Jacob, H.J., Sigmund, C.D., Shockley, T.R., Gross, K.W., and Dzau, V.J.,** Renin promoter SV40 T-antigen transgenic mouse, *Hypertension,* 17, 1167, 1991.

185. **Sigmund, C.D., Okuyama, K., Ingelfinger, J., Jones, C.A., Mullins, J.J., Kane, C., Kim, U., Wu, C., Kenny, L., Rustum, Y., Dzau, V.J., and Gross, K.W.,** Isolation and characterization of renin-expression cell lines from transgenic mice containing a renin-promoter viral oncogene fusion construct, *J. Biol. Chem.,* 265, 19916, 1990.
186. **Botteri, F.M., van der Putten, H., Wong, D.F., Sauvage, C.A., and Evans, R.M.,** Unexpected thymic hyperplasias in transgenic mice harboring a neuronal promoter fused to simian virus 40 large T-antigen, *Mol. Cell. Biol.,* 7, 3178, 1987.
187. **Suda, Y., Aizawa, S., Hirai, S.I., Inoue, T., Furuta, Y., Suzuki, M., Hirohashi, S., and Ikawa, Y.,** Driven by the same Ig enhancer and SV40 promoter ras induced lung adenomatous tumors, myc induced pre-B cell lymphomas and SV40 large T-gene a variety of tumors in transgenic mice, *EMBO J.,* 6, 4055, 1987.
188. **Behringer, R.R., Peschon, J.J., Messing, A., Gartside, C.L., Hauschka, S.D., Palmiter, R.D., and Brinster, R.L.,** Heart and bone tumours in transgenic mice, *Proc. Natl. Acad. Sci. U.S.A.,* 85, 2648, 1988.
189. **Miller, C.C., Samani, N.J., Carter, A.T., Brooks, J.I., and Brammar, W.J.,** Modulation of mouse renin gene expression by dietary sodium chloride intake in one-gene, two-gene and transgenic animals, *J. Hypertens.,* 7, 861, 1989.
190. **Brinster, R.L., Allen, J.M., Behringer, R.R., Gelinas, R.E., and Palmiter, R.D.,** Introns increase transcriptional efficiency in transgenic mice, *Proc. Natl. Acad. Sci. U.S.A.,* 85, 836, 1988.
191. **Oliver, W.J. and Gross, F.,** Unique specificity of mouse angiotensinogen to homologous renin, *Proc. Soc. Exp. Biol. Med.,* 122, 923, 1966.
192. **Klett, Ch., Hellmann, W., Müller, F., Suzuki, F., Nakanishi, S., Ohkubo, H., Ganten, D., and Hackenthal, E.,** Angiotensin II controls angiotensinogen secretion at a pretranslational level, *J. Hypertens.,* 6 (4), S442, 1988.
193. **Seo, M.S., Fukamizu, A., Nomura, T., Yokoyama, M., Katsuki, M., and Murakami, K.,** The human renin gene in transgenic mice, *J. Cardiovasc. Pharmacol.,* 16 (Suppl. 4), S8, 1990.
194. **Fukamizu, A., Uehara, S., Sugimura, K., Kon, Y., Sugimura, M., Hasegawa, T., Yokoyama, M., Nomura, T., Katsuki, M., and Murakami, K.,** Cell-type specific expression of the human renin gene, *J. Biol. Regul. Homeost. Agents,* 5, 112, 1991.
195. **Fukamizu, A., Hatae, T., Kon, Y., Sugimura, M., Hasegawa, T., Yokoyama, M., Nomura, T., Katsuki, M., and Murakami, K.,** Human renin in transgenic mouse kidney is localized to juxtaglomerular cells, *Biochem. J.,* 278, 601, 1991.
196. **Takahashi, S., Fukamizu, A., Hasegawa, T., Yokoyama, M., Nomura, T., Katsuki, M., and Murakami, K.,** Expression of the human angiotensinogen gene in transgenic mice and transfected cells, *Biochem. Biophys. Res. Commun.,* 180, 1103, 1991.
197. **Johns, D.W., Peach, M.J., Gomez, R.A., Inagami, T., and Carey, R.M.,** Angiotensin II regulates renin gene expression, *Am. J. Physiol.,* 259, F882, 1990.
198. **Steinhelper, M.E., Cochrane, K.L., and Field, L.J.,** Hypotension in transgenic mice expressing atrial natriuretic factor fusion genes, *Hypertension,* 16, 301, 1990.
199. **Helene, C. and Toulme, J.-J.,** Specific regulation of gene expression by antisense, sense and antigene nucleic acids, *Biochim. Biophys. Acta,* 1049, 99, 1990.
200. **Cotten, M. and Birnstiel, M.L.,** Ribozyme mediated destruction of RNA in vivo, *EMBO J.,* 8, 3861, 1989.
201. **Cameron, F.H. and Jennings, P.A.,** Specific gene suppression by engineered ribozymes in monkey cells, *Proc. Natl. Acad. Sci. U.S.A.,* 86, 9139, 1989.
202. **Katsuki, M., Sato, M., Kimura, M., Yokoyama, M., Kobayashi, K., and Nomura, T.,** Conversion of normal behavior to shiverer by myelin basic protein antisense cDNA in transgenic mice, *Science,* 241, 593, 1988.
203. **Watt, F.M.,** Cell culture models of differentiation, *FASEB J.,* 5, 287, 1991.
204. **Nichols, J., Evans, E.P., and Smith, A.G.,** Establishment of germ-line-competent embryonic stem (ES) cells using differentiation inhibiting activity, *Development,* 110, 1341, 1990.

205. **Rossant, J. and Joyner, A.L.,** Towards a molecular-genetic analysis of mammalian development, *Trends Genet.*, 5, 277, 1989.
206. **Frohman, M.A. and Martin, G.R.,** Cut, paste, and save: new approaches to altering specific genes in mice, *Cell,* 56, 145, 1989.
207. **Capecchi, M.,** Altering the genome by homologous recombination, *Science,* 244, 1288, 1991.
208. **Le Mouellic, H., Lallemand, Y., and Brulet, P.,** Targeted replacement of the homeobox gene Hox-3.1 by the Escherichia coli lacZ in mouse chimeric embryos, *Proc. Natl. Acad. Sci. U.S.A.,* 87, 4712, 1990.

Section II

Angiotensinogen

Chapter 4

TISSUE RENIN-ANGIOTENSIN SYSTEMS

M.I. Phillips, E.A. Speakman, and B. Kimura

TABLE OF CONTENTS

0-8493-4622-3/93/$0.00 + $.50

I. INTRODUCTION

The existence of tissue renin-angiotensin systems (RAS), independent of the circulating RAS, became apparent in the early 1970s. During this time renin was purified from the mouse submaxillary gland[1] and experiments on the brain pointed to an independent brain RAS with evidence for renin,[2,3] angiotensin converting enzyme,[4] angiotensin,[5] and angiotensinogen.[6] Renin activity was also measurable in the mesenteric and splanchnic vascular beds.[7] However, although measurements of components of the RAS was possible, the question could also be raised whether they were taken up from blood-borne components of peripheral RAS or synthesized independently in the different tissues. Various approaches to answer this question were taken. One was to carry out bilateral nephrectomy to remove the source of circulating RAS. We reported that after nephrectomy, plasma renin activity (PRA) was abolished but spontaneously hypertensive rats (SHR) still responded to an angiotensin antagonist in the brain with a decrease in blood pressure.[8] Aguilera et al.[9] confirmed that 48 h after bilateral nephrectomy, PRA levels were zero, but found that angiotensin levels were still present in plasma in measurable amounts. It was argued that the source of this angiotensin after nephrectomy had to be nonrenal. This approach has recently been revisited by Trolliet and Phillips,[10] who have found that plasma angiotensin II (AngII) persists (10%) in rats maintained by chronic dialysis without kidneys for 5 d. Nussberger et al.[11] have argued that the angiotensin after nephrectomy is due to *in vitro* production during assay procedure. However, we have found that leaving the AngII at 4°C for 0, 10, or 60 min after collection does not alter the levels of AngII measured. The evidence points to an extrarenal source of angiotensin, persisting after bilateral nephrectomy.

Campbell[12] compared the metabolism of AngI and II in peripheral vascular beds by measuring the arterial-venous differences. He found that venous levels were higher than arterial levels. This implied that AngII, which is normally degraded during one pass through the same tissue, must be produced by the perfused tissue. Another approach to show the independence of the tissue RAS from the circulating RAS was the use of cell cultures. Cell cultures in isolated media or in controlled media in which AngII levels could be defined were used in studies with neurons, glial cells, and neuroblastoma. All the components of an independent RAS have been determined in these cultures.[13,14] The most convincing evidence for an independent tissue RAS, however, has come from the use of molecular biology using probes for angiotensinogen[15] and renin mRNA,[16] which has established that a variety of tissues express these mRNAs, indicating that they have the synthetic apparatus to make angiotensin products in those tissues. This approach has been confirmed and extended in several tissues and cell cultures with the mRNAs transcribing the genes for the RAS.[17-24]

The significance of all this evidence is that the RAS functions in different ways.[16,23] In the blood it acts as an endocrine system forming AngII extracellularly and delivering AngII by blood to various target tissues. In tissue, an independent RAS exists by which AngII may have both paracrine and autocrine activities. Further, within the cells the RAS may mediate intracrine actions. In this review evidence for these systems in multiple tissues is discussed in relation to the molecular biology of the system.

II. THE ENDOCRINE RENIN-ANGIOTENSIN SYSTEM

The classic endocrine RAS has become better understood with the new data revealing its molecular basis. This is briefly reviewed.

A. RENIN

The aspartyl protease which cleaves angiotensinogen to form AngI is synthesized as a prezymogen. The nucleotide sequence of the genome is 12,000 bp with 10 exons separated by 9 introns. The 10 exons correspond to a 1500-nucleotide mRNA.[25,26] The renin gene of the endocrine system is expressed in the kidneys in the juxtaglomerular cells (JG) (Figure 1). Regulation of the 5′-flanking region of the renin gene by promoters, enhancers, and regulatory elements have been detailed by Dzau et al.[16] Translation of

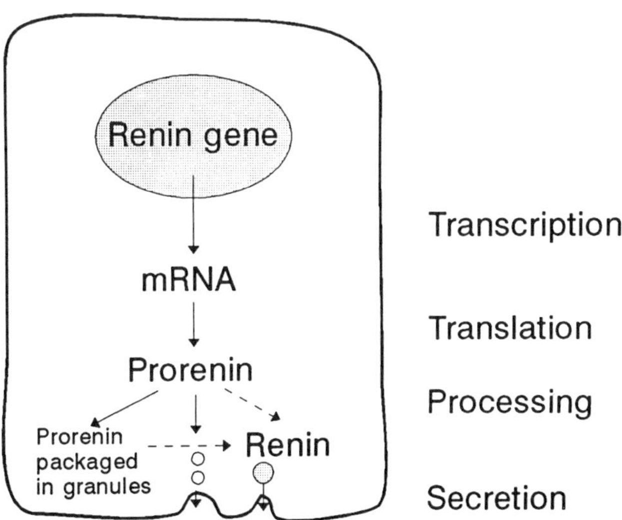

FIGURE 1. Diagram of a renin-producing cell. This would represent the juxtaglomerular cells of the kidney and renin- and prorenin-secreting cells of the tissue renin-angiotensin system. Active renin is secreted from granules by exocytosis under cellular regulation, inactive renin (prorenin) is secreted constitutively. Dotted arrows indicate putative synthetic steps from prorenin to renin.

renin mRNA produces pre-prorenin with a 2-kDa signal peptide. The signal peptide is removed by the endoplasmic reticulum and a 5-kDa oligosaccharide added to produce the zymogen-prorenin.

Prorenin (47 kDa) is packaged in immature storage granules or secreted constitutively directed from the Golgi apparatus as inactive renin. The kidney is the primary source of this inactive renin.[27] The active renin (41 kDa) is packaged in maturing granules in which the prorenin is cleaved and glycosylated. Renin secretion from these granules is regulated. This process occurs inside the JG cells which are derived from smooth muscle cells in the afferent arteriole of the renal glomeruli. Antibodies to renin show dark staining in these cells when applied by immunocytochemistry.[28] The conversion of prorenin to renin is not yet understood. *In vitro,* prorenin can be converted to renin by low temperature (4°C) and by acid, but neither condition would occur physiologically. Prorenin is biologically inactive and the levels of prorenin are not correlated to levels of renin released.[28] Thus, there is a posttranslational inhibition of prorenin conversion to renin. The source of this inhibition and the conditions which regulate conversion of prorenin to active renin are being investigated. Recent evidence points to the kallikrein system and serine proteases such as tissue plasminogen activator (tPA) playing a role in this conversion.[30,16] Recombinant human renin has been produced by transfection of cultured mammalian cells with a human renin gene construct. X-ray crystallography of this renin is being used to study the residues at the active sites of cleavage.[31] Renin is an unusually specific aspartyl protease which specifically acts on angiotensinogen as a substrate.

B. ANGIOTENSINOGEN

Angiotensinogen is the only known precursor of angiotensin peptides and limits the enzymatic reaction of renin by its availability. In plasma the concentration is 1 to 2 mg AngI/ml. The major source of the substrate for the endocrine RAS is the liver.[31] The molecular aspects have been recently reviewed by Lynch and Peach.[32] Angiotensinogen is a 55- to 60-kDa α-2-glycoprotein with a sequence of 14 amino acids[33] and is derived from a preform of 477 amino acids. Rat liver cDNA has been cloned by Ohkubo.[34] It is 12 kbp and has 5 exons, separated by 4 introns. S1 nuclease mapping showed there are several sites of polyadenylation,[34] although the significance for the RAS is not clear. A partial length preangiotensinogen cDNA from rat was cloned by Lynch et al.[21] The angiotensinogen gene has multiple regulatory sites which are responsive to glucocorticoids, estrogens, and cytokines. Cleavage by renin to the product AngI occurs at the leucine[10]-valine[11] bond for which renin is highly specific. The regulation of angiotensinogen has been extensively studied.

Plasma angiotensinogen levels are elevated by estrogens, nephrectomy, and thyroid and glucocorticoid hormones. Angiotensinogen mRNA levels in liver are increased 2- to 5-fold by a 4-h infusion of AngII.[35] Inhibiting AngII with enalapril (3 mg/kg po) decreased angiotensinogen mRNA levels by 50%, while saralasin also decreased the mRNA expression (45%). Thus, AngII regulates the gene expression of liver angiotensinogen. Angiotensinogen mRNA in liver is also increased by Na^+ depletion and decreased by Na^+ repletion.[36] Angiotensinogen is secreted constitutively from the liver into the circulation and the levels of plasma angiotensinogen remain relatively constant despite the multiple fold increase in mRNA produced by these regulating substances.

C. ANGIOTENSIN-CONVERTING ENZYME (ACE)

When angiotensin I is produced it has to be cleaved at the phe^8-his^9 dipeptide bond of his-leu at the carboxyl terminus. This is the action of ACE, a dipeptidyl carboxyl peptidase (E.C.3.4.15.1), which is bound to membranes of endothelial cells. ACE has a molecular weight of 130 to 169 kDa and is a zinc (Zn)-containing metalloprotein exopeptidase. The cDNA for human pulmonary ACE has been cloned[37] and a cDNA sequence corresponding to the 5′ extremity of the mouse kidney ACE mRNA.[38] The DNA structure is similar to other Zn metallopeptidases, collagenase, neutral endopeptidase, and thermolysin. There is variation in the molecular weights of ACE from different tissues. They may arise from different genes or from the same gene by alternate splicing of RNA. In order for AngI to be converted to AngII with sufficient speed and quantity in the circulation, it was calculated that conversion must occur in the lungs where ACE is found in high quantity and where the total surface area of the pulmonary capillaries would allow the requisite speed of conversion. While the lungs are the major site of conversion, other sites exist in the kidney, liver, and systemic vascular beds. The detached carboxyl-terminal dipeptide his-leu does not possess any cardiovascular effects (unpublished observations). The specificity of ACE for angiotensinogen is weak. ACE has low substrate specificity with the major requirements being (1) a precarboxylic acid group and (2) absence of proline which inhibits its action. In addition to the phe-his bond, ACE cleaves a number of dipeptide bonds and also some tripeptide sequences with a free carboxyl terminal. Thus, ACE also acts on bradykinin, metenkephalin, substance P, luteinizing hormone releasing hormone (LHRH), and neurotensin. ACE inactives the nonapeptide bradykinin (arg-pro-pro-gly-phe-ser-pro-phe-arg) by hydrolytic removal of phe-arg. The importance of the kininase II activity of ACE, whereby the vasodilatory effects of bradykinin are terminated, is still an unresolved issue in hypertension research. Inhibition of kininase II by ACE inhibitors (ACEI) would cause a decrease of bradykinin breakdown, enhancing its vasodilatory effects, as well as inhibiting the conversion of AngII from AngI and reducing the vasoconstrictor effects of AngII.

ACE is a membrane-bound peptidase with an anchor peptide in the membrane and a zinc ectoenzyme on the luminal surface of the membrane. Thus, ACE conversion of AngI and AngII occurs extracellularly in the lungs.

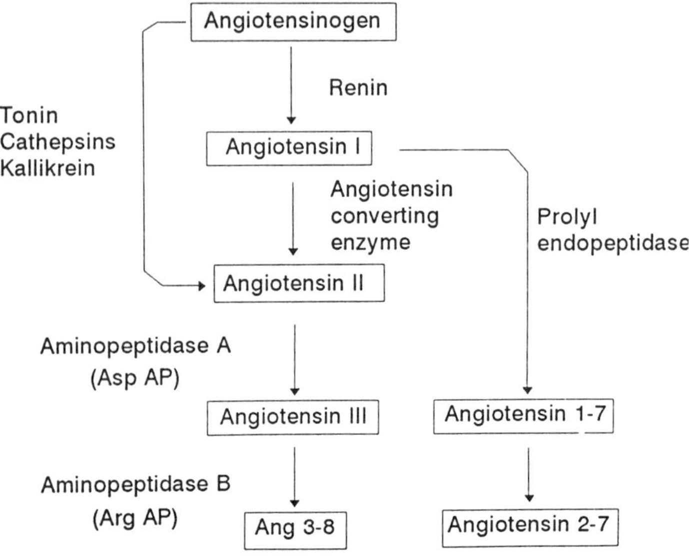

FIGURE 2. Synthetic pathways of the renin-angiotensin system.

D. METABOLISM OF ANGIOTENSIN

Angiotensin II in plasma has a half life of about 30 s. It is metabolized by aminopeptidase A, the aspartyl-aminopeptidase (Asp AP) (EC 3.4.11.) which by removal of the amino-terminal Asp^1 produces des-Asp^1-AngII (AngIII). The AngIII is then metabolized in part by arginyl aminopeptidase (Arg AP) (EC 3.4.11.6).[39] AngIII has physiological effects on aldosterone release and vasoconstriction. The angiotensins are further metabolized to the hexapeptide, Ang 3-8. This metabolite may prove to be an active peptide, as Swanson et al.[40] have proposed specific receptors exist for the hexapeptide. The aminopeptidases are in both the soluble and membrane-bound forms in different tissues. There are alternate routes of angiotensin metabolism (Figure 2). One is via prolyl endopeptidase action on AngI which produces the carboxyl-terminal removal of phe^8 to produce angiotensin 1-7[41] This is further metabolized to Ang 2-7 by aminopeptidase action. In addition to these alternate pathways, there have been several suggestions over the years that angiotensinogen may be cleaved without renin by tonin or cathepsins to form AngII directly. In the endocrine RAS the known pathways appear to be via renin to AngII and then metabolism to AngIII. The more recently discovered carboxyl-terminal deleted Ang 1-7 has been intensively investigated but has not been shown to have a vasoconstrictive action or to mobilize calcium as AngII does.[42] Ang 1-7 may be important for the tissue RAS but not the endocrine RAS. Although these alternate pathways exist, the question is, do they act physiologically? Tonin is a slow-acting enzyme. Cathepsin D works only at low pH since it is a lysosomal enzyme. Cathepsin G, and endothelial cell protease, acts very fast. Other contenders for alternate, nonrenin routes of AngII production include members of the serine protease family which share in common homologies of the serpin gene. These include the elastase, α-1 protease, nixin, antithrombin III, and plasminogen activating factor. They are related to angiotensinogen which raises the thought that while angiotensinogen may be the precursor for AngI and the specific substrate for renin, it lives a double life as a serine protease.

E. RECEPTORS

For the system to have any effect it has to work through specific angiotensin receptors. (These are reviewed in Chapters 9 and 14.) After numerous attempts to synthesize the receptor through solubilization and purification procedures, the use of molecular cloning techniques have produced the isolation of a bovine[43] and the rat vascular smooth muscle[44] cDNA for the AT_1 receptor. The AT_1 is a receptor recognized by the recent introduction of specific AngII antagonists for AT_1 receptors (DuP 753) and AT_2 receptors (PD 123177, CGRP). The AT_1 receptors are distinguished from the AT_2 receptors by being G-protein-coupled receptors and linked to phosphatidyl inositol hydrolysis through phospholipase C. Using primers corresponding to the cloned rat cDNA for AT_1 receptors, Iwai and Inagami[45] have isolated a subtype of the AT_1 receptor in the rat adrenal. Thus, there are two subtypes of the AT_1 receptor, designated AT_1-A and AT_1-B, with a nucleotide sequence that is highly homologous and yet distinct, despite a 96% identity. This subtyping might help to explain the diversity of AngII actions in various tissues. Further work by others has revealed that the AT_1-A receptor is insensitive to PD 123177 and sensitive to nanomolar amounts of DuP 753 (losartan), whereas the AT_1-B receptor is slightly sensitive to PD 123177 (μM) and less sensitive than the AT_1-A to losartan (μM range). The AT_1-B receptor is sensitive to AngIII. The potential sites for posttranslational modification of the receptor subtypes are being studied by the polymerase chain reaction (PCR) method using primers corresponding to regions of the extracellular loops and N-terminus of the putative seven-domain transmembrane molecule.

So far the vasoconstrictive and fluid balance effects of circulating AngII appear to be mediated by the G-protein-coupled receptors. A clear distinction between type AT_1 and AT_2 receptors occurs in the adrenal gland where the adrenal medulla has predominately AT_2 receptors and the adrenal cortex AT_1.[46] The strongest data for second messenger linkage to the AT_2 receptor is inhibition of cyclic GMP.[47] The importance of these receptors for the endocrine RAS are related both to its effective role in cardiovascular function and in its own regulation by hormonal feedback. High angiotensin levels are known to inhibit renin release from the kidney and this process involves both neurogenic and humoral signals. Some of these signals could be stimulated by AngII acting on receptors in the brain and other tissues. The endocrine RAS has multiple tissue target sites of action and delivers AngII, III, 3–8, and 1–7 to the receptors specific for these angiotensins.

III. TISSUE RENIN-ANGIOTENSIN SYSTEMS

The concept of the tissue RAS is that the close proximity of cells in tissue allows angiotensin produced by the cells to act in a paracrine or autocrine fashion.[16,23,24] This concept implied that within each tissue there is a separate synthetic mechanism for the production of AngII. In the case of the paracrine function, one cell produces AngII and delivers it by regulated secretion to a neighboring target effector cell which binds and responds to the AngII. The advantage of this system lies in its proximity. Since the cells are close together, AngII is not degraded and responses can occur rapidly. In the autocrine concept, the cell produces AngII, releases it, and this feeds back via membrane receptors onto the same cell to regulate the rate of synthesis. Intracrine function refers to the intracellular effect of AngII and these may be related to the possibility of AngII receptors in the nucleus regulating DNA pretranscriptional events, as well as mRNA transcription of growth factors and the cell regulatory proteins. Since we now know the structure of the trans-membrane seven domain AT_1 receptor,[43,44] we can also hypothesize that some of the internal loops may serve as receptor sites for intracellular AngII. Preliminary evidence with an antibody to an amino acid peptide, corresponding to the amino acid sequence of the third cytosolic internal loop of the published AT_1 receptor, shows intracellular staining by immunocytochemistry in brain and kidney cells. The following is a brief review of evidence for a tissue paracrine RAS in different organs. In each case we have considered the mRNA, angiotensin levels, and regulation.

A. BRAIN

The separation of the endocrine from the tissue RAS may seem arbitrary, if not wrong, for research on the vascular system. The tissue may receive the products of the endocrine system and act on them or incorporate them into the tissue. In the brain, however, the tissue is protected from circulating AngII by the blood brain barrier except in the limited locations of the circumventricular organs. Thus, an independent brain tissue RAS has been investigated. The brain contains mRNA for renin[19,48] but the levels are low and the distribution is dissimilar to the distribution of angiotensinogen mRNA. Figure 3 shows mRNA for renin in the cerebral cortex. We also found renin mRNA expressed in brain stem in very low abundance. Renin was measurable in cultured brain cells and a surprising finding was strong staining for renin antibody in glial cells.[49] Angiotensinogen mRNA is expressed in the brain, but a debate has arisen whether it is found only in glial cells or also in neurons.

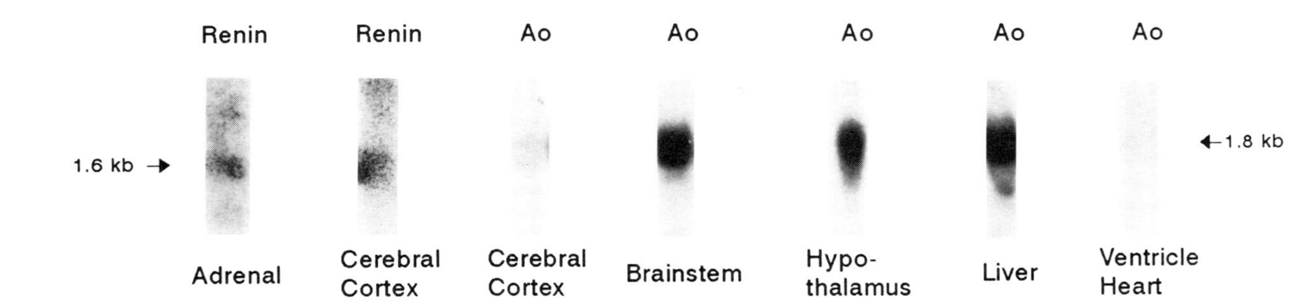

FIGURE 3. Northern blot of renin and angiotensinogen mRNA in rat tissues. RNA was extracted from tissue with guanidine thiocyanate followed by phenol chloroform and precipitated with isopropanol according to the method by Chomczynski and Sacchi.[147] 20 μg total RNA (for hypothalamus, 10 μg) were separated on agarose-formaldehyde gels, transferred to nylon membranes, and hybridized with riboprobes (Promega Gemini System II) for angiotensinogen or renin. (Renin and angiotensinogen cDNA in pGEM-4 was a gift from Dr. K. Lynch.) The strongest signal for angiotensinogen mRNA was in the liver, followed by the hypothalamus and brainstem; cortex and heart ventricle had weak signals. Renin mRNA was detected in the cortex in the brain and also in the adrenal. (From Phillips et al., *Regul. Pept.*, 43, 1, 1992. With permission.)

Stornetta et al.,[50] using *in situ* hybridization, localized angiotensinogen mRNA in cells that stained for glial fibrillary acid protein (GFAP) which is a marker of glial cells. Lynch et al.[51] found extensive preangiotensinogen mRNA in *in situ* hybridization which was not clustered in neuronal groups. Deschepper and Ganong,[52] using angiotensinogen antibodies, reported that angiotensinogen in the hypothalamus was localized in glial cells. Thus, it seems that glial cells possess the substrate to produce AngII. This would fit a paracrine model of tissue RAS where the glial cells are the synthesizer cells and the neurons the target-effector. However, in a preliminary study of glial and neuronal cells in culture, we found both glial cells and neurons expressed angiotensinogen mRNA in the SHR and WKY rats but the signal was extremely low in the neurons of the Sprague Dawley strain. Direct measurements of angiotensinogen in 95% pure neuronal cultures and 100% pure glial cultures by the use of an angiotensinogen antibody with radioimmunoassay and high-performance liquid chromatography (HPLC) also demonstrated quantifiable amounts of angiotensinogen in both neurons and glial cells.[53] In both studies the question of glial contamination in the neurons can be raised. Recently, Thomas et al.[54] have shown the distribution of angiotensinogen in the brain by immunocytochemistry and apart from the hypothalamus, both neurons and glial cells stain distinctly for angiotensinogen. Thus, the claim that angiotensinogen mRNA is only expressed in glial cells may depend on the region of the brain that is investigated. Although it is not clear how the angiotensinogen in glial cells in the hypothalamus interacts with the neuronal cells of the hypothalamus, other evidence points to a complete RAS in neural tissue. Okamura et al.[55] demonstrated renin AngI, AngII, and ACE in neuroblastoma cells. A similar finding was reported by Fishman et al.[13] in the neuroblastoma × glioma hybrid cell line. The experiments on cells in culture led to the first hypothesis of an endogenous, intracellular RAS.[56] ACE mRNA is expressed in choroid plexus, caudate putamen, cerebellum, brainstem, and hippocampus.[57] Therefore, although mRNA for renin, angiotensinogen, and ACE have been found in the brain, their individual location and abundance do not provide evidence for single cell RAS and suggest extracellular synthesis of AngII. Receptors for angiotensin have been amply demonstrated in brain tissue[58-60] and neuronal cultures.[61] Both AT_1 and AT_2 receptor subtypes exist in the adult rat brain. The AT_1 receptors are distributed in areas associated with the cardiovascular effects of central AngII. The AT_2 receptors are distributed in areas associated with movement control. In the fetal brain the predominant receptor is AT_2.[62] In addition to the classical angiotensin receptor, Hanley[63] has argued that the mammalian proto-oncogene mas is an angiotensin receptor which is preferentially sensitive to AngIII over AngII in brain tissue. MAS has been found in the cerebral cortex, hippocampus, cerebellum, and olfactory bulb. It is exclusively expressed in neurons.[64] This distribution is similar to the distribution found by *in situ* hybridization of ACE mRNA.[57] The full range of ligands for this receptor are not known. The production of angiotensin fragments, which have been shown to have central effects, may indicate specific angiotensin receptors and MAS may serve as a receptor for one of these metabolites.

AngII, electrolytes, and sex steroids appear to regulate gene expression leading to AngII production in the brain. We have measured levels of AngII in multiple blocks of brain tissue using radioimmunoassay and HPLC.[65,66] The results are shown in Table 1. The highest levels were found in the hypothalamus, which fits with the earlier picture from immunocytochemistry of AngII in the paraventricular and supraoptic nuclei.[67] There is a gender difference between the levels of AngII in male and female brains. AngII levels in male rat brains, compared to female brains, are consistently higher in all tissues except the olfactory bulbs. Olfactory bulb levels are very low in male rats and not exceptionally high in female rat brains. New evidence for the regulation of the brain RAS being independent of the circulating RAS, comes from the recent study by Trolliet and Phillips[10] showing that in rats chronically dialyzed for 5 d after bilateral nephrectomy, levels of brain AngII are increased above control levels. This regulation appears to depend on the level of K^+ and when K^+ is elevated after nephrectomy, the levels of brain AngII also increase (Figure 4). When K^+ is reduced after nephrectomy by low K^+ diet, brain AngII also decreases. This is confirmed in our recent study with angiotensinogen mRNA expression where low K^+ shows a decrease in angiotensinogen mRNA in the hypothalamus, whereas mesenteric artery angiotensinogen mRNA is increased.[68] Bilateral nephrectomy with chronic dialysis represents a low-angiotensin model. The opposite model is the 2 kidney-1 clip (2K1C) rat which has high renin and high angiotensin. In the 2K1C rat brain, levels of AngII are not increased significantly, suggesting that brain RAS is independent of peripheral RAS. In the genetically hypertensive rat (SHR) the levels of brain AngII are significantly increased compared to the

TABLE 1
Angiotensin II Levels in Brain

Location	Number	Male	Number	Female
Amygdala	17	75.4 ± 11.2	7	35.6 ± 2.7
Brainstem	44	89.2 ± 16.1	13	52.5 ± 7.3
Cerebellum	17	65.9 ± 8.2	13	50.8 ± 9.0
Cortex	68	75.2 ± 15.3	15	81.3 ± 30.6
Hippocampus	14	36.1 ± 10.5	7	50.4 ± 12.5
Hypothalamus	74	102.1 ± 9.8	15	69.3 ± 12.7
Olfactory bulbs	5	6.0 ± 4.0	7	63.1 ± 16.5
Pituitary	2	190.5 ± 74.5	20	22.6 ± 66.0
Spinal cord	30	90.4 ± 16.2	13	44.8 ± 8.0
Striatum	1	26	6	35.2 ± 18.0

Note: Levels are expressed as picograms AngII per gram tissue (mean ± standard error).

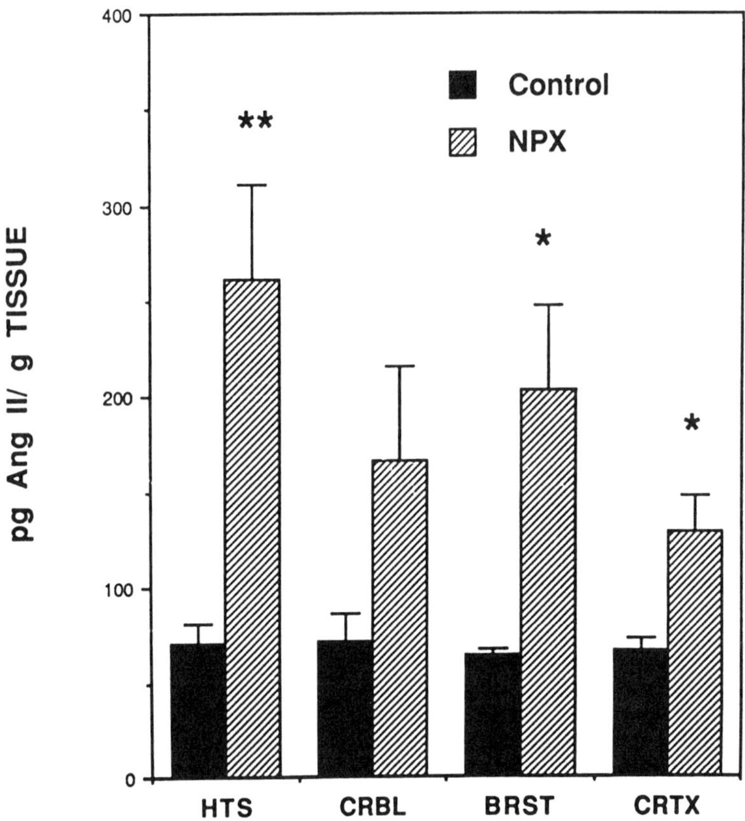

FIGURE 4. Levels of brain angiotensin measured 5 d after bilateral nephrectomy with chronic intraperitoneal dialysis to maintain rats in excellent health. HTS, Hypothalamus; CRBL, cerebellum; BRST, brainstem; CRTX, cortex. (From Trolliet, M.R. and Phillips, M.I., *J. Hypertens.*, 10, 29, 1992. With permission.)

normotensive (WKY) control rats.[69] These differences are maintained in both the prehypertensive and the hypertensive phases of hypertension.[69] A study of mRNA expression for angiotensinogen in these rats revealed that the highest level appeared in the prehypertensive phase at 4 weeks of age.[70] Thus, we hypothesize that early exposure to increased expression of angiotensinogen in the prehypertensive phase may contribute to the development of hypertension in the hypertensive phase. A test of this is with central infusion of ACE inhibitors (ACEI) which shows that hypertension does not develop if ACE is inhibited early in life. There is evidence that ACEIs given chronically in rats 4 to 8 weeks of age crosses the blood brain barrier and could inhibit the central synthesis of AngII (see Chapter 8).

B. PITUITARY

Angiotensin II levels in the pituitary are higher per gram of tissue than in the brain (Table 1). The concentration is approximately 190 pg AngII/g tissue in male rats.[65] In female rat brains, levels are 200 pg AngII/g and can rise during proestrus induced by high estrogen and progesterone levels. A similar level is found in sheep pituitary and pig pituitary by our radioimmunoassay for AngII (unpublished observations). The problem with measurements in rat pituitary is that the low protein level in rats requires that several pituitaries be collected for accurate measurement. Renin mRNA is found in the rat pituitary by *in situ* hybridization in both the anterior and intermediate lobes but not in the posterior lobe.[17,20,71] Angiotensinogen mRNA, however, although it is abundant in brain, was low or undetectable in the pituitary gland.[15,21] Immunocytochemical staining with renin antibody and angiotensin antibody showed that they were colocated in gonadotroph cells of the anterior pituitary.[20,71] Imboden et al.[72] have shown immunocytochemical localization of fibers containing AngII-like immunoreactivity in the neurohypophysis and measurements of AngII in the neurohypophysis have yielded moderate levels slightly than those of the anterior pituitary.[73] AngII-like protein is rich in the median eminence.[67] This might imply that AngII produced in the hypothalamus where the levels are high, is transported to the median eminence, and either travels to the neural lobe or is released into the portal system to arrive at the anterior pituitary. Efforts to show release of angiotensin through the portal vessel, however, have not met with success. Therefore, the angiotensin and renin found in anterior pituitary cells appears to be endogenous to the pituitary gland. Since angiotensinogen mRNA is apparently lacking from the pituitary, the possibility that angiotensin is taken up from the blood remains.

However, if uptake occurs, it is selectively located in the gonadotrophs which appear to be the only location of angiotensin in the anterior pituitary. The pituitary RAS fits with a paracrine functioning model. It has been suggested that AngII mediates LHRH stimulation of prolactin.[74] In this paracrine model the synthesizing cells are gonadotrophs and the AngII released stimulates inositol phosphate and prolactin secretion in lactotrophs.

C. VASCULAR

Re[75] showed that there was renin in the dog aorta, using an antibody to renin. Since then the components of the RAS have been discovered but some of the components are in different layers of the vessel wall. Angiotensinogen mRNA and renin mRNA are found in the endothelium and the medial and advential layers. ACE is also found in the endothelium but does not appear in the deeper layers of the vessel wall. ACE mRNA is expressed in endothelial cells only[36,76] and has not been found in the medial or adventitial layers. Angiotensinogen mRNA, however, is expressed only in adventitia, and surprisingly in the fatty tissue surrounding the blood vessels.[22,77] Measurements in our laboratory of mesenteric artery and endothelial cells, show levels of angiotensin of 184 ± 10 pg AngII/g. Unger et al.[78] have raised the question whether it is necessary to postulate a separate tissue RAS for vascular cells. They point out that the uptake of renin could occur from plasma.[79] Since ACE is an ectoenzyme anchored in the membrane of the endothelial cells,[76] its active site is lumenal and they argue that ACE could act on lumenally delivered AngI and produce AngII outside the cell. The AngII would then be taken up by deeper tissues of the vascular wall. Arguments in favor of an endogenous, vascular tissue RAS come from the following experiments. Cultured vascular cells synthesize and secrete AngII.[80] Using a new renin inhibitor for rabbit renin, Higashimori and colleagues[81] have shown that AngII production can be suppressed in isolated mesenteric arteries of the rabbit by the renin inhibitor. This suggests that there is local inhibition of AngII synthesis, independent of blood-borne renin. Trolliet et al.[68] found that angiotensinogen mRNA in the mesenteric artery was increased in the bilaterally nephrectomized, dialyzed rat. Since the nephrectomized rats have greatly reduced circulating angiotensin and an absence of plasma renin activity, this data supports an endogenous, vascular RAS. A vascular RAS does not preclude AngI being produced in the endothelial cell, secreted, and converted extracellularly by ACE to produce AngII. The AngII could either be taken up into the same cell (autocrine action) or delivered to neighboring cells such as vascular smooth muscle (paracrine action).

A case can be made for the paracrine action of AngII between endothelial cells as synthesizers and smooth muscle cells as target cells. AngII has been shown to cause hypertrophy in cultured, quiescent, rat aortic smooth muscle cells. Naftilan et al.[82,83] reported the induction of c-fos expression in smooth muscle which was independent of protein synthesis. Using nuclear runoff transcription assay, the rise in c-fos mRNA was found to be due to an increase in transcription rate. This finding has led to a series of studies elegantly summarized by Dzau et al.[84] on the role of vascular AngII in the mechanisms of growth of vascular smooth muscle. Blood vessel walls must maintain their tone in order to respond to vasoconstrictors. The finding that AngII induces the proto-oncogene c-fos and c-jun[82] indicates a mechanism by which AngII could control growth, tone, hyperplasia, and in pathophysiological states, induce hypertrophy of blood vessels. The c-fos and c-jun proto-oncogene produces Fos and Jun which combine with a leucine zipper to become the AP-1 protein that interacts with the regulatory elements of DNA. The products are growth factors such as platelet derived growth factor (PDGF) which stimulates mitogenesis and the hyperplasia of cells. The idea put forward by Dzau et al.[84] is very intriguing. At the same time the AP-1 induces PDGF, it also induces transforming growth factor beta (TGFβ), which inhibits growth. When the two factors are equal, hyperplasia is prevented. As PDGF begins to predominate, hyperplasia results. Further reduction of the TGFβ inhibition results in polyploidy plus hyperplasia. If PDGF totally dominates as the AngII-stimulated cell product, then hypertrophy will result. By this mechanism, endogenous vascular AngII may be involved in the increased wall thickness associated with hypertension and also with repair and atherosclerotic plaque in vascular injury. While these growth effects could result from circulating AngII in high renin hypertension, the vascular RAS would play the vital role in structural responses to high blood pressure and vascular injury in normal or low renin hypertension.

Injury to the neointima of the blood vessel wall produces a rapid regrowth of the neointimal cells.[85] This is a major problem in the treatment of stenosis by arterial balloon catheterization which is frequently negated by subsequent restenosis. Powell et al.[85] showed that pretreating rats with captopril and other ACEIs prevented the intimal proliferation. This important result was interpreted in terms of lowering the levels of vascular AngII released by injury to the endothelial cells and thereby inhibiting the AngII from stimulating through the c-fos and c-jun-AP-1 pathway, growth factors that lead to the myointimal proliferation.[85] Further studies will demonstrate if the AngII receptors involved in this process are of the AT_1 or AT_2 subtype and if receptor blockade is as effective as ACEIs in preventing post-injury neointimal proliferation. In terms of a paracrine function, the endothelial cell as a synthesizer and the smooth muscle cell as a target effector cell (for growth and contraction), fits quite well except for the mystery of angiotensinogen mRNA.

D. HEART

In the past it was believed that the beneficial effects of ACEI in clinical practice was solely due to their action on reducing the workload of the heart. However, many studies have demonstrated that antihypertensive therapy caused different effects on cardiac mass. Treatment of aortic banded rats with ACEI which did not cause a reduction in blood pressure of afterload[86,87] resulted in the regression of heart hypertrophy. Hence, evidence for a dissociation between changes in blood pressure and cardiac mass gave indirect support for the possibility that heart tissue RAS acts as a stimulator of cardiac growth and causes changes in cardiac mass.

Messenger RNA for components of the cardiac RAS have been detected in cardiac tissue using methods of Northern hybridization, S1 nuclease protection assay, and *in situ* hybridization,[86,88,89] but the exact location of the system is still unknown. In view of the diversity of cell type in cardiac tissue, the RAS could be present in cardiomyocytes, vascular smooth muscle cells, or endothelial cells.

Angiotensinogen mRNA is expressed in the atria,[15,86,90-96] in the ventricles,[86,90,92,93] and whole heart[89,93] in rat and mouse. Estimates of angiotensinogen mRNA levels in the rat heart were approximately 5% of that in the rat liver.[19]

Renin mRNA has been isolated from heart tissue of both mice and rats. mRNA for renin is expressed in the ventricles, atria, and whole heart.[74,89,91,94] Dzau et al.[89] reported that the level of heart renin mRNA was 2% of the level in the kidney. Renin activity has been detected in isolated left ventricular rat cardiomyocytes[95] and in cardiac tissue of mice.[19,89] Using the method of *in situ* hybridization, renin and angiotensinogen mRNA have been found to be evenly distributed in all four cardiac chambers of neonatal rat heart[88] but are expressed to different extents in the cardiac chambers of adult rat hearts,[92] with angiotensinogen mRNA higher in the atria than in the ventricles.

Recently, mRNA for the converting enzyme (ACE) has been located in left ventricular tissue of rats.[98] Converting enzyme activity has been reported in atria and ventricles[57,96,99,100] of rats. The location of ACE in the rat heart was studied by *in vivo* autoradiography and found to be at many sites in the heart, including the coronary vasculature, atrial and ventricular myocardium, and the valve leaflets in the rat heart.[101] Northern blot analysis of ACE mRNAs showed a double band, 4 to 5 kb in size when using a pulmonary ACE mRNA probe to hybridize heart tissue but a testis ACE mRNA probe was not effective.[57]

Therefore, mRNA for all components of RAS synthesis have been identified in mammalian heart tissue and there is evidence that these mRNAs are translated to proteins. Figure 5 summarizes the reported location and relative mRNA levels of the components of the RAS in the heart.

There is evidence that the components of the RAS found in the heart interact to form AngII. AngII levels have been measured in the rhesus monkey heart, and were highest in the right atrium, with decreasing levels in the right ventricle, left atrium, interventricular septum, and left ventricle.[102] There are moderately low but detectable levels of AngII in whole rat heart and in specific areas of the heart: right atrium, 71.6 ± 25.8; left atrium, 31.7 ± 5.8; right ventricle, 67.8 ± 5.9; and left ventricle, 54.5 ± 6.1 pg AngII per gram of tissue.[97]

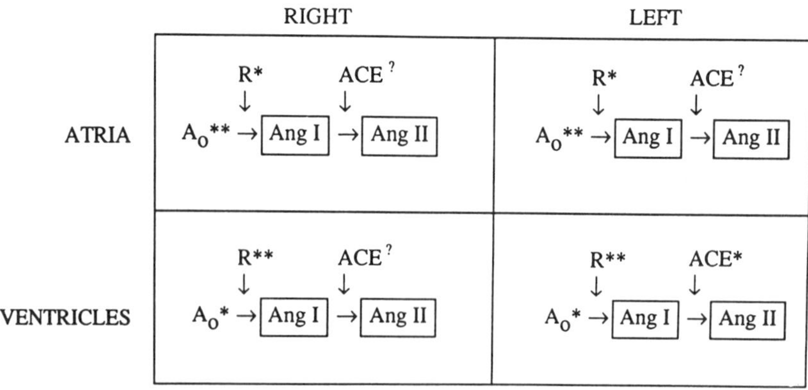

FIGURE 5. Diagram of the distribution of the renin-angiotensin system in the heart. A_o, angiotensinogen; R, renin; ACE, angiotensin converting enzyme; AngI, angiotensin I; AngII, angiotensin II; □, peptide; *<** indicates relative levels of mRNA for a component of the RAS.

Several methods have been used to differentiate between peptides synthesized in the heart and those taken up from the circulation *in vivo*. In nephrectomized rabbits with low levels of circulating hormone, atrial AngII levels have been shown to be reduced by treatment with ACE inhibitors,[102] suggesting the existence of a local RAS. The use of the isolated perfused heart[102] demonstrates that AngI is converted to AngII by an independent functional RAS in the heart.

Regulation of RAS has been demonstrated at the molecular level under various physiological conditions and stimuli. Glucocorticoid, estrogen, and thyroid hormone increase angiotensinogen mRNA levels in atria and ventricles.[15,92] A reduced dietary sodium intake resulted in an increase in angiotensinogen mRNA and an increase in renin mRNA expression in the heart.[19] Lindpainter et al.[102] demonstrated a fall in angiotensinogen mRNA expression when bilaterally nephrectomized rabbits were treated with ACE inhibitors. Pressure overload on the heart resulting in heart hypertrophy is associated with a rise in ACE content and mRNA levels for ACE and angiotensinogen.[86,98,100]

Evidence for AngII receptors in heart tissue gives support for a physiological role of the RAS and AngII peptide in cardiac tissue. Lory et al.[103] reported that AngII receptors were expressed in *Xenopus* oocyte following the injection of rat heart mRNA. Binding sites for AngII have been found on the cell membrane of cardiac muscle (sarcolemmal membrane preparations) of rabbit,[104,105] rat,[106,107] and human[108] myocardium. Both high-[101,104,105,107] and low-affinity[104,107] binding sites have been reported. High-affinity receptors have been reported in the rat conduction system.[106] Both AT_1 and AT_2 receptors are equally distributed in the atria and ventricular myocardium[109] of the rat and the ventricular myocardium of the rabbit.[110] Autoradiographic data from human heart show that receptors are localized in the myocardium, coronary vessels, and sympathetic nerves.[108]

Many years ago it was proposed that AngII may have a direct effect on the nucleus of the cell, since it has been found that labeled AngII injected into the left ventricles of rats was taken up and localized in the nuclei of cardiac muscle cells.[111] Support for this idea comes from the report that AngII can bind nuclear chromatin and cause solubilization.[112] AngII may therefore act directly on the nucleus and cause the subsequent stimulation of transcription. This could be a mechanism for intracrine effects of AngII if the source of the AngII were from the same cell.

Evidence for the involvement of AngII and the cardiac RAS in growth and hypertrophy of the heart has accumulated from several studies. There is indirect evidence for AngII as a growth factor by the identification of the mas proto-oncogene product as an angiotensin receptor.[63,113] SHR with cardiac hypertrophy show increased expression of renin mRNA[102] when compared to normotensive control rats. Increased cardiac ACE activity[114,115] and angiotensinogen gene expression[86] in hypertrophic left ventricles of aortic banded rats, and converting enzyme activity[100] in the left ventricle after coronary occlusion stress suggest a role for the RAS in the pathogenesis of cardiac hypertrophy. Significantly higher ACE activity in left hypertrophied heart of rats with chronic aortic stenosis[98] and higher angiotensinogen mRNA and ACE activity were found in ventricular myocardium of rats with tachypacing heart failure than those of untreated rats.[96] In studies of cultured chick embryonic cardiomyocytes *in vitro,* AngII induces cellular hypertrophy, shown as an increase in protein synthesis which can be blocked by an AngII antagonist.[116] The proposed AngII receptors in the nucleus and by mas oncogene would promote protein synthesis and cell growth in myocardial cells.

The local physiologic effects of a cardiac RAS are a positive inotropic action in rabbit aorta[105] increased contractility[117] and a positive chronotropic effect in rat cardiomyocytes.[118] Therefore, in addition to the growth promoting action of AngII, it can also cause direct positive tropic actions. The cardiac RAS may play an important role in modulating cardiac function under normal and changing conditions. What is lacking is more information on the cellular location of the cardiac RAS since there are several cell types in heart.

E. KIDNEY

In a series of studies, Celio and Inagami,[119] using polyclonal antibodies, demonstrated that angiotensin and renin coexist in the kidney. They found angiotensin-like and renin-like immunostaining located in the JG cells of the kidney. The appropriate mRNAs have been demonstrated in kidney tissue of mice and rats. Mouse submandibular gland renin cDNA was used to demonstrate the presence of renin mRNA in mouse kidney.[89,120-122] In the rat, nick-translated angiotensinogen cDNA (derived from partial length rat liver angiotensinogen cDNA) was hybridized with specific mRNA sequences in rat kidney. Rat kidney angiotensinogen mRNA was estimated to be 5% of the rat liver angiotensinogen mRNA concentration. In mice the amount was 20%.[122] Renin mRNA was also found in the kidney at about 20% of the level found in the submandibular gland in mouse.[89,122] Thus, both renin and angiotensinogen mRNA are detectable in kidney at moderate levels compared to the high levels of liver. Although the renin was derived from mouse renin and expressed in the rat kidney, the size was the same (1.6 kb).[89] Studies in our laboratory have generally found kidney angiotensinogen mRNA to be low compared to other tissue angiotensinogen mRNA. This may represent a difference in our procedure. Ingelfinger et al.[120] use tissue that is snap-frozen in liquid nitrogen. We use tissue collected on dry ice. However, the difference might imply that kidney angiotensinogen mRNA abundance is lower or not as stable as the angiotensinogen mRNA is in other tissues. In the study by Campbell and Habener[15] of angiotensinogen mRNA, kidney levels were low compared to the levels in the liver (approximately 3%) and about 10% of levels in the mesentery, spinal cord, and brain. The kidney RAS is regulated by several factors. Kidney angiotensinogen mRNA was increased by dexamethasone, ethylestradiol, and triiodothyronine compared to controls. The kidney RAS is regulated by Na^+ volume reduction and high AngII. Ingelfinger et al.[120] showed that renin mRNA and angiotensin mRNA in the kidney are increased by Na^+ depletion compared to levels under Na^+ repletion. The effect with Na^+ depletion was also found by Iwao et al.,[36] but under conditions of Na^+ repletion, they found decreases in mRNA levels. Pratt et al.[123] found a 50% increase in Na^+-depleted WKY but no change in SHR. It is known that AngII suppresses plasma renin activity by inhibiting renin synthesis or release.[124] Johns et al.,[125] using a rat renin cDNA, showed that the renin gene expression in the kidneys increased in density as demonstrated by Northern blot analysis and also showed a spread of renin immunostaining along the efferent arteriole when an ACEI was given. The spread of renin mRNA was inhibited by AngII. It was concluded that ACEI stimulates renin gene expression and AngII suppresses renin gene induction.

The control of transcription may be more complicated than that. In a model of hypertension, the 2K 1C rat, renin mRNA was elevated in the clipped kidney but not the unclipped kidney during the first 4 weeks and significantly elevated in both nonclipped and clipped kidney in the chronic phase.[126] The results in the acute stage indicate that there is a posttranscriptional inhibitory event which decreases the renin activity. Using enzymatic dispersion and ultracentrifugation in Percoll to separate a population of renin-containing rat kidney cells, Johns et al.[127] showed that the strongest signal for renin mRNA was in the cortical extract of the kidney. The authors imply that this renin-rich population of kidney cells may include cells which are not JG cells. A physiological study with angiotensin infusions and the ACEI, enalapril, showed that renin mRNA in the kidney is regulated by AngII plasma levels.[34] Gomez et al.[128] did not find significant changes of angiotensinogen mRNA in the kidneys after treatment with ACEI. In the study by Nakamura et al.,[35] AngII for 4 h decreased the renin mRNA level but had no effect on angiotensinogen mRNA. Thus, there seems to be a modulation of gene regulation of the kidney RAS by levels of AngII. ACE mRNA (based on pulmonary ACE) was strongly expressed in kidney.[57] ACE has been localized to the brush border of the proximal tubule[28] but the localization of ACE mRNA has not yet been localized in the kidney tubules.

From reviewing these studies it appears that the kidney tissue RAS genes are independently regulated and angiotensinogen expression may be independent of the circulating RAS. The possibility of an intercellular paracrine system exists but data are fragmentary and proof is lacking for kidney RAS without more knowledge of the cellular molecular biology of the RAS.

F. ADRENAL

The adrenal gland contains the highest level of angiotensin that we have measured in tissue. The mean level was 3320 ± 281 pg AngII/g (Table 2). Further analysis showed that the majority of this AngII came from the cortex, although the medulla contains AngII. Within the cortex the majority of AngII is in the zona glomerulosa (ZG) but it is also distributed unevenly in the zona fasciculata (ZF). Immunocytochemical staining with an AngII antibody

TABLE 2
Angiotensin II Concentration in Different Tissues[66]

Tissue	pg Ang II/g (±SE)	Number
Aorta	85±22	2
Mesenteric artery	184±10	16
Kidney	142±6	7
Left ventricle	54±6	4
Right ventricle	68±6	4
Adrenal	3320±281	9
Brown fat	440±5	2
White fat	52±7	3
Ovary	259±92	6
Uterus	147±12	6
Spleen	2134±1330	5

confirms this distribution (Figure 6). Doi et al.[129] and Naruse et al.[130] had earlier shown that immunoreactive renin was present in the rat adrenal cortex and the levels of adrenal renin are independent of plasma or renal renin. The amount of renin mRNA in the adrenal is low compared to that in kidney and heart.[17] Angiotensinogen mRNA in the adrenal cortex that we have observed has a low abundance (Figure 3). The signal is not strong, especially considering the high amounts of AngII compared to the angiotensinogen mRNA density in other tissues with lower AngII levels. Published Northern blots of angiotensinogen mRNA in mouse adrenal gland and rat adrenal is low compared to liver.[90] Adrenal renin is regulated by AngII.[131] It seems surprising that such a high level of AngII would be so close to the dense distribution of AngII receptors. The AngII receptors in the adrenal cortex are distributed throughout the ZG and also in the ZF. These receptors are of the AT_1 type.[46] In the medulla of the adrenal gland the receptors are of the AT_2 subtype.[46] The assumption is that the AngII is involved in aldosterone secretion. However, this function can be adequately handled by circulating AngII. The presence of high concentrations of AngII close to high concentrations of AngII receptors warrants further consideration from a physiological point of view. Are the ZF cells synthesizers for the ZG cells, and do ZG cells take up AngII by internalization of their dense receptor distribution?

G. SALIVARY GLAND

The mouse maxillary gland is unique in having high levels of renin and it was from this gland that renin was first purified.[1] Even among mice, only a few strains have the high renin in the submaxillary gland (SMG). Those mice with high renin have two renin genes, the *ren-1* and *ren-2*. The mice with low renin or normal renin in salivary glands, have a single renin gene (*ren-1*). The 5′ sequences of *ren-1* and *ren-2* are homologous. *Ren-1* has a

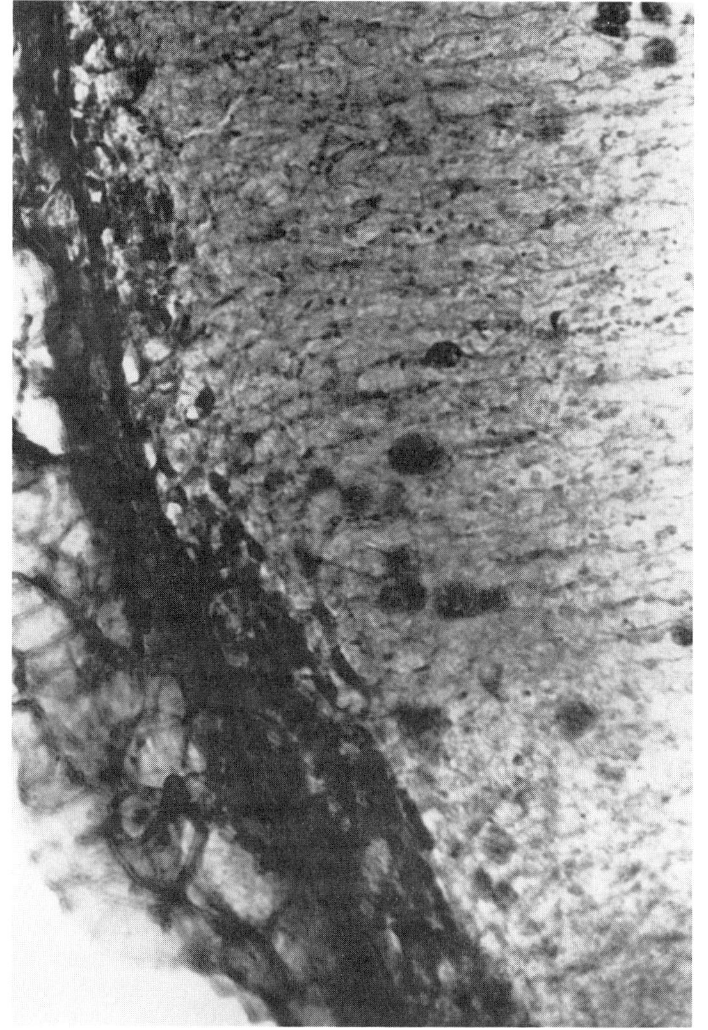

FIGURE 6. Immunocytochemical staining for AngII-like protein in the zona glomerulosa and zona fasciculata of the adrenal gland.

cAMP regulatory region which is interrupted in the *ren-2* gene. (A full description of these genes can be found in Chapters 1 and 2.) The purification of mouse SMG renin was critical in the development of new methodologies that led to the purification of renal renin.[29,132] The two allelic designations for the renin regulator locus (Rnr) were based on whether the mice had high levels of renin (Swr strain) and/or low SMG renin, (C57PBL/6 strain).[133] In both strains the renin levels were regulated by testosterone. The high renin levels do not appear in the salivary glands in other species and renin mRNA has not been shown in rat.[39] The mouse salivary gland has not been pursued as a model for the tissue RAS. It has been pursued as a model for studying renin cDNA sequence, the amino acid sequence, and as a cDNA probe for other cDNA genomic clones for other species.[122,132] Renin expression is in the granula convoluted tubule cells which make up the glandular epithelium of the SMG. In males the gene is regulated at puberty by testosterone and also by thyroxine.[122] It would seem that the SMG is the Cinderella of the tissue RAS story and might be used as a model for the study of tissue RAS in culture to test whether the granular cells of the tubules have paracrine and autocrine function involving AngII.

H. TESTES

The testes have cells and the developing sperm protected from blood by a blood-testes barrier. Therefore, the access of blood-borne renin is restricted to those parts of the Sertoli cells protected by the barriers. Renin and renin mRNA have been detected in the Leydig cells of a mouse tumor cell line, cloned from the C57B1/6J strain of mice[122] and rat.[17] Angiotensinogen mRNA was not detected in rat testes.[34] The secretion of renin and prorenin is low in testes.[132] Angiotensin I, II, and III were measured in the Leydig cells and confirmed by HPLC and radioimmunoassay.[134] The majority of angiotensins were AngI, followed by AngII, and a small amount of AngIII. AngII was released into the culture medium by treatment with luteinizing hormone, while AngI was only minimally released by gonadotropins.[134] Thus, the testes may be an example of a tissue RAS independent of plasma renin and angiotensin, which is regulated by extracellular peptides, the gonadotropins, and causes the release of AngII preferentially over the release of AngI and renin. The Leydig cells also contain cDNAs encoding rabbit testicular ACE which have been cloned and sequenced.[135] Oligonucleotide probes derived from testicular ACE cDNA demonstrated expression of mRNA only in the testes and not in heart, brain, kidney, or lung.[57] Thus, the testicular ACE is highly specific. It is distributed in the epididymis on the lumen wall.[135] cDNA to mouse SMG renin mRNA has been used as a hybridization probe for *in situ* hybridization in the rat testes. A positive signal was detected in the cytoplasm of Leydig cells.[20]

Since Leydig cells are part of a paracrine system, namely the interaction of Leydig cells with the Sertoli cells, the testes may offer a model of tissue RAS in a paracrine process whereby AngII is synthesized in one type of cell, released, and transported to the nearby Sertoli cells, where it would produce a functional response. What that function is — androgen release or spermatogenesis, androgen binding protein release — is not known.

I. SPLEEN

In the spleen we find a very high, though variable, concentration of AngII (2134 ± 1330 pg AngII/g, n = 5). Renin mRNA has not been found in spleen.[22] Suzuki et al.[17] measured renin mRNA, derived from a rat renin cRNA, by solution hybridization with Northern blotting. They detected high quantities of renin in kidney, atria, ventricle, and testes and modest levels in tissue known to have low renin i.e., the whole brain and hypothalamus, yet they could not detect any mRNA in spleen or liver.[17] Angiotensinogen mRNA, however, has been detected in spleen, although it is in low abundance.[135] Given the lack of renin mRNA in the spleen, it is possible that angiotensin synthesis occurs by one of the alternate pathways referred to earlier (Figure 2). The spleen is high in cathepsin D, which under the right conditions (low pH), could cleave angiotensinogen directly to AngII. The AngII that we have measured from spleen is predominately AngII (1–8) as shown by HPLC. AngII receptors have been localized in the red pulp in the spleen of rats and mice[136] and these are specific since they are antagonized by the AngII antagonist, saralasin. The role of AngII in the spleen may be local, on splenic volume and blood flow or in the modulation of lymphocyte function. The connection between tissue RAS and components of the immune system is gradually unfolding. Binding sites for AngII have been localized on lymphocytes[137] and SHR rats which are believed to have overactive AngII expression have depressed T-lymphocyte populations in the prehypertensive and posthypertensive phases of hypertension.[138]

J. PANCREAS

Campbell and Habener,[22] using angiotensinogen mRNA based on the rat angiotensinogen gene, were unable to detect any signal by Northern blotting of rat pancreas RNA. More recently, Chappell et al.[139] have investigated dog pancreas and reported evidence of angiotensinogen mRNA expression using a human angiotensinogen cDNA as a hybridization probe. They detected amounts of angiotensinogen mRNA that they calculated to be approximately 2% of that in the liver. In addition, they measured AngII levels and found moderately high quantities (524 ± 74 fmol/g tissue for AngII, 221 ± 54 fmol/g for AngIII, and 156 ± 21 fmol/g for Ang 1–7). There was no regional difference in the distribution of the peptide within the pancreas. These levels are considerably higher than blood levels, which would imply that they represent endogenously synthesized angiotensins. However, neither AngI nor renin activity were found in the pancreas. The significance of this finding for tissue RAS awaits further evidence before it can be categorically stated that AngII is not taken up from the plasma but synthesized through one of the angiotensin alternate pathways in cells of the pancreas.

K. OVARIES

In the rat, *in situ* hybridization identified renin mRNA in the corpus luteum.[140] The production of renin in humans is from thecal, stromal, and luteal cells.[141] Thus, there is a discrepancy between the results. Prorenin, as well as renin, angiotensinogen, and AngII immunoreactivity, have been demonstrated in human ovarian follicular fluid.[142,143] The main source of renin in the ovary is the theca cells.[144] Angiotensin receptors have been studied in the ovarian follicles of rat ovaries by autoradiography.[145] The receptors were localized exclusively to a subpopulation of follicles occurring on the granulosa cells. AngII is present in the whole ovary and levels in the ovary, as we have measured them, are 259 ± 92 pg/AngII/g (Table 2). The tissue is highly vascularized and the AngII and receptors may be associated with blood vessels. Granulosa and theca cells are primarily regulated by pituitary follicular stimulating hormone (FSH), luteinizing hormone (LH), and prolactin. Their function is to secrete estrogens. The human ovary synthesizes and secretes prorenin and the concentration of prorenin in the ovarian follicular fluid is much higher than renin in the plasma.[141] Prorenin is secreted continuously by the ovary during pregnancy.[143] This is in contrast to the male, where renin and prorenin secretion are extremely low in testes. Although the details are far from complete, the paracrine model for the ovary may exist between the theca cells as the synthetic cell which secrete prorenin and renin, and the granulosa cells which have AngII binding sites and could be the target cells. The production of AngII by theca cells or by extracellular synthesis needs to be determined.

L. ADIPOSE TISSUE

When angiotensinogen mRNA was closely studied in the rat aorta by dissecting apart the vessel wall and dispersing cells, it was discovered that the major source of angiotensinogen mRNA was in the periaortic brown adipose tissue and to a lesser extent in the adventitia.[22,77] Both hybridization *in situ* to brown adipose tissue and Northern blot analysis confirm the presence of angiotensinogen mRNA equivalent to 5% of the content in control liver. The levels of angiotensinogen mRNA could be raised by treatment with dexamethasone, ethylestradiol, or T_3. This important finding by two groups independently, casts some doubt on previous measurements of angiotensinogen mRNA in rat aorta and shows the need for *in situ* hybridization in this research. Brown fat contains a high level of AngII (440 ± 5 pg/AngII/g) and white fat a lesser, but modest, level of 104 ± 9 pg/g (Table 2). The presence of brown adipose tissue, which is highly vascularized, raises the possibility of a paracrine function between vascular cells and adipose cells. In this model the adipose cells are the synthesizers in which angiotensinogen generates angiotensin via a converting enzyme and the angiotensin is released to act on the nearby target cell — the vascular smooth muscle cells. The associated vascular smooth muscle contract in response to AngII by phospholipase C stimulation and calcium entry. Such a paracrine system would have consequence for obesity and hypertension. In obese patients, the larger amount of higher vascularized fat tissue would be a perfect setting for the secretion of locally formed AngII onto adjacent blood vessels causing vasoconstriction and raising total peripheral resistance. Another function of fat, particularly brown fat, is its vital role in thermogenesis. The presence of norepinephrine fibers innervating the fat would fit well with the known action of AngII in facilitating norepinephrine release. Again, this would be an appropriate example of paracrine action between an adipocyte producing AngII and an adjacent neurite as a target cell, containing norepinephrine vesicles but not containing angiotensinogen mRNA or angiotensins. Brown fat may have high levels of catecholamines because of this paracrine interaction between nerves and adipose AngII.

IV. SUMMARY AND FUTURE TRENDS

While the story is by no means finished, a picture is emerging of tissue RAS which is independent of the classical endocrine RAS and the fragmentary evidence, so far, fits with the idea of a paracrine function in many tissues. In a paracrine system one cell is the synthesizer and another cell is the target. Synthesizer cells may be target cells for the same and other hormones and target cells may be synthesizers of peptides. However, in the theoretical sense we may conceive of two neighboring cells in tissue, one of which has the genes for synthesizing AngII and the evidence for the transcription of these genes is expression of mRNA and posttranslational events which allow the processing, packaging, and secretion of the RAS components (Figure 7). The cell is subject to regulating influences. The target cell has angiotensin receptors on the membrane and possibly in the nucleus, linked to second messenger and third messenger systems.

In summary, evidence for high levels of renin mRNA have been found in the SMG of certain strains of mice (but not in rat or low renin strains of mice), the heart, and testes (Leydig cells). Moderate to very low levels of renin mRNA have been found in the brain, adrenals, kidney, and pituitary. Absence of renin mRNA has been reported in small intestine, spleen, liver, and pancreas. Angiotensinogen mRNA has high levels in brain, spinal cord, mesenteric arteries, adipose tissue, and medium to low levels in the lung, atria, kidney, adrenal gland, large intestine, and stomach. It is absent in the testes and rat SMG. ACE mRNA, while not as well studied, is high in choroid plexus striatum, kidney, and lung, low in cerebellum, hippocampus, and testes, except when testicular ACE is tested. ACE is present in liver. AngII concentration is high in the adrenal gland and spleen, moderately high in brown fat and ovary, moderate in mesenteric artery, kidney, brain tissue, pituitary, and uterus, and low in left ventricle, right ventricle, and testes. Regulators of the tissue RAS include stimuli which increase components of

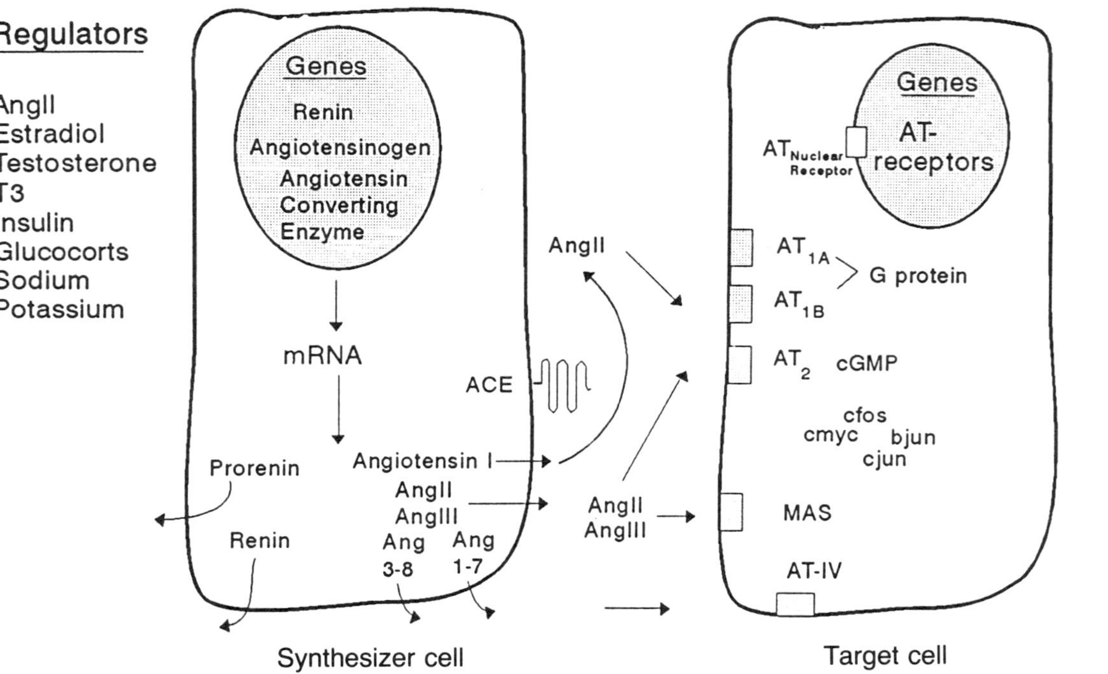

FIGURE 7. Model of the tissue RAS paracrine system. One cell is the synthesizer cell in which prorenin, renin, and angiotensins are produced. The AngII may be synthesized inside the cell or may be converted extracellularly by extracellular ACE and then delivered to the target cell via the interstitial fluid. The target cell, which may also synthesize angiotensin, receives the angiotensin from the nearby synthesizer cells to receptor subtypes which stimulate second and third messengers that regulate the cellular and nuclear processing of the target cell. Some of the AngII may be endocytosed and transported to nuclear receptors. For examples of this paracrine arrangement see text. (From Phillips et al., *Regul. Pept.*, 43, 1, 1992. With permission.)

the RAS, estradiol, nephrectomy, T-3, low sodium, high potassium, and those which decrease the system include glucocorticoids, insulin, low AngII, high sodium, and high potassium.

The localization of mRNA expressing RAS components in different tissues has established tissue RAS, but to understand its local actions, more cellular specificity needs to be defined. Further understanding of the cellular physiology of tissue RAS may come with the development of isolated cell lines which express RAS components. Sigmund and Gross[122] have proposed using SV40-T antigen to develop a library of renin expressing cells with different tissue and gene identifications. The strategy is to use the SV40-T antigen to express tissue- and cell-specific renin regulatory regions. In this way a variety of specific cell types will be perpetuated through multiple passages with renin secretion intact. Another development is the use of transgenic mice[121] and transgenic rats.[146] The expression of an inserted gene product or products of the RAS can be studied in these animals for their pathophysiological effects. For example, the renin gene has been inserted in the transgenic rat with activation of their RAS and hypertension is a consequence. The question of whether hypertension requires overexpression of one gene, such as renin, or multiple genes can be investigated in this model. Also the tissue- or cell-type specificity, which ultimately depends on the gene, can be tested. By the same logic, treatment for pathophysiological states induced through transgenic insertion can be studied by deletion of the appropriate gene.

REFERENCES

1. **Cohen, S., Taylor, J.M., Murakami, K., Michelakis, A.M., and Inagami, T.,** Isolation and characterization of renin-like enzymes from mouse submaxillary glands, *Biochemistry,* 11, 4286, 1972.
2. **Ganten, D., Minnich, J.L., Granger, P., Hayduk, K., Brecht, H.M., Barbeau, A., Boucher, R., and Genest, J.,** Angiotensin forming enzyme in brain tissue, *Science,* 173, 64, 1971.
3. **Hirose, S., Yokosawa, H., and Inagami, T.,** Immunochemical identification of renin in rat brain and distinction from acid proteases, *Nature,* 274, 392, 1978.
4. **Yang, H.T. and Neff, N.H.,** Distribution and properties of angiotensinogen converting enzyme of rat brain, *J. Neurochem.,* 19, 2243, 1972.
5. **Phillips, M.I.,** Angiotensin in the brain, *Neuroendocrinology,* 25, 354, 1978.
6. **Lewicki, J.A., Fallon, J.H., and Printz, M.P.,** Regional distribution of angiotensinogen in rat brain, *Brain Res.,* 159, 359, 1978.
7. **Ganten, D., Fuxe, J., Phillips, M.I., and Mann, J.F.E.,** The brain isorenin angiotensin system: biochemistry, localization and possible role in drinking and blood pressure regulation, in *Frontiers in Endocrinology,* Ganong, W.F. and Martini, L., Eds., Raven Press, New York, 1978, 61.

8. **Phillips, M.I., Mann, J.F.E., Haebara, H., Hoffman, W.E., Dietz, R., Schelling, P., and Ganten, D.,** Lowering of hypertension by central saralasin in the absence of plasma renin, *Nature,* 270, 445, 1977.
9. **Aguilera, G., Schivar, A., Bankal, A., and Catt, K.J.,** Circulating angiotensin II and adrenal receptors after nephrectomy, *Nature (London),* 289, 507, 1981.
10. **Trolliet, M.R. and Phillips, M.I.,** The effect of chronic bilateral nephrectomy on plasma and brain angiotensin, *J. Hypertens.,* 10, 29, 1992.
11. **Nussberger, J., Fluckiger, J.-P., Hui, K.Y., Evequoz, D., Waeber, B., and Brunner, H.R.,** Angiotensin I and II disappear completely from circulating blood within 48 hours after binephrectomy: improved measurement of angiotensins in rat plasma, *J. Hypertens.,* 9 (Suppl. 6), S230, 1991.
12. **Campbell, D.J.,** The site of angiotensin production, *J. Hypertens.,* 3, 199, 1985.
13. **Fishman, M.C., Zimmerman, E.A., and Slater, E.E.,** Renin and angiotensin: the complete system within the neuroblastoma × glioma cells, *Science,* 24, 921, 1981.
14. **Raizada, M.K., Phillips, M.I., Crews, F.T., and Sumners, C.,** Distinct angiotensin II receptor in primary cultures of glial cells from rat brain, *Proc. Natl. Acad. Sci. U.S.A.,* 84, 4655, 1987.
15. **Campbell, D.J. and Habener, J.F.,** Angiotensinogen gene is expressed and differentially regulated in multiple tissues of the rat, *J. Clin. Invest.,* 78, 31, 1986.
16. **Dzau, V.J., Burt, D., and Pratt, R.E.,** Molecular biology of the renin-angiotensin system, *Am. J. Physiol.,* 255, F563, 1988.
17. **Suzuki, F., Ludwig, G., Hellmann, W., Paul, M., Lindpainter, K., Murakami, K., and Ganten, D.,** Renin gene expression in rat tissues: a new quantitative assay method for rat renin mRNA using synthetic cRNA, *Clin. Exp. Hypertens. A,* 10, 345, 1988.
18. **Dzau, V.J., Brody, T., Ellison, K.E., Pratt, R.E., and Ingelfinger, J.R.,** Tissue-specific regulation of renin expression in the mouse, *Hypertension,* 9 (Suppl. 3), 36, 1987.
19. **Dzau, V.J., Ingelfinger, J.R., Pratt, R.E., and Ellison, K.E.,** Identification of messenger RNA sequences in mouse and rat brains, *Hypertension,* 8, 544, 1986.
20. **Deschepper, C.F., Mellon, S.H., Cumin, F., Baxter, J.D., and Ganong, W.F.,** Analysis by immunocytochemistry and *in situ* hybridization of renin and its mRNA in kidney, testis, adrenal and pituitary of the rat, *Proc. Natl. Acad. Sci. U.S.A.,* 83, 7552, 1986.
21. **Lynch, K.R., Simnad, V.T., Ben-Ari, E.T., Maniatis, T., Zinn, K., and Garrison, J.C.,** Localization of preangiotensinogen messenger RNA sequences in the rat brain, *Hypertension,* 8, 540, 1986.
22. **Campbell, D.J. and Habener, J.G.,** Cellular localization of angiotensinogen gene expression in brown adipose tissue and mesentery: quantification of messenger ribonucleic acid abundance using hybridization in situ, *Endocrinology,* 121, 1616, 1987.
23. **Campbell, D.J.,** Tissue renin-angiotensin system: sites of angiotensin formation, *J. Cardiol. Pharmacol.* Suppl., 7, S1, 1987.
24. **Phillips, M.I.,** Functions of brain angiotensin, *Annu. Rev. Physiol.,* 49, 413, 1987.
25. **Hardman, J.A., Hort, Y., and Catanzaro, D.F.,** Primary structure of human renin gene, *DNA,* 3, 457, 1984.
26. **Hobart, P.M., Fogliano, M., O'Connor, B.A., Schaefer, I.M., and Chirgwin, J.M.,** Human renin gene: structure and sequences analysis, *Proc. Natl. Acad. Sci. U.S.A.,* 81, 5026, 1984.
27. **Kim, S., Hosoi, M., Nakajima, K., and Yamamoto, K.,** Immunological evidence that kidney is primary source of circulating inactive prorenin in rats, *Am. J. Physiol.,* 260, E526, 1991.
28. **Taugner, R., Hackenthal, E., Rix, R., Nobiling, R., and Poulsen, K.,** Immunocytochemistry of the renin-angiotensin system, *Kidney Int.* Suppl., 12, 533, 1982.
29. **Corvol, P. and Menard, J.,** Renin purification and cloning, *Hypertension,* 18(3), 252, 1991.

30. **Hsueh, W.A., Carlson, E.J., and Israel-Hagelman, M.,** Mechanism of acid activation of renin: role of kallikrein in renin activation, *Hypertension,* 3 (Suppl. I), 22, 1981.
31. **Sielecki, A.R., Hayakawa, K., Fujinaga, M., Murphy, M.E., Fraser, M., Muir, A.K., Carilli, C.T., Lewicki, J.A., Baxter, J.D., and James, M.N.,** Structure of recombinant human renin, a target for cardiovascular-active drugs, at 2.5 Å resolution, *Science,* 243, 1346, 1989.
32. **Lynch, K.R. and Peach, M.J.,** Molecular biology of angiotensinogen, *Hypertension,* 17, 3, 1991.
33. **Hilgenfeldt, U.,** Half-life of rat angiotensinogen: influence of nephrectomy and lipopolysaccharide stimulation, *Mol. Cell Endocrinol.,* 56, 91, 1988.
34. **Okhubo, S.O.,** Tissue distribution of rat angiotensinogen mRNA and structural analysis of its heterogeneity, *J. Biol. Chem.,* 261 (1), 319, 1986.
35. **Nakamura, A., Iwao, H., Fukui, K., Kimura, S., Tamaki, T., Nakanishi, S., and Abe, Y.,** Regulation of liver angiotensinogen and kidney renin mRNA levels by angiotensin II, *Am. J. Physiol.,* 258, E1, 1990.
36. **Iwao, H., Fukui, K., Kim, S., Nakayama, K., Ohkubo, S., Nakansishis, S., and Abe, Y.,** Sodium balance effects on renin angiotensinogen and atrial natriuretic peptide mRNA levels, *Am. J. Physiol.,* 255, E219, 1988.
37. **Bernstein, K.E., Martin, B.M., Bernstein, E.A., Linton, J., Striker, L., and Striker, G.,** The isolation of angiotensin converting enzyme cDNA, *J. Biol. Chem.,* 163, 11021, 1988.
38. **Soubrier, F., Alhenc-Gelas, F., Hubert, C., Allegrini, J., Treager, J.M., and Corvol, P.,** Two putative active centers in human angiotensin I converting enzyme revealed by molecular cloning, *Proc. Natl. Acad. Sci. U.S.A.,* 85, 9386, 1988.
39. **Abhold, R.H. and Harding, J.W.,** Metabolism of angiotensin II and III by membrane bound peptides from rat brain, *J. Pharmacol. Exp. Ther.,* 245, 171, 1988.
40. **Swanson, G.N., Hanesworth, J.M., Sardinia, M.F., Coleman, J.K.M., Wright, J.W., Hall, K.L., Miller-Wing, A.V., Stobb, J.W., Cook, V.I., Harding, E.C., and Harding, J.W.,** Discovery of a distinct binding site for angiotensin II (3 – 8), a putative angiotensin IV receptor, *Regul. Pept.,* 40(3), 409, 1992.
41. **Chappell, M.C., Brosnihan, K.B., Diz, D., and Ferrario, C.M.,** Identification of angiotensin 1 – 7 in rat brain, *J. Biol. Chem.,* 164, 16518, 1989.
42. **Ferrario, C.M., Brosnihan, K.B., Diz, D.I., Jaiswal, N., Khosla, M.C., Milsted, A., and Tallent, E.A.,** Angiotensin (1 – 7): a new hormone of the angiotensin system, *Hypertension,* 17 (Suppl III), 126, 1991.
43. **Sasaki, K., Yamano, Y., Bardhan, S., Iwai, N., Murray, J.J., Hasegawa, M., Matsuda, Y., and Inagami, T.,** Cloning and expression of a complementary DNA encoding a bovine adrenal angiotensin II type-1 receptor, *Nature,* 351, 230, 1991.
44. **Murphy, T., Alexander, R.W., Griendling, K.K., Runge, M.S., and Bernstein, K.E.,** Isolation of a cDNA encoding the vascular $type_1$ angiotensin II receptor, *Nature,* 351, 233, 1991.
45. **Iwai, N. and Inagami, T.,** Identification of two subtypes in the rat type 1 angiotensin II receptor, *FEBS Lett.,* 298 (2,3), 257, 1992.
46. **Chiu, A., Herblin, W.F., McCall, D.E. et al.,** Identification of angiotensin II receptor subtypes, *Biochem. Biophys. Res. Commun.,* 165, 196, 1989.
47. **Sumners, C., Tang, W., Zelenza, B., and Raizada, M.K.,** Angiotensin II receptor subtypes are coupled with distinct signal-transduction mechanisms in neurons and astrocytes from rat brain, *Proc. Natl. Acad. Sci. U.S.A.,* 88, 7567, 1991.
48. **Ekker, M., Tronik, D., and Rengeon, F.,** Extrarenal transcription of the renin genes in multiple tissues of mice and rats, *Proc. Natl. Acad. Sci. U.S.A.,* 86, 5155, 1989.
49. **Hermann, K., Phillips, M.I., Hilgenfeldt, U., and Raizada, R.,** Biosynthesis of angiotensinogen and angiotensin by brain cells in primary culture, *J. Neurochem.,* 51, 398, 1988.

50. **Stornetta, R.L., Hawelyn-Johnson, C.L., Guyenet, P.G., and Lynch, K.R.,** Astrocytes synthesize angiotensinogen in brain, *Science,* 242, 1444, 1988.
51. **Lynch, K.R., Hawelyn-Johnson, C.L., and Guyenet, P.G.,** Localization of brain angiotensinogen mRNA by hybridization histochemistry, *Mol. Brain Res.,* 2, 149, 1987.
52. **Deschepper, C.F. and Ganong, W.F.,** Renin and angiotensin in endocrine glands, in *Frontiers in Neuroendocrinology,* Martini, L. and Ganong, W.F. Eds., Raven Press, New York, 1988, 79.
53. **Hermann, K., Raizada, M.K., Sumners, C., and Phillips, M.I.,** Presence of renin in primary neuronal and glial cells from rat brain, *Brain Res.,* 437, 205, 1987.
54. **Thomas, W.G., Greenland, K.J., Shinkel, T.A., and Sernia, C.,** Angiotensinogen is secreted by pure rat neuronal cell cultures, *Brain Res.,* in press.
55. **Okamura, T., Clemens, D.L., and Inagami, T.,** Renin, angiotensin and angiotensin converting enzyme in neuroblastoma cells. Evidence for intracellular formation of angiotensins, *Proc. Natl. Acad. Sci., U.S.A.,* 78, 6940, 1981.
56. **Inagami, T.,** Renin in the brain and neuroblastoma cells: an endogenous and intracellular system, *Neuroendocrinology,* 35, 475, 1982.
57. **Whiting, P., Nava, S., Mozley, L., Eatham, H., and Poat, J.,** Expression of angiotensin converting enzyme mRNA in rat brain, *Mol. Brain Res.,* 11, 93, 1991.
58. **Song, K., Allen, A.M., Paxinos, G., and Mendelsohn, F.A.O.,** Mapping of angiotensin II receptor subtype heterogeneity in rat brain, *J. Comp. Neurol.,* 316, 467, 1992.
59. **Tsutsumi, K. and Saavedra, J.M.,** Quantitative autoradiography reveals different angiotensin II receptor subtypes in selected rat brain nuclei, *J. Neurochem.,* 56, 348, 1991.
60. **Speth, R.C., Rowe, B.P., Grove, K.L., Center, M.R., and Saylor, D.,** Sulfhydryl reducing agents distinguish two subtypes of angiotensin II receptors in the rat brain, *Brain Res.,* 548, 1, 1990.
61. **Raizada, M.K., Yang, J.W., Phillips, M.I., and Fellows, R.E.,** Rat brain cells in primary culture: characterization of angiotensin II binding sites, *Brain Res.,* 207, 343, 1981.
62. **Cook, V.I., Grove, K., McManamin, K.M., Carter, M.R., Harding, J.W., and Speth, R.C.,** The AT_2 angiotensin receptor subtype in the 18 day gestation fetal rat brain, *Brain Res.,* 560, 334, 1991.
63. **Hanley, M.R.,** Molecular and cell biology of angiotensin receptors, *J. Cardiol. Pharmacol.,* 18 (Suppl. 2), S7, 1991.
64. **Bunneman, B., Fuxe, K., Metzger, R., Mullin, J., Jackson, T.R., Hanley, M.R., and Ganten, D.,** Autoradiographic localization of the MAS proto-oncogene mRNA in adult rat bran, *Neurosci. Lett.,* 114, 147, 1990.
65. **Phillips, M.I. and Stenstrom, B.,** Angiotensin II in rat brain comigrates with authentic angiotensin II in HPLC, *Circ. Res.,* 56, 212, 1985.
66. **Phillips, M.I., Kimura, B., and Raizada, M.K.,** Measurement of brain peptides: angiotensin and ANP in tissue and cell culture, *Methods Neurosci.,* 6, 177, 1991.
67. **Phillips, M.I., Weyhenmeyer, J.A., Felix, D., and Ganten, D.,** Evidence for an endogenous brain renin angiotensin system, *Fed. Proc.,* 38, 2260, 1979.
68. **Trolliet, M., Rydzewski, B., Raizada, M.K., and Phillips, M.I.,** Effect of electrolyte balance on steady-state levels of angiotensin mRNA in chronically nephrectomized rats, *FASEB J.,* 5 (Abstr.), 1818, 1992.
69. **Phillips, M.I. and Kimura, B.,** Brain angiotensin in the developing spontaneously hypertensive rat, *J. Hypertens.,* 6, 607, 1988.
70. **Phillips, M.I. and Kimura, B.,** Gene expression of brain angiotensinogen mRNA during the development of hypertension in the spontaneously hypertensive rat (SHR), *Hypertension,* 18 (3), 376, 1991.
71. **Naruse, K., Naruse, M., Obana, K., Demura, R., Demura, H., Inagami, T., and Shizume, K.,** Renin in the rat pituitary coexists with angiotensin II and depends on testosterone, *Endocrinology,* 118, 2470, 1986.

72. **Imboden, H., Harding, J.W., and Felix, D.,** Hypothalamic angiotensinergic fiber systems terminate in the neurohypophysis, *Neurosci. Lett.*, 96, 42, 1989.
73. **Galli, S.M., Phillips, M.I., and Aguilera, G.,** Sodium intake modulates angiotensin II content in hypothalamus and pituitary gland but not in the brainstem, *FASEB J.*, 5 (Abstr.), 949, 1991.
74. **Jones, T.H., Brown, B.L., and Dobson, P.R.M.,** Evidence that angiotensin II is a paracrine agent mediating gonadotrophin-releasing hormone-stimulated inositol phosphate production and prolactin secretion in the rat, *J. Endocrinol.*, 116, 367, 1988.
75. **Re, R.N.,** Cellular mechanisms of growth in cardiovascular tissue, *Am. J. Cardiol.*, 60, 1041, 1987.
76. **Alhenc-Gelas, F., Soubrier, C., Allegrini, H.J., Lattion, A.L., and Corvol, P.,** The angiotensin I-converting enzyme (kininase II): progress in molecular and genetic structure, *J. Cardiol. Pharmacol.*, 15 (Suppl. 6), S25, 1990.
77. **Cassis, L.A., Lynch, K.R., and Peach, M.J.,** Localization of angiotensinogen mRNA in rat aorta, *Circ. Res.*, 62, 1259, 1988.
78. **Unger, T., Gohlke, P., Paul, M., and Rettig, R.,** Tissue renin-angiotensin systems: fact or fiction, *J. Cardiol. Pharmacol.*, 18 (Suppl. 2), S20, 1991.
79. **Louden, M., Bind, R.F., Thurston, H., and Swales, J.D.,** Atrial wall uptake of renal renin and blood pressure control, *Hypertension,* 5, 629, 1983.
80. **Kifor, I. and Dzau, V.J.,** Endothelial renin-angiotensin pathway: evidence for intracellular synthesis and secretion of angiotensins, *Circ. Res.*, 60, 422, 1987.
81. **Higashimori, K., Gante, J., Holzemann, G., and Inagami, T.,** Significance of vascular renin for local generation of angiotensins, *Hypertension,* 17, 270, 1992.
82. **Naftilan, A.J., Gilliland, G.K., Eldridge, C.S., and Kraft, A.S.,** Induction of the proto-oncogene c-jun by angiotensin II, *Mol. Cell. Biol.*, 10 (10), 5536, 1990.
83. **Naftilan, A.J., Pratt, R.E., Eldridge, C.S., Lin, H.L., and Dzau, V.J.,** Angiotensin II induces c-fos expression in smooth muscle via transcriptional control, *Hypertension,* 13 (6), 706, 1989.
84. **Dzau, V.J., Gibbons, G.H., and Pratt, R.E.,** Molecular mechanisms of vascular renin-angiotensin system in myointimal hyperplasia, *Hypertension,* 18 (4), II-100, 1991.
85. **Powell, J.S., Muller, R.K., and Baumgartner, H.R.,** Suppression of the vascular response to injury: the role of angiotensin-converting enzyme inhibitors, *J. Am. Coll. Cardiol.*, 17, 137B, 1991.
86. **Baker, K.M., Chernin, M.I., Wixson, S.K., and Aceto, J.F.,** Renin-angiotensin system involvement in pressure-overload cardiac hypertrophy in rats, *Am. J. Physiol.*, 259, H324, 1990.
87. **Linz, W., Scholkens, B.A., and Ganten, D.,** Converting enzyme inhibition specifically prevents the development and induces regression of cardiac hypertrophy in rats, *Clin. Exp. Hypertens. Theory Pract.*, A11 (7), 1325, 1989.
88. **Baker, K.M., Chernin, M.I., and Cooper, G.R.,** Localization of heart angiotensinogen and renin mRNA by hybridization histochemistry, *Physiologist,* 33, A96, 1990.
89. **Dzau, V.J., Ellison, K.E., Brody, T., Ingelfinger, J., and Pratt, R.E.,** A comparative study of the distributions of renin and angiotensinogen messenger ribonucleic acids in rat and mouse tissues, *Endocrinology,* 120, 2334, 1987.
90. **Hellman, W., Suzuki, F., Ohkubo, H., Nakanishi, S., Ludwig, G., and Ganten, D.,** Angiotensinogen gene expression in extrahepatic rat tissues: application of a solution hybridization assay, *Naunyn-Schmiedeberg's Arch. Pharmacol.*, 338, 327, 1988.
91. **Paul, M., Wagner, D., Metzger, R., Ganten, D., Lang, R.E., Suzuki, F., Murakami, K., Burbach, J.H.P., and Ludwig, G.,** Quantification of renin mRNA in various mouse tissues by a novel solution hybridization assay, *J. Hypertens.*, 6, 247, 1988.
92. **Lindpainter, K., Wilhelm, M.J., Jin, M., Unger, T., Lang, R.E., Schoelkens, B.A., and Ganten, D.,** Tissue renin-angiotensin systems: focus on the heart, *Circ. Res.*, 67, 564, 1990.

93. **Kunapuli, S.P. and Kumar, A.,** Molecular cloning of human angiotensinogen cDNA and evidence for the presence of its mRNA in rat heart, *Circ. Res.,* 60, 786, 1987.
94. **Samani, N.J., Morgan, K., Brammer, W.J., and Swales, J.D.,** Detection of renin messenger RNA in rat tissues: increased sensitivity using an RNAse protection technique, *J. Hypertens.,* 10 (Suppl. 2), S19, 1987.
95. **Re, R.,** The myocardial intracellular renin-angiotensin system, *Am. J. Cardiol.,* 59, 56A, 1987.
96. **Finckh, M., Hellman, W., Ganten, D., Furtwangler, A., Allgeier, J., Boltz, M., and Holtz, J.,** Enhanced cardiac angiotensinogen gene expression and angiotensin converting enzyme activity in tachypacing-induced heart failure in rats, *Basic Res. Cardiol.,* 86, 303, 1991.
97. **Speakman, E.A. and Phillips, M.I.,** Increased angiotensin II levels in the heart of the hypertensive rat, *Physiologist,* 33 (4), A53, 1990.
98. **Schunkert, H., Dzau, V.J., Tang, S.S., Hirsch, A.T., Apstein, C.S., and Lorell, B.H.,** Increased rat cardiac angiotensin converting enzyme activity and mRNA expression in pressure overload left ventricular hypertrophy. Effects on coronary resistance, contractility and relaxation, *J. Clin. Invest.,* 86, 1913, 1990.
99. **Rosenthal, J., von Lutterotti, N., Thurnreiter, G.S., Rothemund, J., Reiter, W., Kazda, S., Garthoff, B., Jacob, I., and Dahlheim, H.,** Suppression of renin-angiotensin system in the heart of spontaneously hypertensive rats, *J. Hypertens.,* 5 (Suppl. 2), S23, 1987.
100. **Fabris, B., Jackson, B., Kohzuki, M., Perich, R., and Johnston, C.I.,** Increased cardiac angiotensin-converting enzyme in rats with chronic heart failure, *Clin. Exp. Pharmacol. Physiol.,* 17, 309, 1990.
101. **Yamada, H., Fabris, B., Allen, A.M., Jackson, B., Johnston, C.I., and Mendelsohn, F.A.O.,** Localization of angiotensin converting enzyme in rat heart, *Circ. Res.,* 68, 141, 1991.
102. **Lindpainter, K., Jin, M., Nidermaier, N., Wilhelm, M.J., and Ganten, D.,** Cardiac angiotensinogen and its local activation in the isolated perfused beating heart, *J. Hypertens.,* 5 (Suppl. 2), S33, 1987.
103. **Lory, P., Richard, S., Rassendren, F.A., Tiaho, F., and Nargeot, J.,** Electrophysiological expression of endothelin and angiotensin receptors in Xenopus oocytes injected with rat heart mRNA, *FEBS Lett.,* 258 (2), 289, 1989.
104. **Wright, G.B., Alexander, R.W., Ekstein, L.S., and Gimbrone, M.A.,** Characterization of the rabbit ventricular myocardial receptor for angiotensin II. Evidence for two sites of different affinities and specificities, *Mol. Pharmacol.,* 24, 213, 1983.
105. **Baker, K.M., Campanile, C.P., Trachte, G.J., and Peach, M.J.,** Identification and characterization of the rabbit angiotensin II myocardial receptor, *Circ. Res.,* 54, 286, 1984.
106. **Saito, K., Gutkind, J.S., and Saavedra, J.M.,** Angiotensin II binding sites in the conduction system of rat hearts, *Am. J. Physiol.,* 253, H1618, 1987.
107. **Rogers, T.B., Gaa, S.T., and Allen, I.S.,** Identification and characterization of functional angiotensin II receptors on cultured heart myocytes, *J. Pharmacol. Exp. Ther.,* 236, 438, 1986.
108. **Urata, H., Healy, B., Stewart, R.W., Bumpus, F.M., and Husain, A.,** Angiotensin II receptors in normal and failing hearts, *J. Clin. Endocrinol.,* 69, 66, 1989.
109. **Sechi, L.A., Grady, E.F., Griffin, C.A., Kalinyak, J.E., and Schambelan, M.,** Characterization of angiotensin II receptor subtypes in the rat kidney and heart using the non-peptide antagonists DuP753 and PD123177, *J. Hypertens.,* 9, S224, 1991.
110. **Rogg, H., Schmid, A., and de Gasparo, M.,** Identification and characterization of angiotensin II receptor subtypes in rabbit ventricular myocardium, *Biochem. Biophys. Res. Commun.,* 173, 416, 1990.
111. **Robertson, A.L. and Khairallah, P.A.,** Angiotensin II: rapid localization in nuclei of smooth and cardiac muscle, *Science,* 172, 1138, 1971.

112. **Re, R.N., Vizard, D.L., Brown, J., and Bryan, S.E.,** Angiotensin II receptors in chromatin fragments generated by micrococcal nuclease, *Biochem. Biophys. Res. Commun.,* 119, 220, 1984.
113. **Jackson, T.R., Blair, L.A.C., Marshall, J., Goedert, M., and Hanley, M.R.,** The mas oncogene encodes an angiotensin receptor, *Nature,* 335, 437, 1988.
114. **Lorell, B.H., Schunkert, H., Grice, W.N., Tang, S.S., Apstein, C.S., and Dzau, V.J.,** Alteration in cardiac angiotensin converting enzyme activity in pressure overload hypertrophy, *Circulation,* 80 (Suppl. 2), 297, 1989.
115. **Drexler, H., Lindpainter, K., Lu, W., Schieffer, B., and Ganten, D.,** Transient increase in the expression of cardiac angiotensinogen in a rat model of myocardial infarction and failure, *Circulation,* 80 (Suppl.), 459, 1989.
116. **Aceto, J.F. and Baker, K.M.,** [Sar^1]angiotensin II receptor-mediated stimulation of protein synthesis in chick heart cells, *Am. J. Physiol.,* 258, H806, 1990.
117. **Neyses, L. and Vetter, H.,** Action of atrial natriuretic peptide and angiotensin II on the myocardium: studies in isolated rat ventricular cardiomyocytes, *Biochem. Biophys. Res. Commun.,* 163, 1435, 1989.
118. **Allen, S.S., Cohen, N.M., Dhallan, R.S., Gaa, S.T., Lederer, W.J., and Rogers, T.B.,** Angiotensin II increases spontaneous contractile frequency and stimulates calcium current in cultured neonatal rat heart myocytes: insights into the underlying biochemical mechanisms, *Circ. Res.,* 62, 524, 1988.
119. **Celio, M.R. and Inagami, T.,** Renin in human kidney, *Hypertension,* 72, 453, 1981.
120. **Ingelfinger, J.R., Pratt, R.E., Ellison, K.E., and Dzau, V.J.,** Sodium regulation of angiotensinogen mRNA expression in rat kidney cortex and medulla, *J. Clin. Invest.,* 78, 1311, 1986.
121. **Field, L.J., McGowan, R.A., Dickinson, D.P., and Gross, K.W.,** Tissue and gene specificity of mouse renin expression, *Hypertension,* 6, 597, 1984.
122. **Sigmund, C.D. and Gross, K.W.,** Structure, expression and regulation of the murine renin genes, *Hypertension,* 18(4), 446, 1991.
123. **Pratt, R.E., Zou, W.M., Naftilan, A.J., Ingelfinger, J.R., and Dzau, V.J.,** Altered sodium regulation of renal angiotensinogen mRNA in the spontaneously hypertensive rat, *Am. J. Physiol.,* 256, F469, 1989.
124. **Menard, J., Guyene, T.-T., Chatellier, G., Kleinbloesem, C.H., and Bernadet, P.,** Renin release regulation during acture renin inhibition in normal volunteers, *Hypertension,* 18, 257, 1991.
125. **Johns, D.W., Carey, R.M., Gomez, R.A., Lynch K., Inagami, T., Saye, J., Geary, K., Farnsworth D.E., and Peach, M.J.,** Isolation of renin-rich rat kidney cells, *Hypertension,* 10, 488, 1987.
126. **Morishita, R., Higaki, J., Okunishi, H., Tanaka, T., Ishii, K., Nagano, M., Mikami, H., Ogihara, T., Murakami, K., and Miyazaki, M.,** Changes in gene expression of the renin-angiotensin system in two-kidney, one clip hypertensive rats, *J. Hypertens.,* 9, 187, 1991.
127. **Johns, D.W., Lynch, K.R., Gomez, A., and Carey, R.M.,** Angiotensin II suppresses renin gene expression and redistributes renal renin content, *Hypertension,* 10 (Abstr.), 358, 1991.
128. **Gomez, R.A., Chevalier, R.L., Everett, A.D., Elwood, J.P., Peach, M.J., Lynch, K.R., and Carey, R.M.,** Recruitment of renin gene expressing cells in adult rat kidneys, *Am. J. Physiol.,* 259, F660, 1990.
129. **Doi, Y., Atarashi, K., Franco-Saenz, R., and Mulrow, P.,** Adrenal renin: a possible regulator of aldosterone production, *Clin. Exp. Hypertens.,* 8, 1019, 1983.
130. **Naruse, M., Naruse, K., and Inagami, T.,** Immunoreactive renin in mouse adrenal: localization in inner cortical region, *Hypertension,* 6, 275, 1984.
131. **Yamaguchi, T., Franco-Saenz, R., and Mulrow, P.J.,** Effect of angiotensin II on renin production by rat adrenal glomerulosa cells in culture, *Hypertension,* 19(3), 263, 1992.

132. **Inagami, T.,** Purification of renin and prorenin, *Hypertension,* 18(3), 241, 1991.
133. **Wilson, C.M., Erdos, E.G., Dunn, S.F., and Wilson, W.D.,** Genetic control of renin activity in the submaxillary gland of the mouse, *Proc. Natl. Acad. Sci. U.S.A.,* 74, 1185, 1977.
134. **Pandey, K.N. and Inagami, T.,** Regulation of renin-angiotensins by gonadotropic hormones in cultured Leydig cells: release of angiotensin but not renin, *J. Biol. Chem.,* 261, 3934, 1986.
135. **Kumar, R.S., Kusari, J., Roy, S.N., Soffer, R.L., and Ganes, C.,** Structure of testicular angiotensin converting enzyme, *J. Biol. Chem.,* 164, 16754, 1989.
136. **Castren, E., Kurhiara, M., and Saavedra, J.M.,** Autoradiographic localization and characterization of angiotensin II binding sites in the spleen of rats and mice, *Peptides,* 8 (4), 737, 1987.
137. **Shimada, K. and Yazaki, Y.,** Binding sites for angiotensin II in human mononuclear leucocytes, *J. Biochem.,* 84, 1013, 1978.
138. **Fannon, L.D., Braylan, R.C., and Phillips, M.I.,** Alterations of lymphocyte populations during development in the spontaneously hypertensive rat, *J. Hypertens.,* 10, 629, 1992.
139. **Chappell, M.C., Millsted, A., Diz, D.I., Brosnihan, K.B., and Ferrario, C.M.,** Evidence for an intrinsic angiotensin system in the canine pancreas, *J. Hypertens.,* 9, 751, 1992.
140. **Lightman, A., Tarlatzis, B.C., and Rzasa, P.J.,** The ovarian renin-angiotensin system: renin-like activity and angiotensin II/III immunoreactivity in gonadotropin-stimulated and unstimulated human follicular fluid, *Am. J. Obstet. Gynecol.,* 156, 808, 1987.
141. **Hsueh, W.A. and Baxter, J.D.,** Human prorenin, *Hypertension,* 17, 469, 1991.
142. **Glorioso, N., Atlas, S.A., and Laragh, J.H.,** Prorenin in high concentrations in human ovarian follicular fluid, *Science,* 233, 1422, 1986.
143. **Sealey, J.E.,** A separate prorenin-angiotensin system in reproductive and renal tissues? Evidence for renin gene expression in extrarenal vascular tissues is unconvincing, *Excerpta Med.,* 29, 29, 1989.
144. **Do, Y.S., Sherrod, A., and Lobo, R.A.,** Human ovarian theca cells are a source of renin, *Proc. Natl. Acad. Sci. U.S.A.,* 85, 1957, 1988.
145. **Husain, A., Bumpus, F.M., DeSilva, P., and Speth, R.C.,** Localization of angiotensin II receptors in ovarian follicles and the identification of angiotensin II in rat ovaries, *Proc. Natl. Acad. Sci. U.S.A.,* 84, 2489, 1987.
146. **Mullins, J.J., Peters, J., and Ganten, D.,** Fulminant hypertension in transgenic rats having the mouse Ren-2 gene, *Nature (London),* 344, S41, 1990.
147. **Chomcynski, P. and Saachi, N.,** Single-step method of RNA isolation by acid quanidium thiocyanate phenol chloroform extraction, *Anal. Biochem.,* 162, 156, 1987.
148. **Phillips, M.I., Speakman, E.A., and Kimura, B.,** Levels of angiotensin and molecular biology of the tissue renin-angiotensin system, *Regul. Pept.,* 42, 1, 1992.

Chapter 5

MOLECULAR, BIOCHEMICAL, AND FUNCTIONAL BIOLOGY OF ANGIOTENSINOGEN

K.R. Lynch and D.P. O'Connell

TABLE OF CONTENTS

0-8493-4622-3/93/$0.00 + $.50

I. INTRODUCTION

Angiotensinogen is a 55,000- to 60,000-Da glycoprotein that is moderately abundant in blood and cerebrospinal fluid. The amino-terminal decapeptide of angiotensinogen, angiotensin I (AngI), is released by the action of renin and angiotensinogen is the only known naturally occurring renin substrate. Angiotensinogen is synthesized by a variety of cell types, most prominently hepatocytes, adipocytes, and astrocytes. Angiotensinogen is predominantly extracellular and its constitutive secretion would appear to exclude rapid changes in local concentrations of this protein. Its gene, which is single copy, is regulated by several systemic hormones (e.g., glucocorticoid, estrogen, and thyroid hormone). Angiotensinogen is an acute-phase protein and a member of the serpin gene superfamily. The available evidence indicates that angiotensinogen functions solely as an extracellular reservoir of angiotensin peptides.

II. PROTEIN STRUCTURE AND SYNTHESIS LOCATIONS

The single angiotensinogen gene encodes one protein product as shown by analysis of numerous cDNA and genomic clones. This protein is nominally 477 amino acids in rodents[1,2] and 485 amino acids in the human.[3] It contains a 24-amino acid leader peptide that is cotranslationally removed to reveal the mature protein (calculated mol wt [from rat cDNA] is 49,548 Da for the mature protein). The invariant AngI decapeptide sequence resides at the amino terminus. Beyond this amino-terminal decapeptide, the rodent and human amino acid sequences differ, offering a possible explanation for the observed low activity of primate renin with nonprimate angiotensinogen.[4] This explanation was supported by Quinn and Burton,[5] who used synthetic tetradecapeptides to mimic renin substrates. In this manner, the four amino acids immediately beyond the scissile bond (between amino acids ten and eleven) were found to impart substrate specificity.

Angiotensinogen contains several potential sites for N-linked glycosylation (at [rat] Asn-23, Asn-271, Asn-295). Differential glycosylation may be responsible for the isoelectric point and size variants of this protein.[6-8] The glycosylation is a cotranslational event that, together with cleavage of the leader peptide, occurs during passage of the polypeptide through the endoplasmic reticulum membrane.[9] Some angiotensinogen circulates as a high-molecular-weight form;[10] but this is normally only a minor component of plasma angiotensinogen. However, the larger form of angiotensinogen exists at higher plasma levels during the last trimester of pregnancy and is occasionally the predominate form in the plasma of hypertensive, pregnant women.[11] This higher-molecular-weight angiotensinogen is an effective renin substrate and consists of angiotensinogen that is noncovalently linked to another, unidentified serum protein.[12] The physiologic relevance of these multiple forms of angiotensinogen remains to be defined.

In contrast to the rat and cow AngII receptors, which are highly conserved proteins (>95% identical[13,14]), it is only the amino-terminal decapeptide of angiotensinogen that is highly conserved. The rat and human angiotensinogen amino acid sequences are 60% identical overall, while the amino termini (i.e., AngI) are 100% identical. Following conceptualization of the rat angiotensinogen amino acid sequence from its cDNA,[1] Doolittle pointed out that angiotensinogen is a serpin-like protein because it is significantly related to α_1-antitrypsin (23% identity over 404 amino acids), antithrombin III (20%, 405 amino acids), and ovalbumin (18%, 378 amino acids).[15] This relatedness led to speculation that angiotensinogen might have serine protease inhibitor activity as do many members of the serpin superfamily (but see below).

Angiotensinogen is predominantly extracellular and for this reason its site of synthesis can be readily determined only by measuring the local accumulation of its mRNA. These RNA sequences are detected (i.e., by Northern blots) in liver, fat, and brain extracts and have been found also, albeit in lesser amounts, in a bewildering variety of tissue extracts.[16-18] Indeed, it is difficult to find a tissue that does not contain some angiotensinogen mRNA! In some sites, the source of angiotensinogen may not be parenchymal cells but rather associated adipocytes or fibroblasts. Angiotensinogen mRNA sequences were identified in fibroblast-like cells by *in situ* hybridization performed on sections of rat mesenteric vessels[19] and in extracts of clonal lines of mouse 3T3 fibroblasts.[20] However, kidney angiotensinogen mRNA has been localized also to the proximal convoluted tubule,[21] thus it would appear that not all low-level angiotensinogen mRNA accumulation can be attributed to fibroblasts or adipocytes. An additional complication is that angiotensinogen may be synthesized in the bloodstream, since its mRNA has been detected in human and rat leukocytes.[98]

The detection of angiotensinogen mRNA in liver confirmed earlier studies showing renin substrate release from perfused liver[22] and dispersed hepatocytes.[23] Immunohistochemical localization of hepatic angiotensinogen revealed diffuse cytoplasmic immuno-deposits in periportal hepatocytes consistent with a presence in the Golgi apparatus and rough endoplasmic reticulum.[24] Immunoreactivity of this angiotensinogen varies in a diurnal fashion and it seems likely that hepatocytes actively synthesize and secrete angiotensinogen from the late dark to the early light periods.[24] Adipocyte angiotensinogen RNA was detected first in rat periaortic adipocytes by *in situ* hybridization[19] and later noted in both brown and white fat extracts including perirenal, epididymal, and interscapular depots.[25,26] Furthermore, white and brown fat pads isolated from rats were shown to secrete this protein.[26] Angiotensinogen RNA accumulates also in cultured Swiss mouse 3T3-L1 fibroblasts, and the level of the mRNA increases as these cells differentiate to adipocytes.[20] The reason for angiotensinogen synthesis (and secretion) by adipocytes remains speculative. Developmental regulation of angiotensinogen gene expression (discussed later) suggests that angiotensinogen or the angiotensin peptides might have a functional role in cell differentiation. The predominant source of circulating angiotensinogen is assumed to be the liver, but the extent to which adipose tissues contribute to the blood angiotensinogen levels has not been determined.

Renin-like activity was demonstrated in the lysates of wild-type 3T3 cells.[27] AngI-like immunoreactivity was detected in the culture media of 3T3-L1 and 3T3-F442A cells in the absence of exogenous renin.[99] (Saye and Peach, unpublished data). Therefore, adipocytes may have the capability of processing angiotensinogen into its constitutive peptides, but the nature of the protease is not known at present. The K_m of renin for angiotensinogen is about equal to the physiologic concentration of plasma angiotensinogen (1 μM).[28,29] This rate constant, if relevant to the *in vivo* situation, implies that changes in local concentration of *either* renin or angiotensinogen would affect the amount of the released AngI. To this end, AngII stimulates the production of prostaglandins in isolated epididymal adipocytes, suggesting a paracrine regulatory function for the locally produced and processed angiotensinogen.[30] In addition, adipocyte-derived angiotensinogen may serve as a local source for angiotensin peptides that enhance norepinephrine release at sympathetic nerve terminals.[25] Since AngII is a potent physiologic regulator of arterial constriction, adipocyte-derived angiotensinogen may modulate local blood flow.

Some astrocytes synthesize angiotensinogen in brain as demonstrated by several methods, including combined *in situ* hybridization and immunocytochemistry and testing for angiotensinogen RNA in cultured astrocytes and neurons. Examination of brainstem and hypothalamic sections revealed a striking congruence of silver grains (labeled oligonucleotide probe) and glial fibrillary acidic protein (GFAP, a standard astrocyte marker) immunoreactivity, but not with microtubule-associated protein II (MAPII, a neuronal marker) immunoreactivity.[31] Angiotensinogen RNA sequences were found in extracts of astrocyte-enriched, but not in neuron-enriched, cultures from neonatal rat brains.[32] Likewise, angiotensinogen protein was associated largely with the astrocytic, but not neuronal, culture media.[32] Interestingly, several other serpins, including antithrombin III[33] and plasminogen activator inhibitor type I[34] are synthesized in these astrocytic cultures. Neither data from cultured cells nor brain slices exclude the possibility of some low-level accumulation of angiotensinogen mRNA in neurons. Astrocytic angiotensinogen synthesis was not entirely unexpected as immunocytochemical analyses had mapped immunoreactive material mostly to astrocytes, but neuronal localization has also been described.[35,36] There are inconsistencies in published reports concerning the neuronal localization of angiotensinogen immunoreactivity both with respect to whether neurons stain with anti-angiotensinogen antisera and, if so, the anatomical localization of immunopositive neurons. Studies by Campbell et al.[37] have demonstrated clearly that the presence or absence of neuronal staining for angiotensinogen is dependent on how the tissue is processed. For paraffin sections a significant variation among animals can be seen in the location of immunopositive neurons. It is suggested that paraffin embedding could be unmasking, or generating, antibody-binding sites in molecules other than angiotensinogen that are recognized by anti-angiotensinogen antibodies. This hypothesis is supported by the apparent lack of correlation in brain location between AngII-positive neurons and putatively angiotensinogen-positive neurons.[35,36] Further indications for a predominantly astrocytic source of angiotensinogen include: (1) the temporal relationship between angiotensinogen mRNA appearance[38] and astrocyte development in the late fetal/neonatal rat and (2) the presence of angiotensinogen mRNA in cultures developed from human stage IV astrocytomas[39] and astrocytic cell lines such as NG108 and C6 glioma.[40] Thus, the preponderance of evidence suggests that most, in not all, brain angiotensinogen is released from astrocytes.

The astrocytes that contain angiotensinogen mRNA are unevenly distributed in rat brain with clustering in numerous hypothalamic, midbrain, and brainstem nuclei.[41,42] Quantitative analysis of these brain regions showed that individual brain nuclei differ in their proportion of angiotensinogen mRNA positive astrocytes. Furthermore, individual astrocytes differ in their individual content of angiotensinogen mRNA.[100] There is excellent concordance between angiotensinogen mRNA accumulation and [^{125}I]-AngII (or AngII analogs) binding to both AT_1 and AT_2 sites in adult brain sections.[43,44] There are brain regions that show a disparity between angiotensin binding and angiotensinogen mRNA levels (i.e., high binding, low mRNA), but these are outside the blood brain barrier (e.g., area postrema). The angiotensinogen in cerebrospinal fluid (0.2 μM)[28] is synthesized within the central nervous system (CNS), because circulating angiotensinogen does not gain access to the brain parenchyma.[45]

III. ANGIOTENSINOGEN mRNA STRUCTURE

The 1431-nucleotide translational open reading frame that encodes rat angiotensinogen is embedded in a mRNA molecule of about 1800 nucleotides. The reading frame is preceded by 61 nucleotides of untranslated sequence. Angiotensinogen mRNA is not a single entity because at least four different polyadenylation sites are used in the posttranscriptional modification of the precursor. These sites are similarly located among different molecules at 1650, 1785, 1800, and 1840 nucleotides and were identified by S_1 nuclease mapping.[17] Size heterogeneity of rat angiotensinogen mRNA results also from an unusual feature of its gene. This is illustrated by experiments wherein treatment of primary hepatocyte cultures with glucocorticoids elicits the use of two additional, upstream transcription start sites along the angiotensinogen gene (at -386 and -328 nucleotides relative to the uninduced cap site).[46] These extended rat angiotensinogen mRNAs apparently have the same coding potential as the predominant mRNA species, i.e., they have extended 5′

nontranslated regions. For this reason, these larger mRNAs have no obvious bearing on the functioning of the renin-angiotensin system. There is no evidence to date for alternate splicing of the five-exon angiotensinogen gene RNA transcript.

IV. GENE STRUCTURE AND PROMOTER FUNCTION

The angiotensinogen gene is single copy and has been isolated from rat,[47] in which it has been mapped to chromosome 19,[48] from mouse[2] (chromosome 8),[49] and from human[50] (chromosome 1).[51] The rat gene remains the most thoroughly characterized, but the structure and function of the mouse and human genes are not significantly different. The rat angiotensinogen gene is roughly 12 kbp long and contains four introns. The first exon encodes only a portion of the 5′ untranslated region of the angiotensinogen mRNA and is separated from the remaining exons by a 5-kbp intron. The intron/exon boundaries conform to the consensus sequence of pre-mRNA introns[52] and are fully conserved among rat, mouse, and human genes. The upstream flanking region of the rat gene contains several recognizable promoter elements, including a "TATA" element at position −30 bp (relative to the capping site) and two putative glucocorticoid responsive elements (GREs, 5′-AGAACA-3′) at positions −586 bp and −477 bp.[47]

Several laboratories have reported studies on the rodent and human angiotensinogen gene promoters. These studies have been directed predominantly at identifying the elements, and their corresponding transcription factors, that confer hormone responsiveness on the gene. Deletions of the rat angiotensinogen gene flanking DNA, which was coupled to a luciferase reporter gene, identified a minimal promoter region of 688 bp that lies immediately 5′ to the angiotensinogen gene. This DNA fragment supported expression of the fusion gene construct in both placental JEG-3 and hepatoma HepG2 cells.[53] DNAse I footprinting of this fragment indicated that it contains six transcriptionally active elements including an enhancer element. The latter element acts to suppress transcription in its normal position (−108 to −60 nucleotides), but enhances transcription when moved upstream. This element can function also to inhibit the activity of heterologous promoters such as those found in the SV-40 early region. Another cis-acting element at −545 bp was identified by DNAse I footprinting[54] and has sequence similarity with the NFκB transcription factor binding site. This element appears to function in the acute-phase response, i.e., transcriptional activation driven by exposure to cytokines. APRE (acute phase response element), when fused upstream of a cytokine-unresponsive gene, confers cytokine responsiveness on that gene. The APRE binds both NFκB (previously known as BPi, i.e., binding protein, inducible) and members of the C/EBP transcription factor family (previously known as BPc [constitutive]).[55] Interestingly, NFκB gene expression is enhanced by cytokine (i.e., interleukin-6) treatment.[55]

It was noted that interleukin-6 treatment of rat hepatoma H35 cells was ineffective in increasing angiotensinogen mRNA accumulation without prior addition of glucocorticoids.[56] This observation was confirmed and extended to the cytokine interleukin-1 by Ron et al. working with the same H35 cell line.[57] The human hepatoma line HepG2, which is spontaneously deficient in glucocorticoid receptor, is unresponsive to interleukin-1 induction of angiotensinogen gene expression. However, if the glucocorticoid receptor is introduced subsequently via transfection of a functional glucocorticoid receptor gene, the cells become cytokine inducible with regards to angiotensinogen synthesis.[57] It is noteworthy that one (of two) glucocorticoid response elements overlaps the APRE. In a recent study from Habener's group,[55] rat liver cDNA expression library was screened with a labeled APRE DNA to identify clones encoding a potential APRE binding protein. In this manner, a portion of the sequence of a so-called AGIE-BP1 (angiotensinogen gene-inducible enhancer-binding protein 1) was conceptualized. This sequence, which was previously unknown, contains two zinc finger motifs and is related to other transcription factors (e.g., MBP-1/PRDII-BF1).[58] The DNA binding domain of AGIE-BP1 (expressed in bacteria) binds to the APRE in a fashion indistinguishable from purified NFκB. Finally, the distribution of AGIE-BP1 mRNA has been described and varies considerably in abundance among tissues, the greatest amounts being located in heart, brain, spleen, skeletal muscle, and lung, and the least amounts in kidney, liver, testes, and prostate. Thus, there is no obvious quantitative relationship between angiotensinogen and AGIE-BP1 expression. When poly(A)$^+$ RNA from brains of developing rat embryos was analyzed on Northern blots, expression of both transcripts peaked on embryonic day 17 (E17). This observation prompted the authors to speculate that AGIE-BP1 may modulate the expression of brain angiotensinogen in the rat embryo.

An equivalent minimal promoter region was identified immediately upstream of the mouse angiotensinogen gene.[2] Likewise, an analogous minimal promoter region of the human angiotensinogen gene has been identified also.[59] The mouse minimal promoter region, which is contained within a 750-bp fragment immediately flanking the gene, is sufficient to confer glucocorticoid, estrogen, and bacterial endotoxin (the latter induces an acute-phase response) inducibility on a mini (i.e., internally deleted for ease of detection)-transgene. This transgene was expressed also in a tissue-specific manner in the recipient mice. Human angiotensinogen gene promoter function was examined by fusing a 1.3-kbp flanking DNA fragment to a chloramphenicol acetyl transferase reporter gene. This fusion gene construct was tested following its introduction into both the human hepatoma cell line HepG2 and in the human astrocytoma line T98G. Activity was observed only in the HepG2 cell line.[59] Furthermore, a 14-kbp human genomic DNA fragment encoding angiotensinogen and containing the same 1.3 kbp-fragment was introduced into a transgenic mouse. The transgene was expressed in a tissue-specific fashion as regards high-level human angiotensinogen mRNA accumulation in liver, but this expression was quantitatively aberrant in the sense that high levels of the transgene mRNA were found also in kidney extracts.[60] The conclusion that can be drawn from these studies is that located within about a kilobase of the major start site of transcription, there are multiple regulatory elements including those ultimately responsive to steroids (i.e., glucocorticoids and estrogens) and some cytokines. This region also contains an enhancer element that confers tissue specificity on angiotensinogen gene expression. However, ''tissue specificity'' viz a vis angiotensinogen gene expression is equated often with the level of mRNA accumulation in only a few large, well-defined tissues (liver, adipose tissue, kidney, and brain) but the plethora of other organs/cell groups that contain angiotensinogen mRNA (see above), are often not examined.

V. ONTOGENY AND CONTROL OF ANGIOTENSINOGEN GENE EXPRESSION

The angiotensinogen gene is subject to a temporal progression of developmental expression that is influenced positively by a variety of hormones/ cytokines. Low levels of angiotensinogen RNA sequences were detected on embryonic day 11 (E11) in rat fetuses (bodies) and yolk sacs. By Northern analysis yolk sac placenta angiotensinogen mRNA was found to be 20-fold more abundant, and about 200 bases longer, than that extracted from the fetal rat bodies. These RNA levels increased from day E13 to E17, while the increase in angiotensinogen RNA in rat fetus heads did not occur until days E20 to E21,[38] when the angiotensinogen mRNA content is threefold higher in brain than in liver.[61] That angiotensinogen RNA is detectable only barely in fetal (E15 to E20) rat liver[62] suggests the bulk of fetal angiotensinogen RNA is derived from nonhepatic sources. Functional AngII receptors are expressed transiently at unique sites in the rat fetus. This observation has led to speculation that AngII may play a role in fetal development.[63] Liver angiotensinogen RNA levels rise rapidly after birth and achieve adult levels within 24 h postpartum.[61,62] A number of factors may contribute to this rapid increase in hepatic angiotensinogen production, including the postpartum increases in the circulating concentrations of glucocorticoids and/or thyroid hormone. With respect to factors regulating angiotensinogen levels in the fetus, it has been suggested that glucocorticoid hormones may also play an important role. Recent studies in chronically instrumented fetal sheep have demonstrated that a physiologic increase in fetal plasma cortisol concentration near term decreases both liver angiotensinogen mRNA and albumin mRNA expression.[64] Because the 5′-flanking regions of both the albumin and angiotensinogen genes contain sequences that are significantly similar to the GRE,[17] it may imply a common mechanism of negative regulation. However, this interpretation is paradoxical in the sense that the GREs enhance transcription along these genes in adult rats and in a variety of cultured cells.

Perhaps the inhibition of RNA accumulation in fetal sheep as a function of cortisol concentration is an indirect event.

In the aging rat (3 to 20 months) there is a 50% diminution in liver angiotensinogen mRNA levels.[61] Hepatic angiotensinogen mRNA concentrations are reduced in the pregnant rat when compared with the adult nonpregnant female or male rat, and are regulated during gestation.[65] Brain angiotensinogen levels do not show a dramatic change at birth, rather they increase gradually during the first three weeks of life in concert with the development of astrocytes.[61] During phenotypic modulation of 3T3 fibroblasts (3T3-L1 and 3T3-F442A lines) to an adipocyte-like cell, angiotensinogen mRNA accumulation increases dramatically. This increase is similar to changes in glycerol phosphate dehydrogenase mRNA expression, an established marker of adipocyte maturation, and parallels the morphologic changes associated with adipocyte differentiation. This would suggest that, in the adipocyte at least, the expression of angiotensinogen mRNA is controlled by a developmental regulatory mechanism.[20,66]

There are numerous reports in the literature describing hormonal modulation of angiotensinogen gene expression. Prominent among the hormones studied are glucorticoids, estrogens, thyroid hormone, cytokines, and AngII. Herein, we do not attempt a thorough critical review of this literature, but instead refer the reader to Chapter 6. Below, a brief summary of this literature is given, in an attempt to highlight what we consider to be the salient features of the hormonal modulations.

The most studied regulatory phenomenon regarding angiotensinogen is that exerted by glucocorticoids. For the most part, the glucocorticoid responsiveness of the angiotensinogen gene follows the pattern of numerous other such genes. Multiple GREs, which bind the activated glucocorticoid receptor to enhance transcription, occur upstream of the transcription initiation site. However, there are several unusual features of glucocorticoid activity regarding modulation of the angiotensinogen gene. These were alluded to previously and include: (1) the unmasking of at least two additional transcription start sites,[46] (2) the permissive effect with acute-phase response mediators,[57] and (3) the suppression of angiotensinogen RNA accumulation in fetal sheep in response to cortisol.[64] The synergism between glucocorticoids and cytokines is similar to the effect on angiotensinogen mRNA accumulation observed in dispersed hepatocytes treated with a cAMP analog (S_p-cAMP) and glucocorticoids.[67] Angiotensinogen synthesizing tissues respond differentially to glucocorticoid regimens. For example, treatment of rats with dexamethasone resulted in a marked increase in liver angiotensinogen RNA accumulation, a modest increase in whole brain angiotensinogen RNA, and no detectable increase in kidney angiotensinogen RNA.[68] Likewise, a single dose of estradiol was found to enhance liver angiotensinogen RNA accumulation, increased this RNA in the brain to a lesser extent, but had no effect on the heart angiotensinogen RNA accumulation.[69]

Thyroid hormone and estrogens have been observed to increase angiotensinogen mRNA accumulation in a number of tissues. AngII has been shown also to increase angiotensinogen release when the peptide was infused continuously over a period of several hours into rat livers,[70,71] but attempts to reproduce this stimulation in hepatocytes[67,72] have produced mixed results. More recent studies have confirmed that endogenous AngII regulates hepatic angiotensinogen production independent of adrenal glucocorticoid.[73,74] The regulation of angiotensinogen gene expression by insulin has been reported in studies on cultured cells where this hormone exerts an inhibitory effect.[75]

Renal cortical angiotensinogen mRNA along with renin mRNA levels are increased in normotensive (WKY) rats that are sodium depleted but no increase in renal angiotensinogen mRNA was demonstrable in spontaneously hypertensive rats (SHRs).[76,77] In addition, SHRs appear to have lower renal angiotensinogen mRNA but substantially higher brain angiotensinogen mRNA levels compared to WKY and these differences are independent of sodium intake.[77] Interestingly, Naftilan and colleagues[78] have shown also that increased salt intake in the WKY strain results in increased angiotensinogen mRNA accumulation in the vessel (aorta) wall proper, while these sequences are difficult to demonstrate in vascular smooth muscle of rats not fed a high-salt diet.[25] The enhanced expression in the brains of SHR was localized to the anteroventral portion of hypothalamus, preoptic area, and medial septum.[79] These effects appear to be tissue specific, as liver angiotensinogen mRNA levels do not differ between the two strains. No significant changes in either renal or brain angiotensinogen mRNA accumulation, but a marked enhancement of renal renin mRNA levels, have been described during chronic angiotensin-converting enzyme inhibition in both WKY and SHR rats.[80,81] In slowly progressing, low-output cardiac failure induced in rats by tachypacing, both heart angiotensinogen mRNA accumulation and angiotensin converting enzyme activity are enhanced.[82] In this type of cardiac failure, neither increased systolic wall stress or myocardial ischemia are prerequisite factors required for this increase in angiotensinogen mRNA.

VI. FUNCTIONS OF ANGIOTENSINOGEN

The widely agreed on function of angiotensinogen is to act as a circulating renin substrate with the products of the reaction being the (amino-terminal derived) decapeptide AngI and the 443 (rat)-amino acid des(AngI)-angiotensinogen. While it is obvious that angiotensinogen is the sole source of the angiotensin peptides (be they angiotensin 1-10, 1-8, or 1-7), there is debate regarding other possible functions of angiotensinogen. One wonders what function, if any, the larger renin product has in normal physiology and why so many tissues, in addition to liver and brain, synthesize and release angiotensinogen? If angiotensinogen is simply and only an extracellular reservoir of angiotensin peptides, then why is the precursor so large (45 times the AngI peptide) and why cannot all the peripheral tissues be serviced by liver synthesis?

A feature of the angiotensin system that must be considered is the extracellular nature of angiotensinogen. It is present at low levels intracellularly; for example, immunocytochemical detection of the hepatic protein is difficult without prior colchicine treatment.[83] Likewise, brain angiotensinogen is predominantly extracellular.[84] Hepatocytes, adipocytes, and astrocytes do not possess any known means of concentrating secretory proteins, but rather constitutively export such proteins. Furthermore, Deschepper and Reudelhuber[85] have reported that angiotensinogen (in contrast to renin) is not directed to secretory granules when expressed in cultured mouse corticotroph-like AtT-20 cells, but rather released constitutively. Thus, there is no evidence in favor of, and much evidence opposed to, the hypothesis that angiotensin peptides can be generated and concentrated intracellularly. Local angiotensin systems might consist of a locally released (or activated) protease that cleaves the extracellular angiotensinogen present in probably all extracellular fluids. Currently the only physiologically relevant processing enzyme recognized is renin; however, several serine proteases such as tonin[86] and other tissue kallikreins[87] and cathepsin G[88] are capable of releasing AngII directly from angiotensinogen, at least *in vitro*.

A plausible hypothesis proposed by Campbell[89] to explain why the hepatic output of angiotensinogen is insufficient to supply all non-CNS tissues is that individual angiotensinogen-secreting tissues would have higher levels of angiotensinogen than those that receive this protein only from circulation. Depending on the concentration and activity of renin (or renin-like) enzymes, the elevated angiotensinogen could result in higher levels of angiotensin peptide released locally. It is worth noting that the local concentrations of renin-angiotensin system components are not known in any tissue angiotensin system. We reiterate that the K_m for the renin angiotensinogen reaction is nearly the concentration of circulating angiotensinogen (about 1 μM).[29]

Finally, functions of angiotensinogen unrelated to its role as a prohormone need to be considered. The hypothesis that angiotensinogen is a precursor to erythropoietin,[90] which was based largely on cross-reacting antisera, is apparently incorrect as the erythropoietin precursor has been described subsequently by molecular cloning and is not significantly similar to angiotensinogen.[91] The suggestion that des(AngI)-angiotensinogen functions as a renin inhibitor[92] has not been confirmed experimentally.[93] Angiotensinogen is a serpin, being most closely related to α_1-antitrypsin and antithrombin III.[15] Although this large gene family is named after its *ser*ine *p*rotease *in*hibitor members, there are noninhibitor members[94] as well. Angiotensinogen contains an amino-terminal extension relative to most other serpins and the putative active sites of the most closely related serpins are not conserved between human and rodent angiotensinogens. Proteolytic cleavage of α_1-antitrypsin, and other serine protease inhibitors, at the reactive center loop generates a protein characterized by increased heat stability associated with the conversion of the stressed to relaxed form. Cleavage of angiotensinogen at the analogous site fails to bring about a similar conversion of protein forms.[95] We know of no direct evidence supporting the contention that angiotensinogen is a physiological serine protease inhibitor.

At least four groups have reported the breeding of mice that contain angiotensinogen transgenes. Two groups[96,97] have reported on transgenic mice expressing rat angiotensinogen. In both cases, the transgenic mice expressed high levels of the transgene in the liver and the plasma levels of angiotensinogen were about threefold higher than in normal animals. However, one group[97] generated a (single) transgenic mouse line that was hypertensive (mean arterial pressure: 158 mmHg, male; 132 mmHg, female) while the other group[96] developed ten transgenic mice expressing the angiotensinogen gene; all of the latter were normotensive. A notable difference was that the Nakanishi group[96] used an angiotensinogen gene driven by a metallothionein promoter leading to a high level of expression, as expected, in the liver but not the brain. The Kaling group[97] used the natural rat angiotensinogen promoter (i.e., 1.6 kbp of 5′-flanking sequence). In their single hypertensive mouse line, the rat gene was expressed highly in both brain and liver. Furthermore, the pattern of brain expression was that predicted from *in situ* hybridization studies conducted with rat brain tissue.[42] (The Kaling group[97] reported a second, normotensive, transgenic mouse line wherein the rat transgene was expressed only in brain and in an aberrant pattern.) Thus, elevated angiotensinogen levels are not the sole cause of the hypertension in these mice. An obvious interpretation, as suggested by Kimura et al.,[97] is that the vigorous expression of the transgene in normal brain angiotensinogen expression regions underlies the development of hypertension in their transgenic mouse line. Interestingly, transgenic mouse lines expressing *both* angiotensinogen and renin were found to be hypertensive.[96]

VII. SUMMARY

The only firm conclusion that we draw from the data presented above is that angiotensinogen functions as an extracellular reservoir of angiotensin peptides. Hepatocytes, astrocytes, and adipocytes are most active in the synthesis of angiotensinogen, but its mRNA is detectable in a wide variety of tissues/cell types; presumably, these are synthesizing angiotensinogen also. The reason for the widespread synthesis of this protein remains speculative; perhaps local angiotensinogen concentrations are significantly higher in the extracellular spaces of these tissues. The expression of the angiotensinogen gene is controlled by glucocorticoids, estrogens, thyroid hormone, cytokines, angiotensin II and is developmentally regulated. The angiotensinogen gene has been well characterized regarding a number of cis-acting regulatory elements and with respect to some of their corresponding transcription factors.

Increases in the concentration of this constitutively secreted protein are expected to be gradual as compared to, for example, renin. There is no direct evidence for intracellular processing of angiotensinogen. On the contrary, a coherent body of evidence suggests angiotensinogen is processed entirely extracellularly. Angiotensinogen, although clearly a member of the serpin gene superfamily, is probably devoid of serine protease inhibitor activity. No role has been shown for the larger product of the renin reaction, i.e., des(AngI)-angiotensinogen. The role of circulating (or local) angiotensinogen concentrations in the development and/or maintenance of hypertension remains speculative.

REFERENCES

1. **Ohkubo, H., Kageyama, R., Ujihara, M., Hirose, T., Inayama, S., and Nakanishi, S.,** Cloning and sequence analysis of cDNA for rat angiotensinogen, *Proc. Natl. Acad. Sci. U.S.A.*, 80, 2196, 1983.
2. **Clouston, W.M., Evans, B.A., Haralambidis, J., and Richards, R.I.,** Molecular cloning of the mouse angiotensinogen gene, *Genomics,* 2, 240, 1988.
3. **Kageyama, R., Ohkubo, H., and Nakanishi, S.,** Primary structure of human preangiotensinogen deduced from the cloned cDNA sequence, *Biochemistry,* 23, 3603, 1984.
4. **Peach, M.J.,** Renin-angiotensin system: biochemistry and mechanisms of action, *Physiol. Rev.*, 57, 313, 1977.
5. **Quinn, T. and Burton, J.,** Amino acid sequence of angiotensinogen as a basis for species specificity of renin, *Chem. Abst.*, 97, 276, 1982.
6. **Hilgenfeldt, U. and Hackenthal, E.,** Separation and characterization of two different species of rat angiotensinogen, *Biochim. Biophys. Acta,* 708, 335, 1982.
7. **Tewksbury, D.A.,** Angiotensinogen, *Fed. Proc.*, 42, 2724, 1983.
8. **Sernia, C. and Mowchanuk, M.D.,** Brain angiotensinogen: in vitro and chromatographic characterization, *Brain Res.*, 259, 275, 1983.
9. **Lennarz, W.,** Overview: role of intracellular membrane systems in glycosylation of proteins, *Meth. Enzymol.*, 98, 91, 1983.
10. **Gordon, D.B., and Sachin, I.N.,** Chromatographic separation of multiple renin substrates in women: effect of pregnancy and oral contraceptives, *Proc. Soc. Exp. Biol. Med.*, 156, 461, 1977.
11. **Tewksbury, D.A. and Dart, R.A.,** High molecular weight angiotensinogen levels in hypertensive pregnant women, *Hypertension,* 4, 729, 1982.
12. **Tewksbury, D.A. and Tyron, E.S.,** Immunochemical comparison of high molecular weight angiotensinogen from amniotic fluid, plasma of men, and plasma of pregnant women, *Am. J. Hypertens.*, 2, 411, 1989.

13. **Murphy, T.J., Alexander, R.W., Griendling, K.K., Runge, M.S., and Bernstein, K.E.,** Isolation of a cDNA encoding the vascular type-1 angiotensin II receptor, *Nature,* 351, 233, 1991.
14. **Sasaki, K., Yamano, Y., Bardhan, S., Iwai, N., Murray, J.J., Hasegawa, M., Matsuda, Y., and Inagami, T.,** Cloning and expression of a complementary DNA encoding a bovine adrenal angiotensin II type-1 receptor, *Nature,* 351, 230, 1991.
15. **Doolittle, R.F.,** Angiotensinogen is related to the antitrypsin-antithrombin-ovalbumin family, *Science,* 222, 417, 1983.
16. **Campbell, D.J. and Habener, J.F.,** Angiotensinogen gene is expressed and differentially regulated in multiple tissues of the rat, *J. Clin. Invest.,* 78, 31, 1986.
17. **Ohkubo, H., Nakayama, K., Tanaka, T., and Nakanishi, S.,** Tissue distribution of rat angiotensinogen mRNA and structural analysis of its heterogeneity, *J. Biol. Chem.,* 261, 319, 1986.
18. **Chappell, M.C., Millsted, A., Diz, D.I., Brosnihan, K.B., and Ferrario, C.M.,** Evidence for an intrinsic angiotensin system in the canine pancreas, *J. Hypertens.,* 9, 751, 1991.
19. **Campbell, D.J. and Habener, J.F.,** Cellular localization of angiotensinogen gene expression in brown adipose tissue and mesentery: quantification of messenger ribonucleic acid abundance using hybridization in situ, *Endocrinology,* 121, 1616, 1987.
20. **Saye, J.A., Cassis, L.A., Sturgill, T.W., Lynch, K.R., and Peach, M.J.,** Angiotensinogen gene expression in mouse 3T3-L1 cells, *Am. J. Physiol.,* 256, C448, 1989.
21. **Inglefinger, J.R., Zuo, W.M., Fon, E.A., Ellison, K.E., and Dzau, V.J.,** In situ hybridization evidence for angiotensinogen messenger RNA in the rat proximal tubule, *J. Clin. Invest.,* 85, 417, 1990.
22. **Nasjletti, A. and Masson, G.M.C.,** Studies on angiotensinogen formation in a perfused liver system, *Circ. Res.,* 30, S187, 1972.
23. **Weigand, K., Wernze, H., and Falge, C.,** Synthesis of angiotensinogen by rat liver cells and its regulation in comparison to serum albumin, *Biochem. Biophys. Res. Commun.,* 75, 102, 1977.
24. **Waguri, S., Watanabe, T., Kominami, E., and Uchiyana, Y.,** Variations in immunoreactivity of angiotensinogen and cathepsins B and H in rat hepatocytes over 24 hours, *Am. J. Anat.,* 187, 175, 1990.
25. **Cassis, L.A., Lynch, K.R., and Peach, M.J.,** Localization of angiotensinogen mRNA in rat aorta, *Circ. Res.,* 62, 1259, 1988.
26. **Cassis, L.A., Saye, J., and Peach, M.J.,** Location and regulation of rat angiotensinogen messenger RNA, *Hypertension,* 11, 591, 1988.
27. **Schelling, P., Ganten, D., Speck, G., and Fischer, H.,** Effects of angiotensin II and angiotensin II antagonist saralasin on cell growth and renin in 3T3 and SV3T3 cells, *J. Cell. Physiol.,* 98, 503, 1979.
28. **Genain, C., Bouhnik, J., Tewksbury, D., Corvol, P., and Menard, J.,** Characterization of plasma and cerebrospinal fluid human angiotensinogen and des-angiotensin I-angiotensinogen by direct radioimmunoassay, *J. Clin. Endocrinol. Metab.,* 59, 478, 1984.
29. **Do, Y.S., Shinagawa, T., Tam, H., Inagami, T., and Hsueh, W.A.,** Characterization of pure human renal renin, *J. Biol. Chem.,* 262, 1037, 1987.
30. **Richelson, B.,** Factors regulating the production of prostaglandin E_2 and prostacylin (prostaglandin I_2) in rat and human adipocytes, *Biochem. J.,* 247, 389, 1987.
31. **Stornetta, R.L., Hawelu-Johnson, C.L., Guyenet, P.G., and Lynch, K.R.,** Astrocytes synthesize angiotensinogen in brain, *Science,* 242, 1444, 1988.
32. **Intebi, A.D., Flaxman, M.S., Ganong, W.F., and Deschepper, C.F.,** Angiotensinogen production by rat astroglial cells in vitro and in vivo, *Neuroscience,* 34, 545, 1990.
33. **Deschepper, C.F., Birgornia, V., Berens, M.E., and Lapointe, M.C.,** Production of thrombin and antithrombin III by brain and astroglial cell cultures, *Mol. Brain Res.,* 11, 355, 1991.

34. **Rydzewski, B., Zelezna, B., Tang, W., Sumners, C., and Raizada, M.K.,** Angiotensin II stimulation of plasminogen activator inhibitor I gene expression in astroglial cells from the brain, *Endocrinology,* 130, 1255, 1992.
35. **Deschepper, C.F., Bouhnik, J., and Ganong, W.F.,** Co-localization of angiotensinogen and glial fibrillary acidic protein in astrocytes in the rat brain, *Brain Res.,* 374, 195, 1986.
36. **Thomas, W.G., and Sernia, C.,** Immunocytochemical localization of angiotensinogen in the rat brain, *Neuroscience,* 25, 319, 1988.
37. **Campbell, D.J., Sernia, C., Thomas, W.G., and Oldfield, B.J.,** Immunocytochemical localization of angiotensinogen in rat brain: dependence of neuronal immunoreactivity on method of tissue processing, *J. Neuroendocrinol.,* 3, 653, 1991.
38. **Lee, H.U., Campbell, D.J., and Habener, J.F.,** Developmental expression of the angiotensinogen gene in rat embryos, *Endocrinology,* 121, 1335, 1987.
39. **Milsted, A., Barna, B.P., Ransohoff, R.M., Brosnihan, K.B., and Ferrario, C.M.,** Astrocyte cultures derived from human brain tissue express angiotensinogen mRNA, *Proc. Natl. Acad. Sci. U.S.A.,* 87, 5720, 1990.
40. **Fatigati, V., Washington, J.M., Lynch, K.R., Peach, M.J., and Miller, E.D.,** Presence of angiotensinogen mRNA in various cultured cell lines, *Hypertension,* 9, 25, 1987.
41. **Lynch, K.R., Simnad, V.I., Ben-Ari, E.T., and Garrison, J.C.,** Localization of preangiotensinogen mRNA sequence in the rat brain, *Hypertension,* 8, 540, 1986.
42. **Lynch, K.R., Hawelu-Johnson, C.L., and Guyenet, P.G.,** Localization of brain angiotensinogen mRNA by hybridization histochemistry, *Mol. Brain Res.,* 2, 149, 1987.
43. **Mendelsohn, F.A.O., Quirion, R., Saavedra, J.M., Aguilera, G., and Catt, K.J.,** Autoradiographic localization of angiotensin II receptors in rat brain, *Proc. Natl. Acad. Sci. U.S.A.,* 81, 1575, 1984.
44. **Healy, D.P., Maciejewski, A.R., and Printz, M.P.,** Localization of central angiotensin II receptors with ^{125}I- sar^{1} ile^{8}-angiotensin II: periventricular sites of the anterior third ventricle, *Neuroendocrinology,* 44, 15, 1986.
45. **Lewicki, J.A., Printz, J.M., and Printz, M.P.,** Clearance of rabbit plasma angiotensinogen and relationship to CSF angiotensinogen, *Am. J. Physiol.,* 244, H577, 1983.
46. **Ben-Ari, E.T., Lynch, K.R., and Garrison, J.C.,** Glucocorticoids induce the accumulation of novel angiotensinogen gene transcripts, *J. Biol. Chem.,* 264, 13074, 1989.
47. **Tanaka, T., Ohkubo, H., Nakanishi, S., and Kageyama, R.,** Common structural organization of the angiotensinogen and the α_1-antitrypsin genes, *J. Biol. Chem.,* 259, 8063, 1984.
48. **Mori, M., Ishizaki, K., Yamada, T., Chen, H., Sugiyama, T., and Serikawa, T.,** Restriction fragment length polymorphisms of the angiotensinogen gene in inbred rat strains and mapping of the gene on chromosome 19q, *Cytogenet. Cell Genet.,* 50, 42, 1989.
49. **Clouston, W.M., Fournier, R.E., and Richards, R.I.,** The angiotensinogen gene is located on mouse chromosome 8, *FEBS Lett.,* 255, 419, 1989.
50. **Gaillard, I., Clauser, E., and Corvol, P.,** Structure of human angiotensinogen gene, *DNA,* 8, 87, 1989.
51. **Gaillard-Sanchez, I., Mattei, M.G., Clauser, E., and Corvol, P.,** Assignment by in situ hybridization of the angiotensinogen gene to chromosome band 1q4, the same region as the human renin gene, *Hum. Genet.,* 84, 341, 1990.
52. **Green, M.R.,** Pre-mRNA introns, *Annu. Rev. Genet.,* 20, 671, 1986.
53. **Brasier, A.R., Tate, J.E., Ron, D., and Habener, J.F.,** Multiple cis-acting DNA regulatory elements mediate hepatic angiotensinogen gene expression, *Mol. Endocrinol.,* 3, 1022, 1989.
54. **Ron, D., Brasier, A.R., Wright, K.A., Tate, J.E., and Habener, J.F.,** An inducible 50-kilodalton NFκB-like protein and a constitutive protein both bind the acute-phase response element of the angiotensinogen gene, *Mol. Cell. Biol.,* 10, 1023, 1990.

55. **Ron, D., Brasier, A.R., and Habener, J.F.,** Angiotensinogen gene-inducible enhancer-binding protein1, a member of a new family of large nuclear proteins that recognize nuclear factor B-binding sites through a zinc finger motif, *Mol. Cell. Biol.*, 11, 2887, 1991.
56. **Itoh, N., Matsuda, T., Ohtani, R., and Okamoto, H.,** Angiotensinogen production by rat hepatoma cells is stimulated by B cell stimulatory factor 2/interleukin-6, *FEBS Lett.*, 244, 6, 1989.
57. **Ron, D., Brasier, A.R., Wright, K.A., and Habener, J.F.,** The permissive role of glucocorticoids on interleukin-1 stimulation of angiotensinogen gene transcription is mediated by an interaction between inducible enhancers, *Mol. Cell. Biol.*, 10, 4389, 1990.
58. **Baldwin, A.S., LeClair, K.P., Singh, H., and Sharp, P.A.,** A large protein containing a zinc finger domain binds to related sequence elements in the enhancers of the class 1 major histocompatibility complex and kappa immunoglobulin genes, *Mol. Cell. Biol.*, 10, 1406, 1990.
59. **Fukamizu, A., Takahashi, S., Seo, M.S., Tada, M., Tanimoto, K., Uehara, S., and Murakami, K.,** Structure and expression of the human angiotensinogen gene, *J. Biol. Chem.*, 265, 7576, 1990.
60. **Takahashi, S., Fukamizu, A., Hasegawa, T., Yokoyama, M., Nomura, T., Katsuki, M., and Murakami, K.,** Expression of the human angiotensinogen gene in transgenic mice and transfected cells, *Biochem, Biophys. Res. Commun.*, 180, 1103, 1991.
61. **Kalinyak, J.E., Hoffman, A.R., and Perlman, A.J.,** Ontogeny of angiotensinogen messenger RNA and angiotensin-II receptors in rat brain and liver, *J. Endocr. Invest.*, 14, 647, 1991.
62. **Gomez, R.A., Cassis, L., Lynch, K.R., Chevalier, R.L., Wilfong, N., Carey, R.M., and Peach, M.J.,** Fetal expression of the angiotensinogen gene, *Endocrinology*, 123, 2298, 1988.
63. **Millan, M.A., Carvallo, P., Izumi, S.I., Zemel, S., Catt, K.J., and Aguilera, G.,** Novel sites of expression of functional angiotensin II receptors in late gestation fetus, *Science*, 244, 1340, 1989.
64. **Olson, A.L., Robillard, J.E., Kisker, C.T., Smith, B.A., and Perlman, S.,** Negative regulation of angiotensinogen gene expression by glucocorticoids in fetal sheep liver, *Pediatr. Res.*, 30, 256, 1991.
65. **Everett, A.D., Chevalier, R.L., and Gomez, R.A.,** Hepatic angiotensinogen gene regulation in the fetal and pregnant rat, *Pediatr. Res.*, 30, 252, 1991.
66. **Saye, J.A., Lynch, K.R., and Peach, M.J.,** Changes in angiotensinogen messenger RNA in differentiating 3T3-F442A adipocytes, *Hypertension*, 15, 867, 1990.
67. **Ben-Ari, E.T. and Garrison, J.C.,** Regulation of angiotensinogen mRNA accumulation in rat hepatocytes, *Am. J. Physiol.*, 255, E70, 1988.
68. **Kalinyak, J.E. and Perlman, A.J.,** Tissue-specific regulation of angiotensinogen mRNA accumulation by dexamethasone, *J. Biol. Chem.*, 262, 460, 1987.
69. **Kunapuli, S.P., Benedict, C.R., and Kumar, A.,** Tissue specific hormonal regulation of the rat angiotensinogen gene expression, *Arch. Biochem. Biophys.*, 254, 642, 1987.
70. **Khayyall, M., MacGregor, J., Brown, J.J., Lever, A.F., and Robertson, J.I.S.,** Increase of plasma renin substrate concentration after infusion of angiotensin in the rat, *Clin. Sci. London*, 44, 87, 1973.
71. **Nasjletti, A. and Masson, G.M.C.,** Stimulation of angiotensinogen formation by renin and angiotensin, *Proc. Soc. Exp. Biol. Med.*, 142, 307, 1973.
72. **Stuzmann, M., Radziwill, R., Komischke, K., Klett, C., and Hackenthal, E.,** Hormonal and pharmacological alteration of angiotensinogen secretion from rat hepatocytes, *Biochim. Biophys. Acta*, 886, 48, 1986.
73. **Iwao, H., Nakamura, A., Fukui, K., Kimura, S., Tamaki, T., and Abe, Y.,** Endogenous angiotensin II regulates hepatic angiotensinogen production, *Life Sci.*, 47, 2343, 1990.

74. **Nakamura, A., Iwao, H., Fukui, K., Kimura, S., Tamaki, T., and Abe, Y.,** Effect of angiotensin II on angiotensinogen production in adrenalectomized rats, *Life Sci.*, 46, 1657, 1990.
75. **Chang, E. and Perlman, A.J.,** Angiotensinogen mRNA: regulation by cell cycle and growth factors, *J. Biol. Chem.*, 263, 5480, 1988.
76. **Inglefinger, J.R., Pratt, R.E., Ellison, K., and Dzau, V.J.,** Sodium regulation of angiotensinogen mRNA expression in rat kidney cortex and medulla, *J. Clin. Invest.*, 78, 1311, 1986.
77. **Pratt, R.E., Zou, W.M., Naftilan, A.J., Inglefinger, J.R., and Dzau, V.J.,** Altered sodium regulation of renal angiotensinogen mRNA in the spontaneously hypertensive rat, *Am. J. Physiol.*, 256, F469, 1989.
78. **Naftilan, A.J., Zuo, W.M., Inglefinger, J., Ryan, T.J., Pratt, R.E., and Dzau, V.J.,** Localization and differential regulation of angiotensinogen mRNA expression in the vessel wall, *J. Clin. Invest.*, 87, 1300, 1992.
79. **Yongue, B.G., Angulo, J.A., McEwen, B.S., and Myers, M.M.,** Brain and liver angiotensinogen messenger-RNA in genetic hypertensive and normotensive rats, *Hypertension*, 17, 485, 1991.
80. **Gomez, R.A., Lynch, K.R., Chevalier, R.L., Everett, A.D., Johns, D.W., Wilfong, N., Peach, M.J., and Carey, R.M.,** Renin and angiotensinogen gene expression and intrarenal renin distribution during angiotensin converting enzyme inhibition, *Am. J. Physiol.*, 254, F900, 1988.
81. **Morishita, R., Higaki, J., Okunishi, H., Kawamoto, T., Ishii, K., Hakamura, F., Katahira, K., Nagano, M., Mikami, H., Miyazkai, M., and Ogihara, T.,** Effect of long-term treatment with an angiotensin-converting enzyme-inhibitor on the renin-angiotensin system in spontaneously hypersensitive rats, *Clin. Exp. Pharmacol. Physiol.*, 18, 685, 1991.
82. **Finckh, M., Hellmann, W., Ganten, D., Furtwangler, A., Allegeier, J., Boltz, M., and Holtz, J.,** Enhanced cardiac angiotensinogen gene-expression and angiotensin converting enzyme activity in tachypacing-induced heart-failure in rats, *Basic Res. Cardiol.*, 86, 303, 1991.
83. **Richoux, J.P., Cordonnier, J.L., Bouhnik, J., Clauser, E., Corvol, P., Menard, J., and Grignon, G.,** Immunocytochemical localization of angiotensinogen in rat liver and kidney, *Cell Tissue Res.*, 233, 439, 1983.
84. **Morris, B.J. and Reid, I.A.,** The distribution of angiotensinogen in dog brain studied by cell fractionation, *Endocrinology*, 103, 492, 1978.
85. **Deschepper, C.F. and Reudelhuber, T.L.,** Rat angiotensinogen is secreted only constitutively when transfected into AtT-20 cells, *Hypertension*, 16, 147, 1990.
86. **Boucher, R., Asselin, J., and Genest, J.,** A new enzyme leading to the direct formation of angiotensin II, *Circ. Res.*, 34 (Suppl. I), 2, 1974.
87. **Araujo, G.W., Pesquero, J.B., Lindsey, C.J., Paiva, A.C.M., and Pesquero, J.L.,** Identification of serine proteinases with tonin-like activity in the rat submandibular and prostate glands, *Biochim. Biophys. Acta*, 1074, 167, 1991.
88. **Tonnesen, M.G., Klempner, M.S., Austen, K.F., and Wintroub, B.U.,** Identification of a human neutrophil angiotensin II-generating protease as cathepsin G, *J. Clin. Invest.*, 69, 25, 1982.
89. **Campbell, D.J.,** Tissue renin-angiotensin system: sites of angiotensin formation, *J. Cardiovasc. Pharmacol.*, 10, S1, 1987.
90. **Fyhrquist, F., Rosenlof, K., Gronhagen-Riska, C., Hortling, L., and Tikkanen, I.,** Is renin substrate an erythropoietin precursor?, *Nature*, 308, 649, 1984.
91. **Jacobs, K., Shoemaker, C., Rudersdorf, R., Neill, S.D., Kaufman, R.J., Mufson, A., Seehra, J., Jones, S.S., Hewick, R., Fritsch, E.F., Kawakita, M., Shimizu, T., and Miyake, T.,** Isolation and characterization of genomic and cDNA clones of human erythropoietin, *Nature*, 313, 806, 1985.

92. **Poulsen, K. and Jacobsen, J.,** Is angiotensinogen a renin inhibitor and not the substrate for renin?, *J. Hypertens.,* 4, 65, 1986.
93. **Hackenthal, E., Hackenthal, R., and Hofbauer, K.G.,** No evidence for product inhibition of the renin-angiotensinogen reaction in the rat, *Circ. Res.,* 41, 49, 1977.
94. **Doolittle, R.F.,** Similar amino acid sequences revisited, *Trends Biochem. Sci.,* 14, 244, 1989.
95. **Stein, P.E., Tewksbury, D.A., and Carrell, R.W.,** Ovalbumin and angiotensinogen lack serpin S-R conformational change, *Biochem. J.,* 262, 103, 1989.
96. **Ohkubo, H., Kawakami, H., Kakehi, Y., Takumi, T., Arai, H., Yokota, Y., Iwai, M., Tanabe, Y., Masu, M., Hata, J., Iwao, H., Okamoto, H., Yokoyama, M., Nomura, T., Katsuki, M., and Nakanishi, S.,** Generation of transgenic mice with elevated blood pressure by introduction of the rat renin and angiotensinogen genes, *Proc. Natl. Acad. Sci. U.S.A.,* 87, 5153, 1990.
97. **Kimura, S., Mullins, J.J., Bunnemann, B., Metzger, R., Hilgenfeldt, U., Zimmermann, F., Jacob, H., Fuxe, K., Ganten, D., and Kaling, M.,** High blood pressure in transgenic mice carrying the rat angiotensinogen gene, *EMBO J.,* 11, 821, 1992.
98. **Gomez, R.A., Lynch, K.R. et al.,** submitted.
99. **Saye, J.A. and Peach, M.J.,** unpublished data.
100. **Stornetta, R.L. and Lynch, K.R.,** unpublished data.

Chapter 6

HORMONAL REGULATION OF THE ANGIOTENSINOGEN GENE IN LIVER AND OTHER TISSUES

C.F. Deschepper and L.Q. Hong-Brown

TABLE OF CONTENTS

0-8493-4622-3/93/$0.00 + $.50

I. INTRODUCTION

Angiotensinogen (AOG) is the glycoprotein precursor of the angiotensin peptides. It is cleaved by renin to yield the decapeptide angiotensin I (AngI); angiotensin-converting enzyme (ACE) then removes a dipeptide from the C-terminal end of AngI to form the physiologically active octapeptide angiotensin II (AngII). Circulating AngII increases vascular tone and promotes salt and water retention by a combined action on many target organs, including vascular smooth muscles, kidneys, adrenals, brain, and sympathetic nervous system. Both renin and AOG are rate-limiting factors in the generation of this important peptide.[1-3] The importance of maintaining adequate levels of AOG is underscored by the fact that its concentration in circulating plasma is close to the K_m of the proteolytic activity of renin;[1,2,4] as a consequence, renin generates amounts of AngI that are directly proportional to the amounts of available substrate.[4,5] This observation has prompted many investigations aimed at delineating which factors control the production of AOG.

All available evidence indicates that AOG is secreted only constitutively and cannot be stored within secretory granules.[6] This implies that the production of AOG is controlled mostly at the level of synthesis (instead of secretion). In agreement with this requirement, the hepatic production of AOG is affected primarily by hormones that act at the genomic level, i.e., steroids and thyroid hormones. The effects of such hormones on circulating levels of AOG were recognized more than 40 years ago.[7,8] However, the AOG cDNA and its corresponding gene have been cloned and characterized only much more recently.[9-12] The availability of these reagents has made it possible to investigate the molecular mechanisms responsible for the regulation of AOG in further detail. In addition to liver, AOG is produced by a variety of tissues, including brain, adipose tissue, and kidneys.[13,14] Similarly to hepatocytes, cells responsible for the production of AOG in these tissues (astrocytes, adipocytes, fibroblasts, etc.) lack secretory granules.[14] Many of the same hormones that affect the production of AOG by liver also regulate the AOG gene in tissues, but there are tissue-specific differences.[2,13]

Recent and comprehensive reviews have been devoted to the topic of AOG.[14-16] In the present chapter, we will devote special attention to particular aspects of the regulation of AOG. First, we will compare the mechanisms of action of different hormones. Indeed, the rate of production of AOG is regulated primarily at the level of its mRNA concentration, but this may be achieved by different mechanisms. Second, we will discuss tissue-specific differences in the effects of hormones on the AOG gene. Finally, we will put special emphasis on discussing the effect of each hormone in the light of their physiopathological importance. Indeed, a controversy has recently arisen concerning the exact physiological impact of the regulation of AOG;[14,17] therefore, we will examine whether or when variations in the production of AOG result in changes in the concentration or activity of other components of the renin-angiotensin system.

II. SEX STEROIDS

The stimulatory effects of estrogens on circulating levels of AOG were first reported by Helmer and Griffith.[7] This has been confirmed in many subsequent reports;[18-20] in addition, synthetic steroids were found to be more effective than naturally occurring steroids.[21] Administration of estrogens to male rats increases the synthesis and secretion of AOG by liver slices *in vitro*.[22] Estrogens probably act directly on the liver, since they stimulate the production of AOG by either a perfused liver system or confluent hepatoma cells when added to the perfusion[23] or incubation medium.[24] Likewise, we have observed that administration of ethynyl estradiol (3mg/kg) to male rats increased the concentration of AOG mRNA in liver tissue extracts; the effect was maximal 24 h after administration, and changes in mRNA concentrations or in plasma AOG were of the same magnitude.[25] These hormones probably increase the transcriptional rate of the AOG gene, since they upregulate the expression of a minigene containing 0.75 kb of 5′-flanking sequence of the AOG gene in cognate transgenic mice.[26]

Some AOG-producing models have been found to be nonresponsive to estrogens. These include freshly dispersed hepatocytes,[27] certain rat hepatoma cell lines,[28,29] and the livers of immature or hypophysectomized rats.[7,30,31] In each of these models, it is possible that a lack of estrogen receptors accounts for the nonresponsitivity.[28,30] Conversely, it has been shown that there are higher levels of liver estrogen receptors in female than in male rats, and that plasma AOG was stimulated to a greater extent by low doses of estrogens in female rats.[31] It therefore appears that the levels of estrogen receptors are important modulators of the effect of estrogens on the production of AOG.

The effects of estrogens are also tissue specific. Administration of ethynyl estradiol to male rats had no effect on the concentration of AOG mRNA in brown adipose tissue; in the brain, estrogens did not increase, but decreased AOG mRNA concentration in diencephalon, and they had no effect in pons or in cerebellum.[25] Our results are in agreement with those of Printz et al., who reported a decrease of AOG protein in specific brain regions after administration of estrogens.[32] However, they are hard to reconcile with those of Kunapuli et al., who reported an increase of whole brain AOG mRNA 4 h after administration of estradiol.[33] Much less information is available concerning the effects of androgens. Helmer and Griffith[7] reported that the effect of diethylstilbestrol on plasma AOG was inhibited by testosterone; however, this observation could not be confirmed by others.[19] Ellison et al. reported that expression of the AOG gene in kidney was greatly dependent upon the presence of androgens.[34] Testosterone also had a small effect on AOG mRNA concentration in liver, but it was much less pronounced than the effect on kidneys; in addition there was no effect of testosterone on plasma AOG.[34]

In order to discuss the effects that increased levels of AOG may have on the renin-angiotensin system, it is necessary to describe briefly the two kinds of enzymatic assays that are used to measure renin. Plasma renin concentration

(PRC) corresponds to the actual amounts of active renin in plasma; this measurement is independent of the availability of substrate. Plasma renin activity (PRA) is an index of the capacity of plasma to generate AngI, and depends both on PRC and plasma AOG. *In vitro,* PRA increases linearly with the concentration of AOG;[4,5] in contrast, upregulation of AOG rarely affects PRA or plasma AngII in normal *in vivo* conditions, because renin secretion is simultaneously downregulated (presumably by the negative feedback action of AngII on the juxtaglomerular cells).[14] However, the negative feedback control of renin may be altered in some situations, like exposure to estrogens. For instance, the drop in PRC is delayed and occurs only 5 d after administration of estrogens to rats.[20] Many studies indicate that in humans, plasma AOG is increased by administration of estrogen-containing medications.[5,35-38] This is usually accompanied by some suppression of PRC, but not always to a sufficient extent to prevent a rise in PRA.[5,35,36,38] A rise in PRA and plasma AngII during contraceptive treatment has not been observed in all studies;[37] however, this may be explained by the fact that sodium and water retention make it sometimes difficult to detect a high renin state.[39] When plasma AngII and PRA are plotted v. daily sodium excretion, these two parameters are clearly higher in women taking contraceptive agents than in controls, along with an AngII-dependent reduction in renal blood flow.[36] Therefore, it appears that increased levels of plasma AOG are responsible for small but significant increases in PRA in most women taking contraceptives; in a smaller percentage of these patients, an escape from the normal feedback mechanisms results in more important increases in PRA and in arterial blood pressure.

Another condition where estrogens have meaningful consequences is pregnancy: both plasma AOG and PRA are increased. This leads to increased levels of circulating AngII,[37,40] which may play a special role in the last trimester of pregnancy by improving the perfusion of the uteroplacental unit.[40] However, mild forms of hypertension develop in some cases; the increase in plasma AngII may be partly responsible for these disorders.[40]

III. THYROID HORMONES

In rats, induction of hypothyroidism results in a decrease in circulating levels of AOG[1,41,42] and in the rate of production of AOG by the liver;[22] these levels are restored to normal by administration of T_3.[41,43] When animals are made hyperthyroid by daily administration of T_4, plasma levels of AOG are elevated.[1,44] Some authors reported that thyroid hormones (TH) increase the production of AOG by primary hepatocytes[45] and rat hepatoma cells,[28] while others found no such effect.[29] However, we have determined that the effect of TH on hepatoma cells depends on which particular cell line is used.[44] In addition, the effect was dependent on the growth status of the cells; thus, T_3 increased the production of AOG in confluent, but not in subconfluent, cells.[15]

In all reported studies, there was an excellent correlation between the magnitudes of changes in AOG secretion and liver AOG mRNA concentrations.[42,44,46] This implies that the effect of TH is either at the level of transcription of the AOG gene or stability of its mRNA. However, the difference in responsiveness between confluent and subconfluent cells indicates that the action of TH may be dependent on the induction of secondary genes within these cells.[44] To verify this point, we have examined the effect of cycloheximide on the response of confluent H35 hepatoma cells. This drug blocked the effect of T_3, showing that ongoing protein synthesis is a requirement in order for TH to stimulate the production of AOG (Figure 1); in contrast, cycloheximide did not block the effect of dexamethasone (discussed further below).

TH also effect the production of AOG in at least one other tissue, i.e., the brain. *In vitro,* primary cultures of astrocytes produce more AOG in T_3-containing than in T_3-deficient serum-free medium.[15] *In vivo,* chronic hypothyroidism lowers the concentration of AOG mRNA in all brain regions we have examined. However, as opposed to what was observed in liver, brain AOG mRNA was not increased in hyperthyroid vs. control animals.[15]

In contrast to other hormonal deficiencies, hypothyroidism is the only one of such conditions that leads to decreased levels of production of AOG

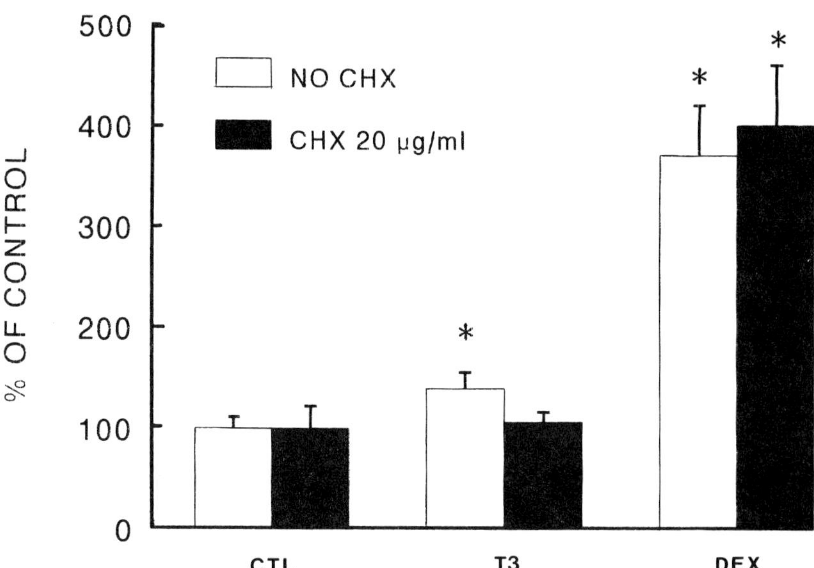

FIGURE 1. Changes in concentrations of AOG mRNA (as determined by Northern blot analysis) in confluent H35 hepatoma cells after 4-h exposure to either T3 10^{-6} *M* or dexamethasone (DEX) 10^{-7} *M*. Values are expressed as percent of the mean values in the control groups (CTL; cells not exposed to hormones) and are means ± SD, n = 6. Protein synthesis was blocked by adding cycloheximide (CHX) 20 μg/ml to the incubation medium, 1 h prior to and during the exposure to the hormones (black bars). $*p \leq 0.05$.

by the liver. Circulating levels of AOG are also decreased after hypophysectomy or hypothalamic lesions, but this can be completely accounted for by a deficit in thyroid function.[47] TH therefore appear to be the main endocrine factor involved in the maintenance of *basal* levels of plasma AOG. When plasma levels of AOG are altered as a result of a thyroid disorder, there is also an impairment of the normal feedback mechanism on renin secretion; as a consequence, PRA may be low during hypothyroidism, and high during hyperthyroidism.[1,41,48,49]

IV. GLUCOCORTICOIDS

Since the initial observation of Helmer and Griffith,[8] numerous reports have confirmed that glucocorticoids (GC) increase the rate of production of AOG by liver.[22,50,51] Likewise, GC increase the concentration of AOG mRNA in liver,[42,52,53] primary hepatocytes,[27] and hepatoma cell lines.[28,53] Earlier studies suggested that GC exert their effects at the level of transcription of the AOG gene, since their action was blocked by actinomycin D.[51] This has since been confirmed by four types of studies:

1. Dexamethasone was found to increase AOG gene transcription fivefold in a rat pancreatic islet cell line.[54]
2. By sequence analysis, it was determined that GC-responsive elements (GRE) were present in the 5′-flanking region of the AOG gene.[55]
3. A 0.75-kb portion of the 5′-flanking region was found to be sufficient to confer appropriate specificity of expression and regulation to a minigene in transgenic mice.[26]
4. A construct containing a reporter gene, driven by a 688 portion of the 5′-flanking region, was found to be appropriately expressed and regulated when transfected into HepG2 hepatoma cells, and specific liver nuclear proteins were found to bind to the GREs in the promoter region.[56]

Unlike thyroid hormones, the effect of GC on AOG mRNA concentration does not require ongoing protein synthesis, since it is unaffected by preincubation of cells with cycloheximide (Figure 1).

Certain features of the action of GC appear to differentiate the response of the AOG gene from that of other genes. Two distinct "GREs" have been identified within the 5′-flanking sequence of the AOG gene.[15,57] These two sequences are located on both sides of an "acute-phase response element" (APRE),[15] and GC-induced transcription of AOG gene was found to be dependent upon occupancy of the APRE by specific binding proteins.[58] In addition, GC induce the accumulation of novel and longer transcripts by the use of two new transcription sites in the AOG gene.[57] However, the longer mRNAs do not appear to code for novel forms of the AOG protein.

Adrenalectomy results in a decrease in circulating levels of AOG.[59] However, this merely reflects consumption of plasma AOG by high levels of

TABLE 1

	Control	Acute DEX	Chronic DEX
PAC (fmol/ml)	702 ± 19	1905* ± 65	875 ± 87
LIV AOG mRNA (O.D. units)	7.3 ± 1.3	26.1* ± 2.9	32.7* ± 5.5

Note: Values for plasma AOG concentration (PAC) and concentration of AOG mRNA in liver (LIV AOG mRNA) are means ± SE, n = 5. Rats were injected with either saline (Control) or dexamethasone 1 mg/kg s.c. 6 h prior to sacrifice (Acute DEX), or were given dexamethasone 40 μg/d for 5 d (through an implanted osmotic minipump).

*$p \leq 0.05$

circulating renin, instead of a decrease in the rate of production of AOG by the liver.[50,60] Thus, the concentration of AOG mRNA in liver is not decreased after adrenalectomy.[52,53] In contrast, the concentration of AOG mRNA in brain was significantly reduced after adrenalectomy; this suggests that glucocorticoids participate in the maintenance of basal levels of AOG in brain, but not in liver. There are other tissue-specific differences in the effects of GC in nonhepatic tissues; thus, acute administration of dexamethasone has only small or no effect on the concentration of AOG mRNA in brain and kidney (in contrast to liver).[52,53] We have observed that the magnitude of the effect of dexamethasone on AOG mRNA in brown adipose tissue was at least as high as what was obtained in liver.[61]

Previous studies have generally examined the effects of acute administration of high doses of dexamethasone; it is therefore difficult to predict from these results whether GC are physiological regulators of AOG. More recently, we have administered lower doses of dexamethasone (40 μg/d) to rats chronically, using osmotic minipumps. Both chronic and acute dexamethasone treatments increased liver AOG mRNA concentration to about the same extent. However, plasma AOG was not elevated significantly in the group of rats receiving dexamethasone chronically (see Table 1); in contrast to what we had observed with estrogens and TH, there was a poor correlation between liver AOG mRNA concentration and plasma AOG during chronic exposure to GC. *In vitro,* we have made similar observations when incubating rat hepatoma cells with GC: AOG was stimulated to a lesser extent in cells preincubated for 5 d with dexamethasone than in cells freshly exposed to the drug. Taken together, these data suggest that chronic exposure to GC has differential effects on transcription of the AOG gene (which is increased) and translation of its mRNA (which is decreased). These findings may provide an explanation of why plasma levels of AOG were much higher in humans treated with estrogens than in patients treated with GC or afflicted with Cushing's syndrome.[30] More studies are required to elucidate the molecular mechanisms underlying these effects.

The findings listed above provide clues to understanding the relative importance of GC in the regulation of AOG. As opposed to thyroid hormones, they are not involved in the maintenance of basal levels of production of AOG by the liver, although they may play such a role in the brain. Acute injections of GC result in a two- to threefold increase of plasma AOG;[51,53,62] however, this effect is buffered by compensatory decreases in renin secretion. As a result, PRA is affected only marginally (although sometimes significantly).[62] Chronic exposure to GC increases transcription of the AOG gene, but plasma AOG is not elevated because of an opposite effect upon translation. In this context, there is no evidence that GC are primary regulators of the hepatic production of AOG; their main role appears to be limited to a permissive and synergistic effect to the action of cytokines, as discussed below.

V. CYTOKINES

AOG has been found to be a structural member of the superfamily of serine proteases inhibitors (serpins).[63] The fact that most serpins behave as acute-phase proteins prompted some investigators into looking at the effect of acute inflammatory reactions on the production of AOG. Peripheral administration of *Escherichia coli* lipopolysaccharide (LPS) was found to increase AOG mRNA fivefold in rat liver, but had not effect in brain.[64] The effect of LPS is mediated by leukocytes-derived cytokines, and GC have an essential permissive role.[65] Likewise, interleukin-6 was found to increase the production of AOG by H35 hepatoma cells, and the presence of GC was prerequisite;[66] under similar conditions, interleukin-1 and interferon-α had no effect on the production of AOG.[66]

Some of the molecular mechanisms underlining the mechanisms of action of cytokines have been recently analyzed in detail.[15,58,67,68] A sequence of nucleotides bearing resemblance with the consensus sequence for binding by NFκB (APRE) has been identified in the promoter region of the rat AOG gene.[15,64] This APRE is located between a tandem of GRE, and can bind to nuclear proteins belonging to two main classes (NFκB and the C/EBP-like family). Such a modular arrangement explains the essential permissive role of GC in the action of interleukins; thus, induction of angiotensinogen gene transcription involves interactions between the GC receptor and the specific proteins that bind to the APRE.[58] Differences have been found in the pattern of APRE-binding proteins in various tissues;[15] this observation provides a possible molecular explanation for some of the tissue-specific differences in the effects of GC.

Two classes of acute-phase proteins have been distinguished on the basis of their response to cytokines: some respond primarily to interleukin-1 or tumor necrosis factor (TNF)-α, while others respond to either interleukin-6 or LIF.[69] It is still unclear which cytokines mediate the effects of acute inflammation on AOG. Experiments with rat hepatoma cells suggest that

AOG responds mainly to interleukin-6.[66] Ron and Brasier have shown that the AOG promoter was induced by interleukin-1 and TNF-α when transfected into Hep G2 cells.[15] However, these cells are of human origin, and they may respond different from cells of rat origin.

Taken together, the data suggest that a coordinate interaction between GC and some cytokines is responsible for the induction of the AOG gene during acute phase situations. *In vivo,* circulating AngII is elevated after LPS injection at the same time that plasma AOG reaches its peak.[70] However, it is not known whether AngII is a physiologically important component of the acute-phase response. Alternatively, it is possible that AOG plays a yet-undefined role during inflammation.

VI. ANGIOTENSIN II

Administration of AngII increases plasma AOG in dogs and rats.[71,72] More recently, it has been shown that the concentration of AOG mRNA in liver is increased by infusion of exogenous AngII, and decreased either by infusion of saralasin (an AngII antagonist) or administration of captopril (an inhibitor of converting enzyme).[73] In rats fed with diets containing different amounts of sodium, the high correlation between plasma AngII and liver AOG mRNA concentration suggests that endogenous AngII plays an active role in the regulation of AOG synthesis.[74] AngII may also regulate the production of AOG in other tissues, since kidney AOG mRNA is affected by maneuvers that effect circulating AngII.[75]

In vitro, AngII increases both the rate of secretion of AOG and AOG mRNA accumulation in rat hepatocytes, without affecting the synthesis of total proteins or albumin secretion.[76,77] The effect of AngII is not affected by modulators of intracellular calcium, it is blocked by pertussis toxin, and it appears to be mediated by inhibition of adenylate cyclase.[77] Two observations suggest that AngII exerts its effect by stabilizing AOG mRNA: (1) AngII attenuates the actinomycin-induced decrease in AOG mRNA, and (2) the effect of AngII is more rapid than that of dexamethasone, indicating that it acts at a different level than upon transcription.[78]

The role of AngII may best be illustrated by recent data from Iwao et al.[79] These authors have looked at the effect of administration of Dup753 (a drug that blocks the AT_1 AngII receptors selectively) to adrenalectomized rats. They report that PRC is higher in adrenalectomized than in control rats, and that it increases even further after administration of Dup753 (see Table 2). Despite the huge increase in PRC, PRA is similar in adrenalectomized rats and Dup753-treated adrenalectomized rats, because consumption of AOG prevents PRA from rising any further. From these data, one can appreciate that some stimulation in the production of AOG is prerequisite in order to obtain higher levels of PRA. AngII is therefore a perfect physiological regulator of AOG, since it increases its production in those very situations when the needs for substrate are increased. Since it may act on AOG mRNA stability

TABLE 2

	PRA (ng AngI/ml/h)	PRC (ng AngI/ml/h)	PAC (ng AngI/ml)
Intact	7.4 ± 0.7	58.9 ± 4.6	1,015 ± 23
ADX	31 ± 5.1	376 ± 95	549 ± 44
ADX + Dup753	35 ± 2.7	10,795 ± 708	42 ± 4.1

Note: Values for plasma renin activity (PRA), plasma renin concentration (PRC), and PAC are mean ± SE. The experimental groups are intact rats, 7 d post-adrenalectomy (ADX), and adrenalectomized animals receiving 10 mg/kg of Dup753 daily in the drinking water (ADX + Dup753).

Adapted from Iwao, H. et al., *Hypertension,* 18 (Abstr.), 411, 1991. With permission.

instead of transcription, it is also the only hormone that can achieve this effect in a relatively short period of time.

VII. OTHER REGULATORY MECHANISMS

Numerous reports indicate that after binephrectomy or renal tubular damage, there is an increase in plasma AOG, in the rate of production of hepatic AOG, and in the concentration of hepatic AOG mRNA in liver.[22,23,51,59,62,80,81] Although this condition is one of the most potent stimulators of liver AOG, the operative mechanisms have still not been elucidated. The effect of nephrectomy appears to be mediated by an induction of transcription, since it is blocked by actinomycin D.[51,82] Stimulation of AOG is not maximal after nephrectomy, since the effect of other hormones (GC, estrogens) can still be superimposed.[62,81,82] In animals that have been adrenalectomized 7 d prior to removal of the kidneys, the effect of nephrectomy is greatly attenuated, but not abolished; the effect is therefore not totally dependent on steroids.[51,62] Iwao et al. have argued that GC are permissive and synergistic for the action of a yet-unidentified factor.[62] Indeed, others have suggested that an AOG-stimulating factor is present in the gammaglobulin fraction of plasma from nephrectomized animals.[83] The effect of nephrectomy is also tissue specific: in addition to liver, it increases AOG mRNA concentration in adipose tissue.[84] However, it has no effect on AOG mRNA in brain and it lowers the transcript in other tissues, including aorta, adrenal, mesentery, and lung.[13,62]

In vitro experiments have determined that the production of AOG by hepatoma cells is decreased by agents that stimulate proliferation; these include insulin and TPA in H35 cells[85] and high concentrations of mestranol in HepG2 cells.[24] The relevance of these observations to the regulation of AOG *in vivo* has not been determined yet.

VIII. PHYSIOPATHOLOGICAL CONSEQUENCES OF ANGIOTENSINOGEN REGULATION

While most hormones that regulate the production of AOG operate by increasing the accumulation of AOG mRNA, they achieve this effect by different mechanisms. Glucocorticoids, cytokines, and estrogens all appear to upregulate transcription of the AOG gene directly. Thyroid hormones appear to require the induction of secondary genes, while AngII probably stabilizes the AOG mRNA transcript. With the exception of AngII, the effects of all of these hormones is not apparent before several hours. It is therefore obvious that the regulation of AOG will have little to do with the short-term regulation of blood pressure. Renin, which can respond swiftly to a range of stimuli, is the most important factor in determining how much angiotensins will be produced. However, its ability to do so varies according to the availability of its substrate. The importance of appropriate levels of substrate is illustrated in recent experiments by Ohkubo et al.[86] Thus, elevated levels of renin in transgenic mice increase blood pressure only in the presence of concomitant increases in AOG. In that context, it appears that the primary role of hormones in the regulation of AOG is to maintain levels of substrate that are optimal for renin, either during basal or stimulated conditions.

Each hormone appears to play a role that is most important in specific physiological circumstances: *estrogens* are especially important in the last trimester of pregnancy, when both renin and AngII are elevated; *thyroid hormones* participate in the maintenance of basal levels of AOG; *AngII* is mainly responsible for maintaining adequate levels of substrate when demand is higher; *glucocorticoids* may have some importance in subacute situations, but are mainly permissive for the action of *interleukins,* which stimulate the production of AOG in acute-phase situations. These regulatory processes are not the primary mechanisms controlling the concentration of AngII, and do not lead to large changes in the levels of plasma AOG. Instead, they operate rather unobtrusively, providing a continuous fine-tuning of the renin-angiotensin system by adjusting the availability of the substrate. Alterations in the rate of production of AOG have more obvious consequences in some of the particular conditions listed below.

1. Most studies on the regulation of AOG have investigated whether hormones do increase the production of AOG. Little attention has been paid to the effects of reduced levels of AOG. However, the maintenance of basal levels of AOG is clearly important, as exemplified by the fact that injections of antiserum to AOG lead to a decrease in blood pressure.[87] As discussed above, reduced levels of AOG are responsible for a decrease in PRA during TH insufficiency. Similarly, reductions in circulating AOG have been shown to be responsible for the reduction

in arterial pressure observed during liver cirrhosis or portocaval anastomosis.[79] Thus, failure to maintain appropriate basal levels of renin substrate will limit the ability of circulating renin to generate angiotensins.

2. Under conditions of rapid conversion of AOG to AngII, it is important to increase the production of AOG to keep pace with its consumption. One such situation is hemorrhage, where a stimulation of AOG has been shown to be an essential element that allows renin to mount an efficient response.[88] AngII (and possibly, GC) are the physiological regulators of AOG under these circumstances; however, the net effect of these hormones is to maintain ''normal'' levels of AOG in the face of continuous consumption, rather than increasing the concentration of AOG in plasma.
3. Most often, increased levels of plasma AOG are not accompanied by changes in PRA or circulating AngII, because PRC varies in the opposite direction (presumably as a consequence of a negative feedback action of AngII). However, this feedback mechanism does not operate optimally in certain conditions, including abnormally high levels of TH or estrogens. When the feedback mechanism is impaired, increases in plasma AOG are accompanied by increases in PRA, which may be responsible for high blood pressure (as discussed above).
4. In addition to being secreted in the circulation by the juxtaglomerular cells from the kidney, renin is produced by a variety of other tissues, including adrenals, blood vessels, or other cells within the kidney.[89-91] Just as in the circulation, the ability of renin (or renin-like enzymes) to generate angiotensins in these locations will be partly determined by the availability of substrate. For instance, it has been recently shown that in kidneys (where the concentration of AngII in glomerular filtrate is about 1000 times higher than in plasma[92]), the local concentration of AngI was reduced after administration of a converting-enzyme inhibitor (which lowers plasma AOG).[93] This observation is in contrast to what happens to plasma AngI (which increases because of high PRC and diminished conversion into AngII), and suggests that AOG is a major rate-limiting factor in the local generation of angiotensin peptides in kidneys.[93] In addition, tissular renin-like enzymes may contribute to the levels of angiotensins in the circulation as well. For instance, it is noteworthy that there are very little changes in plasma AngI and AngII after nephrectomy, despite a rapid decline in PRC.[82,94,95] This suggests that AOG is still processed despite the absence of renin in the circulation, presumably by tissular renin-like enzymes. In this context, high levels of AOG may be one factor fostering the generation of angiotensins, especially if there is no negative feedback mechanism of AngII on the production of renin (or renin-like enzymes) in tissues.

AOG is produced in tissues other than liver as well.[13] The importance of extrahepatic production of AOG has not been determined, but it is conceivable that the concentration of AOG is higher in tissues producing this protein than in those that receive the protein only from the circulation.[90] There are multiple examples of tissue-specific regulation of AOG, and these may represent mechanisms involved in regulating the availability of substrate at the level of specific tissues.

IX. CONCLUSIONS

More than 40 years have passed since Helmer and Griffith first reported that the production of AOG was under hormonal control.[7,8] However, details of the mechanisms of actions of these hormones have been revealed only much more recently. As discussed above, these findings have been helpful in defining the exact physiopathological importance of each respective hormone. In addition, recent studies on the AOG gene have yielded results whose relevance goes beyond the immediate role of AOG in the control of blood pressure; these include findings on the molecular arrangement of cis-acting elements in cytokine-responsive genes, or experiments unmasking differences in the effects of acute and chronic exposure to glucocorticoids. Our understanding of the molecular biology of the AOG gene has progressed rapidly, to the point that Lynch and Peach wondered recently whether this gene had "already yielded its innermost secrets".[14] However, the recent studies mentioned above illustrate the usefulness of ongoing investigations, and promise that future studies on the regulation of the AOG gene will keep yielding pertinent results. A recent study by Jeunemaitre et al. (*Cell,* 71, 169, 1992) has established that there is a genetic linkage between the AOG gene and hypertension; this constitutes additional evidence for the importance of this gene.

ACKNOWLEDGMENTS

We thank Drs. D.J. Campbell, P. Corvol, W.F. Ganong, I.A. Reid, M.K. Steele, and J. Sealy for helpful comments and discussions. This work was partly supported by United States Public Health Service Grants HL38774 and HL29714.

REFERENCES

1. **Dzau, V.J. and Herrmann, H.C.,** Hormonal control of angiotensinogen production, *Life Sci.,* 30, 577, 1982.
2. **Reid, I.A., Morris, B.J., and Ganong, W.F.,** The renin-angiotensin system, *Annu. Rev. Physiol.,* 40, 377, 1978.

3. **Poulsen, K.,** Kinetics of the renin system, *Scand. J. Clin. Lab. Invest.*, 31 (Suppl. 132), 186, 1973.
4. **McDonald, W.J., Cohen, E.L., Lucas, C.P., and Conn, J.W.,** Renin-renin substrate kinetic constants in the plasma of normal and estrogen-treated humans, *J. Clin. Endocrinol. Metab.*, 45, 1297, 1977.
5. **Skinner, S.L., Lumbers, E.R., and Symonds, E.M.,** Alterations by oral contraceptives of normal menstrual changes in plasma renin activity, concentration and substrate, *Clin. Sci.*, 36, 67, 1969.
6. **Deschepper, C.F. and Reudelhuber, T.L.,** Rat angiotensinogen is secreted only constitutively when transfected into AtT-20 cells, *Hypertension*, 16, 147, 1990.
7. **Helmer, O.M. and Griffith, R.S.,** The effect of administration of estrogens on the renin-substrate content of the rat plasma, *Endocrinology*, 51, 421, 1952.
8. **Helmer, O.M. and Griffith, R.S.,** Biological activity of steroids as determined by assay of renin-substrate, *Endocrinology*, 49, 154, 1951.
9. **Ohkubo, H., Kageyama, R., Ujihara, M., Hirose, T., Inayama, S., and Nakanishi, S.,** Cloning and sequence analysis of cDNA for rat angiotensinogen, *Proc. Natl. Acad. Sci. U.S.A.*, 80, 2196, 1983.
10. **Tanaka, T., Ohkubo, H., and Nakanishi, S.,** Common structural organization of the angiotensinogen gene and the α_1-antitrypsin genes, *J. Biol. Chem.*, 259, 8063, 1984.
11. **Clouston, W.M., Evans, B.A., Haralambidis, J., and Richards, R.I.,** Molecular cloning of the mouse angiotensinogen gene, *Genomics*, 2, 240, 1988.
12. **Gaillard, I., Clauser, E., and Corvol, P.,** Structure of human angiotensinogen gene, *DNA*, 8, 87, 1989.
13. **Campbell, D.J. and Habener, J.F.,** Angiotensinogen gene is expressed and differentially regulated in multiple tissues of the rat, *J. Clin. Invest.*, 78, 31, 1986.
14. **Lynch, K.R. and Peach, M.J.,** Molecular biology of angiotensinogen, *Hypertension*, 17, 263, 1991.
15. **Ron, D., Brasier, A.R., and Habener, J.F.,** Transcriptional regulation of hepatic angiotensinogen gene expression by acute-phase response, *Mol. Cell. Endocrinol.*, 74, C97, 1990.
16. **Tewksbury, D.A.,** Angiotensinogen: biochemistry and molecular biology, in *Hypertension: Pathophysiology, Diagnosis and Management*, Laragh, J.H. and Brenner, B.M., Eds., Raven Press, New York, 1990, 1197.
17. **Ménard, J., El Amrani, A.-I.K., Savoie, F., and Bouhnik, J.,** Angiotensinogen: an attractive and underrated participant in hypertension and inflammation, *Hypertension*, 18, 705, 1991.
18. **Ménard, J., Malmejac, A., and Milliez, P.,** Influence of diethylstilbestrol on the renin-angiotensin system of male rat, *Endocrinology*, 86, 774, 1970.
19. **Nasjletti, A., Matsunaga, M., and Masson, G.M.C.,** Effect of sex hormones on the renal pressor system, *Can. J. Physiol. Pharmacol.*, 49, 292, 1971.
20. **Ménard, J. and Catt, K.J.,** Effects of estrogen treatment on plasma renin parameters in the rat, *Endocrinology*, 92, 1382, 1973.
21. **Ménard, J., Corvol, P., Foliot, A., and Raynaud, J.P.,** Effects of estrogens on renin substrate and uterine weights in rats, *Endocrinology*, 93, 747, 1973.
22. **Clauser, E., Bouhnik, J., Coezy, E., Corvol, P., and Ménard, J.,** Synthesis and release of immunoreactive angiotensinogen by rat liver slices, *Endocrinology*, 112, 1188, 1983.
23. **Nasjletti, A. and Masson, G.M.C.,** Studies on angiotensinogen formation in a liver perfusion system, *Circ. Res.*, 30, II187, 1972.
24. **Coezy, E., Auzan, C., Lonigro, A., Philippe, M., Ménard, J., and Corvol, P.,** Effect of mestranol on cell proliferation and angiotensinogen production in HepG2 cells: relation with the cell cycle and action of tamoxifen, *Endocrinology*, 120, 133, 1987.

25. **Hong-Brown, L.Q. and Deschepper, C.F.,** unpublished data.
26. **Clouston, W.M., Lyons, I.G., and Richards, R.I.,** Tissue-specific and hormonal regulation of angiotensinogen minigenes in transgenic mice, *EMBO J.*, 8, 3337, 1989.
27. **Ben-Ari, E.T. and Garrison, J.C.,** Regulation of angiotensinogen mRNA accumulation in rat hepatocytes, *Am. J. Physiol.*, 255, E70, 1988.
28. **Chang, E. and Perlman, A.J.,** Multiple hormones regulate angiotensinogen messenger ribonucleic acid levels in a rat hepatoma cell line, *Endocrinology,* 121, 513, 1987.
29. **Togami, M., Blazka, D., and Hayashi, J.,** Control of angiotensinogen production by H4 rat hepatoma cells in serum-free medium, *In Vitro Cell. Dev. Biol.*, 24, 699, 1988.
30. **Krakoff, L.R.,** Measurement of plasma renin substrate by radioimmunoassay of angiotensin I: concentration in syndromes associated with steroid excess, *J. Clin. Endocrinol. Metab.*, 37, 110, 1973.
31. **Ignatenko, L.L., Mataradze, G.D., Bunyatyan, A.F., and Rozen, V.B.,** Sex differences in estrogen receptor accumulation in liver cell nuclei and increase of blood plasma angiotensinogen concentration after injection of low doses of synthetic estrogens in rats, *Bull. Exp. Biol. Med.*, 110, 1636, 1990.
32. **Printz, M.P., Hawkins, R.L., Wallis, C.J., and Mia Chen, F.,** Steroid hormones as feedback regulators of brain angiotensinogen and catecholamines, *Chest,* 83S, 308S, 1983.
33. **Kunapuli, S.P., Benedict, C.R., and Kumar, A.,** Tissue specific hormonal regulation of the rat angiotensinogen gene expression, *Arch. Biochem. Biophys.*, 254, 642, 1987.
34. **Ellison, K.E., Ingelfinger, J.R., Pivor, M., and Dzau, V.J.,** Androgen regulation of rat renal angiotensinogen messenger RNA expression, *J. Clin. Invest.*, 83, 1941, 1989.
35. **Cain, M.D., Walters, W.A., and Catt, K.J.,** Effects of oral contraceptive therapy on the renin-angiotensin system, *J. Clin. Endocrinol. Metab.*, 33, 671, 1971.
36. **Hollenberg, N.K., Williams, G.H., Burger, B., Chenitz, W., Hoosmand, I., and Adams, D.F.,** Renal blood flow and its response to angiotensin II. An interaction between oral contraceptive agents, sodium intake, and the renin-angiotensin system in healthy young women, *Circ. Res.*, 38, 35, 1976.
37. **Derckx, F.H.M., Stuenkel, C., Schalekamp, M.P.A., Visser, W., Huisveld, I.H., and Schalekamp, M.A.D.H.,** Immunoreactive renin, prorenin, and enzymatically active renin in plasma during pregnancy and in women taking oral contraceptives, *J. Clin. Endocrinol. Metab.*, 63, 1008, 1986.
38. **Newton, M.A., Sealey, J.E., Ledingham, J.G.G., and Laragh, J.H.,** High blood pressure and oral contraceptives: changes in plasma renin and renin substrate and in aldosterone secretion, *Am. J. Obstet. Gynecol.*, 101, 1037, 1968.
39. **Streeten, D.H.P., Anderson, G.H., and Dalakos, T.G.,** Angiotensin blockade: its clinical significance, *Am. J. Med.*, 60, 817, 1976.
40. **Broughton Pipkin, F.,** The renin-angiotensin system in normal and hypertensive pregnancies, in *Handbook of Hypertension, Vol. 10, Hypertension in Pregnancy,* Rubin, P.C., Ed., Elsevier, Amsterdam, 1988, 118.
41. **Bouhnik, J., Galen, F.X., Clauser, E., Ménard, J., and Corvol, P.,** The renin-angiotensin system in the thyroidectomized rats, *Endocrinology,* 108, 647, 1981.
42. **Sernia, C., Clements, J.A., and Funder, J.W.,** Regulation of liver angiotensinogen mRNA by glucocorticoids and thyroxine, *Mol. Cell. Endocrinol.*, 61, 147, 1989.
43. **Faust, P.L., Chirgwin, J.M., and Kornfeld, S.,** Renin, a secretory glycoprotein, acquires phosphomannosyl residues, *J. Cell Biol.*, 105, 1947, 1987.
44. **Hong-Brown, L.Q. and Deschepper, C.F.,** Effects of thyroid hormones on angiotensinogen gene expression in rat liver, brain, and cultured cells, *Endocrinology,* 130, 1231, 1992.
45. **Ruiz, M., Montiel, M., Jimenez, E., and Morell, M.,** Effect of thyroid hormones on angiotensinogen production in the rat in vivo and in vitro, *J. Endocrinol.*, 115, 311, 1987.

46. **Kimura, S., Iwao, H., Fukui, K., Abe, Y., and Tanaka, S.,** Effects of thyroid hormone on angiotensinogen and renin messenger RNA levels in rats, *Jpn. J. Pharmacol.,* 52, 281, 1990.
47. **Kjos, T., Gotoh, E., Tkcas, N., Shackelford, R., and Ganong, W.F.,** Neuroendocrine regulation of plasma angiotensinogen, *Endocrinology,* 129, 901, 1991.
48. **Murakami, E., Hiwada, K., and Kokubu, T.,** Effects of angiotensin II, thyroxine and estrogen on plasma renin substrate concentration and renin substrate production by the liver, *Jpn. Circ. J.,* 45, 1078, 1981.
49. **Resnick, L.M. and Laragh, J.H.,** Plasma renin activity in syndromes of thyroid hormone excess and deficiency, *Life Sci.,* 30, 585, 1982.
50. **Nasjletti, A. and Masson, M.C.,** Effects of corticosteroids on plasma angiotensinogen and renin activity, *Am. J. Physiol.,* 217, 1396, 1969.
51. **Freeman, R.H. and Rostorfer, H.H.,** Hepatic changes in renin substrate biosynthesis and alkaline phosphatase activity in the rat, *Am. J. Physiol.,* 223, 364, 1972.
52. **Kalinyak, J.E. and Perlman, A.J.,** Tissue-specific regulation of angiotensinogen mRNA accumulation by dexamethasone, *J. Biol. Chem.,* 262, 460, 1987.
53. **Deschepper, C.F. and Flaxman, M.S.,** Glucocorticoid regulation of rat diencephalon angiotensinogen production, *Endocrinology,* 126, 963, 1990.
54. **Brasier, A.R., Philippe, J., Campbell, D.J., and Habener, J.F.,** Novel expression of the angiotensinogen gene in a rat pancreatic islet cell line, *J. Biol. Chem.,* 261, 16148, 1986.
55. **Ohkubo, H., Nakayama, K., Tanaka, T., and Nakanishi, S.,** Tissue distribution of rat angiotensinogen mRNA and structural analysis of its heterogeneity, *J. Biol. Chem.,* 261, 319, 1986.
56. **Brasier, A.R., Tate, J.E., Ron, D., and Habener, J.F.,** Multiple cis-acting regulatory elements mediate hepatic angiotensinogen gene expression, *Mol. Endocrinol.,* 3, 1022, 1989.
57. **Ben-Ari, E.T., Lynch, K.R., and Garrison, J.C.,** Glucocorticoids induce the accumulation of novel angiotensinogen gene transcripts, *J. Biol. Chem.,* 264, 13074, 1989.
58. **Brasier, A.R., Ron, D., Tate, J.E., and Habener, J.F.,** Synergistic enhansons located within an acute phase responsive enhancer modulate glucocorticoid induction of angiotensinogen gene transcription, *Mol. Endocrinol.,* 4, 1921, 1990.
59. **Carretero, O. and Gross, F.,** Renin substrate in plasma under various experimental conditions in the rat, *Am. J. Physiol.,* 213, 695, 1967.
60. **Ménard, J., Bouhnik, J., Clauser, E., Richoux, J.P., and Corvol, P.,** Biochemistry and regulation of angiotensinogen, *Clin. Exp. Hypertens. (A),* 5, 1005, 1983.
61. **Deschepper, C.F.,** unpublished data.
62. **Iwao, H., Kimura, S., Fukui, K., Nakamura, A., Tamaki, T., Ohkubo, H., Nakanishi, S., and Abe, Y.,** Elevated angiotensinogen mRNA levels in rat liver by nephrectomy, *Am. J. Physiol.,* 258, E413, 1990.
63. **Carrell, R. and Travis, J.,** Alpha 1-antitrypsin and the serpins: variation and countervariation, *Trends Biochem. Sci.,* 10, 20, 1985.
64. **Kageyama, R., Ohkubo, H., and Nakanishi, S.,** Induction of rat liver angiotensinogen mRNA following acute inflammation, *Biochem. Biophys. Res. Commun.,* 129, 826, 1985.
65. **Okamoto, H., Ohashi, Y., and Itoh, N.,** Involvement of leukocyte and glucocorticoid in acute-phase response or angiotensinogen, *Biochem. Biophys. Res. Commun.,* 145, 1225, 1987.
66. **Itoh, N., Matsuda, T., Ohtani, R., and Okamoto, H.,** Angiotensinogen production by rat hepatoma cells is stimulated by B cell stimulatory factor 2/interleukin-6, *FEBS Lett.,* 244, 6, 1989.
67. **Ron, D., Brasier, A.R., Wright, K.A., Tate, J.E., and Habener, J.F.,** An inducible 50-kilodalton NFkappaB-like protein and constitutive protein both bind the acute-phase response element of the angiotensinogen gene, *Mol. Cell. Biol.,* 10, 1023, 1990.

68. **Ron, D., Brasier, A.R., Wright, K.A., and Habener, J.F.,** The permissive role of glucocorticoids on interleukin-1 stimulation of angiotensinogen gene transcription is mediated by an interaction between inducible enhancers, *Mol. Cell. Biol.*, 10, 4389, 1990.
69. **Baumann, H.,** Hepatic acute phase reaction in vivo and in vitro, *In Vitro Cell. Dev. Biol.*, 25, 115, 1989.
70. **Ohtani, R., Ohashi, Y., Muranaga, K., Itoh, N., and Okamoto, H.,** Changes in activity of the renin-angiotensin system of the rat by induction of acute inflammation, *Life Sci.*, 44, 237, 1989.
71. **Reid, I.A.,** Effect of angiotensin II and glucocorticoids on plasma angiotensinogen concentration in the dog, *Am. J. Physiol.*, 232, E234, 1977.
72. **Nasjletti, A. and Masson, G.M.C.,** Stimulation of angiotensinogen formation by renin and angiotensin, *Proc. Soc. Exp. Biol. Med.*, 142, 307, 1973.
73. **Nakamura, A., Iwao, H., Fukui, K., Kimura, S., Tamaki, T., Nakanishi, S., and Abe, Y.,** Regulation of liver angiotensinogen and kidney renin mRNA levels by angiotensin II, *Am. J. Physiol.*, 258, E1, 1990.
74. **Iwao, H., Nakamura, A., Fukui, K., Kimura, S., Tamaki, T., and Abe, Y.,** Endogenous angiotensin II regulates hepatic angiotensinogen production, *Life Sci.*, 47, 2343, 1990.
75. **Iwao, H., Fukui, K., Kim, S., Nakayama, K., Ohkubo, H., Nakanishi, S., and Abe, Y.,** Sodium balance effects on renin, angiotensinogen, and atrial natriuretic polypeptide mRNA levels, *Am. J. Physiol.*, 255, E129, 1988.
76. **Stuzmann, M.M., Radziwill, R., Komischke, K., Klett, C., and Hackenthal, E.,** Hormonal and pharmacological alteration of angiotensinogen secretion from rat hepatocytes, *Biochim. Biophys. Acta,* 886, 48, 1986.
77. **Klett, C., Muller, F., Gierschik, P., and Hackenthal, E.,** Angiotensin II stimulates angiotensinogen synthesis in hepatocytes by a pertussis toxin-sensitive mechanism, *FEBS Lett.*, 259, 301, 1990.
78. **Klett, C., Hellman, W., Müller, F., Suzuki, F., Nakanishi, S., Ohkubo, H., Ganten, D., and Hackenthal, E.,** Angiotensin II controls angiotensinogen secretion at a pretranslational level, *J. Hypertens.*, 6 (Suppl. 4), S442, 1988.
79. **Iwao, H., Tamaki, T., Yasuhara, A., Reid, I.A., and Abe, Y.,** Effect of DUP753 on renal renin and hepatic angiotensinogen mRNA levels in intact and adrenalectomized rats, *Hypertension,* 18 (Abstr.), 411, 1991.
80. **Hirasawa, K., Yamamoto, H., Matsui, A., Shinozaki, K., Kobayashi, S., Yagi, Y., Morimoto, S., Takeda, R., and Murakami, M.,** The effect of mercuric chloride, of bilateral ureteral ligation and of bilateral nephrectomy on plasma renin substrate concentration in rats, *Jpn. Circ. J.*, 32, 1591, 1968.
81. **Eggena, P. and Barrett, J.D.,** Renin substrate release in response to perturbations of renin-angiotensin system, *Am. J. Physiol.*, 254, E389, 1988.
82. **Voigt, J. and Köster, H.,** Induction of plasma angiotensin by steroid hormones in nephrectomized rats, *Eur. J. Biochem.*, 110, 57, 1980.
83. **Hasegawa, H., Tateishi, H., and Masson, G.M.C.,** Evidence for an angiotensinogen-stimulating factor after nephrectomy, *Can. J. Physiol. Pharmacol.*, 51, 563, 1973.
84. **Cassis, L.A., Saye, J.A., and Peach, M.J.,** Location and regulation of rat angiotensinogen messenger RNA, *Hypertension,* 11, 591, 1988.
85. **Chang, E. and Perlman, A.J.,** Angiotensinogen mRNA. Regulation by cycle cell and growth factors, *J. Biol. Chem.*, 263, 5480, 1988.
86. **Ohkubo, H., Kawakami, H., Kakehi, Y., Takumi, T., Arai, H., Yokota, Y., Iwai, M., Tanabe, Y., Masu, M., Hata, J., Iwao, H., Okamoto, H., Yokoyama, M., Nomura, T., Katsuki, M., and Nakanishi, S.,** Generation of transgenic mice with elevated blood pressure by introduction of the rat renin and angiotensinogen genes, *Proc. Natl. Acad. Sci. U.S.A.*, 87, 5153, 1990.

87. **Gardes, J., Bouhnik, J., Clauser, E., Corvol, P., and Ménard, J.,** Role of angiotensinogen in blood pressure homeostasis, *Hypertension,* 4, 185, 1982.
88. **Beaty, O., III, Sloop, C.H., Schmid, H.E., Jr., and Buckalew, V.M., Jr.,** Renin response and angiotensinogen control during graded hemorrhage and shock in the dog, *Am. J. Physiol.,* 231, 1300, 1976.
89. **Deschepper, C.F. and Ganong, C.F.,** Renin and angiotensin in endocrine glands, in *Frontiers in Neuroendocrinology,* Vol. 10, Martini, L. and Ganong, W.F., Eds., Raven Press, New York, 1988, 79.
90. **Campbell, D.J.,** Tissue renin-angiotensin system: sites of angiotensin formation, *J. Cardiovasc. Pharmacol.,* 10, S1, 1987.
91. **Lenz, T. and Sealey, J.E.,** Tissue renin systems as a possible factor in hypertension, in *Hypertension: Pathophysiology, Diagnosis, and Management,* Laragh, J.H. and Brenner, B.M., Eds., Raven Press, New York, 1990, 1319.
92. **Seikaly, M.G., Arant, B.S., and Seney, S.D., Jr.,** Endogenous angiotensin concentrations in specific fluid compartments of the rat, *J. Clin. Invest.,* 86, 1352, 1990.
93. **Campbell, D.J., Lawrence, A.C., Towrie, A., Kladis, A., and Valentijn, A.J.,** Differential regulation of angiotensin peptide levels in plasma and kidney of the rat, *Hypertension,* 18, 763, 1991.
94. **Aguilera, G., Shirar, A., Baukal, A., and Catt, K.J.,** Circulating angiotensin II and adrenal receptors after nephrectomy, *Nature,* 289, 507, 1981.
95. **Suzuki, H., Smeby, R.R., Mikami, H., Brosnihan, K.B., Husain, A., De Silva, P., Saruta, T., and Ferrario, C.M.,** Nonrenal factors contribute to plasma and cerebrospinal fluid angiotensin II, *Hypertension,* 8 (Suppl. I), 5, 1986.

Chapter 7

ANGIOTENSINOGEN GENE EXPRESSION AND HYPERTENSION: TRANSGENIC ANIMAL MODELS

J.E. Kalinyak

TABLE OF CONTENTS

0-8493-4622-3/93/$0.00 + $.50

I. HYPERTENSION AND THE RENIN-ANGIOTENSIN SYSTEM

The renin-angiotensin system has always been viewed as an endocrine system involved in the feedback regulation of blood pressure and fluid-electrolyte homeostasis.[1-9] It has also been viewed as an atypical endocrine system, because the biologically active peptide, angiotensin II (AngII), is generated in the plasma. Angiotensinogen, predominantly synthesized and secreted into the circulation by the liver, is cleaved by renin, producing angiotensin I (AngI). AngI is subsequently cleaved by angiotensin-converting enzyme (ACE), producing AngII. AngII is the major biologically active peptide that binds to its receptors to stimulate the secretion of aldosterone or cause vasoconstriction of the vasculature.[1,3,10] Both of these actions are important in the maintenance of blood pressure routinely and in cases of hemorrhage.[11] AngII also increases arterial pressure by activating the sympathetic nervous system and by potentiating the norepinephrine responsiveness or release from nerve endings in blood vessels.[12-16]

There is a good correlation between the plasma concentration of AngII and plasma renin activity under normal physiologic circumstances.[17] This is due to AngII decreasing the secretion of renin from the juxtaglomerular cells.[18] AngII also regulates the synthesis of its precursor protein via a positive feedback regulatory mechanism. In addition, it modulates AngII receptor abundance in a tissue-specific fashion.[19] Therefore, the renin-angiotensin system involves the complex interaction of five different proteins. The role of the renin-angiotensin system in the development of hypertension is clear only in the infrequent situations of renovascular occlusion,[20] while its role in either the initiation and/or maintenance of essential hypertension is an area of active research.

Interest in defining the role the renin-angiotensin system plays in essential hypertension increased because of two major findings. The first was the discovery of a new class of antihypertensive agents, the ACE inhibitors,[21] and the second was the observation that many tissues, previously ignored, express angiotensinogen and all the proteins in the enzymatic cascade needed for the generation of AngII.[10,19,22-25] This latter finding changed our view of the renin-angiotensin system from an endocrine system to a system with paracrine, and possibly autocrine, functions. These local renin-angiotensin systems have been identified in a number of organs important in cardiovascular homeostasis, i.e., brain, heart, arteries, kidney, and adrenal gland.[26-39] It is clear that the brain renin-angiotensin system contributes to blood pressure homeostasis by stimulating adrenergic outflow,[40,41] stimulating thirst,[42] and vasopressin release.[41,43]

How important is the concentration of angiotensinogen in the generation of AngII and blood pressure regulation? Circulating levels of angiotensinogen are elevated in hypercortisolism[44] and contraceptive therapy.[45] Both situations are also associated with elevated blood pressure. Low levels of angiotensinogen are associated with adrenal insufficiency[46] and converting-enzyme inhibition.[47] In addition, a direct role for angiotensinogen in blood pressure regulation has been demonstrated by the drop in blood pressure induced by the administration of angiotensinogen-specific antibody.[48] This decrease is greater in rats on a salt-free diet than those on a normal diet and is absent in binephrectomized animals. These data clearly show a role for angiotensinogen in blood pressure regulation, but many people believe that the concentration of renin is the most important determinant in controlling the amount of circulating AngI and subsequently, AngII. In fact, early studies suggested that the action of renin on substrate could best be described by zero-order kinetics.[49,50] Contrary to this statement, studies by Gould and Green[51] demonstrated that the Michaelis-Menten constant (K_m) for the hydrolysis of human angiotensinogen by renin is numerically equal to the concentration of angiotensinogen that will permit the reaction to proceed at half maximal velocity. More recently, Cumin et al.[52] showed that the K_m for the human renin-angiotensinogen reaction was $1.25 \pm 0.1\ \mu M$. They claimed that this level indicates that under physiological conditions, a 10-fold increase in circulating angiotensinogen concentration would be required to obtain zero-order kinetics. Further support for the importance of angiotensinogen levels in blood pressure regulation comes from two reported cases of angiotensinogen-secreting hepatic tumors associated with hypertension.[53,54]

Menard et al.[55] recently demonstrated that intravenous infusion of 250 μg of rat angiotensinogen into anesthetized sodium-deplete, sodium-replete, or binephrectomized sodium-deplete rats resulted in similar plasma angiotensinogen levels, but an increase in blood pressure was only found in the sodium-deplete animals with intact kidneys. This suggests first, that the enzymatic activity of renal renin is necessary to see the pressor effects. The plasma renin concentration, a measure of circulating protein, was increased 70-fold in the sodium-deplete animals. Following the angiotensinogen infusion, plasma renin concentration decreased 2.5-fold, while plasma renin activity, a measure of AngI generation, increased 3-fold. This reflects the rapid enzymatic cleavage of angiotensinogen into AngI and suggests that the subsequent rapid generation of AngII probably resulted in a decrease in renal renin secretion. Although the amount of angiotensinogen infused was not physiologic, this experiment does demonstrate the rapidity with which any alteration in the circulating AngII level activates a precise feedback regulatory system in an attempt to maintain a set equilibrium. One could speculate that to see the pressor effects of physiologic doses of angiotensinogen in normal animals might require the ability to block the feedback of AngII on renal renin secretion and AngII receptor abundance. It is also because of this precise feedback regulatory system that multiple components of the renin-angiotensin system must be investigated in transgenic experiments to fully understand either positive or negative findings.

II. REGULATION OF GENE EXPRESSION

The cell and organ phenotype is due to the selective expression of a subset of available genes in the genome. This cell-specific regulation of gene expression is to a large extent regulated at the point of gene transcription. One of the uses of transgenic technology is to investigate how the 5′- and 3′-flanking regions of DNA modulate hormone-regulated, developmental, and tissue-specific gene expression.

A prototypic eukaryotic gene[56,57] would consist structurally of multiple coding regions of DNA, called exons, interspersed with noncoding pieces of DNA, called introns.[58] The structural regions contain the information that will be transcribed into mRNA and subsequently translated into protein. In addition to this structural gene, regulatory sequences exist at the 5′- and 3′-flanking regions of DNA. Although the regulatory regions appear to be quite complex, some general conserved patterns have emerged. In particular, there is a sequence called a TATA box located 25 to 30 nucleotides upstream of the 5′ end of the structural gene. This A-T-rich region is where the RNA polymerase II begins its interaction with the DNA to begin transcription. Other important regulatory regions upstream are cis-acting DNA sequences that bind and interact with specific nuclear and cytoplasmic trans-acting factors. Two well-characterized sequences are the CAAT box and GC-rich boxes that bind CAAT-binding proteins and Sp1 respectively. Another cis-DNA element is termed an enhancer. This DNA sequence of 6 to 20 bp binds to specific proteins within the cells and results in a dramatic increase in transcription.[59,60] An enhancer can also mediate a negative effect on transcription, in which it is termed a silencer sequence. These enhancer and silencer sequences are position independent. A subset of these enhancer elements, termed hormone-regulatory elements, interacts with specific steroid hormones such as glucocorticoid, estrogen, progesterone, testosterone, vitamin D, and others.[61,62] It is important to note that enhancer sequences may be present at variable distances either 5′ upstream or 3′ downstream from the transcription initiation site and even may be located in introns.

Gene transcription produces a heteronuclear RNA transcript that includes both introns and exons of the structural gene. Rapidly following transcription, the 5′ end of the RNA is modified by the addition of a 7-methylguanine residue. This reaction is referred to as 5′-capping and is essential for optimal translational efficiency and RNA stability.[63] The 3′ end of the heteronuclear RNA contains a poly-A tract. This sequence lies 2 to 25 nucleotides downstream from an AAUAAA sequence. The addition of the Poly-A tail is thought to be important in increasing RNA stability.[64] The last process necessary to produce a mature, functional mRNA prior to movement into the cytoplasm is the elimination of the introns by a process referred to as splicing.

III. TRANSGENIC ANIMALS: TESTING THE ANGIOTENSINOGEN GENE *IN VIVO*

Transgenic technology has numerous applications. Some of these involve generating animal models of human diseases. Other uses of transgenic technology include investigating the functions of developmentally regulated genes, producing large quantities of economically important biological materials, and mapping regulatory regions of specific genes. The challenge in hypertension research is to determine the genetic cause of elevated blood pressure and to apply this new-found knowledge to the development of new pharmacologic interventions. The major problem in understanding the genetics of hypertension is that several genes are thought to be involved. Therefore, it is the polygenetic nature of this disease that has prompted the use of transgenic animals to identify and dissect out the contribution of various genes important in the development of high blood pressure. This technology involves the introduction of genes into the germline of mammals, thereby providing scientists with new insights into the actions of these genes. The methods for introducing genes into animals will be briefly reviewed to facilitate understanding of the potential problems in data interpretation.

A. TRANSGENIC TECHNIQUES: METHODS FOR INTRODUCING GENES INTO ANIMALS

1. Microinjection of DNA into Pronucleus

Microinjection of recombinant DNA into the pronucleus of a fertilized mouse egg has been the most widely and successfully used technique for obtaining transgenic animals.[65,66] One-cell fertilized eggs are removed and the pronucleus is microinjected with several hundred molecules of recombinant DNA. The embryos are then transferred to the oviducts of pseudopregnant females. The progeny are evaluated by DNA hybridization for retention of recombinant DNA. The principal advantage of this technique is the relative simplicity and efficiency in generating transgenic mice that express most genes in a predictable manner. Following microinjection, the foreign DNA is typically first built up into a concatemer, called an array, and then is typically inserted into a single site[67,68] on one of the chromosomes by homologous recombination.[67,68] To date, there does not appear to be any preferred sites for insertion of the DNA. Therefore, each transgenic line, consisting of the recipient mouse (G_0) and its offspring, is the product of a unique series of molecular events. The results of these unique molecular events are evident when several transgenic lines are produced by introducing the same foreign DNA construct. Some G_0 mice show no expression and their offspring are equally expressionless.[69] In the expressing lines, the tissues expressing the gene can vary in the level of gene expression.[70] The cause of this variation remains uncertain, but the arrangement of the genes in the array, gene mutation or rearrangement, and possible position effects, such as ill-defined influences of the chromosomal region around the inserted array may potentially influence the transgene expression.[67]

One must be aware that frequently rearrangements, deletions, duplications,[71] or translocations[72] of the recipient sequences occur at the insertion sites. In addition, the injected DNA does not always integrate into the recipient genome, but can remain as an episome.[73] Finally, one disadvantage of this method is that it cannot be used to introduce genes into cells at a later developmental stage.

2. Retrovirus Infection

Retroviral gene transfer is accomplished by infection of preimplantation embryos (day 8 to 12) with concentrated recombinant virus.[74] The main advantage of this approach is that only a single copy of the retrovirus integrates by a precisely defined mechanism into the genome of the infected cell. In addition, this approach is technically easier and one can introduce genes at various developmental stages. Finally, it is easier to isolate the flanking host sequences of a proviral insert when compared to the flanking DNA derived from pronuclear injection. This technique allows infection of cells from many somatic tissues, while germ cells are infected with a low frequency. The main disadvantage of this approach is the size limitation for transduced DNA.

3. Embryonic Stem Cells

Embryonic stem cells are pluripotent cells obtained from the blastocyst stage of embryogenesis that can be cultured *in vitro*. The foreign gene is introduced into the cell by a variety of transfer techniques such as electroporation, calcium-phosphate precipitation, microinjection, or retroviral infection.[66] An advantage of this technique is that the investigator can screen or select for cells with specific phenotypic characteristics prior to reintroduction into other blastocysts. This approach also provides an opportunity to use homologous recombination to disrupt genes and the transgenic progeny can then be studied.

B. TRANSGENIC ANIMALS: ADDITION OF AN ANGIOTENSINOGEN GENE

Three laboratories have generated transgenic mice by introducing truncated mouse,[75] full length rat,[76] and full length human[77] angiotensinogen genes. All of these laboratories used the technique of microinjection of DNA into the pronucleus. Therefore, each transgenic line generated, even within the same laboratory, reflects the products of a unique series of molecular events in that varying arrays of genes have been integrated into unique sites within the recipient genome. The questions being asked were different in each of these laboratories.

The laboratory of Clauston et al.[75] generated transgenic mice as a way to map the tissue-specific and inducible enhancers of the angiotensinogen gene. Two angiotensinogen minigenes were constructed from a mouse angiotensinogen gene in which the coding region was deleted to allow size discrimination of the mRNA of the introduced gene from the mRNA produced by the endogenous angiotensinogen gene. The translational reading frame was preserved so that the same termination signal would be used, thereby maintaining identical 3′-untranslated regions and preserving similar RNA stability. The normal initiation methionine and leader sequence was also preserved to avoid intracellular accumulation of this truncated protein. This construct allowed the investigators to directly compare the expression of both the endogenous and minigene mRNAs using the same Northern blot and the same radiolabeled probe. The constructs differed only in the amount of 5′-flanking DNA. The longer minigene (4.0 A/2) contained 4 kb of DNA 5′ upstream from the cap-site. The shorter minigene (0.75 A/2) contained 0.75 kb of 5′-flanking DNA.

Using these two constructs they were able to demonstrate that both the short and long minigenes were expressed at similar levels in all the tissues known to express angiotensinogen, thereby suggesting that the 0.75 A/2 minigene contained sufficient information to direct appropriate tissue-specific expression. In addition, they showed that the hormone and acute-phase response elements mapped within the 5′-flanking DNA 0.75 kb upstream from the start site. Following hormone and acute-phase induction, the mRNA levels were greater in the minigenes than in the endogenous gene, suggesting that multiple copies of the minigene were transcriptionally active. However, because there was not a direct correlation between copy number and mRNA levels, this suggests that some insertion position-dependent influences are also influencing minigene transcription.

The lack of minigene expression in the testis, a tissue known to have very high levels of endogenous angiotensinogen mRNA, suggested that the BALB/c minigene constructs lacked a testis-specific enhancer element. In addition, the expression of the BALB/c minigenes was consistently higher than the endogenous angiotensinogen gene in the salivary gland of the Swiss mice. On further investigation, they found that the angiotensinogen mRNA levels were variable but higher in the random-bred Swiss mice than in the inbred BALB/c mice. Therefore, the overexpression of the BALB/c minigene in the Swiss mice was paradoxical. Studies defined the strain polymorphisms affecting angiotensinogen expression in the testis and salivary gland. Many transgenic laboratories use outbred or F2 hybrid zygotes to make transgenic mice. Therefore, it may be found that strain polymorphisms may be an important cause of the variable expression of other genes in transgenic mice which to date has been attributed to effects due to the site of integration.

The 5′-flanking regulatory elements on the human angiotensinogen gene were mapped by Takahashi et al.,[77] using transgenic mice. Two lines of transgenic mice (hAG2-5 and hAG3-2) were generated from the same DNA construct containing the 14-kb human angiotensinogen gene with 1.3-kb 5′-flanking DNA. The *in vivo* tissue-specific expression of these lines were examined by Northern blot hybridization using a human specific angiotensinogen probe. Expression of the transgenes were highest in the livers of both mouse lines. In addition, the transcripts showed a normal start site, transcript processing, and polyadenylation. High levels of transgene transcripts were, surprisingly, also found in the kidneys of both mouse lines. Lower levels of transcripts were easily detected in the hearts of both lines, while much lower transcript levels were found in the brains, lungs, spleens, and submaxillary glands. They also demonstrated that the human angiotensinogen transgene did not interfere with the normal regulation of the mouse angiotensinogen gene expression.

The finding of high levels of kidney expression of the transgene could result from positional effects due to its insertion site, increased copy number, or the lack of negative regulatory sequences in the transgene. The authors chose to investigate the last potential explanation by attaching the 1.3-kb 5′-flanking regulatory sequences to the reporter gene, CAT. Three different constructs were transfected into a hepatoma, an embryonic kidney, and a glioma cell line and CAT activity assayed. High levels of CAT expression were found in the renal cells, again suggesting that the region containing the putative negative regulatory sequences was not within the 1.3-kb 5′-flanking DNA or the 14-kb angiotensinogen structural gene. Investigation of 5′-flanking DNA regions further upstream or 3′-flanking regions will be necessary to confirm the existence of a putative negative regulatory sequence.

An alternative use of transgenic techniques is to study the relative importance of a gene in a pathologic state, such as how the renin-angiotensin system is involved in the development of hypertension. This was the question being asked by Ohkubo et al.[76] Therefore, to answer this question, they generated transgenic mice that were capable of expressing endogenous rat renin or angiotensinogen or both genes under the control of the mouse metallothionein I (MT-I) promoter. The use of this promoter allows the investigator to induce the expression of the transgenes by supplying the mice with water containing 25 mM $ZnSO_4$. The mouse MT-I/rat angiotensinogen fusion gene (MAG-16) consists of 1.7 kbp of the MT-I promoter and 70 bp of the 5′-untranslated region of the MT-I gene and the entire rat angiotensinogen gene extending from 3 bp of the 5′-flanking region up to ≈900 bp of the 3′-flanking region. Of 664 eggs injected and 54 resulting rat pups, 10 transgenic mice were capable of expression appreciable, albeit varying, amounts of transgene mRNAs. Neither these F1 mice nor F2 mice made homozygous for the angiotensinogen transgene showed any elevation in blood pressure even when placed on water containing 25 mM $ZnSO_4$. This result is difficult to explain, because it has been reported that mouse renin can cleave rat angiotensinogen.[78] Thus, one would have expected elevated levels of AngII and an increase in blood pressure. Measurement of AngII levels or plasma renin concentration would have been helpful in determining if the increase in angiotensinogen levels was sufficient to result in increased generation of AngII, or if the feedback regulation of the renin-angiotensin system was sufficient to compensate and establish a new, but normotensive, equilibrium. Renin transgenic mice also were normotensive, but it has been shown that mouse angiotensinogen is not cleaved by rat renin,[78] thus one would have predicted these findings. Hypertension was demonstrable when an angiotensinogen and renin transgenic lines were mated. All but four animals carrying both transgenes exhibited elevated systolic blood. Northern blot and primer extension studies showed that the rat angiotensinogen transgene was expressed to the same extent as the endogenous gene. In addition, the circulating angiotensinogen levels were higher in the mice possessing both transgenes than in the mice with only the rat angiotensinogen transgene. Finally, captopril, a potent inhibitor of ACE activity, was shown to be effective in decreasing blood pressure in all the hypertensive mice.

The polygenetic nature of essential hypertension has prompted the use of transgenic animal models as a means to define the contribution of individual genes in the development of elevated blood pressure. The studies have been limited to the introduction of foreign DNA constructs into the germline. When positive results are found, the interpretation is straightforward. However the finding that a gene does not alter the phenotype of the transgenic animals must be interpreted more cautiously. One must exclude outbred strain polymorphisms. In addition, it is important to examine the feedback regulatory mechanisms which might establish a new steady state without changing the end phenotype. With the eventual development of gene deletion in transgenic rats, this technology could be applied to the spontaneous hypertensive rat and may facilitate our investigation into the polygenetic etiology of essential hypertension.

REFERENCES

1. **Bengis, R.G., Coleman, T.G., Young, D.B., and McCaa, R.E.,** Long-term blockade of angiotensin formation in various normotensive and hypertensive rat models using converting enzyme inhibitor (SQ 14,225), *Circ. Res.*, 43, 145, 1978.
2. **Watkins, L., Jr., Burton, J.A., Haber, E., Cant, J.R., Smith, F.W., and Barger, A.C.,** The renin-angiotensin-aldosterone system in congestive failure in conscious dogs, *J. Clin. Invest.*, 57, 1607, 1976.
3. **Brown, J.J., Casals-Stenzel, J., Cumming, A.M.M., Davies, D.L., Fraser, R., Lever, A.F., Morton, J.J., Semple, P.F., Tree, M., and Robertson, J.I.S.,** Angiotensin II, aldosterone and arterial pressure: a quantitative approach, *Hypertension*, 1, 159, 1979.

4. **Hall, J.E., Guyton, A.C., Smith, M.J., Jr., and Coleman, T.G.,** Chronic blockade of angiotensin II formation during sodium deprivation, *Am. J. Physiol.*, 237, F424, 1979.
5. **Laragh, J.H. and Sealey, J.E.,** The renin-angiotensin-aldosterone hormonal system and regulation of sodium, potassium, and blood pressure homeostasis, in *Handbook of Physiology*, Orloff, J. and Berliner, R.W., Eds., The American Physiological Society, Washington, D.C., 1973, 831.
6. **Freeman, R.H., Davis, J.O., Williams, G.M., Deforest, J.M., Seymour, A.A., and Rowe, B.P.,** Effects of the converting enzyme inhibitor, SQ14225, in a model of low cardiac output in dogs, *Circ. Res.*, 45, 540, 1979.
7. **Hall, J.E.,** Regulation of glomerular filtration rate and sodium excretion by angiotensin II, *Fed. Proc.*, 45, 1431, 1986.
8. **Kimbrough, H.M., Jr., Vaughan, E.D., Jr., Carey, R.M., and Ayers, C.R.,** Effect of intrarenal angiotensin II blockade on renal function in conscious dogs, *Circ. Res.*, 40, 174, 1977.
9. **Haber, E.,** The role of renin in normal and pathological cardiovasculat hemostasis, *Circulation*, 54, 849, 1976.
10. **Campbell, D.J.,** Circulating and tissue angiotensin systems, *J. Clin. Invest.*, 79, 1, 1987.
11. **Brough, R.B., Cowley, A.W., Jr., and Guyton, A.C.,** Quantitative analysis of the acute response to haemorrhage of the renin-angiotensin vasoconstrictor feedback loop in areflexic dogs, *Cardiovasc. Res.*, 9, 722, 1975.
12. **Ferrario, C.M., Gildenberg, P.L., and McCubbin, J.W.,** Cardiovascular effects of angiotensin mediated by the central nervous system, *Circ. Res.*, 30, 257, 1972.
13. **Peach, M.J.,** Renin-angiotensin system: biochemistry and mechanisms of action, *Physiol. Rev.*, 57, 313, 1977.
14. **Zimmerman, B.G.,** Adrenergic facilitation by angiotensin: does it serve a physiological function?, *Clin. Sci.*, 60, 343, 1981.
15. **Reid, I.A., Brooks, V.L., Rudolph, C.D., and Keil, L.A.,** Analysis of the actions of angiotensin on the central nervous system of conscious dogs, *Am. J. Physiol.*, 243, R82, 1982.
16. **Zimmerman, B.G., Sybertz, E.J., and Wong, P.C.,** Interaction between sympathetic and renin-angiotensin system, *J. Hypertens.*, 2, 581, 1984.
17. **Brunner, H.R., Waeber, B., and Nussberger, J.,** Angiotensin-converting enzyme inhibition versus blockade of the renin-angiotensin system, *Am. J. Med.*, 87 (Suppl. 6B), 1989.
18. **Ganong, W.F., Davis, J.O., and Sambhi, M.P.,** Symposium on control of renin secretion, *J. Hypertens.*, 2 (Suppl. I), 1, 1984.
19. **Johnston, C.I.,** Biochemistry and pharmacology of the renin-angiotensin system, *Drugs*, 39 (Suppl. 1), 21, 1990.
20. **Pickering, T.G.,** Diagnosis and evaluation of renovascular hypertension. Indications for therapy, *Circulation*, 83 (Suppl 2), I147, 1991.
21. **Ondetti, M.A., Rubin, B., and Cushman, D.W.,** Design of specific inhibitors of angiotensin-converting enzyme: new class of orally effective antihypertensive agents, *Science*, 4196, 441, 1977.
22. **Dzau, V.J., Burt, D.W., and Pratt, R.E.,** Molecular biology of the renin-angiotensin system, *Am. J. Physiol.*, 255, F563, 1988.
23. **Frolich, E.D., Iwata, T., and Saske, O.,** Clinical and physiologic significance of local tissue renin-angiotensin systems, *Am. J. Med.*, 87, 6B, 1989.
24. **Ferrario, C.M.,** Importance of the renin-angiotensin-aldosterone system in the physiology and pathology of hypertension, an overview, *Drugs*, 39 (Suppl. 2), 1, 1990.
25. **Campbell, D.J.,** The site of angiotensin production, *J. Hypertens.*, 3, 199, 1985.
26. **Field, L.J., McGowan, R.A., Dickenson, D.P., and Gross, K.W.,** Tissue and gene specificity of mouse renin expression, *Hypertension*, 6, 597, 1984.

27. **Ohkubo, H., Nakayama, K., Tanaka, T., and Nakanishi, S.,** Tissue distribution of rat angiotensinogen mRNA and structural analysis of its heterogeneity, *J. Biol. Chem.*, 261, 319, 1986.
28. **Campbell, D.J. and Habener, J.F.,** Angiotensin gene is expressed and differentially regulated in multiple tissues of the rat, *J. Clin. Invest.*, 78, 31, 1986.
29. **Kalinyak, J.E. and Perlman, A.J.,** Tissue specific regulation of angiotensinogen by dexamethasone, *J. Biol. Chem.*, 262, 460, 1987.
30. **Dzau, V.J., Ellison, D.E., Brody, R., Ingelfinger, J., and Pratt, R.E.,** A comparative study of the distribution of renin and angiotensinogen messenger ribonucleic acids in rat and mouse tissues, *Endocrinology,* 120, 2334, 1987.
31. **Cassis, L.A., Lynch, K.R., and Peach, M.J.,** Localization of angiotensinogen messenger RNA in rat aorta, *Circ. Res.*, 62, 1259, 1988.
32. **Cohen, M.L. and Kurz, K.D.,** Angiotensin converting enzyme inhibition in tissues from spontaneously hypertensive rats after treatment with captopril or MK-421, *J. Pharmacol. Exp. Ther.*, 220, 63, 1982.
33. **Strittmatter, S.M., de Souza, E.B., Lynch, D.R., and Snyder, S.H.,** Angiotensin-converting enzyme localized in the rat pituitary and adrenal glands by [3H]captopril autoradiography, *Endocrinology,* 118, 1690, 1986.
34. **Rosenthal, J., von Lutterotti, N., Thurnreiter, M., Gomba, S., Rothemund, J., Reiter, W., Kazda, S., Garthoff, B., Jacob, I., and Dahlheim, H.,** Suppression of renin-angiotensin system in the heart of spontaneously hypertensive rats, *J. Hypertens.*, 5 (Suppl. 2), S23, 1987.
35. **Rosenthal, J.H., Pfeifle, B., Michailov, M.L., Pschorr, J., Jacob, I.C.M., and Dahlheim, H.,** Investigations of components of the renin-angiotensin system in rat vascular tissue, *Hypertension,* 6, 383, 1984.
36. **Mizuno, K., Nakamaru, M., Higashimori, K., and Inagami, T.,** Local generation and release of angiotensin II in peripheral vascular tissue, *Hypertension,* 22, 223, 1988.
37. **Lindpaintner, K., Wehelm, M.J., Jin, M., Unger, T., Lang, R.E., Schoelkens, B.A., and Ganten, D.,** Tissue renin-angiotensin systems: focus on the heart, *J. Hypertens.*, 5 (Suppl. 2), S33, 1987.
38. **Unger, R., Badoer, E., Ganten, D., Lang, R.E., and Rettig, R.,** Brain angiotensin: pathways and pharmacology, *Circulation,* 77 (Suppl. I), 40, 1988.
39. **Ganten, D., Hermann, K., Bayer, C., Unger, R., and Lang, R.E.,** Angiotensin synthesis in the brain and increased turnover in hypertensive rats, *Science,* 221, 869, 1983.
40. **Unger, T., Badoer, E., Ganten, D., Lang, R.E., and Retig, R.,** Brain angiotensin pathways and pharmacology, *Circulation,* 77 (Suppl. I), 40, 1988.
41. **Reid, J.L. and Rubin, P.C.,** Peptides and central neural regulation of the circulation, *Physiol. Rev.*, 67, 725, 1987.
42. **Fitzsimmons, J.T.,** Angiotensin stimulation of the central nervous system, *Rev. Physiol. Biochem. Pharmacol.*, 87, 117, 1980.
43. **Keil, L.C., Summy-Long, J., and Severs, W.B.,** Release of vasopressin by angiotensin II, *Endocrinology,* 96, 1063, 1975.
44. **Krakoff, L.W.,** Measurement of plasma renin substrate by radioimmunoassay of angiotensin I: concentration in syndromes associated with steroid excess, *J. Clin. Endocrinol. Metab.*, 37, 110, 1973.
45. **Skinner, S.L., Lumbers, E.R., and Symonds, E.M.,** Alterations by oral contraceptives of normal menstrual changes in plasma renin activity, concentration and substrate, *Clin. Sci.*, 36, 67, 1969.
46. **Genain, C., Bouhnik, J., Tewksbury, D., Corvol, P., and Menard, J.,** Characterization of plasma and cerebrospinal fluid human angiotensinogen and des-angiotensin I-angiotensinogen by direct radioimmunoassay, *J. Clin. Endocrinol. Metab.*, 59(3), 478, 1984.

47. **Rasmussen, S., Nielsen, M.D., and Giese, J.,** Captopril combined with thiazide lowers renin substrate concentration: implication for methodology on renin assays, *Clin. Sci.*, 60, 591, 1981.
48. **Gardes, J., Bouhnik, J., Clauser, E., Corvol, P., and Menard, J.,** Role of angiotensinogen in blood pressure homeostasis, *Hypertension*, 4, 185, 1982.
49. **Boyd, G.W., Adamson, A.R., Fitz, A.E., and Peart, W.S.,** Radioimmunoassay determination of plasma-renin activity, *Lancet, I,* 213, 1969.
50. **Haber, E.,** Recent developments in pathophysiologic studies of the renin-angiotensin system, *N. Engl. J. Med.*, 280, 148, 1969.
51. **Gould, A.B. and Green, D.,** Kinetics of the human renin and human substrate reaction, *Cardiovasc. Res.*, 5, 86, 1971.
52. **Cumin, F., Le-N'guyen, D., Castro, B., Menard, J., and Corvol, P.,** Comparative enzymatic studies of human renin acting on pure natural or synthetic substrates, *Biochim. Biophys. Acta*, 913, 10, 1987.
53. **Ueno, N., Yoshida, K., Hirose, S., Yokoyama, H., Uehara, H., and Murakami, K.,** Angiotensinogen-producing hepato cellular carcinoma, *Hypertension*, 6, 931, 1984.
54. **Kew, M.C., Leckie, B.J., and Greef, M.C.,** Arterial hypertension as a paraneoplastic phenomenon in hepato cellular carcinoma, *Arch. Intern. Med.*, 149, 2111, 1989.
55. **Menard, J., El Amrani, A.-I. K., Savoie, F., and Bouhnik, J.,** Angiotensinogen: an attractive and underrated participant in hypertension and inflammation, *Hypertension*, 18, 705, 1991.
56. **Chin, W.W.,** Hormonal regulation of gene expression, in *Endocrinology*, DeGroot, L.J., Ed., W.B. Saunders, Philadelphic, 1989.
57. **Nevins, J.R.,** The pathway of eukaryotic mRNA formation, *Annu. Rev. Biochem.*, 52, 441, 1983.
58. **Sharp, P.A.,** On the origin of RNA splicing and introns, *Cell*, 42, 397, 1985.
59. **Darnell, J., Lodish, H., and Baltimore, D.,** *Molecular Cell Biology*, Scientific American Books, New York, 1986.
60. **Dynan, W.S. and Tjian, R.,** Control of eukaryotic messenger RNA synthesis by sequence-specific DNA-binding proteins, *Nature*, 316, 774, 1985.
61. **Yamamoto, K.R.,** Steroid receptor regulated transcription of specific genes and gene networks, *Annu. Rev. Genet.*, 19, 209, 1985.
62. **Chambon, P., Dierich, A., and Guab, M.-P.,** Promoter elements of genes coding for proteins and modulation of transcription by estrogens and progesterone, *Recent Prog. Horm. Res.*, 40, 1, 1984.
63. **Shatkin, A.J.,** mRNA cap binding proteins: essential factors for initiation translation, *Cell*, 40, 223, 1985.
64. **Platt, T.,** Transcription termination and the regulation of gene expression, *Annu. Rev. Biochem.*, 55, 339, 1986.
65. **DePamphillis, M.L., Herman, S.A., Martinez-Salas, E., Chalifour, L.E., Wirak, D.O., Cupo, D.Y., and Miranda, M.,** Microinjecting DNA into mouse ova to study DNA replication and gene expression and to produce transgenic animals, *Biotechniques*, 6(7), 662, 1988.
66. **Jaenisch, R.,** Transgenic animals, *Science*, 240, 1468, 1988.
67. **Bishop, J.O. and Smith, P.,** Mechanism of chromosomal integration of microinjected DNA, *Mol. Biol. Med.*, 6, 283, 1989.
68. **Gordon, J.W. and Ruddle, F.H.,** DNA-mediated genetic transformation of mouse embryos and bone marrow — a review, *Gene*, 33, 456, 1985.
69. **Palmiter, R.D. and Brinster, R.L.,** Germline transformation of mice, *Annu. Rev. Genet.*, 20, 465, 1986.
70. **Al-Shawi, R., Burke, J., Jones, C.T., Simons, J.P., and Bishop, J.O.,** A MUP promoter-thymidine kinase reporter gene shows relaxed tissue-specific expression and confers male sterility upon transgenic mice, *Mol. Cell. Biol.*, 8, 4821, 1988.

71. **Covarrubias, L., Nishida, Y., and Mintz, B.,** Early postimplantation embryo lethality due to DNA rearrangements in a transgenic mouse strain, *Proc. Natl. Acad. Sci. U.S.A.*, 83, 6020, 1986.
72. **Mahon, K.A., Overbeek, P.A., and Westphal, H.,** Prenatal lethality in a transgenic mouse line is the result of a chromosomal translocation, *Proc. Natl. Acad. Sci. U.S.A.*, 85, 1165, 1988.
73. **Lacey, M., Alpert, S., and Hanahan, D.,** Bovine papillomavirus genome elicits skin tumours in transgenic mice, *Nature*, 322, 609, 1986.
74. **Harbers, K., Jahner, D., and Jaenisch, R.,** High frequency of unequal recombination in pseudoautosomal region shown by proviral insertion in transgenic mouse, *Nature*, 293, 540, 1981.
75. **Clauston, W.M., Lyons, I.G., and Richards, R.I.,** Tissue-specific and hormonal regulation of angiotensinogen minigenes in transgenic mice, *EMBO J.*, 8, 3337, 1989.
76. **Ohkubo, H., Kawakami, H., Kakehi, Y., Takumi, T., Arai, H., Yokota, Y., Iwai, M., Tanabe, Y., Masu, M., Hata, J., Iwao, H., Okamoto, H., Yokoyama, M., Nomura, T., Katsuki, M., and Nakanishi, S.,** Generation of transgenic mice with elevated blood pressure by introduction of the rat renin and angiotensinogen genes, *Proc. Natl. Acad. Sci. U.S.A.*, 87, 5153, 1990.
77. **Takahashi, S., Fukamizu, A., Hasegawa, R., Yokoyama, M., Nomura, R., Katsuki, M., and Murakami, K.,** Expression of the human angiotensinogen gene in transgenic mice and transfected cells, *Biochem. Biophys. Res. Commun.*, 180, 1103, 1991.
78. **Oliver, W.J. and Gross, F.,** Unique specificity of mouse angiotensinogen to homologous renin, *Proc. Soc. Exp. Biol.*, 122, 923, 1981.

Section III

Angiotensin Converting Enzyme

Chapter 8

ANGIOTENSIN-CONVERTING ENZYME AND CONVERTING ENZYME INHIBITORS

K.H. Berecek, S.J. King, and J.N. Wu

TABLE OF CONTENTS

0-8493-4622-3/93/$0.00 + $.50

I. INTRODUCTION

Angiotensin I (AngI), an inactive decapeptide generated by action of the enzyme renin on a glycoprotein substrate angiotensinogen, is converted to the active pressor octapeptide angiotensin II (AngII). The exopeptidase responsible for this conversion was first identified and isolated in plasma by Skeggs et al.,[1] who accordingly named it angiotensin-converting enzyme (ACE). This enzyme was later found to be the same enzyme as kininase II, and is able to hydrolyze bradykinin and various other peptides.[2,3] The conversion of AngI to AngII was assumed to take place in the circulation until Ng and Vane[4] determined that the enzyme activity present in the plasma was insufficient to account for the rapidity of the *in vivo* conversion and demonstrated that most of the conversion of circulating AngI to AngII occurred during passage through the lungs. The importance of the lung as a site of ACE activity was confirmed by Stanley and Biron,[5] who demonstrated that conversion of AngI to AngII in dogs on cardiopulmonary bypass was markedly lowered. However, the amount of conversion was still higher than that in blood alone, suggesting that ACE was present in vascular beds other than the lungs. Indeed, this dipeptidyl carboxypeptidase has been found to be widely distributed throughout the body as a membrane-bound ectoenzyme on the surface of vascular endothelial cells and epithelial cells of many organs. Since AngII is a potent vasoconstrictor shown to be involved in normal blood pressure regulation as well as in the pathogenesis of hypertension, it is believed that ACE plays a role in blood pressure regulation. In fact, over the past 10 years, inhibitors of ACE have become highly effective agents in the treatment of hypertension and heart failure.[10-13] In spite of the effectiveness of converting enzyme inhibitors in lowering blood pressure, the antihypertensive mechanisms of these agents are not fully understood.

In this chapter, we will review the biochemistry and cell biology of ACE and ACE inhibitors. Since we are limited in the amount of information we can provide in this chapter, we refer the readers to several recent, excellent monographs and books which have been published on ACE and ACE inhibitors.[14-20]

II. PROPERTIES OF ACE

ACE (kininase II, dipeptidyl carboxypeptidase, peptidyl carboxyhydrolase, EC 3.4.15.1) facilitates the removal of dipeptides and, in some cases, tripeptides from the carboxy-terminal of compatible substrates.[1,4,21-23] The enzyme is a glycoprotein containing 8 to 32% carbohydrate and is heavily sialated. It has a molecular weight between 130 and 160 kDa. Using the predicted amino acid sequence for the human vascular form of the enzyme, the peptide backbone has an estimated molecular weight of 146.6 kDa.

The enzyme acts on the decapeptide AngI (Asp-Arg-Val-Tyr-Ile-His-Pro-Phe-His-Leu) by cleaving a dipeptide (His-Leu) from the carboxy-terminal end of AngI to form the octapeptide AngII (Asp-Arg-Val-Tyr-Ile-His-Pro-Phe). ACE is an unusual zinc-containing exopeptidase[24] that cleaves dipeptide not only from the C-terminal end of AngI but also a variety of peptide substrates including substance P, enkephalins, neurotensin, and luteinizing hormone releasing hormone (LHRH).[25-28] Table 1 indicates the peptide substrates of ACE. It is evident that peptides with widely differing sequences can serve as substrates for the enzyme. Note that tripeptide and dipeptide products are cleaved by ACE from substance P, LHRH, and enkephalins. This same enzyme, often designated as kininase II, inactivates the vasodepressor nanopeptide, bradykinin (Arg-Pro-Pro-Gly-Phe-Ser-Pro-Phe-Arg) by the hydrolytic removal of its C-terminal dipeptide Phe-Arg.[29] Early studies indicated that ACE was a peptidase similar to pancreatic carboxypeptidase A and that both enzymes were zinc-containing metalloproteins since both were inhibited by cyanide and ethylene diamine tetraacetate (EDTA).[23,24]

Enzyme activity of ACE requires the presence of chloride or other monovalent anions such as bromide, fluoride, and nitrate.[1] The enzyme has low requirements of substrate specificity: (1) a free carboxylic acid group at the C-terminal end of the polypeptide and (2) absence of proline in the penultimate position. ACE will not hydrolyze a peptide unless it contains a free C-terminal carboxyl group, e.g., the terminal Leu of AngI. This free carboxyl group forms an ionic bond with an amino acid receptor site (Arg) of ACE. The distance between the zinc ion receptor site and the amino acid receptor is greater in ACE than in carboxypeptidase A, resulting in a cleavage of dipeptides from the substrate rather than monopeptides, as is the case of carboxypeptidase A. Hence, ACE converts the decapeptide AngI to the octapeptide AngII, whereas, carboxypeptidase A does not. Similarly to carboxypeptidase A, ACE will not hydrolyze a peptide that lacks a C-terminal carboxyl group, one with a proline-donating amino group for the susceptible peptide bond,[31,32] or one with a C-terminal decarboxylic amino acid.[33] ACE appears to have high affinity for peptide substrates with an aromatic amino acid in the antepenultimate position.[24,29] Also, tripeptides with a free amino group on the N-terminal amino acid are hydrolyzed.[21] The catalytic site of ACE like that of carboxypeptidase A involves functional Arg, Lys, Tyr, glutamic acid, and zinc.[34] One zinc atom per molecule is necessary for enzymatic activity.

III. DISTRIBUTION AND CELLULAR LOCALIZATION OF ACE

As previously stated, ACE was first discovered by Skeggs et al.,[1] in horse plasma. Tissue ACE was first detected in a microsomal (membrane) fraction

TABLE 1
Active Peptide Substrates of ACE

Peptide	Amino acid sequence
Angiotensin-I	Asp - Arg - Val - Tyr - Ile - His - Pro - Phe -↑ His - Leu
Bradykinin	Arg - Pro - Pro - Gly - Phe - Ser - Pro -↑ Phe - Arg
des-Arg^9-bradykinin	Arg - Pro - Pro - Gly - Phe -↑ Ser - Pro - Phe
Enkephalines	Tyr - Gly - Gly -↑ Phe - Met
Heptapeptide	Tyr - Gly - Gly -↑ Phe - Met -↑ Arg - Phe
Octapeptide	Tyr - Gly - Gly - Phe -↑ Met - Arg -↑ Gly - Leu
Chemotactic Peptide	f - Met -↑ Leu - Phe
Neurotensin	<Glu - Leu - Tyr - Glu - Asn - Lys - Pro - Arg - Arg - Pro - Tyr -↑ Ile - Leu
Substance-P	Arg - Pro - Lys - Pro - Gln - Gln - Phe - Phe -↑ Gly -↑ Leu - Met - NH_2 (4 : 1)
LHRH	<Glu - His - Trp -↑ Ser - Tyr - Gly - Leu -↑ Arg - Pro - Gly - NH_2

Note: Arrows denote primary (↑) and secondary (↑) sites of cleavage by human ACE.

After Skidgel, R.A. and Erdos, E.G., Novel activity of human AI converting enzyme: release of the NH_2- and COOH-terminal tripeptides from the luteinizing hormone-releasing hormone, *Proc. Natl. Acad. Sci. U.S.A.*, 82, 1025, 1985.

demonstrated in most tissues and fluids, although the content of the enzyme varies from tissue to tissue (Table 2).[28,37,38] The first specific and quantitative measurements of levels of ACE activity in various tissues of the rat utilized the fluorometric assay with Z-Phe-His-Leu as substrate.[39] Similar results were obtained in the same tissues using the spectrophotometric assay employing Hip-His-Leu as substrate.[40] By measuring enzyme activity or by using immunofluorescence or radioinhibitor binding assays, ACE was found to be widely distributed in the vascular tissue of large and small arteries and veins of man and various animal species.[41-43] The enzyme was found predominately in endothelium, but may also be found in the adventitia.[43,44] Velleteri and Bean[45] localized the enzyme in tunica media of the rat aorta, suggesting AngII

TABLE 2
Major Distribution of ACE

Endothelial cells (vasculature)
Epithelial cells
  Lung
  Kidney
  Gastrointestinal tract
  Reproductive tract
  Choroid plexus
  Placenta
  Salivary glands
Central nervous system cells
  Neurons
  Dendrites
Heart
  Cardiac valves
  Atrial cells
  Ventricular myocardium
Male genital tract
  Testis (germinal cells and spermatozoa)
  Prostate gland
  Epididymides
  Seminal plasma
Body fluids
  Blood
  Urine
  Lung edema
  Amniotic fluid
  Cerebrospinal fluid
  Lymph

Data from References 9 and 37.

may be generated in the vascular smooth muscle. Functional studies by Saye et al.,[46] showing that AngI causes contraction of endothelium-denuded rings of aortic tissue from rabbits, suggested that the conversion of AngI to AngII occurs in extra-endothelial layers of the vascular wall. ACE is also synthesized by epithelial cells of many tissues, including the kidney, the gastrointestinal tract, the reproductive tract, the choroid plexus, the placenta, and salivary glands. Its concentration in some epithelial cells has been found to be much higher than in endothelial cells, e.g., in the brush border of the kidney, the small intestine, and the placenta.[28,38] Vascular ACE is an intrinsic membrane protein bound within the membrane by a small nonantigenic hydrophobic anchor near the C-terminal end.

Localization of ACE in the brain has been demonstrated by biochemical measurements of enzyme activity and by quantitative autoradiography.[47-49] ACE is present in both neurons and glial cells. The choroid plexus, the ependyma, the blood vessels of the brain, the subfornical organ, and the organ vasculosum of the lamina terminalis are the richest sources of the enzyme in the brain, followed by basal ganglia, the neural secretory nuclei of the hypothalamus (the paraventricular nucleus and supraoptic nucleus, the median eminence, and the posterior pituitary gland).[50-54] There is a close correlation between the distribution of ACE and AngII in the various brain regions, with the exception of the basal ganglia. ACE has been found in cultured neuronal and glial cells from rats and mice[55] and in cultured neuroblastoma cells.[56]

Both *in vitro* autoradiographic studies and detailed ligand binding studies in homogenates from chambers of the heart have shown that ACE is widely distributed in the heart. The highest concentrations were found in cardiac valves, followed by the aorta, pulmonary and coronary arteries. Lower concentrations of ACE were found in the atria, and very small amounts were found in ventricular myocardial cells. ACE was absent in the conduction system of the heart.[57,58] Very high concentrations of ACE have also been found in various tissues of the male reproductive tract.[59] Although in most tissues ACE is membrane bound, the enzyme is also found in soluble form in body fluids, including blood, urine, lung edema, amniotic fluid, cerebral spinal fluid, and lymph.[28,51,59-61]

Structural studies of ACE purified from several species indicate that the same protein is responsible for enzymatic activity in many tissues. However, adult testis appears to contain large quantities of a different protein. Testicular ACE is also a glycoprotein but its molecular mass is approximately 105 kDa. Although pulmonary ACE and testicular ACE are different sizes, they share extensive biochemical and immunological similarities.[62-64] The pulmonary and testicular isozymes are encoded by two different mRNAs.[65-68] ACE mRNA from mature, but not immature, testis leads to the synthesis of testicular ACE *in vitro,* whereas mRNA ACE from lungs of both mature and immature animals can produce the synthesis of pulmonary ACE *in vitro*.[62,63] Recently, the complete amino acid sequence of the testicular isozymes from rabbit[68,72] and human[69,70] and the complete structures of pulmonary ACE from human[65] and mouse[66] have been deduced through cloning and sequencing of the respective cDNAs. Studies comparing the sequence of the isozymes suggest that the two mRNAs are transcribed from the same gene[62,71,72] but an alternate start site is used in the testis. The transcription of ACE is known to be regulated by several factors, although the details of transcriptional regulation remain to be determined.

IV. MEASUREMENT OF ACE

Many methods have been utilized to identify and quantify ACE and ACE activity. The first methods developed utilized whole animal bioassay, using the pressor response to AngI as an indication of ACE activity. Isolated smooth muscle contraction has also been used extensively as an end point. These methods had inherent difficulties with standardization and only limited numbers of samples could be run. Two classes of chemical assays were then developed. One was based on the release of the terminal His-Leu from AngI. This assay system used either radiolabeled AngI as the substrate and followed the release of radiolabeled products[71,72] or unmodified AngI and followed peptide release by fluorometry[73,74] or ninhydrin-reactive product.[75] A major disadvantage with these assay systems was that they were best suited for broken cell preparations and could not be used to assess ACE activity in crude samples. We have successfully used a method using ^{125}I-AngI as the substrate and follow the generation of radiolabeled AngII by high-voltage paper electrophoresis.[76,77] Combining this technique with the use of active site-directed ACE inhibitors allowed us to assess reaction rates in whole cells, broken cells, and soluble samples such as serum and conditioned culture medium. A second class of chemical assays uses model, N-terminus blocked tripeptides as the substrate. Some examples of blocking groups that have been successfully used include benzyloxycarbonyl (Z), hippuryl (Hip), *tert*-butyloxycarbonyl (*t*-BOC), and 2-furanacrylic acid (FA). The release of His-Leu from Z-Phe-His-Leu can be followed directly by fluorometry.[73] Using Hip-His-Leu as the substrate, the release of hippuric acid is followed by spectrophotometry after organic extraction from the reaction mixture.[78]

With the advent of active site-directed ACE inhibitors, several assay systems have been developed using the inhibitors to assess the numbers of active ACE molecules. Several of the inhibitors have been successfully employed to determine the tissue distribution of ACE as well as the relative abundance of the enzyme in a given tissue.

V. REGULATION OF ACE ACTIVITY AND RELEASE

ACE does not seem to be a rate-limiting step in the generation of AngII. ACE is in abundant supply both in the plasma and on the luminal surface of endothelial cells and it rapidly and effectively converts all of the AngI produced from angiotensinogen by renin to AngII. Testicular ACE, unlike the pulmonary isozyme, is developmentally regulated. Expression of the testis-specific isozyme does not occur until puberty. With the onset of puberty, there is a profound upregulation of ACE mRNA protein and activity.[45,63] Furthermore, the increase in ACE at puberty does not occur in the hypophysectomized rat.[78]

Developmental factors also appear to regulate ACE activity in cultured endothelial cells.[79] ACE activity is very low in cultured endothelial cells until confluence is reached, whereupon there is a rapid rise in ACE activity reaching maximum levels within two days post-confluence and remaining at this level for several days. Krulewitz and Fanburg[80] have given evidence that ACE activity of cultured bovine endothelial cells is under hormonal control. ACE activity is increased by glucocorticoids and thyroid hormone and decreased by insulin. It was further shown that ACE activity is responsive to cyclic AMP-related agents such as dibutyryl cyclic AMP, theophylline, and isoproterenol, all of which were shown to increase ACE activity.

Exposure to chronic normobaric hypoxia is associated with alterations in ACE activity in intact animals and cultured pulmonary artery endothelial cells.[77,81-83] Jackson et al.[81] have shown that 21-d exposure of rats to normobaric hypoxia (10% O_2) was associated with a significant depression in the single-pass conversion of AngI to AngII in the lung. These studies were performed in an isolated perfused rat lung preparation in which perfusate flow rates were held constant, thus minimizing the possibility that the decreased conversion seen in hypoxic lungs was related to altered kinetics of substrate exposure. Cultured bovine pulmonary artery endothelial cells exposed to hypoxia (3% O_2) for 24 to 48 h demonstrated increased ACE activity associated with the cells but no increase in the medium.[83] We reported that exposure of porcine pulmonary artery endothelial cells to chronic normobaric hypoxia (2.5% O_2) for 24 to 72 h induced a time-dependent increase in ACE protein synthesis without a concomitant increase in cell-associated ACE activity.[77] No change in either ACE protein content or activity was detected in the medium. We have demonstrated that the increased ACE protein synthesis was associated with increased ACE mRNA (unpublished results). The nonconcomitant increase in ACE activity following increased ACE mRNA and protein seems to indicate either that ACE is deactivated by an endogenous inhibitor that is induced by hypoxia or that ACE is synthesized as an inactive enzyme and is activated at some point prior to expression on the cell surface.

ACE has also been reported to be induced by exposure to active site-directed ACE inhibitors both *in vivo*[84] and *in vitro*.[85] We have recently shown that ACE mRNA is induced in neuronal enriched primary cultures prepared from 1-d-old rat pups exposed *in utero* to captopril. In addition, we found that neuronal cultures obtained from unexposed pups showed an increase in ACE mRNA expression when exposed to lisinopril *in vitro*.[86] Although ACE mRNA was increased in the neuronal cultures, ACE activity was decreased; in contrast, both ACE mRNA and activity were increased in whole brains from rats that had been exposed to lifetime captopril. Since ACE in the brain is expressed by neurons, glial and endothelial cells, we suggest that in neurons there may be an uncoupling of ACE activity from ACE synthesis. This uncoupling of ACE activity and ACE synthesis is apparently specific to neurons, as both glial and endothelial cells express increased ACE activity and synthesis after exposure to ACE inhibitor.

Recently, Schunkert et al. indicated that there may be feedback inhibition of AngII on ACE mRNA and activity.[87] Intravenous AngII infusion for 3 d to male Sprague-Dawley (SD) rats decreased the levels of ACE mRNA and ACE activity in the lung and testis with no change in ACE activity in serum. They also demonstrated an up-regulation of ACE mRNA expression and ACE activity after ACE inhibition by quinapril. We have also shown both dose- and time-dependent decreases in ACE mRNA levels in neuronal cultures exposed to AngII *in vitro*. These decreases can be blocked with AngII type 1 receptor blockade; in contrast, blockade with type 2 blocker potentiated the decrease in ACE mRNA. These data suggest that the AngII type 2 receptor in the brain may serve as a feedback receptor to decrease expression of ACE.

VI. DEVELOPMENT OF ACE INHIBITORS

The development of the first class of ACE inhibitors originated from the observation that various polypeptides isolated from the venom of the Brazilian snake *Bothrops jararaca* could potentiate bradykinin[88,89] and inhibit ACE.[90] Among these polypeptides, a nonapeptide inhibitor snake venom (teprotide — BPP_{5qa}) SQ20881 was eventually synthesized in the laboratory[90] and was found to be one of the most potent and selective ACE inhibitors. This agent was used extensively in animal experimentation and clinical research studies. Utilization of this ACE inhibitor proved the importance of the renin-angiotensin system in normal cardiovascular homeostasis,[91,92] as well as the various forms of hypertension.[93-97] This peptide reversed or prevented the development of renal vascular or malignant hypertension in experimental animals.[95,98,99] This converting enzyme inhibitor was also used to show that the renin-angiotensin system is important in the regulation of blood flow in various vascular beds, in particular, the regulation of renal blood flow.[100,101] SQ20881 was also used to show that inhibition of the renin-angiotensin system could be used as a therapeutic tool in accelerated or malignant hypertension,[102,103] in congestive heart failure,[104,105] and was found to reverse the delayed cerebral arteriospasm which follows subarachnoid hemorrhage.[106] This ACE inhibitor was found to have an immediate onset and a long duration of action (4 to 16 h) and was found to be nontoxic. However, its therapeutic potential was limited by the fact that it could only be administered intravenously and was very costly. The key idea in the development of oral inhibitors of ACE was the hypothesis that ACE is similar to carboxypeptidase A and the construction of a hypothetical model of ACE with the zinc atom appropriately positioned to induce vulnerability of the penultimate peptide bond. These efforts culminated in the development of captopril (SQ14225, D-3-mercapto-2-methylpropanyl-proline), the first orally active inhibitor, by Cushman, Ondetti and colleagues[107,108] and enalapril by Patchett and colleagues in 1980[109] (Figure 1).

The structure of captopril and the key hypothesized binding sites with ACE are illustrated in Figure 1.[110] Addition of a sulfhydryl group into captopril increased the inhibitory potency of this compound many times over that of its chemical precursor, which contained a carboxyl function, since the sulphur moiety binds tightly with the zinc ion of ACE. The onset of action of captopril is apparent within 15 min after oral administration, and the compound has a short (approximately two hours) plasma half-life. There is, however, evidence that its pharmacodynamic half-life is longer due to dynamic interconversion of captopril and its disulfide metabolites *in vivo*.[111] The clinical effectiveness of captopril in the treatment of hypertension and congestive heart failure has

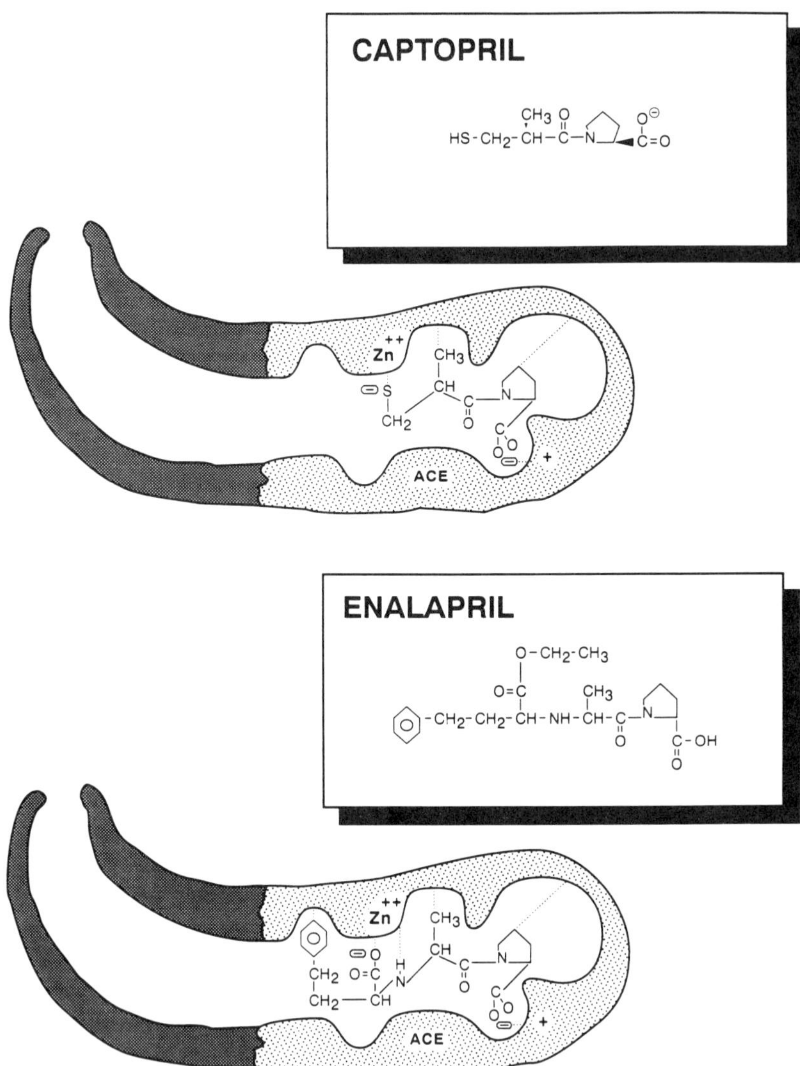

FIGURE 1. A schematic representation of the active site of ACE. The left-hand panel shows the structure of captopril and the putative binding of captopril to the active site of ACE. The right-hand panel shows the structure of enalapril with, and the putative binding of, enalaprilat to the active site of ACE. Note enalapril is a pro-drug metabolized by the liver by ester hydrolysis to form the active drug enalaprilat.

been well established.[112-118] However, its sulfhydryl moiety which contributes to its potency is believed to be responsible for a number of adverse effects such as proteinuria, loss of taste, and skin rash.[115-118] Therefore, an effort was made to develop a nonthiol-containing ACE inhibitor.

Patchett and collaborators[109] developed a new class of potent ACE inhibitors, the carboxyalkyl dipeptides, which includes enalapril (MK 421). Enalapril is an orally active pro-drug that is well absorbed after oral administration and is rapidly metabolized in the liver by ester hydrolysis to the active diacid parent form, enalaprilat (MK-422). This ACE inhibitor is four to nine times more potent than captopril.[120,121] The plasma level of enalapril reaches its peak within an hour after oral ingestion and disappears within 4 h. In contrast, enalaprilat reaches a peak level at 4 h and is slowly eliminated in the urine over a 24-h period without further metabolism. The duration of action of this compound varies between 12 and 24 h. Enalaprilat is not absorbed after oral administration, but can be given intravenously. Given intravenously, it has a very rapid onset and short duration of action, approximately 6 h. The difference in duration of action between oral and intravenous administration is due to the fact that the liver slowly de-esterifies enalapril into enalaprilat and slowly releases this compound into the blood stream, acting as a depot for the active compound. In contrast, following intravenous injection, the compound is rapidly eliminated. Detailed structure-activity studies indicate that both the sulfhydryl-containing inhibitors and carboxyalkyl dipeptides bind to the active site of ACE in a similar manner.[121-125] The zinc-binding moiety of enalaprilat is a carboxyl group that is relatively weak. However, the addition of more binding sites (seven for enalaprilat vs. five for captopril) (Figure 1) overcame the weakness of the carboxyl group binding to the zinc ion and produced an ACE inhibitor that was more potent than captopril.[126]

A number of newer ACE inhibitors are now available for clinical use. The new sulfhydryl-containing compounds are alacepril, phentiapril, pivalopril, and zofenopril (Tables 3 and 4). The newer carboxyalkyl inhibitors are

TABLE 3
Classification of ACE Inhibitors

I. Natural
   A. Peptides
      1. Snake venom polypeptides (teprotide, BBP_{5a})
      2. Human (des-pro^3 bradykinin or converstatin, enkephalins, endorphins, substance P)
      3. Microbial (EDTA-like aspergillomanasmine bicyclic lactams, phenacein)

II. Synthetic
   A. Peptides (val-trp; phe-ala-pro-)
   B. Peptide analogs (di- or tripeptides)
      1. Sulfhydryl group as zinc ligand (captopril, alacepril, phentiapril, pivalopril, zotenopril)
      2. Carboxyl group as zinc ligand (enalapril, perindopril, ramipril, quinapril, delapril, pentopril, lisinopril, cilazapril)
      3. Phosphinyl group as zinc ligand (fosenopril)

After Kostis, J. et al., Comparative clinical pharmacology of ACE inhibitors, *Angiotensin Converting Enzyme Inhibitors,* Kostis, J.B. and DeFelice, E.A., Eds., Alan R. Liss, New York, 1987, 19.

TABLE 4
Comparative Features of New ACE Inhibitors

	Pro-drug	Sulfur moiety	Duration of ACE inhibition	Antihypertensive action
Alacepril (Du-1219)	Y	Y	L	L
Cilazapril (RO-31-2848)	Y	N	L	L
Delapril (CV-3317)	Y	N	L	L
Fentiapril (SA-446)	N	Y	I	I
Fosenopril (SQ-28,555)	Y	N	L	L
Lisinopril (MK-521)	N	N	L	L
MC-838	Y	Y	?	?
Pentopril (CGS-13945)	Y	N	I	I
Perindopril (SA-9490-3)	Y	N	L	L
Pivalopril (RHC-3659)	Y	Y	S	S
Quinapril (CI-906)	Y	N	L	L
Ramapril (Hoe-498)	Y	N	L	L
Spirapril (TI-211-950/SCH-33844)	Y	Y	L	L
Zofenopril (SQ-26,991)	Y	Y	L	L

Note: Y, Yes; L, long duration (24 h or greater); N, no; I, intermediate duration; S, short duration (less than 12 h).

Adapted from DeFelice, E.A. and Kostis, J.B., *Angiotensin Converting Enzyme Inhibitors*, Kostis, J.B. and DeFelice, E.A., Eds., Alan R. Liss, New York, 1987, 213.

perindopril, ramipril, quinapril, delapril, pentopril, lisinopril, and cilazapril. A third class of ACE inhibitors contains a phosphinic acid group as the zinc-binding moiety; a prototype of this group is fosenopril. Like the carboxyalkyl dipeptides, the phosphinic acids must be administered as esters in order to obtain sufficient oral bioavailability. Comparative features of the new ACE inhibitors are outlined in Table 4. Of the 14 new ACE inhibitors listed in this table, 12 are pro-drugs. Eleven of these new drugs have a longer duration of ACE inhibition and/or antihypertensive effects than captopril. There are numerous publications characterizing these various ACE inhibitors.[14-19,118,124,125]

VII. MECHANISMS OF ACTION OF ACE INHIBITORS

Within a few years after their introduction the ACE inhibitors became established as highly effective therapeutic agents for all grades of hypertension and congestive heart failure. In addition, ACE inhibitors were also found to be beneficial in other cardiovascular diseases such as cardiac hypertrophy, myocardial ischemia, myocardial infarction, and arrhythmias. Echocardiographic measurements have shown that cardiac hypertrophy can be controlled or actually reversed by use of ACE inhibitors and that this decrease in left ventricular mass occurred without an effect on cardiac contractility or conductivity.[128-131] These agents were found to be effective in most patients with hypertension, irrespective of etiology or renin status. In addition, these inhibitors were shown to be as effective in hypertensive patients with diabetes mellitus. In spite of the great clinical efficacy of these compounds, their mechanisms of action are not fully understood. It was first thought that the major antihypertensive mechanism of action of ACE inhibition was due to blockade of the generation of AngII in the pulmonary capillary bed. However, attempts to correlate the magnitude of the blood pressure-lowering effects of these agents with their inhibitory action on ACE in the plasma and the lung endothelium yielded conflicting results, suggesting that the initial concept of ACE inhibition was overly simplistic.[132-135] Recent studies have shown that a reduction in circulating plasma AngII is probably not the sole mechanism involved in the antihypertensive action of ACE inhibitor. Table 5 outlines the mechanisms of action that have been suggested for the ACE inhibitors.

A. INHIBITION OF THE CIRCULATING RENIN-ANGIOTENSIN SYSTEM

It has been difficult to accurately assess the quantitative effect of ACE inhibitors on the renin-angiotensin system. One approach has been to assess the effect of ACE inhibitor on the pressor response to exogenous AngI. This technique provides direct evidence of the efficacy of ACE inhibitors, but has disadvantages in that pharmacological doses of AngI have to be administered. The measurement of plasma ACE activity has been problematic in that, in the case of captopril, enzyme activity has to be determined immediately following blood sampling because the captopril-ACE complex dissociates *in vitro* and could lead to an underestimation of the extent of ACE inhibition.[134,135] Another difficulty with measurement of plasma ACE activity as an index of ACE inhibition is that the ACE in this pool is not responsible for the bulk of conversion of AngI to AngII. Thus, determinations of plasma ACE should be corroborated by measurements of plasma AngII levels. Measurement of plasma AngII levels has also been difficult due to cross-reacting

TABLE 5
Mechanisms of Action of ACE Inhibitors

1. Inhibition of the renin-angiotensin system
   - Circulating
   - Tissue (i.e., cardiac, vascular, brain)
2. Potentiation of the kallikrein-kinin system
   - Circulating
   - Tissue
   - Via prostaglandins
3. Prostaglandins
4. Changes in aldosterone-sodium homeostasis
5. ACE inhibition and vasopressin release
6. Alterations in central and peripheral nervous system function
7. Changes in the angiotensin receptor binding

precursors and metabolites, production of AngII *in vitro* and enzymatic degradation of AngII during sample processing.[136-139]

The magnitude of the initial blood pressure reduction induced by ACE inhibitors was found to be related to pretreatment plasma renin activity and plasma AngII levels.[140-143] With long-term ACE inhibition, however, the relationship between the decrease in blood pressure and plasma levels of components of the renin-angiotensin system is weak.[144,145] Studies in animals with various forms of experimental hypertension and in hypertensive patients have demonstrated that blood pressure can be lowered by chronic administration of ACE inhibitors, regardless of whether the plasma renin-angiotensin system was stimulated prior to treatment. Moreover, blood pressure can even be lowered in anephric animals.[146] The lack of correlation between the antihypertensive effect and expected changes in components of the plasma renin-angiotensin following ACE inhibition led to the hypothesis that inhibition of the renin-angiotensin system in the tissues may be a more important antihypertensive mechanism than inhibition of circulating ACE.

B. INHIBITION OF THE TISSUE RENIN-ANGIOTENSIN SYSTEM(S)

After chronic administration of ACE inhibitors, induction of new ACE synthesis has been shown to occur.[147] This is probably the explanation for the failure of some studies to demonstrate long-term plasma ACE inhibition despite the continual administration of ACE inhibitors.[148] In order to study whether or not induction of the enzyme occurred in tissues, Johnston and coworkers[148] developed new techniques to measure induction and inhibition of ACE simultaneously in a variety of tissues. This was achieved by dissociation of the inhibitor from the enzyme by chelating zinc in tissue slices with EDTA and then exposing tissue to ^{125}I-MK351A. MK351A is a thiouracil derivative of the potent ACE inhibitor, lisinopril. Comparison of autoradiographs with or without EDTA together with tissue from untreated animals allowed simultaneous calculation of the degree of induction or inhibition of tissue ACE after chronic therapy. Utilizing this technique, Jackson et al.[155] reported that chronic oral administration of ACE inhibitors (lisinopril, 1 mg/kg/d $\times$ 2 weeks) produced a variable degree of ACE inhibition in plasma and tissues. ACE inhibition was greatest in plasma and lung. Plasma ACE was decreased in plasma samples before dialysis against buffer-containing EDTA and increased in plasma samples dialyzed against EDTA containing buffer.

^{125}I-MK351A has also been used successfully by Mendelsohn et al.[149] to anatomically map ACE in the central nervous system and peripheral tissues.[149,150] Fyhrquist et al. and Jackson et al. have also used this compound to develop a radioinhibitor binding assay to quantitate ACE levels in plasma and tissue.[151-154] There is a great deal of evidence for an independent renin-angiotensin system in various tissues, most notably the vascular wall, heart, adrenal gland, kidney, and brain.[156-164] Discovery of the components in the renin-angiotensin system in these various tissues gave rise to the idea that functional tissue renin-angiotensin systems exist independent of the circulating renin-angiotensin system. Recently, the messenger RNAs for the protein components of the renin-angiotensin system have been identified in these same organs.[157,164]

1. Renin-Angiotensin System in Vascular Tissue

Direct evidence has been provided[164-168] for local generation and release of AngII by peripheral vascular tissues of SD rats, spontaneously hypertensive rats (SHR), and humans. These investigators[191] also demonstrated suppressed AngII release from isolated mesenteric arteries in SHR treated with three different ACE inhibitors. Elevations in vascular ACE activity have been reported in renal hypertensive dogs and in various rat models of hypertension. Studies in rats showed that ACE levels varied among vascular beds and among the various models of hypertension.[164-168] ACE levels were highest in the lung, aorta, and mesenteric arteries of two-kidney, one-clip (2K1C) renal hypertensive rats; intermediate in lung, aorta, and mesenteric arteries in one-kidney, one-clip renal hypertensive rats; and showed the lowest in DOCA-salt hypertensive rats. ACE levels in these vascular tissues and in the adrenal medulla correlated positively with plasma renin activity, whereas in the kidney, pituitary gland, testis, and chorioid plexus of the brain ACE activity levels correlated negatively with plasma renin activity. Increases in vascular ACE in the SHR have also been reported in the aorta and its branches, but not in the hepatic, pulmonary, or basilar arteries compared to normotensive Wistar Kyoto (WKY) rats.[167]

Locally generated AngII may exert a number of different actions to influence vascular tone and distensibility/structure. These include a direct vasoconstrictor effect by stimulation of AngII receptors on the smooth muscle cells of the vascular media,[170] facilitation of adrenergic transmission, thus producing vasoconstriction by increased vascular tone,[171] and stimulation of sodium and calcium transport systems across the cell membranes in cultured vascular smooth muscle cells.[172] AngII may also exert effects on vascular tone through stimulation of endothelial prostacyclin synthesis.[173] The physiological significance of this effect is not clear. An important recent finding is that pressor hormones such as AngII may actually be growth factors in cardiac and vascular tissue and promote hypertrophy and/or hyperplasia in cells of these tissues by pressure-independent actions.[174] One of the mechanisms whereby AngII may promote tissue growth is via a stimulation of proto-oncogenes. These elements may be the signals which trigger cell growth and differentiation. The products of proto-oncogenes make up a series of proteins that transmit the growth signals across the plasma membrane to the nucleus.[175] The proto-oncogene c-*sis* encodes for one chain of the mitogen, plasma-derived growth factor (PDGF), and the H-ras oncogene may encode for G-proteins that link receptors to a possible phospholipase C that hydrolizes inositol lipids. Alterations in either one of these proto-oncogenes could result in persistent growth signals. For cell growth to occur or to proceed, it is thought that mitogens require intracellular proteins encoded by the proto-oncogenes c-*myc* and c-*fos*. The fos protein is thought to act as a cofactor by controlling gene expression by interacting physically with regulatory DNA sequences flanking particular target genes.[176] The major sequelae in response to an increase in blood pressure is hypertrophy in the left ventricle, the aorta, and other vascular beds. Pressure overload may induce myocardial hypertrophy by improving the efficiency of existing components of the protein synthetic pathway of the smooth muscle cells as well as expanding the maximal capacity of the cell for protein synthesis.[177] The cellular processes that contribute to hypertrophy are still unclear. Work in cultured cells suggest that AngII may be involved in the development of tissue hypertrophy. In cardiac myocytes, AngII is known to induce a contractile response via inositol lipid hydrolysis;[178] however, there is no conclusive proof that AngII can induce growth in these cells. Moreover, in vascular smooth muscle cells, it has been shown that AngII can induce a contractile response via phosphoinositide lipid breakdown.[203] Along with the induction of contraction, it has been demonstrated that when AngII is added to aortic vascular smooth muscle cells in culture, there is an increase in protein and DNA synthesis and cell size but not an increase in the number of cells.[180] Berk et al.[181] showed that AngII was not mitogenic for vascular smooth muscle cells; but after 24-h exposure to 100 n*M* AngII, there was an 80% increase in protein synthesis as measured by ^{3}H-leucine incorporation. Naftilan et al. have studied the effects of AngII on the expression of the c-*fos* and c-*jun* proto-oncogenes in rat vascular smooth muscle cells.[182,183] AngII was found to produce a dose-dependent increase in c-*fos* mRNA expression which was abolished by the AngII receptor antagonist, saralasin. Recently, Rainier et al.[184] have treated aortic smooth muscle cells with an oligonucleotide spanning the translation initiation site for both the sense and anti-sense orientation for the c-*fos* gene and demonstrated that AngII-induced protein synthesis was blocked in the cells. In addition to alterations in proto-oncogene expression, AngII has also been shown to induce genetic expression of PDGF, which is a powerful vasoconstrictor and mitogen. Itoh and colleagues have shown that antisense oligonucleotide to PDGF mRNA attenuated the hypertrophy induced by AngII.[185] These aforementioned studies suggest that AngII is able to induce hypertrophy by direct hydrolysis of inositol lipids and stimulation of nuclear proto-oncogene expression, as well as by induction of other genes encoding for powerful mitogens such as PDGF. In contrast to studies performed with cells cultured from aorta of rat, Lyall et al.[186] treated cells cultured from rat mesenteric arteries with increasing doses of AngII for 4 to 6 d and performed cell counts. These investigators reported a dose-dependent increase in cell number in vascular smooth muscle cells cultured from rat mesenteric artery. Similar findings have been reported in cells cultured from human aorta.[187] The differences between the findings in human aortic cells and rat aortic cells may be due to species variation, differences in the serum that is used to culture the cells, or in the fact that the human cells were prepared from explants which may have different growth properties than the rat cells, which were enzymatically prepared. The studies by Lyall et al.[186] in the rat mesenteric arteries suggest that there may be differences in the response of vascular smooth muscle cells from various vascular beds to AngII, with this peptide being able to induce a hyperplastic response in smaller arteries but not in larger arteries such as the aorta. It should be pointed out that there are problems in studies where cells are put into culture. In the case of vascular smooth muscle cells, it has been well known that once these cells were put in culture they change from contractile types of cells to synthetic types of cells.[188,189] Furthermore, cells in culture are exposed to mitogens which are contained in the serum that is used to raise the cells and they are devoid of the modulating influences of nerves and endothelial cells.

For studies in intact vascular tissues, most investigators have used a pressure overload model which is produced by coarctation of the aorta to raise the afterload placed on the heart as well as the pressure load placed on arteries above the coarctation. Utilizing a model of this nature, Komuro et al.[190] reported an increase in the expression of c-*fos* and c-*myc* in thoracic aorta from animals who had had a clip placed around the abdominal aorta above the origins of both renal arteries. The increase in expression of these proto-oncogenes occurred fairly rapidly. There was increased expression of c-*fos* oncogene at 30 min and c-*myc* at 2 h in the myocardium of the rat.[190] The levels of these oncogenes peaked at 8 h and returned to normal by 48 h. In contrast to the c-*myc* and c-*fos* oncogenes, H-ras expression showed a gradual increase. In using a model with coarctation of the aorta between the renal arteries, which creates a less severe hypertensive state, it was shown that there was increased expression of the H-ras, c-*myc,* and c-*fos* proto-oncogenes in the left ventricle and the proximal aorta in rats at 72 h after placing the ligature.[174,191] These studies, compared to the studies reported above, suggest that following a return to baseline at 48 h after ligature, there must be a further rise in the activity of these cellular signals. Regardless of which oncogene or mitogenic growth factor is altered by AngII, evidence strongly supports the hypothesis that the local AngII system in the blood vessels and the heart may exert important trophic and growth-regulating influences on the blood vessels and the heart in hypertension.

2. Intracardiac Renin-Angiotensin System

The components of the renin-angiotensin system, including renin and angiotensinogen mRNA, have been detected in heart. Studies in isolated cardiac myocytes and the isolated perfused heart have demonstrated local production of AngII.[162] Local AngII may influence coronary and myocardial function. Furthermore, enhanced local vascular AngII production in the heart in areas of injury or inflammation may result in increased vasoconstriction or vasospasm. In addition to its vasoconstrictor effects, AngII also has a positive ionotropic effect and may contribute to regulation of cardiac contractility.[292] Since AngII has a myocyte growth-stimulating property, it may play an important role in cardiac hypertrophy and/or in remodeling. Cardiac AngII may also adversely affect myocardial metabolism and provoke ventricular arrhythmias during ischemia and reperfusion-induced myocardial injury. In studies in isolated perfused hearts, infusion of AngII was shown to produce reperfusion arrhythmia, an effect which was attenuated by treatment with ACE inhibitor.[162]

Numerous groups have demonstrated that chronic oral antihypertensive treatment of SHR with the ACE inhibitor, captopril, increased renin concentrations in the aortic wall.[193,194] In addition, there is much evidence for inhibition of ACE in the vascular wall after oral treatment with ACE inhibitor.[195-197] Captopril and other ACE inhibitors are known to produce a marked depression in reactivity to exogenous vasoconstrictors.[198-200] This is probably related to attenuation in sympathetic vasoconstrictor tone, but may also be attributable to captopril-induced attenuation in the development of structural changes in the blood vessels of the hypertensive animal. Long-term treatment of SHR with captopril in the drinking water prevented the development of hypertension and also caused long-lasting decreases in myocardial and aortic wall hypertrophy.[201] In the mesenteric resistance vessels, the same treatment caused a reduction of wall-to-lumen ratio, an increase in the compliance of the vessel, and a decrease in contractile activity. Furthermore, perindopril, another converting enzyme inhibitor, prevented the development of hypertension and attenuated the structural and functional alterations in resistance vessels from the mesenteric vascular bed in SHR.[202] Recently, we have shown that treatment of SHR with captopril throughout life prevented the development of hypertension and of the functional and structural alterations in mesenteric arteries and cardiac hypertrophy.[203] Ertl et al.[204] demonstrated a reduction in the size of infarction after coronary artery ligation in dogs treated with ACE inhibitors. ACE inhibitors have also been found to increase coronary blood flow and improve myocardial metabolism.[205] The observations that the effects of ACE inhibitors on cardiac hypertrophy were seen in *in situ* isolated heart, and that the effects of ACE inhibitors on vascular and cardiac hypertrophy could be obtained if subhypertensive doses of ACE inhibitor were given, suggest that factors beyond blood pressure reduction and load changes may contribute to the cardiovascular benefits of the ACE inhibitors. This can come about through interference with local paracrine and/or autocrine effects of the renin-angiotensin renin angiotensin system and/or participation of kinins.[206-208]

3. The Brain Renin-Angiotensin System

There is a great deal of evidence for an intrinsic renin-angiotensin system and local AngII generation in the brain.[161,209-211] Components of the renin-angiotensin system, including the mRNA for angiotensinogen and renin, have been localized in the brain. AngII-immunoreactive cell bodies and fibers, as well as specific angiotensin receptors, have been identified in areas of the brain that participate in cardiovascular regulation, including the circumventricular organs, paraventricular nucleus in the hypothalamus, locus coeruleus, nucleus tractus solitaris, area postrema, and the ventral lateral medulla.[212-214] The presence of components of the renin-angiotensin system in neuroblastoma cells, and in primary neuronal enriched cell cultures,[215-217] and the finding that these cells synthesize immunoprecipitable AngII which co-migrates with authentic AngII on high-performance liquid chromatography (HPLC), add further support to the concept of local synthesis of AngII in the brain. Furthermore, brain AngII has been shown to participate in blood pressure and fluid, and electrolyte regulation through mechanisms distinct from those in the periphery. The mechanisms through which brain AngII affects blood pressure involve enhancement of sympathetic outflow, stimulation of vasopressin and ACTH release, and increased catecholamine biosynthesis and turnover.[210,218,219] In addition, brain AngII elicits two behaviors, thirst and salt appetite, which are critical in body fluid and electrolyte regulation.[220] Increasing evidence suggests that AngII may act as a neurotransmitter in the central nervous system. Radioreceptor assays have detected specific high-affinity AngII receptors in nerve terminals (synaptosomes).[221] Iontophoresis of AngII was found to enhance the rate of discharge of 75% of the neurons in the subfornical organ. This effect was dose dependent and antagonized by the angiotensin receptor antagonist, saralasin.[222] Other studies have shown that AngII activates neurons of the supraoptic and paraventricular nuclei, areas that synthesize and release vasopressin.[223] These observations correlate with the known enhancement of vasopressin release by AngII.

Several lines of anatomical and functional evidence support the hypothesis that brain AngII may play a role in the pathogenesis of hypertension. Most of the studies have been carried out in SHR. In comparison to WKY rats, SHR show increased sympathetic activity[224] and vasopressin and ACTH secretion.[225,226] SHR also show increased levels of AngII-like material in brain and cerebrospinal fluid,[209,210] increased turnover of AngII in the brain,[209,216] increased numbers of AngII receptors in some brain areas,[227,228] and increased pressor and neuronal responsiveness to central administration of AngII.[229] Findings in cell culture confirm previous data obtained from adult brain tissue.[216,217] Comparison of primary neuronal cultures from SHR and WKY has revealed differences in AngII stores, metabolism, and binding between strains. The finding of altered AngII mechanisms in brain cells from SHR neonates suggests that these alterations may antedate the onset of hypertension, and therefore, may contribute to the development of hypertension rather than being a consequence of it.

The strongest evidence that brain AngII may play a role in the pathogenesis of hypertension in SHR has come from the findings of numerous laboratories that intracerebroventricular administration of saralasin, an AngII receptor antagonist, or captopril, in doses that are ineffective or much less effective when given systemically, significantly attenuated the development of hypertension.[6,7,207,231-239] Administration of captopril into the cerebrospinal fluid has also been found to lower blood pressure in renal hypertensive rats[234] and

DOCA-salt hypertensive rats.[235] This blood pressure-lowering effect of captopril does not appear to be due to the chemical structure of the compound, since Phillips and Kimura[211] confirmed a reduction in blood pressure in SHR using enalaprilat and ramipril. Although earlier studies reported that captopril did not cross the blood brain barrier,[236,237] there is increasing functional and biochemical evidence that both acute and chronic peripheral administration of captopril and other ACE inhibitors inhibit brain ACE activity and alter the brain renin-angiotensin system.[238,239] The various ACE inhibitors differ with respect to their ability to gain access to structures within the blood brain barrier and inhibit brain ACE upon systemic administration. Differences in lipid solubility among the various ACE inhibitors may partly account for variability in the access to the brain. Several groups have shown inhibition of brain ACE in structures within the blood brain barrier using more lipophilic ACE inhibitors such as ramipril, HOE288, SA446, captopril, or prendipril. On the other hand, hydrophilic agents such as enalapril and lisinopril did not produce detectable inhibition of central ACE.[239-241] An additional factor governing the access of pro-drug ACE inhibitors such as ramipril, enalapril, and prendipril to structures inside the blood brain barrier may be the degree and site of metabolic activation of these drugs after oral administration. For instance, the rat brain has very little capacity to hydrolyze enalapril to the active enalaprilat.[240,241] The degree to which enalapril can inhibit brain ACE is related to the amount of enalaprilat that can enter the brain from the blood. Inhibition of ACE in human cerebrospinal fluid was reported by Geppetti et al.[242] after individuals were given a single dose of captopril. If this observation can be confirmed with other ACE inhibitors, it will strengthen the concept of the effect of ACE inhibitors on the central renin-AngII system. The lack of effect of enalapril and lisinopril on brain ACE does not necessarily mean that penetration of the central nervous system is essential for decreasing blood pressure. Even the most hydrophilic agents, such as enalapril and lisinopril, produce a decrease in ACE in the circumventricular organs such as the subfornical organ (SFO) and the organum vasculosum of the lamina terminalis (OVLT). These are the sites where AngII evokes central actions of vasopressin release, drinking, and pressor response.[61] In addition, there are AngII-containing pathways from the circumventricular organs to the paraventricular nucleus and other areas involved in cardiovascular regulation.[212] The effects of these hydrophilic agents have been found to be even greater than those elicited by lipophilic agents. Therefore, all ACE inhibitors may have central nervous effects. The more lipophilic ACE inhibitor may have additional central effects at doses sufficient for penetration of the blood brain barrier.

C. CONVERTING ENZYME INHIBITION AND THE KALLIKREIN-KININ SYSTEM

Plasma kallikrein produces bradykinin, a potent vasodilator nonapeptide, from a high-molecular-weight substrate. Kallikrein of the glandular subtype is also present in renal tissue and in the urine. Renal prekallikrein may be activated by phospholipase A, in the arachidonic acid cascade, suggesting a role for prostaglandins in the regulation of the renal kallikrein-kinin system.[243] Renal kallikrein cleaves both a low- and a high-molecular-weight kininogen to produce kallidin, a decapeptide. Kallidin is then processed to bradykinin by an aminopeptidase. Kallidin and bradykinin are rapidly metabolized by kininases, kininase II being identical with ACE. The question of whether ACE inhibition results in accumulation of bradykinin and whether such accumulation produces a substantial contribution to the antihypertensive effect of ACE inhibitors is still much debated. Some investigators have reported an increase in blood kinin levels with acute administration of teprotide and captopril.[244,245] In a majority of studies, however, no consistent changes in circulating bradykinin could be detected either during short- or long-term inhibition of converting enzyme activity.[246,247] It should be pointed out, however, that kinins are autocoids and changes in biological activity and degradation may not necessarily be reflected by changes in their plasma levels.[248] Studies in hypertensive animals have shown that specific anti-kinin antibodies acutely attenuated the antihypertensive effect of captopril,[248] while competitive antagonists of bradykinin gave conflicting results.[249,250] Scholkens and Linz[251] recently reported that the cardioprotective effects of ramipril was mimicked by bradykinin and abolished by coadministration of a bradykinin antagonist. Some preliminary evidence also suggests that bradykinin may also be involved in the increased glucose utilization that has been reported with ACE inhibition.[252] It is likely that the kallikrein-kinin system does participate in the antihypertensive action of ACE inhibitors. However, the nature of the participation, be it in the cardioprotective function, the reduction in vascular tone, or in the metabolic effects of ACE inhibition, is far from clear. Hence, a great deal more research is needed in this area.

D. ACE INHIBITION AND PROSTAGLANDINS

The prostaglandins represent a group of 20-carbon fatty acids that are synthesized from arachidonic acid, a component of phospholipids contained in cell membranes.[253] A complex interaction is known to exist between prostaglandins and the renin-angiotensin system. AngII enhances the release of arachidonic acid, thereby triggering the formation of prostaglandins. Arachidonic acid is converted to prostaglandin G_2 (PGG_2) by cyclooxygenase. PGG_2 is then transformed into either prostaglandin E_2 (PGE_2, a vasodilator), prostaglandin F_2 ($PGF_2\alpha$, a vasoconstrictor), thromboxane A_2 (TXA_2, a vasoconstrictor which also produces aggregation of platelets), or prostaglandin I_2 (PGI_2, a vasodilator which prevents platelet aggregation). Vasodilator prostaglandins produce direct relaxation of vascular smooth muscle, and also attenuate the vasoconstrictor effect of AngII.[254] PGE_2 and PGI_2 also stimulate the release of renin.[254] Increased kinin levels cause vasodilation and stimulate prostaglandins synthesis,[254,255] thus potentiating the depressor response. Reports regarding changes in blood prostaglandin levels in response to ACE inhibition are conflicting.[255,256] Coadministration of indomethacin, a nonsteroidal anti-inflammatory agent known to reduce prostaglandin synthesis, with ACE inhibitors reduced the antihypertensive efficacy of ACE inhibitors in some, but not all, studies. This may not be specific for the antihypertensive action of ACE inhibitors.[256] There is evidence that indomethacin also reduces the antihypertensive effect of other agents, such as beta blockers.[257,258] The contribution of prostaglandins to the antihypertensive effect of ACE inhibitors needs to be clarified.

E. CHANGES IN ALDOSTERONE-SODIUM HOMEOSTASIS WITH ACE INHIBITORS

AngII is one of the most potent stimuli of aldosterone secretion. Numerous laboratories have reported a decrease in aldosterone production during ACE inhibition.[259,260] This effect should produce a natriuresis and help to prevent the development of sodium retention when blood pressure is lowered by the ACE inhibitors. In some studies, plasma aldosterone has been found to be decreased even after long-term ACE inhibition; however, this has not been a universal finding.[262-264] In some studies, an increase in total body sodium content was reported in patients after prolonged ACE inhibition.[260,261] Sanchez et al.[262] demonstrated a reduction in total exchangeable sodium 16 weeks after enalapril treatment. In contrast, other investigators have found no significant change in other markers or volume states such as body weight, plasma volume, and extra- or intracellular fluid volumes.[265]

F. ACE INHIBITION AND VASOPRESSIN RELEASE

Vasopressin secretion has been reported to be reduced during treatment with ACE inhibitors in patients with severe and malignant hypertension.[266,267] Whether this effect contributes to the antihypertensive action of ACE inhibitors in such patients is unclear. Recent studies from our laboratory in SHR, treated *in utero* with captopril, demonstrated decreased vasopressin-like immunoreactivity in the paraventricular and supraoptic nuclei of the hypothalamus,[269] and decreased plasma vasopressin levels compared to control SHR.[268] Since AngII is a major stimulus for the release of vasopressin, part of the pressor effect of AngII is due to a release of vasopressin; part of the antihypertensive effect of ACE inhibitors may come about by a decrease in vasopressin synthesis and release. More research is necessary in this area.

G. ALTERATIONS IN CENTRAL AND PERIPHERAL NERVOUS FUNCTION WITH ACE INHIBITION

A large body of evidence supports the view that the sympathetic nervous system is hyperactive in hypertensive patients and animals.[269-271] There is a great deal of interaction between the sympathetic nervous system and the renin-angiotensin system, particularly in the brain and the peripheral vasculature. In the brain central, AngII produces its pressor effect in large part by increasing sympathetic outflow.[272,273] AngII has also been reported to increase the synthesis of catecholamines, facilitate neurotransmitter release, and inhibit the reuptake of norepinephrine.[61] Circulating AngII can gain access to the brain through areas which lack blood brain barrier, namely the circumventricular organs.[274] Ablation of some of these regions, particularly the subfornical organ, can prevent or blunt the pressor effect of intravenous administration of AngII. In the periphery, AngII interacts with the sympathetic system at both the pre- and postjunctional receptors.[275-277] Angiotensin inhibits the reuptake of norepinephrine released from nerve terminals[275] and facilitates the release of norepinephrine from the sympathetic boutons.[276] AngII potentiates the vasoconstrictor effects of norepinephrine postjunctionally[277] and causes the release of epinephrine from the adrenal medulla.[278] The influence of ACE inhibitor on cardiovascular responses to sympathetic activation has been studied in both animals and humans. Captopril diminishes the pressor effect of both sympathetic nerve stimulation and alpha-adrenergic receptor stimulation with exogeneous norepinephrine.[279-281] Captopril also potentiates baroreflex control of heart rate in normotensive and hypertensive humans[282] and this baroreflex potentiation may be an important part of its antihypertensive action. A similar increase in baroreflex sensitivity has been reported in SHR after chronic intracerebroventricular or lifetime oral administration of captopril.[283,284] Furthermore, this effect can be reversed by intracerebroventricular administration of AngII to captopril-treated rats.[284] In addition, bilateral injections of an AngII antagonist to the nucleus tractus solitarii enhances baroreflex control of heart rate.[285] These studies suggest that captopril may potentiate baroreflex sensitivity in hypertensive humans and rats by inhibiting brain AngII. Thus, the antihypertensive actions of ACE inhibitor may be mediated in part through a neurogenic AngII-related mechanism.

H. ACE INHIBITION AND THE ANGIOTENSIN RECEPTOR

Numerous studies have suggested that AngII receptors in brain may be regulated differently from those in the periphery and that AngII receptors in SHR may be regulated differently from those in WKY rats.[286-292] In support of this hypothesis, an earlier study of Felix and Schelling[222] reported an increase in septal neuronal firing rate in response to microiontophoresis of AngII in SHR compared to normotensive rats. The increase in septal AngII receptor sensitivity in SHR was abolished when the rats were placed on oral captopril treatment in the early weaning period. These data suggest that captopril does indeed cross the blood brain barrier and can decrease the responsiveness to AngII by altering receptor number and/or affinity. Many studies have shown that oral captopril affects the brain renin-AngII system. Rats that have been subjected to chronic captopril therapy show significantly lower drinking and pressor responses to intracerebroventricular administration of AngI, suggesting a chronic inhibition of converting enzyme. Furthermore, the pressor and drinking responses to intracerebroventricular administration of AngII are also decreased, suggesting a decrease in the number and/or affinity of AngII receptors. Radioligand binding and autoradiographic studies have shown that treatment of animals with ACE inhibitors produces a decrease in the number of AngII receptors. Recently, we compared ^{125}I-angiotensin binding in neuronal-enriched primary cultures from whole brains of one-day-old SHR pups treated *in utero* with captopril vs. control SHR.[292] We found that ^{125}I-AngII binding in primary neuronal cultures from captopril-treated SHR was decreased compared to cultures from control SHR. Scatchard analysis revealed no differences in K_d, but B_{max} was less in the captopril-treated SHR than in controls. Both chronic treatment with ACE inhibitors *in vivo* and brief periods of incubation with ACE inhibitors *in vitro* decreased the number of AngII receptors expressed in the neuron-enriched cultures. These findings suggest a potential new mechanism of antihypertensive action of ACE inhibitors: these agents may produce downregulation of brain AngII receptors, which may lead to a decrease in blood pressure. This hypothesis remains to be tested in other models of hypertension, as well as in human subjects.

ACKNOWLEDGMENTS

The authors would like to thank Betty Little, April Sandlin, Sherry Crittenden, and Peg White for typing the manuscript and Dr. Suzanne Oparil for her critical review of the manuscript and suggestions for changes. Work presented in this manuscript by the authors was supported by United States Public Health Service grant HL 31515.

REFERENCES

1. **Skeggs, L.T., Kahn, J.R., and Shumway, N.P.,** The preparation and function of the hypertensin-converting enzyme, *J. Exp. Med.*, 103, 295, 1956.
2. **Yang, H.Y.T., Erdos, E.G., and Levin, Y.,** A dipeptide carboxypeptidase that converts AI and inactivates bradykinin, *Biochim. Biophys. Acta,* 214, 374, 1970.
3. **Erdos, E.G. and Skidgel, R.A.,** Structure and functions of human AI converting enzyme (kininase II), *Biochem. Soc. Trans.*, 13, 42, 1985.
4. **Ng, K.K.F. and Vane, J.R.,** Conversion of AI to AII, *Nature,* 216, 762, 1967.
5. **Stanley, P. and Biron, P.,** Pressor response to AI during cardiopulmonary bypass, *Experientia,* 26, 46, 1969.
6. **Caldwell, P.R.B., Seegal, B.C., Hsu, K.C., Das, M., and Soffer, R.L.,** Angiotensin-converting enzyme: vascular endothelial localization, *Science,* 191, 1050, 1976.
7. **Ryan, U.S., Ryan, J.W., Whitaker, C., and Chiu, A.,** Localization of ACE (kininase II). II. Immunocytochemistry and immunofluorescence, *Tissue Cell,* 8, 125, 1976.
8. **Soffer, R.L.,** Angiotensin-converting enzyme and the regulation of vasoactive peptides, *Annu. Rev. Biochem.*, 45, 73, 1976.
9. **Erdos, E.G. and Skidgel, R.A.,** The AI converting enzyme, *Lab. Invest.*, 56, 345, 1987.
10. **Cushman, D.W. and Ondetti, M.A.,** Inhibitors of ACE for treatment of hypertension, *Biochem. Pharmacol.*, 29, 1871, 1980.

11. **Johnston, C.I., Jackson, B.J., Larmour, I. et al.,** Plasma enalapril levels and hormonal effects after short- and long-term administration in essential hypertension, *Br. J. Clin. Pharmacol.,* 18, 2335, 1984.
12. **Schwartz, J.B., Taylor, A., Abernathy, D. et al.,** Pharmacokinetics and pharmacodynamics of enalapril in patients with congestive heart failure and patients with hypertension, *J. Cardiovasc. Pharmacol.,* 7, 767, 1985.
13. **Gavras, I. and Gavras, H.,** The use of ACE inhibitors in hypertension, in *ACE Inhibitors,* Kostis, J.B. and Felice, E.A., Eds., Alan R. Liss, New York, 1987, 93.
14. **Kostis, J.B. and Felice, E.A., Eds.,** *ACE Inhibitors,* Alan R. Liss, New York, 1987.
15. **Ferguson, R.K. and Vlasses, P.H., Eds.,** *Angiotensin CEI,* Futura Publishing, Mount Kisco, NY, 1987.
16. **MacGregor, G.A. and Sever, P.S., Eds.,** *Current Advances in ACE Inhibition,* Churchill-Livingston, Edinburgh, 1989.
17. **Reid, J.L. and Zanchetti, A.,** ACE Inhibition in the 1990's, *J. Hypertens.,* 7 (Suppl. 5), S1, 1989.
18. **Wood, J.M.,** The renin-angiotensin system as a therapeutic target, *J. Cardiovasc. Pharmacol.,* 16 (Suppl. 4), S1, 1990.
19. **Kaneko, Y. and Laragh, J.H.,** New developments in ACE inhibitory action, *Am. J. Hypertens.,* 4 (Suppl.), 1S, 1991.
20. **Clozel, J.P. and Owens, G.K.,** The renin angiotensin system and the vascular wall from experimental models to man, *Hypertension,* 18 (Suppl. II), 2, 1991.
21. **Skeggs, L.T., Dorer, F.E., Kahn, J.R., Lintz, K.E., and Levine, M.,** Experimental renal hypertension: the discovery of the renin-AII system, in *Biochemical Regulation of Blood Pressure,* Soffer, R.L., Ed., John Wiley & Sons, New York, 1981, 3.
22. **Kostls, J.B., DeFelice, E.A., and Pianko, L.J.,** The renin-angiotensin system, in *ACE Inhibitors,* Kostis, J.B. and Felice, E.H., Eds., Alan R. Liss, New York, 1987, 1.
23. **Ondetti, M.A. and Cushman, D.W.,** Enzymes of the renin-angiotensin system and their inhibitors, *Annu. Rev. Biochem.,* 31, 283, 1982.
24. **Cushman, D.W. and Ondetti, M.A.,** Inhibitors of angiotensin-converting enzyme, *Prog. Med. Chem.,* 17, 41, 1980.
25. **Skidgel, R.A., Engelbrecht, S., Johnson, A.R., and Erdos, E.G.,** Hydrolysis of substance P and neurotensin by converting enzyme and neutral endopeptides, *Peptides,* 5, 769, 1984.
26. **Skidgel, R.A. and Erdos, E.G.,** Novel activity of human AI converting enzyme: release of the NH_2- and COOH-terminal tripeptides from the luteinizing hormone-releasing hormone, *Proc. Natl. Acad. Sci. U.S.A.,* 82, 1025, 1985.
27. **Strittmatter, S.M., Thiele, E.A., Kapiloff, M.S., and Synder, S.H.,** A rat brain isoenzyme of angiotensin-converting enzyme: unique specificity for amidated peptide substrates, *J. Biol. Chem.,* 260, 9825, 1985.
28. **Skidgel, R.A. and Erdos, E.A.,** The broad substrate specificity of human AI converting enzyme, *Clin. Exp. Hypertens.,* A9 (2 & 3), 243, 1987.
29. **Erdos, E.G.,** The AI converting enzyme, *Fed. Proc.,* 36, 1760, 1977.
30. **Skeggs, L., Marsh, W., Kahn, J., and Shumwag, N.,** The existence of two forms of hypertension, *J. Exp. Med.,* 99, 275, 1954.
31. **Ferreira, S.H.,** A bradykinin-potentiating factor (BPF) present in the venom of Bothrops jararaca, *Br. J. Pharmacol.,* 24, 163, 1965.
32. **Bakhle, Y.S.,** Conversion of AI to AII by cell-free extracts of dog lung, *Nature,* 220, 929, 1968.
33. **Ferreira, S.H., Bartelt, D.C., and Greene, L.J.,** Isolation of bradykinin-potentiating peptides from Bothrops jararaca venom, *Biochemistry,* 9, 2583, 1970.
34. **Bunning, P., Holmquist, B., and Riordan, J.F.,** Functional residues at the active site of ACE, *Biochem. Biophys. Res. Commun.,* 83, 1442, 1978.

35. **Erdos, E.G. and Yang, H.Y.T.,** An enzyme in microsomal fraction of kidney that inactivates bradykinin, *Life Sci.,* 6, 569, 1967.
36. **Erdos, E.G. and Yang, H.Y.T.,** Inactivation and potentiation of the effects of bradykinin, in *Hypotensive Peptides,* Erdos, E.G., Back, N., and Sicuteri, F., Eds., Springer-Verlag, New York, 1966, 235.
37. **Johnston, C.I. and Kohzuki, M.,** ACE: localization and inhibition, in *Current Advances in ACE Inhibition,* MacGregor, G.A. and Sever, P.S., Eds., Churchill-Livingston, Edinburgh, 1989, 3.
38. **Kokubu, T. and Takada, Y.,** Biochemistry of human converting enzyme, *Clin. Exp. Theory Pract.,* A9 (2 & 3), 217, 1987.
39. **Roth, M., Weitzman, A.F., and Piquilloud, Y.,** Converting enzyme content of different tissues of the rat, *Experientia,* 25, 1247, 1969.
40. **Cushman, D.W. and Cheung, H.S.,** Concentration of angiotensin-converting enzyme in tissues of the rat, *Biochim. Biophys. Acta,* 250, 261, 1971.
41. **Miyazaki, M., Okunishi, H., Mishimura, K., and Toda, N.,** Vascular angiotensin-converting enzyme activity in man and other species, *Clin. Sci.,* 66, 39, 1984.
42. **Jackson, B., Cubela, R., and Johnston, C.,** Angiotensin converting enzyme (ACE) characterization by ^{125}I-MK351A binding studies of plasma and tissue ACE during variation of salt status in the rat, *J. Hypertens.,* 4, 759, 1986.
43. **Wilson, S.K., Lynch, D.R., and Snyder, S.H.,** Angiotensin-converting enzyme labelled with ^{3}H-captopril: tissue localization and changes in different models of hypertension in the rat, *J. Clin. Invest.,* 80, 841, 1987.
44. **Okunishi, H., Miyazaki, M., Okamura, T., and Toda, N.,** Different distribution of two types of AII-generating enzymes in the aortic wall, *Biochem. Biophys. Res. Commun.,* 149, 1186, 1987.
45. **Velletri, P. and Bean, B.L.,** The effect of captopril on rat aortic angiotensin-converting enzyme, *J. Cardiovasc. Pharmacol.,* 4, 315, 1982.
46. **Saye, J.A., Singer, H.A., and Peach, M.J.,** Role of endothelium in conversion of AI to AII in rabbit aorta, *Hypertension,* 6, 216, 1984.
47. **Strittmatter, S.T., Lo, M.M.S., Javitch, J.A., and Synder, S.H.,** Autoradiographic visualization of angiotensin-converting enzyme in rat brain with (^{3}H) captopril: localization to a striatonigral pathway, *Proc. Natl. Acad. Sci. U.S.A.,* 81, 1599, 1984.
48. **Correa, F.A., Plunkett, L.M., Saavedra, J.M., and Hichens, M.,** Quantitative autoradiographic determination of angiotensin-converting enzyme (kininase II) binding in individual rat brain nuclei with ^{124}I-351A, a specific enzyme inhibitor, *Brain Res.,* 347, 192, 1985.
49. **Chai, S.Y., Mendelsohn, F.A.O., and Paxinos, G.,** Angiotensin converting enzyme in rat brain visualized by quantitative in vitro autoradiography, *Neuroscience,* 2, 615, 1987.
50. **Yang, H.Y.T. and Neff, N.H.,** Distribution and properties of ACE of rat brain, *J. Neurochem.,* 19, 2243, 1972.
51. **Schelling, P., Ganten, U., Sponer, G., Unger, T., and Ganten, D.,** Components of the renin-angiotensin system in the cerebrospinal fluid of rats and dogs with special consideration of the origin and the fate of angiotensin II, *Neuroendocrinology,* 31, 297, 1980.
52. **Rix, E., Ganten, D., Schull, B., Unger, T., and Taugner, R.,** Converting enzyme in the chorioid plexus, brain and kidney: immunocytochemical and biochemical studies in rats, *Neurosci. Lett.,* 22, 125, 1981.
53. **Paul, M., Printz, M.P., Harms, E., Unger, T., Lang, R.E., and Ganten, D.,** Localization of renin (ED 3.4.23) and converting enzyme (EC 3.4.15.1) in nerve endings of rat brain, *Brain Res.,* 334, 315, 1985.
54. **Unger, T. and Gohlke, P.,** Tissue renin-angiotensin systems in the heart and vasculature: possible involvement in the cardiovascular actions of CEI, *Am. J. Cardiol.,* 65 (19), 31, 1990.

55. **Koshiya, K., Okada, M., Imai, K., Kato, T., Tanka, R., Hatanaka, H., and Kato, T.,** Localization of angiotensin-converting enzyme, prolylendopeptidase and other peptidase in cultured neuronal or glial cells, *Neurochem. Int.,* 7, 125, 1985.
56. **Clemens, D.L., Okamura, T., and Inagami, T.,** Subcellular localization of angiotensin-converting enzyme in cultured neuroblastoma cells, *J. Neurochem.,* 47, 1837, 1986.
57. **Johnston, C.I., Fabris, B., Yamada, H., Mendelsohn, F.A.O. et al.,** Comparative studies of tissue inhibition by ACE inhibitors, *J. Hypertens.,* 7 (Suppl. 5), S11, 1989.
58. **Yamada, H., Fabris, B., Allen, A.M., Jackson, B., Johnston, C.I., and Mendelsohn, F.A.O.,** Localization of ACE in rat heart, *Circ. Res.,* 1989.
59. **Erdos, E.G., Schulz, W.W., Gafford, J.T., and Defendini, R.,** Neutral metalloendopeptidase in human male genital tract. Comparison to angiotensin I-converting enzyme, *Lab. Invest.,* 52, 437, 1985.
60. **Yasui, T., Alhenc-Gelas, F., Corvol, P., and Menard, J.,** Angiotensin I-converting enzyme in amniotic fluid, *J. Lab. Clin. Med.,* 104, 741, 1984.
61. **Ganten, D., Printz, M., Phillips, M.I., and Scholkens, B.A., Eds.,** *The Renin Angiotensin System in the Brain,* Springer-Verlag, Heidelberg, 1982.
62. **Sen, G.C., Thekkamkara, T.J., and Kumar, R.S.,** ACE: structural relationship of the testicular and pulmonary forms, *J. Cardiovasc. Pharmacol.,* 16 (Suppl. 4), 3, 1990.
63. **El-Dorry, H., Bull, H.G., Iwata, K., Thornberry, N.A., Cordes, E.H., and Soffer, R.L.,** Molecular and catalytic properties of rabbit testicular dipeptidyl carboxypeptidase, *J. Biol. Chem.,* 257, 14128, 1982.
64. **El-Dorry, H., Pickett, C.B., MacGregor, J.S., and Soffer, R.L.,** Tissue-specific expression of mRNAs for dipeptidyl carboxypeptidase isozymes, *Proc. Natl. Acad. Sci. U.S.A.,* 79, 4295, 1982.
65. **Soubrier, F., Alhenc-Gelas, F., Hubert, C. et al.,** Two putative active centers in human angiotensin I-converting enzyme revealed by molecular cloning, *Proc. Natl. Acad. Sci. U.S.A.,* 85, 9386, 1988.
66. **Bernstein, K.E., Martin, B.M., Bernstein, E.A., Linton, J., Striker, L., and Striker, G.,** The isolation of angiotensin-converting enzyme cDNA, *J. Biol. Chem.,* 263(23), 11021, 1988.
67. **Roy, S.N., Kusari, J., Soffer, R.L., Lai, C.Y., and Sen, G.C.,** Isolation of cDNA clones of rabbit ACE: identification of two distinct mRNAs for the pulmonary and the testicular isozymes, *Biochem. Biophys. Res. Commun.,* 155, 678, 1988.
68. **Kumar, R.S., Kusari, J., Roy, S.N., Soffer, R.L., and Sen, G.C.,** Structure of testicular ACE: a segmental mosaic isozyme, *J. Biol. Chem.,* 264, 16754, 1989.
69. **Lattion, A.L., Soubrier, F., Allegrini, J., Hubert, C., Corvol, P., and Alhenc-Gelas, F.,** The testicular transcript of the AI-converting enzyme encodes for the ancestral, non-duplicated form of the enzyme, *FEBS Lett.,* 252, 99, 1989.
70. **Ehlers, M.R.W., Fox, E.A., Strydom, D.A., and Riordan, J.F.,** Molecular cloning of human testicular angiotensin-converting enzyme: the testis isozyme is identical to the C-terminal half of endothelial angiotensin-converting enzyme, *Proc. Natl. Acad. Sci. U.S.A.,* 86, 7741, 1989.
71. **Huggins, C.G. and Thampi, N.S.,** A simple method for determination of angiotensin I converting enzyme, *Life Sci.,* 7, 633, 1968.
72. **Lee, H.J., Larue, J.N., and Wilson, I.B.,** Angiotensin-converting enzyme from porcine plasma, *Biochim. Biophys. Acta,* 235, 521, 1971.
73. **Piquilloud, Y., Reinharz, A., and Roth, M.,** Studies on angiotensin converting enzyme with different substrates, *Biochim. Biophys. Acta,* 206, 136, 1970.
74. **Cheung, H.S. and Cushman, D.W.,** Inhibition of homogenous angiotensin-converting enzyme of rabbit lung by synthetic venom peptides of Bothrops Tararaed, *Biochim. Biophys. Acta,* 293, 451, 1973.
75. **Dorer, F.E., Kahn, J.R., Lentz, K.E., and Skeggs, L.T.,** Purification and properties of angiotensin-converting enzyme from hog lung, *Circ. Res.,* 31, 356, 1972.

76. **Oparil, S., Koemer, T., and O'Donoghue, J.K.,** Structural requirements for substrates and inhibitors of angiotensin I converting enzyme in vivo and in vitro, *Circ. Res.*, 34, 19, 1973.
77. **King, S.J., Booyse, F.M., Lin, P.H., Traylor, M., Markates, A.J., and Oparil, S.,** Hypoxia stimulates endothelial cell angiotensin converting enzyme antigen synthesis, *Am. J. Physiol. (256) Cell Physiol.*, 25, C1231, 1989.
78. **Cushman, D.W. and Cheung, H.S.,** Spectrophotometric assay and properties of the angiotensin-converting enzyme of rabbit lung, *Biochem. Pharmacol.*, 20, 1637, 1971.
79. **DelVecchio, P.J. and Smith, J.R.,** Synthesis of angiotensin converting enzyme by endothelial cells in culture in vivo, *J. Cell Physiol.*, 108, 308, 1981.
80. **Krulewitz, A.H. and Fanburg, B.L.,** Stimulation of bovine endothelial cell angiotensin I converting enzyme activity by cyclic AMP-related agents, *J. Cell Physiol.*, 129, 147, 1986.
81. **Jackson, R.M., Narkates, A.J., and Oparil, S.,** Impaired pulmonary conversion of angiotensin I to angiotensin II in rats exposed to chronic hypoxia, *J. Appl. Physiol.*, 60, 1121, 1986.
82. **Keane, P.M., Kay, J.J., Suyama, K.L., Gauthier, D., and Andrew, K.,** Lung angiotensin converting enzyme activity in rats with pulmonary hypertension, *Thorax*, 37, 198, 1982.
83. **Krulewitz, A.H. and Fanburg, B.L.,** The effect of oxygen tension on the in vitro production and release of angiotensin converting enzyme by bovine pulmonary artery endothelial cells, *Am. Rev. Respir. Dis.*, 130, 866, 1984.
84. **Forstuma, T., Tikkanen, I., Gronhagen-Riska, C., and Fyhrquist, F.,** Dissociation of the effect of captopril on blood pressure and angiotensin converting enzyme in serum and lungs of spontaneously hypertensive rats, *Acta Pharmacol. Toxicol.*, 49, 416, 1981.
85. **Fyhrquist, F., Gronhagen-Riska, C., Hortling, L., Forslund, T., Tikkanen, I., and Klockars, M.,** The induction of angiotensin converting enzyme by its inhibitors, *Clin. Exp. Hypertens.*, A5, 1319, 1983.
86. **King, S.J., Oparil, S., and Berecek, K.H.,** Neuronal angiotensin converting enzyme (ACE) gene expression is increased by converting enzyme inhibitors (CEI), *Mol. Cell. Neurosci.*, 2, 13, 1991.
87. **Schunkert, H., Hirsch, A.T., Pinto, Y., Pelletier, P., Jacob, H., Remme, W.J., Ingelfinger, J.R., and Dzau, V.J.,** Feedback regulation of angiotensin converting enzyme mRNA and activity by angiotensin II, *Circulation*, 82, 230, 1990.
88. **Ferreira, S.H.,** A bradykinin-potentiating factor (BPF) present in the venom of Bothrops jararaca, *Br. J. Pharmacol.*, 24, 163, 1965.
89. **Ferreira, S.H., Bartelt, D.C., and Greene, L.J.,** Isolation of bradykinin-potentiating peptides from Bothrops jararaca venom, *Biochemistry*, 9, 2583, 1970.
90. **Ondetti, M.A., Williams, N.J., Sabo, E.F., Pluscec, J., Weaver, E.R., and Kocy, O.,** Angiotensin-converting enzyme inhibitors from the venom of Bothrops jararaca. Isolation, elucidation of structure, and synthesis, *Biochemistry*, 10, 4033, 1971.
91. **Engel, S.L., Schaeffer, T.R., Gold, B.I., and Rubin, B.,** Inhibition of pressor effects of angiotensin I and augmentation of depressor effects of bradykinin by synthetic peptides, *Proc. Soc. Exp. Biol. Med.*, 140, 240, 1972.
92. **Greene, L.J., Camargo, A.C.M., Krieger, E.M., Stewart, J.M., and Ferreira, S.H.,** Inhibition of the conversion of angiotensin I to II and potentiation of bradykinin by small peptides present in Bothrops jararaca venom, *Circ. Res.*, 30, 1162, 1972.
93. **Engel, S.L., Schaeffer, T.R., Waugh, M.H., and Rubin, B.,** Effects of the nonapeptide SQ 20,881 on blood pressure of rats with experimental renovascular hypertension, *Proc. Soc. Exp. Biol. Med.*, 143, 483, 1973.
94. **Muirhead, E.E., Brooks, B., and Arora, K.K.,** Prevention of malignant hypertension by the synthetic peptide SQ 20,881, *Lab. Invest.*, 30, 129, 1974.

95. **Gavras, H., Brunner, H.R., Laragh, J.H., Sealey, J.E., Gavras, I., and Vukovich, R.A.,** An ACE inhibitor to identify and treat vasoconstrictor and volume factors in hypertensive patients, *N. Engl. J. Med.,* 291, 817, 1974.
96. **Samuels, A.I., Miller, E.D., Jr., Fray, J.C.S., Haber, E., and Barger, A.C.,** Renin-angiotensin antagonists and the regulation of blood pressure, *Fed. Proc.,* 35, 2512, 1976.
97. **Case, D.B., Wallace, J.M., Keim, H.J. et al.,** Possible role of renin in hypertension as suggested by renin-sodium profiling and inhibition of converting enzyme, *N. Engl. J. Med.,* 296, 641, 1977.
98. **Miller, E.D., Samuels, A.I., Haber, E., and Barger, A.C.,** Inhibition of angiotensin conversion in experimental renovascular hypertension, *Science,* 177, 1108, 1972.
99. **Ayers, R.C., Vaughan, E.D., Yancey, M.R. et al.,** Effect of 1-sarcosin-8-alanine AII and converting enzyme inhibitor on renin release in dog acute renovascular hypertension, *Circ. Res.,* 34 – 35 (1), 27, 1974.
100. **Gavras, H., Liang, C.S., and Brunner, H.R.,** Redistribution of regional blood flow after inhibition of the angiotensin-converting enzyme, *Circ. Res.,* 43 (Suppl. 1), 1, 1978.
101. **Hollenberg, N.K., Swartz, S.L., Passan, D.R., and Williams, G.H.,** Increased glomerular filtration rate after converting-enzyme inhibition in essential hypertension, *N. Engl. J. Med.,* 301, 9, 1979.
102. **Johnson, J.G., Black, W.D., Vukovich, R.A., Hatch, F.E., Jr., Friedman, B.I., Blackwell, C.F., Shenouda, A.W., Share, L., Shade, R.E., Acchiardo, S.R., and Muirhead, E.E.,** Treatment of patients with severe hypertension by inhibition of angiotensin-converting enzyme, *Clin. Sci. Mol. Med.,* 48, 53S, 1975.
103. **Tifft, C.P., Gavras, H., Kershaw, G.R., Gavras, I., Brunner, H.R., Liang, C.S., and Chobanian, A.V.,** Converting enzyme inhibition in hypertensive emergencies, *Ann. Intern. Med.,* 90, 1, 1979.
104. **Gavras, H., Faxon, D.P., Berkoben, J. et al.,** Angiotensin-converting enzyme inhibition in patients with congestive heart failure, *Circulation,* 58, 770, 1978.
105. **Curtiss, C., Cohn, J.N., Vrobel, T., and Franciosa, J.A.,** Role of the renin-angiotensin system in the systemic vasoconstriction of chronic congestive heart failure, *Circulation,* 58, 763, 1978.
106. **Gavras, H., Andrews, P., and Papadakis, N.,** Reversal of experimental delayed cerebral vasospasm by angiotensin-converting enzyme inhibition, *J. Neurosurg.,* 55, 6, 1981.
107. **Ondetti, M.A., Rubin, B., and Cushman, D.W.,** Design of specific inhibitors of angiotensin-converting enzyme: new class of orally active antihypertensive agents, *Science,* 196, 441, 1977.
108. **Cushman, D.W., Cheung, H.S., Sabo, E.F., and Ondetti, M.A.,** Design of potent competitive inhibitors of angiotensin-converting enzyme, *Biochemistry,* 16, 5484, 1977.
109. **Patchett, A.A., Harris, E., Tristram, E.W. et al.,** A new class of angiotensin-converting enzyme inhibitors, *Nature,* 228, 280, 1980.
110. **Antonaccio, M. and Cushman, D.,** Drugs inhibiting the renin-angiotensin system, *Fed. Proc.,* 40, 2275, 1981.
111. **Migdalof, B.H. et al.,** Evidence for dynamic interconversion of captopril and its disulfide metabolites in vivo, *Fed. Proc.,* 39, 757, 1980.
112. **Case, D.B., Atlas, S.A., Laragh, J.H., Sealey, J.E., Sullivan, P.A., and McKinstry, D.N.,** Clinical experience with blockade of the renin-angiotensin-aldosterone system with an oral converting enzyme inhibitor (SQ 14,225; captopril) in hypertensive patients, *Prog. Cardiovasc. Dis.,* 21, 195, 1978.
113. **Jenkins, A.C. and McKinstry, D.N.,** Review of clinical studies of hypertensive patients treated with captopril, *Med. J. Aust. (Spec. Suppl.),* 32, 1979.
114. **Gavras, H., Brunner, H.R., Turini, G.A., Kershaw, G.R., Tifft, C.P., Cuttelod, S., Gavras, I., Vukovich, R.A., and McKinstry, D.N.,** Antihypertensive effect of the oral ACE inhibitor SQ 14,225 in man, *N. Engl. J. Med.,* 298, 991, 1978.
115. **Bauer, J.H.,** ACE inhibitors, *Am. J. Hypertens.,* 3, 331, 1990.

116. **Duchin, K.L., Singhvi, S.M., Willard, D.A., Migdalo, F.E.B.H., and McKinstry, D.N.,** Captopril kinetics, *Clin. Pharmacol. Ther.,* 31, 452, 1982.
117. **Todd, P.A. and Hell, R.C.,** Enalapril. A review of its pharmacodynamic and pharmacokinetic properties and therapeutic use in hypertension and congestive heart failure, *Drugs,* 31, 198, 1986.
118. **Frohlich, E.D.,** ACE inhibitor: present and future, *Hypertension,* 13, I125, 1989.
119. **Vlasses, P.H., Larijani, G.E., Conner, D.P. et al.** Enalapril, a nonsulfhydryl angiotensin-converting enzyme inhibitor, *Clin. Pharmacol.,* 4, 27, 1985.
120. **Sweet, C.S.,** Pharmacological properties of the converting enzyme inhibitor, enalapril maleate (MK-421), *Fed. Proc.,* 42, 167, 1983.
121. **Petrillo, E.W. and Ondetti, M.A.,** Angiotensin-converting enzyme inhibitor: medicinal chemistry and biological actions, *Med. Res. Rev.,* 2, 1, 1982.
122. **Ondetti, M.A. and Cushman, D.W.,** Angiotensin-converting enzyme inhibitor: biochemical properties and biological action, *Crit. Rev. Biochem.,* 16, 381, 1984.
123. **Wyvratt, M.J. and Patchett, A.A.,** Recent developments in the design of angiotensin-converting enzyme inhibitor, *Med. Res. Rev.,* 5, 483, 1985.
124. **Kostis, J., Raia, J.J., Jr., DeFelice, E.A., Barone, J.A., and Letter, R.G.,** Comparative clinical pharmacology of ACE inhibitors, in *Angiotensin Converting Enzyme Inhibitors,* Kostis, J.B. and DeFelice, E.A., Eds., Alan R. Liss, New York, 1987, 19.
125. **DeFelice, E.A. and Kostis, J.B.,** New ACE inhibitors, in *Angiotensin Converting Enzyme Inhibitors,* Kostis, J.B. and DeFelice, E.A., Eds., Alan R. Liss, New York, 1987, 213.
126. **Cushman, D.W., Ondetti, M.A., Gordon, E.M., Natarajean, S., Karanewsky, D.S., Krapcho, J., and Petrillo, E.W., Jr.,** Rational design and biochemical utility of specific inhibitors of ACE, *J. Cardiovasc. Pharmacol.,* 10 (Suppl. 7), S17, 1987.
127. **Gravras, H., Brunner, H.R., Laragh, J.H., Sealey, J.E., Garvas, I., and Vukovich, R.A.,** An angiotensin converting enzyme inhibitor to identify and treat vasoconstrictor and volume factors in hypertensive patients, *N. Engl. J. Med.,* 291, 817, 1974.
128. **Jenkins, A.C., Dreslinski, G.R., Tadors, S.S., Groel, J.T., Fand, R., and Herczeg, S.A.,** Captopril in hypertension; seven years later, *J. Cardiovasc. Pharmacol.,* 7 (Suppl. 1), S96, 1985.
129. **Messerli, F.H., Orens, S., and Grossman, E.,** Left ventricular hypertrophy and antihypertensive therapy, *Drugs,* 35 (Suppl. 5), 27, 1988.
130. **Sullivan, P.A., Kellerher, M., Twomey, M., and Dineen, M.,** Effects of converting enzyme inhibitor on blood pressure, plasma-renin activity and plasma-aldosterone in hypertensive diabetics compared to patients with essential hypertension, *J. Hypertens.,* 3, 359, 1985.
131. **Zannad, F.,** The emerging role of ACE inhibitors in the treatment of cardiovascular disease, *J. Cardiovasc. Pharmacol.,* 15 (Suppl. 2), S1, 1990.
132. **Swartz, S.L. and Williams, G.H.,** Angiotensin-converting enzyme inhibition and prostaglandins, *Am. J. Cardiol.,* 29 (6), 1405, 1982.
133. **Cushman, D.W., Wang, F.L., Fung, W.C. et al.,** Comparison in vitro, ex vitro and in vivo of the actions of seven structurally diverse inhibitors of angiotensin converting enzyme (ACE), *Br. J. Clin. Pharmacol.,* 28, 115S, 1989.
134. **Cohen, M.L., Kurz, K.D., and Schenck, K.W.,** Tissue angiotensin-converting enzyme inhibition as an index of the disposition of enalapril (MK-421) and metabolite MK-422, *J. Pharmacol. Exp. Ther.,* 226 (1), 192, 1983.
135. **Unger, T.H., Ganten, D., and Lang, R.E.,** Pharmacology of CEI: new aspect, *Clin. Exp. Hypertens. A,* 5 (7/8), 1333, 1983.
136. **Waeber, B., Nussberger, J., and Brunner, H.R.,** Angiotensin-converting inhibitors in hypertension, in *Hypertension: Pathophysiology, Diagnosis, and Management,* Larage, J.H. and Brenner, B.M., Eds., Raven Press, New York, 1990, chap. 140.

137. **De Silva, P.E., Husain, A., Smeby, R.R., and Khairallah, P.A.,** Measurement of immunoreactive angiotensin peptides in rat tissues: some pitfalls in angiotensin II analysis, *Anal. Biochem.,* 174, 80, 1988.
138. **Nussberger, J., Brunner, D.B., Waeber, B., and Brunner, H.R.,** True versus immunoreactive angiotensin II in human plasma, *Hypertension,* 7 (Suppl. I), 1, 1985.
139. **Nussberger, J., Brunner, D.B., Waeber, B., and Brunner, H.R.,** Specific measurement of angiotensin metabolites and in vitro generated angiotensin II in plasma, *Hypertension,* 8, 476, 1986.
140. **Unger, T., Gohlke, P., Paul, M., and Rettig, R.,** Tissue renin-angiotensin systems: fact or fiction?, *J. Cardiovasc. Pharmacol.,* 18 (Suppl. 2), S20, 1991.
141. **Gavras, H., Brunner, H.R., Turini, G.A., Kershaw, G.R., Fiffi, C.P., Cuttelod, S., Vukovich, R.A., and McKinstry, D.N.,** Antihypertensive effect of oral angiotensin converting enzyme inhibitor SQ 14225 in man, *N. Engl. J. Med.,* 298, 991, 1978.
142. **Case, D.B., Atlas, S.A., Laragh, J.H., Sealey, J.E., Sullivan, P.A., and McKinstry, D.N.,** Clinical experience with blockade of the renin-angiotensin-aldosterone system by an oral converting-enzyme inhibitor (SQ 14,225 or captopril) in hypertensive patients, *Prog. Cardiovasc. Dis.,* 21, 195, 1978.
143. **Vukovich, R.A., McKinstry, D.N., and Gavras, I.,** Oral angiotensin-converting enzyme inhibitor in long term treatment of hypertensive patients, *Ann. Intern. Med.,* 90, 19, 1979.
144. **Case, D.B., Atlas, S.A., Laragh, J.H., Sullivan, P.A., and Sealey, J.E.,** Use of first dose response or plasma renin activity to predict the long-term effect of captopril: identification of triphasic pattern of blood pressure response, *J. Cardiovasc. Pharmacol.,* 2, 239, 1980.
145. **Waeber, B., Gavras, I., Brunner, H.R., Cook, C.A., Charocopos, F., and Gavras, H.,** Prediction of sustained antihypertensive efficacy of chronic captopril therapy: relationships to immediate blood pressure response and control plasma renin activity, *Am. Heart J.,* 103, 384, 1982.
146. **Unger, T. and Gohlke, P.,** Tissue renin-angiotensin systems in the heart and vasculature: possible involvement in the cardiovascular actions of converting enzyme inhibitors, *Am. J. Cardiol.,* 65, 31, 1990.
147. **Fyhrquist, F., Forslund, T., Tikkanen, I., and Gronhagen-Riska, C.,** Induction of angiotensin I converting enzyme in rat lung with captopril SQ 14285, *Eur. J. Pharmacol.,* 67, 473, 1980.
148. **Johnston, C.I., Mendelsohn, F.A.O., Cubela, R.B., Jackson, B., Kihzuki, M., and Fabris, B.,** Inhibition of angiotensin converting enzyme (ACE) in plasma and tissues: studies ex vivo after administration of ACE inhibitors, *J. Hypertens.,* 6 (Suppl. 3), S17, 1988.
149. **Mendelsohn, F.A.O., Chai, S.Y., and Dunbar, M.,** In vitro autoradiographic localization of angiotensin-converting enzyme in rat brain using ^{125}I-labelled MK 351A, *J. Hypertens.,* 2 (Suppl. 3), 41, 1984.
150. **Chai, S.Y., Mendelsohn, F.A.O., and Paxinos, G.,** Angiotensin converting enzyme in rat brain visualized by quantitative in vitro autoradiography, *Neuroscience,* 20, 615, 1987.
151. **Fyhrquist, F., Tikkanen, I., Gronhagen-Riska, C., Hurtling, L., and Hichens, M.,** Inhibitor binding assay for angiotensin converting enzyme, *Clin. Chem.,* 30, 696, 1984.
152. **Jackson, B., Cubela, R., and Johnston, C.I.,** Angiotensin converting enzyme measurement in human serum using radioinhibitor ligand binding, *Aust. J. Exp. Biol. Med. Sci.,* 64, 149, 1986.
153. **Jackson, B., Cubela, R., and Johnston, C.I.,** Angiotensin converting enzyme (ACE) characterization by ^{125}I-MK351A binding studies of plasma and tissue ACE during variation of salt status in the rat, *J. Hypertens.,* 4, 759, 1986.

154. **Johnston, C.I., Cubela, R., and Jackson, B.,** Inhibitory potency and plasma drug levels of angiotensin converting enzyme inhibitors in the rat, *Clin. Exp. Pharmacol. Physiol.,* 15, 123, 1988.
155. **Jackson, B., Cubela, R., and Johnston, C.I.,** Tissue angiotensin converting enzyme (ACE) during changes in the renin-angiotensin system, *J. Cardiovasc. Pharmacol.,* 10 (Suppl. 70), S137, 1987.
156. **Campbell, D.J.,** Circulating and tissue angiotensin systems, *J. Clin. Invest.,* 79, 1, 1987.
157. **Dzau, V.J.,** Vascular renin-angiotensin: a possible autocrine or paracrine system in control of vascular function, *J. Cardiovasc. Pharmacol.,* 6 (Suppl. 3), S377, 1984.
158. **Mizuno, K., Nakamura, M., Higashimori, K., and Higashimori, K.,** Local generation and release of angiotensin II in peripheral vascular tissue, *Hypertension,* 11, 223, 1988.
159. **Mizuno, K., Tani, M., Niimura, S. et al.,** Direct evidence for local generation and release of angiotensin II in human vascular tissue, *Biochem. Biophys. Res. Commun.,* 165, 457, 1989.
160. **Kifor, I. and Dzau, V.J.,** Endothelial renin-angiotensin pathway: evidence for intracellular synthesis and secretion of angiotensin, *Circ. Res.,* 60, 422, 1987.
161. **Unger, T., Badoer, E., Ganten, D., Lang, R.E., and Rettig, R.,** Brain angiotensin: pathways and pharmacology, *Circulation,* 77 (Suppl. I), 40, 1988.
162. **Lindpainter, K., Jin, M., Wilhalm, M.J., Suzuki, F., Linz, W., Scholkens, B.A., and Ganten, D.,** Intracardiac generation of angiotensin and its physiologic role, *Circulation,* 77 (Suppl. I), 118, 1988.
163. **Inagami, T., Ilamura, T., Clemens, D., Celio, M.R., Naruse, K., and Naruse, M.,** Local generation of angiotensin in the kidney and in tissue culture, *Clin. Exp. Hypertens. A,* 5, 1137, 1983.
164. **Campbell, D.J. and Habener, J.F.,** Cellular localization of angiotensinogen gene expression in brown adipose tissue and mesentery: quantification of messenger ribonucleic acid abundance using hybridization in situ, *Endocrinology,* 121, 16, 1987.
165. **Nakamura, M., Inagami, T., Ogihara, T., and Kumahara, Y.,** Effect of captopril on angiotensin II release from vascular tissues in rats, *Clin. Exp. Hypertens.,* 9 (2-3), 477, 1987.
166. **Mizuno, K., Tani, M., Niimura, S., Hashimoto, S., Satoh, A., Shimamoto, K., Inagami, T., and Fukuchi, S.,** Direct evidence for local generation and release of AII in human vascular tissue, *Biochem. Biophys. Res. Commun.,* 165(1), 457, 1989.
167. **Higashimori, K., Nakamura, M., Tabuchi, Y., Nagano, M., Mikami, H., Ogihara, T., and Inagami, T.,** Angiotensin converting enzyme inhibitors suppress the vascular renin angiotensin system of spontaneously hypertensive rats, *Am. J. Hypertens.,* 4, 56S, 1991.
168. **Miyazaki, M., Ikunishi, H., Mishimura, K., and Toda, N.,** Vascular angiotensin converting enzyme activity in man and other species, *Clin. Sci.,* 66, 39, 1984.
169. **Wilson, S.K., Lynch, D.R., and Snyder, S.H.,** Angiotensin converting enzyme labelled with ^{3}H-captopril. Tissue localization and changes in different models of hypertension in the rat, *J. Clin. Invest.,* 30, 841, 1987.
170. **Oliver, J.A. and Sciacca, R.R.,** Local generation of angiotensin II as a mechanism of regulation of peripheral vascular tone in the rat, *J. Clin. Invest.,* 74, 1247, 1984.
171. **Malik, K.U. and Nasjletti, A.,** Facilitation of adrenergic transmission by locally generated AII in rat mesenteric arteries, *Circ. Res.,* 38, 26, 1976.
172. **Kuriyama, S., Nakamura, A., Hopp, L. et al.,** Angiotensin II effect on $^{22}Na^+$ transport in vascular smooth muscle cells, *J. Cardiovasc. Pharmacol.,* 11, 136, 1988.
173. **Toda, N.,** Endothelium-dependent relaxation induced by angiotensin II and histamine in isolated arteries of dog, *Br. J. Pharmacol.,* 81, 301, 1984.
174. **Heagerty, A.M.,** AII: vasconstrictor or growth factor?, *J. Cardiovasc. Pharmacol.,* 18 (Suppl. 2), S14, 1991.

175. **Mulvagh, S.L., Roberts, R., and Schneider, M.D.,** Cellular oncogenes in cardiovascular disease, *J. Mol. Cell. Cardiol.*, 20, 657, 1988.

176. **Rauscher, F.J., III, Sambucetti, L.C., Curran, T., Distel, R.J., and Spiegelman, B.M.,** A common DNA binding site for fos complexes and transcription factor AP-1, *Cell,* 52, 471, 1988.

177. **Marban, E. and Koretsune, Y.,** Cell calcium oncogenes and hypertrophy, *Hypertension,* 15, 652, 1990.

178. **Allen, I.S., Cohen, N.M., Dhalian, R.S., Gaa, S.T., Lederer, W.J., and Rogers, T.B.,** AII increases contractile frequency and stimulates calcium current in cultured neonatal rat heart myocytes: insights into the underlying biochemical mechanisms, *Circ. Res.*, 62, 524, 1988.

179. **Griendling, K.K., Berk, B.C., Ganz, P., Gimrone, M.A., and Alexander, R.W.,** Angiotensin II stimulation of vascular smooth muscle phosphoinositide metabolism. State of the art lecture. *Hypertension,* 9 (Suppl. III), 181, 1987.

180. **Geisterfer, A.A.T., Peach, M.J., and Owens, G.K.,** Angiotensin II induces hypertrophy, not hyperplasia of cultured rat aortic smooth muscle cells, *Circ. Res.*, 62, 749, 1988.

181. **Berk, B.C., Vekshtein, V., Gordon, H.M., and Tsuda, T.,** Angiotensin II-stimulated protein synthesis in cultured vascular smooth muscle cells, *Hypertension,* 13, 305, 1989.

182. **Naftilan, A.J., Pratt, R.E., Eldridge, C.S., Lin, H.L., and Dzau, V.J.,** Angiotensin II induces c-fos expression in smooth muscle via transcriptional control, *Hypertension,* 13, 706, 1989.

183. **Naftilan, A.J., Gilliland, G.K., Eldridge, C.S., and Kraft, A.S.,** Induction of the protooncogene c-jun by angiotensin II, *Mol. Cell. Biol.*, 10 (10), 5536, 1990.

184. **Rainer, R.S., Eldridge, C.S., Gilliland, G.K., and Naftilan, A.J.,** Antisense oligonucleotide to c-fos blocks the angiotensin II-induced stimulation of protein synthesis in rat aortic smooth muscle cells, *Hypertension,* 16 (Abstr.), 326, 1990.

185. **Itoh, H., Pratt, R.E., and Dzau, V.J.,** Antisense oligonucleotides complementary to PDGF mRNA attenuate angiotensin II-induced vascular hypertrophy, *Hypertension,* 16 (Abstr.), 325, 1990.

186. **Lyall, F., Morton, J.J., Lever, A.F., and Cragoe, E.J.,** Angiotensin II activated Na^+-H^+ exchange and stimulates growth in cultured vascular smooth muscle cells, *J. Hypertens.*, 6 (Suppl. 4), S438, 1988.

187. **Campbell-Boswell, M., and Robertson, L.A., Jr.,** Effects of angiotensin II and vasopressin on human smooth muscle cells in vitro, *Exp. Mol. Pathol.*, 35, 265, 1981.

188. **Chamley-Campbell, J.H., Campbell, G.R., and Ross, R.,** Phenotype dependent response of cultured aortic smooth muscle to serum mitogens, *J. Cell Biol.*, 89, 379, 1981.

189. **DeMey, J.G.R., Uitendaal, M.P., Boonen, H.C.M., Vrijdag, M.J.J.F., Daemen, M.J.A.P., and Struyker-Boudier, H.A.J.,** Acute and long-term effects of tissue culture on contractile reactivity in renal arteries of the rat, *Circ. Res.*, 65, 1125, 1989.

190. **Komuro, I., Kurabayashi, M., Takaku, F., and Yazaki, Y.,** Expression of cellular oncogenes in the myocardium during the developmental stage and pressure-overloaded hypertrophy of the rat heart, *Circ. Res.*, 62, 1075, 1988.

191. **Offerenshaw, J.D., Heagerty, A.M., West, K.P., and Swales, J.D.,** The effects of coarctation hypertension upon vascular inositol phospholipid hydrolysis in Wistar rats, *J. Hypertens.*, 6, 733, 1988.

192. **Ahmed, S.S., Levinson, G.E., Weisse, A.B., and Regan, T.J.,** The effect of angiotensin on myocardial contractility, *J. Clin. Pharmacol.*, 15, 276, 1975.

193. **Assad, M.M. and Antonaccio, M.J.,** Vascular wall renin in spontaneously hypertensive rats, *Hypertension,* 4, 487, 1982.

194. **Unger, T., Hubner, D., Schull, B., Lang, R.E., Rascher, W., Retting, R., and Ganten, D.,** Effect of chronic oral captopril treatment on tissue renin concentration and converting enzyme activity in stroke-prone spontaneously hypertensive rats, in *Hypertensive Mechanisms,* Rascher, W., Clough, D., and Ganten, D., Eds., Schattauer Verlag, New York, 1982, 768.

195. **Cohen, M.L. and Kutz, K.D.,** Angiotensin converting enzyme inhibition in tissues from spontaneously hypertensive rats after treatment with captopril or MK-421, *J. Pharmacol. Exp. Ther.,* 220, 63, 1982.
196. **Cohen, M.L., Wiley, K.S., and Kurz, K.D.,** Effect of acute oral administration of captopril and MK-421 on vascular angiotensin converting enzyme activity in the spontaneously hypertensive rat, *Life Sci.,* 32, 565, 1983.
197. **Unger, T., Gohlke, P., Ganten, D., and Lang, R.E.,** Converting enzyme inhibitors and their effects on the renin-angiotensin system of the blood vessel wall, *J. Cardiovasc. Pharmacol.,* 13 (Suppl. 3), S8, 1989.
198. **Berecek, K.H., Okuno, T., Nagahama, T., and Oparil, S.,** Altered vascular reactivity and baroreflex sensitivity induced by chronic central administration of captopril in the spontaneously hypertensive rat, *Hypertension,* 6, 689, 1983.
199. **Berecek, K.H. and Shier, D.N.,** Alterations in renal vascular reactivity induced by chronic central administration of captopril in the spontaneously hypertensive rat, *Clin. Exp. Hypertens.,* A8 (7), 1081, 1986.
200. **Berecek, K.H., Kirk, K.A., Nagahama, S., and Oparil, S.,** Alterations in sympathetic function in spontaneously hypertensive rats after chronic administration of captopril, *Am. J. Physiol.,* 252, H796, 1987.
201. **Freslon, J.L. and Giudicelli, J.F.,** Compared myocardial and vascular effects of captopril and dihydralazine during hypertension development in spontaneously hypertensive rats, *Br. J. Pharmacol.,* 80 (3), 533, 1983.
202. **Christensen, K.L., Jespersen, L.T., and Mulvany, M.J.,** Development of blood pressure in spontaneously hypertensive rats after withdrawal of long-term treatment related to vascular structure, *J. Hypertens.,* 7 (2), 83, 1989.
203. **Lee, R.M., Berecek, K.H., Tsoporis, J., McKenzie, R., and Triggle, C.R.,** Prevention of hypertension and vascular changes by captopril treatment, *Hypertension,* 17 (2), 141, 1991.
204. **Ertl, G., Kloner, R.A., Alexander, R.W., and Braunwald, E.,** Limitation of experimental infarct size by an angiotensin-converting enzyme inhibitor, *Circulation,* 65, 40, 1982.
205. **Linz, W., Scholkens, B.A., and Han, Y.F.,** Beneficial effects of the converting enzyme inhibitor, ramipril, in ischemic rat hearts, *J. Cardiovasc. Pharmacol.,* 8 (Suppl. 10), S91, 1986.
206. **Scholkens, B.A., Linz, W., and Martorana, P.A.,** Experimental cardiovascular benefits of angiotensin converting enzyme inhibitors: beyond blood pressure reduction, *J. Cardiovasc. Pharmacol.,* 18 (Suppl. 2), S26, 1991.
207. **Martorana, P.A., Kettenbach, B., Breipohl, G., Linz, W., and Scholkens, B.A.,** Reduction of infarct size by local angiotensin-converting enzyme inhibition is abolished by a bradykinin antagonist, *Eur. J. Pharmacol.,* 182, 395, 1990.
208. **Scholkens, B.A. and Linz, W.,** Cardioprotective effects of angiotensin converting enzyme inhibitors: experimental proof and clinical perspectives, *Clin. Physiol. Biochem.,* 8 (Suppl. 1), 33, 1990.
209. **Ganten, D., Hermann, K., Bayer, C., Unger, T., and Lang, R.E.,** Angiotensin synthesis in the brain and increased turnover in hypertensive rats, *Science,* 221, 869, 1983.
210. **Ganten, D., Fuxe, K., Phillips, M.I., Mann, J.F.E., and Ganten, U.,** The brain isorenin-angiotensin system: biochemistry, localization and possible role in drinking and blood pressure regulation, in *Frontiers in Neuroendocrinology,* Vol. 5, Raven Press, New York, 1978, 61.
211. **Phillips, M.I. and Kimura, B.,** Converting enzyme inhibitors and brain angiotensin, *J. Cardiovasc. Pharmacol.,* 8 (Suppl. 10), S82, 1986.
212. **Lind, R.W., Swanson, L.W., and Ganten, D.,** Organization of angiotensin II immunoreactive cells and fibers in the rat nervous system, *Neuroendocrinology,* 40, 2, 1985.

213. **Campbell, D.J., Bouhnik, J., Menard, J., and Corvol, P.,** Identity of angiotensinogen precursors of rat brain and liver, *Nature,* 308, 208, 1984.
214. **Mendelsohn, F.A.O., Quirion, R., Saavedra, J.M., Aguilera, G., and Catt, K.J.,** Autoradiographic localization of angiotensin II receptors in rat brain, *Proc. Natl. Acad. Sci. U.S.A.,* 81, 1575, 1984.
215. **Fishman, M., Zimmerman, E., and Slater, E.,** Renin and angiotensin: the complete system within the neuroblastoma × glial cell, *Science,* 214, 921, 1981.
216. **Hermann, K., Raizada, M.K., Sumners, C., and Phillips, M.I.,** Immunocytochemical and biochemical characterization of angiotensin I and II in cultured neuronal and glial cells from rat brain, *Neuroendocrinology,* 47, 125, 1988.
217. **Hermann, K., Phillips, M.I., Hilgenfeldt, U., and Raizada, M.K.,** Biosynthesis of angiotensinogen and angiotensins by brain cells in primary culture, *J. Neurochem.,* 51, 398, 1988.
218. **Peach, M.J.,** Renin-angiotensin system: biochemistry and mechanisms of action, *Physiol. Rev.,* 57, 313, 1977.
219. **Vallotton, M.B.,** The renin-angiotensin system, *Trends Pharmacol. Sci.,* 8, 69, 1987.
220. **Epstein, A.N., Fitzsimons, J.T., and Rolls, B.J.,** Drinking induced by injection of angiotensin into the brain of the rat, *J. Physiol. (London),* 210, 457, 1970.
221. **Bradford, H.F.,** *Chemical Neurobiology,* W.H. Freeman, New York, 1986, 265.
222. **Felix, D. and Schelling, P.,** Angiotensin converting enzyme blockade by captopril changes antiotensin II receptors and angiotensinogen concentrations in the brain of SHR-SP and WKY rats, *Neurosci. Lett.,* 34, 45, 1982.
223. **Sladek, C.D. and Armstrong, W.E.,** Effect of neurotransmitters and neuropeptides on vasopressin release, in *Vasopressin: Principles and Properties,* Gash, D.M. and Boer, G.J., Eds., Plenum Press, New York, 1987, 275.
224. **Schomig, A., Dietz, R., Rascher, W., Luth, J.B. et al.,** Sympathetic vascular tone in spontaneously hypertensive rats, *Klin. Wochenschr.,* 56 (Suppl. I), 131, 1978.
225. **Crofton, J.T., Share, L., Shade, R.E., Allen, C., and Tarnowski, D.,** Vasopressin in the rat with spontaneous hypertension, *Am. J. Physiol.,* 235 (4), H361, 1978.
226. **Dietz, R., Schomig, A., Haebara, H., Mann, J.F. et al.,** Studies on the pathogenesis of spontaneous hypertension of rats, *Circ. Res.,* 43 (Suppl. I), 98, 1978.
227. **Plunkett, L.M. and Saavedra, J.M.,** Increased angiotensin II binding affinity in the nucleus tractus solitarius of spontaneously hypertensive rats, *Proc. Natl. Acad. Sci. U.S.A.,* 82, 7721, 1985.
228. **Saavedra, J.M., Correa, F.M.A., Kurihara, M., and Shigematsu, K.,** Increased number of angiotensin II receptors in the subfornical organ of spontaneously hypertensive rats, *J. Hypertens.,* 4 (Suppl. 5), S27, 1986.
229. **Castro, R. and Phillips, M.I.,** Neuropeptide action in nucleus tractus solitarius: angiotensin specificity and hypertensive rats, *Am. J. Physiol.,* 249, R341, 1985.
230. **Schelling, P. and Felix, D.,** Influence of captopril treatment on angiotensin II receptors and angiotensinogen in the brain of spontaneously hypertensive rats, *Hypertension,* 5, 935, 1983.
231. **Hutchinson, J.S., Mendelsohn, F.A.O., and Doyle, A.E.,** Blood pressure responses of conscious normotensive and spontaneously hypertensive rats to intracerebroventricular and peripheral administration of captopril, *Hypertension,* 2, 546, 1980.
232. **Okuno, T., Nagahama, S., Lindheimer, M.D., and Oparil, S.,** Attenuation of the development of spontaneous hypertension in rats of chronic central administration of captopril, *Hypertension,* 5, 653, 1983.
233. **Unger, T., Kaufmann-Buhler, I., Scholkens, B., and Ganten, D.,** Brain converting enzyme inhibition: a possible mechanism for the antihypertensive action of captopril in spontaneously hypertensive rats, *Eur. J. Pharmacol.,* 70, 467, 1981.
234. **Suzuki, H., Kondo, K., Hand, M., and Saruta, T.,** Role of the brain isorenin-angiotensin system in experimental hypertension in rats, *Clin. Sci.,* 61, 175, 1981.

235. **Pochiero, M., Nicoletta, P., Losi, E., Bianchi, A., and Caputi, A.P.,** Cardiovascular responses of conscious DOCA-salt hypertensive rats to acute intracerebroventricular and intravenous administration of captopril, *Pharmacol. Res. Commun.*, 15, 173, 1983.
236. **Heald, A.F. and Ita, C.E.,** Distribution in rats of an inhibitor of angiotensin-converting enzyme, SQ 14225, as studied by whole body autoradiography and liquid scintillation counting, *Pharmacologist,* 19, 129, 1977.
237. **Vollmer, R.R. and Boccagno, J.A.,** Central cardiovascular effects of SQ 14225, an angiotensin-converting enzyme inhibitor in chloralose-anesthetized cats, *Eur. J. Pharmacol.*, 45, 117, 1977.
238. **Cohen, M.L. and Kurz, K.,** Angiotensin converting enzyme inhibition in tissue from spontaneously hypertensive rats after treatment with captopril or MK 421, *J. Pharmacol. Exp. Ther.*, 220, 63, 1982.
239. **Unger, T., Ganten, D., Lang, R.E., and Scholkens, B.A.,** Is tissue converting enzyme inhibition a determinant of the antihypertensive efficacy of converting enzyme inhibitors? Studies with two different compounds Hoe 498 and MK 421 in spontaneously hypertensive rats, *J. Cardiovasc. Pharmacol.*, 6, 872, 1984.
240. **Unger, T., Schull, B., Rascher, W., Lang, R.E., and Ganten, D.,** Selective activation of the converting enzyme inhibitor MK 421 and comparison of its active diacid form with captopril in different tissues of the rat, *Biochem. Pharmacol.*, 19, 3063, 1982.
241. **Cohen, M.L. and Kurz, K.,** Captopril and MK 421; stability on storage, distribution to the central nervous system, and onset of activity, *Fed. Proc.*, 42, 171, 1983.
242. **Geppetti, P., Spillantini, M.G., Frilli, S., Petrini, U., Fanciullaci, M., and Sicuteri, F.,** Acute oral captopril inhibits angiotensin converting enzyme activity in human cerebrospinal fluid, *J. Hypertens.*, 5, 151, 1987.
243. **Nishimura, K., Alhenc-Gelas, F., White, A., and Erdos, E.G.,** Activation of membrane-bound kallikrein and renin in the kidney, *Proc. Natl. Acad. Sci. U.S.A.*, 77, 4975, 1980.
244. **Swartz, S.L., Williams, G.H., Hollenberg, N.K., Levine, L., Dluhy, R.G., and Moore, T.J.,** Captopril-induced changes in prostaglandin production. Relationship to vascular responses in normal man, *J. Clin. Invest.*, 65, 1257, 1980.
245. **Crantz, F.R., Swartz, S.L., Hollenberg, N.K., Moore, T.J., Dluhy, R.G., and Williams, G.H.,** Differences in response to the peptidyldipeptidase hydrolase inhibitors SQ 20,881 and SQ 14,225 in normal-renin essential hypertension, *Hypertension,* 2, 604, 1980.
246. **Ogihara, T., Maruyama, A., Hata, T., Mikami, H., Nakamura, M., Naka, T., Ohde, H., and Kumahara, Y.,** Hormonal responses to long-term converting enzyme inhibition in hypertensive patients, *Clin. Pharmacol. Ther.*, 30, 328, 1981.
247. **Johnston, C.I., Clappison, B.H., Anderson, W.P., and Yasujima, M.,** Effect of angiotensin-converting enzyme inhibition on circulating and local kinin levels, *Am. J. Cardiol.*, 49, 1401, 1982.
248. **Sancho, J., Re, R.N., Burton, J., Barger, C.A., and Haber, E.,** Role of the renin angiotensin system in cardiovascular homeostasis in normal human subjects, *Circulation,* 53, 400, 1976.
249. **Carretero, O.A., Miyazaki, S., and Scicli, A.G.,** Role of kinin in the acute antihypertensive effect of the converting enzyme inhibitor, captopril, *Hypertension,* 2, 18, 1981.
250. **Beneto, A., Gavras, H., Stewart, J.M., Vavrek, R.J., Hatinoglou, S., and Gavras, I.,** Vasodepressor role of endogenous bradykinin assessed by a bradykinin antagonist, *Hypertension,* 8, 971, 1986.
251. **Scholkens, B.A. and Linz, W.,** ACE inhibition in mechanisms of ''cardioprotection'' in acute myocardial ischemia, *Klin. Wochenschr.*, 69 (24), 1, 1991.
252. **Dietz, G.J., Rett, K., Jauch, K.W., Wicklmayr, M., Fink, E., Har Guenther, B., Fritz, H., and Mehnert, H.,** Captopril bei Hyperton mit diabetes mellitus type II, *Herz,* 12 (I), 16, 1987.

253. **Ramwell, P.W., Leovey, E.M., and Sintetos, A.L.,** Regulation of the arachidonic cascade, *Biol. Reprod.*, 16, 70, 1977.

254. **Lonigro, A.J., Itskovitz, H.D., Crowshaw, K., and McGiff, J.C.,** Dependency of renal blood flow on prostaglandin synthesis in the dog, *Circ. Res.*, 32, 712, 1973.

255. **Moore, T.J., Crantz, F.R., Hollenberg, N.K., Koletsky, R.J., Leboff, M.S., Swartz, S.L., Levine, L., Podolsky, S., Dluhy, R.G., and Williams, G.H.,** Contribution of prostaglandins to the antihypertensive action of captopril in essential hypertension, *Hypertension,* 3, 168, 1981.

256. **McGiff, J.C., Terragno, N.A., Malik, K.U., and Lonigro, A.J.,** Release of prostaglandin E-like substance from canine kidney by bradykinin, *Circ. Res.*, 31, 36, 1972.

257. **Needelman, P., Wyche, A., Bronson, S.D., Holmberg, M., and Morrison, A.R.,** Specific regulation of peptide-induced renal prostaglandin biosynthesis, *J. Biol. Chem.*, 254, 9772, 1979.

258. **Swartz, S.L.,** The role of prostaglandins in mediating the effects of angiotensin converting enzyme inhibitors and other antihypertensive drugs, *Cardiovasc. Drugs Ther.*, 1 (1), 39, 1987.

259. **Brunner, H.R., Gavras, H., Waeber, B., Kershaw, G.R., Turini, G.A., Vukovich, R.A., McKinstry, D.N., and Gavras, I.,** Oral angiotensin-converting enzyme inhibitor in long term treatment of hypertensive patients, *Ann. Intern. Med.*, 90, 19, 1979.

260. **Case, D.B., Atlas, S.A., Laragh, J.H., Sullivan, P.A., and Sealey, J.E.,** Use of first dose response or plasma renin activity to predict the long-term effect of captopril: identification of triphasic pattern of blood pressure response, *J. Cardiovasc. Pharmacol.*, 2, 339, 1980.

261. **Atlas, S.A., Case, D.B., Sealey, J.E., Laragh, J.H., and McKinstry, D.N.,** Interruption of the renin-angiotensin system in hypertensive patients by captopril induces sustained reduction in aldosterone secretion, potassium retention and natriuresis, *Hypertension,* 1, 274, 1979.

262. **Sanchez, R.A., Marco, E., Gilbert, H.B., Raffaele, G.P., Brito, M., Gimenez, M., and Moledo, L.I.,** Natriuretic effect and changes in renal hemodynamics induced by enalapril in essential hypertension, *Drugs,* 30 (I), 49, 1985.

263. **De Zeeuw, D., Navis, G.J., Donker, A.J.M., and De Jong, P.E.,** The angiotensin converting enzyme inhibitor enalapril and its effects on renal function, *J. Hypertens.*, 1(I), 93, 1983.

264. **Biollaz, J., Brunner, H.R., Gavras, I., Waeber, B., and Gavras, H.,** Antihypertensive therapy with MK 421: angiotensin II-renin relationships to evaluate efficacy of converting enzyme blockade, *J. Cardiovasc. Pharmacol.*, 4, 966, 1982.

265. **Brunner, H.R., Waeber, B., Nussberger, J., Schaller, M.D., and Gomez, J.H.,** Long-term clinical experience with enalapril in essential hypertension, *J. Hypertens.*, 1 (1), 103, 1983.

266. **Thibonnier, M., Aldigier, J.C., Soto, M.E., Sassano, P., Menard, J., and Corvol, P.,** Abnormalities and drug-induced alterations of vasopressin in human hypertension, *Clin. Sci.*, 61, 149, 1981.

267. **Santucci, A., Luparini, R.L., Ferri, C., Ficara, C., Giarrizzo, C., and Balsano, F.,** Relationship between vasopressin and the renin-angiotensin aldosterone system in essential hypertension: effect of converting enzyme inhibitor on plasma vasopressin, *J. Hypertens.*, 3 (2), 133, 1985.

268. **Berecek, K.H., Wyss, J.M., and Swords, B.H.,** Alterations in vasopressin mechanisms in captopril treated spontaneously hypertensive rats, *Clin. Exp. Hypertens. A,* 13(5), 1019, 1991.

269. **Ammann, F.W., Bolli, P., Kiowski, W., and Buhler, F.R.,** Enhanced α-adrenoceptor mediated vasoconstriction in essential hypertension, *Hypertension,* 3 (I), 119, 1981.

270. **Goldstein, D.S.,** Plasma catecholamines and essential hypertension. An analytical review, *Hypertension,* 5, 86, 1983.

271. **Dietz, R., Schomig, A., Haebara, H., Mann, J.F., Rascher, W., Luth, J.B., Grunherz, N., and Gross, F.,** *Circ. Res.,* 43 (I), 98, 1978.
272. **Judy, W.V., Wantanabe, A.M., Henry, D.P., Besch, H.R., Jr., Murphy, W.R., and Hockel, H.,** *Circ. Res.,* 38 (II), 21, 1976.
273. **Bickerton, R.K. and Buckley, J.P.,** Evidence for a central mechanism in angiotensin induced hypertension, *Proc. Soc. Exp. Biol. Med.,* 106, 834, 1961.
274. **Simpson, J.B.,** The circumventricular organs and the central action of angiotensin, *Neuroendocrinology,* 32, 248, 1981.
275. **Khairallah, D.A.,** Action of angiotensin in adrenergic nerve endings: inhibition of norepinephrine uptake, *Fed. Proc.,* 31, 1351, 1972.
276. **Zimmerman, B.G.,** Actions of angiotensin on adrenergic nerve endings, *Fed. Proc.,* 37, 199, 1978.
277. **Zimmerman, B.G.,** Blockade of adrenergic potentiating effects of angiotensin by 1-sar,8-ala angiotensin, *J. Pharmacol. Exp. Ther.,* 183, 486, 1973.
278. **Feldberg, W. and Lewis, G.P.,** The action of peptides on the adrenal medulla. Release of adrenaline by bradykinin and angiotensin, *J. Physiol.,* 171, 98, 1964.
279. **Antonaccio, M.J. and Kerwin, L.,** Pre- and post-junctional inhibition of vascular sympathetic function by captopril in spontaneously hypertensive rats, *Hypertension,* 3 (I), 154, 1981.
280. **Hatton, R. and Clough, D.P.,** Captopril interferes with neurogenic vasoconstriction in the pithed rat by angiotensin-dependent mechanisms, *J. Cardiovasc. Pharmacol.,* 4, 116, 1982.
281. **Imai, Y., Abe, K., Seino, M., Haruyama, T., Tajima, J., Yoshinaga, K., and Sekino, H.,** Captopril attenuates pressor responses to norepinephrine and vasopressin through depletion of endogenous angiotensin II, *Am. J. Cardiol.,* 49, 1537, 1982.
282. **Mancia G., Parati, G., Pomidossi, G. et al.,** Modification of arterial baroreflexes by captopril in essential hypertension, *Am. J. Cardiol.,* 6 (10), 791, 1988.
283. **Cheng, S.W.T., Swords, B.H., Kirk, K.A., and Berecek, K.H.,** Effects of lifetime captopril treatment on baroreceptor function in the spontaneously hypertensive rat, *Hypertension,* 13, 63, 1989.
284. **Cheng, S.W.T., Kirk, K.A., Robertson, J., and Berecek, K.H.,** Brain angiotensin II and baroreflex function in spontaneously hypertensive rats, *Hypertension,* 14, 274, 1989.
285. **Casto, R. and Phillips, I.,** Angiotensin attenuates baroreflexes at nucleus solitarius of rats, *Am. J. Physiol.,* R193, 1986.
286. **Catt, K.J., Harwood, J.P., Aguilera, G., and Dufau, M.L.,** Hormonal regulation of peptide receptors and target cell responses, *Nature,* 280, 109, 1979.
287. **Thomas, W.G. and Sernia, C.,** Regulation of rat brain angiotensin II (AII) receptors by intravenous AII and low dietary Na^+, *Brain Res.,* 345, 54, 1985.
288. **Bradshaw, B. and Moore, T.J.,** Abnormal regulation of adrenal angiotensin II receptors in spontaneously hypertensive rats, *Hypertension,* 11, 49, 1988.
289. **Messenger, E.A., Stonier, C., and Aber, G.M.,** Differences in glomeruli binding and response to angiotensin II between normotensive and spontaneously hypertensive rats, *Clin. Sci.,* 75, 191, 1988.
290. **Matsushima, Y., Kawamura, M., Akabane, S., Imanishi, M., Kuramochi, M., Ito, K. and Omae, T.,** Increase in renal angiotensin II content and tubular AII receptors in prehypertensive spontaneously hypertensive rats, *Hypertension,* 6 (10), 791, 1988.
291. **Grammas, P., Diglio, C., Giacommeli, F., and Weiner, J.,** Cerebrovascular angiotensin II receptors in spontaneously hypertensive rats, *J. Cardiovasc. Pharmacol.,* 13, 227, 1989.
292. **Berecek, K.H., Swords, B.H., Lo, S., and Kirk, K.A.,** Effect of angiotensin converting enzyme inhibitors on brain angiotensin II binding, *J. Hypertens.,* in press.

Chapter 9

ANGIOTENSIN I-CONVERTING ENZYME (ACE) GENE STRUCTURE AND POLYMORPHISM: RELATION TO ENZYME FUNCTION AND GENE EXPRESSION

F. Soubrier, L. Wei, C. Hubert, E. Clauser, F. Alhenc-Gelas, and P. Corvol

TABLE OF CONTENTS

0-8493-4622-3/93/$0.00 + $.50

I. ACE IS A WIDELY DISTRIBUTED ENZYME WITH A DIVERSE SUBSTRATE SPECIFICITY

Angiotensin I-converting enzyme (ACE; Kininase II, dipeptidyl-carboxypeptidase I; EC 3.4.15.1) acts primarily as a dipeptidyl carboxypeptidase and is involved in the metabolism of two major vasoactive peptides, angiotensin II and bradykinin. ACE generates the potent vasopressor angiotensin II (AII) by cleaving the C-terminal dipeptide from angiotensin I (AI) and inactivates the vasodepressor bradykinin (BK), by the sequential removal of two C-terminal dipeptides.[1] The most favorable ACE substrate is BK (Km: 0.2 μM) for which the Km is approximately 80 times lower than for AI (Km: 16 μM).

ACE can hydrolyze a wide range of oligopeptides *in vitro* via its dipeptidyl carboxypeptidase action. These substrates include neurotensin,[2] the enkephalins and C-terminal extended pro-enkephalins.[3] However, the Km values of ACE for these substrates are substantially higher than for BK or AI. ACE also displays some endopeptidase activity in the cleavage of certain substrates which are blocked at the C-terminus. For example, ACE can release the C-terminal tripeptide amide from substance P[2,4] and from luteinizing hormone-releasing hormone (LH-RH).[5] In addition, and perhaps more surprisingly, ACE can cleave the tripeptide from the blocked N-terminus of LH-RH.[5] However, the tripeptide amide is not invariably released from C-terminal blocked substrates: cholecystokinin-8 and various gastrin analogs are hydrolyzed by ACE to release the amidated C-terminal dipeptides as the initial and major products.[6] While these enzymatic activities on natural peptides have been observed *in vitro,* there is no convincing evidence that they bear any physiological relevance, except perhaps for substance P in the brain and the upper respiratory tract.[3,7] To establish a role for ACE in the metabolism of a given peptide, the co-localization of enzyme and substrate must be demonstrated, the presence of other competing enzymes should be considered, and pharmacological experiments using a specific ACE blocker should clearly result in substrate accumulation and/or potentiation of action.

Monovalent anions, especially chloride, enhance the activity of ACE towards all known substrates[8] although the presence of chloride is not always essential; for example, bradykinin is hydrolyzed in the absence of chloride. This activation is thought to be a consequence of a change in protein conformation occurring on chloride binding to a putative lysine residue located in the region of the active site.[9]

ACE is a monomeric zinc metallopeptidase which exists predominantly as a single molecular form of 170 kDa in man. It is an extensively glycosylated protein: the pulmonary enzyme contains 30% carbohydrate which includes fucose, mannose, N-acetylglucosamine, galactose, and sialic acid.[10] The pulmonary enzyme contains a higher proportion of sialic acid than kidney ACE.[11]

ACE is a membrane-bound enzyme and is orientated such that the catalytic sites are exposed at the extracellular surface of the cell. Due to the association of ACE with the plasma membrane of vascular endothelial cells, the enzyme displays an ubiquitous tissue distribution. In addition to an endothelial location, high levels of ACE are also found in the brush borders of absorptive epithelia, such as the apical microvillosities of the small intestine and the kidney proximal convoluted tubule.[12] Other epithelial locations of ACE include the choroid plexus, where the enzyme is found at a high concentration and is probably the source of ACE in the cerebrospinal fluid,[13,14] the prostate, and epididymis of the male genital tract.[15,16] ACE is also found in mononuclear cells, such as monocytes after macrophage differentiation and T-lymphocytes.[17,18] *In vitro* autoradiography, employing radiolabeled specific ACE inhibitors, and immunohistochemical studies have mapped the locations of ACE in brain. ACE was found primarily in the choroid plexus, ependyma, subfornical organ, basal ganglia (caudate-putamen and globus pallidus) but, most notably, the enzyme was found to be highly concentrated in the striatonigral neuronal pathway.[19-21] This suggests a role for ACE in the metabolism of a neuropeptide present in the same pathway. Substance P co-localizes with ACE to the same neuronal pathway; however, neutral endopeptidase, also highly abundant in this area, is more likely to be involved in the *in vivo* degradation of substance P.[22,23] Neuronal ACE is slightly smaller in size compared to the endothelial enzyme and this observation has been attributed to differences in the extent of N-glycosylation.[24] However, both forms of the enzyme display the same neuropeptide specificities and inhibition profiles and, therefore, do not represent distinct isoforms.[25]

A soluble form of ACE also exists and has been detected in biological fluids such as plasma,[26] semen,[27] amniotic fluid,[28] and cerebrospinal fluid.[14] Soluble ACE appears to originate predominantly from endothelial cells on the basis of similar glycosylation contents and the secretion of ACE by endothelial cell cultures.

In the testis, in addition to the endothelial form of ACE, a distinct isoform is present in germinal cells. This germinal form of ACE is smaller, with a molecular weight of 100 kDa in humans, and its expression is specifically regulated. The presence of germinal ACE is dependent on the maturation of germinal cells occurring at the onset of puberty, since ACE is only present in spermatids and not in immature forms of germinal cells.[16,29]

The molecular cloning of the endothelial and germinal ACE cDNAs and of the ACE gene from different species allowed the elucidation of the primary structure of the two ACE isoforms. The determination of the primary structure of the enzyme displayed an unexpected organization of the molecule with two active sites and raised new questions concerning ACE function. Polymorphic genetic markers of the ACE gene were developed and used for investigating the effect of the ACE gene on the control of ACE expression.

II. ACE STRUCTURE RESULTS FROM THE DUPLICATION OF AN ANCESTRAL GENE

Molecular cloning of the endothelial ACE cDNA revealed that the enzyme is composed of two highly homologous domains, thus indicating a gene duplication event in the course of evolution. This hypothesis was further substantiated by the elucidation of the structure of the germinal form of ACE: this form is encoded by a mRNA which corresponds to the putative ancestral form of the ACE gene and contains only one of the two homologous domains. The discovery of two initiation transcription sites and the determination of the ACE gene structure allowed the elucidation of the mechanism of transcription of the two ACE mRNAs.

A. SOMATIC ACE

Affinity-chromatography of ACE, employing the specific ACE inhibitor lisinopril as the affinity ligand together with other appropriate steps for solubilization and selection of the inhibitor-bound form of ACE, enabled the purification of the enzyme to homogeneity.[30,31] Oligonucleotide probes designed on the basis of the N-terminal sequence and of several short internal peptides of the enzyme were used to screen cDNA libraries. Complete ACE cDNAs were isolated from human endothelial and mouse kidney cDNA libraries. The human endothelial ACE mRNA is 4.3 kb long and there appears to be some species differences: two kidney ACE mRNAs were detected in mice (4.9 and 4.15 kb) in contrast to the rabbit where the pulmonary ACE mRNA is 5.0 kb long.[32,33] Human endothelial ACE, which comprises 1306 amino acids, possesses 17 potential N-glycosylation sites and 13 cysteyl residues (Figure 1). Comparison of the amino-terminal end of the human ACE precursor, as deduced from the cDNA sequence, with the amino-terminal sequence of the purified human enzyme enabled us to propose a model for the intracellular processing of ACE and its mode of membrane anchorage.[31] The sequence of a 29 amino acid hydrophobic peptide at the N-terminus was absent in the mature enzyme, and therefore constitutes the cleaved signal peptide. A second sequence of 17 hydrophobic amino-acids is located near the carboxy-terminal end of the molecule. Computerized prediction of the secondary structure revealed that this C-terminal hydrophobic sequence corresponds to a membrane-associated alpha-helix, and is, therefore, the candidate region for anchoring the enzyme to the cell membrane. *In vitro* expression of wild type and mutated recombinant ACE allowed us to define the mechanism of anchorage, as described below.

One surprising feature of the enzyme, disclosed by molecular cloning of the cDNA, is the presence of two large highly homologous domains, referred to here as N and C domains. This suggests that the ancestral gene was duplicated during evolution. This duplication accounts for the higher molecular weight of ACE, when compared to other zinc metallopeptidases, such

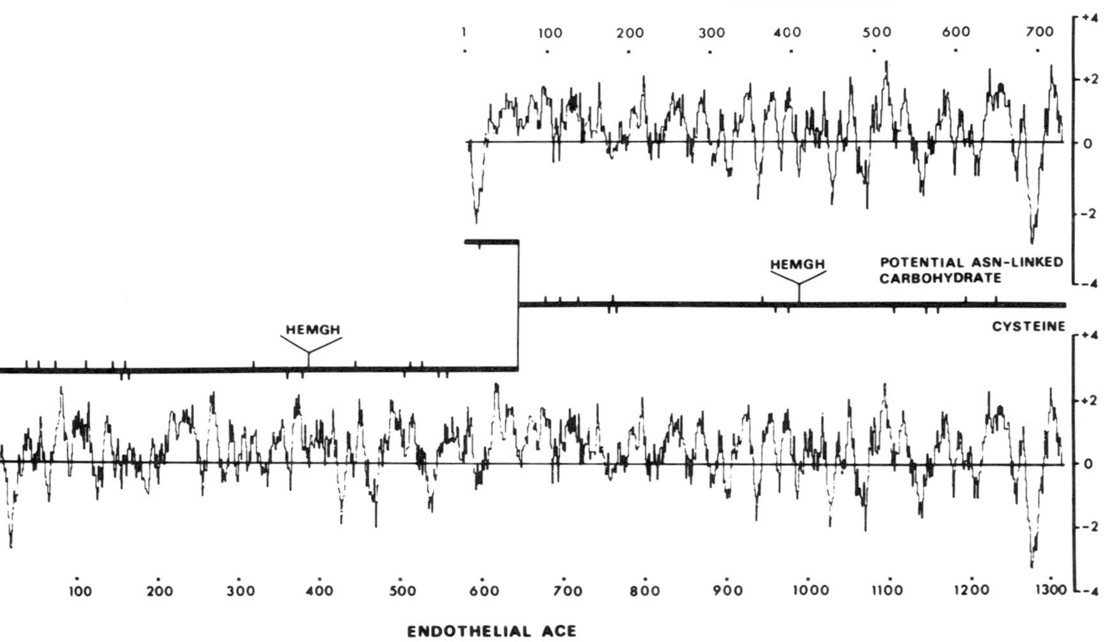

FIGURE 1. Schematic representation of the testicular and endothelial ACE enzymes.[39,41] (Middle) Diagram showing the cysteine positions, the potential asparagine-linked glycosylation sites, and the positions of the putative residues of the active site of the two enzymes. Beyond the point of divergence, the testicular enzyme is figured on the upper line and the endothelial enzyme on the lower line. Hydropathy plots of the predicted testicular (top) and endothelial (bottom) amino acid sequences. Amino acid numbering is presented above and under the hydropathy plots. (From Lattion, A.-L. et al., *FEBS Lett.*, 252, 99, 1989. With permission.)

hACE	T	V	H	H	E	M	G	H	365
hACE	V	A	H	H	E	M	G	H	963
THERM	V	V	A	H	E	L	T	H	146
rNEP	V	I	G	H	E	I	T	H	588

FIGURE 2. Consensus sequence of the active site of zinc-metallopeptidases. hACE: first and second domain of human ACE.[31] THERM: *bacillus thermoproteolyticus.*[36] rNEP: rabbit neutral endopeptidase.[34] Numbers refer to the last amino acid position in each protein segment. Identical residues are boxed.

as neutral endopeptidase.[34] The overall sequence similarity between the two domains of human ACE is more than 60% at the nucleotide level and at the amino acid level; however, in the latter case this homology increases to 89% in the region containing essential residues of the active site.[31] There is no sequence homology at the carboxy- and amino-terminal extremities of the molecule, and the central region which separates the two homologous domains is also unique. Each domain contains a pentapeptide sequence which complies with the consensus sequence His-Glu-Xaa-Xaa-His for the active site of zinc metallopeptidases.[35] The two histidine residues provide two of the three zinc-coordinating ligands and the glutamic acid is the base donor in the catalytic reaction, as suggested from crystallographic studies of another zinc metallopeptidase, thermolysin[36] and from amino acid comparisons with this and other metallopeptidases (Figure 2). The presence of two putative active sites in the ACE molecule was unexpected since previous studies had suggested that ACE contained a single zinc atom per molecule[37] and a single high affinity inhibitor site per molecule of ACE.[38] However, by *in vitro* mutagenesis studies, we have demonstrated that ACE in fact contains two functional catalytic sites (see below).

Other residues have a key role in catalysis by zinc metallopeptidase such as His 231 of thermolysin, which stabilizes the transition state of the substrate, and Glu 166, which provides the third zinc-coordinating ligand.[36] Although we proposed histidine 404 and 1002, and glutamic acid 389 and 987 as candidate residues for these functions, respectively, their roles have not been confirmed. A tyrosine and a lysine residue have also been identified as essential residues in rabbit lung ACE by inactivation of the purified enzyme with 1-fluoro-2,4,-dinitrobenzene (Dnp-F),[39] confirming previous findings.[40] By peptide mapping and sequencing these amino acids have been identified as lysine 694 and tyrosine 776 in the C-domain. Similarly, Dnp-F reacts with lysine 122 and tyrosine 204 of rabbit testicular ACE, which correspond to lysine 118 and tyrosine 200 of human testicular ACE.[41] A mutant human testicular ACE, in which tyrosine 200 was replaced by a phenylalanine, was expressed in CHO cells.[42] The mutant ACE exhibited a 15-fold decrease of the Kcat for the hydrolysis of the synthetic substrate furanacryloyl-phe-gly-gly and a seven-fold decrease for angiotensin I hydrolysis. The Km values of the mutant enzyme for these substrates were slightly increased, but the specific ACE inhibitor, lisinopril, was 100-fold less tightly bound to the mutant enzyme compared to the wild-type enzyme. These results may suggest that the tyrosine 200 residue is involved in stabilizing the transition state. The two Dnp-F reactive residues, which are absent in the corresponding sequence of the N-domain of ACE, could be responsible, at least in part, for the increased catalytic activity of the C-domain, as described below.

A high degree of sequence identity was found between the different mammalian ACE cDNA sequences cloned to date. Human[31] and mouse[32] ACE coding sequences are 83% homologous at both the nucleotide and amino acid levels, and the identity is even higher between human and rabbit[43] ACE sequences. The most conserved sequences are in the regions of the homologous domains. The bacterial dipeptidyl carboxypeptidase shares common synthetic and natural substrates with ACE and is also inhibited by captopril,[44] with an IC_{50} of 40 n*M* as compared with 18 n*M* for the pig endothelial enzyme.[25] The *Salmonella typhimurium* gene, which codes for this enzyme, does not display any sequence similarity with mammalian ACE except for the consensus zinc-binding motif.[45] This is in contrast to the clear homology between the *Escherichia coli* and human genes coding for another zinc metallopeptidase, aminopeptidase N.[46]

B. GERMINAL ACE

Early experiments, using *in vitro* cell-free translation of rabbit testis mRNA followed by immunoprecipitation of ^{35}S-labeled proteins with an ACE antiserum showed that the testicular ACE isoform was translated as a shorter precursor (90 kDa) compared to the lung ACE precursor (140 kDa),[47] thus implying that the smaller MM did not result from a post-translational modification. Immunohistochemical studies showed that testicular ACE is recognized by antibodies raised against the somatic enzyme, demonstrating the presence of common epitopes.[48]

Using the endothelial ACE cDNA as a probe in northern blot experiments, a single 3 kb hybridizing mRNA species was detected in the human testis, in contrast to the 4.3 kb mRNA detected in endothelial cells.[31]

Various strategies were used for cloning the testicular ACE cDNA from humans[49,50] and rabbits.[51] We used the endothelial ACE cDNA probe to isolate clones from a human testis cDNA library,[49] whereas Roy et al.[51] used ACE antibodies raised against the rabbit enzyme to isolate cDNA clones by expression screening of lambda GT11 libraries.

The primary structure of testicular ACE was deduced from the sequences of cDNA clones covering the entire coding sequence.[49,50] As expected from the size of the mRNA, testicular ACE contains only 732 amino acids, as compared to the 1306 amino acids present in endothelial ACE. A schematic representation of endothelial and testicular ACE is shown in Figure 2.

The testicular ACE sequence corresponds to the C domain of endothelial ACE, and therefore contains only one of the two putative catalytic sites identified in endothelial ACE. The absolute sequence identity downstream from residue 1944 of endothelial ACE suggests transcription of both mRNAs from a single gene. The 5′ end of the testicular ACE mRNA, however, contains a 227 base pair sequence, which is absent in endothelial ACE, suggesting differential splicing of a common primary RNA transcript or the use of an alternative promoter. This N-terminal sequence of 67 residues encodes the signal peptide and a serine- and threonine-rich region presumed to be O-glycosylated.[49] By site-directed mutagenesis, it was established that a segment of 36 amino acid residues encoded by the testicular ACE specific 5′ sequence contained more than 90% of all O-linked sugars and was not required for enzyme stability and activity.[52]

The complete sequence similarity includes, at the 3′ end of testicular ACE, the hydrophobic segment which represents the anchoring domain (see below), suggesting a similar mechanism of membrane anchorage for the germinal and somatic enzymes. The rabbit pulmonary and testis ACE mRNAs use different 3′ polyadenylation sites located within the same exon.[33] The pulmonary mRNA polyadenylation site is 628 bp 3′ to the testis polyadenylation site, as demonstrated by hybridization with specific probes. It is not known whether the cell-specific use of different polyadenylation signals is important for gene expression or mRNA stability.

C. STRUCTURE OF THE ACE GENE

By Southern blot experiments and gene cloning, it was demonstrated that the somatic and germinal ACE mRNAs are transcribed from a unique gene[53]. The somatic and germinal ACE mRNAs could be generated by either differential splicing of a common primary RNA transcript or by transcription from alternative initiation sites under the control of separate promoters. Several studies have demonstrated the presence of two functional promoters in the ACE gene. The somatic promoter is located on the 5′ side of the first exon of the gene, and the germinal promoter is located on the 5′ side of the specific 5′ end of germinal ACE mRNA. Primer extension and RNase protection assays on testicular RNAs were performed in mice, rabbits and humans.[53-55] In all species, transcription of the germinal ACE mRNA was initiated inside the gene. Therefore, intron 12, corresponding to the genomic sequence flanking the 5′ region of the testicular-specific exon 13, as deduced from the complete analysis of the ACE gene in humans, was proposed as the putative germinal ACE promoter.[53]

The promoter function of this sequence was firmly established by using intron 12 to drive the transcription of a reporter gene in a germinal specific fashion.[56] A mouse genomic fragment, 689 bp long, containing intron 12, exon 12 and a part of intron 11 of the mouse ACE gene was fused to the beta-galactosidase coding sequence and microinjected into pro-nuclei of one-cell stage embryos. A histochemical analysis of the transgenic mice revealed that beta-galactosidase was only expressed, together with ACE, in elongating spermatozoa. Thus, the 689 bp sequence, driving the transcription of the reporter gene, corresponds to an intragenic promoter with a strong cell specificity. The two alternate promoters of the ACE gene exhibit highly contrasted cell specificities, as the somatic promoter is active in several cell types, whereas the germinal promoter is only active in a stage-specific manner in male germinal cells. To our knowledge, this is the only example of an intragenic, alternative promoter, present inside a duplicated gene and driving transcription of the ancestral nonduplicated form of the gene.

The complete intron-exon structure of the human ACE gene was determined by restriction mapping of genomic clones and by sequencing of the intron-exon boundaries.[53] The human ACE gene contains 26 exons. The somatic ACE mRNA is transcribed from exon 1 to 26, but exon 13 is spliced during maturation of the somatic ACE transcript. The germinal mRNA is transcribed from exon 13 to exon 26.

The structure of the human ACE gene provides further support for the duplication of an ancestral ACE gene (Figure 3). Exons 4-11 and 17-24, encoding the two homologous domains of the ACE molecule, are highly similar both in size and in sequence.[53] In contrast, intron size separating homologous exons is not conserved. The ACE gene duplication appears to have occurred early in evolution. In all mammalian species where the ACE gene has been cloned, i.e., rabbits, mice and humans, the ACE gene appears to be duplicated. A dipeptidyl carboxypeptidase, affinity-purified from the electric organ of *Torpedo marmorata,* was recognized by a polyclonal antiserum raised against pig kidney ACE and was activated by chloride.[57] The molecular weight of *torpedo* ACE is 190,000, and therefore this enzyme would also appear to be transcribed from a duplicated ACE gene. If this is the case, then the duplication of the ACE gene must have occurred more than 600 million years ago. Conservation in remote species of the two transcription units inside the ACE duplicated gene reflects their physiological significance, although a local and specific substrate of testicular ACE, which would definitively signify its function, is not yet known. A gene duplication has also been postulated for another zinc ectopeptidase, aminopeptidase N, although there is a single active site. A duplicated structure with two active sites has been found in other epithelial ectoenzymes, such as lactase-phlorizin hydrolase[58] and sucrase-isomaltase.[59] However, in contrast to ACE, these two brush-border hydrolases contain two active sites with less homology and a clear difference in substrate specificity.

III. IS ACE A BIFUNCTIONAL ENZYME?

The primary structure of human endothelial ACE indicated the presence of two putative active sites. A series of studies were performed to establish whether both domains of ACE were indeed enzymatically active and whether both active sites had similar or different kinetic properties and substrate specificities.

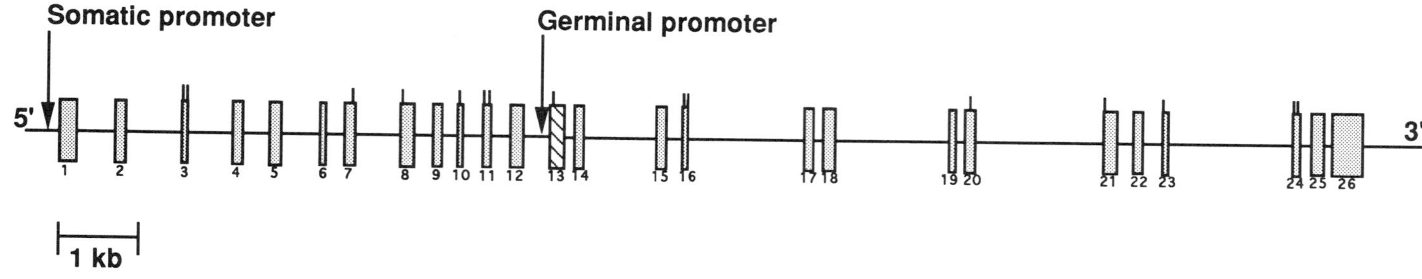

FIGURE 3. Structure of the human ACE gene and presence of two alternate promoters. Location of the 26 numbered exons (vertical bars). Exon 13 (open bar) is specific to the testicular ACE mRNA. The two promoters are indicated by vertical arrows. Vertical bars above the exon boxes indicate the location of the cysteine residues. (Adapted from Hubert, C. et al., *J. Biol. Chem.*, 266, 15377, 1991.)

A. N AND C DOMAINS OF HUMAN ACE ARE BOTH CATALYTICALLY ACTIVE

The sequence identity between the N- and C-domains is more than 60% over a stretch of 357 amino acids. Each of the two homologous domains contain the consensus sequence found at the active sites of thermolysin, neutral endopeptidase and other zinc peptidases.[31] A similar structure, with two domains, has also been determined for the mouse,[32] rabbit,[33] and bovine[60] somatic enzymes. However, until recently, two lines of evidence favored the existence of a single active site in the ACE molecule: the analysis of the zinc content of somatic ACE indicated the presence of a single zinc atom per molecule of ACE and studies with radiolabeled ACE inhibitors detected a single apparent high affinity binding site per ACE molecule. The elucidation of the primary structure of testicular ACE showed that it corresponds to the C-terminal part of endothelial ACE, thus suggesting that the C-domain of somatic ACE contains the zinc atom and the binding site for competitive inhibitors.

To establish whether both putative active sites of the somatic enzyme were functional, a series of ACE mutants, each containing only one intact domain, was constructed by deletion or point mutations of putative critical residues of the other domain.[61] Wild-type ACE and the different mutants were expressed in stable heterologous Chinese hamster ovary (CHO) cell lines. As negative controls, ACE constructs were designed in which either the zinc-binding histidines or the catalytic glutamic acid of the two domains were mutated. Figure 4 shows a schematic representation of these different constructions.

The recombinant wild-type ACE was structurally, immunologically, and enzymatically identical to affinity-purified human kidney ACE.[62] Conversely, the two mutants which possessed critical mutations in both N- and C- domains were devoid of enzymatic activity. The mutants possessing only an intact N-domain were able to hydrolyze a synthetic substrate which constitutes the C-terminus of angiotensin I (Hip-His-Leu) and angiotensin I. For these mutants, the Km values for both substrates, Hip-His-Leu and angiotensin I, were similar to the wild-type enzyme but the Kcat values were 10 and 4 times lower, respectively. Similar results were obtained with the truncated mutants containing only the C-domain or the full-length mutants containing mutations which inactivated the N-domain, although, in this case, the Kcat value for angiotensin I was three times higher than for the N-domain active site. Both the N- and C-domains had an absolute zinc requirement for activity (Table 1). In all these experiments, both domains appeared to function independently since the activity of the wild-type enzyme was equal to the sum of the activities of the N- and C-domains.[61] These studies established that ACE possesses two functional catalytic sites, which were both dependent on a zinc cofactor. In agreement with these findings, the zinc content of somatic ACE has been reinvestigated and found to be two atoms of zinc per molecule.[63,64]

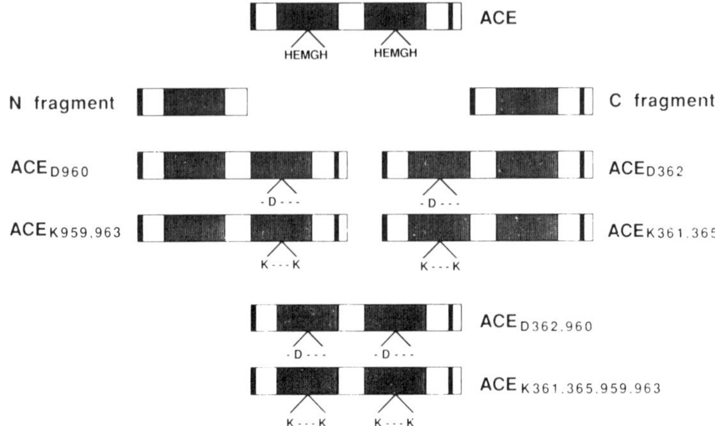

FIGURE 4. Diagram of ACE constructions. The wild type ACE (ACE) contains the signal peptide (left black box), the two large homologous domains (hatched box), and the transmembrane domain (right black box) near the C-terminus. The sequence containing the two putative zinc-binding histidines and the putative catalytic glutamic acid is shown under each domain. The truncated mutant, N fragment, contains the signal peptide and the N domain. The truncated mutant, C fragment, contains the signal peptide, the C domain and the transmembrane domain. For all full-length mutants, the mutated sequence domain. Unchanged amino acids are represented by dashes. The position of mutated amino acids is indicated in the mutant name. (From Wei, L. et al., *J. Biol. Chem.*, 266, 5540, 1991. With permission.)

It is well established that chloride has a marked influence on ACE activity[65] and it is interesting to note that the catalytic activity of the two domains was affected differently by chloride.[61] For AI hydrolysis, the N active site exhibited low levels of activity in the absence of chloride and optimal activity at a concentration of 10 m*M* chloride. In contrast, the C-active site was essentially inactive in the absence of chloride and required much higher chloride concentrations for optimal activity ($\geq$800 m*M*) (Table 2). Bradykinin was hydrolyzed efficiently by both active sites with similar Kcat values in the presence of NaCl, but again, different chloride activation profiles. Therefore, the two domains both act as dipeptidyl carboxypeptidases towards AI and bradykinin but display different catalytic constants for these substrates and different chloride activation profiles.

The interaction of the two active sites of ACE with competitive ACE inhibitors has recently been investigated.[66] Both the N- and C- domains contain a high affinity binding site for [^{3}H]-trandolaprilat, a potent ACE inhibitor. Chloride stabilizes the enzyme-inhibitor complex and slows the dissociation rate, an effect which is more marked for the C-domain than for the N-domain. In addition, various ACE inhibitors show a difference in their potency towards the N- and the C-domains when chloride concentration varies providing further evidence for the existence of structural differences between the two active

TABLE 1
Kinetic Parameters and Chloride Dependence for Hydrolysis

	Hip-His-Leu[a]			Angiotensin I[b]		
Enzyme	**Km μM**	**Kcat s^{-1}**	**Optimal $[Cl^-]$[c] mM**	**Km μM**	**Kcat s^{-1}**	**Optimal $[Cl^-]$[c] mM**
Wild type	1540	408	800	16	40	30
N-domain	2000	40	10	16	12	10
C-domain	2000	360	800	16	35	30

[a] Assay conditions for hydrolysis of Hip-His-Leu were: 100 m*M* potassium phosphate, pH 8.3, 300 m*M* NaCl, 10 μ*M* $ZnSO_4$, 37°C.

[b] Assay conditions for angiotensin I hydrolysis were: 50 m*M* HEPES, pH 7.5, 50 m*M* NaCl, 1 μ*M* $ZnSO_4$, 37°C.

[c] Optimal $[Cl^-]$: optimal chloride concentration for hydrolysis of Hip-His-Leu and angiotensin I.

TABLE 2
Comparison of the Inhibitor Binding Properties of the N- and C-Domains in Presence of 300 m*M* NaCl

	Trandolaprilat				
Enzyme	**K_D[a] × 10^{10} M**	**K_{+1}[b] × 10^5 M^{-1} s^{-1}**	**K_{-1}[c] × 10^5 s^{-1}**	**Ki[c] × 10^{10} M**	**Relative[d] inhibitory potency[e]**
N-domain	2.9	8.6	24	3.1	T>C>E>L
C-domain	0.3	4.3	1.4	0.3	T>L>E>C

[a] K_D was determined by equilibrium dialysis in 100 m*M* HEPES, pH 7.

[b] K_{+1} and K_1 are the rat constants of [^{3}H] transdolaprilat binding.

[c] Ki is the inhibitory constant using Hip-His-Leu as a substrate, as in Table 1 legend.

[d] Relative inhibitory potency of Trandolapril, Captopril, Enalapril, and Lysinopril on binding properties of [^{3}M] transdolaprilat on various ACE molecules.

[e] T: Trandolapril; C: Captopril; E: Enalapril; L: Lisinopril.

sites of ACE (Table 2). Further studies with various ACE inhibitors may help to elucidate the structure/function of these two active sites and possibly lead to the design of specific inhibitors for each active site.

B. DOES EACH ACTIVE SITE DISPLAY A DISTINCT SUBSTRATE SPECIFICITY?

ACE is the only zinc metallopeptidase known to date to possess two active sites. Some other enzymes contain two active sites and are truly bifunctional since they exhibit distinct functions as, for example, the brush-border hydrolases sucrase-isomaltase[59] and lactase-phlorizin hydrolase[58]. In these cases, a duplication of an ancestral gene occurred during evolution forming two active sites; subsequently a divergent evolution resulted in a different catalytic activity for each domain. In the case of ACE, the two active sites display highly similar kinetic parameters for the most well-known natural substrates. However, studies with natural and synthetic peptide substrates, chloride and competitive inhibitors, as discussed above, suggest structural and functional differences between subsites raising the possibility of different substrate specificities of the N and C active sites. The demonstration of another natural substrate with more favorable kinetic parameters for the N-domain compared to the C-domain would allow us to define ACE as a bifunctional enzyme. Ehlers et al.[63] reported that the somatic enzyme hydrolyzed LH-RH and cleaved the Trp_3-Ser_4 bond of this peptide more efficiently compared to the testicular enzyme, suggesting that the N-domain may be preferentially involved in this cleavage. This would be a strong argument for a distinct specificity of the N-domain, although it is not firmly established that ACE inactivates this peptide in the pituitary. Our own studies also indicate that LH-RH is hydrolysed by both domains, releasing a carboxy-terminal tripeptide. However, this primary cleavage between Trp_3 and Ser_4 is performed more efficiently by the N-domain (unpublished results). These results indicate the occurrence of functional and structural differences between the two active sites, despite the high degree of sequence homology. However, a truly specific substrate for each active site has not yet been identified.

In addition to the possible bifunctionality of ACE, alternative hypotheses can be proposed to account for the specific properties of each active site: first, the N-domain of ACE could be more efficient under particular physicochemical conditions (for example, in the intracellular medium where the chloride concentration is low); second, a lower selection pressure on the N-domain may have been responsible for mutational changes of key residues involved in the catalytic mechanism.

IV. ANCHORAGE AND MECHANISM OF SOLUBILIZATION

Conventional studies on the purification of ACE revealed that the enzyme was bound to the membrane of the cell and that a solubilization step, either by proteolytic cleavage with trypsin or by detergent, was required to purify the enzyme.[67] Immunohistochemical studies localized the enzyme to the surface of the ceii on various epithelia and endothelium.[19,68] In addition to the membrane-bound form of the enzyme, ACE is also found as a soluble form in plasma, semen, cerebrospinal fluid and other body fluids. The relationship between the membrane bound and the soluble forms of the enzyme are not clear, although their molecular sizes are similar, indicating that the solubilization process does not greatly affect the size of the enzyme.

A. THE MOLECULAR MECHANISM OF ACE ANCHORAGE

The mode of anchorage of ACE to the plasma membrane and the mechanism of its solubilization was first studied by Hooper et al.[67] who demonstrated that ACE was inserted via its C-terminus, as the trypsin-solubilized (hydrophilic) and the detergent-solubilized (amphipathic) enzymes had identical N-terminal sequences.

The study by Hooper et al. was further supported by the determination of the primary structure of ACE which revealed the presence of a highly hydrophobic sequence located at the C-terminal extremity of the molecule (residues 1231 – 1247). This sequence almost certainly constitutes the membrane anchor since this is the only region, with the exception of the cleaved N-terminal signal sequence, of high hydrophobicity capable of spanning the lipid bilayer.

Further evidence for C-terminal anchorage was obtained from site-directed mutagenesis studies on human endothelial ACE[62] and human testicular ACE.[69] Transfection of CHO cells with the wild-type recombinant ACE cDNA resulted in 95% of newly synthesized enzyme bound to the cell membrane whereas 5% was secreted into the cell supernatant. However, when a stop codon was introduced into the cDNA on the 5′ side immediately before the sequence encoding the putative membrane anchor, ACE was only recovered in the cell supernatant. Among zinc ectopeptidases, ACE is the only enzyme to be inserted via its C-terminus in contrast to neutral endopeptidase and aminopeptidase N which retain the membrane inserted signal peptide.

B. MECHANISM AND PHYSIOLOGICAL SIGNIFICANCE OF ACE SECRETION

The relationship between the membrane bound form of ACE and its circulating form has been elucidated. The plasma enzyme is probably derived from vascular endothelial cells and lacks the C-terminus of ACE, as it is not recognized by a specific antibody directed against this part of the enzyme.[62] Therefore it seems that the secreted, circulating form of ACE is derived from the membrane bound form by a post-translational event. Such a mechanism has been described for the solubilization of several ectoproteins. A post-translational processing of the membrane bound form of ACE into a soluble form, probably similar to that occurring in endothelial cells, has been observed during the culture of CHO cells which express recombinant ACE.[62] The enzymatic cleavage, which generates soluble ACE from the membrane bound form, occurs intracellularly or at the plasma membrane. An endogenous EDTA-sensitive protease, identified in lung and kidney, was able to release a soluble form of the enzyme from pig kidney membranes.[67] The cellular localization, the identification of this putative proteolytic enzyme, and the exact site(s) of ACE cleavage are unknown.

The functional significance of endothelial ACE vs. circulating plasma ACE was first investigated by Ng and Vane[70] who showed that most of the conversion of AI into AII occurred in the lung. It is interesting to note that many investigators failed to correlate the antihypertensive effect of ACE inhibitors with the degree of plasma ACE inhibition.[71] Alterations in the tissue levels of ACE as opposed to plasma ACE may be of importance for vasoactive peptide metabolism.

V. EFFECT OF THE ACE GENE POLYMORPHISM ON CIRCULATING ACE LEVELS

Plasma serum ACE (SACE) concentration is stable in a given individual but varies widely between individuals.[72] In healthy humans, no hormonal or environmental factors affecting SACE were identified; however, in some pathological conditions, such as granulomatous disease, hyperthyroidism, and diabetes, elevated ACE levels are observed.[73]

A family study of serum ACE levels was initially performed in normal nuclear families which showed intrafamilial correlations of SACE. The segregation analysis of SACE showed that the familial resemblance was predominantly due to a major gene effect which modulates SACE.[74]

Cloning of the human ACE cDNA allowed the detection of a polymorphism of the ACE gene, which consists of the presence or absence of a 250 bp DNA fragment (Figure 4).[75] The polymorphic insertion is located inside intron 16 of the ACE gene and corresponds to an *Alu* repetitive sequence.[76] We analyzed this polymorphism in a group of 80 healthy subjects, and measured the serum ACE levels. The frequency of the I allele (insertion) was 0.44 whereas the frequency of the D allele (deletion) was 0.56. In this study, the I/D polymorphism accounted for 47% of the total SACE variance. Using this polymorphism as a DNA marker, we showed a strong association between alleles of the ACE gene polymorphism and the levels of SACE. Individuals who were homozygote for the D allele had a mean serum ACE level nearly two-fold higher compared to individuals with the II genotype[75] (Figure 5). Therefore, the ACE gene locus itself could be responsible for variations in the genetic control of SACE. Another study combining segregation and linkage analysis showed that the I/D polymorphism was in fact only a neutral marker in strong linkage disequilibrium with the putative variant.[77] The measured genotype effect of the I/D polymorphism accounted for 28% of the total variance of SACE. These studies did not detect a relationship between the ACE genotype and blood pressure, although a weak correlation between serum ACE and blood pressure was observed in offspring, after adjustment for age. These observations have been extended to the membrane bound form of ACE on the basis of the expression of ACE in human circulating T lymphocytes. In these cells, ACE levels are also genetically determined and associated with the DD polymorphism,[18] indicating a genetic determinant of the biosynthesis of ACE probably occurring in many, if not all, tissues which express the enzyme.

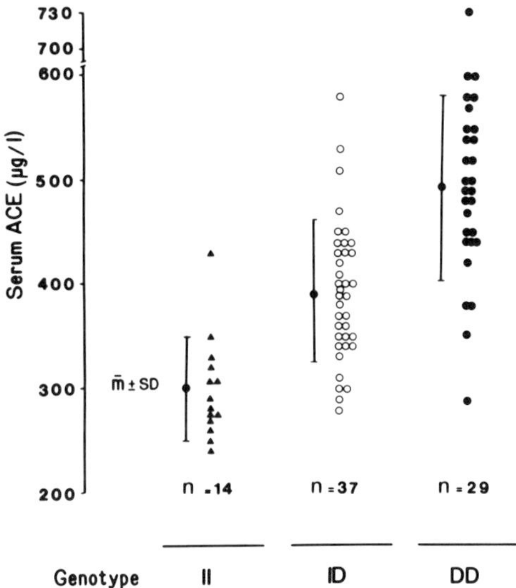

FIGURE 5. Serum immunoreactive ACE concentrations (μg/l) for individual with the II, ID and DD genotypes, respectively shown in the left, middle and right panels. Solid vertical bars indicate mean concentration and standard deviation for each group. (From Rigat, B. et al., *J. Clin. Invest.*, 86, 1343, 1990. With permission.)

VI. CONCLUSION

The molecular cloning of the somatic and germinal forms of ACE, and structure/function studies of the protein have radically modified our knowledge of the structure, biosynthesis, and expression of ACE. However, important questions still remain to be answered. Among these questions are the possible substrate specificity of each active site, the nature, and the localization of the enzyme involved in the solubilization of ACE. The regulatory mechanisms allowing a cell- and stage-specific transcription of the germinal ACE as well as its function are unknown. Identification of the ACE variant responsible for the genetic variation of SACE is one of the most difficult and interesting future challenges of the molecular genetics of ACE, with regards to its possible implication in angiotensin II modulation. It is clear that ACE is not only an important enzyme for blood pressure regulation and neuropeptide metabolism but also represents a fascinating model from a biochemical, evolutionary, and genetic point of view.

ACKNOWLEDGMENTS

We are grateful to Dr. T.A. Williams for her invaluable help during the preparation of this manuscript. We also wish to thank Mrs. N. Braure for secretarial assistance, and Mrs. A. Depardieu for artwork. This work was supported by INSERM, Collège de France, CNRS, and by a grant from Bristol-Myers-Squibb Institute for Medical Research.

REFERENCES

1. **Erdös, E.G.,** Angiotensin I converting enzyme and the changes in our concepts through the years (Lewis K. Dahl memorial lecture), *Hypertension,* 16, 363, 1990.
2. **Skidgel, R.A., Engelbrecht, S., Johnson, A.R., and Erdös, E.G.,** Hydrolysis of substance P and neurotensin by converting enzyme and neutral endopeptidase, *Peptides,* 5, 769, 1984.
3. **Skidgel, R.A. and Erdös, E.G.,** Angiotensin I converting enzyme and its role in neuropeptide metabolism, in *Neuropeptides and their Peptidases,* Turner, A.J., Ed., Ellis Horwood, Chichester, 1967, p. 165.
4. **Yokosawa, H., Endo, S., Ogura, Y., and Ishii, S.,** A new feature of angiotensin-converting enzyme in the brain: hydrolysis of substance P, *Biochem. Biophys. Res. Commun.,* 116, 735, 1983.
5. **Skidgel, R.A. and Erdös, E.G.,** Novel activity of human angiotensin I converting enzyme: release of the NH2- and COOH-terminal tripeptides from the luteinizing hormone-releasing hormone, *Proc. Natl. Acad. Sci. U.S.A.,* 82, 1025, 1985.
6. **Dubreuil, P., Fulcrand, P., Rodriguez, M., Fulcrand, H., Laur J., and Martinez, J.,** Novel activity of angiotensin-converting enzyme hydrolysis of cholecystokinin and gastrin analogues with release of the amidated carboxy-terminal dipeptide, *Biochem. J.,* 262, 125, 1989.
7. **Subissi, A., Guelfi, M., and Criscuoli, M.,** Angiotensin converting enzyme inhibitors potentiate the bronchoconstriction induced by substance P in the guinea-pig., *Br. J. Pharmacol.,* 100, 502, 1990.
8. **Bünning, P. and Riordan, J.F.,** Activation of angiotensin converting enzyme by monovalent anions, *Biochemistry,* 22, 110, 1983.
9. **Shapiro, R., Holmquist, B., and Riordan, J.F.,** Anion activation of angiotensin-converting enzyme: dependence on nature of substrate, *Biochemistry,* 22, 3850, 1983.
10. **Das, M. and Soffer, R.L.,** Pulmonary angiotensin-converting enzyme: structural and catalytic properties, *J. Biol. Chem.,* 250, 6762, 1975.
11. **Weare, J.A., Gafford, J.T., Lu, H.S., and Erdös, E.G.,** Purification of human kidney angiotensin I-converting enzyme using reverse-immunoadsorption chromatography, *Anal. Biochem.,* 123, 310, 1982.
12. **Bruneval, P., Hinglais, N., Alhenc-Gelas, F., Tricottet, V., Corvol, P., Ménard, J., Camilleri, J.P., and Bariety, J.,** Angiotensin I converting enzyme in human intestine and kidney. Ultrastructural immunohistochemical localization, *Histochemistry,* 85, 73, 1986.
13. **Arregui, A. and Iversen, L.L.,** Angiotensin-converting enzyme: presence of high activity in choroid plexus of mammalian brain, *Eur. J. Pharmacol.,* 52, 147, 1978.
14. **Schweisfurth, H., Schioberg-Schiegnitz, S.,** Assay and biochemical characterization of angiotensin I-converting enzyme in cerebrospinal fluid, *Enzyme,* 32, 12, 1984.
15. **Yokoyama, M., Takada, Y., Iwata, H., Ochi, K., Takeuchi, M., Hiwada, K., and Kokubu, T.,** Correlation between angiotensin-converting enzyme activity and histologic patterns in benign prostatic hypertrophy tissue, *J. Urol.,* 127, 368, 1982.

16. **Cushman, D.W. and Cheung, H.S.,** Concentrations of angiotensin-converting enzyme in tissues of rat, *Biochim. Biophys. Acta,* 250, 261, 1971.
17. **Friedland, J., Setton, C., and Silverstein, E.,** Induction of angiotensin-converting enzyme in human monocytes in culture, *Biochem. Biophys. Res. Commun.,* 83, 843, 1978.
18. **Costerousse, O., Allegrini, J., Lopez, M., and Alhenc-Gelas, F.,** Angiotensin I-converting enzyme in human circulating mononuclear cells. Genetic polymorphism of expression in T-lymphocytes, submitted.
19. **Defendini, R., Zimmerman, E.A., Weare, J.A., Alhenc-Gelas, F., and Erdös, E.G.,** Angiotensin-converting enzyme in epithelial and neuroepithelial cells, *Neuroendocrinology,* 37, 32, 1983.
20. **Strittmatter, S.M., Lo, M.M.S., Javitch, J.A., and Snyder, S.H.,** Autoradiographic visualization of angiotensin-converting enzyme in rat brain with [3H]captopril: localization to a striatonigral pathway, *Proc. Natl. Acad. Sci. U.S.A.,* 81, 1599, 1984.
21. **Barnes, K., Matsas, R., Hooper, N.M., Turner, A.J., and Kenny, A.J.,** Endopeptidase 24.11 is striosomally ordered in pig brain and, in contrast to aminopeptidase N and dipeptidyl dipeptidase, an angiotensin converting enzyme, is a marker for a set of striatal efferent fibers, *Neuroscience,* 27, 799, 1988.
22. **Matsas, R, Kenny, A.J., and Turner, A.J.,** The metabolism of neuropeptides. The hydrolysis of peptides including enkephalins, tachynins and their analogues by endopeptidase 24.11 (E.C. 3.4.24.11), *Biochem. J.,* 223, 433, 1984.
23. **Oblin, A., Danse, M.J., and Zivkovic, B.,** Degradation of substance P by membrane peptidases in the rat substantia nigra: effect of selective inhibitors, *Neurosci. Lett.,* 84, 91, 1988.
24. **Hooper, N.M. and Turner, A.J.,** Isolation of two differentially glycosylated forms of peptidyl dipeptidase A (angiotensin converting enzyme) from pig brain: a re-evaluation of their role in neuropeptide metabolism, *Biochem. J.,* 241, 625, 1987.
25. **Williams, T.A., Hooper, N.M., and Turner, A.J.,** Characterization of neuronal and endothelial forms of angiotensin converting enzyme in pig brain, *J. Neurochem.,* 57, 193, 1991.
26. **Das, M., Hartley, J.L., and Soffer, R.L.,** Serum angiotensin-converting enzyme. Isolation and relationship to the pulmonary enzyme, *J. Biol. Chem.,* 252, 1316, 1977.
27. **El-Dorry, H.A., MacGregor, J.S., and Soffer, R.L.,** Dipeptidyl carboxypeptidase from seminal fluid resembles the pulmonary rather than the testicular isoenzyme, *Biochem. Biophys. Res. Commun.,* 115, 1096, 1983.
28. **Yasui, T., Alhenc-Gelas, F., Corvol, P., Ménard, J.,** Angiotensin I-converting enzyme in amniotic fluid, *J. Lab. Clin. Med.,* 104, 741, 1984.
29. **Brentjens, J.R., Matsuo, S., Andres, G.A., Caldwell, P.R.B., and Zamboni L.,** Gametes contain angiotensin converting enzyme (kininase II), *Experientia,* 42, 399, 1986.
30. **Bull, H.G., Thornberry, N.A., and Cordes, E.S.,** Purification of angiotensin-converting enzyme from rabbit lung and human plasma by affinity chromatography, *J. Biol. Chem.,* 260, 2963, 1985.
31. **Soubrier, F., Alhenc-Gelas, F., Hubert, C., Allegrini, J., John, M., Tregear, G., and Corvol, P.,** Two putative active centers in human angiotensin I-converting enzyme revealed by molecular cloning, *Proc. Natl. Acad. Sci. U.S.A.,* 85, 9386, 1988.
32. **Bernstein, K.E., Martin, B.M., Edwards, A.S., and Bernstein, E.A.,** Mouse angiotensin I-converting enzyme is a protein composed of two homologous domains, *J. Biol. Chem.,* 264, 11945, 1988.
33. **Thekkumkara, T.J., Livingston III, W., Kumar, R.S., and Sen, G.C.,** Use of alternative polyadenylation sites for tissue-specific transcription of two angiotensin-converting enzyme mRNAs, *Nucl. Acids Res.,* 20, 683, 1992.

34. **Devault, A., Lazure, C., Nault, C., Le Moual, H., Seidah, N.G., Chrétien M., Kahn, P., Powell, J., Mallet J., Beaumont A., Roques, B., Crine P., and Boileau G.,** Amino acid sequence of rabbit kidney neutral endopeptidase 24.11 (enkephalinase) deduced from a complementary DNA, *EMBO J.,* 6, 1317, 1987.
35. **Vallee, B.L. and Auld, D.S.,** Zinc coordination, function, and structure of zinc enzymes and other proteins, *Biochemistry,* 29, 5647, 1990.
36. **Kester, W.R. and Matthews, B.W.,** Crystallographic study of the binding of dipeptide inhibitors to thermolysin: implications for the mechanism of catalysis, *Biochemistry,* 16, 2506, 1977.
37. **Bünning, P., Riordan, J.F.,** The functional role of zinc in angiotensin converting enzyme. Implication for the enzyme mechanism, *J. Inorg. Biochem.,* 24, 183, 1985.
38. **Cumin, F., Vellaud, V., Corvol, P., Alhenc-Gelas, F.,** Evidence for a single active site in the human angiotensin I-converting enzyme from inhibitor binding studies with [3H] RU44403: role of chloride, *Biochem. Biophys. Res. Commun.,* 163, 718, 1989.
39. **Bünning, P., Kleemann, S.G., and Riordan, J.F.,** Essential residues in angiotensin converting enzyme: modification with 1-fluoro-2,4-dinitrobenzene, *Biochemistry,* 29, 10488, 1990.
40. **Bünning, P., Holmquist, B., and Riordan, J.F.,** Functional residues of the active site of angiotensin-converting enzyme, *Biochem. Biophys. Res. Commun.,* 83, 1442, 1978.
41. **Chen, Y. N-P. and Riordan, J.F.,** Identification of essential tyrosine and lysine residues in angiotensin converting enzyme: evidence for a single active site, *Biochemistry,* 29, 10493, 1990.
42. **Chen, Y.N.-P., Ehlers, M.R.W., and Riordan, J.F.,** The functional role of tyrosine-200 in human testis angiotensin-converting enzyme, *Biochem. Biophys. Res. Commun.,* 184, 306, 1992.
43. **Kumar, R.S., Kusari, J., Roy, S.N., Soffer, R.L., and Sen G.C.,** Structure of testicular angiotensin-converting enzyme. A segmental mosaic isozyme, *J. Biol. Chem.,* 264, 16754, 1989.
44. **Deutch, C.E. and Soffer, R.L.,** Escherichia Coli mutants defective in dipeptidyl carboxypeptidase, *Proc. Natl. Acad. Sci. U.S.A.,* 75, 5998, 1978.
45. **Hamilton, S. and Miller, C.G.,** Cloning and nucleotide sequence of the *Salmonella typhimurium* dcp gene encoding dipeptidyl carboxypeptidase, *J. Bacteriol.,* 174, 1626, 1992.
46. **Olsen, J., Cowell, G.M., Konigshofer, E., Danielsen, E.M., Moller, J., Laustsen, L., Hansen, O.C., Welinder, K.G., Engberg, J., Hunziker, W. Spiess, M., Sjöström, H., and Norèn, O.,** Complete amino acid sequence of human intestinal aminopeptidase N as deduced from cloned cDNA, *FEBS Lett.,* 238, 307, 1988.
47. **El-Dorry, H.A., Pickett, C.B., McGregor, J.S., and Soffer, R.L.,** Tissue-specific expression of mRNAs for dipeptidyl carboxypeptidase isoenzymes, *Proc. Natl. Acad. Sci. U.S.A.,* 79, 4295, 1982.
48. **El-Dorry, H.A., Bull, H.G., Iwata, K., Thornberg, N.A., Cordes, E.H., and Soffer, R.L.,** Molecular and catalytic properties of rabbit testicular dipeptidyl carboxypeptidase, *J. Biol. Chem.,* 257, 14128, 1982.
49. **Lattion, A.-L., Soubrier, F., Allegrini, J., Hubert, C., Corvol, P., and Alhenc-Gelas, F.,** The testicular transcript of the angiotensin I-converting enzyme encodes for the ancestral, non-duplicated form of the enzyme, *FEBS Lett.,* 252, 99, 1989.
50. **Ehlers, M.R.W., Fox, E.A., Strydom, D.J., and Riordan, J.F.,** Molecular cloning of human testicular angiotensin-converting enzyme: The testis isozyme is identical to the C-terminal half of endothelial angiotensin-converting enzyme, *Proc. Natl. Acad. Sci. U.S.A.,* 86, 7741, 1989.
51. **Roy, S.N., Kusari, J., Soffer, R.L., Lai, C.Y., and Sen, G.C.,** Isolation of cDNA clones of rabbit angiotensin converting enzyme: identification of two distinct mRNAs for the pulmonary and the testicular isoenzymes, *Biochem. Biophys. Res. Commun.,* 155, 678, 1988.

52. **Ehlers, M.R.W., Chen Y.-N.P., and Riordan, J.F.,** The unique N-terminal sequence of testis angiotensin-converting enzyme is heavily O-glycosylated and unessential for activity or stability, *Biochem. Biophys. Res. Commun.*, 183, 199, 1992.
53. **Hubert, C., Houot, A.-M., Corvol, P., and Soubrier, F.,** Structure of the Angiotensin I-converting enzyme gene. Two alternate promoters correspond to evolutionary steps of a duplicated gene, *J. Biol. Chem.*, 266, 15377, 1991.
54. **Howard, T.E., Shai, S.-Y., Langford, K.G., Martin, B.M., and Bernstein, K.E.,** Transcription of testicular angiotensin-converting enzyme (ACE) is initiated within the 12th intron of the somatic ACE gene, *Mol. Cell. Biol.*, 10, 4294, 1990.
55. **Kumar, R.S., Thekumkara, T.J., and Sen G.,** The mRNAs encoding the two angiotensin-converting isozymes are transcribed from the same gene by a tissue-specific choice of alternative transcription initiation sites, *J. Biol. Chem.*, 266, 3854, 1991.
56. **Langford, K.G., Shai, S.-Y., Howard, T.E., Kovac, M.J., Overbeek, P.A., and Bernstein, K.E.,** Transgenic mice demonstrate a testis-specific promoter for angiotensin-converting enzyme, *J. Biol. Chem.*, 266, 15559, 1991.
57. **Turner, A.J., Hryszko, J., Hooper, N.M., Dowdall, and M.J.,** Purification and characterization of a peptidyl dipeptidase resembling angiotensin converting enzyme from the electric organ of torpedo marmorata, *J. Neurochem.*, 48, 910, 1987.
58. **Mantei, N., Villa, M., Enzler, T., Wacker, H., Boll, W., James, P., Hunziker, W., and Semenza, G.,** Complete primary structure of human and rabbit lactase-phlorizin hydrolase: implication for biosynthesis, membrane anchoring and evolution of the enzyme, *EMBO J.*, 7, 2705, 1988.
59. **Hunziker, W., Spiess, M., Semenza, G., and Lodish, H.F.,** The sucrase isomaltase complex: Primary structure, membrane orientation and evolution of stalked, intrinsic brush border protein, *Cell,* 46, 227, 1986.
60. **Shai, S.-Y., Fishel, R.S., Martin, B.M., Berk, B.C., and Bernstein, K.B.,** Bovine angiotensin converting enzyme cDNA cloning and regulation. Increased expression during endothelial cell growth arrest, *Circ. Res.*, 70, 1274, 1992.
61. **Wei, L., Alhenc-Gelas, F., Corvol, P., and Clauser, E.,** The two homologous domains of the human angiotensin I-converting enzyme are both catalytically active, *J. Biol. Chem.*, 266, 9002, 1991.
62. **Wei, L., Alhenc-Gelas, F., Soubrier, F., Michaud, A., Corvol, P., and Clauser, E.,** Expression and characterization of recombinant human angiotensin I-converting enzyme. Evidence for a C-terminal transmembrane anchor and for a proteolytic processing of the secreted recombinant and plasma enzymes, *J. Biol. Chem.*, 266, 5540, 1991.
63. **Ehlers, M.R.W. and Riordan, J.F.,** Angiotensin-converting enzyme: Zinc- and inhibitor binding stoichiometries of the somatic and testis isozymes, *Biochemistry,* 30, 7118, 1991.
64. **Williams, T.A., Barnes, K., Kenny, A.J., Turner, A.J., and Hooper, N.M.,** A comparison of the zinc content and substrate specificity of the endothelial and testicular forms of porcine angiotensin converting enzyme, and the isolation of isoenzyme specific antisera, *Biochem. J.*, in press.
65. **Shapiro, R., Holmquist, B., and Riordan, J.F.,** Anion activation of angiotensin-converting enzyme: dependence on nature of substrate, *Biochemistry,* 22, 3850, 1983.
66. **Wei, L., Clauser, E., Alhenc-Gelas, F., and Corvol, P.,** The two homologous domains of human angiotensin I-converting enzyme interact differently with competitive inhibitors, *J. Biol. Chem.*, 267, 13398, 1992.
67. **Hooper, N.M., Keen, J., Pappin, D.J.C., and Turner, A.J.,** Pig kidney angiotensin-converting enzyme. Purification and characterization of amphiphatic and hydrophilic forms of the enzyme establishes C-terminal anchorage to the plasma membrane, *Biochem. J.*, 247, 85, 1987.
68. **Ryan, J.W., Ryan, U.S., Schultz, D.R., Whitaker, C., Shung, A., and Dorer, F.E.,** Subcellular localization of pulmonary angiotensin convering enzyme (kininase II), *Biochem. J.*, 146, 497, 1975.

69. **Ehlers, M.R.W., Chen, Y.-N.P., and Riordan, J.F.,** Spontaneous solubilization of membrane-bound human testis angiotensin-converting enzyme expressed in chinese hamster ovary cells, *Proc. Natl. Acad. Sci. U.S.A.,* 88, 1009, 1991.
70. **Ng, K.K.F. and Vane, J.R.,** Conversion of angiotensin I to an angiotensin II, *Nature,* 216, 762, 1967.
71. **Waeber, B., Nussberger, J., Juillerat, L., and Brunner, H.R.,** Angiotensin converting enzyme inhibitor: discrepancy between antihypertensive effect and suppression of enzyme activity, *J. Cardiovasc. Pharmacol.,* 14, S53, 1989.
72. **Alhenc-Gelas, F., Weare, J.A., Johnson, R.L., and Erdös, E.G.,** Measurement of human converting enzyme level by direct radioimmunoassay, *J. Lab. Clin. Med.,* 101, 83, 1983.
73. **Lieberman, J.,** Elevation of serum angiotensin-converting enzyme (ACE) level in sarcoidosis, *Am. J. Med.,* 59, 365, 1975.
74. **Cambien, F., Alhenc-Gelas, F., Herbeth, B., André, J.L., Rakotovao, R., Gonzales, M.F., Allegrini, J., and Bloch, C.,** Familial resemblance of plasma angiotensin-converting enzyme level: the Nancy study, *Am. J. Hum. Genet.,* 43, 774, 1988.
75. **Rigat, B., Hubert, C., Alhenc-Gelas, F., Cambien, F., Corvol, P., and Soubrier, F.,** An insertion/deletion polymorphism in the angiotensin I-converting enzyme gene accounting for half the variance of serum enzyme levels, *J. Clin. Invest.,* 86, 1343, 1990.
76. **Rigat, B., Hubert, C., Corvol, P., and Soubrier, F.,** PCR detection of the insertion/deletion polymorphism of the human angiotensin converting enzyme gene (DCP1) (dipeptidyl carboxypeptidase 1), *Nucl. Acids Res.,* 20, 1433, 1992.
77. **Tiret, L., Rigat, B., Visvikis, S., Breda, C., Corvol, P., Cambien, F., and Soubrier, F.,** Evidence, from combined segregation and linkage analysis, that a variant of the angiotensin I-converting enzyme (ACE) gene controls plasma ACE levels, *Am. J. Hum. Genet.,* 51, 197, 1992.

Section IV

Angiotensin Receptors

Chapter 10

DEFINING ANGIOTENSIN RECEPTOR SUBTYPES

A.T. Chiu, K.H. Leung, R.D. Smith, and P.B.M.W.M. Timmermans

TABLE OF CONTENTS

0-8493-4622-3/93/$0.00 + $.50

I. HISTORICAL PERSPECTIVES

As with all other hormonal systems, our understanding of the angiotensin receptors has also evolved and refined with the passage of time. Ever since the elucidation of the structure of the octapeptide angiotensin II (AngII),[1] the search for its molecular mechanism of action had stimulated a fury of studies describing and conceptualizing the existence of heterogeneous receptive components called receptors on the cell membrane that mediate its numerous biological responses. Angiotensin II, a primary effector hormone of the renin-angiotensin system, exerts a wide range of biological and biochemical effects in tissues associated with the cardiovascular, renal, endocrine, and neuronal functions.[2]

A. STRUCTURE-ACTIVITY RELATIONSHIP APPROACH

Historically, the characteristics of angiotensin receptors have been defined by studies examining the structure-activity relationship (SAR) of a large series of synthetic peptide homologs and analogs of both agonists and antagonists for a particular response. The majority of these studies, however, were focused primarily on the contractile responses of isolated vascular or nonvascular tissues and on vasoconstrictor responses in intact animals. A rather limited SAR was obtained for most other biological effects.

Culminated from numerous SAR studies, Peach and Levens[3] in 1980 suggested that there were at least two different types of angiotensin receptor, differing mainly in their requirements for phenylalanine in position 8 of the peptide (see Table 1). They designated that type 1 receptors were those mediating the pressor and myotropic responses which had an absolute requirement for phenylalanine as the eighth amino acid of the octapeptide and a free C-terminal carboxylic functionality for full activity. The rank order of potency of agonists was characterized as AngII > AngIII >> AngI. In fact, AngI was considered inactive in these systems. As for the type 2 receptors, aliphatic substitutions could be made for phenylalanine-8 without greatly attenuating the activity of the molecule on the sympathetic nerve endings, adrenal medullary catecholamine release, or intestinal salt and water transfer. The rank order of potency of agonists in these type 2 systems was portrayed as AngII > AngI > AngIII. Importantly, AngI was considered to be directly active. Results obtained from the adrenal cortex also suggested the existence of another angiotensin receptor subtype, since the order of potency of agonists (AngIII > AngII >> AngI) was presumably different from the other two subtypes.[4,5]

The use of SAR and the affinity (pA_2) of competitive peptide antagonists (generally partial agonists) in defining receptor subtypes has provided a greater appreciation of the complexity of the interaction between the polypeptide, angiotensin, and a biological target.[3,6] However, it is rather cumbersome using the agonistic potency of an extensive series of peptide analogs to determine

TABLE 1
Classification of Subtypes of Angiotensin Receptors Based on SAR

	Receptor type 1	Receptor type 2
System Response		
	Pressor systems[78]	Catecholamine release
		Adrenal medulla[79]
	Myotropic systems[78]	Sympathetic nerve[80]
	Vasopressin release[81]	Intestinal salt and H_2O transfer[82]
	Thirst[83]	Renal tubules[84]
SAR Requirements		
	Positions 4 and 6 necessary for binding	Positions 4 and 6 necessary for binding
	Phe[8] and a free –COOH are necessary for activity	Aliphatic substitution at position 8 and no free –COOH are permissible
	AngI acts via AngII	AngI acts directly

Adapted from Peach, M.J. and Levens, N.R., *Adv. Exp. Med.*, 130, 171, 1980.

the type of angiotensin receptor in any tissue or response being examined. There are numerous limitations which might affect the interpretation of the results. For example, the potency or activity of a compound, particularly a polypeptide, is dependent on many factors, such as chemical purity, metabolic stability, intrinsic efficacy, the microenvironment surrounding the receptor, the nature of transduction mechanisms available in each tissue, and the possible presence of receptor subtypes. As Peach and Dostal[7] pointed out, one cannot safely assume that a peptide fully characterized in the aorta or uterus will have the same activity profile in all other effector organs or with the same organ in different species. Because of these complexities, controversies derived from this approach remain unresolved.[6]

B. RADIOLIGAND-RECEPTOR BINDING APPROACH

Another approach to characterizing receptors is by using radioligand binding, which has been widely used to obtain the kinetics of binding, the number of receptor sites in a given tissue, and the apparent affinities of competitive ligands.[8] This approach, if used in conjunction with the measurement of biologic responses in the same tissue under investigation, may uncover subtle differences in receptor characteristics that can be overlooked by either approach used alone. However, in most studies, angiotensin receptors were characterized by binding alone, using homologs and analogs of AngII whose biologic responses were reported in different organs or in the same tissue evaluated under totally different experimental conditions.

Radiolabeled AngII binding has been characterized in numerous target tissues, such as vascular smooth muscles, adrenal, kidney, and brain from various species.[3] In the literature, either a single population of high-affinity binding sites or sometimes two populations of binding sites has been noted in the same or different tissues, depending on the conditions of the studies. The assignment of high- and low-affinity sites was quite arbitrary. The high-affinity sites may have a K_d value ranging from .07 to 20 n*M* and the low-affinity sites may span from 2 to 60 n*M* (see Reference 3 for details). It is not apparent whether the differences among the high-affinity values signify the presence of receptor subtypes or just reflect the technical artifacts produced under different experimental procedures or conditions. In general, there is also inadequate indication that the low-affinity binding sites might actually represent a different type of functional receptors. Furthermore, the SAR studies were carried out in conditions generally indiscriminative between possible receptor subtypes.[9,10]

The radioligand-receptor binding approach encounters quite the same problems as described for the SAR approach. This technique, in most cases, does neither discriminate agonists with different intrinsic activity nor an agonist from an antagonist. In addition, AngII, being a linear polypeptide, can assume multiple conformational states that are compatible with different receptor subtypes. As a result, the presence of these receptor subtypes cannot be easily detectable with any certainty when using this peptide as the primary ligand. Devynck and co-workers,[11] however, have reported the detection of two populations of specific binding sites in rat adrenal cortical membranes using [^{3}H]AngIII as the radioligand. They suggested that the one with a higher affinity (K_d = 0.2 n*M*) but with smaller capacity was specific for AngIII, whereas the other with lower affinity (K_d = 4 n*M*) but with greater capacity was the receptor for AngII that was distinct from the AngIII site. These interesting findings, however, were not confirmed by other investigators.[12,13]

C. BIOCHEMICAL AND RECEPTOR COUPLING APPROACH

More recently, Douglas[14] proposed a classification of two types of AngII receptors, in either rat or rabbit kidneys, based on the differences not only in the binding affinity for AngII and AngIII, but also in the nature of receptor regulation and in the mechanisms of signal transduction (see Table 2). The receptor of the rat glomerular mesangium, classified as type A, was characterized by high affinity (K_d = 0.8 to 2 n*M*) for AngII and AngIII, ''down-regulation'' with high ambient concentrations of AngII, and signal transduction mediated by phospholipase C-induced Ca^{2+} transients. The rabbit tubular epithelial AngII receptor, designated as type B, had lower affinity (K_d = 2 to 6 n*M*) for AngII, had a 10-fold lower affinity for AngIII, was ''upregulated'' by high levels of AngII, and mediated inhibition of adenylate cyclase by coupling to an inhibitory GTP-binding protein. The latter characteristic (i.e., the difference in coupling) of this proposal raises the issue of the definition of a receptor. What constitutes a receptor? Should the receptive component in the cell membrane having a unique structural recognition site represent a receptor type regardless of its potential of being coupled to different

TABLE 2
Classification of Kidney AngII Receptors by Subtypes

Feature	Type A	Type B
AngII affinity (n*M*)	0.8–2	2–6
AngIII affinity	AngIII = AngII	AngIII < AngII
Signal transduction		
Primary	(+) Phospholipase C	(−) Adenyl cyclase
Secondary	(−) Adenyl cyclase	(+) Phospholipase A_2
Example	Rat mesangial cells	Rabbit proximal tubular epithelial cells

Note: (+), Stimulation; (−), inhibition.

Adapted from Douglas, J.G., *Am. J. Physiol.*, 253, F1, 1987.

intracellular events? Or should the same receptive macromolecule coupled to different transduction mechanisms be regarded as different receptor subtypes? These questions may be resolved by either the cloning of angiotensin receptors and/or by the discovery of novel and receptor subtype-specific antagonists. Since the verdict is not yet available, the use of the above criteria in defining the receptor subtypes in different target organs and in different species under varied conditions may not lead to an unambiguous and conclusive assignment.

There is one biochemical criterion reported so far which appeared capable of discriminating two types of angiotensin receptors. In 1984, Gunther[15] reported the identification of two classes of [^{125}I]AngII binding sites in rat liver membranes. The high-affinity binding sites (K_d = 0.35 n*M*) were inactivated by dithiothreitol (DTT, 0.1 to 10 m*M*), whereas the lower-affinity binding sites (K_d = 3.1 n*M*) were relatively unaffected by this treatment. Furthermore, AngII stimulation of glycogen phosphorylase in isolated rat hepatocytes (EC_{50} = 0.4 n*M*) was completely abolished by 10 m*M* DTT. In contrast, AngII inhibition of glucagon-stimulated adenylate cyclase activity (EC_{50} = 3 n*M*) in hepatocytes was unaffected by the same treatment. These data provided the strongest evidence for AngII receptor subtypes with different biochemical characteristics and mechanism of actions. Although the discriminatory effect of DTT on radiolabeled binding of angiotensin in other tissues, such as brain and ovary, has been confirmed by a number of investigators,[16,17] no further attempts on the functional consequences of DTT treatment have been reported.

For more details, there are numerous excellent reviews describing the physiology and biochemical pharmacology of the angiotensin receptors derived from the peptide approach.[3,5,7,14] There was little doubt that heterogeneity of AngII receptor did exist, although the evidence was inconclusive. The foregoing perspectives were intended only to highlight the most relevant historical observations which ushered the advent of nonpeptide AngII receptor antagonists and the cloning of angiotensin receptors into a new era of receptor research.

II. THE ADVENT OF NONPEPTIDE ANGII RECEPTOR ANTAGONISTS

In late 1982, at the height of the clinical success of angiotensin-converting enzyme inhibitors and the early stage of renin-inhibitor development, the scientists at Du Pont Pharmaceuticals embarked on a program aimed at developing nonpeptide AngII antagonists acting at the receptor level. The successful development of potent and specific nonpeptide AngII receptor antagonists has now been extensively reviewed.[18,19] The discovery of highly specific and potent nonpeptide AngII antagonists has provided another vantage point and specific tools for a better and in-depth understanding of the angiotensin receptor heterogeneity.

The rat adrenal cortex was initially chosen as the primary source of AngII receptors on which the polypeptide structure-affinity requirements were established.[20] The choice of selecting this tissue was based on a number of practical reasons: (1) it contained the highest density of AngII binding sites per mg of protein; (2) it appeared to have AngII receptors similar to those found in the vascular tissues and in the kidneys; and (3) the rat was the animal model used to determine the antihypertensive properties of potential receptor antagonists.

In this preparation, the rat adrenal cortical microsomes, a single population of high-affinity [^{3}H]AngII binding sites with K_d = 1.2 n*M* and a B_{max} of 2.6 pmol/mg protein was found.[21] A traditional SAR was established for the angiotensin receptors in this tissue using various angiotensin homologs and analogs. The rank order of affinity with increasing IC_{50} values (n*M*) was shown as [Sar^1,Ala^8] (2) $\geq$ AngII (4) = AngIII (4) $>>$ AngI (530) $>$ Ang 5–8 (800). The structural requirements for high-affinity binding have been previously described.[20,22] These affinity results with peptide analogs, together with the usual monophasic inhibition curves, provided no indication for the presence of heterogeneous populations of AngII receptors. Our current concept of receptor subtypes, however, was derived from the observation that our initial series of nonpeptide AngII receptor antagonists such as EXP6155, EXP6803, EXP7711, and DuP 753 (losartan), only inhibited the total specific binding of labeled AngII maximally up to 70 to 80%, whereas peptide analogs displayed a complete elimination.[21,23,24] This was the first indication of receptor heterogeneity present in a seemingly homogeneous preparation. Subsequently, several lines of biochemical evidence have been reported by Chiu et al.[24] and Whitebread et al.[25] that led to the current classification of AngII receptors.

A. CLASSIFICATION OF ANGIOTENSIN RECEPTORS

Angiotensin receptors have been divided into two subtypes based on the binding characteristics for their respective selective nonpeptide receptor ligands, losartan and PD123177 or peptide analog, CGP42112A.[24,25] Table 3 summarizes the various biochemical properties of these receptors, AT_1 and AT_2, which is the unified nomenclature recommended by the Ad Hoc Nomenclature Committee.[26]

Biochemically, AT_1 receptors or binding sites are characterized by their high affinity for the octapeptide AngII and its peptide antagonist, [Sar^1,Ala^8]AngII, typically effective at a low nanomolar range. The rank order of affinity is typified as [Sar^1,Ala^8]AngII $\geq$ AngII $>>$ AngIII $>>$ AngI $>$ Ang 5 – 8. Coincidentally, this rank order is the same whether the AT_1 receptor is localized in vascular tissue or in nonvascular tissue such as the adrenal cortex. Most distinctively, losartan has high affinity for this receptor, with a K_d value in the midnanomolar range and it is highly selective, with a 10,000-fold preference for AT_1 over the AT_2 receptor. As a result of this selectivity, losartan has been used as a prototypic marker for the AT_1 receptor.[27] There is now a host of nonpeptide AngII receptor antagonists which are selective for AT_1 receptor (Figures 1 and 2). Further details will be provided in the next section.

TABLE 3
Biochemical Properties of Angiotensin Receptor Subtypes

	AT_1	AT_2
Previous names	AngII-1; AngII-B; $AngII_\alpha$	AII-2; AII-A; AII_β
Affinity order	[Sar^1,Ala^8]AngII $\geq$ AngII $>$ AngIII $>>$ AngI	AngIII $\geq$ AngII $\geq$ [Sar^1,Ala^8]AngII $>>$ AngI
Selective antagonists	Losartan (DuP 753); EXP3174; DuP 532; L-158,809; GR117289; SK&F108566	PD123177; PD123319; PD121981; PD124125; CGP42112A
Selective radioligands	[^{3}H]Losartan, [^{125}I]EXP985	[3-(^{125}I)Tyr]CGP42112A
Sensitivity to $-$SH reagents	Inactivation	Enhancement
Coupling to G-proteins	Likely	Unlikely
Signal transduction	$\uparrow$ Ca^{2+}/$\uparrow$ IP_3 $\downarrow$ Adenylyl cyclase $\uparrow$ Prostaglandins	$\downarrow$ cGMP/$\uparrow$ cGMP $\uparrow$ Prostaglandins
Structure	359 amino acids 7 transmembrane domains	Unknown
Molecular size	~65–80 kDa	~120 kDa

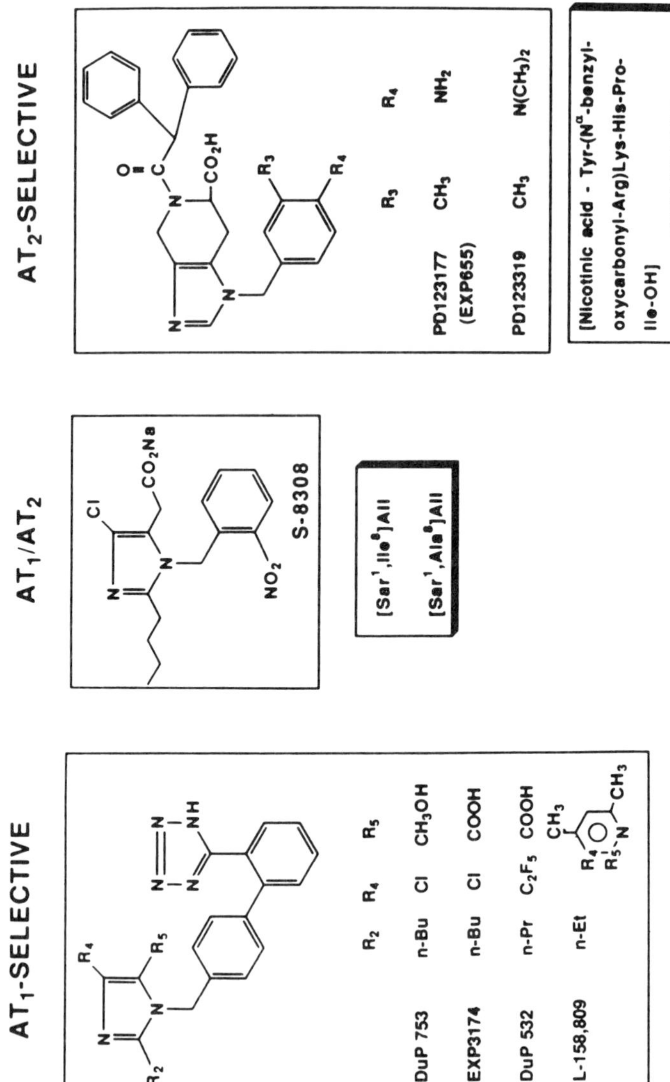

FIGURE 1. Angiotensin receptor subtype-selective ligands or antagonists.

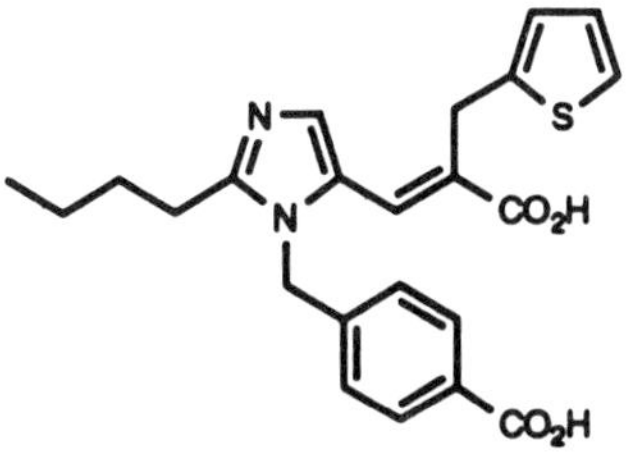

SK&F 108,566

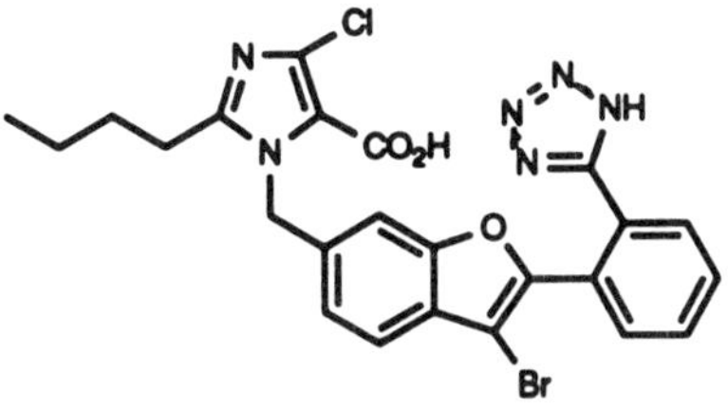

GR 117,289

FIGURE 2. Newer additions to the list of AT_1-selective receptor antagonists.

On the other hand, the AT_2 receptor subtype or, more appropriately, AT_2 binding sites, exhibit a subtle difference in its preference for the Ang peptides. The rank order of affinity is characterized as AngIII > AngII > [Sar[1],Ala[8]]AngII >> AngI >> Ang 1 – 7 > Ang 5 – 8. This order of preference has also been observed in numerous nonvascular tissues.[28-30] However, the AT_2 receptor is more uniquely defined as those angiotensin receptors which are inhibited by either CGP42112A or PD123177 or its congeners.[25,27,31] These ligands show a 1- to 4000-fold higher affinity for AT_2 than AT_1 receptors. PD123177 is effective and selective in the concentration range between 10 to 1000 n*M,* whereas CGP42112A is more potent but selective between 1 to 100 n*M*. These agents have been used as the prototypic markers for the AT_2 receptors.[25,27] Other selective AT_2 ligands have now been identified and are described in a later section.

B. RECEPTOR SUBTYPE-SELECTIVE LIGANDS OR AGENTS

The list of peptidic and nonpeptidic ligands available for studies of the biochemistry and physiology of angiotensin receptor subtypes is continuously expanding. Figures 1 and 2 illustrate the structures of some of the better-known compounds that have gone through extensive characterization.[32-36] These compounds have been shown to be specific for the AT receptors and are grouped according to their selectivity for each receptor subtype.

1. Nonselective AT_1/AT_2 Ligands

In general, peptide AngII homologs and analogs have been shown to be nonselective ligands for the AT_1/AT_2 receptors. However, AngIII has slightly higher affinity for the AT_2 than the AT_1 receptors (15-fold). [Sar[1],Ala[8]]AngII and [Sar[1],Ile[8]]AngII are potent nonselective antagonists for both receptor subtypes with affinity constants in the low nanomolar range. These peptide antagonists should be used as reference standards for characterizing Ang receptors.[27] However, caution must be exercised because of their intrinsic partial agonistic properties, metabolic instability, and the lack of oral bioavailability. There are two less potent nonpeptide AngII receptor antagonists,

S-8307 and S-8308, which show a balanced affinity for both receptor subtypes. The *in vitro* and *in vivo* pharmacology has been previously described.[37,38]

2. AT_1-Selective Antagonists

Beside losartan, there are a number of highly potent and selective AT_1 receptor antagonists that have been described.[33-36] Structurally, they bear a great deal of resemblance to losartan, having the imidazobenzylphenyltetrazole as the nucleus for structural modification. As a class of AngII antagonists, these losartan-like compounds show a selectivity of about 10,000-fold for the AT_1 receptor over the AT_2 receptor (Table 4). Losartan is the least potent among this group, having an affinity constant (K_B) of 3×10^{-9} *M*, whereas EXP3174, DuP 532, L-158,809, and GR117289 are the most potent nonpeptide AngII antagonists so far reported, possessing K_B values of about 10^{-10} *M*. These highly potent AngII antagonists are characterized by their noncompetitive or insurmountable inhibitory action, contrasting with the competitive antagonists exemplified by losartan.[18,32,39] The mechanism(s) of this noncompetitive action is presently unknown. There are suggestions that this may be related to the pseudoirreversible type of mechanism by which these antagonists may dissociate from the receptor or its surrounding complex very slowly.[40] An example is given contrasting losartan with EXP3174 in Figure 3. EXP3174 appears to dissociate from the rat adrenal cortical membranes at a rate 20 times slower than that of losartan. However, in the usual conditions

TABLE 4
Relative Inhibitory Potency of Specific Antagonists on AngII Receptor Binding and Contractile Response

Ligand/blocker	IC_{50} (n*M*)		Kinetics	K_B[c] n*M*	Ref.
	AT_1[a]	AT_2[b]			
DuP 753	5.5	>10,000	S	3.3	85
EXP3174	1.3	>10,000	I	0.1	41
DuP 532	1.8	4,000	I	0.11	40
L-158,809	0.6	ND	I	0.4	86
SK&F108566	3.0	>10,000	S	2.6	36
GR117289	1.0	>10,000	I	0.16	35
PD123177	>10,000	30	ND	ND	24,28
CGP42112A	700	0.5	ND	ND	25

Note: I, Insurmountable; S, surmountable; ND, not determined.

[a] [^{125}I]AngII binding to rat adrenal cortical AT_1 receptors.
[b] [^{125}I]AngII binding to rat adrenal medullary AT_2 receptors.
[c] Inhibitory constant for AngII-induced rabbit aortic contraction.

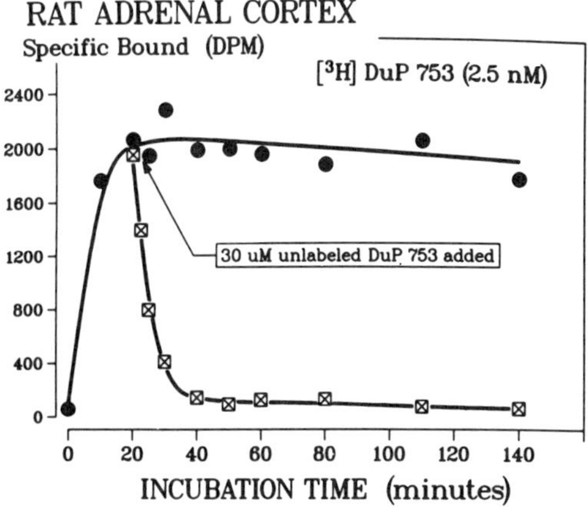

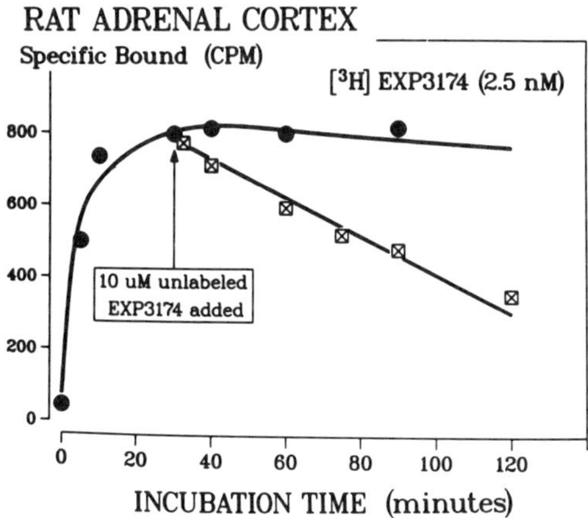

FIGURE 3. Comparison of the dissociation characteristics of [^{3}H]DuP 753 (losartan), a competitive antagonist, and [^{3}H]EXP3174, a noncompetitive antagonist. The experimental conditions have been described by Chiu et al.[45]

carried out for the competitive ligand-receptor binding studies where competing ligand is mixed with the radioligand AngII at the same time, one cannot detect the difference between a competitive and a noncompetitive antagonist. There was only a fourfold increase in affinity for the most potent compound over that of losartan in contrast to the 33-fold difference in K_B values.[41,42] To resolve this intriguing observation of insurmountable inhibition, further studies determining the biochemical or biophysical mechanisms are highly recommended.

All of these AT_1-selective agents, listed in Table 5, are direct-acting, potent receptor antagonists. However, in the case of losartan, a more potent

TABLE 5
AT_1 Receptor-Mediated Signal Transduction Pathways Sensitive to Blockade by Losartan

Signal pathway	Species	Tissue	Losartan (*M*)	Ref.
(+) Ca^{2+} mobilization	Rat	Aortic smooth muscle cells	2×10^{-8} [a]	85
	Rat	Hepatocytes	10^{-5}	54
	Mouse/rat	NG108-15 cells	10^{-6}	87
	Human	Omental endothelial cells	1.9×10^{-8} [a]	42
(−) Ionic channels	Rat	Fetal neuronal cells	10^{-7}	88
(+) Inositol metabolism	Rat	Mesangial cells	3.8×10^{-9} [a]	59
	Rat	Hepatocytes	10^{-5}	54
	Rat	Hepatocytes	10^{-6}	89
	Rat	Liver clone 9 cells	1.9×10^{-8} [a]	
	Rat	Neonatal astrocytic glia	10^{-6}	67
	Rat	Adult astrocytic glia	10^{-6}	90
(±) cAMP	Rat	(−) Liver membranes	10^{-5}	54
	Rat	(+) Fetal fibroblasts	10^{-6}	91
	Rat	(−) Adrenal glomerulosa cells	10^{-5}	65
(+) PGE_2/PGI_2	Rat	C_6 glioma cells	10^{-7}	64
(+) PGE_2	Human	Astrocytes	10^{-6}	71
(+) PGE_2/PGI_2	Pig	Aortic smooth muscle cell	10^{-6}	62

Note: (+), Stimulation; (−), inhibition.

[a] IC_{50} value.

metabolite (EXP3174) has been found in the *in vivo* and also the *in vitro* preparations such as liver slices from rat, Rhesus monkeys, and humans.[43] Oxidation of the imidazole-5-alcoholic moiety to the carboxylate EXP3174 is the major pathway of metabolism in rat, whereas tetrazole glucuronidation is the major route in monkey liver slices. Obviously, the pharmacokinetics of drug action will be altered in preparations where these metabolisms are actively involved.

Nonspecific protein bindings of these nonpeptide AT_1 receptor antagonists may also contribute to confusion in interpreting the experimental results. Even in simple *in vitro* preparations such as ligand receptor binding or Ca^{2+} mobilization in aortic smooth muscle cells, the presence of 0.25% bovine serum albumin can render the inhibitory potency of losartan 3 times less effective, 20 times less for EXP3174, and 1500 times less for DuP 532.[40] Under the same conditions, the inhibitory potency of saralasin was unaffected. In general, the noncompetitive nonpeptide antagonists, which are diacidic in nature, are more affected by this nonspecific, high-capacity protein binding. However, the severity of this problem may vary with individual chemical structure.[35,44]

The use of these nonpeptide AT_1 receptor antagonists and their radioactive ligands,[45,46] [^{3}H]DuP 753 and [^{125}I]EXP985, as markers for the AT_1 receptors has become an important pharmacological tool in renin-angiotensin receptor research. However, when comparing the inhibitory potencies and binding affinities among different preparations, the above complications or other unknown factors must be taken into consideration. Differences in potency or affinity observed may not necessarily be indicative of the presence of different AT_1 receptor subtypes.

3. AT_2-Selective Ligands

Two peptide AT_2-selective ligands, CGP42112A and pNH_2Phe^6-AngII, have been reported.[25,30] They are characterized as AT_2-selective primarily by their distinctive receptor-binding profile. The original claim that CGP42112A was an antagonist was based on it ability to block AngII-induced contraction at micromolar concentrations.[25] In fact, at these concentrations, it also acts on the AT_1 receptors. It is not certain if this is indeed an antagonist for the AT_2 receptors. Similarly, the alleged agonist effect of pNH_2Phe^6-AngII was also due to AT_1 activation.[30] Therefore, the use of these agents must be carefully scrutinized before any conclusion can be made. Based on receptor-ligand binding studies, these agents are about 1000-fold more selective for the AT_2 than for AT_1 receptors. A radioiodinated CGP42112A is currently available for receptor studies.[47]

There are a number of AT_2-selective nonpeptide ligands currently available from the Parke-Davis Pharmaceutical Company.[28] The better-known compounds are PD123177 (formerly known as EXP655) and PD123319. Structurally, they bear some resemblance to losartan and its predecessor, S-8308 (see Figure 1). PD123177 has been extensively characterized both biochemically and pharmacologically.[27,48] It displays a selectivity of 3500-fold higher affinity for the AT_2 receptor than for the AT_1 receptor. It demonstrates neither agonistic nor antagonistic properties on biological responses associated with the AT_1 receptors. However, at this time, it is not conclusive if PD123177 or its congeners is a pure antagonist or a partial agonist for the AT_2-mediated processes that still remain to be defined and substantiated. More studies are warranted, especially in defining the AT_2 functions, before reaching a verdict on this class of chemical agents.

C. BIOPHYSICAL PROPERTIES OF AT_1 AND AT_2 RECEPTORS

There are a number of biophysical properties that uniquely distinguish the two receptor subtypes. AT_1 receptors are quite sensitive to inactivation by millimolar concentrations of a sulfhydryl-reducing agent, DTT, suggesting an importance of disulfide bond(s) for its structural integrity.[25,27,49] The binding of radiolabeled AngII or [^{3}H]losartan to rat aortic smooth muscle cells containing exclusively AT_1 receptors is abolished by pretreatment of the tissue with 5 m*M* DTT. However, in rat adrenal cortical membranes where there is a mixture of both receptor subtypes, the binding of labeled AngII is increased in a concentration-dependent manner by DTT. By Scatchard analysis of the binding data, it was shown that the maximal binding capacity (B_{max}) was reduced, whereas the affinity for AngII was increased by several folds. The enhancement of AT_2 affinity for AngII by DTT is partly due to elimination of AT_1 receptor and partly due to increase in receptor affinity. Based on the results described, it is clear that AT_2 receptor is not a product of AT_1 modified by exposure to the reducing agent. Therefore, DTT or other reducing agents may be used as an additional biochemical tool for differentiating angiotensin receptor subtypes. However, caution must be exercised because the degree of AT_1 inactivation may depend on the duration of exposure, the concentration of reducing agent, the nature of tissue, and also the ionic composition of bathing solution. To ascertain if the remaining AngII binding subsequent to DTT treatment is due to the presence of AT_2 receptor or other subtypes, losartan should be used in confirming the lack of an effect on the residual bindings.

In terms of the molecular size, AT_2 appears to be slightly larger than the AT_1 receptor.[29,50] By covalent chemical crosslinking of AngII to its binding sites in tissue expressing exclusively AT_2 receptors, a protein with apparent molecular weight of about 79 kDa was found in rat ovarian granulosa cell[29] or one with about 100 kDa was noted in R3T3 cells.[28] The apparent discrepancy between the two findings is not clear but the possibility of heterogeneity of AT_2 receptors may exist. We have also used the chemical crosslinker disuccinimidyl suberate to crosslink AngII to its binding sites in rat adrenal membranes from the cortex and medulla, where AT_1 predominates in the former and AT_2 in the latter tissue. The method described by Paglin and Jamieson[51] was adopted for our experiment, in which radiolabeled protein samples were subjected to SDS-PAGE under reduced and denatured conditions. Our results (unpublished) indicate that the presence of 5 m*M* DTT in the reaction mixture enhanced the covalent linking of labeled AngII to only one protein, with an apparent molecular weight of about 106 kDa from the adrenal cortical microsomes. In the absence of DTT, the same size protein was noted, although weakly labeled. The affinity labeling was drastically reduced by pretreatment with 10^{-8} *M* AngII. In contrast, crosslinking of labeled AngII to the medulla membranes was readily achieved, even in the absence of DTT, although its presence could further enhance the coupling. However, a protein with an apparent molecular weight of about 120 kDa was detected. These results are quite similar to the earlier findings[51] where the membranes from whole adrenals were used. Most interestingly, in our studies, the crosslinking in both tissues was blocked by 10^{-6} *M* PD123177 but not by 10^{-6} *M* losartan, whether in the presence or absence of DTT. Based on the above findings, it appears that this crosslinker selectively coupled those AngII molecules that bind to the AT_2 binding sites. One can further deduce that the conformation of AngII that binds to AT_1 appears to be different from that bound to AT_2 and is less susceptible to activation by disuccinimidyl suberate. On the other hand, when the rat adrenal cortical microsomes were subjected to photoactivation with [^{125}I-Sar^1,N$_3$DPhe^8]AngII (Du Pont-NEN), a protein with an apparent molecular weight of about 80 kDa was found and this labeling was blocked by 10^{-8} *M* AngII and 10^{-6} *M* losartan but not by 10^{-6} *M* PD123177 (unpublished observations). Before the recognition of AT_1 and AT_2 receptor subtypes, the size of AngII receptors was known to be either 116 to 120 kDa or 63 to 68 kDa, depending on the type of labeling reagent or tissue used.[51,52] In view of the current findings, these data are consistent with the interpretation that the 120-kDa protein is likely to be the AT_2 receptor, whereas the 68-kDa protein is likely to be the AT_1 receptor. Further studies are encouraged in order to establish if such a differential chemical reactivity does exist between various AT receptor subtypes.

D. RECEPTOR-SECOND MESSENGER COUPLING

Angiotensin receptor subtypes may be distinguished by the presence or absence of coupling to a GTP-binding protein. Most AT_1 receptors are thought to involve GTP-binding proteins for the receptor-second messenger couplings whereas AT_2 may not. The binding proteins (G_i, G_s, G_q), when bound to receptors, are thought to confer a high-affinity state to the receptor for the agonist ligand.[53] Consequently, GTP or its nonhydrolyzable analog [GTPγS or Gpp(NH)p], has a marked inhibitory effect on the binding of AngII to receptors described above. The marked GTP effect on binding of AngII was most fully characterized in rat liver membranes which contains primarily AT_1 receptors.[15,54,55] In contrast, the binding of AngII to tissues (human myometrium, bovine cerebellar cortex, or PC12W cells) expressing exclusively AT_2 receptors was not affected by these GTP analogs.[30,56] Furthermore, Bottari et al.[56] found that AngII was unable to induce GTPγ[^{35}S] incorporation in human myometrium and bovine cerebella cortex, suggesting an absence of a G-protein in AT_2 receptor coupling. Based on these results, it was concluded that AT_2 receptors are not G-protein linked and do not belong to the G-protein linked superfamily of receptors. Although the foregoing evidence is strongly supportive, it remains prudent at this time not to take the lack of a GTP effect as a marker for receptor differentiation. It is reasonable to expect that not all AT_1 receptors are coupled to G-protein at all times and in every tissue. There are some AT_1 receptors which are not functionally coupled, such as those on platelets and those considered as spare receptors, desensitized receptors, or receptors in the "low-affinity agonist conformation." For receptor-ligand binding, these AT_1 receptor preparations may not be sensitive to GTP analogs.[57] We have also examined the effect of GTPγS on the binding of AngII to rat liver membranes and rat adrenal cortical membranes. The marked effect of GTPγS is readily detectable in rat liver but not in adrenal cortical membranes, even when AT_2 receptors have been masked by PD123177 (Figure 4) (unpublished data). The adrenal AT_1 receptors in this preparation still exhibit a high-affinity state for AngII and other agonists. It should be noted that even if the AT_2 is not a G-protein-linked

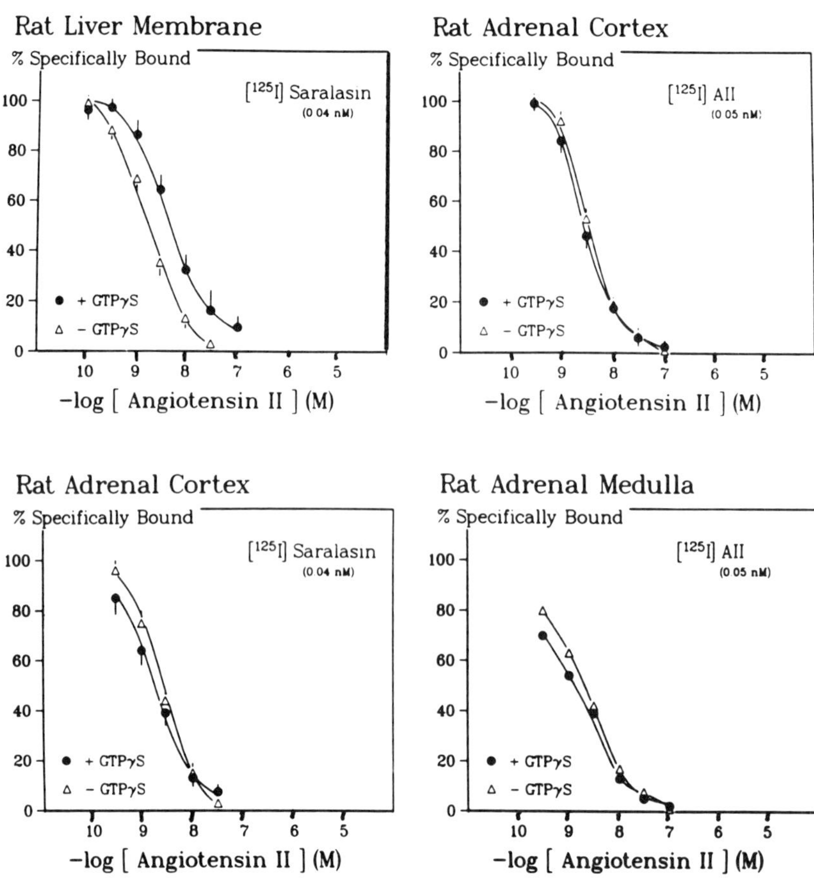

FIGURE 4. Effects of GTPγS (10^{-4} *M*) on the inhibition of either [^{125}I]saralasin or [^{125}I]AngII binding to rat liver membranes, rat adrenal cortical, and medullary microsomes. Details of the experimental conditions have been described by Chiu et al.[27,42]

receptor, it maintains a higher affinity state for both AngIII and AngII. The question whether AT_2 receptor is truly independent of a G-protein linkage may be resolved either through receptor cloning or by identification of its biochemical function.

In most cases, AngII receptors were shown to couple to many of the traditional signal transduction pathways in various sensitive target tissues.[58] For example, in adrenal cortex, liver, and vascular smooth muscle, angiotensin receptors are coupled to phospholipase C, activating the conversion of phosphatidylinositol to 1,4,5-inositol *tris*phosphate and diacylglycerol.[7,54,59] In some of these cells, AngII may directly couple to the Ca^{2+} channel, transducing an increased influx of extracellular Ca^{2+},[60,61] or to phospholipase A_2, leading to the release of arachidonate and its metabolic products.[62-64] Furthermore, AngII receptors found in liver, adrenal, and kidney are shown also to inhibit adenylate cyclase.[14,15,65] In spite of the multiplicity of biochemical actions, all of these AngII receptors are blocked by losartan. Therefore, they are all of the AT_1 receptor subtype. Various specific examples are listed in Table 5. The fact that AT_1 receptors are capable of interacting with multiple biochemical pathways suggests possible heterogeneity may exist among the AT_1 receptors. Losartan may well be a nonselective antagonist for the different AT_1 subtypes. On the other hand, one could invoke the possibility that all of these AT_1 receptors may share one common (or similar) extracellular recognition site but differ only in the multiple interactive potentials in the intracellular domains for coupling to various second messenger systems, the availability of which may be dependent on intrinsic properties of each particular tissue. These issues will be discussed in Chapter 10. In contrast to the findings, none of these traditional transducing pathways (PLC, PLA_2, Ca^{2+}, adenyl cyclase) was found to be coupled to the AT_2 receptor subtype which is exclusively expressed in Swiss R3T3 cells, rat ovarian granulosa cells, PC12W cells, and some other cell types.[29,50,66] In these cultured cell lines, AngII did not affect the basal or stimulated inositol phosphate production, intracellular Ca^{2+} mobilization, adenyl cyclase or guanyl cyclase activity, and prostaglandins production. Leung et al.[66] observed that in intact PC12W cells, atrial naturetic peptide at $10^{-7}\,M$ stimulated a 10-fold increase in cGMP. This effect of ANP was not modified even in the presence of a broad range of concentrations of AngII, indicating that AT_2 receptors do not interfere with the biochemical response of ANP receptors. On the other hand, there are a number of studies reporting a functional linkage between AT_2 receptors and cGMP or prostaglandin (PG) production, or ionic channel activation in fetal neuronal cell cultures, in human astrocytes, or in N1E-115 neuroblastoma cells. Sumners and co-workers[67] observed that neuronal cultures from neonatal rat brain contain predominantly the AT_2 receptors which are coupled to a reduction in basal cGMP levels. In another preparation using tetrodotoxin-treated neurons cultured from rat hypothalamus and brain stem, they found two distinct effects of AngII on the net outward current (I_{no}). AngII acts on AT_2 receptors to stimulate an increase in I_{no} and on AT_1 receptors to induce a reduction in I_{no}.[68] The biological consequences of these cellular events have yet to be determined. In N1E-115 neuroblastoma cells, Zarahn and co-workers[69] found that AngII stimulated an increase in cGMP production which was mediated predominantly through AT_1 and partly through AT_2 receptor subtype. Interestingly, this effect was attenuated by the nitric oxide synthetase inhibitor, *N*-monomethyl-L-arginine, suggesting an involvement of NO in AngII-induced cGMP production. Chen and Re[70] also observed that, in insulin-treated human SHSY5Y neuroblastoma cells, AngII-induced increase in [^{3}H]thymidine incorporation can be partially blocked by either losartan or PD123177, suggesting that both AT receptor subtypes may have a functional role in cell growth. In human astrocytes, Jaiswal and co-workers[71] reported that AT_1 receptors are involved in the release of PGs and in mobilization of calcium, whereas the AT_2 receptors are coupled to the release of PGs only through a calcium-independent mechanism. More recently, Bottari et al.[72] suggested that the AT_2 receptors found in membrane preparation of rat adrenal glomerulosa or PC12W cells may signal through activation of a phosphotyrosine phosphatase resulting in an inhibition of a particulate guanylate cyclase activity that was effected through ANP receptors. This is an interesting example of cross-talk between different receptors on the plasma membrane level. However, the significance of such modulation is rather unclear, especially in view of the lack of overall effects on the intact cell system.[29,50,66]

Judging from these disparate observations, one must question why AT_2 receptors in certain cell types are apparently inert while in other cell types, they appear to be functional. It seems that those cells or cell mixtures which express the AT_2 function, may also possess AT_1 receptors either on the same cells or on other co-culturing cells. It is plausible that the AT_2 function is a secondary event subsequent to AT_1 activation. In other words, AT_2 receptors may play a modulatory role directly on AT_1 receptors or their activated biochemical processes.[69] Additionally, crosstalk between AT_2 receptor and other hormonal receptors may be another possibility for consideration.[72] Further studies are necessary in order to translate the biochemical events into system responses subsequent to AT_2 receptor activation.

E. INTEGRATED FUNCTIONAL CORRELATES

The biochemical functions of AT_1 receptors, as discussed above, can be integrated into overall system responses. Virtually all the robust acute or long-term effects of AngII are mediated through the AT_1 receptors. Coincidentally, the biological responses listed in Table 6 are those that have been well characterized for AngII even before the discovery of any receptor subtypes. These responses that include vasoconstriction, vascular or nonvascular smooth muscle contraction, hormone secretions, renin release, and cellular hypertrophy have been shown to be effetively blocked by losartan. For more details on the pharmacology of losartan on these biological effects, an excellent compendium has been written by Smith et al.[19]

In spite of the few biochemical events that have been linked to the AT_2 receptors, the overall integrated functional responses remain elusive. The challenge ahead is to translate the cell culture observations into system responses that can be monitored in an *in vitro* or *in vivo* model of normal or pathological states.

F. RECEPTOR CLONING AND NOVEL AT "RECEPTOR" SUBTYPES

The AngII receptor has now been successfully cloned and expressed.[73-75] The receptor cDNA encodes a protein of 359 amino acids with the pharma-

TABLE 6
Integrated Functional Correlates

AT_1-mediated response	Species	Ref.
(+) Vasoconstriction	Rat, monkey, dog, human	92,93
(+) Vascular contraction	Rat, rabbit	28,94
(+) Nonvascular contraction	Rat, guinea pig, human	42,94
(+) Aldosterone secretion	Rat, dog, ovine, human	95–98
(+) Catecholamine release	Rat, dog	48,99
(+) Cell growth	Rat, human	61,100
Protein synthesis		
Thymidine incorporation		
(+) Drinking	Rat	101
(−) Renin release	Rat, monkey, dog, human	97,102–104
(+) Vasopressin release	Rat	105

Note: (+), Stimulation; (−), inhibition.

cological specificity of a AT_1 receptor. As receptor cloning continues to expand, various AT_1 receptor subtypes are being proposed[76,77] based on the differences primarily in the sequence of the isolated complimentary DNA. It may be more prudent at this time to delay further assignment of new nomenclature for any potential AT_1 receptor subtypes until their pharmacological properties are fully described. More details on this subject will be described in Chapter 10.

At the time of this writing, no successful cloning and expression of AT_2 receptors has been reported. The multiplicity of AT receptors are becoming evident from both the molecular cloning and from traditional biochemical and pharmacological characterization. Table 7 summarizes the various AT "receptor" subtypes or binding sites which are biochemically or pharmacologically distinct from AT_1 and AT_2. These receptors or binding sites can be activated by or bind to AngII with high affinity and possibly selectivity, but they cannot be blocked by losartan, CGP42112A, or PD123177 (or congeners). Careful examination of these "receptor" subtypes may reveal subtle differences which may help to establish their identities. The biological relevance of these "receptor" subtypes in relationship to the organism's homeostasis is unclear. Undoubtedly, this area of research will continue to expand and may broaden our horizon in the evolutionary development of the AT receptors or their related proteins.

III. FUTURE DIRECTIONS AND CONCLUDING REMARKS

Since the discovery of potent nonpeptide AngII receptor antagonists and the successful clonings of AngII receptors, the realization of heterogeneity

TABLE 7
Novel Angiotensin "Receptors" or Binding Sites

Occurrence	Response	Blocked by	Insensitive	Ref.
Mas oncogene	(+) Ca^{2+} Transients	[D-Arg, D-Pro^2, D-Trp^{7,9},Leu^{11}] Substance P	Losartan PD123177	106
Xenopus laevis oocytes	(+) $[Ca^{2+}]i$	Saralasin CG42112A	Losartan PD123177	107
Turkey adrenal cortical cells	(+) Aldosterone secretion	Sarile	Losartan PD123177	108
Mycoplasma hyorhinis	AngII binding K_d = 5 n*M*	AngI, aprotinin Bacitracin	Losartan CGP42112A Saralasin	109
Mouse neuroblastoma, Neuro-2A	AngII binding K_d = 12 n*M*	Unknown	Losartan PD123177 AngIII	110
Rabbit/rat liver cytosolic protein	AngII binding K_d = 6.7 n*M*	Saralasin	Losartan PD123177	111 112

Note: (+), Stimulation.

of angiotensin receptors, especially among species, has reached a level far beyond our original expectation. We now have some rather specific tools to dissect the anatomy of these receptor subtypes and more new tools will continue to be developed as a result of further revelations. The path forward has been suggested in each of the foregoing sections. Many of these studies are ongoing and even before the appearance of this publication, we may have answers to many of these questions: What is the biological function(s) of AT_2 receptors? Does each subtype of AT_1 receptor couple selectively to a single transduction pathway? Will there be more subtype-selective AngII blockers or agonists discovered for facilitating pharmacological evaluation? From our vantage point, the future holds many great promises both for research and therapeutic treatments.

ACKNOWLEDGMENTS

The authors wish to thank the following individuals: Dr. E. Escher, University of Sherbrooke, for a gift of [Sar^1,N_3DPhe^8]AngII; Dr. G. Brown, Du Pont-NEN, for the radioiodination of this peptide; Drs. P. K. Chakraverty, R. Rivero, and R. Simpson, all from Merck Sharp & Dohme Research Laboratories at Rahway, New Jersey, for synthesis of [^{3}H]EXP3174; Messrs. D. E. McCall, T. Nguyen, and W. Roscoe for their technical contribution; and Ms. A. Y. K. Best for the preparation of this manuscript.

REFERENCES

1. **Skeggs, L.T., Lentz, K.E., Shumway, N.P., and Woods, K.R.,** The amino acid sequence of hypertensin II, *J. Exp. Med.*, 104, 193, 1956.
2. **Peach, M.J.,** Pharmacology of angiotensin II, *Kidney Horm.*, 3, 273, 1986.
3. **Peach, M.J. and Levens, N.R.,** Molecular approaches to the study of angiotensin receptors, *Adv. Exp. Med.*, 130, 171, 1980.
4. **Chiu, A.T. and Peach, M.J.,** Inhibition of induced aldosterone biosynthesis with a specific antagonist of angiotensin II, *Proc. Natl. Acad. Sci. U.S.A.*, 71, 341, 1974.
5. **Peach, M.J. and Chiu, A.T.,** Stimulation and inhibition of aldosterone biosynthesis in vitro by angiotensin II and analogs, *Circ. Res.*, 1 (Suppl. 1), 34, 1974.
6. **Regoli, D.,** Receptors for angiotensin: a critical analysis, *Can. J. Physiol. Pharmacol.*, 57, 129, 1979.
7. **Peach, M.J. and Dostal, D.E.,** The angiotensin II receptor and the action of angiotensin II, *J. Cardiovasc. Pharmacol.*, 16(4), S25, 1990.
8. **McQueen, J. and Semple, P.F.,** Angiotensin receptor assay and characterization, *Meth. Neurosci.*, 5, 312, 1991.
9. **Glossman, H., Baukal, A.J., and Catt, K.J.,** Properties of angiotensin II receptors in the bovine and rat adrenal cortex, *J. Biol. Chem.*, 249, 825, 1974.
10. **Douglas, J., Aguilera, G., Kondo, T., and Catt, K.,** Angiotensin II receptors and aldosterone production in rat adrenal glomerulosa cells, *Endocrinology*, 102, 685, 1978.
11. **Devynck, M.A., Pernollet, M.G., Matthews, P.G., Khosla, M.C., Bumpus, F.M., and Meyer, P.,** Specific receptors for des-Asp[1]-angiotensin II (''angiotensin III'') in rat adrenals, *Proc. Natl. Acad. Sci. U.S.A.*, 74, 4029, 1977.
12. **Aguilera, G., Capponi, A., Baukal, A., Fujita, K., Hauger, R., and Catt, K.,** Metabolism and biological activities of angiotensin II and des-Asp[1]-angiotensin II in isolated adrenal glomerulosa cells, *J. Endocrinol.*, 104, 1279, 1979.
13. **Douglas, J.G., Khosla, M.C., and Bumpus, F.M.,** Efficacy of octa- and heptapeptide antagonists of angiotensin II as inhibitors of angiotensin II binding in the rat adrenal glomerulosa, *Endocrinology*, 116, 1598, 1985.
14. **Douglas, J.G.,** Angiotensin receptor subtypes of the kidney cortex, *Am. J. Physiol.*, 253, F1, 1987.
15. **Gunther, S.,** Characterization of angiotensin receptor subtypes in rat liver, *J. Biol. Chem.*, 259, 7622, 1984.
16. **Miyazaki, H., Kondoh, M., Ohnishi, J., Masuda, Y., Hirose, S., and Murakami, K.,** High-affinity angiotensin II receptors in bovine ovary are different from those previously identified in other tissues, *Biomed. Res.*, 9, 281, 1988.
17. **Speth, R.C., Rowe, B.P., Grove, K.L., Carter, M.R., and Saylor, D.L.,** Sulfhydryl reducing agents distinguish two subtypes of angiotensin II receptors in the rat brain, *Brain Res.*, 548, 1, 1991.
18. **Timmermans, P.B.M.W.M., Wong, P.C., Chiu, A.T., and Herblin, W.F.,** Nonpeptide angiotensin II receptor antagonists, *Trends Pharmacol. Sci.*, 12, 55, 1991.
19. **Smith, R.D., Chiu, A.T., Wong, P.C., Herblin, W.F., and Timmermans, P.B.M.W.M.,** Pharmacology of nonpeptide angiotensin II receptor antagonists, *Annu. Rev. Pharmacol. Toxicol.*, 32, 135, 1992.
20. **Chiu, A.T., McCall, D.E., Duncia, J.V., Johnson, A.L., Stump, J.M., Taber, R.I., and Timmermans, P.B.M.W.M.,** Structural-affinity relationship (SAR) of angiotensin II in binding to rat adrenal cortical receptors, *FASEB J.*, 3, A732, 1989.
21. **Chiu, A.T., Duncia, J.V., McCall, D.E., Wong, P.C., Price, W.A., Thoolen, M.J.M.C., Carini, D.J., Johnson, A.L., and Timmermans, P.B.M.W.M.,** Nonpeptide angiotensin II receptor antagonists. III. Structure-function studies, *J. Pharmacol. Exp. Ther.*, 250, 867, 1989.

22. **Chiu, A.T., Duncia, J.V., Wong, P.C., Price, W.A., Carini, D.J., and Timmermans, P.B.M.W.M.,** Discovery and utilities of nonpeptide angiotensin II receptor antagonist, in *Applied Cardiovascular Biology 1990 – 91,* Intl. Soc. Applied Cardiovasc. Biol., Zilla, P., Fasol, R., and Callow, A., Eds., S. Karger, Basel, 1992, p. 45.
23. **Chiu, A.T., Duncia, J.V., McCall, D.E., Wong, P.C., Price, W.A., Thoolen, M.J.M.C., Carini, D.J., Johnson, A.L., and Timmermans, P.B.M.W.M.,** Nonpeptide angiotensin II (AII) receptor antagonists. Structure-function studies, *Pharmacologist,* 30(3), A165, 1988.
24. **Chiu, A.T., Herblin, W.F., Ardecky, R.J., McCall, D.E., Carini, D.J., Duncia, J.V., Pease, L.J., Wexler, R.R., Wong, P.C., Johnson, A.L., and Timmermans, P.B.M.W.M.,** Identification of angiotensin II receptor subtypes, *Biochem. Biophys. Res. Commun.,* 165(1), 196, 1989.
25. **Whitebread, S., Mele, M., Kamber, B., and DeGasparo, M.,** Preliminary biochemical characterization of two angiotensin II receptor subtypes, *Biochem. Biophys. Res. Commun.,* 163, 284, 1989.
26. **Bumpus, F.M., Catt, K.J., Chiu, A.T., DeGasparo, M., Goodfriend, T., Husain, A., Peach, M.J., Taylor, D.G., Jr., and Timmermans, P.B.M.W.M.,** Nomenclature for angiotensin receptors, *Hypertension,* 17(5), 720, 1991.
27. **Chiu, A.T., McCall, D.E., Ardecky, R.J., Duncia, J.V., Nguyen, T.T., and Timmermans, P.B.M.W.M.,** Angiotensin II receptor subtypes and their selective nonpeptide ligands, *Receptor,* 1, 33, 1990.
28. **Dudley, D.T., Panek, R.L., Major, T.C., Lu, G.H., Burns, R.F., Klinkefus, B.A., Hodges, J.C., and Weishaar, R.E.,** Subclasses of angiotensin II binding sites and their functional significance, *Mol. Pharmacol.,* 38, 370, 1990.
29. **Pucell, A.G., Hodges, J.C., Sen, I., Bumpus, F.M., and Husain, A.,** Biochemical properties of the ovarian granulosa cell type 2-angiotensin II receptor, *Endocrinology,* 128(4), 1947, 1991.
30. **Speth, R.C. and Kim, K.H.,** Discrimination of two angiotensin II receptor subtypes with a selective agonist analogue of angiotensin II, p-aminophenylalanine angiotensin II, *Biochem. Biophys. Res. Commun.,* 169(3), 997, 1990.
31. **Weishaar, R.E., Panek, R.L., Major, T.C., Lu, G.H., Hodges, J.C., and Dudley, D.T.,** Evidence for subclasses of angiotensin II binding sites and their functional significance, *Am. J. Hypertens.,* 3(5, Pt 2), 98A, 1990.
32. **Timmermans, P.B.M.W.M., Chiu, A.T., Smith, R.D., and Wong, P.C.,** Peptide receptors I — nonpeptide angiotensin II receptor antagonists, a novel way to inhibit the renin-angiotensin system with therapeutic potential, in *Recent Advances in Receptors,* IBC Technical Services Ltd., 1991, p. 1.
33. **Smith, R.D., Chiu, A.T., Wong, P.C., Herblin, W.F., Siegl, P.K.S., and Timmermans, P.B.M.W.M.,** Nonpeptide angiotensin II receptor antagonists: a new class of antihypertensive agents, in *1992 Hypertension Annual,* Hansson, L., Ed., Current Science Ltd., London, 1992, p. 35.
34. **Siegl, P.K.S., Chang, R.S.L., Greenlee, W.J., Lotti, V.J., Mantlo, N.B., Patchett, A.A., and Sweet, C.S.,** In vivo pharmacology of a highly potent and selective nonpeptide angiotensin II (AII) receptor antagonist: L-158,809, *FASEB J.* 5 (No. 6, III), A1576, 1991.
35. **Middlemiss, D., Drew, G.M., Ross, B.C., Robertson, M.J., Scopes, D.I.C., Dowle, M.D., Akers, J.S., Cardwell, K., Clark, K.L., Coote, S., Eldred, C.D., Hamblett, J., Hilditch, A., Hirst, G.C., Jack, T., Montana, J., Panchal, T.A., Paton, J.M.S., Shah, P., Stuart, G., and Travers, A.,** Bromobenzofurans: a new class of potent, nonpeptide antagonists of angiotensin II, *Bioorg. Med. Chem. Lett.,* 1(12), 711, 1991.

36. **Weinstock, J., Keenan, R.M., Samanen, J., Hempel, J., Finkelstein, J.A., Franz, R.G., Gaitanopoulos, D.E., Girard, G.R., Gleason, J.G., Hill, D.T., Morgan, T.M., Peishoff, C.E., Aiyar, N., Brooks, D.P., Fredrickson, T.A., Ohlstein, E.H., Ruffolo, R.R., Jr., Stack, E.J., Sulpizio, A.C., Weidley, E.F., and Edwards, R.M.,** 1-(Carboxybenzyl)imidazole-5-acrylic acids: potent and selective angiotensin II receptor antagonists, *J. Med. Chem.,* 34, 1514, 1991.
37. **Chiu, A.T., Carini, D.J., Johnson, A.L., McCall, D.E., Price, W.A., Thoolen, M.J.M.C., Wong, P.C., Taber, R.I., and Timmermans, P.B.M.W.M.,** Nonpeptide angiotensin II receptor antagonists. II. Pharmacology of S-8308, *Eur. J. Pharmacol.,* 157, 13, 1988.
38. **Wong, P.C., Chiu, A.T., Price, W.A., Thoolen, M.J.M.C., Carini, D.J., Johnson, A.L., Taber, R.I., and Timmermans, P.B.M.W.M.,** Nonpeptide angiotensin II receptor antagonists. I. Pharmacological characterization of 2-n-butyl-4-chloro-1-(2-chlorobenzyl)imidazole-5-acetic acid, sodium salt (S-8307), *J. Pharmacol. Exp. Ther.,* 247, 1, 1988.
39. **Wong, P.C. and Timmermans, P.B.M.W.M.,** Nonpeptide angiotensin II receptor antagonists: insurmountable angiotensin II antagonism of EXP3892 is reversed by the surmountable antagonist DuP 753, *J. Pharmacol. Exp. Ther.,* 252, 49, 1991.
40. **Chiu, A.T., Carini, D.J., Duncia, J.V., Leung, K.H., McCall, D.E., Price, W.A., Wong, P.C., Smith, R.D., Wexler, R.R., and Timmermans, P.B.M.W.M.,** DuP 532: a second generation of nonpeptide angiotensin II receptor antagonists, *Biochem. Biophys. Res. Commun.,* 177(1), 209, 1991.
41. **Wong, P.C., Price, W.A., Chiu, A.T., Duncia, J.V., Carini, D.J., Wexler, R.R., Johnson, A.L., and Timmermans, P.B.M.W.M.,** Nonpeptide angiotensin II receptor antagonists. XI. Pharmacology of EXP3174, an active metabolite of DuP 753 — an orally active antihypertensive agent, *J. Pharmacol. Exp. Ther.,* 255(1), 211, 1990.
42. **Chiu, A.T., McCall, D.E., Price, W.A., Wong, P.C., Carini, D.J., Duncia, J.V., Wexler, R.R., Yoo, S.E., Johnson, A.L., and Timmermans, P.B.M.W.M.,** In vitro pharmacology of DuP 753, a nonpeptide AII receptor antagonist, *Am. J. Hypertens.,* 4(4), 282S, 1991.
43. **Stearns, R.A., Miller, R.R., Doss, G.A., Chakavarty, P.K., Gatto, G.J., and Chiu, S.H.L.,** The metabolism of DuP 753, a non-peptide angiotensin II receptor antagonist, by rat, monkey, and human liver slices, *Drug Metabolism and Disposition,* 20(2), 281, 1992.
44. **Smith, R.D., Duncia, J.V., Lee, R.J., Christ, D.P., Chiu, A.T., Carini, D.J., Herblin, W.F., Timmermans, P.B.M.W.M., Wexler, R.R., and Wong, P.C.,** The nonpeptide angiotensin II-receptor antagonist, losartan, in Neuropeptide Analogs, Conjugates, and Fragments. *Methods in Neurosciences,* Vol. 13, Cohn, M.L., Ed., Academic Press, Orlando, in press.
45. **Chiu, A.T., McCall, D.E., Aldrich, P.E., and Timmermans, P.B.M.W.M.,** [^{3}H]DuP 753, a highly potent and specific radioligand for the angiotensin II-1 receptor subtype, *Biochem. Biophys. Res. Commun.,* 172(3), 1195, 1990.
46. **Chiu, A.T., McCall, D.E., Carini, D.J., Smith, R.D., and Timmermans, P.B.M.W.M.,** [^{125}I]EXP985, a high potent and specific radioligand antagonist for the AT_1 receptor, *J. Hypertens.,* 10(4), S110, 1992.
47. **Whitebread, S., Taylor, V., Bottari, S.P., Kamber, B., and DeGasparo, M.,** Radioiodinated CGP42112A: a novel high affinity and highly selective ligand for the characterization of angiotensin AT_2 receptors, *Biochem. Biophys. Res. Commun.,* 181(3), 1365, 1991.
48. **Wong, P.C., Hart, S.D., Zaspel, A., Chiu, A.T., Smith, R.D., and Timmermans, P.B.M.W.M.,** Functional studies of nonpeptide angiotensin II receptor subtype-specific ligands: DuP 753 (AII-1) and PD123177 (AII-2), *J. Pharmacol. Exp. Ther.,* 255(2), 584, 1990.

49. **Chiu, A.T., McCall, D.E., Nguyen, T.T., Carini, D.J., Duncia, J.V., Herblin, W.F., Wong, P.C., Wexler, R.R., Johnson, A.L., and Timmermans, P.B.M.W.M.,** Discrimination of angiotensin II receptor subtypes by dithiothreitol, *Eur. J. Pharmacol.*, 170, 117, 1989.
50. **Dudley, D.T., Hubbell, S.E., and Summerfelt, R.M.,** Characterization of angiotensin II binding sites (AT_2) in R3T3 cells, *Mol. Pharmacol.*, 40, 360, 1991.
51. **Paglin, S. and Jamieson, J.,** Covalent crosslinking of angiotensin II to its binding sites in rat adrenal membranes, *Proc. Natl. Acad. Sci. U.S.A.*, 79, 3739, 1982.
52. **Marie, J., Seyer, R., Lombard, C., Desarnaud, F., Aumelas, A., Jant, S., and Bonnafous, J.C.,** Affinity chromatography purification of angiotensin II receptor using photoactivatable biotinylated probes, *Biochemistry*, 29, 8943, 1990.
53. **Graziano, M.P. and Gilman, A.G.,** Guanine nucleotide-binding regulatory proteins: mediators of transmembrane signaling, *Trends Pharmacol. Sci.*, 8, 478, 1987.
54. **Bauer, P.H., Chiu, A.T., and Garrison, J.C.,** DuP 753 can antagonize the effects of angiotensin II in rat liver, *Mol. Pharmacol.*, 39, 579, 1991.
55. **Crane, J.K., Campanile, C.P., and Garrison, J.C.,** The hepatic angiotensin II receptor. II. Effect of guanine nucleotides and interaction with cyclic AMP production, *J. Biol. Chem.*, 257, 4959, 1982.
56. **Bottari, S.P., Taylor, V., King, I.N., Bogdal, Y., Whitebread, S., and DeGasparo, M.,** Angiotensin II AT_2 receptors do not interact with guanine nucleotide binding proteins, *Eur. J. Pharmacol.*, 207, 157, 1991.
57. **Sumners, C., Myers, L.M., Kalberg, C.J., and Raizada, M.K.,** Physiological and pharmacological comparisons of angiotensin II receptors in neuronal and astrocyte glial cultures, *Prog. Neurobiol.*, 34, 355, 1990.
58. **Peach, M.J.,** Molecular actions of angiotensin, *Biochem. Pharmacol.*, 30, 2745, 1981.
59. **Pfeilschifter, J.,** Angiotensin II B-type receptor mediates phosphoinositide hydrolysis in mesangial cells, *Eur. J. Pharmacol.*, 184, 201, 1990.
60. **Kojima, I., Shibata, H., and Ogata, E.,** Pertussis toxin blocks angiotensin II-induced calcium influx but not inositol triphosphate production in adrenal glomerulosa cell, *FEBS Lett.*, 204, 347, 1986.
61. **Chiu, A.T., Roscoe, W.A., McCall, D.E., and Timmermans, P.B.M.W.M.,** Angiotensin II-1 receptors mediate both vasoconstrictor and hypertrophic responses in rat aortic smooth muscle cells, *Receptor*, 1(3), 133, 1991.
62. **Leung, K.H., Chang, R.S.L., Lotti, V.J., Roscoe, W.A., Smith, R.D., Timmermans, P.B.M.W.M., and Chiu, A.T.,** AT_1 receptors mediate the release of prostaglandins in porcine smooth muscle cells and rat astrocytes, *Am. J. Hypertens.*, 5, 648, 1992.
63. **Leung, K.H., Roscoe, W.A., Smith, R.D., Timmermans, P.B.M.W.M., and Chiu, A.T.,** DuP 753, a nonpeptide angiotensin II receptor antagonist, does not have a direct stimulatory effect on prostacyclin and thromboxane synthesis, *FASEB J.*, 5(6), A1767, 1991.
64. **Jaiswal, N., Diz, D.I., Tallant, E.A., Khosla, M.C., and Ferrario, C.M.,** Characterization of angiotensin receptors mediating prostaglandin synthesis in C6 glioma cells, *Am. J. Physiol.*, 260(5 Pt. 2), 1000R, 1991.
65. **Balla, T., Baukal, A.J., Eng, S., and Catt, K.J.,** Angiotensin II receptor subtypes and biological responses in the adrenal cortex and medulla, *Mol. Pharmacol.*, 40, 401, 1991.
66. **Leung, K.H., Roscoe, W.A., Smith, R.D., Timmermans, P.B.M.W.M., and Chiu, A.T.,** Angiotensin II AT_2 binding sites are not functionally coupled in the rat pheochromocytoma PC12W cells, *Eur. J. Pharmacol.*, 227, 63, 1992.
67. **Sumners, C., Tang, W., Zelezna, B., and Raizada, M.K.,** Angiotensin II receptor subtypes are coupled with distinct signal transduction mechanisms in cultured neurons and astrocyte glia from rat brain, *Proc. Natl. Acad. Sci., U.S.A.*, 88, 7567, 1991.

68. **Kang, J., Sumners, C., and Posner, P.,** Modulation of net outward current in cultured neurons by angiotensin II: involvement on AT_1 and AT_2 receptors, *Brain Res.,* 580, 317, 1992.
69. **Zarahn, E.D., Ye, X., Ades, A.M., Reagan, L.P., and Fluharty, S.J.,** Angiotensin-induced cGMP production is mediated by multiple receptor subtypes and nitric oxide in N1E-115 neuroblastoma cells, *J. Neurochem.,* 58, 1960, 1992.
70. **Chen, L. and Re, R.N.,** Angiotensin and the regulation of neuroblastoma cell growth, *Am. J. Hypertens.,* 4(5), 82A, 1991.
71. **Jaiswal, N., Tallant, E.A., Diz, D.I., Khosla, M.C., and Ferrario, C.M.,** Subtype 2 angiotensin receptors mediate prostaglandin synthesis in human astrocytes, *Hypertension,* 17, 1115, 1991.
72. **Bottari, S.P., King, I.N., Reichlin, S., Dahlstroem, I., Lydon, N., and DeGasparo, M.,** The angiotensin AT_2 receptor stimulates protein tyrosine phosphatase activity and mediates inhibition of particulate guanylate cyclase, *Biochem. Biophys. Res. Commun.,* 183(1), 206, 1992.
73. **Murphy, T.J., Alexander, R.W., Griendling, K.K., Runge, M.S., and Bernstein, K.E.,** Isolation of a cDNA encoding the vascular type-1 angiotensin II receptor, *Nature,* 351, 233, 1991.
74. **Iwai, N., Yamano, Y., Chaki, S., Konishi, F., Bardhan, S., Tibbetts, C., Sasaki, K., Hasegawa, M., and Inagami, T.,** Rat angiotensin II receptor: cDNA sequence and regulation of the gene expression, *Biochem. Biophys. Res. Commun.,* 177 (1), 299, 1991.
75. **Sasaki, K., Yamano, Y., Bardhan, S., Iwai, N., Murray, J.J., Hasegawa, M., Matsuda, Y., and Inagami, T.,** Cloning and expression of a complementary DNA encoding a bovine adrenal angiotensin II type-1 receptor, *Nature,* 351, 230, 1991.
76. **Furuta, H., Guo, D.F., and Inagami, T.,** Molecular cloning and sequencing of the gene encoding human angiotensin II type 1 receptor, *Biochem. Biophys. Res. Commun.,* 183(1), 8, 1992.
77. **Iwai, N. and Inagami, T.,** Identification of two subtypes in the type 1 angiotensin II receptor, *FEBS Lett.,* 298, 257, 1992.
78. **Khosla, M.C., Smeby, R.R., and Bumpus, F.M.,** Structure-activity relationship in angiotensin II analogs, in *Handbook of Experimental Pharmacology,* Vol. 37, Angiotensin, Page, I.H. and Bumpus, F.M., Eds., Springer-Verlag, Berlin, 1974, 126.
79. **Peach, M.J. and Ober, M.,** Inhibition of angiotensin-induced catecholamine release by 8-substituted analogs of angiotensin II, *J. Pharmacol. Exp. Ther.,* 190, 49, 1974.
80. **Peach, M.J.,** Physiological roles of angiotensin, in *Chemistry and Biology of Peptides,* Meinhofer, J., Ed., Ann Arbor Science Publishers, Ann Arbor, Michigan, 1972, 471.
81. **Gagnon, D.J., Cousineau, D., and Boucher, P.J.,** Release of vasopressin by angiotensin II and prostaglandin E_2 from the rat neurohypophysis in vitro, *Life Sci.,* 12, 487, 1973.
82. **Bumpus, F.M., Levens, N.R., Munday, K.A., and Poat, J.A.,** Structural activity requirements for the action of angiotensin II on fluid transport by rat jejunum, *J. Physiol. (London),* 257, 32, 1976.
83. **Fitzsimons, J.T.,** Renin, angiotensin, and drinking, in *Control Mechanisms of Drinking,* Peters, G., Fitzsimons, J.T., and Peters-Haefeli, L., Eds., Springer-Verlag, Berlin, 1975, 97.
84. **Freedlender, A.E. and Goodfriend, T.L.,** Angiotensin receptors and sodium transport in renal tubules, *Fed. Proc.,* 36, 481, 1977.
85. **Chiu, A.T., McCall, D.E., Price, W.A., Wong, P.C., Carini, D.J., Duncia, J.V., Wexler, R.R., Yoo, S.E., Johnson, A.L., and Timmermans, P.B.M.W.M.,** Non-peptide angiotensin II receptor antagonists. VII. Cellular and biochemical pharmacology of DuP 753, an orally active antihypertensive agent, *J. Pharmacol. Exp. Ther.,* 252, 711, 1990.
86. **Chang, R.S.L., Siegl, P.K.S., Mantlo, N.B., Greenlee, W.J., and Lotti, V.J.,** In vitro pharmacology of a highly potent and selective nonpeptide angiotensin II (AII) receptor antagonist: L-158,809, *FASEB J.,* 5(6), A1575, 1991.

87. **Tallant, E.A., Diz, D.I., Khosla, M.C., and Ferrario, C.M.**, Identification and regulation of angiotensin II receptor subtypes on NG108-15 cells, *Hypertension,* 17(6, Pt. 2), 1135, 1991.
88. **Kang, J., Posner, P., and Sumners, C.**, Angiotensin II type 1 (AT_1)- and angiotensin II type 2 (AT_2)-receptor-mediated changes in potassium currents in cultured neurons: role of intracellular calcium, *FASEB J.*, 6(4), A443, 1992.
89. **Garcia-Sainz, J.A. and Macias-Silva, M.**, Angiotensin II stimulates phosphoinositide turnover and phosphorylase through AII-1 receptors in isolated rat hepatocytes, *Biochem. Biophys. Res. Commun.*, 172(2), 780, 1990.
90. **Raizada, M.K., Zelezna, B., Tang, W., and Sumners, C.**, Astrocytic glial cultures from the brains of adult normotensive and hypertensive rats predominantly express angiotensin II-1 (AII-1) receptors, *FASEB J.*, 5 (4, Pt. I), A871, 1991.
91. **Johnson, C. and Aguilera, G.**, Angiotensin-II receptor subtypes and coupling to signaling systems in cultured fetal fibroblasts, *Endocrinology,* 129(3), 1266, 1991.
92. **Wong, P.C., Price, W.A., Chiu, A.T., Duncia, J.V., Carini, D.J., Wexler, R.R., Johnson, A.L., and Timmermans, P.B.M.W.M.**, Nonpeptide angiotensin II receptor antagonists. VIII. Characterization of functional antagonism displayed by DuP 753, an orally active antihypertensive agent, *J. Pharmacol. Exp. Ther.*, 252, 719, 1990.
93. **Wong, P.C., Hart, S.D., Duncia, J.V., and Timmermans, P.B.M.W.M.**, Nonpeptide angiotensin II receptor antagonists. XIII. Studies with DuP 753 and EXP3174 in dogs, *Eur. J. Pharmacol.*, 202, 323, 1991.
94. **Rhaleb, N.E., Rouissi, N., Nantel, F., D'Orleans-Juste, P., and Regoli, D.**, DuP 753 is a specific antagonist for the angiotensin receptor, *Hypertension,* 17, 480, 1991.
95. **Chang, R.S.L. and Lotti, V.J.**, Two distinct angiotensin II receptor binding sites in rat adrenal revealed by new selective nonpeptide ligands, *Mol. Pharmacol.*, 29, 347, 1990.
96. **Clark, K.L., Robertson, M.J., and Drew, G.M.**, Effects of the non-peptide angiotensin receptor antagonist, DuP 753, on basal renal function and on the renal effects of angiotensin II in the anaesthetized dog, *Br. J. Pharmacol.*, 104 (Suppl.), 78, 1991.
97. **Christen, Y., Waeber, B., Nussberger, J., Porchet, M., Lee, R., Maggon, K., Shum, L., Timmermans, P.B.M.W.M., Brunner, H.R., and Borland, R.M.**, Oral administration of DuP 753, a specific angiotensin II antagonist, to normal male volunteers: inhibition of pressor response to exogenous angiotensin I and II, *Circulation,* 83(4), 1333, 1991.
98. **Fitzpatrick, M.A., Rademaker, M.T., and Espiner, E.A.**, Acute effects of the angiotensin II antagonist, DuP 753, in heart failure, *J. Am. Coll. Cardiol.*, 19(3), 146A, 1992.
99. **Wong, P.C., Hart, S.D., and Timmermans, P.B.M.W.M.**, Effect of angiotensin II antagonism on canine renal sympathetic nerve function, *Hypertension,* 17 (6, Pt. 2), 1127, 1991.
100. **Bakris, G.L., Akerstrom, V., and Re, R.N.**, Insulin, angiotensin II antagonism and converting enzyme inhibition: effect on human mesangial cell mitogenicity and endothelin, *Hypertension,* 3, 326, 1991.
101. **Wong, P.C., Price, W.A., Chiu, A.T., Carini, D.J., Duncia, J.V., Johnson, A.L., Wexler, R.R., and Timmermans, P.B.M.W.M.**, Nonpeptide angiotensin II receptor antagonists: studies with EXP9270 and DuP 753, *Hypertension,* 15, 823, 1990.
102. **Koepke, J.P., Bovy, P.R., McMahon, E.G., Olins, G.M., Reitz, D.B., Salles, K.S., Schuh, J.R., Trapani, A.J., and Blaine, E.H.**, Central and peripheral actions of a nonpeptidic angiotensin II receptor antagonist, *Hypertension,* 15(6), 841, 1991.
103. **Gibson, R.E., Thorpe, H.H., Cartwright, M.E., Frank, J.D., Schorn, T.W., Bunting, P.B., and Siegl, P.K.S.**, Angiotensin II receptor subtypes in the renal cortex of rat and rhesus monkey, *Am. J. Physiol.*, 261(3, Pt. 2), F512, 1991.
104. **Iwao, H.**, Effects of semotiadil on renal hemodynamics and function in dogs, *Eur. J. Pharmacol.*, in press.

105. **Hogarty, D.C. and Phillips, M.I.,** Vasopressin release by central angiotensin II is mediated through an angiotensin type-1 receptor and the drinking response is mediated by both AT-1 and AT-2 receptors, *Soc. Neurosci.*, 17, 1188, 1991.
106. **Hanley, M.R.,** Molecular and cell biology of angiotensin receptors, *Cardiovasc. Pharmacol.*, 18(2), S7, 1991.
107. **Hong, J., Sandberg, K., and Catt, K.J.,** Novel angiotensin II antagonists distinguish amphibian from mammalian angiotensin II receptors expressed in xenopus laevis oocytes, *Mol. Pharmacol.*, 39, 120, 1990.
108. **Kocsis, J.F., Carsia, R.V., Chiu, A.T., and McIlroy, P.J.,** Properties of angiotensin II receptors of domestic turkey adrenocortical cells, *Am. J. Physiol.*, (in press.)
109. **Bergwitz, C., Madoff, S., Abou-Samra, A.B., and Juppner, H.,** Specific, high-affinity binding sites for angiotensin II on mycoplasma hyorhinis, *Biochem. Biophys. Res. Commun.*, 179(3), 1391, 1991.
110. **Chaki, S. and Inagami, T.,** Identification and characterization of a new binding site for angiotensin II in mouse neuroblastoma neuro-2A cells, *Biochem. Biophys. Res. Commun.* 182(1), 388, 1992.
111. **Bandyopadhyay, S.K., Rosenberg, E., Kiron, R.M.A., and Soffer, R.L.,** Purification and properties of an angiotensin-binding protein from rabbit liver particles, *Arch. Biochem. Biophys.*, 263, 272, 1988.
112. **Schelhorn, T.M., Burkard, M.R., Rauch, A.L., Mangiapane, M.L., Murphy, W.R., and Holt, W.F.,** Differentiation of the rat liver cytoplasmic AII binding protein from the membrane angiotensin II receptor, *Hypertension,* 16, 36, 1990.

Chapter 11

CLONING, EXPRESSION, AND REGULATION OF ANGIOTENSIN II RECEPTORS

T. Inagami, N. Iwai, K. Sasaki, Y. Yamano, S. Bardhan, S. Chaki, D.-F. Guo, and H. Furuta

TABLE OF CONTENTS

0-8493-4622-3/93/$0.00 + $.50

I. INTRODUCTION

Angiotensin II (AngII) elicits multiple responses in a variety of tissues,[1] including both short- and long-term effects such as contraction of (vascular) smooth muscle and mesangial cells, stimulation of aldosterone secretion from the adrenal cortex, facilitation of norepinephrine release from nerve endings, stimulation of sodium reabsorption in the renal proximal tubules, inhibition of renin release from the juxtaglomerular cells, induction of hypertrophy of cardiac myocytes and vascular smooth muscle cells. (See Chapter 9.) Administered into the brain either vascularly or intracerebraventricularly, this peptide induces neurogenic hypertension, intense dipsogenesis, and stimulation of the release of vasopressin. (See Chapter 14.) These responses are believed to be mediated by angiotensin receptor(s) which elicits a complex set of intracellular responses that consist of Gq-mediated phospholipase $C_{\beta 1}$ activation which results in transient production of inositol 3,4,5,-*tris*phosphate (IP_3) and diacylglycerol, inducing transient release of Ca^{2+} from sarcoplasmic reticulum. In addition, AngII induces partial inhibition of adenylyl cyclase mediated by a G_i protein and calcium channel opening. Either the elevated intracellular calcium concentration or the activation of protein kinase C by the diacylglycerol formed by the phospholipase C-mediated mechanism causes activation of the phospholipase A_2 and D which initiate the arachidonate cascade. Whether these complex reactions are mediated by multiple AngII receptors or a single multifunctional receptor is not clear.

The AngII receptor has never been isolated in a pure and stable form because it is one of the most unstable receptors when solubilized from plasma membrane by detergents. Thus, it was not clear whether the multiple responses elicited by this peptide hormone are mediated by different types of the receptor or a single-type receptor.

Pharmacological application of subtype-specific inhibitors to various tissues[2-4] and observation of differential stability of AngII receptors against sulfhydryl reagents[2,5,6] revealed that there exists at least two types of AngII receptors. In addition, a soluble AngII binding protein was reported in the cytosol of various tissues.[7,8] The *mas* oncogene product has been postulated as a possible receptor of AngII.[9]

Examination of effects of the subtype-specific inhibitors on various AngII-induced responses revealed that most of the long- and short-term effects of AngII hitherto recognized are mediated by the losartan-sensitive AT_1 receptors.[24] However, it has not been clear whether the AT_1 exists in a single form or its related subtypes exist. Further, it has not been clear whether these responses are elicited by the promiscuous actions of the single-type receptor or specific actions of various subtypes of AT_1. Coexistence of AT_1 and AT_2 subtypes in different cells and tissues limited the scope of studies aimed at the identification of specific actions of receptor subtypes. Moreover, the possible presence of subtypes among the AT_1 receptor made it impossible to use various cells and tissues for the precise identification and characterization of the signalling mechanism of these receptors.

II. CLONING OF ANGIOTENSIN II RECEPTOR cDNA

In order to clarify the biochemical basis for the multiple responses to AngII and multiple AngII receptor/binding proteins, it is essential to purify individual AngII receptors. In view of the exceeding instability of detergent-solubilized AngII receptors, and in view of controversy regarding the nature of *mas* oncogene product as AngII receptor, conventional cloning procedures could not be used for the isolation of AngII receptor cDNA. We employed the expression cloning methods developed by Seed[10] using a shuttle vector pCDM8 as a cloning vehicle and obtained a cDNA encoding the subtype AT_1 from cultured bovine adrenal zona glomerulosa cells.[11] Using the bovine adrenal AT_1 receptor cDNA, we immediately cloned a rat AT_1 cDNA from a cDNA library constructed from the renal mRNA of spontaneously hypertensive rats (SHR).[12] Murphy et al., using a similar method and the pCDM8 vector, was able to clone a cDNA for AT_1 from cultured rat aortic vascular smooth muscle cells.[13]

cDNA (>2.5 kb) inserted into the shuttle vector (pCDM8) were transfected into mammalian cells (Cos-7) and were allowed to express in Cos-7 cells. Cells harboring cDNA for AngII receptor expressed AngII receptor on their plasma membrane, which were detected by the binding of the ^{125}I-labeled angiotensin analog [^{125}I-Sar^1,Ile^8]AngII by autoradiography or binding assay with a small number of subpooled cells.

A very high level of AT_1 mRNA expressed in the primary culture and second passage of bovine adrenocortical zona glomerulosa cells permitted us to detect AT_1 cDNA clones from transfected Cos-7 cells without subpooling. Sixty-seven thousand clones amplified in *Escherichia coli,* were used to transfect one million Cos-7 cells which were then subjected to autoradiography in 12 wells. More than 10 positive clones were isolated. A similar method was used by Murphy et al.[13] in cloning cDNA for an AT_1 receptor from cultured rat aortic vascular smooth muscle cells. These cells also expressed the AT_1 cDNA at a high level but cDNA (43,000 clones) were amplified in 144 subpools of approximately 300 clones each.

Tissue-dependent differences in the functions and developmental expression of AT_1 receptor made us suspect the presence of organ-dependent subtypes of AT_1-type receptors. However, comparison of AT_1s from different species did not permit us to identify such subtypes with possible functional differences. For example while bovine adrenal AT_1 and rat renal or vascular AT_1 showed approximately 8% amino acid nonidentity as shown in Figure 1, it was not possible to determine how much of this difference is due to species difference and how much is due to organ difference.

It was necessary to determine the number of related genes in the same species. Southern blot analysis using rat renal AT_1 cDNA as probe suggested

```
                 ▼                                       I
Rat AT1A      MALNSSAEDG  IKRIQDDCPK  AGRHSYIFVM  IPTLYSIIFV  VGIFGNSLVV   50
Rat AT1B      -T----T---------------------------------------------------
Bovine AT1    -I----T---------------------N---I-------------------------
Human AT1     -I----T---------------------------------------------------

                                         II
Rat AT1A      IVIYFYMKLK  TVASVFLLNL  ALADLCFLLT  LPLWAVYTAM  EYRWPFGNHL  100
Rat AT1B      ----------------------------------------------------------
Bovine AT1    --------------------------------------------------------Y-
Human AT1     --------------------------------------------------------Y-

                             III
Rat AT1A      CKIASASVTF  NLYASVFLLT  CLSIDRYLAI  VHPMKSRLRR  TMLVAKVTCI  150
Rat AT1B      --------S-------------------------------------------------
Bovine AT1    --------S-------------------------------------------------
Human AT1     --------S-------------------------------------------------

                        IV                  ▼             ▼
Rat AT1A      IIWLMAGLAS  LPAVIHRNVY  FIENTNITVC  AFHYESRNST  LPIGLGLTKN  200
Rat AT1B      -----------------Y------------------------Q---------------
Bovine AT1    ----L---------TI-----F--------------------Q-------V-------
Human AT1     ----L----------I-----F--------------------Q---------------

                 V
Rat AT1A      ILGFLFPFLI  ILTSYTLIWK  ALKKAYEIQK  NKPRNDDIFR  IIMAIVLFFF  250
Rat AT1B      ----V-------------------------K------T--------------------
Bovine AT1    ------------------------T---------------K----K----L-------
Human AT1     ---------------------------------------------K------------
```

```
                 VI                                       VII
              ________________                  ___________________________
Rat AT1A      FSWVPHQIFT  FLDVLIQLGV  IHDCKISDIV  DTAMPITICI  AYFNNCLNPL   300
Rat AT1B      ---------------------I---R--E-A---------------------------
Bovine AT1    -------------M-------L---R----E--------------L------------
Human AT1     ---I-----------------I---R--R-A---------------------------

              ___
Rat AT1A      FYGFLGKKFK  KYFLQLLKYI  PPKAKSHSSL  STKMSTLSYR  PSDNMSSSAK   350
Rat AT1B      --------------------------T----AG-------------------------
Bovine AT1    --------------------------------N-----------------E-GN--T-
Human AT1     ------------R-------------------N-------------------V---T-

Rat AT1A      KPASCFEVE*                                                   359
Rat AT1B      -S--F-----
Bovine AT1    ---P-I----
Human AT1     ---P------
```

FIGURE 1. Amino acid sequences of AngII type 1 receptors (AT_1) from various sources, including subtypes A and B from rat. The amino acid sequences were deduced from the base sequences in the coding regions of cDNAs for rat AT_{1A},[12,13] rat AT_{1B},[14] bovine AT_1,[11] human AT_1,[16] and the open reading frame for human genomic DNA.[15] Thick bars with Roman numerals (above amino acid sequences) indicate seven putative transmembrane domains deduced by hydropathy analysis. Broken lines indicate amino acid residues identical with rat AT_{1A}. Only those that are different are indicated by alphabets. Three arrow heads indicate potential N-glycosylation sites.

that rat genome could contain between two and four subtypes of the AT_1 gene. From 30,000 clones of a rat adrenal cDNA (>2 kb) library in pSPORT a new AT_1 cDNA was cloned by colony hybridization using rat kidney AT_1 cDNA which covered almost all of the coding region.[14] Conditions for the hybridization were nonstringent (35% formamide, 6 × SSC, and 1% sodium dodecyl sulfate at 42°C for 12 h). The new rat AT_1 cDNA (AT_{1B}) thus cloned was different from the rat renal or vascular cDNA (AT_{1A}) both in the coding and noncoding regions. In the coding region only 16 amino acid residues (4.5%) were different out of a total of 359 amino acid residues encoded by the cDNA (Figure 1), whereas 16% of base sequences were different. However, in the noncoding region no base sequence homology was identifiable between rat AT_{1A} and AT_{1B}. Thus, it is clear that AT_{1A} and AT_{1B} represent two different genes.

The presence of two subtypes (AT_{1A} and AT_{1B}) in rat suggests similar subtypes in other species. Although the bovine adrenal cortical receptor we cloned could be AT_{1B} in view of its adrenal origin and partial sequence similarity to rat AT_{1B}, no definitive evidence for such a contention is apparent until another subtype is cloned and compared with the existing one.

As discussed later, the AT_1 receptor gene has an exon in which the entire coding region is contained. Based on the assumption that this property extends to the human AT_1 gene, we have cloned a genomic DNA encoding human AT_1 receptor. A human lymphocyte genomic library in a lambda phage (1.2 × 10^6 plaques) was screened with rat AT_1 cDNA fragment, which covered almost all the open reading frame of rat AT_{1A} (identical with that used in the cloning of rat AT_{1B} cDNA). Hybridization conditions were again nonstringent: 30% formamide, 6 × SSC, 2.5 × Denhardt's solution, and 1% sodium dodecyl sulfate at 42°C. Base sequence analysis of a 6.1-Kb *Bam*H1 fragment thus cloned contained an open reading frame consisting of 1080 base pairs encoding 359 amino acid residues.[15] This sequence showed 91% base sequence identity and 95% amino acid sequence identity with bovine AT_1. Compared with rat AT_1, the amino acid sequence of the human AT_1 showed 95% identity to rat AT_{1A} and 97% with rat AT_{1B}, the latter being somewhat more closely related to the human AT_1, suggesting possible identity of this human sequence with the AT_{1B} subtype.

A human AT_1 cDNA was also cloned from a human cDNA library prepared from human liver poly(A)$^+$ RNA.[16] The base sequence of this cDNA and the amino acid sequence deduced thereof were identical with those of human genomic DNA.[15]

A. THE STRUCTURE OF THE AT_1 RECEPTOR

The base sequence of cDNA for bovine adrenocortical AT_1[11] is shown in Figure 2. The initiation codon is preceded by a nucleotide sequence in agreement with the Kozak's consensus sequence. The polypeptide sequence encoded by the open reading frame consists of 359 amino acid residues with a relative molecular mass (M_r) of 41,093. This M_r is in reasonable agreement with that of the solubilized and deglycosylated form of the AngII receptor (35,000) determined by polyacrylamide gel electrophoresis (PAGE) in sodium dodecyl sulfate.[17,18] The total number of amino acid residues (359) is conserved throughout subtypes (AT_{1A}[12] and AT_{1B}[14]) and species, encompassing bovine,[11] rat,[12-14] and humans.[15,16]

Hydropathy analysis (Figure 3) of the amino acid sequences suggested the presence of seven transmembrane domains which are again located in the same sequences of AT_1s from different species and subtypes,[11-16] as indicated by lines above the sequences in Figures 1 and 2. The amino acid sequence showed 20 to 30% sequence identity with those of other G-protein-coupled receptors. The protein showed only 9% sequence identity with human *mas* oncogene product, which had been proposed to be an AngII receptor,[9] but failed to show binding to AngII peptide or its analogs. These observations negate the eligibility of the *mas* oncogene as the AngII receptor gene.

The protein contains several potential sites for posttranslational modification, which include three consensus sites for potential N-glycosylation,[19] one of which is in the N-terminal region preceding the first transmembrane domain and the other two in the third extracellular loop (indicated by arrow heads in Figure 2). Several serine and threonine residues are in the second and C-terminal cytoplasmic domains for possible regulatory phosphorylation although neighboring amino acid sequences are not exactly congruous with sites for cAMP-dependent protein kinase, or protein kinase C.

Other features are a short third intracellular loop in contrast to very long ones for monoaminergic and muscarinic receptors, and the presence of a cysteine residue in the C-terminal (cytosolic) region, a possible palmitoylation site.[20,21] One cysteine residue is present in each of the four extracellular loops. Those in the second and third extracellular domains are conserved throughout most of the receptors with the seven transmembrane domain motif.[5] However, those in the N-terminal (first) and fourth extracellular domain are unique to AngII receptor. It is likely that these two pairs of cysteine residues form disulfide bridges which confer a conformation essential for AngII binding. The AT_1 receptor is known for sensitivity to reducing agents such as dithiothreitol.[17,18] This sensitivity may be due to these disulfide bridges, particularly the one unique to AT_1 receptor, presumably formed between the first and fourth extracellular domains.

The above features are commonly found both in the rat AT_{1A} and AT_{1B} receptors, as well as in the bovine and human AT_1 receptors. However, it was noted that important differences exist between rat AT_{1A} and AT_{1B}. The latter was found to lack a cysteine residue in the carboxy-terminal region and a potential site for phosphorylation probably by protein kinase C is present in ^{232}Thr.[14] This site is not present in AT_{1A}.

Based on these studies, a model of AT_1 with seven transmembrane regions was constructed as shown in Figure 3. Two tentative disulfide bridges are indicated on the basis of the above discussion. The disulfide bridge linking Cys^{18} in the first extracellular (N-terminal) segment and Cys^{274} in the fourth

```
                                                                  -409   GGAGGAGAGAGTG -397
CAAAACACAGCCTCGCCCCTGGAACCCTCGAAGAAGCAACGTCCCTCGCTATAAATTTGGAGCTGCCTCAGAGAGGACGATTCCAGCGCCAGTCAGCCA -298
AGGGCCTGAGGCGACTGAGCGCCCGGGCAGACGCGCGACTAGCAGCTCTGCCTACCCGAGACGGCGATCCACGAGGAGGAACCCAAGCGGACCCCAGGA -199
GTGAGGGCGCACCCTGAGCCGGCGCCGCCAGCTCGGCCCGACGCCGGCGCAGCGTCCCGAGTGAGAGCGGACTGATGTATGATAAGCTGATCTAAAATG -100
ACAGGTTTTATCTGACTAAGTCATTGAAGCCATTCCAGAAGGTCTGCATCCAGGTGCATTTGGCATAGTGTCTGAGACTGACTAACCCAAGAGATCAAA   -1

Met Ile Leu Asn Ser Ser Thr Glu Asp Gly Ile Lys Arg Ile Gln Asp Asp Cys Pro Lys Ala Gly Arg His Asn  25
ATG ATC CTC AAC TCT TCC ACT GAA GAT GGT ATT AAA AGA ATC CAA GAT GAT TGT CCC AAA GCT GGA AGG CAC AAT  75
                                                    I
Tyr Ile Phe Ile Met Ile Pro Thr Leu Tyr Ser Ile Ile Phe Val Val Gly Ile Phe Gly Asn Ser Leu Val Val  50
TAC ATA TTT ATC ATG ATC CCT ACT TTA TAC AGT ATT ATC TTT GTG GTG GGG ATA TTT GGA AAC AGC TTG GTG GTG 150

Ile Val Ile Tyr Phe Tyr Met Lys Leu Lys Thr Val Ala Ser Val Phe Leu Leu Asn Leu Ala Leu Ala Asp Leu  75
ATT GTC ATT TAC TTT TAC ATG AAA CTG AAG ACT GTG GCC AGT GTT TTT CTT TTG AAT TTA GCT CTG GCT GAC TTA 225
                 II
Cys Phe Leu Leu Thr Leu Pro Leu Trp Ala Val Tyr Thr Ala Met Glu Tyr Arg Trp Pro Phe Gly Asn Tyr Leu 100
TGC TTT TTA CTG ACT TTG CCA CTG TGG GCT GTC TAC ACT GCT ATG GAA TAC CGC TGG CCC TTC GGC AAT TAC CTA 300
                                                     III
Cys Lys Ile Ala Ser Ala Ser Val Ser Phe Asn Leu Tyr Ala Ser Val Phe Leu Leu Thr Cys Leu Ser Ile Asp 125
TGT AAG ATC GCT TCA GCC AGT GTC AGT TTC AAC CTC TAT GCC AGC GTG TTT CTA CTT ACA TGT CTA AGC ATT GAC 375

Arg Tyr Leu Ala Ile Val His Pro Met Lys Ser Arg Leu Arg Arg Thr Met Leu Val Ala Lys Val Thr Cys Ile 150
CGC TAC CTG GCT ATT GTT CAC CCA ATG AAG TCC CGC CTC CGG CGC ACA ATG CTT GTG GCC AAA GTC ACC TGC ATC 450
                 IV
Ile Ile Trp Leu Leu Ala Gly Leu Ala Ser Leu Pro Thr Ile Ile His Arg Asn Val Phe Phe Ile Glu Asn Thr 175
ATT ATT TGG CTG CTG GCA GGT TTG GCC AGT TTG CCA ACT ATA ATC CAC CGC AAC GTA TTT TTC ATC GAG AAT ACC 525

Asn Ile Thr Val Cys Ala Phe His Tyr Glu Ser Gln Asn Ser Thr Leu Pro Val Gly Leu Gly Leu Thr Lys Asn 200
AAT ATC ACC GTT TGC GCT TTC CAT TAC GAA TCC CAA AAT TCT ACC CTC CCG GTA GGG CTG GGC CTC ACC AAG AAT 600
                 V
Ile Leu Gly Phe Leu Phe Pro Phe Leu Ile Ile Leu Thr Ser Tyr Thr Leu Ile Trp Lys Thr Leu Lys Lys Ala 225
ATA TTG GGA TTC TTG TTT CCT TTT CTG ATC ATT CTT ACA AGC TAC ACT CTT ATT TGG AAG ACC CTC AAG AAG GCT 675

Tyr Glu Ile Gln Lys Asn Lys Pro Arg Lys Asp Asp Ile Phe Lys Ile Ile Leu Ala Ile Val Leu Phe Phe Phe 250
TAT GAA ATT CAG AAG AAC AAA CCA AGA AAA GAT GAT ATT TTC AAG ATA ATT TTG GCA ATC GTG CTT TTC TTT TTC 750
```

```
                                   VI
Phe Ser Trp Val Pro His Gln Ile Phe Thr Phe Met Asp Val Leu Ile Gln Leu Gly Leu Ile Arg Asp Cys Lys  275
TTT TCC TGG GTT CCC CAC CAG ATA TTC ACT TTT ATG GAT GTG TTA ATT CAG TTG GGC CTC ATC CGT GAC TGT AAA  825
                                                              VII
Ile Glu Asp Ile Val Asp Thr Ala Met Pro Ile Thr Ile Cys Leu Ala Tyr Phe Asn Asn Cys Leu Asn Pro Leu  300
ATT GAA GAT ATT GTT GAC ACT GCC ATG CCC ATC ACT ATT TGC TTG GCT TAT TTT AAC AAT TGC CTG AAT CCT CTC  900

Phe Tyr Gly Phe Leu Gly Lys Lys Phe Lys Lys Tyr Phe Leu Gln Leu Leu Lys Tyr Ile Pro Pro Lys Ala Lys  325
TTT TAT GGC TTT CTA GGA AAA AAA TTT AAA AAA TAT TTT CTA CAG CTT CTG AAA TAC ATT CCC CCA AAG GCC AAA  975

 ●       ●       ●                   ●
Ser His Ser Asn Leu Ser Thr Lys Met Ser Thr Leu Ser Tyr Arg Pro Ser Glu Asn Gly Asn Ser Ser Thr Lys  350
TCC CAC TCA AAC CTG TCG ACA AAA ATG AGC ACA CTC TCC TAC CGC CCC TCA GAA AAT GGA AAC TCC TCT ACC AAG 1050

Lys Pro Ala Pro Cys Ile Glu Val Glu TER                                                               359
AAG CCT GCC CCA TGC ATT GAG GTT GAG TAA CACACTTGAAACCTGTCTGTGAAGTCACCTTAAAGAAGGAGAAGAAAAACATTCTTGTA 1139

CAACACCTCACTACCAAATGAGCATGGGCTGCATTTCAGAATTAAAGGAGAAAATAGATTATGTGGACTGAACTGACTTGTCTACAGCTCTGAACAAAA 1238
GCTCTTTCCCTTTGCAACTCAACAAAGCAAAGCCACATTCTCCATGTTTTGCATTAGGTAGATGATGCCCACTCAAAGAACTGATGTCAGAAACTGGAT 1337
GAAAGTGTTGATTTGGAAAATTTTACTGGGAAAAATGTCCTCTCCCTAGTCTGCTCTTGTTCTGTTATTTTTGATTTCCACATGAAAGTACTTCAGTTC 1436
AGTTCAGTCACTCAGTCGTGTCCAACTCTTTGCGACCCCATGAATCGCAGCATGCCAGGCTTCCTGTCCATCACCAACTCCCGGAGTTTACCCAAACTC 1535
ATGTGCTCTGAGTCGGTGATGCCATCCAGCCATCTCATCCTCTGTCGTCCCCTTCTCCTCCTGCCCCCAATCCCTCTCAGCATCAGGGTCTTTTCCAAT 1634
GAGTCAACTCTTCGCATGAGGTGGCCAAAGTATTGGAGTTTCAGCTTCAGCATCAGTCCTTCCAATGAACACCCAGGACTGATCTCCTTTAGGATGGAC 1733
TGGTTGGATCTCCTTGCAGTCCAAGGGATGCTCAAGTCTTCTCCAACACCACAGTTCAAAAGCATCAATTCTTTGGCACTCAGCTTTCTTCACAGTCCA 1832
ACTCTCACATCCATACATGACCACAGGAAAAACCATAGCCTTGACTAGATGGACCTTTGTTGGCAAAGTAATGTCTCTGCTTTTGAATATTCTATCTAG 1931
GCTGGTTATAACTTTCCTTCCAAGGAGTAAGCGTCTTTTAATTTCATGGCTGCAGTCACCATCTGCAGTGATTTTGGAGCCCAAAAAAAATAAAATCTG 2030
ACACTGTTTCCATGTTTCCCCATCTATTTCCCATGAGGTACTTAGAATATATTAAACCTCAAGAGGAGCAACAGGAGAGAAGAGCTTAAGACTGCCATG 2129
CCCAATTTCCAAAGGGCAGCAAAGCTTTCGTGCCTATTTAGCTATTAGCAACTGCGGCCCACTTGTACCTGCTACTGCGCATTTCATACAAAGACTCGC 2228
TAAGCAGTAGTTGTCAGGTTCCAGAAGCTGTTGTGAAACCCAACCCGTGTCTTATAGATTCACACTGCCGAAATATGCCAGCTGAGTGGCCTATTTGTA 2327
TAATGATGTCTCTGTAGTCATGTATAAAAATTATATTACTTGTAAAGGTGTTATTCCAGTCCCAGGTAGTCATGTTACTTTTATTTCAGATCTGAGAAG 2426
TATATATATTTTGTGGTGAAAGATTATATATCATAA[illegible]ATATTCCTTACTGTTTTAAAAAGTATATATTCAACCTGTGCATGTATCTCTCTAAACTGCTA 2525
TCTTATTAAAATTTGGCAAAGTTATATTCACATTAAATAATGTTATTGCAATGTATTTAATCTTCATTACTTAAAATAAATGTTGGTTTATTTTT      2620
```

FIGURE 2. The base sequence of the cDNA for bovine adrenocortical AngII type 1 (AT_1) receptor and deduced amino acid sequence.[11] Bars with roman numerals indicate seven putative transmembrane regions in the amino acid sequence as revealed by hydropathy analysis; three arrow heads on asparagine residues are potential N-glycosylation sites. Serine and threonines indicated by filled circles are potential phosphorylation sites by various protein kinase. The base sequences ATTTA in boxes are signals for labeling mRNA.

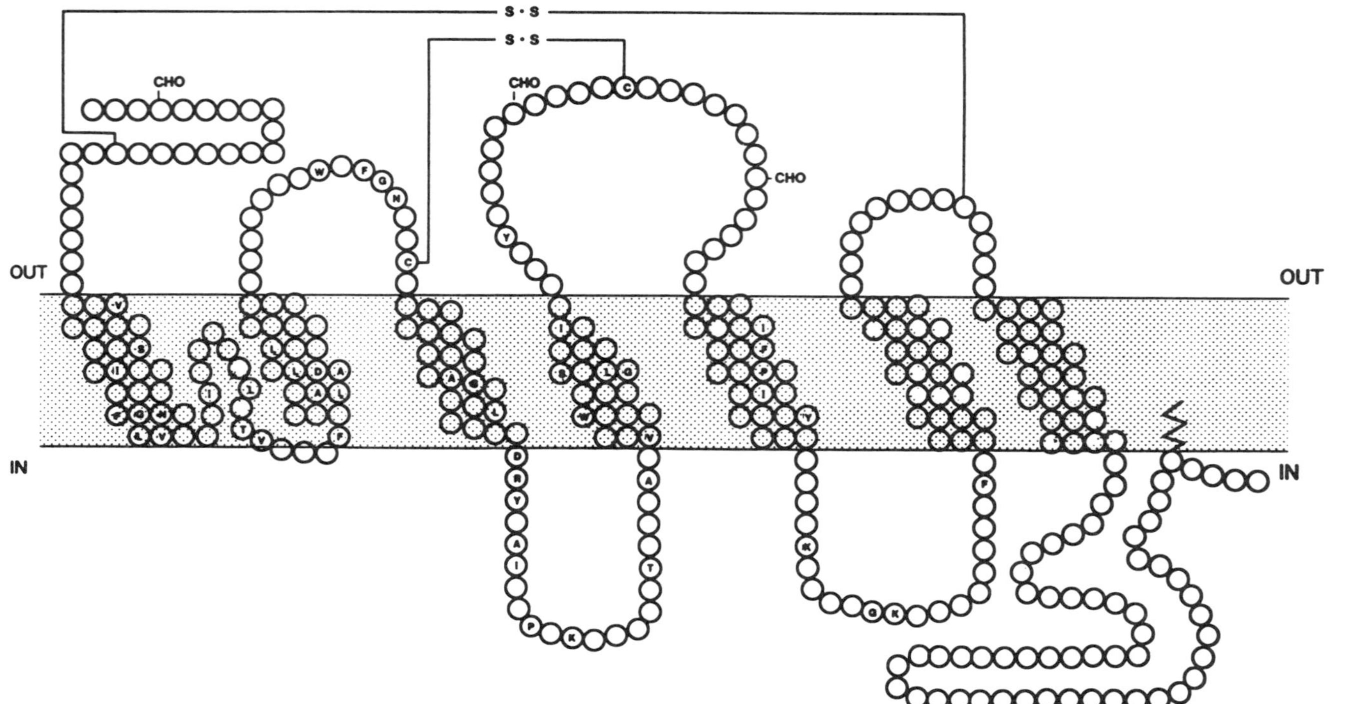

FIGURE 3. A model of bovine adrenocortical AT_1 receptor with seven hydrophobic domains spanning plasma membrane.[11] The positions of the two disulfide bridges are located by inference based on the two conserved cysteine residues in the second and third extracellular domains and two cysteine residues unique to AT_1 receptor located in the first and fourth extracellular domains. CHO indicates carbohydrate chains at potential N-glycosylation sites. Residues indicated by alphabets are those homologous with β_2 receptor. (From Kobilka, B.K. et al., *Proc. Natl. Acad. Sci. U.S.A.*, 84, 46, 1987.)

extracellular loop strongly indicate that the first and seventh transmembrane columns must be located close to each other and that the overall structure of the receptor could very well be such that the seven transmembrane columns are arranged to form a well-like structure across plasma membrane in a tentative configuration given in Figure 4. In constructing further detailed structure of the receptor, attention should be given to the fact that of a total of seven transmembrane regions which may form α-helical columns, five of them contain a proline residue which causes a bend in the α-helical column.

Another feature to note is that the first cytosolic loop is full of hydrophobic amino acid residues, suggesting the possibility that it may be either tucked inside the plasma membrane or interacting with another membrane protein.

B. FUNCTIONS OF CLONED AT_1 RECEPTOR

As discussed above, AT_1 receptors have been cloned in an expression vector pCDM8, which expresses the receptor molecules on plasma membrane when transfected into Cos-7 cells. Cos-7 cells do not contain a detectable amount of endogenous AT_1 receptor or its mRNA; on the other hand it seems to have a good assortment of G-proteins. Thus, it was possible to examine its binding properties and some signal transduction mechanisms without constructing permanently expressing cells.

As shown in Figure 5, bovine adrenocortical AT_1 receptor expressed in Cos-7 cells showed a single high-affinity binding. Plasma membrane preparations from these cells showed a K_d of 0.28 n*M* for the binding of [^{125}I]AngII.[11]

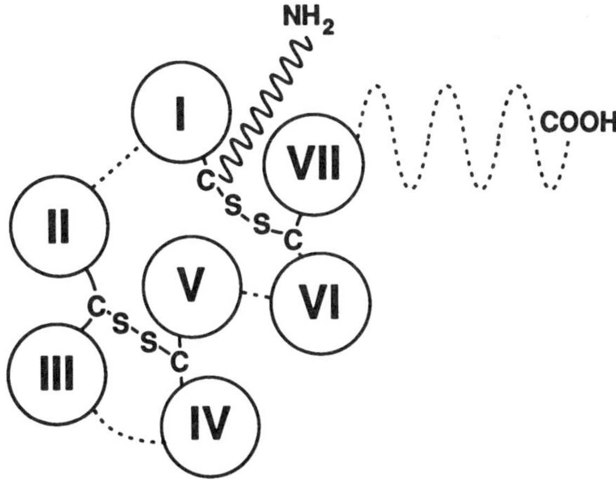

FIGURE 4. Putative arrangement of seven hydrophobic transmembrane columns forming a well-like structure across plasma membrane due to structural constraint imposed by two disulfide bridges (looking down from the outside of the membrane). Roman numerals indicate seven transmembrane domains. Dotted lines are peptide chains inside the plasma membrane and solid lines are peptide chains in the extracellular space. C-S-S-C indicate disulfide bridges as inferred in Figure 3.

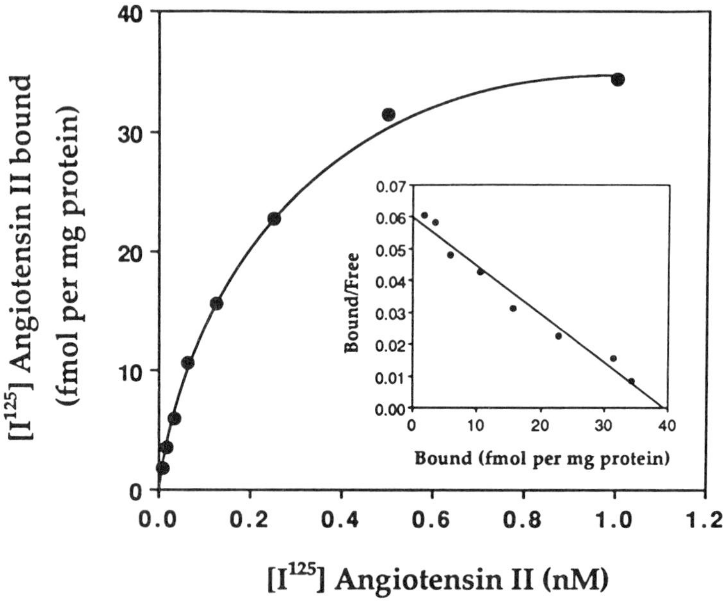

FIGURE 5. Binding isotherm of [^{125}I]AngII to bovine adrenal AT_1 receptor transiently expressed in Cos-7 cells. The inset is a Scatchard plot for the binding isotherm which indicates the presence of a single high-affinity binding site with a K_d of 0.26 n*M*. (From Sasaki, K. et al., *Nature*, 351, 230, 1991. With permission.)

Rat AT_{1A} receptor expressed in Cos-7 cells showed a K_d of 1.9 n*M* for saturable and single-site high-affinity binding of [^{125}I]AngII to unbroken cells.[12] Likewise AT_{1A} cloned from rat vascular smooth muscle cells and expressed in Cos-7 cell membrane showed a saturable binding to [^{125}I]-[Sar[1],Ile[8]]AngII with a K_d of 0.68 n*M*.

Various AngII analogs showed saturable binding as indicated by effective competition with [^{125}I]AngII for the binding as shown in Figure 6, peptidic inhibitors [Sar[1],Ile[8]]AngII binding most strongly, followed by [Sar[1],Ala[8]]AngII, AngII, AngIII, then by losartan (Dup 753). AngI also showed some binding. In contrast to the effective competitive binding of the AT_1-specific inhibitor losartan, the AT_2-specific inhibitor Exp 655 or PD 123177 did not compete, even at 10^{-3} *M*.[11]

AngII receptors expressed in Cos 7 cells from their cDNAs cloned from bovine[11] and rat adrenals,[14] rat vascular smooth muscle cells,[13] rat kidney,[12] and human source showed similar specificity to the AT_1-specific nonpeptidic inhibitor losartan but not to the AT_2-specific PD123177, demonstrating that the cDNA thus far cloned from the rat, bovine, and human tissues are all subtype 1 (AT_1).

A subtle but interesting difference was noted in the binding affinity to losartan between bovine adrenal receptor and receptors from other tissues.

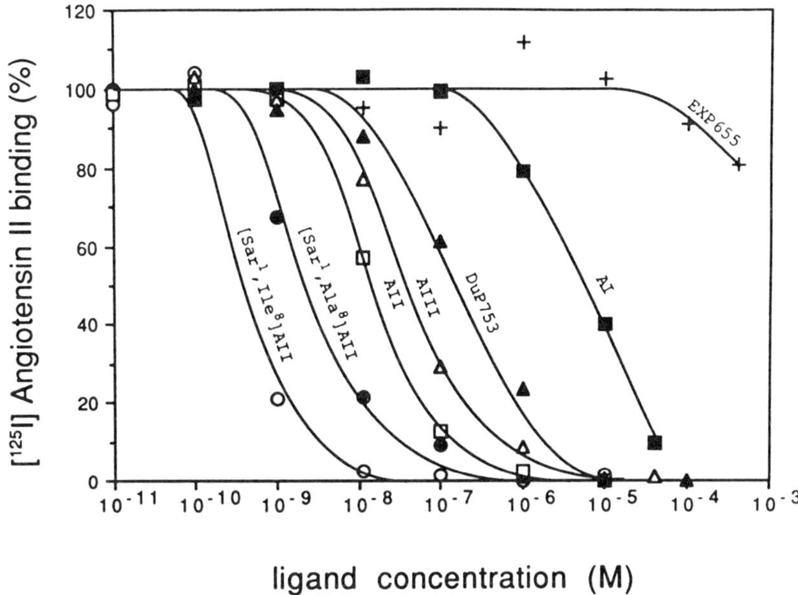

FIGURE 6. Displacement of [^{125}I]AngII binding to bovine adrenocortical AT_1 receptor transiently expressed in Cos-7 cells by various AngII homologs and subtype-specific nonpeptidic antagonists. Note displacement by AT_1-specific DuP 357 (losartan) but not by the AT_2-specific EXP655. (From Sasaki, K. et al., *Nature,* 351, 230, 1991. With permission.)

Losartan showed a binding affinity comparable to AngIII with a K_i of less than 10^{-8} *M* in rat AT_{1A} and AT_{1B}.[12-14] However, its affinity to bovine adrenal AT_1 is approximately 10 times less than AngIII, with a K_i for losartan being higher than 10^{-7} *M*. The mechanistic basis for the species-dependent difference is yet to be clarified. This peculiarity of bovine adrenal AT_1 derived from cloned cDNA is in agreement with a similar peculiarity seen with native bovine adrenal AT_1 receptor.[22] AT_1 receptor either transiently expressed in Cos-7 or permanently expressed in Chinese hamster ovary (CHO) cells responded to AngII by increased production of inositol *tris*phosphate, IP_3, and transient increase in intracellular calcium concentration (Figure 7).

The IP_3 response is congruous with the general concept that receptors possessing the structural feature of the seven transmembrane domains is coupled to G-proteins. In the IP_3 response the AT_1 receptor seems to be coupled to Gq in activating phospholipase β_1 ($PLC_{\beta1}$).

IP_3 is known to elicit the release of Ca^{2+} from intracellular storage. Cells transiently or permanently expressing AT_1 receptor responded to AngII by raising intracellular Ca^{2+}, as shown in Figure 8.[11] These observations ascertain that the AT_1 cDNA cloned from bovine adrenal cortical cells and rat vascular smooth muscle cells and kidney represents AngII receptors coupled to phospholipase C (presumably phospholipase $C_{\beta1}$).

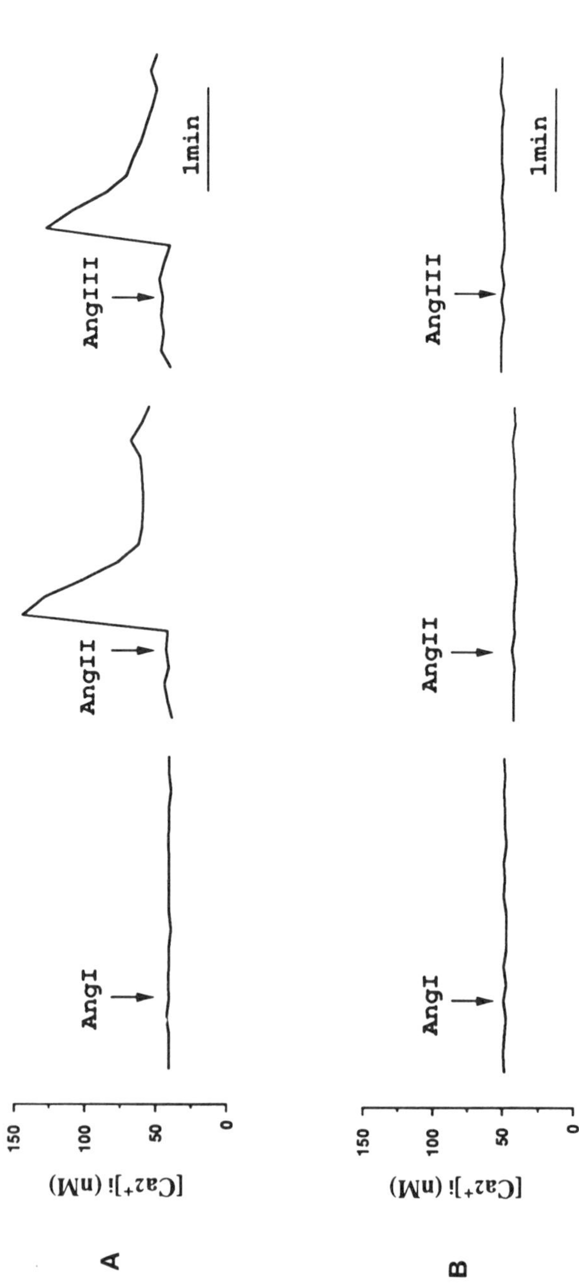

FIGURE 7. (A) Increase in the intracellular calcium concentration induced by 10^{-7} *M* AngII in Cos-7 cells transfected with bovine adrenocortical AT_1 receptor cDNA inserted into the expression plasmid pCDM8. (From Sasaki, K. et al., *Nature*, 351, 230, 1991. With permission.) (B) Control cells transfected with the plasmid not containing cDNA for AT_1.

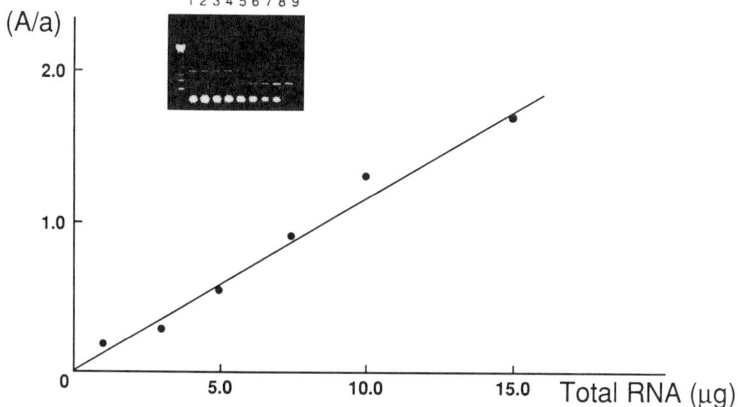

FIGURE 8. A standard curve for competitive polymerase chain reaction (PCR) for rat AT_{1A} receptor. A/a is the ratio of the intensity of PCR product from natural RNA of 479 base pairs to that produced from the truncated reference RNA of 191 base pairs added to the PCR mixture. A good linear relationship is seen over a 15-fold range of total RNA. The insert shows the photograph of the gels used for the construction of the standard curve. (From Iwai, N. and Inagami, T., *Biochem. Biophys. Res. Commun.*, 182, 1094, 1992. With permission.)

AngII has been shown to inhibit adenylyl cyclase through a G-protein (G_i). However, our preliminary observations with permanently expressed (in CHO cells) AT_{1A} show inhibition of adenylyl cyclase. These results indicate that a single subtype of receptor may be capable of interacting with different G-proteins and can mediate at least two different signal transduction pathways.

Studies on the effects of AngII on intracellular calcium concentration have revealed the presence of a tonic phase of the elevation of intracellular calcium concentration which is dependent on the extracellular calcium. While the tonic phase is observed in transiently transfected Cos-7 cells, detailed mechanistic studies are yet to be completed.

The function of the AngII receptor is to bind AngII and transmit the signal generated by the binding of the agonist to the G-protein in such a way as to activate its GTPase activity while promoting the release of the G-protein from the receptor. AngII receptor is particularly noted for its strong tendency for desensitization. Although we do not understand the structure-function mechanism of these receptor-mediated mechanisms at present, the availability of the cloned receptor subtypes will greatly assist our inquiry into these mechanisms. For example, the permanently expressed AT_{1A} seems to be avidly internalized.

III. METHODS FOR THE QUANTIFICATION OF THE AT_1 RECEPTOR

Expression levels of AT_1 receptor mRNA is very low except for liver, kidney, and adrenals. In assessing quantitatively the expression levels of AT_1

receptor, it is necessary to use a sensitive method such as S_1 nuclease-protection assay or polymerase chain reaction (PCR). A competitive PCR method for quantitative assessment of AT_1 mRNA has been developed by Iwai et al.,[12,23] in which a truncated mRNA (191-base) which retains identical primer binding sites is used as a reference for the quantification of the native 479-base mRNA. A good linear relationship between the concentration ratios of the cDNAs from the native mRNA segment and the reference mRNA and the amount of total sample RNA was obtained (Figure 8). This method was capable of reliably detecting a 1.7-fold increment of mRNA over a baseline level.

Now upon the discovery of AT_1 subtypes AT_{1A}, AT_{1B},[14] and possibly AT_{1C}, with small base sequence differences, a quantitative assessment of each of the subtypes in the mixture has become necessary. A more sophisticated competitive PCR specific for each receptor subtype, or S_1 nuclease-protection assay, may become necessary to meet such a demand effectively.

IV. REGULATION OF RECEPTOR

At a molecular or cellular level, the receptor function seems to be regulated by several different mechanisms, which involve transcription, degradation of receptor mRNA, internalization of receptor protein, modification of the intracellular domain, possibly by phosphorylation, in such a way as to modify the interaction of the receptor with G-protein or ligand binding. On the other hand regulation involving the whole body may be mediated by various neurohumoral factors.

In our preliminary studies with rat aortic smooth muscle cells in culture mRNA for AT_{1A}, the receptor is downregulated by AngII. This is in consonance with past observations on the agonist-mediated downregulation of the receptors. However, in *in vivo* experiments in which various manipulations to modulate circulating AngII were used, adrenal AT_1 was found to change in parallel with plasma AngII.[22] The adrenal cortex is one of the areas in which the expression level of AT_1 is among the highest. Bilateral nephrectomy drastically reduced the AT_1 receptor expression (down to 15% of the control), whereas prolonged infusion of AngII at a rate of 20 μg/h for 14 d increased the adrenal AT_1 level by 140%.[23] Although such drastic manipulation as bilateral nephrectomy or prolonged AngII infusion at a hypertensinogenic dose may induce various side effects, prolonged oral administration of losartan at a dose of 15 mg/kg/d for 3 d reduced adrenal AT_1 mRNA by 50%. These observations suggest that AngII upregulates rat adrenal AT_1 receptor at a whole body level.[23] Similar manipulations did not affect the expression levels of AT_1 in rat kidney, liver,[12,23] aorta, or brain stem.[23] Whether these tissue-dependent responses are observed in isolated cells and tissues is an intriguing question. In the whole body experiments changes in plasma AngII levels markedly affect various other factors such as circulating aldosterone, which could affect the expression of AT_1. Thus, in the whole body experiments, a secondary mechanism of receptor regulation cannot be excluded. Nevertheless, adrenal AT_1 receptors seem to respond to various manipulations. For example, low-salt diets also increase AT_1 expression.[12] The effect of the low-salt feeding may be a secondary effect mediated through increased plasma renin and consequent elevation of circulating AngII concentration.

V. FUTURE PROSPECT

The AT_1 receptor is implicated for practically all the harmful effects of AngII as it has been shown that practically all the hypertensinogenic and mitogenic effects of AngII can be blocked by losartan.[24] Abnormality in the functioning and expression of AT_1 receptor may be a strong candidate for pathophysiological bases for abnormality in blood pressure, electrolyte, and fluid volume homeostasis. To address the abnormality in receptor expression, one must identify the regulatory mechanism of each of the subtype genes involving transcriptional regulation by promotor-enhanced function. Cloning of AT_1 genes and studies of their structural and functional features will be of great value for these studies.

Cloned receptor cDNA for each subtype will permit us to clarify detailed pictures of structure-function relationship of the receptor which will involve receptor-ligand interaction, receptor G-protein interaction, and the intricate desensitization mechanism at various stages of receptor function. These studies will be valuable for developing specific and efficient inhibitors of receptors at various levels as antihypertensive drugs.

Our knowledge of the functions, structures, and pathophysiological roles of other receptor subtypes AT_2 and AT_3 are not nearly as developed as that for the AT_1 receptor. These areas may offer exciting areas of research in the future.

REFERENCES

1. **Peach, M.J.,** Molecular actions of angiotensin, *Biochem. Pharmacol.,* 30, 2745, 1981.
2. **Whitebread, S., Mele, M., Kamber, B., and deGaspero, M.,** Preliminary biochemical characterization of two angiotensin II receptor subtypes, *Biochem. Biophys. Res. Commun.,* 163, 284, 1989.
3. **Chiu, A.T., Herblin, W.F., McCall, D.E., Ardecky, R.J., Carini, D.J., Duncia, J.V., Pease, L.J., Wong, P.C., Wexler, R.R., Johnson, A.L., and Timmermans, P.B.M.W.W.,** Identification of angiotensin II receptor subtypes, *Biochem. Biophys. Res. Commun.,* 165, 196, 1989.
4. **Chang, R.S.L. and Lotti, V.J.,** Two distinct angiotensin II receptor binding sites in rat adrenal revealed by new selective nonpeptide ligands, *Mol. Pharmacol.,* 29, 347, 1990.

5. **Chang, R.S.L., Lotti, V., and Keegan, M.E.,** Inactivation of angiotensin II receptors in bovine adrenal cortex by dithiothreitol, *Biochem. Pharmacol.*, 31, 1903, 1982.
6. **Miyazaki, H., Kondoh, M., Ohnishi, J., Masuda, Y., Hirose, S., and Murakami, K.,** High affinity angiotensin II receptors in the bovine ovary are different from those previously identified in other tissues, *Biomed. Res.*, 9, 281, 1988.
7. **Rosenberg, E., Kiron, M.A.R., and Soffer, R.,** Soluble angiotensin II-binding protein from rabbit liver, *Biochem. Biophys. Res. Commun.*, 151, 466, 1988.
8. **Hagiwara, H., Sugiura, N., Wakita, K., and Hirose, S.,** Purification and characterization of angiotensin-binding protein from porcine liver cytosolic fraction, *Eur. J. Biochem.*, 185, 405, 1989.
9. **Jackson, T.R., Blair, L.A.C., Marshal, J., Goedert, M., and Hanley, M.R.,** The mas oncogene encodes an angiotensin receptor, *Nature*, 335, 437, 1988.
10. **Seed, B.,** An LFA-3 cDNA encodes a phospholipid-linked membrane protein homologous to its receptor CD2, *Nature*, 329, 840, 1987.
11. **Sasaki, K., Yamano, Y., Bardhan, S., Iwai, N., Murray, J.J., Hasegawa, M., Matsuda, Y., and Inagami, T.,** Cloning and expression of a complementary DNA encoding a bovine adrenal angiotensin II type 1 receptor, *Nature*, 351, 230, 1991.
12. **Iwai, N., Yamano, Y., Chaki, S., Konishi, F., Bardhan, S., Tibbetts, C., Sasaki, K., Hasegawa, M., Matsuda, Y., and Inagami, T.,** Rat angiotensin II receptor: cDNA sequence and regulation of the gene expression, *Biochem. Biophys. Res. Commun.*, 177, 299, 1991.
13. **Murphy, T.J., Alexander, R.W., Griendling, K.K., Runge, M.S., and Bernstein, K.E.,** Isolation of a cDNA encoding the vascular type-1 angiotensin II receptor, *Nature*, 351, 233, 1991.
14. **Iwai, N. and Inagami, T.,** Identification of two subtypes in the rat type 1 angiotensin II receptor, *FEBS Lett.*, 298, 257, 1992.
15. **Furuta, H., Guo, D.-F., and Inagami, T.,** Molecular cloning and sequencing of the gene encoding human angiotensin II type 1 receptor, *Biochem. Biophys. Res. Commun.*, 183, 8, 1992.
16. **Takayanagi, R., Ohnaka, K., Sakai, Y., Nakao, R., Yanase, T., Haji, M., Inagami, T., Furuta, H., Guo, D.-F., Nakamuta, M., and Nawata, H.,** Molecular cloning, sequence analysis and expression of a cDNA encoding human type-1 angiotensin II receptor, *Biochem. Biophys. Res. Commun.*, 183, 910, 1992.
17. **Carson, M.D., Leach-Harper, C.M., Baukal, A.J., Aguilera, G., and Catt, K.J.,** Physicochemical characterization of photo affinity-labeled angiotensin II receptors, *Mol. Endocrinol.*, 1, 147, 1987.
18. **Rondeau, J.J., McNecoll, N., Meloche, S., Ong, H., and DeLean, A.,** Hydrodynamic properties of the angiotensin II receptor from bovine adrenal zona glomerulosa, *Biochem. J.*, 268, 443, 1990.
19. **Hubbard, S.C. and Watt, R.J.,** Synthesis and processing of asparagine-linked oligosaccharides, *Annu. Rev. Biochem.*, 50, 555, 1981.
20. **O'Dowd, B., Hnatowich, M., Carou, M.G., Lefkowitz, R.J., and Bouvier, M.,** Palmitoylation of the human β_2-adrenergic receptor, *J. Biol. Chem.*, 264, 7564, 1989.
21. **Ovchinnikov, Y., Abdulaev, N., and Bogachuk, A.,** Two adjacent cysteine residues in the C-terminal cytoplasmic fragment of bovine rhodopsin are palmitoylated, *FEBS Lett.*, 230, 1, 1988.
22. **Balla, T., Baukal, A.J., Eng, S., and Catt, K.J.,** Angiotensin II receptor subtypes and biological responses in the adrenal cortex and medulla, *Mol. Pharmacol.*, 40, 401, 1991.
23. **Iwai, N. and Inagami, T.,** Regulation of the expression of the rat angiotensin II receptor mRNA, *Biochem. Biophys. Res. Commun.*, 182, 1094, 1992.

24. **Chiu, A.T., Roscoe, W.A., McCall, D.E., and Timmermans, P.B.M.W.M.,** Angiotensin II-1 receptors mediate both vasoconstrictor and hypertrophic responses in rat aortic smooth muscle cells, *Receptor,* 1, 133, 1991.
25. **Kobilka, B.K., Dixon, R.A., Frielle, T., Dohlman, H.G., Bolanowski, M.A., Sigal, I.S., Yang-Feng, T.L., Francke, U., Caron, M.G., and Lefkowitz, R.J.,** cDNA for the human β_2-adrenergic receptor: a protein with multiple membrane-spanning domains and encoded by a gene whose chromosonal location is shared with that of the platelet-derived growth factor, *Proc. Natl. Acad. Sci. U.S.A.,* 84, 46, 1987.

Chapter 12

ANGIOTENSIN RECEPTOR SUBTYPES (AT_{1A} and AT_{1B}): CLONING AND EXPRESSION

S.S. Kakar and J.D. Neill

TABLE OF CONTENTS

0-8493-4622-3/93/$0.00 + $.50

I. INTRODUCTION

Blood pressure and salt and water metabolism are among the most important homeostatic processes in mammals. The key regulator of these processes is the interaction of angiotensin II (AngII) with its receptor located in the cell membranes of arterioles, adrenal cortex, and the central nervous system.[1,2] The established mechanisms by which interaction of AngII with its receptor effects these homeostatic mechanisms are stimulation of arteriolar vascular smooth muscle constriction which increases blood pressure, of aldosterone secretion by the adrenal cortex which in turn increases renal sodium resorption, and of neurons in the subfornical organ which increases salt and water appetite.[1,2] Two angiotensin receptor types have been described: one binds preferentially to the nonpeptide antagonist, Dup 753, and subserves the functions of AngII described above (type AT_1), and another which binds the nonpeptide antagonist, PD 123319, and whose functions are not established (type AT_2).[3,4] Because calcium is the established cellular second messenger transducing the interaction of AngII with the AT_1 receptor,[2] it was predictable that this receptor type would be a member of the family of seven-transmembrane, G-protein-coupled receptors.[5] Indeed, the recent molecular analysis of rat[6,7] and bovine[8] AT_1 receptor types in vascular smooth muscle and kidney revealed a seven-transmembrane receptor coupled to increases in intracellular calcium.

Although the existence of different AngII receptor types (AT_1 and AT_2)[3,4] was well established prior to the recent cloning of the AT_1 receptor,[6-8] no compelling evidence existed to predict the existence of AT_1 receptor subtypes (AT_{1A} and AT_{1B}). Nevertheless, we recently discovered a second subtype of the AngII receptor (AT_{1B}) in the rat anterior pituitary gland[9] which showed high similarity with the AT_{1A} receptor subtype originally cloned from vascular smooth muscle and kidney.[6,7] The existence of an AT_{1B} receptor also was reported by Iwai and Inagami.[10]

In this chapter, we will review the molecular characteristics of the AT_{1B} receptor, the tissue distribution of its mRNA, the hormonal regulation of its mRNA expression, the predicted functional differences between AT_{1B} and AT_{1A} receptor proteins, and, finally, the evidence that the AT_1 receptor subtypes are encoded by separate genes.

II. AT_{1B} RECEPTOR CLONING AND ITS SEQUENCE ANALYSIS

We cloned the AT_{1B} receptor from a λZAP II cDNA library prepared from rat anterior pituitary RNA; transcribed RNAs from aliquots of the library were screened by injection into *Xenopus* oocytes and by use of the standard two-electrode voltage clamp procedure as described previously in detail.[9] Nucleotide sequencing of the isolated AT_{1B} cDNA clone revealed a typical seven-transmembrane receptor composed of 2153 nucleotides and which ex-

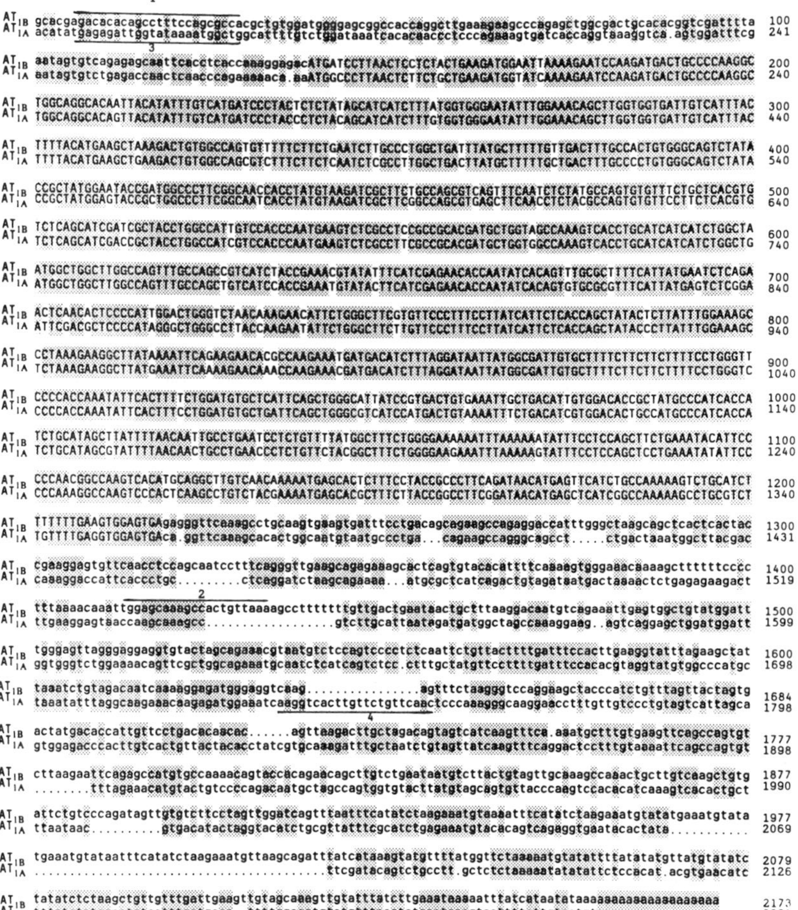

FIGURE 1. Comparison of the nucleotide sequences of AngII receptor subtypes, AT_{1B} and AT_{1A}. Nucleotide sequences that are identical in the AT_{1B} and AT_{1A} cDNAs are shaded; dots are gaps introduced to optimize alignment of the sequences. Nucleotides comprising the open reading frames are capitalized. Over-lines 1 and 2 (AT_{1B}) and under-lines 3 and 4 (AT_{1A}) are the sequences of synthetic oligonucleotides used for the polymerase chain reactions described in Figure 5. (From Kakar, S.S., Sellers, J.C., Devor, D.C., Musgrove, L.C., and Neill, J.D., *Biochem. Biophys. Res. Commun.*, 183, 1090, 1992. With permission.)

hibited an overall 74% identity with the AT_{1A} receptor cDNA and 91% identity between their open reading frames (Figure 1). The open reading frames of both AT_1 receptor subtypes encoded proteins composed of 359 amino acids exhibiting 95% identity (Figure 2). This high level of amino acid identity between subtypes appears to be unprecedented among the large family of seven-transmembrane receptors.[11,12]

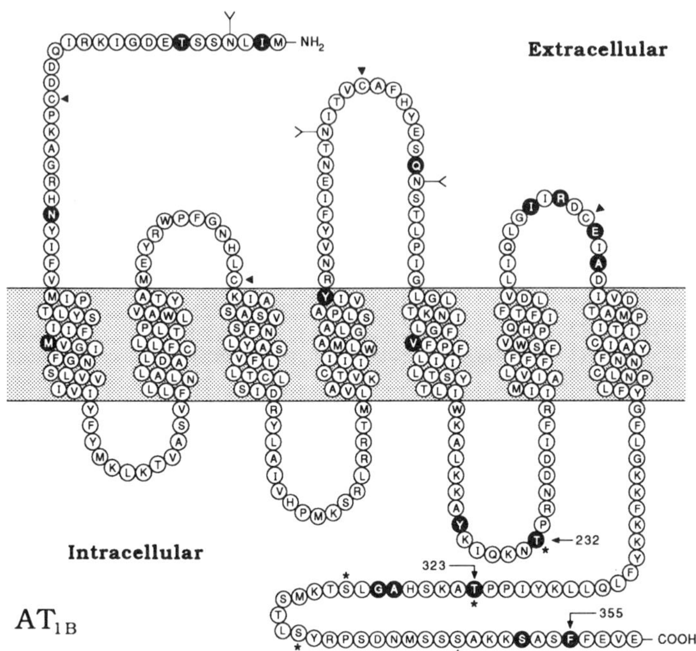

FIGURE 2. Comparison of the proposed primary structures of AT_{1B} and AT_{1A} receptors. Identical residues in the receptors are shown as hollow circles and different residues are shown as solid circles. The shaded areas represent the membrane regions. Also indicated are the canonical sites of N-linked glycosylation (Y), phosphorylation by protein kinase C (*), cysteine residues in the extracellular loops (▼), and the potential palmitoylated cysteine in the C-terminal domain of AT_{1A}.

In addition to the quantitative similarities of the AT_{1B} and AT_{1A} receptor subtypes, they also show several qualitative similarities (Figure 2): (1) three predicted asparagine glycosylation sites (one in the amino-terminal tail and two in the second extracellular loop); (2) four extracellular cysteines (one in the amino-terminal tail and one in each of the three extracellular loops), which probably form the disulfide bridges necessary for ligand binding;[13] and (3) three serines in the carboxy-terminal tail (amino acids 331, 338, and 348) which are potential sites for phosphorylation by protein kinase C.

Important qualitative differences between the AT_{1B} and AT_{1A} receptors are (Figure 2): (1) the presence of threonine as residues 232 (in the third intracellular loop) and 323 (in the carboxy-terminal tail) in the AT_{1B} receptor which are absent in the AT_{1A} receptor — these residues are potential sites for regulatory phosphorylation and may be involved in the desensitization of the receptor after ligand binding;[14] and (2) the absence of the cysteine at residue 355 in the AT_{1B} receptor which is present in the AT_{1A} receptor — this cysteine is a potential site for palmitoylation and may anchor the carboxy-terminal tail

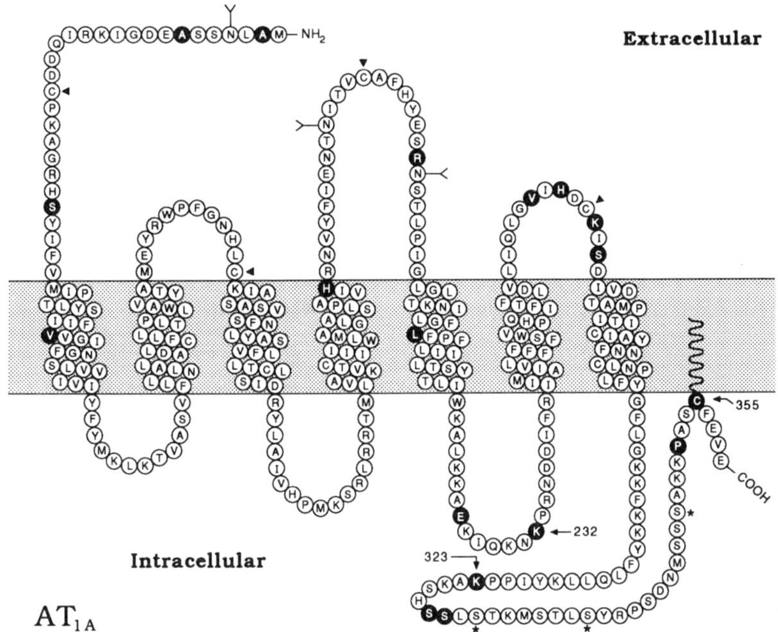

FIGURE 2. (*continued*)

of the receptor to the plasma membrane, thus forming a fourth intracellular loop in the AT_{1A} receptor which is important for binding to its G-protein.[15] These two qualitative differences between the AT_{1B} and AT_{1A} receptors suggest the hypothesis that the AT_{1B} receptor may desensitize more rapidly than does the AT_{1A} receptor after binding to AngII.

III. LIGAND BINDING AND SECOND MESSENGER PRODUCTION BY THE AT_{1B} RECEPTOR

AngII receptor radioassays performed on Cos-7 cells transiently transfected with the AT_{1B} cDNA[9] revealed a pattern of ligand displacement (Figure 3) similar to that observed in a variety of AngII target tissues[3] including the anterior pituitary gland.[16] IC_{50} values were as follows (Figure 3): AngII, $\sim 1 \times 10^{-9}$ *M*; Sar^1-Ile^8-AngII, $\sim 3 \times 10^{-10}$ *M*; AngIII, $\sim 3 \times 10^{-9}$ *M*; AngI, $\sim 4 \times 10^{-7}$ *M*; Dup 753, $\sim 3 \times 10^{-9}$ *M*; and PD 123319, $>1 \times 10^{-6}$ *M*. Thus, our cloned AngII receptor is of the AT_1 type and shows an affinity of binding and pattern of ligand displacement similar with the AT_{1A} receptor.[6,7] Consistent with this finding, we also performed Scatchard analysis of AngII binding to the AT_{1B} receptor and found a K_d (0.7 n*M*)[9] similar to those reported for the cloned AT_{1A} receptor (0.68 and 1.9 n*M*)[6,7] and for the endogenous anterior pituitary AngII receptor (0.95 n*M*).[17] A major exception in the

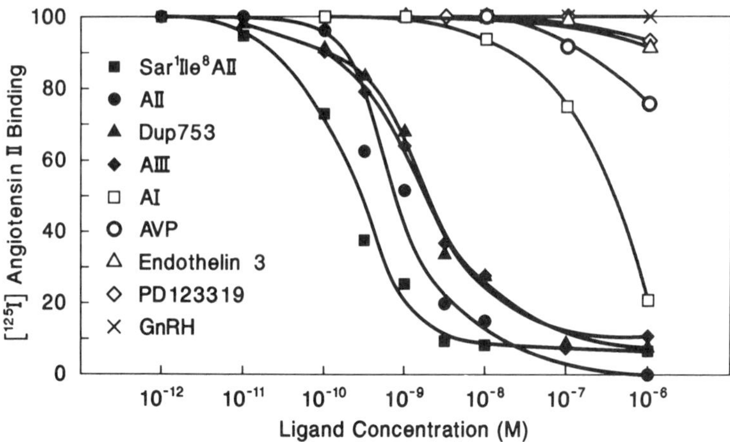

FIGURE 3. DNA transfection and binding assay: Cos-7 cells were transfected with purified pcDNA I plasmid bearing the cDNA encoding the AT_{1B} receptor. Binding of ^{125}I-AngII was measured by incubating the transfected cell membrane in the presence of various concentrations of unlabeled ligands.[9] Results are expressed as the percentage of specific ^{125}I-AngII binding. (From Kakar, S.S., Sellers, J.C., Devor, D.C., Musgrove, L.C., and Neill, J.D., *Biochem. Biophys. Res. Commun.*, 183, 1090, 1992. With permission.)

similarity of ligand binding is that AngI is approximately 10-fold less potent with the AT_{1B} than the AT_{1A} receptor; the physiological significance of this difference is questionable, since levels of AngI are not likely to reach such high concentrations ($\sim 10^{-7}$ *M*) in plasma but might do so in some of the local tissue renin-angiotensin systems.[18]

AT_1 receptors in cells from various tissues,[5] including the anterior pituitary gland,[19] appear to be coupled functionally to increases in intracellular Ca^{2+}. We measured Ca^{2+} levels using video-based fluorescence ratio imaging of the calcium identicator dye, Fura-2, on Cos-7 cells transfected with the AT_{1B} receptor cDNA and found a mean 2.2-fold peak increase in the Fura-2 fluorescence ratio of responding cells.[9] Thus, the AT_{1B} receptor appears to be functionally coupled to increases in intracellular Ca^{2+}, as has been reported also for the AT_{1A} receptor.[6,7]

IV. DIFFERENTIAL TISSUE EXPRESSION OF AT_1 RECEPTOR SUBTYPE mRNA

High-stringency Northern blot analysis of poly(A^+) RNA using [α^{32}P]dCTP-labeled AT_{1B} cDNA as probe revealed 2.3-kb hybridizing transcripts in anterior pituitary, adrenal, vascular smooth muscle cells, liver, and lung (Figure 4). Similarly, Murphy et al.,[6] using rat AT_{1A} cDNA as probe, reported the expression of AT_1 receptor mRNA in various tissues including liver, kidney, aorta, uterus, adrenal, ovary, lung, vascular smooth muscle

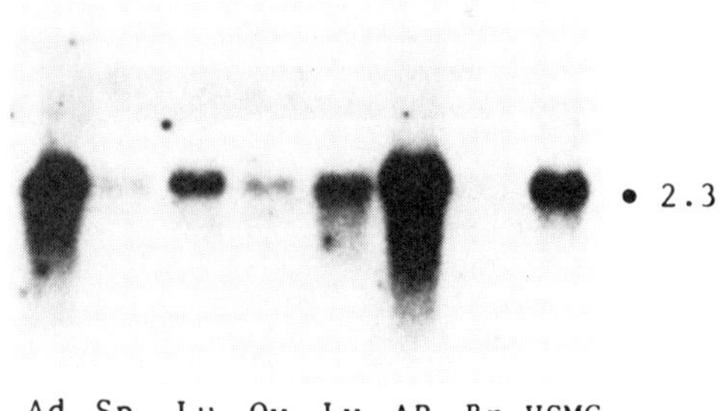

FIGURE 4. Northern blot analysis for AT_1 receptor mRNA in various tissues; 5 μg of poly(A^+) RNA from various tissues was used for Northern blot analysis under high-stringency conditions.[29] The probe used was [α^{32}P]dCTP-labeled full-length AT_{1B} cDNA. Ad, Adrenal; Sp, spleen; Lu, lung; Ov, ovary; Lv, liver; AP, anterior pituitary; Br, brain; and VSMC, vascular smooth muscle cells.

cells, heart, thymus, and spleen. However, the high level of nucleotide identity between the two AT_1 cDNAs resulted in cross-hybridization of AT_{1B} cDNA with AT_{1A} mRNA and vice versa in Northern blots[9] under the hybridization conditions used (Figure 4). Therefore, these results (Figure 4), as well as those of Murphy et al.,[6] do not differentiate the expression of AT_{1A} and AT_{1B} mRNA in these tissues and thus they represent the mixed expression of both forms of the AT_1 receptor.

Therefore, to determine the differential tissue expression of AT_{1B} and AT_{1A} mRNA levels and to assess the ratio of their expression in the various tissues, we[9] used the reverse transcriptase/polymerase chain reaction (PCR) method: dissimilar oligonucleotide primers for the corresponding cDNAs were designed from their 5′ and 3′ nontranslated regions where AT_{1A} and AT_{1B} receptor cDNAs exhibit minimum sequence homology (Figure 1). RNA from various tissues and cultured aortic vascular smooth muscle cells from rat were subjected to first strand cDNA synthesis using oligo(dT) primer and AMV reverse transcriptase. These solutions were then used in a PCR. The reaction mixture was then electrophoresed through a 1.0% agarose gel and stained with ethidium bromide.[9] As shown in Figure 5, anterior pituitary, adrenal, and uterus expressed primarily AT_{1B} mRNA, whereas aortic vascular smooth muscle cells, lung, and ovary expressed primarily AT_{1A} mRNA; spleen, liver, and kidney expressed similar levels of the two RNAs[9] (Figure 5). These results suggest that the AT_{1B} receptor mediates adrenal aldosterone secretion and pituitary adrenocorticotrophic hormone (ACTH) and prolactin secretion, whereas the AT_{1A} receptor mediates vascular smooth muscle constriction.

Using a similar approach, we also examined the expression of AT_{1A} and AT_{1B} mRNAs in various regions of the brain.[20] As shown in Figure 6, both forms of the AngII type 1 receptor mRNAs (AT_{1A} and AT_{1B}) were found in the cerebellum and hypothalamus (Figure 6a), indicating that the two AT_1 receptor subtypes expressed in peripheral tissues also are expressed in the

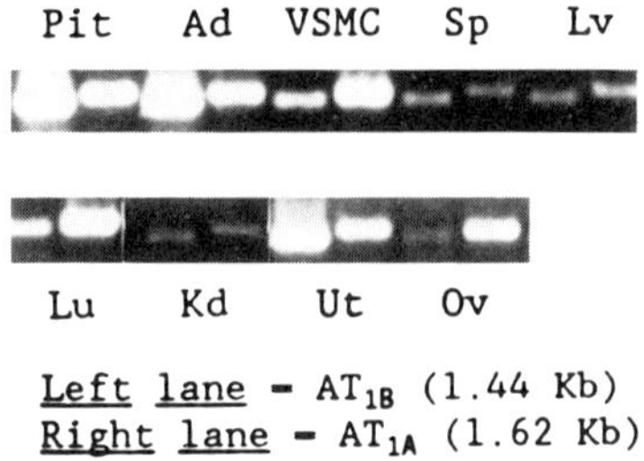

FIGURE 5. Differential expression of AT_{1B} and AT_{1A} mRNA in various tissues. Poly(A^+) RNA from various tissues was used for first strand cDNA synthesis followed by PCR. The PCR reaction mixture was electrophoresed through a 1.0% agarose gel. Ethidium bromide-stained DNA is shown. Pit, Anterior pituitary gland; Ad, adrenal gland; VSMC, vascular smooth muscle cells; Sp, spleen; Lv, liver; Lu, lung; Kd, kidney; Ut, uterus; Ov, ovary. (From Kakar, S.S., Sellers, J.C., Devor, D.C., Musgrove, L.C., and Neill, J.D., *Biochem. Biophys. Res. Commun.*, 183, 1090, 1992. With permission.)

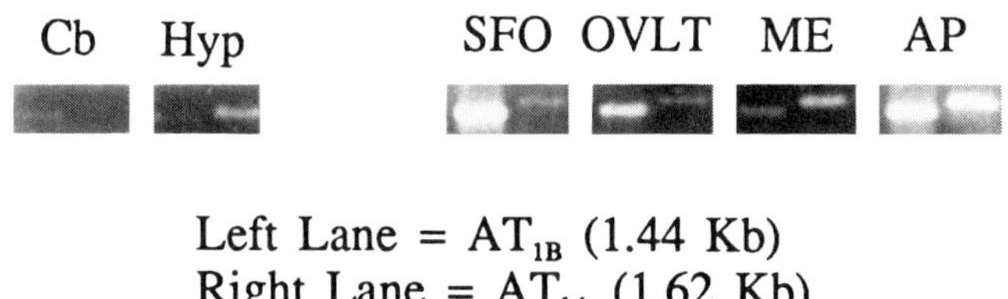

FIGURE 6. AT_{1B} and AT_{1A} mRNA expression in the brain. The experimental conditions were the same as described in Figure 5; 2 μg of total RNA from various regions of the brain was used for first strand cDNA synthesis. Ethidium bromide-stained DNA is shown which was generated from tissue mRNA using reverse transcriptase/polymerase chain reaction. Cb, Cerebellum; Hyp, hypothalamus; SFO, subfornical organ; OVLT, organum vasculosum of the lamina terminalis; ME, median eminence; AP, area postrema. (From Kakar, S.S., Riel, K.K., and Neill, J.D., *Biochem. Biophys. Res. Commun.*, 185, 688, 1992. With permission.)

brain. Cerebellum expressed mostly AT_{1B} mRNA, whereas hypothalamus expressed primarily AT_{1A} mRNA (Figure 6a). Among the circumventricular organs, the subfornical organ (SFO) and organum vasculosum of the lamina terminalis (OVLT) expressed primarily AT_{1B} mRNA with a greater abundance in the former than in the latter; the median eminence expressed mainly AT_{1A} mRNA, whereas the area postrema expressed equal quantities of AT_1 subtype mRNAs (Figure 6b). These results[20] agree with the reported presence of AngII type 1 receptors in the various regions of the brain, including the circumventricular organs, as measured by radioreceptor assays and quantitative autoradiography.[18,21]

AngII stimulates water intake in rats via activation of central AngII type 1 receptors.[22] It is well established that AngII acts on the SFO and the OVLT to induce this behavior.[1,2] Indeed, Dup 753 blocked water intake stimulated by peripheral administration of AngII.[23] In addition, Tsutsumi and Saavedra[21] demonstrated complete inhibition of binding of AngII to the SFO and in the OVLT in the presence of Dup 753, further supporting a role for AT_1 receptors in mediating drinking in rats. As shown in Figure 6b, SFO and OVLT mainly expressed AT_{1B} mRNA, suggesting that the AT_{1B} receptor regulates drinking in female rats.

V. HORMONAL REGULATION OF AT_{1B} RECEPTOR GENE EXPRESSION

Estrogens are reported to potently suppress (75 to 90%) AngII receptor levels in the anterior pituitary as measured by radioreceptor assays.[17] To determine if estrogens exerted a similar effect on AT_1 gene expression, we[9] treated ovariectomized rats with estradiol and then measured AT_{1B} and AT_{1A} mRNA levels in the pituitary gland using the reverse transcriptase/PCR procedure described earlier. As shown in Figure 7, estrogen treatment strongly inhibited AT_{1B} but not AT_{1A} mRNA levels in the pituitary. Progesterone treatment had no effect. In this particular experiment (Figure 7), ovariectomy after two weeks did not result in an increase in AT_{1B} mRNA levels (Figure 7), but in another experiment conducted four weeks after ovariectomy, a large increase was observed (data not shown).[9] Thus, AT_{1B} but not AT_{1A} gene expression appears to be estrogen-regulated.

AT_1 receptors in the SFO and OVLT also are reported to be suppressed by estrogens. Indeed, it has been shown that the volume of fluid intake varied inversely with plasma estrogen levels in rats during the estrous cycle[24] and this apparent inhibitory relationship was confirmed by showing that estrogen treatment of ovariectomized rats decreased AT_1 receptor levels in the SFO

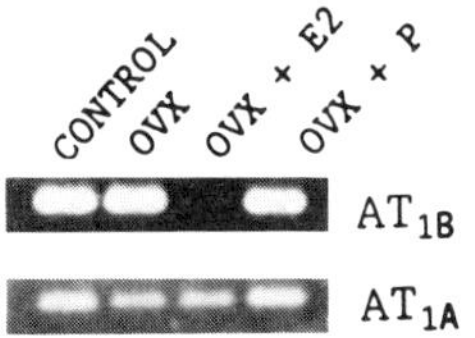

FIGURE 7. Hormonal regulation of AT_{1B} gene expression. Female Sprague-Dawley rats (5 to 6 per group) were used at 14 d after ovariectomy and received two subcutaneous injections at 24-h intervals of 17β estradiol benzoate (25 μg/injection in oil) or progesterone (12.5 mg/injection in oil). The rats were euthanized 24 h after the second injection.[9] The pituitaries were collected, the RNA was prepared, and used for reverse transcriptase/PCR as described in Figure 5; 2 μg of total RNA from anterior pituitaries was used for first strand cDNA synthesis. Ethidium bromide-stained DNA is shown. CONTROL, Intact rats; OVX, ovariectomized rats; OVX + E2, estrogen-treated ovariectomized rats; OVX + P, progesterone-treated ovariectomized rats. (From Kakar, S.S., Sellers, J.C., Devor, D.C., Musgrove, L.C., and Neill, J.D., *Biochem. Biophys. Res. Commun.*, 183, 1090, 1992. With permission.)

and OVLT while suppressing drinking behavior.[25,26] Our demonstration[9] that pituitary AT_{1B} gene expression is suppressed by estrogens (Figure 7), taken together with the evidence that AT_1 receptor levels in SFO and OVLT are also inhibited by estrogens, is consistent with the results shown in Figure 6 that these two circumventricular organs express mostly AT_{1B} mRNA.

VI. SEQUENCE COMPARISONS AMONG RAT, HUMAN, AND BOVINE AT_1 RECEPTORS

The pattern of nucleotide and amino acid sequence differences between AT_{1B} and AT_{1A} receptors (Figures 1 and 2) is not characteristic of alternatively spliced RNA, suggesting that they are encoded by different genes. At least two fragments of rat genomic DNA treated with multiple restriction enzymes (EcoRI, BamHI, HindIII, PstI, and BglII) hybridized under high-stringency conditions with an AT_{1B} cDNA probe representing its open reading frame (Figure 8); this finding is consistent with AT_{1A} and AT_{1B} receptors being encoded by separate genes. Recently, AngII type 1 receptor cDNAs for bovine[8] and human[27,28] also have been isolated and sequenced; however, whether bovine and human AT_1 receptors occur in two forms as in the rat remains unknown. To determine this, we performed Southern blot analysis of restricted genomic DNA from human and bovine species using conditions similar to those described for rat. Only one restriction fragment hybridized in both human and bovine genomic DNA (Figure 8), suggesting the presence of a single AT_1 gene in bovine and human species.

A comparison of the deduced amino acid sequences of AT_1 cDNAs among rat, bovine, and human species revealed that the AT_1 receptor from human and bovine resembles the rat AT_{1A} receptor more than it does the rat AT_{1B} receptor (Figure 9). This hypothesis is based on a number of features: (1) the presence of a number of common amino acids in rat AT_{1A}, and bovine and human AT_1 receptors, which are absent in the rat AT_{1B} receptor; (2) the presence of a potentially palmitoylated cysteine in the C-terminal domain of rat AT_{1A} and bovine and human AT_1 receptors, which is absent in the rat AT_{1B} receptor, and (3) the absence of two potential protein kinase C phosphorylation sites, one in the third intracellular loop and another in the C-terminal portion of rat AT_{1A} and bovine and human AT_1 receptors, which are present in the rat AT_{1B} receptor.

VII. SUMMARY AND CONCLUSIONS

The recent molecular cloning of AT_1 receptors in the rat unexpectedly revealed the existence of two subtypes (AT_{1A} and AT_{1B}) which appear to be encoded by separate genes. These two AT_1 receptor subtypes are related at an unprecedented level, showing 95% identity in their derived amino acid

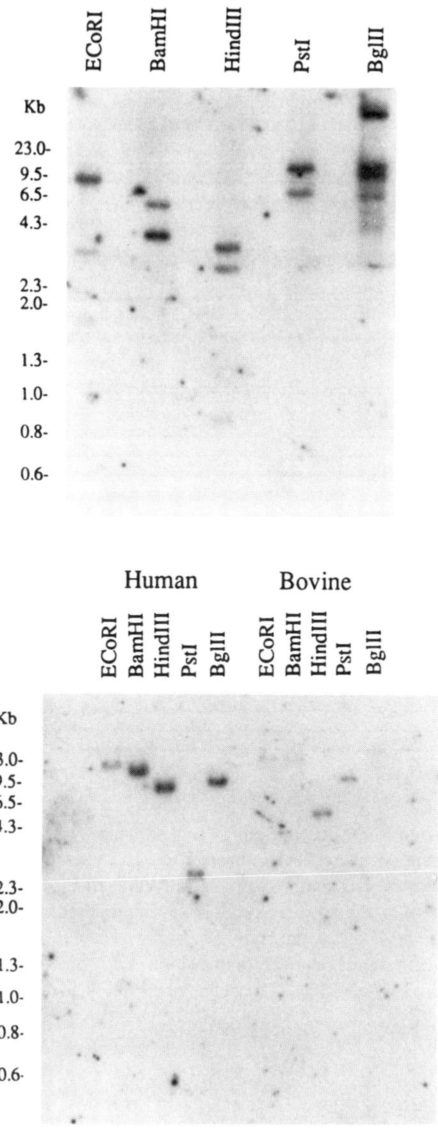

FIGURE 8. Southern blot analysis of rat, human, and bovine genomic DNA: 8 μg of the genomic DNA was completely restricted with EcoRI, BamHI, HindIII, PstI, or BglII. Southern blot analysis was performed under high-stringency conditions[29] using [α^{32}P]dCTP-labeled AT_{1B} cDNA representing its open reading frame.

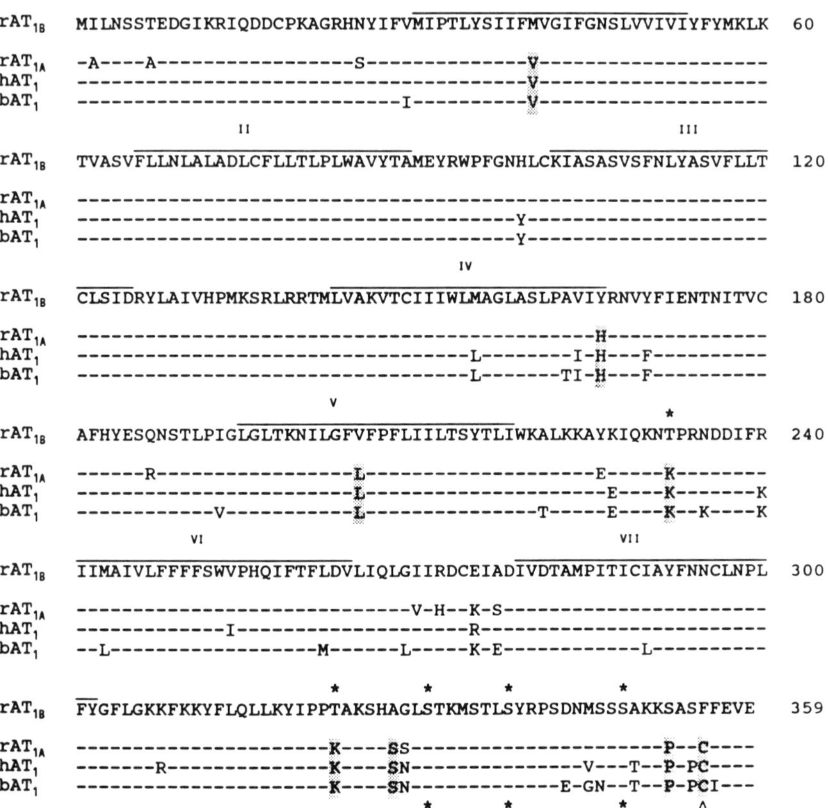

FIGURE 9. Comparison of the deduced amino acid sequences of rat, human, and bovine AT_1 receptors. All amino acids of the rat AT_{1B} receptor are given using the single letter code (top sequence). Dashes indicate exact matches among rat AT_{1B} (rAT_{1B}), rat AT_{1A} (rAT_{1A}), human AT_1 (hAT_1), and bovine AT_1 (bAT_1) receptors. The seven proposed transmembrane domains are indicated by over-lines with Roman numerals. The common amino acids among rAT_{1A}, hAT_1, and bAT_1 receptors, which are different from that in rat AT_{1B}, are shaded. Potential phosphorylation sites are indicated by asterisks. The potential palmitoylated cysteine is indicated by a triangle.

sequences. Nevertheless, three important differences may distinguish them: (1) gene regulation — estrogen regulates AT_{1B} but not AT_{1A} mRNA expression; (2) ligand binding — AngI shows an IC_{50} value that is 10-fold lower with the AT_{1B} receptor than with the AT_{1A} receptor; and (3) desensitization kinetics — the AT_{1B} receptor has two or more protein kinase C phosphorylation sites than the AT_{1A} receptor but lacks the potentially palmitoylated cysteine present in the C-terminal region of AT_{1A} receptor, suggesting the hypothesis that the AT_{1B} receptor desensitizes more rapidly after binding AngII.

The adrenal and anterior pituitary glands and the two circumventricular organs of the brain, subfornical organ, and organum vasculosum of the lamina terminalis, all express primarily AT_{1B} receptor mRNA, whereas vascular smooth muscle cells express primarily AT_{1A} mRNA. These findings suggest that the AT_{1B} receptor is the primary mediator of AngII-induced aldosterone secretion, ACTH and prolactin secretion, and drinking behavior, whereas the AT_{1A} receptor is the primary mediator of AngII-induced arteriolar constriction.

Despite the presence of at least two AT_1 receptor genes in rats, only one appears to exist in the human and bovine genomes. The AT_1 receptor encoded in these two species appears to be more related to the rat AT_{1A} than to the rat AT_{1B} receptor.

REFERENCES

1. **Peach, M.J.,** Renin-angiotensin system: biochemistry and mechanisms of action, *Physiol. Rev.*, 57, 313, 1977.
2. **Vallotton, M.B.,** The renin-angiotensin system, *Trends Pharmacol. Sci.*, 8, 69, 1987.
3. **Timmermans, P.M., Wong, P.C., Chiu, A.T., and Herblin, W.F.,** Nonpeptide angiotensin II receptor antagonists, *Trends Pharmacol. Sci.*, 12, 55, 1991.
4. **Bumpus, F.M., Catt, K.J., Chiu, A.T., De Gasparo, M., Goodfriend, T., Husain, A., Peach, M.J., Taylor, D.G., Jr., and Timmermans, P.B.M.W.M.,** Nomenclature for angiotensin receptors: a report of the nomenclature committee of the Council for High Blood Pressure Research, *Hypertension*, 17, 720, 1991.
5. **Peach, M.J. and Dostal, D.E.J.,** The angiotensin II receptor and the actions of angiotensin II, *J. Cardiovasc. Pharmacol.*, 16 (Suppl. 4), S25, 1990.
6. **Murphy, T.J., Alexander, R.W., Griendling, K.K., Runge, M.S., and Bernstein, K.E.,** Isolation of a cDNA encoding the vascular type-1 angiotensin II receptor, *Nature*, 351, 233, 1991.
7. **Iwai, N., Yamano, Y., Chaki, S., Konishi, F., Bardhan, S., Tibbetts, C., Sasaki, K., Hasegawa, M., Matsuda, Y., and Inagami, T.,** Rat angiotensin II receptor: cDNA sequence and regulation of the gene expression, *Biochem. Biophys. Res. Commun.*, 177, 299, 1991.
8. **Sasaki, K., Yamano, Y., Bardhan, S., Iwai, N., Murray, J.J., Hasegawa, M., Matsuda, Y., and Inagami, T.,** Cloning and expression of a complementary DNA encoding a bovine adrenal angiotensin II type-1 receptor, *Nature*, 351, 230, 1991.
9. **Kakar, S.S., Sellers, J.C., Devor, D.C., Musgrove, L.C., and Neill, J.D.,** Angiotensin II type-1 receptor subtype cDNAs: differential tissue expression and hormonal regulation, *Biochem. Biophys. Res. Commun.*, 183, 1090, 1992.
10. **Iwai, N. and Inagami, T.,** Identification of two subtypes in the rat type 1 angiotensin II receptor, *FEBS Lett.*, 298, 257, 1992.
11. **Dohlman, H.G., Thorner, J., Caron, M.G., and Lefkowitz, R.J.,** Model systems for the study of 7-transmembrane segment receptors, *Annu. Rev. Biochem.*, 60, 653, 1991.
12. **Attwood, T.K., Eliopoulos, E.E., and Findley, J.B.C.,** Multiple sequence alignment of protein families showing low sequence homology: a methodological approach using database pattern-matching discriminators for G-protein-linked receptors, *Gene*, 98, 153, 1991.

13. **Chang, R.S.L., Lotti, V.J., and Keegan, M.,** Inactivation of angiotensin II receptors in bovine adrenal cortex by dithiothreitol: further evidence for the essential nature of disulfide bonds, *Biochem. Pharmacol.*, 31, 1903, 1982.
14. **Leeb-Lundberg, L.M.F., Cotecchia, S., De Blasi, A., Caron, M.G., and Lefkowitz, R.J.,** Regulation of adrenergic receptor function by phosphorylation, *J. Biol. Chem.*, 262, 3098, 1987.
15. **O'Dowd, B.F., Hnatowich, M., Caron, M.G., Lefkowitz, R.J., and Bouvier, M.,** Palmitoylation of the human β_2-adrenergic receptor: mutation of Cys^{341} in the carboxyl tail leads to an uncoupled nonpalmitoylated form of the receptor, *J. Biol. Chem.*, 264, 7564, 1989.
16. **Hauger, R.L., Aguilera, G., Baukal, A.J., and Catt, K.J.,** Characterization of angiotensin II receptors in the anterior pituitary gland, *Mol. Cell. Endocrinol.*, 25, 203, 1982.
17. **Chen, F.M. and Printz, M.P.,** Chronic estrogen treatment reduces angiotensin II receptors in the anterior pituitary, *Endocrinology*, 113, 1503, 1983.
18. **Chen, F.M., Hawkins, R., and Printz, M.P.,** Evidence for a functional, independent brain-angiotensin system: correlation between regional distribution of brain angiotensin receptors, brain angiotensinogen and drinking during the estrous cycle of rats, *Exp. Brain Res. Suppl.*, 4, 157, 1982.
19. **Jones, T.H., Brown, B.L., and Dobson, P.R.M.,** Evidence that angiotensin II is a paracrine agent mediating gonadotrophin-releasing hormone-stimulated inositol phosphate production and prolactin secretion in the rat, *J. Endocrinol.*, 116, 367, 1988.
20. **Kakar, S.S., Riel, K.K., and Neill, J.D.,** Differential expression of angiotensin II receptor subtype mRNAs (AT-1A and AT-1B) in the brain, *Biochem. Biophys. Res. Commun.*, 185, 688, 1992.
21. **Tsutsumi, K. and Saavedra, J.M.,** Angiotensin-II receptor subtypes in median eminence and basal forebrain areas involved in regulation of pituitary function, *Endocrinology*, 129, 3001, 1991.
22. **Wright, J.W., Morseth, S.L., Abhold, R.H., and Harding, J.W.,** Pressor action and dipsogenicity induced by angiotensin II and III in rats, *Am. J. Physiol.*, R514, 1985.
23. **Wong, P.C., Hart, S.D., Zaspal, A.M., Chiu, A.T., Ardecky, R.J., Smith, R.D., and Timmermans, P.B.M.W.M.,** Functional studies of nonpeptide angiotensin II receptor subtype-specific ligands: Dup 753 (AII-1) and PD123177 (AII-2), *J. Pharmacol. Exp. Ther.*, 255, 584, 1990.
24. **Findlay, A.L., Fitzsimons, J.T., and Kucharczyk, J.,** Dependence of spontaneous and angiotensin-induced drinking in the rat upon the oestrous cycle and ovarian hormones, *J. Endocrinol.*, 82, 215, 1979.
25. **Fregly, M.J., Rowland, N.E., Sumners, C., and Gordon, D.B.,** Reduced dipsogenic responsiveness to intracerebroventricularly administered angiotensin II in estrogen-treated rats, *Brain Res.*, 338, 115, 1985.
26. **Jonklaas, J. and Buggy, J.,** Angiotensin-estrogen central interaction: localization and mechanism, *Brain Res.*, 326, 239, 1985.
27. **Bergsma, D.J., Ellis, C., Kumar, C., Nuthulaganti, P., Kerstein, H., Elshourbagy, N., Stadel, J.M., and Aiyar, N.,** Cloning and characterization of a human angiotensin II type 1 receptor, *Biochem. Biophys. Res. Commun.*, 183, 989, 1992.
28. **Takayanagi, R., Ohnaka, K., Sakai, Y., Nakao, R., Yanase, T., Haji, M., Inagami, T., Furuta, H., Gou, D.F., Nakamuta, M., and Nawata, H.,** Molecular cloning, sequence analysis and expression of a cDNA encoding human type-1 angiotensin II receptor, *Biochem. Biophys. Res. Commun.*, 183, 910, 1992.
29. **Sambrook, D., Fritsch, E.F., and Maniatis, T.,** *Molecular Cloning: A Laboratory Manual*, 2nd ed., Cold Spring Harbor Laboratory, Cold Spring Harbor, NY, 1989.

Chapter 13

ANGIOTENSIN II RECEPTORS AND SIGNAL TRANSDUCTION MECHANISMS

K.J. Catt, K. Sandberg, and T. Balla

TABLE OF CONTENTS

0-8493-4622-3/93/$0.00 + $.50

I. INTRODUCTION

Angiotensin II (AngII) was originally characterized as an octapeptide hormone with potent pressor activity in the cardiovascular system. In addition to its marked vasoconstrictor actions, AngII is now recognized as a multifunctional regulatory peptide with actions in numerous target tissues, including adrenal, kidney, brain, liver, and gonads. In each of these locations, AngII elicits responses by binding to high-affinity receptors in the plasma membrane of its target cells. AngII receptors have been characterized by radioligand binding studies and topical autoradiography in a wide variety of tissues, and are abundant in many muscle, epithelial, and neural cells. The regulatory actions of AngII via its receptors in several types of target cells are so widespread that the peptide is comparable to catecholamines and prostaglandins in the breadth of its effects throughout the body.

AngII receptors mediate the expression of phenotypic cellular responses to AngII by coupling via guanyl nucleotide regulatory (G)-proteins to plasma membrane effector enzymes and ion channels. The major enzyme involved in intracellular signaling responses to AngII is phospholipase C; in some tissues, adenylate cyclase, and phospholipases D and A_2, have also been implicated in AngII action. In the majority of its target tissues, AngII stimulates phospholipid turnover and calcium mobilization, which in turn initiate processes leading to muscle contraction, steroidogenesis, secretion, and other cellular responses. The phosphoinositide-calcium signal transduction system is the main intracellular pathway through which AngII exerts its diverse actions on vasoconstriction, aldosterone secretion, renal function, and neural activity. In several of its target tissues, AngII also exerts an inhibitory action on adenylate cyclase, an effect that does not appear to be directly correlated with most of the known cellular responses to AngII. The effects of AngII receptor activation on phospholipase C and adenylate cyclase are mediated by the guanyl nucleotide regulatory proteins, G_q and G_i, respectively. The extents to which other membrane effector enzymes, including phospholipase D and phospholipase A_2, are involved in AngII action are only now being analyzed, and whether they are influenced by G-proteins or indirectly by the products of phosphoinositide hydrolysis is not yet known.

In addition to its peripheral actions on cardiovascular regulation and fluid homeostasis, AngII acts in the central nervous system to influence thirst, blood pressure, behavior, and hormone secretion. AngII affects hormone secretion from the anterior and posterior lobes of the pituitary gland, in part by regulating the production of hypothalamic releasing factors. Its regulatory actions on hormone secretion include stimulation of vasopressin, adrenocorticotrophic hormone (ACTH), and luteinizing hormone (LH) release, and inhibition of prolactin secretion.

The diverse and widespread actions of AngII, and its coupling to multiple signal transduction systems, together with evidence from pharmacological studies in individual target tissues, suggest that the AngII receptor population may include subtypes that mediate specific target cell responses. Previously, competitive binding studies with AngII and its peptide analogs had shown little evidence for the existence of receptor subtypes. However, the presence of receptor subtypes in the kidney and other tissues had been suggested by functional studies and evidence for activation of specific signaling systems in certain target cells.[1,2] In adrenal cortex, heart, liver, and smooth muscle, AngII binds to specific receptors that promote Ca^{2+} mobilization by activating phospholipase C (probably the β isoenzyme) and increasing the formation of inositol 1,4,5-*tris*phosphate [Ins(1,4,5)P_3] (or [IP_3]). The activation of AngII receptors also promotes Ca^{2+} influx by increasing the activity of receptor-operated channels (in adrenal glomerulosa cells, chromaffin cells, cardiocytes, and sympathetic neurons) or by modulating the activity of voltage-sensitive Ca^{2+} channels, as in the heart and sympathetic fibers. Also, although AngII inhibits adenylate cyclase activity in adrenal, liver, kidney, and pituitary, it has no such effect in vascular smooth muscle and adult cardiocytes. In mesangial cells, AngII stimulates diacylglycerol production via activation of phospholipase C and subsequent release of prostaglandin E_2 (PGE_2). However, in distal tubular cells AngII stimulates PGE_2 production without activating phospholipase C. In some tissues, sulfhydryl reagents such as dithiothreitol have been found to inhibit AngII binding to its receptor sites, whereas in others this effect is much less marked. These and other indications of AngII receptor heterogeneity provided indirect evidence for the existence of multiple AngII receptor subtypes that could subserve the several actions of AngII in its numerous target cells.

Direct evidence for the existence of AngII receptor subtypes has become available through the use of newly developed AngII antagonists, and from the molecular cloning of the AngII receptor. Although AngII and its peptide antagonists such as [Sar^1,Ala^8]AngII (saralasin) and [Sar^1,Ile^8]AngII show similar binding affinities for AngII receptors in many tissues, recent studies with novel peptide[3] and nonpeptide[4] antagonists have revealed the existence of two AngII receptor subtypes, termed AT_1 and AT_2. Most of the known biochemical and functional actions of AngII are inhibited by AT_1-specific antagonists, suggesting that such compounds may recognize more than one AT receptor subtype. The function of the AT_2 receptor, which appears to be structurally distinct from the AT_1 subtype, has remained obscure. However, current studies suggest that the AT_2 receptor could mediate hitherto unrecognized actions of AngII in certain target tissues.

II. MECHANISMS OF ANGIOTENSIN II ACTION

A. ANGII RECEPTORS AND SIGNALING MECHANISMS

The concept of second messengers, and their generation at the cytoplasmic surface of the plasma membrane during the interaction of hormones or neurotransmitters with their specific receptors on the cell surface, was introduced by Sutherland and Rall in 1958.[5] The discovery that cyclic AMP acts as an intracellular messenger molecule in mediating the effects of epinephrine on glycogen metabolism in the liver was followed by the recognition that adrenocorticotropin stimulates adrenal steroidogenesis via cell surface receptors, and that its effects are also mediated by cAMP.[6] Subsequently, cAMP became the primary candidate as a messenger to mediate the effects of peptide hormones in nonexcitable tissues. However, since 1975 it has become increasingly obvious that many hormones and transmitters do not increase cAMP levels but act through other second messenger mechanisms(s), many of which are directly or indirectly related to changes in cytoplasmic Ca^{2+} concentration ($[Ca^{2+}]_i$). Hormone-induced increases in Ca^{2+} release from internal stores, and enhanced Ca^{2+} influx through voltage-gated or receptor-operated channels, are important aspects of this signaling process. Such changes are initiated by hydrolysis of plasma membrane phosphoinositides, and are often accompanied by changes in eicosanoid production and cyclic GMP levels.[7]

1. AngII Receptor Properties and Subtypes

Specific binding sites for AngII are present in the adrenal cortex and represent the cell surface receptors through which the peptide stimulates steroidogenesis.[8] In early studies, guanine nucleotides were found to decrease the affinity of AngII for its binding sites,[9,10] an effect which suggested that AngII receptors are coupled to guanine nucleotide binding proteins. The roles of specific guanine nucleotide binding proteins (G_s and G_i) in coupling β-adrenergic and other receptors to adenylate cyclase have been extensively studied and clarified, but the G-proteins responsible for the coupling of AngII and other calcium-mobilizing receptors to phospholipid hydrolysis have only recently been identified (see below).

AngII receptors are widely distributed in numerous tissues, including vascular smooth muscle, adrenal, kidney, liver, and various brain areas.[11,12] As shown in Figure 1, the AngII receptors from various target tissues show

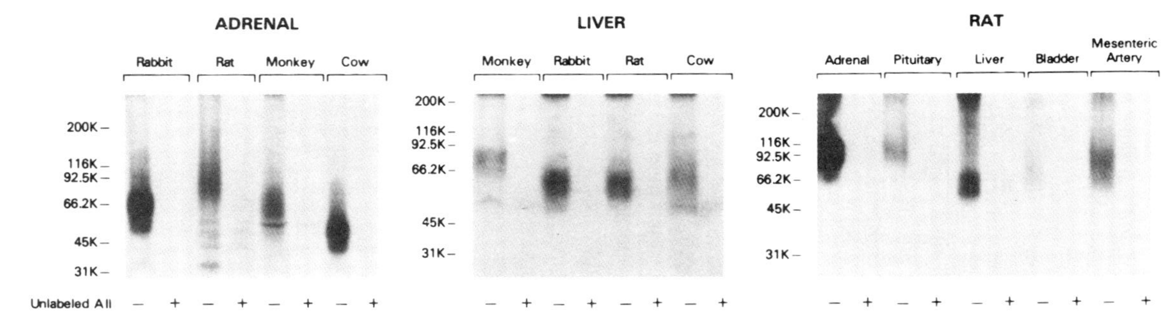

FIGURE 1. Angiotensin II receptors in specific target tissue and different species. After photoaffinity labeling of particulate fractions with ^{125}I-N$_3$-[Sar^1]AngII in the absence (−) or presence (+) of 1 μ*M* AngII, the membranes were solubilized and analyzed by sodium dodecyl sulfate gel electrophoresis and autoradiography.

significant size difference when visualized by photoaffinity labeling and denaturing gel electrophoresis. Much of this variation in molecular size is attributable to differences in glycosylation of the receptor sites present in individual tissues. The AngII receptors in specific tissues also vary in the manner in which their density and affinity are sometimes regulated during changing physiological conditions. For example, AngII receptors are increased in the adrenal zona glomerulosa during sodium deficiency,[13] but are decreased in vascular smooth muscle under the same conditions.[14] The pharmacological properties of AngII receptors, as well as their sensitivity to reducing agents, show significant tissue differences.[1,15] These features suggested the existence of more than one receptor subtype, but ligand binding studies utilizing AngII or its peptide antagonists provided little support for this proposal. As noted above, two pharmacologically distinct AngII receptor subtypes,[3,4] termed AT_1 and AT_2, have been defined by binding studies with new AngII antagonists. Among these new compounds, DuP 753 has about 1000-fold higher affinity for AT_1 than AT_2 receptors, whereas compounds such as PD123177 and CGP42112A have much higher affinity for AT_2 than AT_1 receptors. In addition, AngII binding to AT_1, but not AT_2, receptors is inhibited by guanine nucleotides and sulfhydryl reagents. The known second messenger mechanisms that are activated by AngII, and its numerous effects on its target tissues, appear to be mediated by the AT_1 receptor subtype. However, recent studies have begun to provide insights into the potential function of the AT_2 receptor subtype. The major properties of the AT_1 and AT_2 subtypes of the AngII receptor are shown in Table 1.

TABLE 1
Properties of the AT_1 and AT_2 Angiotensin II Receptor Subtypes

Properties	AT_1	AT_2
Binding affinity		
Antagonists		
[Sar^1,Ile^8]AngII	High	High
DuP 753	High	Low
PD 123177	Low	High
CGP 42112A	Low	High
Agonists		
AngII, AngIII	High	High
[*p*NH_2-Phe^8]AngII	Low	High
Sensitivity to -SH reagent	High	Low
Coupling to G-protein(s)	Yes	Unlikely
Signal transduction	↑ IP_3/Ca^{2+} ↓ Adenylate cyclase	↓ cGMP ↓ TyrPase
Endocytosis	Yes	Doubtful

2. AngII and the cAMP Messenger System

Although early reports suggested that the adrenal actions of AngII are mediated by the cAMP messenger system,[16] subsequent studies demonstrated that AngII does not increase cAMP levels in the adrenal gland,[17-19] and in many tissues causes a fall in cAMP levels. Thus, AngII was found to inhibit adenylate cyclase in adrenal glomerulosa membranes[20] and ACTH-stimulated cAMP formation in intact glomerulosa cells.[21,22] Inhibitory effects of AngII on adenylate cyclase have been described in many tissues including the liver,[23] pituitary gland,[24] kidney,[25] aorta,[26] and testis.[27]

3. AngII Belongs to the Family of Ca^{2+}-Mobilizing Hormones

The importance of Ca^{2+} in AngII-stimulated steroidogenesis was revealed by the demonstration that removal of extracellular Ca^{2+},[28] as well as the addition of Ca^{2+} channel antagonists, inhibits AngII-stimulated steroidogenesis.[29] Furthermore, inhibitors of the Ca^{2+}-dependent regulator protein, calmodulin, reduce the stimulatory effects of AngII on aldosterone production.[30-32] AngII also affects Ca^{2+} fluxes across the plasma membrane; in adrenal glomerulosa cells, it rapidly increases the efflux of $^{45}Ca^{2+}$[33-36] and stimulates $^{45}Ca^{2+}$ influx.[37-39] Calcium ions were also implicated in the actions of AngII in hepatocytes[40] and in vascular and uterine smooth muscle.[41,42]

These observations suggested that AngII activates a Ca^{2+} influx mechanism, and that the resultant increase in $[Ca^{2+}]_i$ stimulates Ca^{2+} efflux via activation of the plasma membrane Ca^{2+} ATPase. However, the observations that total cellular Ca^{2+} decreased rather than increased during AngII stimulation,[34,36] and that AngII increased Ca^{2+} efflux even in the absence of extracellular Ca^{2+},[35,36] indicated that Ca^{2+} must also be mobilized from intracellular stores. Subsequent studies with cell-permeant Ca^{2+}-sensitive fluorescent indicators showed that AngII increased $[Ca^{2+}]_i$ in adrenal glomerulosa cells; this effect was not inhibited by nifedipine, a blocker of voltage-sensitive Ca^{2+} channels (VSCC).[43,44] It was also found that a rapid cytoplasmic Ca^{2+} transient could be evoked by AngII in the absence of extracellular Ca^{2+}.[43] An example of the AngII-induced $[Ca^{2+}]_i$ response in bovine adrenal cells is shown in Figure 2. These data supported the view, based on Ca^{2+} signaling studies in many tissues, that Ca^{2+} mobilization from an agonist-sensitive intracellular pool is an important early component of the cytoplasmic Ca^{2+} response to many hormones and neurotransmitters. This group of stimuli, including AngII, became known as Ca^{2+}-mobilizing agonists to describe this characteristic feature of their activation mechanism.[45]

B. ANGIOTENSIN II AND PHOSPHOINOSITIDE METABOLISM

1. Phosphoinositide Breakdown Links AngII Receptors to Ca^{2+} Signaling

In 1975, Robert Michell suggested that certain hormones and transmitters that act on cell surface receptors, but do not activate adenylate cyclase, exert their effects by increasing the turnover of inositol phospholipids.[46] In this

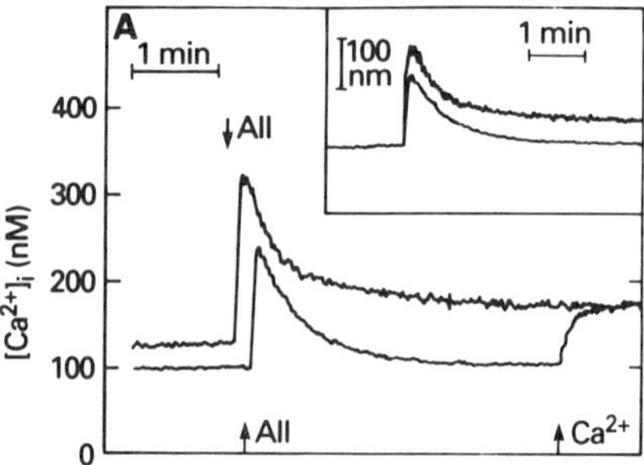

FIGURE 2. Angiotensin II-induced $[Ca^{2+}]_i$ responses in bovine adrenal glomerulosa cells. In the absence of extracellular Ca^{2+} (lower tracings in each pair shown) the peak response was retained but the plateau phase was lost.

hypothesis, the primary signaling event was the receptor-mediated hydrolysis of phosphatidylinositol (PI) to produce diacylglycerol (DAG) and *myo*-inositol 1-phosphate (Ins-1-P). This phospholipase C-catalyzed breakdown of PI was followed by its rapid resynthesis from CDP-DAG and inositol, the former produced from DAG via phosphatidic acid, the latter from Ins-1-P by dephosphorylation. The secondary increase in PI synthesis results in enhanced labeling of PA and PI in cells incubated with $[^{32}P]$phosphate, an effect used in many studies to demonstrate agonist-induced changes in PI turnover. In Michell's scheme, increased PI turnover was an initial event in transmembrane signaling from occupied receptors and was the cause rather than the consequence of the agonist-induced Ca^{2+} signal.

In adrenal glomerulosa cells, AngII was found to elicit a marked increase in the incorporation of $[^{32}P]$phosphate into PA and PI,[47-49] and was also shown to stimulate the breakdown of prelabeled PI.[47,48,50] These effects could be observed in the absence of extracellular Ca^{2+},[51] consistent with the primary role of these changes in Ca^{2+} signal generation. A causal relationship between AngII-stimulated PI turnover and increased aldosterone production was suggested by the ability of Li^+ ions, which interrupt the PI cycle by inhibiting the breakdown of Ins-1-P to inositol,[52,53] to impair the AngII-stimulated steroid response without affecting the stimulatory effects of ACTH or K^+ ions.[50] Further studies on phosphoinositide metabolism revealed that the less abundant inositol phospholipids, phosphatidylinositol 4-phosphate (PtdIns-4-P) and phosphatidylinositol 4,5-bisphosphate (PtdIns-4,5-P_2), are the primary targets of phospholipase C in the plasma membrane. These compounds are formed from PI by sequential phosphorylations and are rapidly broken down upon agonist stimulation to produce inositol 1,4,5-*tris*phosphate [Ins(1,4,5)P_3] and inositol 1,4-bisphosphate [Ins(1,4)P_2], in addition to DAG.[45,54] The effect of AngII on PtdIns-4,5-P_2 hydrolysis was first demonstrated in adrenal glomerulosa cells,[55] with simultaneous formation of water-soluble inositol phosphates.[56,57] Similar effects of AngII were described in liver,[45] vascular smooth muscle,[58,59] and pituitary gland.[60] The rapidity and extent of the breakdown of the polyphosphoinositides in AngII-stimulated glomerulosa cells is shown in Figure 3.

In the adrenal, AngII causes a biphasic increase in the level of Ins(1,4,5)P_3 with a sharp early increase that peaks at 10 to 15 s, followed by a slower secondary rise.[61] The second phase — but not the initial peak — of Ins(1,4,5)P_3 formation is dependent on the presence of extracellular Ca^{2+}.[62,63] The second phase of AngII-stimulated Ins(1,4,5)P_3 production is also impaired when receptor internalization is blocked, while the initial peak increase in Ins(1,4,5)P_3 is not affected.[64] It is not yet clear whether this biphasic Ins(1,4,5)P_3 response to agonist stimulation reflects the breakdown of different pools of PtdIns-4,5-P_2, or results from the stimulatory effect of increased cytoplasmic Ca^{2+} on phospholipase C activity. The existence of multiple forms of phospholipase C in cells[65] also raises the possibility that AngII causes sequential activation of different phospholipase C enzymes.

2. G Proteins Couple AngII Receptors to Phospholipase C

As noted above, guanine nucleotide analogs have long been known to reduce the binding of AngII to its receptors,[9,10] suggesting that G-proteins participate in coupling AngII receptors to phospholipase C. The ability of GTP analogs to stimulate IP_3 or DAG formation in membranes or permeabilized cells[66,67] indicated that phospholipase C can be activated through such

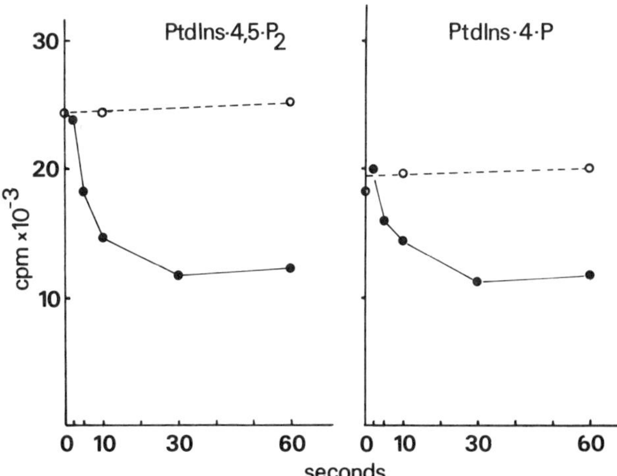

FIGURE 3. Time course of phosphoinositide hydrolysis in [^{3}H]inositol-labeled rat adrenal glomerulosa cells during stimulation by AngII. The rapid and extensive breakdown of PtdIns-4,5-P_2 and PtdIns-4-P was followed within a few minutes by resynthesis of these precursors of Ins(1,4,5)P_3 and DAG and a return to their steady-state level (not shown).

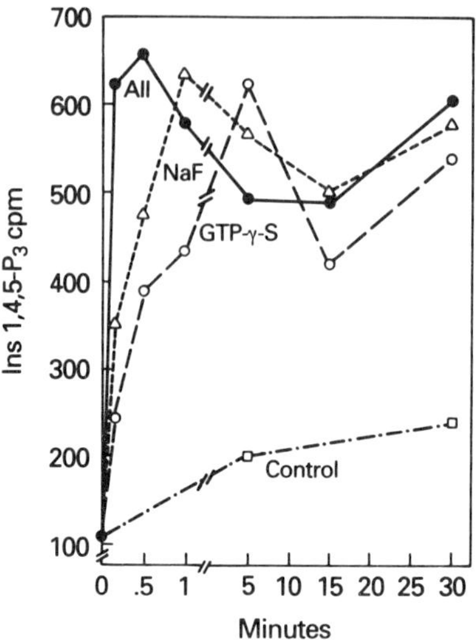

FIGURE 4. Stimulation of Ins(1,4,5)P_3 production in permeabilized glomerulosa cells by AngII, NaF, and GTPγS. Note the more rapid effect of the agonist, which accelerates the exchange of endogenous GTP with GDP bound to the regulatory G-protein that activates phospholipase C.

putative G-proteins. In liver, fluoride ions were shown to stimulate IP_3 formation and Ca^{2+} mobilization; this effect is enhanced by aluminum ions[68] and further supports the involvement of G-proteins in activation of phospholipase C. Similarly, G-protein-regulated formation of IP_3 was demonstrated in adrenal glomerulosa cells.[69,70] An example of the stimulation of IP_3 formation by G-protein activation in permeabilized glomerulosa cells is shown in Figure 4. A potential candidate for coupling AngII receptors to phospholipase C was the inhibitory guanine nucleotide binding protein, G_i. This pertussis toxin-sensitive G-protein is known to mediate the AngII-induced inhibition of adenylate cyclase in several different cell types.[20,23] However, pertussis toxin does not inhibit AngII-stimulated formation of IP_3 in hepatocytes[71] and adrenal glomerulosa cells,[69,70] or the effect of AngII on aldosterone production.[22] Also, GTP analogs continue to inhibit AngII binding in liver membranes when all G_i proteins have been ADP-ribosylated by pertussis toxin.[23] Taken together, these findings indicate that an additional G-protein is involved in the activation of phospholipase C in AngII-treated cells.

Recently, new members of the heterotrimeric G-protein family have been isolated and cloned from the brain[72] and the liver.[73] These proteins (termed G_q and G_{11}) have been shown to activate the β-isoform of phospholipase C[74] and probably include the G-protein that couples AngII receptors to phospholipase C in AngII target tissues.

3. Inositol Phosphate Metabolism in AngII-Stimulated Glomerulosa Cells

In adrenal glomerulosa cells in which the cell membrane phosphoinositides have been labeled with [^{3}H]inositol, stimulation with AngII causes rapid formation of Ins(1,4,5)P_3 and the progressive appearance of several of its metabolic products. An example of the marked increases in inositol phosphates in AngII-stimulated cells is shown in Figure 5. Although the primary product of phospholipase C-mediated PtdIns-4,5-P_2 hydrolysis is Ins(1,4,5)P_3, the majority of the IP_3 in agonist-stimulated cells is the biologically inactive isomer, Ins(1,3,4)P_3.[75] Ins(1,3,4)P_3 is formed by a metabolic pathway in which Ins(1,4,5)P_3 is phosphorylated to Ins(1,3,4,5)P_4, which is then dephosphorylated to Ins(1,3,4)P_3.[76] Thus, Ins(1,4,5)P_3 is rapidly metabolized by two separate mechanisms: it is dephosphorylated to Ins(1,4)P_2 by a membrane-associated 5-phosphatase[77] and is also phosphorylated by a cytosolic 3-kinase; the latter enzyme is activated by Ca^{2+}-calmodulin.[78,79] The dephosphorylation pathway of Ins(1,4,5)P_3 was found to proceed through Ins(1,4)P_2 to yield Ins(4)P rather than Ins(1)P as formerly believed.[80,81] Thus, agonist-induced formation of Ins(4)P provides a specific indicator of polyphosphoinositide breakdown. In contrast, the formation of Ins(1)P is delayed

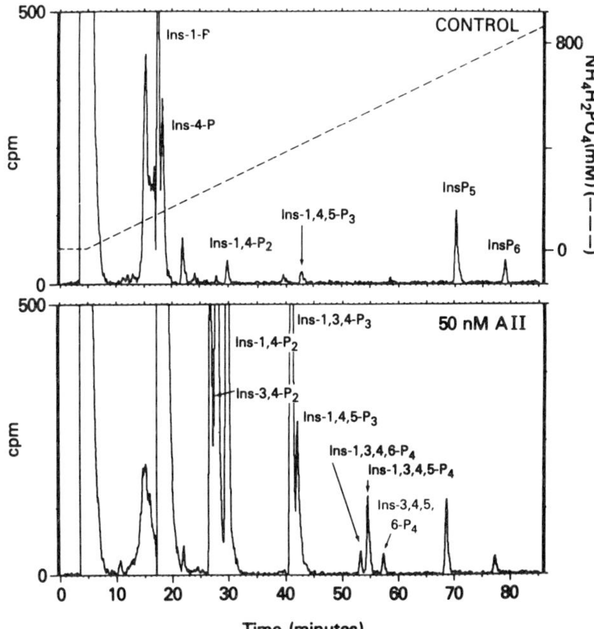

FIGURE 5. Inositol phosphate responses to AngII in [^{3}H]inositol-labeled glomerulosa cells. After incubation for 5 min with 50 n*M* AngII, the [^{3}H]inositol phosphates produced during agonist stimulation were extracted and analyzed by anion exchange HPLC.

in AngII-stimulated adrenal cells and probably reflects the direct hydrolysis of PtdIns.[61,80] It should be noted that Ins(3)P, formed from the degradation of Ins(1,3,4)P_3 via Ins(3,4)P_2 [and to a lesser degree via Ins(1,3)P_2], is an enantiomer of Ins(1)P and therefore co-elutes with the latter on high-performance liquid chromatography systems used for separating inositol phosphates. Many of the inositol phosphate-dephosphorylating enzymes are sensitive to inhibition by Li^+ ions, in particular the inositol polyphosphate 1-phosphatase and the enzyme that dephosphorylates the various Ins-monophosphates,[82] leading to the accumulation of several inositol phosphates including Ins(1,3,4)P_3 [but not Ins(1,4,5)P_3] during agonist stimulation.

More recently, large amounts of highly phosphorylated inositols such as $InsP_5$ and $InsP_6$ have been detected in mammalian cells.[81,83,84] Two $InsP_4$ isomers, Ins(3,4,5,6)P_4 and Ins(1,3,4,6)P_4 [but not Ins(1,3,4,5)P_4] were found to be precursors of Ins(1,3,4,5,6)P_5 in cell-free systems and permeabilized cells,[85,86] and are increased during AngII-stimulation in adrenal glomerulosa cells.[87] One of these potential $InsP_5$ precursors, Ins(1,3,4,6), is formed by the phosphorylation of Ins(1,3,4)P_3 by a distinct 6-kinase.[88,89] The second messenger molecule, Ins(1,4,5)P_3, can be converted to Ins(1,3,4,5,6)P_5 via sequential phosphorylations and dephosphorylations in cytosol preparations from mammalian tissues.[90,91] However, it is not clear whether $InsP_5$ and $InsP_6$ are formed in intact mammalian cells via these pathways, and even less is known about their role(s) in regulating cell functions.

In adrenal glomerulosa cells, prolonged stimulation by AngII results in prominent increases in $InsP_5$ and its putative precursor, Ins(3,4,5,6)P_4. Moreover, both AngII and ACTH (which does not activate phospholipase C) cause a slow but substantial increase in Ins(1,3,4)P_3 6-kinase activity.[87] As shown in Figure 6, there is an early fall in $InsP_5$ levels in AngII-stimulated cells, followed by a plateau and a subsequent increase in $InsP_5$ formation between 5 and 15 h. In contrast, Ins(3,4,5,6)P_4 levels rise continuously during AngII stimulation for up to 15 h. These delayed agonist-induced changes in the levels of the highly phosphorylated inositols suggest their potential involvement in long-term cellular responses such as growth or proliferation. In accordance with this hypothesis, fibroblasts transformed with the *v-src* oncogene showed very high levels of $InsP_5$ and Ins(1,4,5,6)P_4 [an enantiomer of Ins(3,4,5,6)P_4 and a possible product of $InsP_5$ hydrolysis] and express high $InsP_3$ 3-kinase activity.[92]

The purpose of these complex cycles of inositol phosphate metabolism is not yet clear. The only inositol phosphate other than Ins(1,4,5)P_3 to which a messenger function has been attributed is Ins(1,3,4,5)P_4. This molecule has been proposed to act in conjunction with Ins(1,4,5)P_3 to regulate Ca^{2+} entry into agonist-stimulated cells,[93,94] but recent evidence has cast doubt on this suggestion (see below).

4. AngII Stimulates DAG Formation

Diacylglycerol, one of the primary products of phospholipase C-mediated PI hydrolysis, is a potent activator of the Ca^{2+}- and phospholipid-dependent

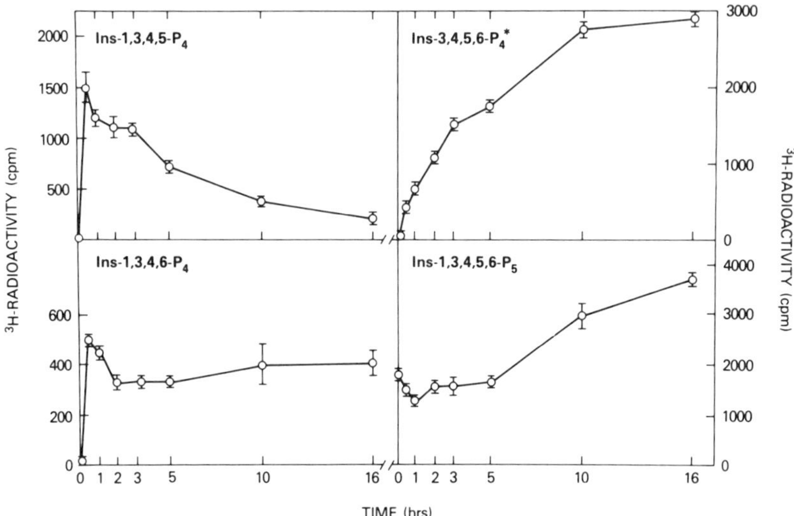

FIGURE 6. Effects of prolonged treatment with AngII on the formation of highly phosphorylated inositols in bovine glomerulosa cells.

enzyme, protein kinase C.[95] It is now accepted that the polyphosphoinositides rather than PI are the primary substrates of phospholipase C, but the importance of DAG formation and the activation of protein kinase C in mediating the effects of Ca^{2+}-mobilizing agonists has become firmly established.[96] The AngII-stimulated incorporation of [^{32}P]phosphate into PA provided an early and indirect indication of the enhanced formation of DAG in the adrenal.[48] Increased labeling of DAG was observed in AngII-stimulated bovine adrenal glomerulosa cells[56] and measurements of DAG content revealed that AngII causes a biphasic response with an early spike-increase, correlating with Ins(1,4,5)P_3 formation, and a subsequent prolonged and larger elevation.[63] In vascular smooth muscle cells, AngII was also found to cause a biphasic increase in DAG levels labeled with [^{3}H]arachidonate. These studies suggested that the early increase in DAG formation resulted from polyphosphoinositide hydrolysis, and the secondary rise from the hydrolysis of PI.[97] The second phase of DAG formation was also found to show a close correlation with AngII receptor internalization.[98]

In hepatocytes, the fatty acid composition of the DAG formed after AngII stimulation indicated that sources other than inositol phospholipids might contribute to DAG production.[99] Also, in bovine adrenal glomerulosa cells, the second phase of AngII-induced DAG formation was only partially inhibited by chelation of extracellular Ca^{2+}, while the sustained hydrolysis of phosphoinositides was almost completely inhibited, again suggesting additional source(s) of DAG production.[63] The changes in DAG and Ins(1,4,5)P_3 production during AngII stimulation are shown in Figure 7, together with the

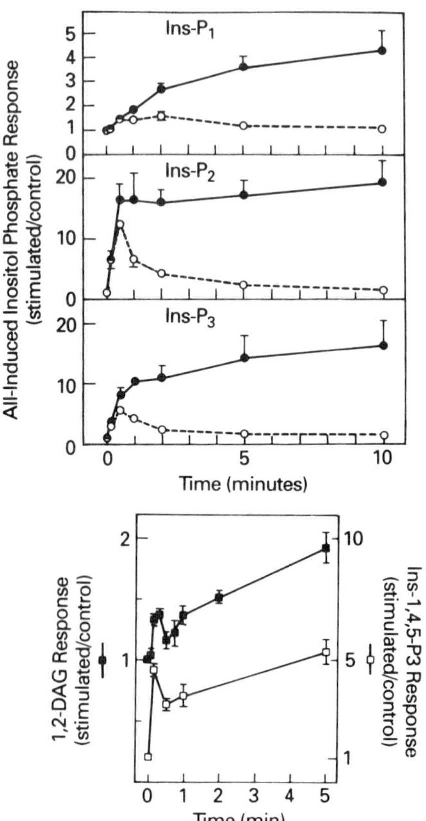

FIGURE 7. Time course of AngII-stimulated inositol phosphate formation and diacylglycerol production in bovine glomerulosa cells. The upper panel shows the Ca^{2+} dependence of the AngII-induced inositol phosphate responses, and the lower panel shows the concomitant changes in Ins(1,4,5)P_3 and 1,2-DAG.

effects of extracellular Ca^{2+} deficiency on the formation of inositol phosphates. Recent evidence has indicated that hydrolysis of phosphatidylcholine by phospholipase D is an important source of the DAG formed during agonist stimulation.[100,101] Consistent with this proposal, a decrease in the [^{32}P]phosphate-[48] and [^{3}H]arachidonate-labeling[102] of phosphatidylcholine was observed in AngII-stimulated rat adrenal glomerulosa cells. In addition, AngII has been shown to activate phospholipase D in bovine adrenal glomerulosa cells.[103]

5. Eicosanoids and AngII Action

Phosphatidylinositol contains a large amount of arachidonic acid (AA) in the 2-position of its glycerol backbone. The release of AA by the direct action of phospholipase A_2 on the phosphoinositides, or of DAG lipase on the DAG liberated from the phospholipids, provides a precursor for the formation of numerous bioactive compounds with possible messenger functions. In the adrenal gland, the ability of prostaglandins to stimulate steroidogenesis[104] raises the possibility that they could mediate the effects of physiological stimuli such as AngII. However, inhibitors of the cyclooxygenase pathway do not inhibit the effect of AngII (or other physiological regulators) on steroid production, and also do not affect the stimulatory action of AA.[105] On the other hand, AngII increases 12-hydroxy-eicosatetraenoic acid (12-HETE) production in human adrenal glomerulosa cells, and inhibition of this pathway interferes with the steroidogenic effect of AngII, indicating the possible importance of the lipoxygenase pathway.[106]

AngII has also been shown to stimulate the formation of AA in human glomerulosa cells, primarily from DAG by the action of DAG lipase.[106] Consistent with these findings, AngII increases the [^{3}H]AA labeling of PI in adrenal glomerulosa cells; this effect is related to the increased PI turnover (see above) and not to a direct deacylation-reacylation cycle involving phospholipase A_2 action on PI.[102] An additional mechanism by which AA can affect cell function is via the action of protein kinase C, since certain of the PKC isoenzymes can be potently activated by AA.

C. ANGIOTENSIN-INDUCED CALCIUM SIGNALING

1. Inositol 1,4,5-*Tris*phosphate Mobilizes Ca^{2+} from Intracellular Stores

Soon after the identification of PtdIns-4,5-P_2 as the primary substrate of phospholipase C in agonist-stimulated cells, one of its hydrolytic products, Ins(1,4,5)P_3 was found to be a potent stimulus of Ca^{2+} release from non-mitochondrial Ca^{2+} pools in permeabilized cells,[107] as well as in subcellular membrane preparations.[108] Since Ins(1,4,5)P_3 is rapidly formed after stimulation, is actively metabolized by both cytoplasmic and plasma membrane-associated enzymes (see above), and has Ca^{2+}-mobilizing activity, it fulfills the criteria of a second messenger.[109] In intact adrenal glomerulosa cells, AngII-stimulated increases in Ins(1,4,5)P_3 levels and cytoplasmic Ca^{2+} concentrations show a very close temporal correlation (Figure 8). Also, the peak increases of Ca^{2+} and Ins(1,4,5)P_3 have a similar dose-response relationship.[110]

Specific binding sites with high affinity for Ins(1,4,5)P_3 have been demonstrated in adrenocortical membrane preparations,[111,112] liver microsomes,[113] and permeabilized hepatocytes,[114] by the use of high specific activity [^{32}P]Ins(1,4,5)P_3. The IP_3 binding activity correlates closely with Ca^{2+}-mobilizing potency in permeabilized hepatocytes,[114] and is clearly different from the enzymes that degrade Ins(1,4,5)P_3.[112] The affinity of the binding site is significantly higher than the Ca^{2+}-mobilizing potency of IP_3 in crude membrane preparations.[112] This discrepancy may result from the different conditions used for binding and Ca^{2+} release studies, but could also reflect heterogeneity of the IP_3 receptors or the requirement for binding of more than

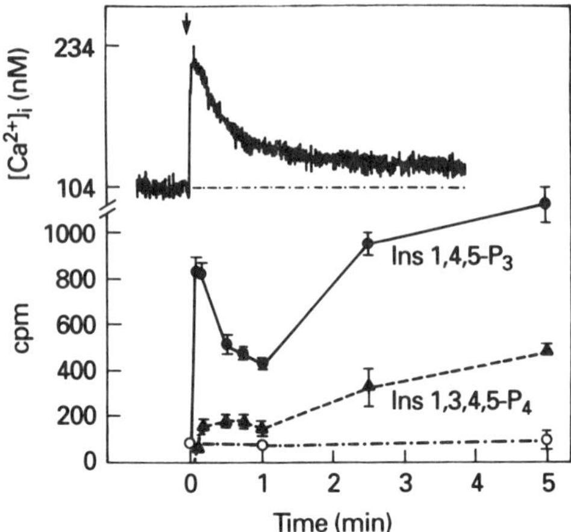

FIGURE 8. Temporal correlations between $[Ca^{2+}]_i$, Ins(1,4,5)P_3, and Ins(1,3,4,5)P_4 in bovine glomerulosa cells during stimulation by AngII.

one IP_3 molecule to open the Ca^{2+} release channel. Such positive cooperativity in IP_3-induced Ca^{2+} release has been observed in leukemia cells.[115] In the liver, IP_3 binding sites are highly enriched in the plasma membrane fraction,[116] suggesting that the IP_3-sensitive organelle has a close association with the plasma membrane. IP_3 receptors have been purified from the cerebellum, in which they are highly abundant in the Purkinje cells.[117] The receptor protein is a homotetramer consisting of four 260-kDa glycoprotein subunits, and exhibits Ins(1,4,5)P_3-regulated Ca^{2+} channel activity in the presence of a small amount of ATP. The IP_3 receptor proved to be identical to a major Purkinje cell protein, termed P_{400}, which had been cloned from the mouse.[118] When expressed, this protein displayed Ins(1,4,5)P_3 binding and Ca^{2+} release activity.[119] The receptor protein has a high degree of homology with the ryanodine receptor/Ca^{2+} channel of the sarcoplasmic reticulum[120] and is localized in subcellular compartments associated with the endoplasmic reticulum.[121,122] IP_3 binding is present in nuclei[123] and is also enriched in organelles, termed "calciosomes",[124] which have been proposed to be IP_3-sensitive Ca^{2+} storage vesicles.[125]

While the Ins(1,4,5)P_3-sensitive Ca^{2+} pool clearly plays a major role in Ca^{2+} signaling, recent evidence suggests that other intracellular Ca^{2+} pools are secondarily recruited during agonist stimulation and contribute to the cytoplasmic Ca^{2+} response. This was first suggested by the finding that GTP greatly potentiates the Ca^{2+}-releasing activity of Ins(1,4,5)P_3 in liver microsomes.[126] This effect requires the presence of PEG and depends on GTP hydrolysis.[127] Similar observations have been made in several membrane preparations[128,129] and permeabilized cells,[130] leading to the proposal that communication between Ins(1,4,5)P_3-sensitive and -insensitive Ca^{2+} pools[131] is mediated by a GTP-dependent process. Another indication that additional pools are involved in Ca^{2+} signaling came from studies in which propagating Ca^{2+} waves were observed in eggs after fertilization[132,133] and in other single cells, and were initiated from a focal site within the agonist-stimulated cell.[134] Based on studies in many different cell types, Ca^{2+}-induced Ca^{2+} release has been proposed to be the most likely mechanism responsible for such propagating Ca^{2+} waves.[135]

2. Regulation of Ca^{2+} Influx in AngII-Stimulated Cells

As mentioned earlier, AngII induces only transient cytoplasmic Ca^{2+} increases in cells incubated in Ca^{2+}-free medium, and the sustained Ca^{2+} response to AngII (also referred to as the "plateau phase") requires the presence of extracellular Ca^{2+}.[39,43,44] The mechanism of the Ca^{2+}-influx component that maintains the agonist-induced Ca^{2+} signal is a matter of intense debate. Since AngII increases Ca^{2+} efflux and decreases the exchangeable Ca^{2+} pool in liver[68] and adrenal glomerulosa cells,[36,136] only very early $^{45}Ca^{2+}$ uptake measurements can detect the AngII-stimulated Ca^{2+} influx.[39,137-139] An example of the AngII-stimulated increase in the rate of $^{45}Ca^{2+}$ influx in bovine glomerulosa cells is shown in Figure 9. Such measurements have shown that ligand-induced Ca^{2+} influx is preceded by IP_3 production, increased cytoplasmic Ca^{2+}, and Ca^{2+} efflux,[39,139] indicating that it might be secondary to activation of intracellular Ca^{2+} release.

AngII-stimulated production of aldosterone can be inhibited by dihydropyridine Ca^{2+} channel antagonists,[140] and is enhanced by the Ca^{2+} channel "agonist", BAY-K 8644,[141] suggesting that voltage-sensitive Ca^{2+} channels are activated by AngII. In the adrenal, such channels are sensitive to minor degrees of depolarization, including those resulting from small increases in potassium concentration (one of the stimuli of aldosterone production). In this regard they resemble the T-type Ca^{2+} channels present in the heart and

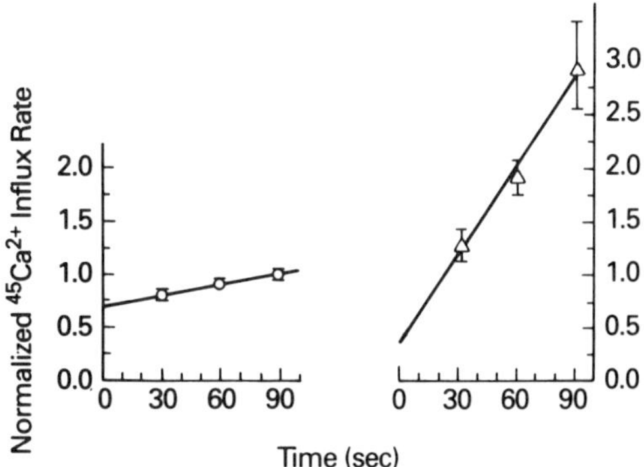

FIGURE 9. $^{45}Ca^{2+}$ influx rates in control (left) and AngII-stimulated (right) bovine glomerulosa cells.

neuronal tissues.[142] Unlike these classical T-channels, the adrenal channels are sensitive to dihydropyridine agonists and antagonists[143,144] and are not sensitive to Ni^+, and in this respect behave more like L-type channels. Patch clamp studies have identified both T-type and L-type Ca^{2+} channels in adrenal glomerulosa cells[143-145] and even N-type channel activities have been described.[144] However, few patch clamp data are available on AngII-stimulated changes in Ca^{2+}-channel behavior. In one study, AngII was shown to stimulate a rapidly inactivating (T-like) but nitrendipine-sensitive Ca^{2+} current in adrenal glomerulosa cells.[143] On the other hand, AngII stimulates the activity of an L-type Ca^{2+} channel in Y-1 adrenocortical tumor cells, and this effect is blocked by pertussis toxin treatment, suggesting the participation of a G_i-like protein.[146] In a third report, AngII was shown to inhibit a Ca^{2+} conductance in rat adrenal glomerulosa cells.[147]

Measurements of cytoplasmic Ca^{2+} and $^{45}Ca^{2+}$ influx rate have also given contradictory results. Thus, AngII-stimulated $^{45}Ca^{2+}$ influx was inhibited by dihydropyridines in bovine[138] but not in rat[139] adrenal glomerulosa cells. AngII-induced sustained Ca^{2+} increases can be observed in adrenal glomerulosa cells treated by dihydropyridine antagonists,[148] but more detailed $[Ca^{2+}]_i$ measurements show only a modest dihydropyridine-sensitive component of the increased cytoplasmic Ca^{2+} in AngII-stimulated bovine and rat adrenal glomerulosa cells.[148a] Furthermore, AngII has also been shown to inhibit depolarization-induced increases in cytoplasmic Ca^{2+} and Ca^{2+} influx rate.[148,149] Thus, AngII appears to both activate and inactivate voltage-sensitive Ca^{2+} influx mechanisms, depending on the prevailing membrane potential and the agonist concentration.

In this regard, it is important to note that AngII influences the membrane potential of adrenal glomerulosa and smooth muscle cells by changing ionic conductances and affecting Na^+/K^+ ATPase activity. AngII causes a rapid and transient hyperpolarization followed by a sustained depolarization in adrenal glomerulosa cells.[143,147,148] The initial hyperpolarization coincides with the rise in cytoplasmic Ca^{2+} and probably results from the opening of Ca^{2+}-activated K^+ channels, as indicated by ^{86}Rb efflux studies in AngII-stimulated glomerulosa cells.[150] AngII also inhibits a major K^+ conductance, leading to more sustained depolarization.[147,151] The inhibitory action of AngII on Na^+/K^+ ATPase activity in adrenal glomerulosa cells[152] would also contribute to the depolarization caused by the peptide. This effect of AngII might be a unique feature of the adrenal glomerulosa cell; in the liver, AngII stimulates Na^+/K^+ pump activity, due to an increase of intracellular Na^+ concentration secondary to the activation of Na^+/H^+ exchange.[153] Stimulation of Na^+/H^+ antiporter activity by AngII has also been demonstrated in vascular smooth muscle cells,[154] but its importance in AngII action is still not understood.

Depolarization of cells by AngII promotes Ca^{2+} influx by activating voltage-gated Ca^{2+} channels, but sustained and greater degrees of depolarization can also cause inactivation of such channels. This could account for the ability of AngII to activate, as well as inactivate, voltage-gated Ca^{2+}

entry depending on the conditions. On the other hand, depolarization can also favor Ca^{2+} entry (or impair Ca^{2+} extrusion) by the Na^{+}/Ca^{2+} exchanger, which is present in both smooth muscle[155] and adrenal glomerulosa cells.[156]

In addition to these discrepant effects of AngII on voltage-gated Ca^{2+} entry, it is clear that AngII enhances Ca^{2+} influx by other mechanisms in adrenal glomerulosa and smooth muscle cells (see above). Also, in the liver, where no voltage-sensitive channels are present, AngII increases Ca^{2+} influx solely by an alternative pathway. This mechanism, which has been extensively studied in cells regulated by Ca^{2+}-mobilizing agonists, is referred to as receptor-operated Ca^{2+} entry. There are two major theories concerning the underlying mechanism(s) of this Ca^{2+} entry process. One proposal is that emptying of the intracellular Ca^{2+} pools by Ins(1,4,5)P_3 leads to enhanced Ca^{2+} entry through the plasma membrane.[157] This ''capacitative'' model of Ca^{2+} entry is based on the finding that when intracellular Ca^{2+} pools are emptied by agonist stimulation in Ca^{2+}-free medium, and the agonist effect is terminated by addition of an antagonist, the readdition of extracellular Ca^{2+} causes a transiently enhanced Ca^{2+} entry even at times when all known messengers have returned to their basal levels.[158,159] Moreover, emptying of the agonist-sensitive Ca^{2+} pools by alternative means, without the addition of hormone, can mimic this effect on Ca^{2+} influx. Thapsigargin, a nonphorbol ester tumor promoter, has been of particular importance in such studies, since it inhibits the Ca^{2+} ATPase of the endoplasmic reticulum and thereby promotes the emptying of these pools via their spontaneous leakage pathway.[160] In bovine adrenal glomerulosa cells, addition of thapsigargin causes a slow increase in $[Ca^{2+}]_i$ and a concomitant enhancement of Ca^{2+} influx. The Ca^{2+} influx response develops in parallel with the depletion of the AngII-sensitive intracellular Ca^{2+} pool. However, both $[Ca^{2+}]_i$ increases and $^{45}Ca^{2+}$ influx rate responses are larger in AngII- than in thapsigargin-stimulated cells.[161] These results indicate that Ca^{2+} pool emptying per se can activate Ca^{2+} entry, but whether AngII-induced Ca^{2+} entry is controlled by the same mechanism is still to be answered.

The other theory suggests that Ca^{2+}-mobilizing agonists regulate Ca^{2+} entry directly by second messengers, of which Ins(1,4,5)P_3 and Ins(1,3,4,5)P_4 are the most favored candidates. Such a combined action of Ins(1,4,5)P_3 and Ins(1,3,4,5)P_4 was observed in the extracellular Ca^{2+}-dependent activation of sea urchin eggs.[93] Also, perfusion studies in single lacrimal gland cells showed that a Ca^{2+}-dependent K^{+} channel was regulated by the combined action of the two second messengers.[162] Based on these observations, and on the ability of Ins(1,3,4,5)P_4 to release Ca^{2+} from microsomes[163] and to increase the amount of Ca^{2+} released by Ins(1,4,5)P_3,[164] the capacitive model was modified to include a role for Ins(1,3,4,5)P_4 in regulating the refilling of the IP_3-sensitive Ca^{2+} stores from the extracellular fluid.[165] Such an effect of Ins(1,3,4,5)P_4 could be mediated by specific Ins(1,3,4,5)P_4 binding sites,[166] which have been identified in the bovine adrenal cortex.[167]

The kinetics of AngII-stimulated formation of Ins(1,3,4,5)P_4 in adrenal glomerulosa cells (shown in Figure 8) are consistent with its proposed role in regulating the Ca^{2+} entry process.[61,110] However, studies in fibroblasts that were transfected with the IP_3 3-kinase enzyme to increase the production of Ins(1,3,4,5)P_4 gave no indication that Ins(1,3,4,5)P_4 plays a role in Ca^{2+} mobilization, entry, or sequestration. Instead, both Ca^{2+} release and entry were closely correlated with the agonist-induced increases in Ins(1,4,5)P_3.[168] Direct regulation of Ca^{2+} entry by Ins(1,4,5)P_3 was demonstrated in the Jurkat T-cell lymphoma cell line by patch clamp studies.[169] Interestingly, the plasma membrane of Jurkat cells contains an Ins(1,4,5)P_3 receptor that differs in its sugar moieties from the Ins(1,4,5)P_3 receptor of the endoplasmic reticulum, and recognizes Ins(1,3,4,5)P_4 almost as well as Ins(1,4,5)P_4.[170] This new form of the IP_3 receptor, if shown to be ubiquitous, might be related to agonist-induced Ca^{2+} entry. Whether such a mechanism is operational in AngII-stimulated cells remains to be elucidated.

3. Ca^{2+} Oscillations in AngII-Stimulated Cells

The development of microfluorometry and high-resolution fluorescent imaging has made it possible to monitor changes in cytoplasmic Ca^{2+} concentration at the single cell level. The improved resolution of such Ca^{2+} measurements has provided additional insights into the signaling process by revealing that many cells display repetitive $[Ca^{2+}]_i$ transients in response to agonists,[171] and that the frequency (rather than the amplitude) of these Ca^{2+}-transients is often regulated by the concentration of the agonist. AngII-stimulated Ca^{2+} oscillations have been described in rat hepatocytes,[172] rat[173] and bovine[174] adrenal glomerulosa cells, and neuroblastoma cells.[175] In all cases, increasing the dose of AngII reduces the lag-time before the first Ca^{2+} transient occurs, and increases the frequency of the oscillations. At higher AngII concentrations, no oscillations are observed and the Ca^{2+} signal consists of a rapid initial spike followed by a lower, sustained plateau phase, as observed in cell suspensions.[172,173] This sequence of changes in single bovine glomerulosa cells during stimulation with increasing doses of AngII is shown in Figure 10. It should be noted that many cell types do not respond to agonist stimulation with Ca^{2+} oscillations but with biphasic (peak and plateau) elevations that are dose dependent.

It is not yet clear whether Ca^{2+} oscillations are attributable to one general mechanism that is characteristic of all cells, or result from different mechanisms that can generate oscillations when the system is working near or at the threshold level of the agonist.[176] However, it appears that Ca^{2+} oscillations in most cells do not require oscillatory changes in Ins(1,4,5)P_3 levels, and that Ca^{2+} oscillations can be provoked without an increase of Ins(1,4,5)P_3, by the release of only a small amount of Ca^{2+} from intracellular stores.[177] It has been suggested that during agonist stimulation, the small amount of Ca^{2+}

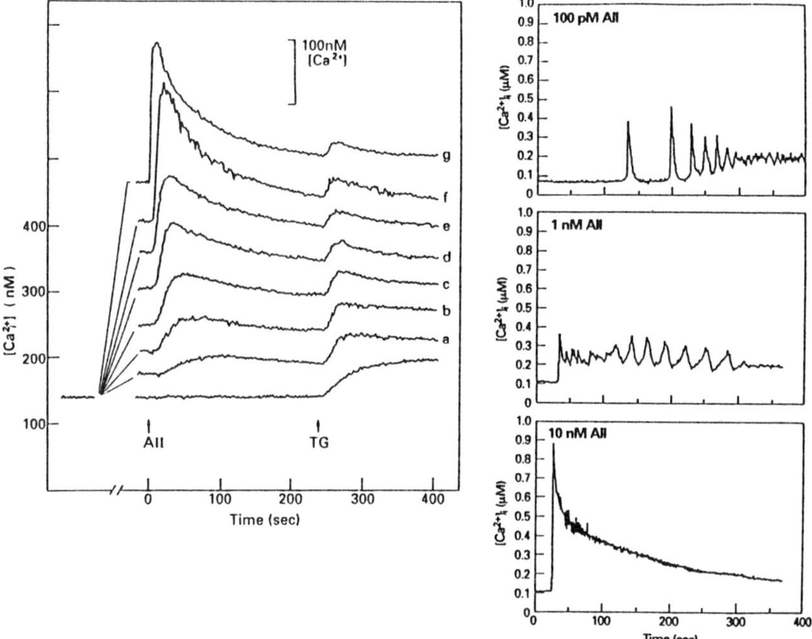

FIGURE 10. Dose-dependent increases in $[Ca^{2+}]_i$ in AngII-stimulated bovine glomerulosa cells. (Left) $[Ca^{2+}]_i$ responses in cell suspensions stimulated by 100 p*M* to 100 n*M* AngII (bottom to top). (Right), Single-cell $[Ca^{2+}]_i$ responses in cells treated with 100 p*M* to 10 n*M* AngII.

released by Ins(1,4,5)P_3 evokes an almost ''all or none''-type Ca^{2+} release from the endoplasmic reticulum by Ca^{2+}-induced Ca^{2+} release.[178] The high ambient Ca^{2+} concentration — or the decrease of the intraluminal Ca^{2+} in the endoplasmic reticulum — would then terminate the effect of Ins(1,4,5)P_3, allowing the Ca^{2+} stores to refill before the next cycle can occur. Many refinements have been made on this theory and several mathematical models have been developed to explain the phenomena observed in oscillating cells, but the physiological relevance of the oscillatory nature of the Ca^{2+} signal and its underlying mechanism are yet to be defined.

One important aspect of the oscillating Ca^{2+} response during AngII action could be related to the voltage-gated Ca^{2+} entry process. If the small depolarization of the adrenal glomerulosa cells by AngII is interrupted by hyperpolarizations caused by the opening of Ca^{2+}-activated K^+ channels (see above), synchronous with the Ca^{2+} transients, this would allow the intermittent reactivation of the voltage-gated Ca^{2+} channels, thereby making Ca^{2+} entry through these channels most active at low doses of AngII. This hypothesis, however, still awaits confirmation.

4. AngII Receptor Internalization and Ca^{2+} Signaling

Many ligands are internalized by receptor-mediated endocytosis after binding to their specific target cells. The ligand-receptor complex is believed to dissociate in the acid milieu within receptosomes and the receptor either recycles to the plasma membrane or is degraded by proteolytic enzymes.[179] In the adrenal gland, rapid internalization of ^{125}I-AngII by glomerulosa cells was demonstrated *in vivo* by electron microscopy.[180] AngII is also internalized in cultured adrenocortical cells[64,181] and vascular smooth muscle.[98] Internalization of AngII agonists occurs rapidly and can be detected within 2.5 min after addition of the agonist ligand (Figure 11). It does not occur with AngII antagonists, and can be prevented by lowering the temperature or by reducing the proton gradient in receptosomes with monensin or K^+ depletion.[64,181] Receptor internalization and recycling is believed to play a role in the regulation of the available receptor sites at the plasma membrane, and to participate in processes such as elimination of the ligand and homologous desensitization. In addition, recent evidence has suggested that receptor-mediated endocytosis could be relevant to AngII-induced second messenger formation.

As mentioned earlier, the second phase of AngII-stimulated DAG formation is closely correlated with receptor internalization in smooth muscle.[98] In adrenal glomerulosa cells, the plateau phase of the cytoplasmic Ca^{2+} signal and the second phase of IP_3 formation[64] are impaired by maneuvers that inhibit receptor internalization (Figure 11). It is possible that AngII receptor complexes continue their signal generation function in internalized vesicles before they dissociate. In this context, it is important to note that the ability of an AngII antagonist to inhibit the plateau phase of the AngII-induced Ca^{2+} signal depends on the time interval between the addition of AngII and the antagonist.[182] Similarly, the sustained phase of AngII-stimulated DAG formation was found to be resistant to the late addition of an AngII antagonist in adrenal glomerulosa cells.[183] These studies indicate that, once activated, AngII receptor complexes maintain their signaling function and are at least partially inaccessable to blockade by specific antagonists. The relationship between AngII receptor internalization and second messenger formation requires further exploration to resolve these issues.

D. MECHANISMS ACTIVATED BY THE ANGII-EVOKED CA^{2+} SIGNAL

In ligand-activated cells, the potential sites at which elevated Ca^{2+} levels can interact with enzymes and processes to affect cell function are so numerous as to be beyond the scope of this review. Therefore, we will only consider those mechanisms which have been explored in some detail during AngII action. The role of the Ca^{2+}-binding regulatory protein, calmodulin, is well established in smooth muscle, where it mediates the stimulatory effect of Ca^{2+} on myosin light chain kinase and promotes contraction. In the liver, although the physiological role of circulating AngII in regulating hepatic glycogenolysis is questionable, AngII-induced $[Ca^{2+}]_i$ increases can activate

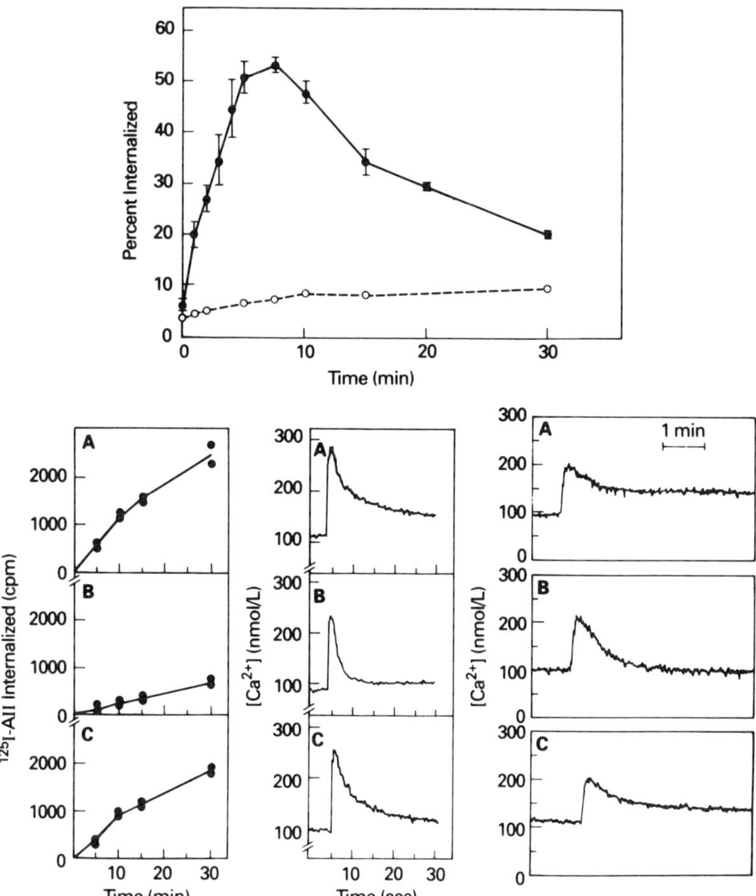

FIGURE 11. Angiotensin II receptor endocytosis and calcium signaling in bovine glomerulosa cells. Above: internalization of receptor-bound ^{125}I-AngII (● — ●) compared with that of the antagonist, ^{125}I-[Sar^1,Ile^8]AngII (○ — ○). Below: left, effects of clathrin depletion (by hypo-osmotic shock and K^+ depletion) and repletion (middle and lower panels) on AngII internalization and $[Ca^{2+}]_i$ responses to AngII. Right, effects of treatment with the endocytosis inhibitor, phenylarsine oxide (B), and its reversal by dithiothreitol (C) on the AngII-induced Ca^{2+} response.

phosphorylase b kinase, an enzyme which contains calmodulin as one of its subunits. Activation of calmodulin converts the inactive phosphorylase b to phosphorylase a and promotes glycogenolysis. A role of calmodulin in AngII-stimulated steroidogenesis has also been suggested,[30,32] but less is known about the possible effectors that are regulated by the Ca^{2+}-calmodulin-dependent protein kinases.

Another Ca^{2+}-dependent kinase that has been implicated in the action of AngII is protein kinase C. One of the earliest suggestions that the Ca^{2+} signal evoked by $Ins(1,4,5)P_3$, and the activation of PKC by DAG, act in concert to regulate responses in agonist-stimulated cells was based on studies of AngII-stimulated aldosterone production.[56] These studies showed that increases in cytoplasmic Ca^{2+}, if not associated with enhanced DAG production (e.g., during treatment with Ca^{2+} ionophores), cause only a transient aldosterone secretory response. Similarly, activation of PKC by phorbol esters without a concomitant $[Ca^{2+}]_i$ increase is a weak stimulus of aldosterone production. However, these stimuli fully mimicked the steroidogenic effect of AngII when operating together, indicating that both limbs of the phosphoinositide/Ca^{2+} messenger system are essential for a full secretory response. The same conclusion was reached when the temporal pattern of AngII-stimulated protein phosphorylations were compared to those evoked by Ca^{2+} ionophores and phorbol esters added alone or in combination.[184] It was also shown that exposure of adrenal glomerulosa cells to AngII for a period of 20 min or more rendered these cells more sensitive to Ca^{2+} entering through the plasma membrane, even after AngII was removed by washing the cells. These data suggested that AngII has the ability to change the sensitivity of the steroidogenic machinery for Ca^{2+} and that these changes leave a "memory" in the cells after the agonists are removed.[185] These and other studies led to the proposal that the AngII-induced Ca^{2+} signal causes translocation and/or modification of PKC (and possibly other Ca^{2+}/calmodulin-dependent kinases) that increases their sensitivity to changes in the sub-plasma membrane Ca^{2+} concentration, which could be significantly higher than the average values revealed by cytoplasmic Ca^{2+} measurements. This model also assumed that the increased recycling of Ca^{2+} in AngII-stimulated cells (due to stimulated influx as well as efflux) represents an important signal to the cells even when the average cytoplasmic Ca^{2+} level is only slightly higher than basal.[186]

Such studies made an important contribution in emphasizing the temporal and spatial aspects of Ca^{2+} signaling in adrenal cells, but the role of protein kinase C in these processes has proven to be more controversial. Translocation of protein kinase C to the plasma membrane by AngII has been observed in bovine,[187] but not in rat,[188] adrenal glomerulosa cells. Moreover, activation of PKC by phorbol esters inhibits AngII-stimulated formation of IP_3 in adrenal glomerulosa[189] and in vascular smooth muscle cells,[190] and therefore represents a negative signal on AngII action. In rat adrenal glomerulosa cells, PKC appears to play an inhibitory rather than a stimulatory role in AngII-stimulated aldosterone secretion.[191] It is not yet clear why such diverse effects are observed in studies on the role of PKC, but it will obviously be necessary to define the roles of the individual PKC isoenzymes in future studies. It is also likely that the use of freshly prepared cells, in which the plasma membranes are damaged to varying degrees, could be accompanied by "basal" activation of PKC. Also, in primary cell cultures the expression of PKC isoenzymes might change as compared to the parent cells *in vivo*. Further studies should reveal which of the above-described effects are relevant to the physiological mechanisms by which AngII activates its target cells.

The effects of increased cytoplasmic Ca^{2+} concentration on adrenal steroidogenesis have revealed new aspects of AngII-stimulated signaling. In permeabilized glomerulosa cells, steroid production was shown to be directly regulated by increased Ca^{2+} levels with half-maximal stimulation at 0.5 μM Ca^{2+}. This effect of Ca^{2+}, however, was only observed in the presence of $NADP^{+}$.[192] Recently, single-cell measurements of endogenous NAD(P)H fluorescence in adrenal glomerulosa cells revealed a Ca^{2+}-dependent increase in the reduced pyridine nucleotides of mitochondrial origin, which are essential components for the steroid hydroxylation steps involved in steroid synthesis.[193] These findings have not only defined the mitochondria as one target of Ca^{2+} action, but also offer a sensitive method through which the steroid production of individual glomerulosa cells could be followed. It is expected that this line of research will provide important information about the flow of information from the AngII receptor to the steroid-synthesizing machinery of the glomerulosa cell.

E. SIGNAL TRANSUCTION BY AT_2 RECEPTORS

Although the signaling systems activated by AngII via the AT_1 receptor are well characterized, much less is known about the function and transduction systems of the AT_2 receptors. Their lack of coupling to G-proteins[194,195] and their failure to undergo internalization[196] suggest that the AT_2 receptors are structurally different entities and have little in common with AT_1 receptors, apart from their identical binding affinities for AngII and most of its peptide analogs. In fact, none of the functional evidence for AngII receptor subtypes has been accounted for by the discovery of the AT_2 receptor. Rather, it seems likely that the AT_2 sites will mediate as-yet-unrecognized aspects of AngII-induced cellular signaling, and may be related to certain of the growth-related actions of AngII.

AT_2 receptors are abundant in the adrenal medulla, PC-12 pheochromocytoma cells, and the uterus, and are also present in other tissues, including the brain and the rat adrenal cortex (Figure 12). They have high affinity for PD123177 and CGP 42112A,[3,4] and for the putative agonist peptide [Phe(*p*-NH_2)^6]AngII.[194] In the rodent and primate fetus, AngII receptors are abundant in connective tissues, skin, and muscle, as well as in the adrenal, kidney, and vascular smooth muscle.[197-199] Although some of the receptors in fetal skin fibroblasts are of the AT_1 subtype and are coupled to the phosphoinositide-Ca^{2+} signaling system, the majority of the fetal skin receptors are of the AT_2 subtype. The ontogeny of the fetal AngII receptors is also of interest, since they increase markedly in late gestation and disappear rapidly after birth.[197,198] In cultured rat fetal fibroblasts, the AngII receptors are initially 90% AT_2 and 10% AT_1, but this ratio is progressively reversed when the cells are subcultured.[200]

AT_2 receptors are also abundant in the brain in immature animals, and decrease progressively with age.[201] In addition to the few regions at which AT_2 receptors are present in the adult brain (lateral septum, subthalamic

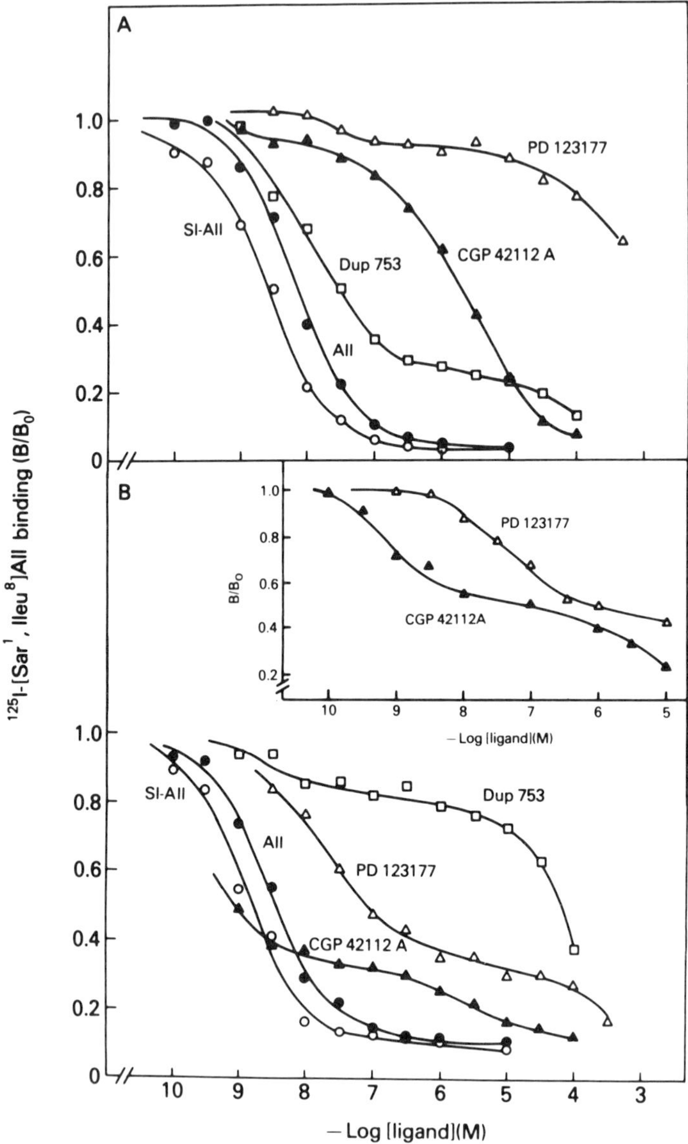

FIGURE 12. Radioligand binding to AT_1 and AT_2 receptors in the rat adrenal gland. In the glomerulosa zone (above), the majority of the AngII receptors detected by binding of ^{125}I-[Sar^1,Ile^8]AngII are of the AT_1 subtype, being displaced by DuP 753 and not by PD 123177. In the medulla (below), the predominance of AT_2 receptors is indicated by the major inhibition of radioligand binding by CGP 42112A and PD 123177, and the minor inhibitory effect of DuP 753.

nucleus, and inferior olive), AT_2 receptors were transiently expressed in several sites in the brains of young animals, including several thalamic nuclei, the 3rd and 4th cranial nerve nuclei, cerebellum, and cingulate gyrus. The AT_1 receptors are present in regions related to the control of blood pressure, water intake, and pituitary hormone secretion. The abundance and location of the AT_2 receptors in young animals, and their decreasing expression with advancing age, suggest that they mediate specific actions of AngII according to the developmental stage of the nervous system.

Although the physiological function of the AT_2 receptor is not yet known, recent studies have begun to elucidate some of the potential signaling pathways that could be relevant to its cellular actions. Whereas astrocyte cultures from neonatal rat brains contain AT_1 receptors that are linked to phosphoinositide hydrolysis, neonatal neuron cultures contain predominantly AT_2 receptors,[202] consistent with the abundance of these sites in the immature brain.[201] The neuronal AT_2 receptors were found to be coupled to a reduction in basal cyclic GMP levels, suggesting the existence of a signal transduction mechanism based on the turnover of this cyclic nucleotide. This could in turn be related to the observation that AngII acts on AT_2 receptors in neurons cultured from the hypothalamus and brain stem to stimulate an increase in net outward current.[203]

Another potential signaling mechanism for the AT_2 receptor has been suggested by a recent report in which AngII was found to inhibit basal and ANP-stimulated guanylate cyclase activity in rat glomerulosa and PC12 cells.[204] This effect of AngII was blocked by orthovanadate, which inhibits phosphotyrosine phosphatase, but not by okadaic acid, which inhibits serine/threonine phosphatases. AngII also induced orthovanadate-sensitive dephosphorylation of phosphotyrosine residues of PC12 cell proteins, suggesting that AngII can signal through a tyrosine phosphatase system that is involved in the regulation of guanylate cyclase activity.

III. ANGIOTENSIN II AS A GROWTH FACTOR

There is increasing evidence for the role of AngII as a growth factor in several of its target tissues. That AngII exerts long-term actions on cell growth and responsiveness, in addition to its rapid regulatory actions on responses such as steroidogenesis and muscle contraction, was indicated by early studies on the hypertrophy and hyperplasia of the adrenal glomerulosa zone during experimental hypertension[205,206] and sodium deprivation[207] in rats. This aspect of the action of angiotensin has been somewhat neglected during the period in which the receptors and signaling mechanisms of AngII were intensively studied, but is now receiving more attention. Such renewed interest in the trophic actions of AngII has been stimulated by the recognition that AngII causes hypertrophy of vascular smooth muscle cells.[208] Also, studies with converting enzyme inhibitors have implicated the renin-angiotensin system in the development of structural changes in the resistance vessels of hypertensive rats.[209] Conventional growth factors such as platelet-derived growth factor (PDGF) and FGF have been implicated in the autocrine control of proliferation of arterial smooth muscle cells in hypertension, but the role of AngII in promoting cell proliferation (as opposed to hypertrophy) remains unclear. AngII increases thymidine incorporation in cultured smooth muscle cells but this evidence of increased DNA synthesis, though usually associated with the induction of cell hyperplasia, is not followed by proliferation of AngII-stimulated muscle cells.

AngII also causes a receptor-mediated increase in protein synthesis in heart cells, as shown by studies in embryonic chick myocytes.[210] This hypertrophic response to AngII is not accompanied by significant changes in DNA and RNA content, again consistent with the absence of a proliferative response to AngII. It has been suggested that the absence of a proliferative response to AngII in the presence of increased thymidine incorporation, as occurs in vascular smooth muscle cells, could result from the development of polyploidy.[208,211] The ability of AngII to promote growth-related responses in smooth muscle cells led to studies on the induction of *c-fos* and other early response genes in AngII-stimulated target cells. In vascular smooth muscle cells, AngII causes a rapid and concentration-dependent increase in the accumulation of *c-fos* mRNA. This effect is dependent on mobilization of intracellular Ca^{2+} and activation of protein kinase C.[212] The effects of AngII on the expression of *c-fos, c-jun,* and other early response genes in bovine adrenal glomerulosa cells are shown in Figure 13. AngII also stimulates the expression of *c-myc* mRNA, an effect that persists for several hours and is accompanied by elevation of mRNA for the PDGF A-chain and increased production of PDGF.[213] It is noteworthy that AngII enhances the stimulatory action of PDGF on DNA synthesis in rabbit vascular smooth muscle cells,[214] suggesting that synergism with the proliferative effects of autocrine or paracrine growth factors could be an important aspect of AngII action in the vascular system.

In addition, AngII and growth factors such as PDGF and transforming growth factor (TGF)-β have been found to induce the expression of mRNA for endothelin in cultured human vascular smooth muscle cells.[215] Endothelin in turn increases the production of these growth factors[216] and promotes mitogenesis and expression of *c-fos* and *c-myc* in smooth muscle cells.[217] Thus, it is possible that an autocrine cycle favoring cell proliferation could be established when the normal proportions of systemic and locally formed vasoactive peptides and growth factors are disturbed by overproduction of AngII or endothelin. It is clear that, in addition to its rapid regulatory actions on target-cell responses such as steroid secretion and muscle contraction, AngII can exert several effects on the expression of specific genes in certain tissues. These include stimulation of the early gene responses and growth factors noted above, as well as stimulation of angiotensinogen gene expression in the liver[218,219] and inhibition of renin gene expression in the kidney.[220]

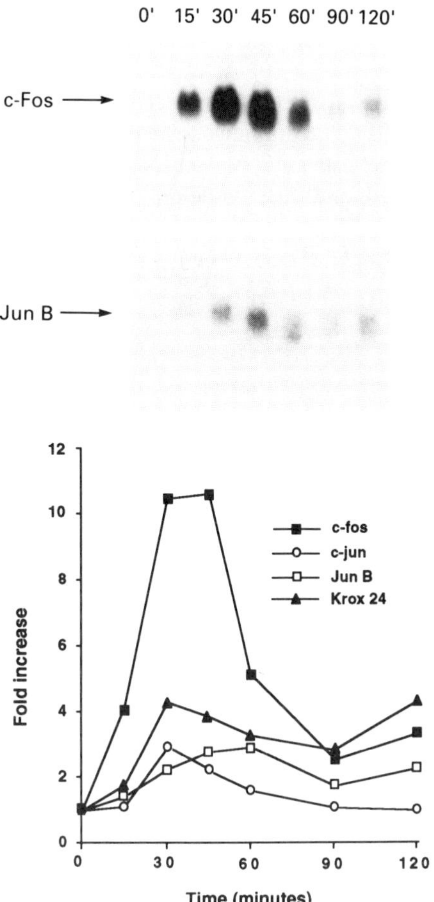

FIGURE 13. Induction of early gene responses by AngII in bovine glomerulosa cells. Above: Northern blots of *c-fos* and *c-jun* mRNA responses to AngII. Below: time course of the changes in four early response gene mRNAs during stimulation by AngII.

Many growth factors, including epidermal growth factor (EGF), PDGF, and FGF, act through receptors that posses a ligand-activated tyrosine kinase domain which is essential for their biological activity.[221] Recently, several peptides and other ligands that act through G-protein-coupled receptors have been found to stimulate tyrosine phosphorylation. These include thrombin, vasopressin, fMetLeuPhe, and muscarinic agonists, as well as AngII. In rat liver epithelial cells, AngII, vasopressin, and epinephrine stimulate rapid and transient increases in tyrosine phosphorylation of several proteins.[222] These effects appear to be secondary to elevation of cytoplasmic Ca^{2+} and were not related to activation of protein kinase C. AngII has also been found to stimulate tyrosine phosphorylation in adrenal glomerulosa cells. These effects of AngII are not due to an intrinsic tyrosine-kinase activity of the receptor itself but appear to be secondary to the effects of AngII on phosphoinositide hydrolysis and Ca^{2+} mobilization. Such increased tyrosine phosphorylation in AngII-stimulated cells could provide clues to the identity of cellular proteins that are involved in the pathways responsible for the long-term actions of AngII on cellular growth and differentiation.

IV. MOLECULAR BIOLOGY OF ANGIOTENSIN II (AT) RECEPTORS

A. CLONING OF AT RECEPTOR SUBTYPES

Recently, cDNAs encoding a bovine and a rat AT receptor (termed bAT and rAT_1, respectively) have been isolated from cDNA libraries synthesized from bovine adrenal glomerulosa cells,[223] rat vascular smooth muscle cells,[224] and rat kidney.[225] Both AT receptors were isolated by expression cloning in COS-7 cells transfected with plasmid DNA prepared from pools of clones, by screening for AT receptor expression by binding assays with ^{125}I-[Sar^1,Ile^8]AngII. Positive pools were subsequently divided until a single positive clone remained. An additional AT receptor subtype which is distinct from the rAT_1 has been recently cloned.[226-228] In one case, this subtype (termed rAT_{1b} or rAT_3) was cloned by probing a rat adrenal cDNA library with a DNA fragment of the rAT_1 receptor under low-stringency hybridization conditions.[226] In another case, the rAT_3 was found serendipitously during an attempt to isolate the receptor for gonadotropin releasing hormone (GnRH).[227] A cDNA library from rat anterior pituitary RNA was screened in *Xenopus* oocytes by following the GnRH-induced Ca^{2+}-dependent chloride current in positive library pools. However, the resulting clone did not bind an ^{125}I-labeled GnRH agonist analog in transfected COS-7 cells. Nucleotide analysis of the cDNA insert revealed a high degree of homology to the rAT_1 receptor, and ^{125}I-[Sar^1,Ile^8]AngII binding to transfected COS-7 cells confirmed that the receptor coded for a unique AT receptor.

In our laboratory, the rAT_3 receptor was isolated by application of nested PCR to cDNA synthesized from adrenal glomerulosa mRNA, using primers based on the sequence of the rAT_1 receptor.[228] PCR yielded a 760-bp product that hybridized to the rAT_1 receptor cDNA on Southern analysis. However, restriction enzyme analysis and DNA sequencing of this subcloned fragment indicated that it was distinct from the cloned rAT_1 receptor. To obtain a full-length receptor cDNA, polymerase chain reaction (PCR) was performed on pools of clones from a rat adrenal glomerulosa library which contained an active AngII receptor as revealed by ligand-induced calcium mobilization in *Xenopus* oocytes injected with RNA transcripts from one of sixteen pools of 50,000 cells. Calcium responses were monitored as light emission from oocytes co-injected with RNA transcripts and the Ca^{2+}-specific photoprotein, aequorin.[229]

A human AT receptor (hAT) has also been cloned by probing lymphocyte genomic[230] and liver cDNA libraries[231,232] with a labeled DNA fragment of the rAT_1 receptor under low-stringency hybridization conditions. All four of the cloned AT receptors (bAT, rAT_1, rAT_3, hAT) code for a 359-amino acid protein with a relative molecular mass (M_r) of approximately 41 kDa. This is similar to the size (~35 kDA) determined by SDS-PAGE of the deglycosylated AngII receptor.[233,234] Hydropathicity profiles[235] of the AT receptors are consistent with seven transmembrane spanning domains connected by three extracellular and three intracellular loops. This pattern is similar to other G-protein-coupled receptors, in which the amino-terminal is extracellular and the carboxy-terminus is located in the cytoplasm.[236] The AT receptor protein shows 20 to 30% sequence identity with other G-protein-coupled receptors and only 20% identity with the human *mas* oncogene product, which has been proposed to be an AngII receptor.[237] Recent evidence suggests that the *mas* oncogene enhances responses to endogenous AT receptors present in the amphibian oocyte[238] and in COS-7 cells.[239,240]

B. COMPARISON OF AT RECEPTOR SEQUENCES

In all four AT receptor cDNAs, the initiation codon is preceded by one ATG and is subsequently followed by in-frame termination codons. It is of interest that less than 10% of 700 vertebrate RNAs have upstream AUG codons; a notable exception is the class of oncogene transcripts, in which 65% have one or more AUGs preceding the start of the major coding region.[241] Upstream AUG codons exert inhibitory effects on translation[242] and could regulate oncogene expression at this level.[243,244] It has been proposed that deletion of 5′ AUG elements contributes to oncogene overexpression in transformed cells. Thus, these upstream AUGs could play a key role in preventing the undesired overexpression of gene products that regulate cell growth.

Another potential control mechanism shared by proto-oncogenes and AT receptors is their unusually long leader sequences (>100 nucleotides). While only 25% of vertebrate mRNAs have leader sequences longer than 100 nucleotides, with most ranging between 20 and 100 nucleotides,[241] the leader sequences for the AT receptors are greater than 149 nucleotides. These long leader sequences provide the possibility that secondary structure can have a regulatory function in translational control.[245] This regulatory mechanism can thus play a critical role in the differential translation of mRNAs that occur during oogenesis and early development. It is interesting that AT receptors, which possess both features of proto-oncogenes (upstream AUG codons and atypically long leader sequences), have been implicated in growth and development. AngII and AngIII stimulate DNA synthesis, cellular proliferation, and protein synthesis.[208,246] AngII also induces the expression of growth-related proto-oncogenes, including *c-fos, c-jun,* and *c-myc* in smooth muscle, cardiac muscle, and adrenal cells[247] (see Figure 12). The expression of growth factors such as PDGF is also increased by AngII.[213] An intriguing finding is the widespread occurrence of AT receptors in the rat fetus.[198] In addition to their typical sites in adult target tissues, AngII receptors were found in fetal skeletal muscle and connective tissue. AT receptors in these novel locations decreased by 80% one day after birth and were nearly undetectable in the adult, suggesting that AngII plays a role in fetal development.

Although the nucleotide identity between the AT receptors is 85 to 91% within the coding region, the overall homology is only 65 to 71% due to significant differences within the 5′ (45 to 48%) and 3′ (56 to 67%) untranslated regions (Table 2). There is one 5′ UTR sequence and two 3′ UTR sequences that are identical in all four receptors. It remains to be seen whether these sequences play any conserved functional or regulatory roles in AT function. The mRNAs of transiently expressed genes including lymphokines, cytokines, and proto-oncogenes frequently contain the destabilizing signal AUUUA. These sequences are also present in AT receptors; two are found in rAT_1, four in rAT_3, four in bAT, and six in hAT. It will be interesting to compare the stabilities of these receptor mRNAs, since these sequences are proposed to be recognition signals for the processing pathways that specifically degrade mRNAs.

The four cloned AT receptors share >90% amino acid sequence homology, and most of the amino acid substitutions occur outside the transmembrane-spanning regions. Several of the potential sites for posttranslational modification are identical (Figure 14). These features include three consensus sites for N-glycosylation; one of these is in the N-terminus preceding the first transmembrane domain (N4) and the other two are in the second extracellular loop (N176 and N188). The N-terminal sequence is highly hydrophilic, and the third transmembrane domain does not contain the aspartate residue that has been implicated in the binding of charged amine moieties in monoaminergic and muscarinic receptors. The signature of G-protein-coupled receptors is found at amino acids 114–130, and the third cytoplasmic loop is relatively short. Cysteine residues are present in each of the four extracellular regions, and probably form the disulfide bridges necessary for ligand binding; dithiothreitol has been shown to dramatically reduce DuP 753-sensitive binding sites.[248]

While most of the structural features of the cloned AT receptors are similar, there are some key differences between the individual receptors. A cysteine residue (C355) in the C-terminal region that might serve as a site

TABLE 2
Nucleotide Homology (%) Among Cloned AngII Receptors

Receptor	5′ UTR	Coding region	3′ UTR
Rat AT_1	100	100	100
Rat AT_3 (= AT_{1b})	48	91	56
Bovine AT	45	85	61
Human AT	45	86	67

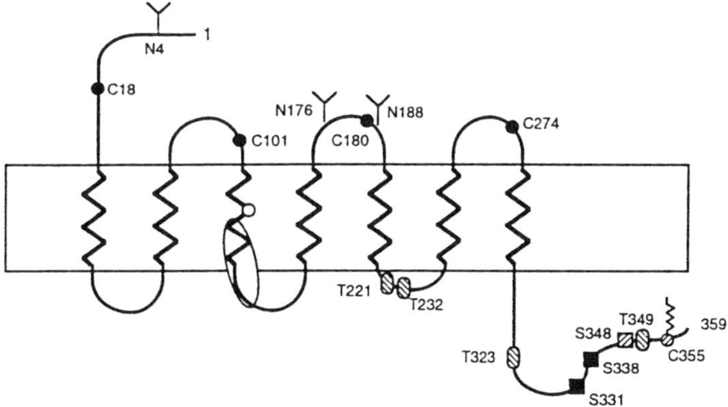

Potential Sites of Post-Translational Modification:
-Glycosylation: N4, N176, N188
-Disulfides: C18, C101, C180, C274
-Palmitoylation: C355 (missing in rAT3)
-Amidation: L305
-Phosphorylation:
Casein Kinase II: S5, S6, T88, T260 (S354 missing in hAT & bAT)
Protein Kinase C: S331, S338
(S348 in rAT1, rAT3 and hAT)
(T232 and T323 in rAT3 only)
(T349 in hAT and bAT only)
(T221 in bAT only)

Other Features:
-No D in TMIII
-Short 3rd IC loop
-G protein signature: 114-130

FIGURE 14. Structural features of the AngII receptor. The potential sites of posttranslational modification and other features of interest are indicated for the following receptors: rat, rAT_1 and rAT_3 (= rAT_{1b}); bovine, bAT; and human, hAT.

for palmitoylation, a posttranslational modification that has been implicated in receptor-effector coupling,[248] is missing in the rAT_3 receptor. Another significant feature of the rAT_3 receptor is the potential for two additional protein kinase C phosphorylation sites at tyrosines 232 and 323. These differences in potential palmitoylation and phosphorylation sites may provide a basis for tissue-specific differential regulation of the AT receptor subtypes in the rat. The hAT and the bAT receptors each have one additional potential protein kinase C phosphorylation site compared with rAT_1 (Figure 1). If other bovine and human AT receptors are cloned, it will be of interest to compare these subtypes within their species.

C. GENOMIC STRUCTURE OF AT RECEPTORS

The rAT_1 receptor is encoded by a single gene that is composed of three exons and two introns.[250] Two of the exons encode 5′ UTR sequences and the third exon contains a small portion of the 5′ UTR and the entire coding region and 3′ UTR. This is consistent with the absence of introns from the coding regions of most other seven-transmembrane domain receptors. The rAT_1 gene contains several sequence motifs that are frequently found in the promoters of other genes.[251] The 5′ sequence adjacent to the coding region includes a putative cap signal, a TATA box sequence, and a GC-rich box.[250] The coding strand from positions 274 to 417 is composed primarily of pyrimidines, suggesting that this sequence might bind histones to form nucleosomes.[252] Another interesting feature of this gene is the presence of minisatellite DNA from positions 2116 to 2147.[250] Although the genomic structure of the rAT_3 receptor is still unknown, PCR with primers based on the 5′ and 3′ untranslated regions of the AT_3 cDNA gave an 1100-bp fragment from both rat adrenal cDNA and rat genomic DNA, suggesting that the gene encoding AT_3 also does not contain introns within the coding region.[228] Furthermore, no evidence for introns was found in a rAT_3 genomic clone which contained the coding region.[253]

D. PHARMACOLOGY OF AT RECEPTORS

Distinct pharmacological differences between the rAT receptor subtypes were observed in comparative binding studies of rAT_1 and rAT_3 receptors expressed in COS-7 cells.[228] AngII and AngIII were more potent at the expressed rAT_3 receptor than at the rAT_1 receptor. While [Sar^1,Ile^8]AngII was equipotent at both receptors, the nonpeptide antagonist, DuP 753, was more potent at the rAT_3 than the rAT_1 receptor. The nonpeptide AT_2 receptor antagonist, PD 123177, did not inhibit binding to rAT_3 receptors even at micromolar concentrations. AngI, which has an IC_{50} of 74 n*M* at the rAT_1 receptor,[224] was reported to be about 10-fold less potent at the rAT_3 receptor.[227] The differences in receptor pharmacology between adrenal and vascular smooth muscle receptors might be related to the specific structural features of the rAT_3 receptor, including the two additional potential sites for phosphorylation. These differences could account for their differential regulation during changes in sodium uptake and the relatively higher potency of des-Asp^1-AngII at the adrenal level. Human astrocytes possess two pharmacologically distinct AT receptors, one of which mediates the release of prostaglandin E_2 (PGE_2) via PLC activation and another that binds AngII[1-7] and activates PGE_2 release by a Ca^{2+}-independent mechanism.[254] This AngII[1-7] receptor is similar in this regard to the renal distal tubule AT receptor in terms of its signaling properties. Recently, a unique rAT pharmacological subtype that binds AngII and AngIII, and the AT_2 antagonist PD 123319, with high affinity and appears to be G-protein coupled has been observed in cultured renal mesangial cells and in proximal tubule epithelial cells.[255] These reports suggest the existence of other AT receptor subtypes that could account for the widespread actions of AngII.

E. COUPLING OF EXPRESSED AT RECEPTORS

In *Xenopus* oocytes, the expression of mRNAs from tissues enriched in AngII receptors has been shown to produce active receptors linked to the phosphoinositide/Ca^{2+} signaling pathway.[229,256] AngII elicits dose-dependent light responses in oocytes co-injected with either rAT_1 or rAT_3 RNAs and aequorin, confirming that these adrenal cDNA clones encode functional AT receptors. Half-maximal responses were evoked by 100 n*M* AngII and the threshold for detectable responses was 5 n*M*. The calcium responses mediated by expressed AT_3 receptors were reduced at high AngII concentrations, in a manner analogous to the decrease in aldosterone responses of adrenal glomerulosa cells to high agonist doses.[22] This finding could be related to the two additional potential protein kinase C phosphorylation sites located in the cytoplasmic domain, an area that is known to be involved in receptor desensitization in other G-protein-mediated receptors.[257]

The AngII-induced Ca^{2+} response mediated by rAT receptors expressed in *Xenopus* oocytes was inhibited by the nonpeptide AngII antagonist, DuP 753 (IC_{50} = 200 ± 35 p*M*), but not by CGP 42112A or PD123177 (IC_{50} >10 μ*M*).[228] The latter finding excludes the possibility of mRNA-induced upregulation of the endogenous amphibian AngII receptor, which is known to recognize CGP 42112A but not DuP 753.[258,259] The hAT receptor was shown to increase Ca^{2+}-dependent chloride current in *Xenopus* oocytes and this effect was blocked by receptor antagonists.[232] rAT_1, rAT_3, and bAT receptors expressed in COS-7 cells mediated ligand-induced increases in intracellular Ca^{2+}, consistent with their coupling to the phosphoinositide signaling system.[223,225,227,228]

F. TISSUE-SPECIFIC EXPRESSION OF AT RECEPTOR SUBTYPES

Northern blot analysis of RNA from known rat AngII target tissues using selective probes based on the 3′ UTR regions of both receptors revealed tissue-specific expression of the two AT receptor subtypes.[228] The rAT_1 receptor probe bound to a 2.4-kb band in rat adrenal glomerulosa, liver, and lung, and also to additional bands in the adrenal (3.5 kb) and liver (1.4 kb) that could reflect unique receptor subtypes or alternative splicing. Under identical conditions, a comparable rAT_3 receptor probe bound solely to a 2.4-kb band in rat adrenal glomerulosa tissue. The expression of rAT_1 and rAT_3 receptor mRNAs in AngII target tissues was also examined by RT-PCR.[227,228,260] One of these studies[228] employed combinations of PCR primers that distinguish between rAT_1 and rAT_3 receptors, and normalized the levels of AT expression to a housekeeping gene (GAPDH). By this method, rAT_3 mRNA was found predominantly in the adrenal cortex and pituitary gland, and was also present in low abundance in several other tissues. rAT_1 receptors were also widely distributed and were most abundant in liver, lung, spleen, pituitary, and adrenal. The molecular size of AngII receptors analyzed by photoaffinity labeling is greater in the adrenal cortex and pituitary glands than in other AngII target tissues,[233] a difference that might be attributable to more extensive glycosylation of the rAT_3 receptor than the rAT_1. These studies demonstrate that most tissues express a mixture of receptor subtypes with different abundances. Tissue-specific regulation of AngII receptor expression and functional variation might account for the differences in the relative tissue levels of the rAT_1 and rAT_3 receptors.

bAT receptors were shown by Northern blot analysis to be expressed in bovine adrenal glomerulosa cells, adrenal medulla, and renal medulla, but not in renal cortex.[223] No other tissues were examined, so it is not clear whether the bAT is more similar in its tissue distribution profile to rAT_1 or rAT_3 receptors. hAT mRNA was detected in human liver, lung, and adrenal[231] and in the heart, placenta, skeletal muscle, and kidney.[232] Assuming that other bovine and human AT receptor subtypes exist, it will be interesting to compare subtype-specific probes to determine whether rAT_1 and rAT_3 have their counterparts in other species. In the brain, the subfornical organ and the organum vasculosum of the lamina terminalis, two of the circumventricular organs associated with AngII-induced drinking behavior, express primarily rAT_3 mRNA.[227] Two other circumventricular organs, the median eminence and the area postrema, expressed primarily rAT_1 and equivalent quantities of rAT_1 and rAT_3, respectively. These data suggest these rAT subtypes mediate distinct functions within the brain.

G. REGULATION OF AT RECEPTORS

In the adrenal gland, the steady-state level of rAT mRNA was found to be positively modulated by sodium restriction, which increases adrenal AngII receptors through activation of the renin-angiotensin system. In rats maintained on low sodium intake for 4 weeks, the level of AT mRNA increased by 2.3-fold in the adrenal but was unchanged in the kidney, liver, and brain.[225] Also, bilateral nephrectomy markedly reduced AT mRNA levels in the adrenal and brain stem (pons and medulla oblongata) by 85 and 60%, respectively.[260] Administration of DuP 753 also reduced the expression levels of the rAT mRNA in the adrenal and brain stem, by 50 and 40%, respectively. The kidney and the aorta were not affected by either treatment. Continuous infusion of AngII increased receptor mRNA by 2.4-fold in the adrenal but not in the brain stem, aorta, or kidney, suggesting that expression of receptor mRNA in the adrenal is dependent on the activity of the renin-angiotensin system.

Renin and AngII-converting enzyme are known to be developmentally regulated in the rat.[261] AngII receptor expression in the liver and kidney is also developmentally regulated, in that rAT mRNA is abundant in adult liver but is not present in the liver of newborn rats. Conversely, AT receptor expression was significantly higher in the kidney of newborn rats than that in adult rats.[261] Estrogens are known to markedly reduce AT receptor binding in the anterior pituitary gland,[262] and estrogen has been shown to inhibit rAT_3

but not rAT_1 mRNA levels in the rat pituitary gland.[227] In addition, ovariectomy was followed by a large increase in rAT_3 mRNA after 4 weeks and this was diminished by subsequent estrogen treatment. It will be interesting to determine whether differences in mRNA tissue distribution and regulation are related simply to the regulation of their expression or whether differential functional properties of the encoded protein will play a key role. In this light, it is interesting that high concentrations of AngII diminish the ligand-induced Ca^{2+} response mediated by rAT_3 but not rAT_1 receptors expressed in *Xenopus* oocytes.[228]

In summary, the recent cloning and expression of AT receptor subtypes has provided a molecular basis for multiple AngII receptors with similar but pharmacologically distinct activities that could mediate diverse cellular responses involved in the physiological actions of AngII. Additional analysis of these subtypes could reveal valuable information about AngII receptor function and regulation.

REFERENCES

1. **Douglas, J.G.,** Angiotensin receptor subtypes of the kidney cortex, *Am. J. Physiol.,* 253, F1, 1987.
2. **Peach, M.J. and Dostal, D.E.,** Angiotensin II receptor and the actions of angiotensin II, *J. Cardiovasc. Pharmacol.,* 4, S25, 1990.
3. **Whitebread, S., Mele, M., Kamber, B., and de Gasparo, M.,** Preliminary biochemical characterization of two angiotensin II receptor subtypes, *Biochem. Biophys. Res. Commun.,* 163, 284, 1989.
4. **Chiu, A.T., Herblin, W.F., McCall, D.E., Ardecky, R.J., Carini, D.J., Duncia, J.V., Pease, L.J., Wong, P.C., Wexler, R.R., Johnson, A.L., and Timmermans, P.B.M.W.M.,** Identification of angiotensin II receptor subtypes, *Biochem. Biophys. Res. Commun.,* 165, 196, 1989.
5. **Sutherland, E.W. and Rall, T.W.,** Fractionation and characterization of a cyclic adenine ribonucleotide formed by tissue particles, *J. Biol. Chem.,* 232, 1077, 1958.
6. **Grahame-Smith, D.G., Buther, R.W., Ney, R.L., and Sutherland, E.W.,** Adenosine 3′,5′-monophosphate as the intracellular mediator of the action of adrenocorticotropic hormone zon the adrenal cortex, *J. Biol. Chem.,* 242, 4435, 1967.
7. **Abdel-Latif, A.A.,** Calcium-mobilizing receptors, phosphoinositides and generation of second messengers, *Pharmacol. Rev.,* 38, 227, 1986.
8. **Glossmann, H., Baukal, A., and Catt, K.J.,** Angiotensin II receptors in bovine adrenal cortex. Modification of angiotensin II binding by guanyl nucleotides, *J. Biol. Chem.,* 249, 664, 1974.
9. **Glossmann, H., Baukal, A.J., and Catt, K.J.,** Properties of angiotensin II receptors in the bovine and rat adrenal cortex, *J. Biol. Chem.,* 249, 825, 1974.
10. **Crane, J.K., Campanile, C.P., and Garrison, J.C.,** The hepatic angiotensin II receptor II. Effect of guanine nucleotides and interaction with cyclic AMP production, *J. Biol. Chem.,* 257, 4959, 1982.
11. **Mendelsohn, F.A., Dunbar, M., Allen, A., Chou, S.T., Millan, M.A., Aguilera, G., and Catt, K.J.,** Angiotensin II receptors in the kidney, *Fed. Proc.,* 45, 1420, 1986.

12. **Catt, K.J., Carson, M.C., Hausdorff, W.P., Leach-Harper, C.M., Baukal, A.J., Guillemette, G., Balla, T., and Aguilera, G.,** Angiotensin II receptors and mechanisms of action in adrenal glomerulosa cells, *J. Steroid Biochem.*, 27, 915, 1987.
13. **Hauger, R.L., Aguilera, G., and Catt, K.J.,** Angiotensin II regulates its receptor sites in the adrenal glomerulosa zone, *Nature*, 271, 176, 1978.
14. **Aguilera, G. and Catt, K.J.,** Regulation of vascular angiotensin II receptors in the rat during altered sodium intake, *Circ. Res.*, 49, 751, 1981.
15. **Douglas, J., Saltman, S., Fredlund, P., Kondo, T., and Catt, K.J.,** Receptor binding of angiotensin II and antagonists. Correlation with aldosterone production by isolated canine adrenal glomerulosa cells, *Circ. Res.*, 38, 108, 1976.
16. **Bing, R.F. and Schulster, D.,** Adenosine 3′:5′-cyclic monophosphate production and steroidogenesis by isolated rat adrenal glomerulosa cells: effects of angiotensin II and [Sar^1,Ala^8]angiotensin II, *Biochem. J.*, 176, 39, 1978.
17. **Saruta, T., Cook, R., and Kaplan, N.M.,** Adrenocortical steroidogenesis: studies on the mechanism of action of angiotensin and electrolytes, *J. Clin. Invest.*, 51, 2239, 1972.
18. **Douglas, J., Saltman, S., Williams, C., Bartley, P., Kondo, T., and Catt, K.J.,** An examination of possible mechanisms of angiotensin II-stimulated steroidogenesis, *Endocr. Res. Commun.*, 5, 173, 1978.
19. **Fujita, K., Aguilera, G., and Catt, K.J.,** The role of cyclic AMP in aldosterone production by isolated zona glomerulosa cells, *J. Biol. Chem.*, 254, 8567, 1979.
20. **Marie, J. and Jard, S.,** Angiotensin II inhibits adenylate cyclase from adrenal cortex glomerulosa zone, *FEBS Lett.*, 159, 97, 1983.
21. **Woodcock, E.A. and Johnston, C.I.,** Inhibition of adenylate cyclase in rat adrenal glomerulosa cells by angiotensin II, *Endocrinology*, 115, 337, 1984.
22. **Hausdorff, W.P., Sekura, R.D., Aguilera, G., and Catt, K.J.,** Control of aldosterone production by angiotensin II is mediated by two guanine nucleotide regulatory proteins, *Endocrinology*, 120, 1668, 1987.
23. **Pobiner, B.F., Hewlett, E.L., and Garrison, J.C.,** Role of N_i in coupling angiotensin II receptors to inhibition of adenylate cyclase in hepatocytes, *J. Biol. Chem.*, 260, 16200, 1985.
24. **Marie, J., Gaillard, R.C., Schoenenberg, P., Jard, S., and Bockaert, J.,** Pharmacological characterization of the angiotensin receptor negatively coupled with adenylate cyclase in rat anterior pituitary gland, *Endocrinology*, 116, 1044, 1985.
25. **Woodcock, E.A. and Johnston, C.I.,** Inhibition of adenylate cyclase by angiotensin II in rat renal cortex, *Endocrinology*, 111, 1687, 1982.
26. **Anand-Srivastava, M.B.,** Angiotensin II receptors negatively coupled to adenylate cyclase in rat aorta, *Biochem. Biophys. Res. Commun.*, 117, 420, 1983.
27. **Khanum, A. and Dufau, M.L.,** Angiotensin II receptors and inhibitory actions in Leydig cells, *J. Biol. Chem.*, 263, 5070, 1988.
28. **Fakunding, J.L., Chow, R., and Catt, K.J.,** The role of calcium in the stimulation of aldosterone production by adrenocorticotropin, angiotensin II, and potassium in isolated glomerulosa cells, *Endocrinology*, 105, 327, 1979.
29. **Fakunding, J.L. and Catt, K.J.,** Dependence of aldosterone stimulation in adrenal glomerulosa cells on calcium uptake: effects of lanthanum and verapamil, *Endocrinology*, 107, 1345, 1980.
30. **Balla, T., Hunyady, L., and Spät, A.,** Possible role of calcium uptake and calmodulin in adrenal glomerulosa cells: effects of verapamil and trifluoperazine, *Biochem. Pharmacol.*, 31, 1267, 1982.
31. **Balla, T. and Spät, A.,** The effect of various calmodulin inhibitors on the response of adrenal glomerulosa cells to angiotensin II and cyclic AMP, *Biochem. Pharmacol.*, 31, 3705, 1982.
32. **Wilson, J.X., Aguilera, G., and Catt, K.J.,** Inhibitory actions of calmodulin antagonists on steroidogenesis in zona glomerulosa cells, *Endocrinology*, 115, 1357, 1984.

33. **Williams, B.C., McDougall, J.G., Tait, J.F., and Tait, S.A.,** Calcium efflux and steroid output from superfused rat adrenal cells: effects of potassium, adrenocorticotropic hormone, 5-hydroxytryptamine, adenosine 3′:5′-cyclic monophosphate and angiotensins II and III, *Clin. Sci.,* 61, 541, 1981.
34. **Elliott, M.E. and Goodfriend, T.L.,** Angiotensin alters $^{45}Ca^{2+}$ fluxes in bovine adrenal glomerulosa cells, *Proc. Natl. Acad. Sci. U.S.A.,* 78, 3044, 1981.
35. **Kojima, I., Kojima, K., and Rasmussen, H.,** Effects of angiotensin II and K^+ on Ca^{2+} efflux and aldosterone production in adrenal glomerulosa cells, *Am. J. Physiol.,* 248, E36, 1985.
36. **Balla, T., Szebeny, M., Kanyár, B., and Spät, A.,** Angiotensin II and FCCP mobilizes calcium from different intracellular pools in adrenal glomerulosa cells; analysis of calcium fluxes, *Cell Calcium,* 6, 327, 1985.
37. **Foster, R., Lobo, M.V., Rasmussen, H., and Marusic, E.T.,** Calcium: its role in the mechanism of action of angiotensin II and potassium in aldosterone production, *Endocrinology,* 109, 2196, 1981.
38. **Kojima, I., Kojima, K., and Rasmussen, H.,** Role of calcium fluxes in the sustained phase of angiotensin II-mediated aldosterone secretion from adrenal glomerulosa cells, *J. Biol. Chem.,* 260, 9177, 1985.
39. **Kramer, R.E.,** Angiotensin II-stimulated changes in calcium metabolism in cultured glomerulosa cells, *Mol. Cell. Endocrinol.,* 60, 199, 1988.
40. **Garrison, J.C., Borland, M.K., Florio, V.A., and Twible, D.A.,** The role of calcium ion as a mediator of the effects of angiotensin II, catecholamines, and vasopressin on the phosphorylation and activity of enzymes in isolated hepatocytes, *J. Biol. Chem.,* 254, 7147, 1979.
41. **Angles d'Auriac, G., Baudouin, M., and Meyer, P.,** Mechanism of action of angiotensin in smooth muscle, *Circ. Res.,* 30, II-151, 1972.
42. **Freer, R.J.,** Calcium and angiotensin tachyphylaxis in rat uterine smooth muscle, *Am. J. Physiol.,* 228, 1423, 1975.
43. **Capponi, A.M., Lew, P.D., Jornot, L., and Vallotton, M.B.,** Correlation between cytosolic free Ca^{2+} and aldosterone production in bovine adrenal glomerulosa cells. Evidence for a difference in the mode of action of angiotensin II and potassium, *J. Biol. Chem.,* 259, 8863, 1984.
44. **Braley, L.C., Menachery, A., Brown, E., and Williams, G.,** The effects of extracellular K^+ and angiotensin II on cytosolic Ca^{++} and steroidogenesis in adrenal glomerulosa cells, *Biochem. Biophys. Res. Commun.,* 123, 810, 1984.
45. **Creba, J.A., Downes, C.P., Hawkins, P.T., Brewster, G., Michell, R.H., and Kirk, C.J.,** Rapid breakdown of phosphatidylinositol 4-phosphate and phosphatidylinositol 4,5-bisphosphate in rat hepatocytes stimulated by vasopressin and other Ca^{2+}-mobilizing hormones, *Biochem. J.,* 212, 733, 1983.
46. **Michell, R.H.,** Inositol phospholipids and cell surface receptor function, *Biochim. Biophys. Acta,* 415, 81, 1975.
47. **Farese, R.V., Larson, R.E., Sabir, M.A., and Gomez-Sanchez, C.,** Effects of angiotensin II and potassium on phospholipid metabolism in the rat adrenal zona glomerulosa, *J. Biol. Chem.,* 256, 11093, 1981.
48. **Hunyady, L., Balla, T., Nagy, K., and Spät, A.,** Control of phosphatidylinositol turnover in adrenal glomerulosa cells, *Biochim. Biophys. Acta,* 713, 352, 1982.
49. **Elliott, M.E., Alexander, R.C., and Goodfriend, T.L.,** Aspects of angiotensin action in the adrenal. Key roles for calcium and phosphatidyl inositol, *Hypertension,* 4, 52, 1982.
50. **Balla, T., Enyedi, P., Hunyady, L., and Spät, A.,** Effects of lithium on angiotensin-stimulated phosphatidylinositol turnover and aldosterone production in adrenal glomerulosa cells: a possible causal relationship, *FEBS Lett.,* 171, 179, 1984.

51. **Hunyady, L., Balla, T., and Spät, A.,** Angiotensin II stimulates phosphatidylinositol turnover in adrenal glomerulosa cells by a calcium-independent mechanism, *Biochim. Biophys. Acta,* 753, 133, 1983.
52. **Hallcher, L.M. and Sherman, W.R.,** The effects of lithium ion and other agents on the activity of myo-inositol 1-phosphatase from bovine brain, *J. Biol. Chem.,* 255, 10896, 1980.
53. **Sherman, W.R., Leavitt, A.L., Honchar, M.P., Hallcher, L.M., and Phillips, B.E.,** Evidence that lithium alters phosphoinositide metabolism: chronic administration elevates primarily D-myo-inositol 1-phosphate in cerebral cortex of the rat, *J. Neurochem.,* 36, 1947, 1981.
54. **Berridge, M.J.,** Rapid accumulation of inositol trisphosphate reveals that agonists hydrolyze polyphosphoinositides instead of phosphatidylinositol, *Biochem. J.,* 212, 849, 1983.
55. **Farese, R.V., Larson, R.E., and Davis, J.S.,** Rapid effects of angiotensin II on polyphosphoinositide metabolism in the rat adrenal glomerulosa, *Endocrinology,* 114, 302, 1984.
56. **Kojima, I., Kojima, K., Kreutter, D., and Rasmussen, H.,** The temporal integration of the aldosterone secretory response to angiotensin occurs via two intracellular pathways, *J. Biol. Chem.,* 259, 14448, 1984.
57. **Enyedi, P., Büki, B., Mucsi, I., and Spät, A.,** Polyphosphoinositide metabolism in adrenal glomerulosa cells, *Mol. Cell. Endocrinol.,* 41, 105, 1985.
58. **Smith, J.B., Smith, L., Brown, E.R., Barnes, D., Sabir, M.A., Davis, J.S., and Farese, R.V.,** Angiotensin II rapidly increases phosphatidate-phosphoinositide synthesis and phosphoinositide hydrolysis and mobilizes intracellular calcium in cultured arterial muscle cells, *Proc. Natl. Acad. Sci. U.S.A.,* 81, 7812, 1984.
59. **Alexander, R.W., Brock, T.A., Gimbrione, M.A., Jr., and Rittenhouse, S.E.,** Angiotensin increases inositol trisphosphate and calcium in vascular smooth muscle, *Hypertension,* 7, 447, 1985.
60. **Enjalbert, A., Sladeczek, F., Guillon, G., Bertrand, P., Shu, C., Epelbaum, J., Garcia-Sainz, A., Jard, S., Lombard, C., Kordon, C., and Bockaert, J.,** Angiotensin II and dopamine modulate both cAMP and inositol phosphate productions in anterior pituitary cells, *J. Biol. Chem.,* 261, 4071, 1986.
61. **Balla, T., Baukal, A.J., Guillemette, G., and Catt, K.J.,** Multiple pathways of inositol polyphosphate metabolism in angiotensin-stimulated adrenal glomerulosa cells, *J. Biol. Chem.,* 263, 4083, 1988.
62. **Woodcock, E.A., Smith, A.I., and White, L.B.S.,** Angiotensin II-stimulated phosphatidylinositol turnover in rat adrenal glomerulosa cells has a complex dependence on calcium, *Endocrinology,* 122, 1053, 1988.
63. **Hunyady, L., Baukal, A.J., Bor, M., Ely, J.A., and Catt, K.J.,** Regulation of 1,2-diacylglycerol production by angiotensin-II in bovine adrenal glomerulosa cells, *Endocrinology,* 126, 1001, 1990.
64. **Hunyady, L., Merelli, F., Baukal, A.J., Balla, T., and Catt, K.J.,** Agonist-induced endocytosis and signal generation in adrenal glomerulosa cells. A potential mechanism for receptor-operated calcium entry, *J. Biol. Chem.,* 266, 2783, 1991.
65. **Rhee, S.G., Suh, P.G., Ryu, S.-H., and Lee, S.Y.,** Studies of inositol phospholipid-specific phospholipase C, *Science,* 244, 546, 1989.
66. **Haslam, R.J. and Davidson, M.M.L.,** Guanine nucleotides decrease the free Ca^{2+} required for secretion of serotonin from permeabilized platelets. Evidence of a role of GTP-binding protein in platelet activation, *FEBS Lett.,* 174, 90, 1984.
67. **Cockcroft, S. and Gomperts, B.D.,** Role of guanine nucleotide binding protein in the activation of polyphosphoinositide phosphodiesterase, *Nature,* 314, 534, 1985.
68. **Blackmore, P.F., Bocckino, S.B., Waynick, L.E., and Exton, J.H.,** Role of guanine nucleotide-binding regulatory protein in the hydrolysis of hepatocyte phosphatidylinositol 4,5-bisphosphate by calcium-mobilizing hormones and the control of cell calcium. Studies utilizing aluminum fluoride, *J. Biol. Chem.,* 260, 14477, 1985.

69. **Enyedi, P., Mucsi, I., Hunyady, L., Catt, K.J., and Spät, A.,** The role of guanyl nucleotide binding proteins in the formation of inositol phosphates in adrenal glomerulosa cells, *Biochem. Biophys. Res. Commun.*, 140, 941, 1986.
70. **Baukal, A.J., Balla, T., Hunyady, L., Hausdorff, W., Guillemette, G., and Catt, K.J.,** Angiotensin II and guanine nucleotides stimulate formation of inositol 1,4,5-trisphosphate and its metabolites in permeabilized adrenal glomerulosa cells, *J. Biol. Chem.*, 263, 6087, 1988.
71. **Lynch, C.J., Prpic, V., Blackmore, P.F., and Exton, J.H.,** Effect of islet-activating pertussis toxin on the binding characteristics of Ca^{2+}-mobilizing hormones and on agonist activation of phosphorylase in hepatocytes, *Mol. Pharmacol.*, 29, 196, 1986.
72. **Smrcka, A.V., Hepler, J.R., Borwn, K.O., and Sternweis, P.C.,** Regulation of polyphosphoinositide-specific phospholipase C activity by purified G_q, *Science*, 251, 804, 1991.
73. **Taylor, S.J., Smith, J.A., and Exton, J.H.,** Purification from bovine liver membranes of a guanine nucleotide-dependent activator of phosphoinositide-specific phospholipase C: immunologic identification as a novel G-protein alpha subunit, *J. Biol. Chem.*, 265, 17150, 1990.
74. **Blank, J.L., Ross, A.H., and Exton, J.H.,** Purification of characterization of two G-proteins that activate the β1 isozyme of phosphoinositide-specific phospholipase C: identification as members of the G_q class, *J. Biol. Chem.*, 266, 18206, 1991.
75. **Irvine, R.F., Letcher, A.J., Lander, D.J., and Downes, C.P.,** Inositol trisphosphates in carbachol-stimulated rat parotid glands, *Biochem. J.*, 223, 237, 1984.
76. **Irvine, R.F., Letcher, A.J., Heslop, J.P., and Berridge, M.J.,** The inositol tris/tetrakis phosphate pathway — demonstration of inositol (1,4,5)trisphosphate-3-kinase activity in mammalian tissues, *Nature*, 320, 631, 1986.
77. **Storey, D.J., Shears, S.B., Kirk, C.J., and Michell, R.H.,** Stepwise enzymatic dephosphorylation of inositol 1,4,5-trisphosphate to inositol in liver, *Nature*, 312, 374, 1984.
78. **Biden, T.J., Comte, M., Cox, J.A., and Wollheim, C.B.,** Calcium-calmodulin stimulates inositol 1,4,5-trisphosphate kinase activity from insulin-secreting RINm5F cells, *J. Biol. Chem.*, 262, 9437, 1987.
79. **Ryu, S.H., Lee, S.Y., Lee, K.Y., and Rhee, S.G.,** Catalytic properties of inositol trisphosphate kinase: activation by Ca^{2+} and calmodulin, *FASEB J.*, 1, 388, 1987.
80. **Balla, T., Baukal, A.J., Guillemette, G., Morgan, R.O., and Catt, K.J.,** Angiotensin-stimulated production of inositol trisphosphate isomers and rapid metabolism through inositol 4-monophosphate in adrenal glomerulosa cells, *Proc. Natl. Acad. Sci. U.S.A.*, 83, 9323, 1986.
81. **Morgan, R.O., Chang, J.P., and Catt, K.J.,** Novel aspects of gonadotropin-releasing hormone action on inositol polyphosphate metabolism in cultured pituitary gonadotrophs, *J. Biol. Chem.*, 262, 1166, 1987.
82. **Majerus, P.W., Connolly, T.M., Bansal, V.S., Inhorn, R.C., Ross, T.S., and Lips, D.L.,** Inositol phosphates: synthesis and degradation, *J. Biol. Chem.*, 263, 3051, 1988.
83. **Heslop, J.P., Irvine, R.F., Tashijan, A.H., Jr., and Berridge, M.J.,** Inositol tetrakis- and pentakisphosphates in GH4 cells, *J. Exp. Biol.*, 119, 395, 1985.
84. **Stephens, L.R., Hawkins, P.T., Carter, N., Chahwala, S.B., Morris, A.J., Whetton, A.D., and Downes, C.P.,** L-myo-inositol 1,4,5,6-tetrakisphosphate is present in both mammalian and avian cells, *Biochem. J.*, 249, 271, 1988.
85. **Stephens, L.R., Hawkins, P.T., Morris, A.J., and Downes, C.P.,** L-myo-inositol 1,4,5,6-tetrakisphosphate (3-hydroxy)kinase, *Biochem. J.*, 249, 283, 1988.
86. **Balla, T., Hunyady, L., Baukal, A.J., and Catt, K.J.,** Structures and metabolism of inositol tetrakisphosphates and inositol pentakisphosphate in bovine adrenal glomerulosa cells, *J. Biol. Chem.*, 264, 9386, 1989.
87. **Balla, T., Baukal, A., Hunyady, L., and Catt, K.J.,** Agonist-induced regulation of inositol tetrakisphosphate isomers and inositol pentakisphosphate in adrenal glomerulosa cells, *J. Biol. Chem.*, 264, 13605, 1989.

88. **Balla, T., Guillemette, G., Baukal, A.J., and Catt, K.J.,** Metabolism of inositol 1,3,4-trisphosphate to a new tetrakisphosphate isomer in angiotensin-stimulated adrenal glomerulosa cells, *J. Biol. Chem.*, 262, 9952, 1987.
89. **Shears, S.B., Parry, J.B., Tang, E.K.Y., Irvine, R.F., Michell, R.H., and Kirk, C.J.,** Metabolism of D-myo-inositol 1,3,4,5-tetrakisphosphate by rat liver, including the synthesis of a novel isomer of myo-inositol tetrakisphosphate, *Biochem. J.*, 246, 139, 1987.
90. **Stephens, L.R., Hawkins, P.T., Barker, C.J., and Downes, C.P.,** Synthesis of myo-inositol 1,3,4,5,6-pentakisphosphate from in ositol phosphates generated by receptor activation, *Biochem. J.*, 253, 721, 1988.
91. **Hunyady, L., Baukal, A.J., Guillemette, G., Balla, T., and Catt, K.J.,** Metabolism of inositol-1,3,4,6-tetrakisphosphate to inositol pentakisphosphate in adrenal glomerulosa cells, *Biochem. Biophys. Res. Commun.*, 157, 1247, 1988.
92. **Johnson, R.M., Wasilenko, W.J., Mattingly, R.R., Weber, M.J., and Garrison, J.C.,** Fibroblasts transformed with v-src show enhanced formation of an inositol tetrakisphosphate, *Science,* 246, 121, 1989.
93. **Irvine, R.F. and Moor, R.M.,** Micro-injection of inositol 1,3,4,5-tetrakisphosphate activates see urchin eggs by a mechanism dependent on external Ca^{2+}, *Biochem. J.*, 240, 917, 1986.
94. **Irvine, R.F.,** Inositol tetrakisphosphate as a second messenger: confusions, contradictions, and a potential resolution, *Bioessays,* 13, 419, 1991.
95. **Nishizuka, Y.,** The role of protein kinase C in cell surface signal transduction and tumour promotion, *Nature,* 308, 693, 1984.
96. **Nishizuka, Y.,** The molecular heterogeneity of protein kinase C and its implications for cellular regulation, *Nature,* 34, 661, 1988.
97. **Griendling, K.K., Rittenhouse, S.E., Brock, T.A., Ekstein, L.S., Gimbrone, M.A., Jr., and Alexander, R.W.,** Sustained diacylglycerol formation from inositol phospholipids in angiotensin II-stimulated vascular smooth muscle cells, *J. Biol. Chem.*, 261, 5901, 1986.
98. **Griendling, K.K., Delafontaine, P., Rittenhouse, S.E., Gimbrone, M.A., Jr., and Alexander, R.W.,** Correlation of receptor sequestration with sustained diacylglycerol accumulation in angiotensin II-stimulated cultured vascular smooth muscle cells, *J. Biol. Chem.*, 262, 14555, 1987.
99. **Bockino, S.B., Blackmore, P.F., and Exton, J.H.,** Stimulation of 1,2-diacylglycerol accumulation in hepatocytes by vasopressin, epinephrine, and angiotensin II, *J. Biol. Chem.*, 260, 14201, 1985.
100. **Bockino, S.B., Blackmore, P.F., Wilson, P.B., and Exton, J.H.,** Phosphatidate accumulation in hormone treated hepatocytes via phospholipase D mechanism, *J. Biol. Chem.*, 2624, 15309, 1987.
101. **Agwu, D.E., McPhail, L.C., Chabot, M.C., Daniel, L.W., and Wykle, R.L.,** Choline-linked phosphoglycerides, *J. Biol. Chem.*, 265, 1405, 1989.
102. **Hunyady, L., Balla, T., Enyedi, P., and Spät, A.,** The effect of angiotensin II on arachidonate metabolism in adrenal glomerulosa cells, *Biochem. Pharmacol.*, 34, 3439, 1985.
103. **Bollag, W.B., Barrett, P.Q., Isales, C.M., Liscovitch, M., and Rasmussen, H.,** A potential role for phospholipase-D in the angiotensin-II-induced stimulation of aldosterone secretion from bovine adrenal glomerulosa cells, *Endocrinology,* 127, 1436, 1990.
104. **Spät A. and Józan, S.,** Effect of prostaglandin E_2 and A_2 on steroid synthesis by the rat adrenal gland, *J. Endocrinol.*, 65, 55, 1975.
105. **Enyedi, P., Spät, A., and Antoni, F.A.,** Role of prostaglandins in the control of the function of adrenal glomerulosa cells, *J. Endocrinol.*, 91, 427, 1981.
106. **Natrarajan, R., Dunn, W.D., Stern, N., and Nadler, J.,** Key role of diacylglycerol-mediated 12-lipoxygenase product formation in angiotensin II-induced aldosterone synthesis, *Mol. Cell. Endocrinol.*, 72, 73, 1990.

107. **Streb, H., Irvine, R.F., Berridge, M.J., and Schulz, I.,** Release of Ca^{2+} from a nonmitochondrial intracellular store in pancreatic acinar cells by inositol-1,4,5-trisphosphate, *Nature,* 306, 67, 1983.
108. **Prentki, M., Janjic, D., Biden, T.J., Blondel, B., and Wollheim, C.B.,** Regulation of Ca transport by isolated organelles of rat insulinoma, *J. Biol. Chem.,* 259, 10118, 1984.
109. **Berridge, M.J. and Irvine, R.F.,** Inositol trisphosphate, a novel second messenger in cellular signal transduction, *Nature,* 312, 315, 1984.
110. **Balla, T., Hausdorff, W.P., Baukal, A.J., and Catt, K.J.,** Inositol polyphosphate production and regulation of cytosolic calcium during the biphasic activation of adrenal glomerulosa cells by angiotensin II, *Arch. Biochem. Biophys.,* 270, 398, 1989.
111. **Baukal, A.J., Guillemette, G., Rubin, R., Spät, A., and Catt, K.J.,** Binding sites for inositol trisphosphate in the bovine adrenal cortex, *Biochem. Biophys. Res. Commun.,* 133, 532, 1985.
112. **Guillemette, G., Balla, T., Baukal, A.J., Spät, A., and Catt, K.J.,** Intracellular receptors for inositol 1,4,5-trisphosphate in angiotensin II target tissues, *J. Biol. Chem.,* 262, 1010, 1987.
113. **Spät, A., Fabiato, A., and Rubin, R.P.,** Binding of inositol trisphosphate by a liver microsomal fraction, *Biochem. J.,* 233, 929, 1986.
114. **Spät, A., Bradford, P.G., McKinney, J.S., Rubin, R.P., and Putney, J.W., Jr.,** A saturable receptor for ^{32}P-inositol-1,4,5-triphosphate in hepatocytes and neutrophils, *Nature,* 319, 514, 1986.
115. **Meyer, T., Holowka, D., and Stryer, L.,** Highly cooperative opening of calcium channels by inositol 1,4,5-trisphosphate, *Science,* 240, 653, 1988.
116. **Guillemette, G., Balla, T., Baukal, A.J., and Catt, K.J.,** Characterization of inositol 1,4,5-trisphosphate receptors and calcium mobilization in a hepatic plasma membrane fraction, *J. Biol. Chem.,* 263, 4541, 1988.
117. **Supattapone, S., Worley, P.F., Baraban, J.M., and Snyder, S.H.,** Solubilization, purification and characterization of an inositol trisphosphate-receptor binding site, *J. Biol. Chem.,* 263, 1530, 1987.
118. **Furuichi, T., Yoshikawa, S., Miyawaki, A., Wada, K., Maeda, N., and Mikoshiba, K.,** Primary structure and functional expression of the inositol 1,4,5-trisphosphate-binding protein P_{400}, *Nature,* 342, 32, 1989.
119. **Miyawaki, A., Furuichi, T., Maeda, N., and Mikoshiba, K.,** Expressed cerebellar-type inositol 1,4,5-trisphosphate receptor, P_{400}, has calcium release activity in a fibroblast L cell line, *Neuron,* 5, 11, 1990.
120. **Mignery, G.A., Sudhof, T.C., Takei, K., and De Camilli, P.,** Putative receptor for inositol 1,4,5-trisphosphate similar to ryanodine receptor, *Nature,* 342, 192, 1989.
121. **Ross, C.A., Meldolesi, J., Milner, T.A., Satoh, T., Supattapone, S., and Snyder, S.H.,** Inositol 1,4,5-trisphosphate receptor localized to endoplasmic reticulum in cerebellar Purkinje neurons, *Nature,* 339, 468, 1989.
122. **Maeda, N., Niinobe, M., Inoue, Y., and Mikoshiba, K.,** Developmental expression and intracellular location of P400 protein characteristic of Purkinje cells in the mouse cerebellum, *Dev. Biol.,* 133, 67, 1989.
123. **Nicotera, P., Orrenius, S., Nilsson, T., and Berggre, P.-O.,** An inositol 1,4,5-trisphosphate-sensitive Ca^{2+} pool in liver nuclei, *Proc. Natl. Acad. Sci. U.S.A.,* 87, 6858, 1990.
124. **Van Delden, C., Favre, C., Spät, A., Cerny, E., Krause, K.H., and Lew, D.P.,** Purification of an inositol 1,4,5-trisphosphate-binding calreticulin-containing intracellular compartment of HL-60 cells, *Biochem. J.,* 281, 651, 1992.
125. **Volpe, P., Krause, K.H., Hashimoto, S., Zorzato, F., Pozzan, T., Meldolesi, J., and Lau, D.P.,** "Calciosome", a cytoplasmic organelle: the inositol 1,4,5-trisphosphate-sensitive Ca^{2+}-store of non-muscle cells, *Proc. Natl. Acad. Sci. U.S.A.,* 85, 1091, 1988.

126. **Dawson, A.P.,** GTP enhances inositol trisphosphate-stimulated Ca^{2+} release from rat liver microsomes, *FEBS Lett.,* 185, 147, 1985.
127. **Dawson, A.P., Comerford, J.G., and Fulton, D.V.,** The effect of GTP on inositol 1,4,5-trisphosphate-stimulated Ca + 2 efflux from a rat liver microsomal fraction: is GTP-dependent protein phosphorylation involved?, *Biochem. J.,* 234, 311, 1986.
128. **Guillemette, G., Balla, T., Baukal, A.J., and Catt, K.J.,** Inositol 1,4,5-trisphosphate binds to a specific receptor and releases microsomal calcium in the anterior pituitary gland, *Proc. Natl. Acad. Sci. U.S.A.,* 84, 8195, 1987.
129. **Nicchitta, C.V., Joseph, S.K., and Williamson, J.R.,** GTP-mediated Ca^{2+} release in rough endoplasmic reticulum. Correlation with a GTP-sensitive increase in membrane permeability, *Biochem. J.,* 248, 741, 1987.
130. **Chueh, S.-H., Mullaney, J.M., Ghosh, T.K., Zachary, A.L., and Gill, D.L.,** GTP- and inositol 1,4,5-trisphosphate-activated intracellular calcium movements in neuronal and smooth muscle cell lines, *J. Biol. Chem.,* 262, 13857, 1987.
131. **Mullaney, J.M., Yu, M., Ghosh, T.K., and Gill, D.L.,** Calcium entry into the inositol 1,4,5-trisphosphate-releasable calcium pool is mediated by a GTP-regulatory mechanism, *Proc. Natl. Acad. Sci. U.S.A.,* 85, 2499, 1988.
132. **Gilkey, J.C., Jaffe, L.F., Ridgeway, E.B., and Reynolds, G.T.,** A free Ca^{2+} wave traverses the activating egg of the medaka *Oryzias latipes, J. Cell Biol.,* 76, 448, 1978.
133. **Busa, W.B. and Nuccitelli, R.,** An elevated free cywolic Ca^{2+} wave follows fertilization in eggs of the frog, *Xenopus laevis, J. Cell Biol.,* 100, 1325, 1985.
134. **Rooney, T.A., Sass, E.J., and Thomas, A.P.,** Agonist-induced cytosolic calcium oscillations originate from a specific locus in single hepatocytes, *J. Biol. Chem.,* 265, 10792, 1990.
135. **Berridge, M.J.,** Calcium oscillations, *J. Biol. Chem.,* 265, 9583, 1990.
136. **Elliott, M.E., Siegel, F.L., Hadjokas, N.E., and Goodfriend, T.L.,** Angiotensin effects on calcium and steroidogenesis in adrenal glomerulosa cells, *Endocrinology,* 116, 1051, 1985.
137. **Mauger, J.P., Poggioli, J., Guesdon, F., and Claret, M.,** Noradrenaline, vasopressin and angiotensin increase Ca^{2+} influx by opening a common pool of Ca^{2+} channels in isolated rat liver cells, *Biochem. J.,* 221, 121, 1984.
138. **Kojima, I., Kojima, K., and Rasmussen, H.,** Characteristics of angiotensin II-, K^+- and ACTH-induced calcium influx in adrenal glomerulosa cells, *J. Biol. Chem.,* 260, 9171, 1985.
139. **Spät, A., Balla, I., Balla, T., Cragoe, E.J., Jr., Hajnóczky, G., and Hunyady, L.,** Angiotensin II and potassium activate different calcium entry mechanisms in rat adrenal glomerulosa cells, *J. Endocrinol.,* 122, 361, 1989.
140. **Aguilera, G. and Catt, K.J.,** Participation of voltage-dependent calcium channels in the regulation of adrenal glomerulosa function by angiotensin II and potassium, *Endocrinology,* 118, 112, 1986.
141. **Hausdorff, W.P. and Catt, K.J.,** Activation of dihydropyridine-sensitive calcium channels and biphasic cytosolic calcium responses by angiotensin II in rat adrenal glomerulosa cells, *Endocrinology,* 123, 2818, 1988.
142. **Fox, A.P., Nowycky, M.C., and Tsien, R.W.,** Kinetic and pharmacological properties distinguishing three types of calcium currents in chick sensory neurons, *J. Physiol.,* 394, 149, 1987.
143. **Cohen, C.J., McCarthy, R.T., Barrett, P.Q., and Rasmussen, H.,** Ca channels in adrenal glomerulosa cells: K^+ and angiotensin II increase T-type Ca channel current, *Proc. Natl. Acad. Sci. U.S.A.,* 85, 2412, 1988.
144. **Durroux, T., Gallo-Payet, N., and Payet, M.D.,** Three components of the calcium current in cultured glomerulosa cells from rat adrenal gland, *J. Physiol.,* 404, 713, 1988.
145. **Matsunaga, H., Maruyama, Y., Kojima, I., and Hoshi, T.,** Transient Ca^{2+}-channel current characterized by low-threshold voltage in zona glomerulosa cells of rat adrenal cortex, *Pfluegers Arch.,* 408, 351, 1987.

146. **Heschler, J., Rosenthal, W., Hinsch, K.D., Wulfern, M., Trautwein, W., and Schultz, G.,** Angiotensin II-induced stimulation of voltage-dependent Ca2+ currents in an adrenal cortical cell line, *EMBO J.*, 7, 619, 1988.
147. **Quinn, S.J., Cornwall, M.C., and Williams, G.H.,** Electrophysiological responses to angiotensin II of isolated rat adrenal glomerulosa cells, *Endocrinology,* 120, 1581, 1987.
148. **Capponi, A.M., Lew, P.D., and Vallotton, M.B.,** Quantitative analysis of the cytosolic-free-Ca^{2+}-dependency of aldosterone production in bovine adrenal glomerulosa cells. Different requirements for angiotensin II and K^+, *Biochem. J.*, 247, 335, 1987.
148a. **Hunyady, L.,** personal communication.
149. **Balla, T., Hollo, Z., Varnai, P., and Spät, A.,** Angiotensin II inhibits K^+-induced Ca^{2+} signal generation in rat adrenal glomerulosa cells, *Biochem. J.*, 273, 399, 1991.
150. **Lobo, M.V. and Marusic, E.T.,** Angiotensin II causes a dual effect on potassium permeability in adrenal glomerulosa cells, *Am. J. Physiol.*, 254, E144, 1988.
151. **Hajnóczky, G., Csordas, G., Bago, A., Chiu, A.T., and Spät, A.,** Angiotensin II exerts its effect on aldosterone production and potassium permeability through receptor subtype AT_1 in rat adrenal glomerulosa cells, *Biochem. Pharmacol.*, 43, 1009, 1992.
152. **Hajnóczky, G., Csordas, G., Hunyady, L., Kalapos, M.P., Balla, T., Enyedi, P., and Spät, A.,** Angiotensin-II inhibits Na^+/K^+ pump in rat adrenal glomerulosa cells: possible contribution to stimulation of aldosterone production, *Endocrinology,* 130, 1637, 1992.
153. **Lynch, C.J., Wilson, P.B., Blackmore, P.F., and Exton, J.H.,** The hormone-sensitive hepatic Na^+-pump, *J. Biol. Chem.*, 261, 14551, 1986.
154. **Berk, B.C., Aronow, M.S., Brock, T.A., Cragoe, E., Jr., Gimbrone, M.A., Jr., and Alexander, R.W.,** Angiotensin II-stimulated Na^+/H^+ exchange in cultured vascular smooth muscle cells, *J. Biol. Chem.*, 262, 5057, 1987.
155. **Smith, J.B., Cragoe, E.J., Jr., and Smith, L.,** Na^+/Ca^{2+} antiport in cultured arterial smooth muscle cells: inhibition by magnesium and other divalent cations, *J. Biol. Chem.*, 262, 11988, 1987.
156. **Hunyady, L., Kayser, S., Cragoe, E.J., Jr., Balla, I., Balla, T., and Spät, A.,** Na^+-H^+ and Na^+-Ca^{2+} exchange in glomerulosa cells: possible role in control of aldosterone production, *Am. J. Physiol.*, 254, C744, 1988.
157. **Putney, J.W.,** A model for receptor-regulated calcium entry, *Cell Calcium,* 7, 1, 1986.
158. **Takemura, H. and Putney, J.W., Jr.,** Capacitative calcium entry in parotid acinar cells, *Biochem. J.*, 258, 409, 1989.
159. **Pandol, S.J., Schoeffield, M.S., Fimmel, C.J., and Muallem, S.,** The agonist-sensitive calcium pool in the pancreatic acinar cell, *J. Biol. Chem.*, 262, 16963, 1987.
160. **Thastrup, O., Foder, B., and Scharff, O.,** The calcium mobilizing and tumor-promoting agent, thapsigargin, elevates the platelet cytoplasmic free calcium concentration to a higher steady state level. A possible mechanism of action for tumor promotion, *Biochem. Biophys. Res. Commun.*, 142, 654, 1987.
161. **Ely, J.A., Ambroz, C., Baukal, A.J., Christensen, S.B., Balla, T., and Catt, K.J.,** Relationship between agonist- and thapsigargin-sensitive calcium pools in adrenal glomerulosa cells, *J. Biol. Chem.*, 266, 18635, 1991.
162. **Morris, A.P., Gallacher, D.V., Irvine, R.F., and Petersen, O.H.,** Synergism of inositol trisphosphate and tetrakisphosphate in activating Ca^{2+}-dependent K^+ channels, *Nature,* 330, 653, 1987.
163. **Ely, J.A., Hunyady, L., Baukal, A.J., and Catt, K.J.,** Inositol 1,3,4,5-tetrakisphosphate stimulates calcium release from bovine adrenal microsomes by a mechanism independent of the inositol 1,4,5-trisphosphate receptor, *Biochem. J.*, 268, 333, 1990.
164. **Spät, A., Lukács, G.L., Eberhardt, I., Kiesel, L., and Runnebaum, B.,** Binding of inositol phosphates and induction of Ca^{2+} release from pituitary microsomal fractions, *Biochem. J.*, 244, 493, 1987.
165. **Irvine, R.F.,** "Quantal" Ca^{2+} release and the control of Ca^{2+} entry by inositol phosphates — a possible mechanism, *FEBS Lett.*, 263, 5, 1990.

166. **Bradford, P.G. and Irvine, R.F.,** Specific binding sites for [^{3}H]inositol (1,3,4,5) tetrakisphosphate on membranes of HL-60 cells, *Biochem. Biophys. Res. Commun.*, 149, 680, 1987.
167. **Enyedi, P. and Williams, G.H.,** Heterogenous inositol tetrakisphosphate binding sites in the adrenal cortex, *J. Biol. Chem.*, 263, 7940, 1988.
168. **Balla, T., Sim, S.S., Iida, T., Choi, K.Y., Catt, K.J., and Rhee, S.G.,** Agonist-induced calcium signaling is impaired in fibroblasts overproducing inositol 1,3,4,5-tetrakisphosphate, *J. Biol. Chem.*, 266, 24719, 1991.
169. **Kuno, M. and Gardner, P.,** Ion channels activated by inositol 1,4,5-trisphosphate in plasma membrane of human T lymphocytes, *Nature*, 326, 301, 1987.
170. **Khan, A.A., Steiner, J.P., and Snyder, S.H.,** Plasma membrane inositol 1,4,5-trisphosphate receptor of lymphocytes: selective enrichment in sialic acid and unique binding specificity, *Proc. Natl. Acad. Sci. U.S.A.*, 89, 2849, 1992.
171. **Woods, N.M., Cuthbertson, K.S.R., and Cobbold, P.H.,** Repetitive transient rises in cytoplasmic free calcium in hormone-stimulated hepatocytes, *Nature*, 319, 600, 1986.
172. **Woods, N.M., Cuthbertson, K.S.R., and Cobbold, P.H.,** Agonist-induced oscillations in cytoplasmic free calcium concentration in single rat hepatocytes, *Cell Calcium*, 8, 79, 1987.
173. **Quinn, S.J., Williams, G.H., and Tillotson, D.L.,** Calcium oscillations in single adrenal glomerulosa cells stimulated by angiotensin II, *Proc. Natl. Acad. Sci. U.S.A.*, 85, 5754, 1988.
174. **Johnson, E.I.M., Capponi, A.M., and Valloton, M.B.,** Cytosolic free calcium oscillates in single bovine adrenal glomerulosa cells in response to angiotensin II stimulation, *J. Endocrinol.*, 122, 391, 1989.
175. **Monck, J.R., Williamson, R.E., Rogulja, I., Fluharty, S.J., and Williamson, J.R.,** Angiotensin II effects on the cytosolic free Ca^{2+} concentration in N1E-115 neuroblastoma cells: kinetic properties of the Ca^{2+} transient measured in single fura-2-loaded cells, *J. Neurochem.*, 54, 278, 1990.
176. **Tsien, R.W. and Tsien, R.Y.,** Calcium channels, stores, and oscillations, *Annu. Rev. Cell Biol.*, 6, 715, 1990.
177. **Rooney, T.A., Renard, D.C., Sass, E.J., and Thomas, A.P.,** Oscillatory cytosolic calcium waves independent of stimulated inositol 1,4,5-trisphosphate formation in hepatocytes, *J. Biol. Chem.*, 266, 12272, 1991.
178. **Wakui, M., Osipchuk, T.V., and Petersen, O.H.,** Receptor-activated cytoplasmic Ca^{2+} spiking mediated by inositol trisphosphate is due to Ca^{2+}-induced Ca^{2+} release, *Cell*, 63, 1026, 1990.
179. **Willingham, M.C. and Pastan, I.,** Endocytosis and membrane traffic in cultured cells, *Recent Prog. Horm. Res.*, 40, 569, 1984.
180. **Bianchi, C., Gutkowska, J., De L'an, A., Ballak, M., Anand-Srivastava, M.B., Genest, J., and Cantin, M.,** Fate of [^{125}I]angiotensin II in adrenal zona glomerulosa cells, *Endocrinology*, 118, 2605, 1986.
181. **Crozat, A., Penhoat, A., and Saez, J.M.,** Processing of angiotensin II (A-II) and (Sar^1,Ala^8)A-II by cultured bovine adrenocortical cells, *Endocrinology*, 118, 2312, 1986.
182. **Ambroz, C. and Catt, K.J.,** Angiotensin II receptor-mediated calcium influx in bovine adrenal glomerulosa cells, *Endocrinology*, in press.
183. **Bollag, W.B., Barrett, P.Q., Isales, C.M., and Rasmussen, H.,** Angiotensin II-induced changes in diacylglycerol levels and their potential role in modulating the steroidogenic response, *Endocrinology*, 128, 231, 1991.
184. **Barrett, P.Q., Kojima, I., Kojima, K., Zawalich, K., Isales, C.M., and Rasmussen, H.,** Temporal pattern of protein phosphorylation after angiotensin II, A23187 and/or 12-O-tetradecanoylphorbol 13-acetate in adrenal glomerulosa cells, *Biochem. J.*, 238, 893, 1986.

185. **Barrett, P.Q., Kojima, I., Kojima, K., Zawalich, K., Isales, C.M., and Rasmussen, H.,** Short term memory in the calcium messenger system, *Biochem. J.*, 238, 905, 1986.
186. **Barrett, P.Q., Bollag, W.B., Isales, C.M., McCarthy, R.T., and Rasmussen, H.,** Role of calcium in angiotensin II-mediated aldosterone secretion, *Endocr. Rev.*, 10, 496, 1989.
187. **Lang, U. and Vallotton, M.B.,** Angiotensin II but not potassium induces subcellular redistribution of protein kinase C in bovine adrenal glomerulosa cells, *J. Biol. Chem.*, 262, 8047, 1987.
188. **Faragó, A., Seprődi, J., and Spät, A.,** Subcellular distribution of protein kinase C in rat adrenal glomerulosa cells, *Biochem. Biophys. Res. Commun.*, 156, 628, 1988.
189. **Kojima, I., Shibata, H., and Ogata, E.,** Phorbol ester inhibits angiotensin-induced activation of phospholipase C in adrenal glomerulosa cells. Its implication in the sustained action of angiotensin, *Biochem. J.*, 237, 253, 1986.
190. **Brock, T.A., Rittenhouse, S.E., Powers, C.W., Ekstein, L.S., Gimbrone, M.A., Jr., and Alexander, R.W.,** Phorbol ester and 1-oleoyl-2-acetylglycerol inhibit angiotensin activation of phospholipase C in cultured vascular smooth muscle cells, *J. Biol. Chem.*, 260, 14158, 1985.
191. **Hajnóczky, G., Varnai, P., Buday, L., Faragó, A., and Spät, A.,** The role of protein kinase-C in control of aldosterone production by rat adrenal glomerulosa cells: activation of protein kinase-C by stimulation with potassium, *Endocrinology,* 130, 2230, 1992.
192. **Capponi, A.M., Rossier, M.F., Davies, E., and Valloton, M.B.,** Calcium stimulates steroidogenesis in permeabilized bovine adrenal cortical cells, *J. Biol. Chem.*, 263, 16113, 1988.
193. **Pralong, W.-P., Hunyady, L., Varnai, P., Wollheim, C.B., and Spät, A.,** Pyridine nucleotide redox state parallels production of aldosterone in potassium-stimulated adrenal glomerulosa cells, *Proc. Natl. Acad. Sci. U.S.A.*, 89, 132, 1992.
194. **Speth, R.C. and Kim, K.H.,** Discrimination of two angiotensin II receptor subtypes with a selective agonist analogue of angiotensin II, p-aminophenylalanine^6 angiotensin II, *Biochem. Biophys. Res. Commun.*, 169, 997, 1990.
195. **Bottari, S.P., Taylor, V., King, I.N., Bogdal, Y., and Whitebread, S.,** Angiotensin II AT_2 receptors do not interact with guanine nucleotide binding proteins, *Eur. J. Pharmacol.*, 207, 157, 1991.
196. **Pucell, A.G., Hodges, J.C., Sen, I., Bumpus, F.M., and Husain, A.,** Biochemical properties of the ovarian granulosa cell type 2-angiotensin II receptor, *Endocrinology,* 128, 1947, 1991.
197. **Jones, C., Millan, M.A., Naftolin, F., and Aguilera, G.,** Characterization of angiotensin II receptors in the rat fetus, *Peptides,* 10, 459, 1989.
198. **Millan, M.A., Carvallo, P., Izumi, S., Zemel, S., Catt, K.J., and Aguilera, G.,** Novel sites of expression of functional angiotensin II receptors in the late gestation fetus, *Science,* 244, 1340, 1989.
199. **Zemel, S., Millan, M.A., Feuillan, P., and Aguilera, G.,** Characterization and distribution of angiotensin II receptors in the primate fetus, *J. Clin. Endocrinol. Metab.*, 71, 1003, 1990.
200. **Johnson, M.C. and Aguilera, G.,** Angiotensin-II receptor subtypes and coupling to signaling systems in cultured fetal fibroblasts, *Endocrinology,* 129, 1266, 1991.
201. **Millan, M.A., Jacobowitz, D.M., Aguilera, G., and Catt, K.J.,** Differential distribution of AT1 and AT2 angiotensin II receptor subtypes in the rat brain during development, *Proc. Natl. Acad. Sci. U.S.A.*, 88, 11440, 1991.
202. **Sumners, C., Tang, W., Zelezna, B., and Raizada, M.K.,** Angiotensin II receptor subtypes are coupled with distinct signal-transduction mechanisms in neurons and astrocytes from rat brain, *Proc. Natl. Acad. Sci. U.S.A.*, 88, 7567, 1991.
203. **Kang, J., Sumners, C., and Posner, P.,** Modulation of net outward current in cultured neurons by angiotensin II: involvement of AT_1 and AT_2 receptors, *Brain Res.*, in press.

204. **Bottari, S.P., King, I.N., Reichlin, S., Dahlstroem, I., Lydon, N., and Gasparo, M.,** The angiotensin AT_2 receptor stimulates protein tyrosine phosphatase activity and mediates inhibition of particulate guanylate cyclase, *Biochem. Biophys. Res. Commun.*, 183, 206, 1992.
205. **Deane, H.W. and Masson, G.M.C.,** Adrenal cortical changes in rats with various types of experimental hypertension, *J. Clin. Endocrinol. Metab.*, 11, 193, 1951.
206. **Hartroft, P.M., Newmark, L.N., and Pitcock, J.A.,** Relationship of renal juxtaglomerular cells to sodium intake, adrenal cortex and hypertension, in *Hypertension*, Moyer, J., Ed., W.B. Saunders, Philadelphia, 1959.
207. **Gross, F., Brunner, H., and Ziegler, M.,** Renin-angiotensin system, aldosterone, and sodium balance, *Recent Prog. Horm. Res.*, 21, 119, 1965.
208. **Geisterfer, A.A.T., Peach, M.J., and Owens, G.K.,** Angiotensin II induced hypertrophy not hyperplasia of cultured rat aortic smooth muscle cells, *Circ. Res.*, 62, 749, 1988.
209. **Korner, P.I., Bobik, A., Angus, J.A., Adam, M.A., and Friberg, P.,** Resistance control in hypertension, *J. Hypertens. Suppl.*, 7, S125, 1989.
210. **Aceto, J.F. and Baker, K.M.,** [Sar^1]angiotensin II receptor-mediated stimulation of protein synthesis in chick heart cells, *Am. J. Physiol.*, 258, H806, 1990.
211. **Berk, B.C. and Alexander, R.W.,** Vasoactive effects of growth factors, *Biochem. Pharmacol.*, 38, 219, 1989.
212. **Taubman, M.B., Berk, B.C., Izumo, S., Tsuda, T., Alexander, R.W., and Nadal-Ginard, B.,** Angiotensin II induces *c-fos* mRNA in aortic smooth muscle: role of Ca^{2+} mobilization and protein kinase C activation, *J. Biol. Chem.*, 264, 526, 1989.
213. **Naftilan, A.J., Pratt, R.E., and Dzau, V.J.,** Induction of platelet-derived growth factor A-chain and *c-myc* gene expressions by angiotensin II in cultured rat vascular smooth muscle cells, *J. Clin. Invest.*, 83, 1419, 1989.
214. **Araki, S., Kawahara, Y., and Sunaki, M.,** Stimulation of platelet-derived growth factor-induced DNA synthesis by angiotensin II in rabbit vascular smooth muscle cells, *Biochem. Biophys. Res. Commun.*, 168, 350, 1990.
215. **Resink, T.J., Hahn, A.W., Scott-Burden, T., Powell, J., Weber, E., and Buhler, F.R.,** Inducible endothelin mRNA expression and peptide secretion in cultured human vascular smooth muscle cells, *Biochem. Biophys. Res. Commun.*, 168, 1303, 1990.
216. **Hahn, A.W., Resink, T.J., Scott-Burden, T., and Powell, J.,** Stimulation of endothelin mRNA and secretion in rat vascular smooth muscle cells: a novel autocrine function, *Cell Regul.*, 1, 649, 1990.
217. **Bobik, A., Grooms, A., Miller, J.A., Mitchell, A., and Grinpukel, S.,** Growth factor activity of endothelin on vascular smooth muscle, *Am. J. Physiol.*, 258, C408, 1990.
218. **Iwao, H., Nakamura, A., Fukui, K., and Kimura, S.,** Endogenous angiotensin II regulates hepatic angiotensinogen production, *Life Sci.*, 47, 2343, 1990.
219. **Klett, C., Muller, F., Gierschik, P., and Hackenthal, E.,** Angiotensin II stimulates angiotensinogen synthesis in hepatocytes, *FEBS Lett.*, 259, 301, 1990.
220. **Johns, D.W., Peach, M.J., Gomez, R.A., Inagami, T., and Carey, R.M.,** Angiotensin II regulates renin gene expression, *Am. J. Physiol.*, 259, F882, 1990.
221. **Druker, B.J., Mamon, H.J., and Roberts, T.M.,** Oncogenes, growth factors, and signal transduction, *N. Engl. J. Med.*, 321, 1383, 1989.
222. **Huckle, W.R., Prokop, C.A., Dy, R.C., Herman, B., and Earp, S.,** Angiotensin II stimulates protein-tyrosine phosphorylation in a calcium-dependent manner, *Mol. Cell. Biol.*, 10, 6290, 1990.
223. **Sasaki, K., Yamano, Y., Bardhan, S., Iwai, N., Murray, J.J., Hasegawa, M., Matsuda, Y., and Inagami, T.,** Cloning and expression of a complementary DNA encoding a bovine adrenal angiotensin II type-1 receptor, *Nature*, 351, 230, 1991.
224. **Murphy, T.J., Alexander, R.W., Griendling, K.K., Runge, M.S., and Bernstein, K.E.,** Isolation of a cDNA encoding the vascular type-1 angiotensin II receptor, *Nature*, 351, 233, 1991.

225. **Iwai, N., Yamano, Y., Chaki, S., Konishi, F., Bardhan, S., Tibbetts, C., Sasaki, K., Hasegawa, M., Matsuda, Y., and Inagami, T.,** Rat angiotensin II receptor cDNA sequence and regulation of the gene expression, *Biochem. Biophys. Res. Commun.*, 177, 299, 1991.
226. **Iwai, N. and Inagami, T.,** Identification of two subtypes in the rat type I angiotensin receptor, *FEBS Lett.*, 298, 257, 1992.
227. **Kakar, S.S., Sellers, J.C., Devor, D.C., Musgrove, L.C., and Neill, J.D.,** Angiotensin II type-1 receptor subtype cDNAs: differential tissue expression and hormonal regulation, *Biochem. Biophys. Res. Commun.*, 183, 1090, 1992.
228. **Sandberg, K., Ji, H., Clark, A.J.L., Shapira, H., and Catt, K.J.,** Cloning and expression of a novel angiotensin II receptor subtype, *J. Biol. Chem.*, 267, 9455, 1992.
229. **Sandberg, K., Markwick, A.J., Trinh, D.P., and Catt, K.J.,** Calcium mobilization by angiotensin II and neurotransmitter receptors expressed in *Xenopus laevis* oocytes, *FEBS Lett.*, 241, 177, 1988.
230. **Furuta, H., Guo, D.-F., and Inagami, T.,** Molecular cloning and sequencing of the gene encoding human angiotensin II type 1 receptor, *Biochem. Biophys. Res. Commun.*, 183, 8, 1992.
231. **Takayanagi, R., Ohnaka, K., Sakai, Y., Nakao, R., Yanase, T., Haji, M., Inagami, T., Furuta, H., Gou, D.-F., Nakamuta, M., and Nawata, H.,** Molecular cloning, sequence analysis and expression of a cDNA encoding human type-1 angiotensin II receptor, *Biochem. Biophys. Res. Commun.*, 183, 910, 1992.
232. **Bergsma, D.J., Ellis, C., Kumar, C., Nuthulaganti, P., Kersten, H., Elshourbagy, N., Griffin, E., Stadel, J.M., and Alyar, N.,** Cloning and characterization of a human angiotensin II type 1 receptor, *Biochem. Biophys. Res. Commun.*, 183, 989, 1992.
233. **Carson, M.C., Leach-Harper, C.M., Baukal, A.J., Aguilera, G., and Catt, K.J.,** Physiochemical characterization of photoaffinity-labeled angiotensin II receptors, *Mol. Endocrinol.*, 1, 147, 1987.
234. **Rondeau, J.-J., McNicoll, N., Escher, E., Meloche, S., Ong, H., and De Lean, A.,** Hydrodynamic properties of the angiotensin II receptor from bovine adrenal zona glomerulosa, *Biochem. J.*, 268, 443, 1990.
235. **Kyte, J. and Doolittle, R.F.,** A simple method for displaying the hydropathic character of a protein, *J. Mol. Biol.*, 157, 105, 1982.
236. **Dohlman, H.G., Caron, M.G., and Lefkowitz, R.J.,** A family of receptors coupled to guanine nucleotide regulatory proteins, *Biochemistry,* 26, 2657, 1987.
237. **Jackson, T.R., Blair, L.A.C., Marshall, J., Goedert, M., and Hanley, M.R.,** The *mas* oncogene encodes an angiotensin receptor, *Nature,* 335, 437, 1988.
238. **Sandberg, K., Bor, M., Ji, H., Markwick, A., Millan, M.A., and Catt, K.J.,** Angiotensin II-induced calcium mobilization in oocytes by signal transfer through gap junctions, *Science,* 249, 298, 1990.
239. **Monnot, C., Weber, V., Stinnakre, J., Bihoreau, C., Teutsch, B., Corvol, P., and Clanser, E.,** Cloning and functional characterization of a novel *mas*-related gene, modulating intracellular angiotensin II actions, *Mol. Endocrinol.*, 5, 1477, 1991.
240. **Ambroz, C., Clark, A.J.L., and Catt, K.J.,** The *mas* oncogene enhances angiotensin-induced $[Ca^{2+}]_i$ responses in cells with pre-existing angiotensin II receptors, *Biochim. Biophys. Acta,* 1133, 107, 1991.
241. **Kozak, M.,** An analysis of 5′-noncoding sequences from 669 vertebrate messenger RNAs, *Nucleic Acids Res.*, 15, 8125, 1987.
242. **Kozak, M.,** Point mutations define a sequence flanking the AUG initiator codon that modulates translation by eukaryotic ribosomes, *Cell,* 44, 283, 1986.
243. **Propst, F., Rosenberg, M.P., Iyer, A., Kaul, K., and Vande Woude, G.F.,** c-*mos* proto-oncogene RNA transcripts in mouse tissues. Structural features, developmental regulation, and localization in specific cell types, *Mol. Cell. Biol.*, 7, 1629, 1987.

244. **Ratner, L., Thielan, B., and Collins, T.**, Sequences of the 5′ portion of the human c-sis gene: characterization of the transcriptional promoter and regulation of expression of the protein product of 5′ untranslated mRNA sequences, *Nucleic Acids. Res.*, 15, 6017, 1987.
245. **Fu, L., Ye, R., Browder, L.W., and Johnston, R.N.**, Translational potentiation of messenger RNA with secondary structure in *Xenopus, Science,* 251, 807, 1991.
246. **Ray, P.E., Aguilera, G., Kopp, J.B., Horikoshi, S., and Klotman, P.E.**, Angiotensin II receptor-mediated proliferation of cultured human mesangial cells, *Kidney Int.*, 40, 764, 1991.
247. **Clark, A.J.L., Balla, T., Jones, M.R., and Catt, K.J.**, Stimulation of early gene expression by angiotensin II in bovine adrenal glomerulosa cells: roles of calcium and protein kinase C, *Mol. Endocrinol.*, 6, 1889, 1992.
248. **Chiu, A.T., McCall, D.E., Nguyen, T., Carini, D.J., Duncia, J.V., Herblin, W.F., Uyeda, R.T., Wong, P.C., Wexler, R.R., Johnson, A.L., and Timmermans, P.B.M.W.M.**, Discrimination of angiotensin II receptor subtypes by dithiothreitol, *Eur. J. Pharmacol.*, 170, 117, 1989.
249. **O'Dowd, B., Hnatowich, M., Caron, M.G., Lefkowitz, R.J., and Bouvier, M.**, Palmitoylation of the human beta$_2$-adrenergic receptor: Mutation of cys^{341} in the carboxyl tail leads to an uncoupled nonpalmitoylated form of the receptor, *J. Biol. Chem.*, 264, 7564, 1989.
250. **Langford, K., Frenzel, K., Martin, B.M., and Bernstein, K.E.**, The genomic organization of the rat AT$_1$ angiotensin receptor, *Biochem. Biophys. Res. Commun.*, 183, 1025, 1992.
251. **Bucher, P.**, Weight matrix descriptions of four eukaryotic RNA polymerase II promoter elements derived from 502 unrelated promoter sequences, *J. Mol. Biol.*, 212L, 563, 1990.
252. **Watson, J., Hopkins, N., Roberts, J., Steitz, J.S., and Weinter, A.**, *Molecular Biology of the Gene,* Benjamin/Cummings, Menlo Park, CA, 1987.
253. **Elton, T.S., Stephan, C.C., Taylor, G.R., Kimball, M.G., Martin, M.M., Durand, J.N., and Oparil, S.**, Isolation of two distinct type 1 angiotensin II receptor genes, *Biochem. Biophys. Res. Commun.*, 184, 1067, 1992.
254. **Jaiswal, N., Tallant, E.A., Diz, D.I., Khosla, M.C., and Ferrario, C.M.**, Subtype 2 angiotensin receptors mediate prostaglandin synthesis in human astrocytes, *Hypertension,* 17, 1115, 1991.
255. **Ernsberger, P., Zhou, J., Damon, T., and Douglas, J.G.**, Angiotensin II receptor subtypes in cultured mesangial cells, *Am. J. Physiol.*, in press.
256. **McIntosh, R.P. and Catt, K.J.**, Coupling of inositol phospholipid hydrolysis to peptide hormone receptors expressed from adrenal and pituitary mRNA in Xenopus laevis oocytes, *Proc. Natl. Acad. Sci. U.S.A.*, 84, 9045, 1987.
257. **O'Dowd, B.F., Hnatowich, M., Regan, J.W., Leader, W.M., Caron, M.G., and Lefkowitz, R.J.**, Site-directed mutagenesis of the cytoplasmic domains of the human β_2-adrenergic receptor, *J. Biol. Chem.*, 263, 15985, 1988.
258. **Ji, H., Sandberg, K., and Catt, K.J.**, Novel angiotensin II antagonists distinguish amphibian from mammalian angiotensin II receptors expressed in Xenopus laevis oocytes, *Mol. Pharmacol.*, 39, 120, 1991.
259. **Sandberg, K., Ji, H., Millan, M.A., and Catt, K.J.**, Amphibian myocardial angiotensin II receptors are distinct from mammalian AT$_1$ and AT$_2$ receptor subtypes, *FEBS. Lett.*, 284, 281, 1991.
260. **Iwai, N. and Inagami, T.**, Regulation of the expression of the rat angiotensin II receptor mRNA, *Biochem. Biophys. Res. Commun.*, 182, 1094, 1992.
261. **Gomez, R.A., Chevalier, R.L., Carey, R.M., and Peach, M.J.**, Molecular biology of the renal renin-angiotensin system, *Kidney Int. Suppl.*, 30, 518, 1990.
262. **Chen, F.-C.M. and Printz, M.P.**, Chronic estrogen treatment reduces angiotensin II receptors in the anterior pituitary, *Endocrinology,* 113, 1503, 1983.

Chapter 14

LOCALIZATION, CHARACTERIZATION, DEVELOPMENT, AND FUNCTION OF BRAIN ANGIOTENSIN II RECEPTOR SUBTYPES

J. M. Saavedra, K. Tsutsumi, C. Strömberg, A. Seltzer, K. Michels, S. Zorad, and M. Viswanathan

TABLE OF CONTENTS

0-8493-4622-3/93/$0.00 + $.50

I. BRAIN ANGIOTENSIN RECEPTORS

Circulating angiotensin II (AngII) does not cross the blood brain barrier.[1,2] However, after injection in the circulation, AngII was shown to produce effects in the brain.[3] For these reasons, it was hypothesized that the brain contained AngII receptors in the circumventricular organs, which are areas outside the blood brain barrier. This hypothesis was confirmed with the use of autoradiographic and fluorescence methods. After intravenous administration of radiolabeled AngII, or after peripheral administration of fluorescent, biologically active AngII derivatives, AngII binding sites were identified in the pituitary gland, the area postrema, the organon vasculosum of the lamina terminalis, the subfornical organ, and the median eminence.[4]

The brain AngII binding sites were later characterized *in vitro* using conventional binding methods with membrane preparations and [^{125}I]AngII.[5,6] There was a striking similarity between the central AngII binding sites and the AngII receptors earlier characterized in peripheral target tissues.[7]

The *in vitro* binding studies revealed that brain AngII binding sites were not restricted to areas containing the circumventricular organs, but were present throughout the brain with a heterogeneous distribution.[8-10] This indicated the possible existence of AngII binding sites in brain structures inside the blood brain barrier. However, the presence of brain AngII binding sites inside the blood brain barrier was only conclusively demonstrated by autoradiography after *in vitro* incubation of brain slices with radiolabeled AngII or AngII derivatives. The first autoradiographic study was that of Mendelsohn et al.[11] This study demonstrated unequivocally that, in addition to the circumventricular organs, many areas of the brain inside the blood brain barrier contained large numbers of AngII binding sites. Since circulating AngII could not have access to these structures, it was only natural to speculate that the AngII binding sites inside the blood brain barrier were connected to AngII endogenously produced in the brain, and therefore to the ''central'' AngII system. Such a hypothesis was also based on the previous demonstration of the presence and distribution of other components of the AngII system in the brain.[12]

A vast amount of experimental evidence has demonstrated unequivocally that AngII, when administered in minute amounts into selective brain structures, resulted in increased blood pressure, drinking behavior, salt appetite, and vasopressin release.[12] The brain structures that responded to AngII injections were those in which autoradiography demonstrated dense AngII binding. Thus, the concept emerged of specific, selective, anatomically discrete, brain areas containing physiologically active AngII receptors, some accessible to circulating AngII and some connected to the brain AngII system.

Once initially demonstrated by autoradiography and the use of a [^{125}I]AngII agonist of high specific activity, the distribution of brain AngII receptors was confirmed by many groups.[13-18] Most mammalian species, including human, showed a distribution of binding sites in general, but sometimes not complete, agreement with that of the rat.[13-18] The cellular localization of brain AngII receptors is for the most part unknown and subject to controversy. A few studies addressed this issue with a combination of autoradiographic methods and relatively specific lesions. The results so far seem to indicate that in the intact brain, AngII receptors are associated with neuronal cell bodies. The best-characterized examples are the AngII receptor localization in neurons of the inferior olivary complex,[19] in presynaptic vagal afferent terminals in the nucleus of the solitary tract (NTS), (which originates in the nodose ganglion and are transported bidirectionally by the vagus nerve),[20,21] and in catecholamine neurons in the locus coeruleus.[22]

Precise, quantitative methods rapidly replaced the initial semiquantitative autoradiographic methods.[18,23-25] These techniques, using varying concentrations of ligand, allowed for construction of full saturation curves in brain nuclei from single rats, and for determination of ligand affinity constants for discrete brain structures. The quantitative autoradiographic methods are the methods of choice to study localization, characterization, and regulation of AngII receptors in brain. Quantification is possible by comparison of optical densities with [^{125}I]-standard curves.[26] The ligand of choice is [^{125}I]Sar^1-AngII,[18,23-25] an AngII agonist with lower degradation and higher affinity than AngII itself.

II. ANGIOTENSIN II RECEPTOR SUBTYPES

A. DISCOVERY

The existence of different AngII receptor subtypes was suspected and had been proposed for receptors located in peripheral tissues.[27] This earlier evidence consisted of differences in agonist and antagonist binding affinity and potency, and differences in sensitivity to disulfide reducing agents.[28,29] However, conclusive proof was obtained only recently, following the development of more selective ligands of AngII receptors.[30-35] These compounds were shown to be able to displace AngII binding from selective receptor populations in a number of peripheral target organs for AngII.

B. NOMENCLATURE

Based on the selective antagonism by nonpeptidic and peptidic ligands, a nomenclature for AngII receptor subtypes has recently been accepted.[35,36] Two receptor subtypes were originally described, AT_1 and AT_2. Binding of AngII to AT_1 receptors is selectively antagonized by the nonpeptidic ligand DuP 753 (losartan potassium); binding to AT_2 receptors is selectively displaced by the nonpeptidic ligands PD 123177 and PD 123319, and by the peptide CGP 42112A.[35,36]

Until recently, the heterogeneity of AngII receptors was a matter of speculation. It is now clear that AngII receptors can at least be subdivided into two subgroups. However, even this classification could turn out to represent an oversimplification. Based on the multiplicity of physiological actions and signal transduction mechanisms, and the differences in regulatory mechanisms of receptor expression and agonist/antagonist potencies, further subdivisions of AT_1 receptors can be expected. For example, in the brain, binding to all AT_1 receptors is sensitive to addition of guanine nucleotides.[37] However, only the AT_1 receptors in the paraventricular nucleus, but not those in the subfornical organ or nucleus of the solitary tract, were sensitive to pretreatment with pertussis toxin, indicating the possibility of different subsets of AT_1 receptors associated with different G-proteins.[37]

A similar case could be made for AT_2 receptors, although much less is known about their function, their second messenger systems, and their biochemical structure. A subclassification of brain AT_2 receptors (AT_{2A} and AT_{2B}), based on their sensitivity to guanine nucleotides and reducing agents, has been recently reported.[37,38]

C. LOCALIZATION

Following the development of the selective receptor competitors, most of the initial studies were performed in peripheral tissues known as target sites for circulating AngII. It was found that in the periphery, AT_1 receptors were present in high concentrations in vasculature, kidney, adrenal cortex, liver, and lung parenchyma.[12,39-43] AT_2 receptors in the periphery have been so far located in the adrenal cortex and medulla, the uterus, the ovaries, the aorta, the heart muscle, and in embryonic connective tissues.[31,44-48]

Soon after the demonstration of AT_1 and AT_2 receptors in the periphery, AT_1 and AT_2 receptors were reported in the brain.[49-56] As was the case in the periphery, binding to brain AT_1 receptors was blocked by losartan.[54] Conversely, binding to brain AT_2 receptors was selectively displaced by CGP 42112 A, PD 123177, or WL19 (Table 1).[49-54,56,57]

Because of large differences in affinity, the proportion of AT_1 and AT_2 receptors in single brain nuclei and discrete areas could be determined using single concentrations of selective AT_1 and AT_2 competitors and quantitative autoradiography.[54] Most brain areas contained a single type of AngII receptor (Table 1).[54]

Of particular interest is the forebrain AngII pathway.[58,59] This pathway extends in continuity from the subfornical organ to median preoptic nucleus, the lamina terminalis, and the organon vasculosum of the lamina terminalis, and contains both AngII-immunopositive nerve terminals and AngII receptors (Figure 1).[12,58,59] The AngII receptor band extends laterally to the paraventricular nucleus, where AngII receptors are concentrated in its parvocellular zone.[60-62] All AngII receptors in this forebrain pathway, including the paraventricular nucleus, and in the median eminence, belong to the AT_1 subtype.[63] The AngII forebrain pathway may be considered as one of the most important neuroanatomical connections between the brain AngII system and its peripheral counterpart.[12]

Another intriguing observation is the localization of AngII receptors in the cerebral arteries.[64] In brain arteries, only AT_2-type AngII receptors have been found (Figure 2).[64]

TABLE 1
Classification of Brain AngII Receptor Subtypes by Displacement of $[^{125}I]Sar^1$-AngII with Specific Antagonists

	2 Week old, % of control		8 Week old, % of control	
	DuP 753 ($10^{-5}M$)	CGP 42112 A ($10^{-7}M$)	DuP 753 ($10^{-5}M$)	CGP 42112 A ($10^{-7}M$)
AT_1				
Group A				
Suprachiasmatic nucleus	0	93 ± 5	0	97 ± 6
Choroid plexus	0	75 ± 25	0	100 ± 17
Nucleus of the solitary tract	0	98 ± 2	0	92 ± 15
Dentate gyrus	ND		0	80 ± 7
Group B				
Nucleus of the lateral olfactory tract	0	100 ± 15	0	117 ± 25
Piriform cortex	0	112 ± 8	0	100 ± 5
Median preoptic nucleus	0	103 ± 6	0	93 ± 12
Paraventricular nucleus	0	101 ± 5	0	100 ± 15
Subiculum	0	94 ± 18	0	115 ± 15
Area postrema	0	100 ± 13	0	112 ± 18
Parasubiculum	0	125 ± 25	ND	
Entorhinal cortex	0	150 ± 17	ND	
Group C				
Subfornical organ	0	107 ± 12	0	111 ± 13
Basolateral amygdaloid nucleus	0	100 ± 18	0	114 ± 14
Retrosplenial granular cortex	0	100 ± 11	ND	
AT_1 and AT_2				
Group D				
Superior colliculus	70 ± 15	35 ± 10[a]	70 ± 10[a,b]	30 ± 10
Cingulate cortex				
Superficial	18 ± 9[a,b]	73 ± 9	ND	
Deep	100 ± 15	15 ± 8[a,b]	ND	
Group E				
Cerebellar cortex (Molecular layer)				
Superficial	31 ± 8[a]	69 ± 8[a]	ND	
Deep	53 ± 6[a]	29 ± 6[a]	ND	
AT_2				
Group F				
Lateral septal nucleus	104 ± 11	0	100 ± 25	0
Ventral thalamic nucleus	108 ± 16	0	ND	
Mediodorsal thalamic nucleus	105 ± 10	0	113 ± 13	0
Locus coeruleus	109 ± 11	0	100 ± 6	0
Principal sensory trigeminal nucleus	108 ± 17	0	ND	
Parasolitary nucleus	100 ± 3	0	82 ± 9	0
Inferior olive	105 ± 7	0	104 ± 20	0
Medial amygdaloid nucleus	103 ± 10	0	100 ± 29	0
Medial geniculate nucleus	93 ± 10	0	87 ± 17	0

TABLE 1 (continued)
Classification of Brain AngII Receptor Subtypes by Displacement of [^{125}I]Sar^1-AngII with Specific Antagonists

	2 Week old, % of control		8 Week old, % of control	
	DuP 753 ($10^{-5}M$)	CGP 42112 A ($10^{-7}M$)	DuP 753 ($10^{-5}M$)	CGP 42112 A ($10^{-7}M$)
Group G				
Anterior pretectal nucleus	85 ± 8	0	ND	
Nucleus of the optic tract	87 ± 7	0	ND	
Ventral tegmental area	100 ± 8	0	ND	
Posterodorsal tegmental nucleus	111 ± 11	0	ND	
Hypoglossal nucleus	109 ± 9	0	ND	
Central medial and paracentral thalamic nuclei	107 ± 11	0	ND	
Laterodorsal thalamic nucleus	104 ± 22	0	ND	
Oculomotor nucleus	92 ± 8	0	ND	

Note: Values are mean ± SE in fmol/mg protein; n = 4 to 5 individually measured animals per group. Concentration of [^{125}I]Sar^1-AngII was 0.5×10^{-9} *M*. 0, Binding was totally displaced at the concentration of antagonist used (binding was not statistically significant when compared to non-specific binding); ND, not detected in control sections (not statistically significant over nonspecific binding).

[a] $p < 0.05$ vs. control (incubated in the absence of AngII receptor antagonists).
[b] $p < 0.05$, DuP 753 vs. CGP 42112 A.

D. SENSITIVITY TO REDUCING AGENTS AND GUANINE NUCLEOTIDES

In peripheral tissues, binding of AngII to some of its receptors has been shown to be sensitive to sulfhydryl reducing agents such as dithiothreitol.[29] It was later revealed that these receptors belong to the AT_1 subtype.[31,32,35] Binding to brain AT_1 receptors was also sensitive to sulfhydryl reducing agents.[38,55,65] Conversely, binding to brain AT_2 receptors was reported to be insensitive to,[55] or enhanced[65,66] by, sulfhydryl reducing agents. Later studies, however, demonstrated that this was not always the case. Binding of AngII to the inferior olive was indeed insensitive to dithiothreitol.[38,55] However, binding of AngII to AT_2 receptors in some other areas, such as those located in the ventral and mediodorsal thalamic, medial geniculate, and oculomotor nuclei, the superior colliculus, and the cerebellar cortex, was sensitive to dithiothreitol.[38]

A further difference between peripheral AT_1 and AT_2 receptors resides in their association to G-proteins. Peripheral AT_1 receptors are coupled to G-proteins, and consequently agonist binding to their receptors in sensitive to the addition of guanine nucleotides.[44,48,67] Conversely, peripheral AT_2 receptors have been reported not to be coupled to G-proteins, and their binding has been found not to be affected by addition of guanine nucleotides.[44,48,68] By analogy to the peripheral AngII receptor subtypes, brain AT_1 receptors have been proposed to be G-protein coupled, whereas brain AT_2 receptors were proposed as not coupled to G-proteins. Again, a more precise study demonstrated that this was not always the case. While AngII binding to brain AT_1 receptors was indeed sensitive to guanine nucleotides,[37] binding to AT_2 receptors was heterogeneous with respect to guanine nucleotide sensitivity. For example, binding to AT_2 receptors in the ventral thalamic and medial geniculate nucleus, and in the locus coeruleus, was sensitive to guanine nucleotides, while binding to the inferior olive was not (Figure 3).[37]

The heterogeneity of brain AT_2 receptors with respect to their nucleotide and reducing agent sensitivity was the basis for the proposal of a subclassification of brain AT_2 receptors into two distinct subgroups. AT_{2A} receptors are sensitive, and AT_{2B} receptors insensitive, to guanine nucleotides and reducing agents.[37] These observations suggest a biochemical, and possibly physiological, difference between AT_2 receptor subtypes (Table 2).[37]

E. SIGNAL TRANSDUCTION MECHANISMS

Peripheral AT_1 receptors have been recently cloned from bovine adrenal zona glomerulosa cells,[69] and from rat aortic vascular smooth muscle cells.[70] AT_1 receptors were identified as a seven-transmembrane domain structure, typical of G-protein-coupled receptors (see above) coupled to phospholipase C and Ca^{2+} mobilization.[69,70] Cysteine residues are present in each of the four extracellular domains and are probably important in the disulfide bridge formation necessary for the ligand-binding conformation, and in the sensitivity of the AT_1 receptor to reducing agents.[69,70]

It is not clear whether a single type of AT_1 receptor is coupled to multiple pathways or multiple receptor subtypes are interacting independently with different signal transducers.[71] In the anterior pituitary, where only AT_1 receptors are expressed,[63] AngII induces phosphatidylinositol metabolism and arachidonate liberation, and also inhibits adenylate cyclase, although the physiological relevance of this last mechanism is not certain.[12]

The AT_2 receptors are probably structurally distinct from the G-protein-coupled receptor superfamily, and their signal transduction mechanisms and physiological functions in peripheral tissues and in brain have not yet been clarified.

Very little is known about mechanisms of signal transduction for brain AT_1 or AT_2 receptors, and very few and contradictory studies have been published. There have been reports of inhibition of phosphatidylinositol metabolism by AngII in the brain.[72] In peripheral sympathetic ganglia, however, stimulation of AT_1 receptors, probably linked to G-proteins, increases phosphatidylinositol metabolism.[67] This further indicates that this signal transduction mechanism, and the G-protein linkage, could be similar for peripheral and central AT_1 receptors.[37,48] It has also recently been reported that AT_1 receptors are involved in the inhibition of brain calmodulin-dependent cGMP phosphodiesterase.[73] There are presently no reports on the nature of the signal transduction mechanisms for AT_2 receptors in brain tissue. However, a recent

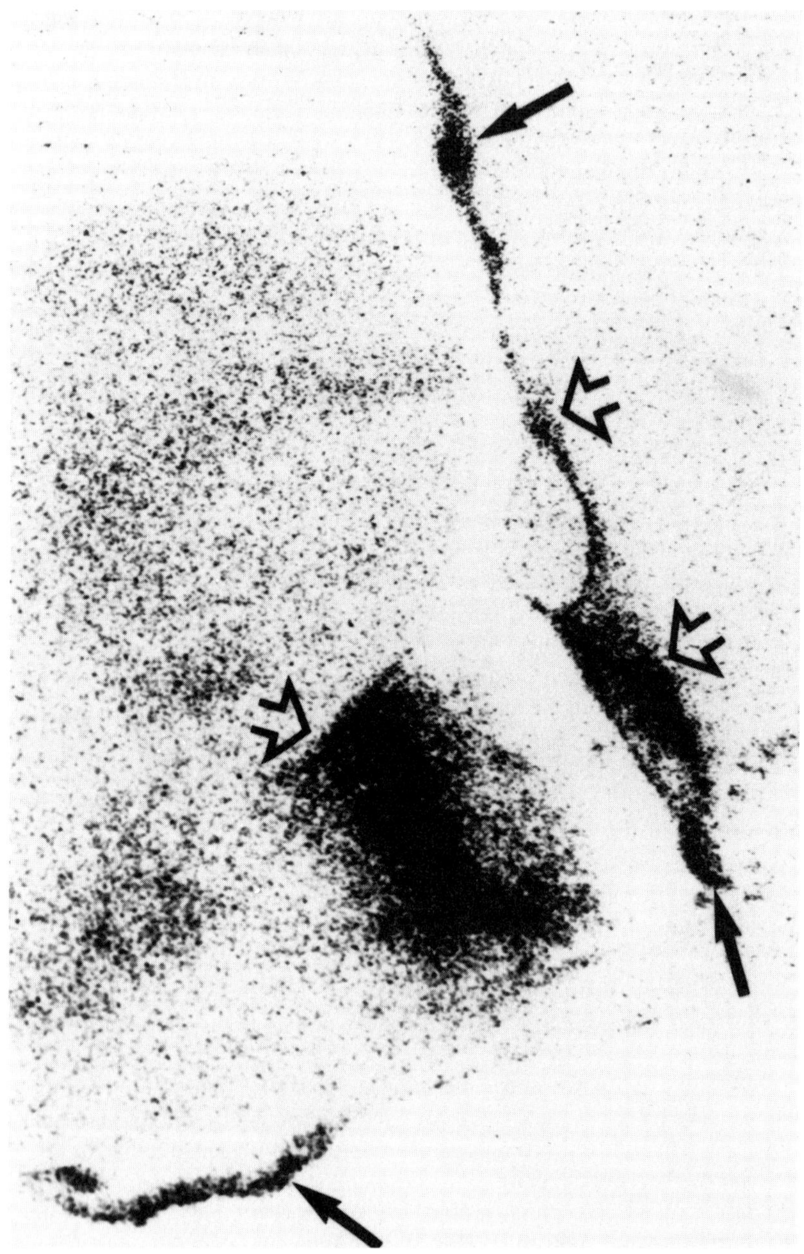

FIGURE 1. The forebrain AngII receptor band. The figure represents parasaggital sections of the rat forebrain. (Left) Autoradiograph after incubation with [^{125}I]Sar^{1}-AngII. (Right) Consecutive section stained with Toluidine blue. Solid arrows point to the subfornical organ (upper right), the organon vasculosum of the lamina terminalis (lower right), and the median eminence

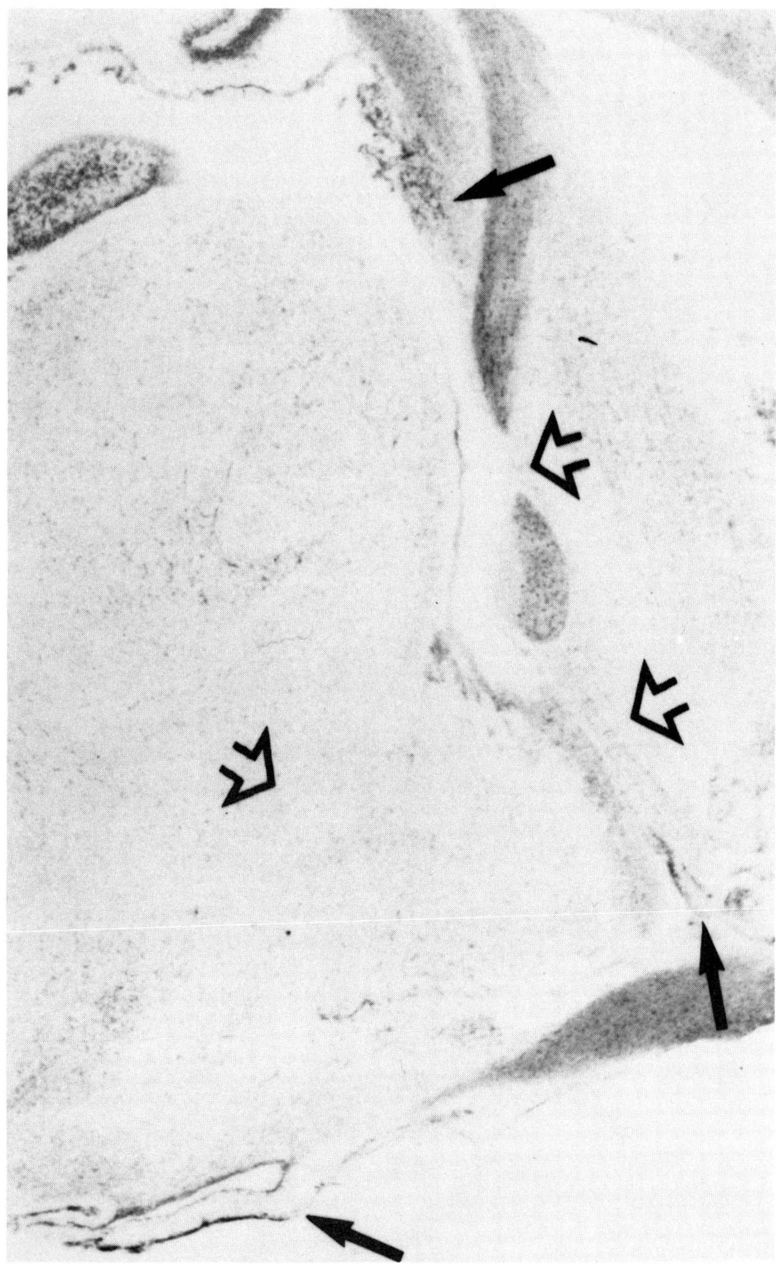

(lower left). All these areas are outside the blood brain barrier. Open arrows point to the median preoptic nucleus (upper right), the lamina terminalis (lower right), and the paraventricular nucleus (lower left).

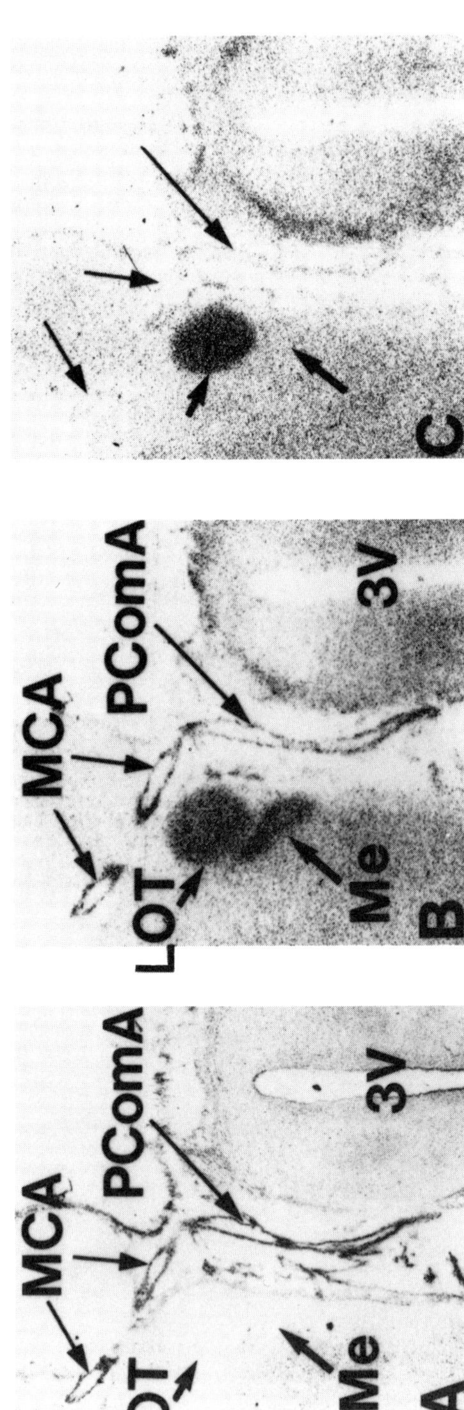

MCA
MCA
PComA
3V
LOT
Me
A
MCA
MCA
PComA
3V
LOT
Me
B
C

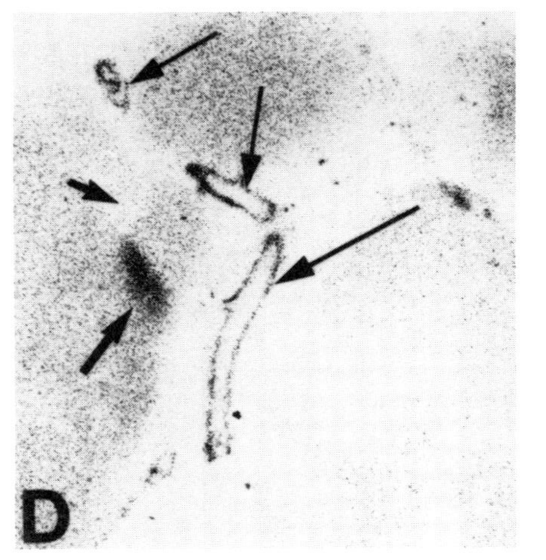

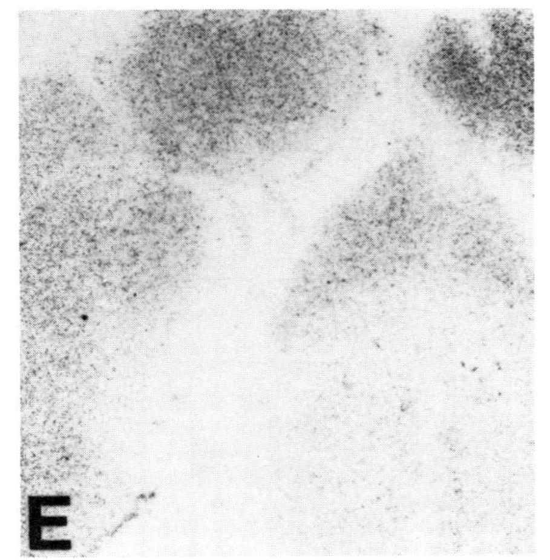

FIGURE 2. Autoradiography of cerebrovascular AngII receptor subtypes. Figures represent horizontal sections of the basal forebrain. (A) Toluidine blue staining; (B) total binding (0.5 n*M* [^{125}I]Sar^1-AngII); (C) AT_1 receptors (incubated as in B in the presence of 10^{-5} *M* CGP 42112A); (D) AT_2 receptors (incubated as in B in the presence of 10^{-5} *M* losartan); (E) nonspecific binding (incubated as in B in the presence of 10^{-6} *M* unlabeled AngII). 3V, Third ventricle, PComA, posterior communicating artery; MCA, medial communicating artery; LOT, nucleus of the lateral olfactory tract; Me, medial amygdaloid nucleus.

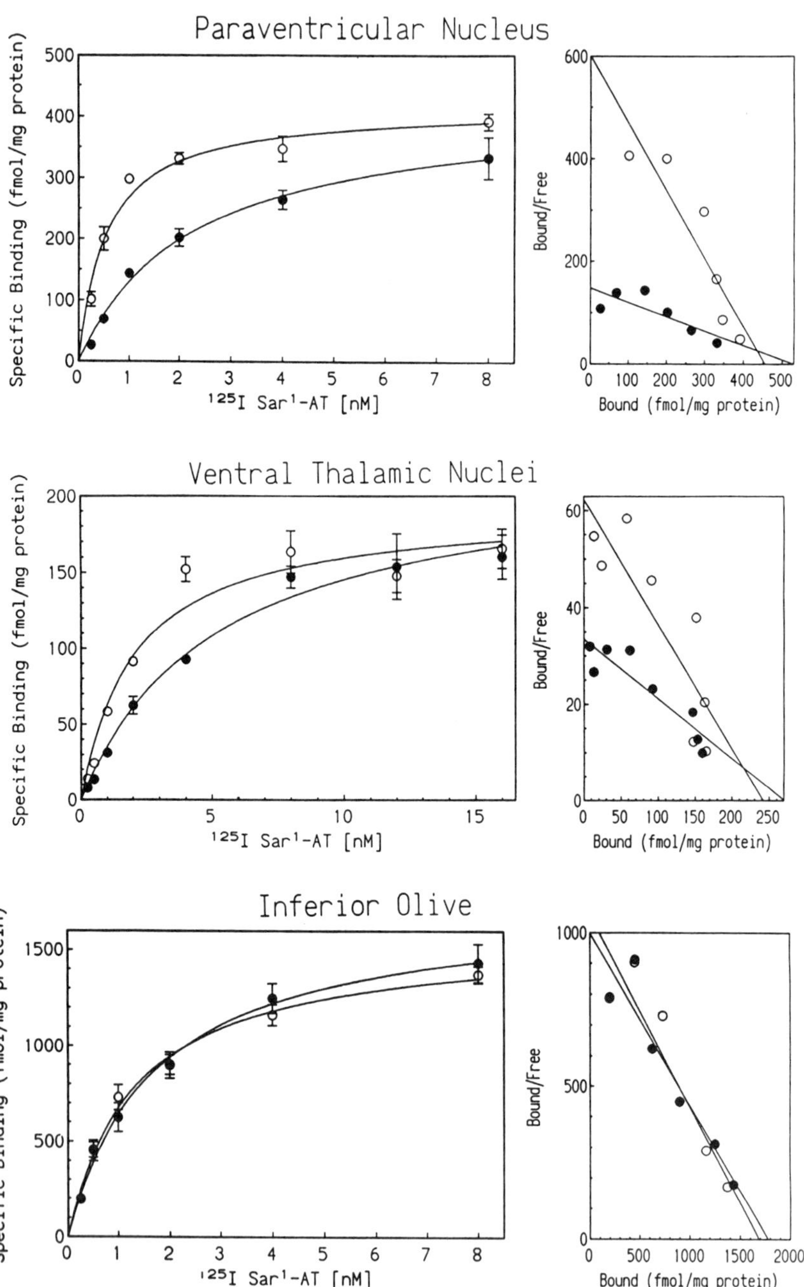

FIGURE 3. Effect of guanine nucleotides on [^{125}I]Sar^1-AngII binding kinetics. The figure represents typical saturation curves and Scatchard plots derived from incubation of consecutive sections without (closed circles) or with (open circles) GTPγS.

TABLE 2
Brain at Receptor Subtypes

AT_1 Receptors

	Subfornical organ	Nucleus of the solitary tract	Paraventricular nucleus
GTPγS sensitivity	Present	Present	Present
Pertussis toxin sensitivity	Absent	Absent	Present

AT_2 Receptors

	AT_{2A}			AT_{2B}
	Ventral thalamic nuclei	**Medial geniculate nucleus**	**Locus coeruleus**	**Inferior olive**
GTPγS sensitivity	Present	Present	Present (higher)	Absent
Pertussis toxin sensitivity	Present	Present	Present	Absent

report demonstrated that in neuronal cultures from neonatal rat brain, which contain mostly AT_2 binding sites, receptor stimulation was coupled to a reduction of basal cGMP levels[74] (Figure 4).

F. DEVELOPMENT

AngII receptors are expressed in large numbers in skeletal muscle and skin during late embryonic life.[75] Most of these peripheral AngII receptors belong to the AT_2 subtype,[48] and their number decreases dramatically after birth.[76] In addition, AT_2 receptors are present in developing aortas in higher numbers than AT_1 receptors.[44]

Primary cultures from fetal rat brain contain AngII receptors.[77] The early autoradiographic studies failed to demonstrate AngII receptors in fetal brain.[75] More detailed autoradiographic analysis, however, clarified that both AT_1 and AT_2 receptors were indeed present in fetal brain (Figure 5).[48,78] During embryonic life, AT_1 receptors are located in the nucleus of the solitary tract and the choroid plexus. AT_2 receptors are expressed in higher numbers in the fetal inferior olive, paratrigeminal, and hypoglossal nuclei, and in the meninges (Figure 5).[78]

The AngII receptor number reached a maximum of 10 times the adult during the first 2 weeks after birth, and the timing of development was different in different brain regions.[79] The number of brain AngII receptors declined after the second postnatal week, to reach relatively lower levels in the adult.[11] These early studies could not address the issue of selective development of receptor subtypes.

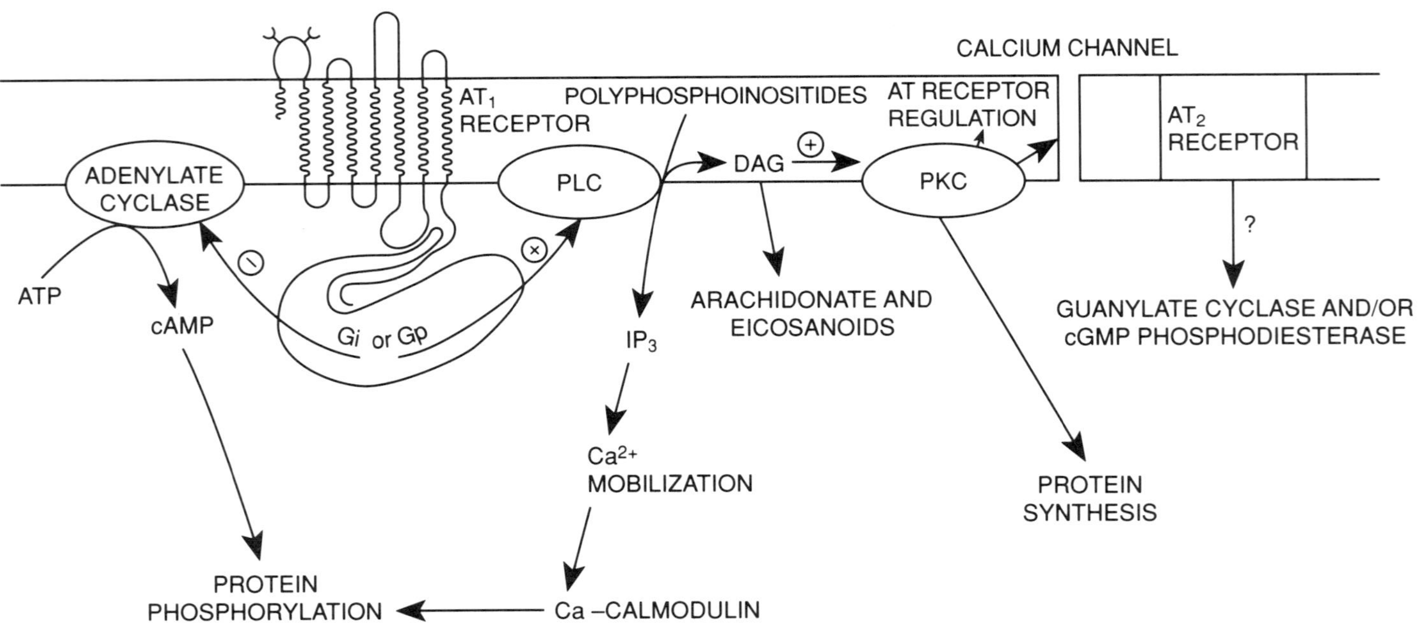

FIGURE 4. Proposed scheme of AngII-regulated cellular events. AT_1 receptor — a G-protein-coupled receptor — can inhibit adenylate cyclase through inhibitory guanine (G_i) proteins, or stimulate phospholipase C through hypothetical G-proteins (G_p). The latter results in hydrolysis of phosphoinositides and production of two second messengers — diacylglycerol (DAG) and inositoltriphosphate (IP_3). IP_3 releases calcium from intracellular stores and DAG increases the affinity of protein kinase C for calcium. AT_2 receptors have been proposed to be linked to guanylate cyclase and/or the cGMP phosphodiesterase system.

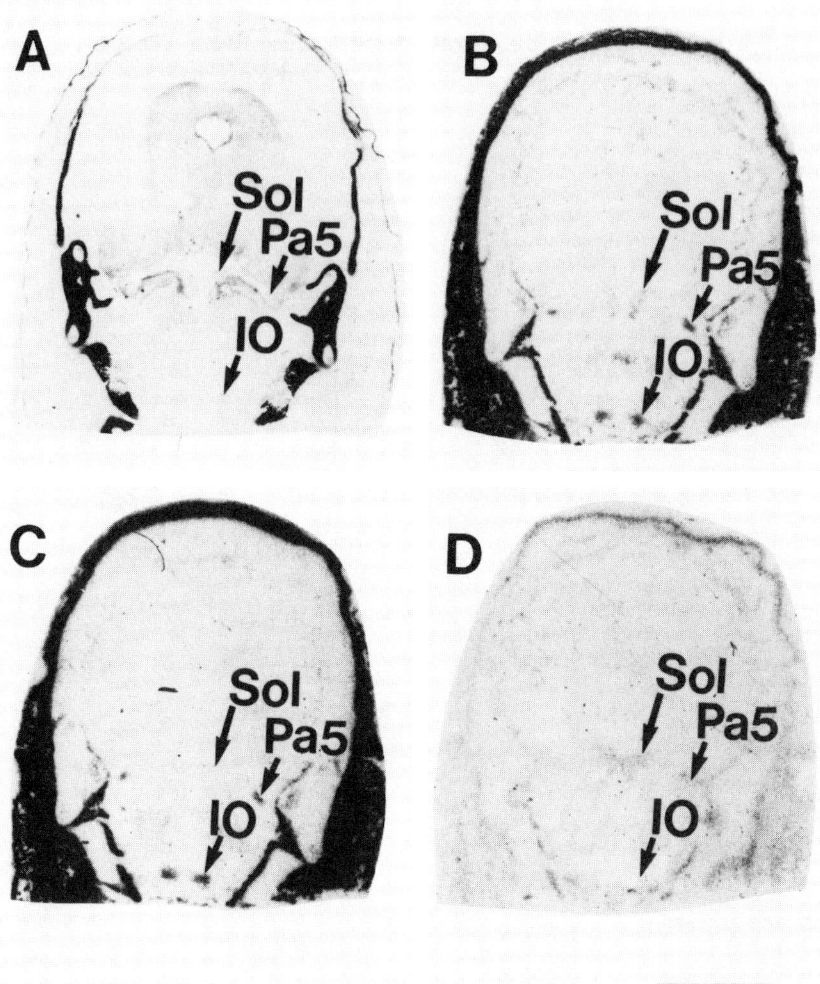

FIGURE 5. Autoradiography of AngII receptors in rat fetal brain. (A) Coronal section of a 18-day-old rat fetus, stained with Toluidine blue. (B, C, and D) Adjacent sections to (A), incubated in the presence of 5×10^{-10} *M* [^{125}I]Sar^1-AngII. (B) Total binding; (C) incubated in the presence of 10^{-5} *M* DuP 753. Binding to Sol (AT_1 receptors) was totally displaced by DuP 753. (D) Incubated in the presence of 10^{-7} *M* CGP 42112 A. Binding to IO and to Pa5 (AT_2 receptors) was totally displaced by CGP 42112A. Sol, Nucleus of the solitary tract; Pa5, paratrigeminal nucleus; IO, inferior olive.

More recently, large numbers of AT_1 and AT_2 AngII receptor subtypes were demonstrated at 2 weeks of postnatal life in the immature rat brain,[54,55] and the receptor subtypes had different developmental patterns. While AT_1 receptor number was similar in immature and adult rats, AT_2 receptors were expressed remarkably in immature animals, and in many areas they were no longer expressed in adults (Table 3).[54]

TABLE 3
AngII Receptor Concentrations in Specific Brain Areas of 2-Week-Old and 8-Week-Old Rats

	2-Week-old	8-Week-old
AT_1		
Group A		
Suprachiasmatic nucleus	128 ± 13	220 ± 14[a]
Choroid plexus	8 ± 2	14 ± 2[a]
Nucleus of the solitary tract	266 ± 12	341 ± 18[a]
Dentate gyrus	18 ± 3	97 ± 7[a]
Group B		
Nucleus of the lateral olfactory tract	56 ± 5	60 ± 3
Piriform cortex	111 ± 5	94 ± 6
Median preoptic nucleus	206 ± 19	183 ± 36
Paraventricular nucleus	205 ± 7	172 ± 14
Subiculum	63 ± 11	66 ± 4
Area postrema	185 ± 13	223 ± 16
Parasubiculum	28 ± 4	22 ± 6
Entorhinal cortex	28 ± 4	20 ± 8
Group C		
Subfornical organ	507 ± 37	348 ± 45[a]
Basolateral amygdaloid nucleus	62 ± 6	30 ± 7[a]
Retrosplenial granular cortex	44 ± 4	ND
AT_1 and AT_2		
Group D		
Superior colliculus	145 ± 5	65 ± 8[a]
Cingulate cortex	19 ± 4	8 ± 4[a]
Group E		
Cerebellar cortex	59 ± 6	ND

G. FUNCTIONS

As is the case in the periphery,[39-43] brain AT_1 receptors mediate all previously described AngII-induced responses, such as blood pressure increase and drinking behavior.[12] Because AT_1 receptors, and not AT_2 receptors, are located in all areas where brain AngII has been earlier shown to mediate its central effects, it is possible to suggest that other centrally mediated, classical effects of AngII are mediated through stimulation of these AT_1 receptors.[12]

The function of the AT_2 receptors, however, whether in the periphery or in the brain, is still a matter of speculation.[12] Because of their association with growing tissues, a participation of AT_2 receptors in organ development has been proposed, but their precise physiological role during development remains a mystery. The localization of brain AT_2 receptors could offer a clue as to their central function. AT_2 receptors are highly expressed in visual pathways, areas related to sensory and motor integration, with limbic system activity, and behavior.[54] Perhaps AT_2 receptors play a role in the development and organization of central structures related to sensory and motor integration.

TABLE 3 (continued)
AngII Receptor Concentrations in Specific Brain Areas of 2-Week-Old and 8-Week-Old Rats

	2-Week-old	8-Week-old
AT_2		
Group F		
Lateral septal nucleus	58 ± 8	18 ± 3[a]
Ventral thalamic nuclei	101 ± 8	24 ± 3[a]
Mediodorsal thalamic nucleus	165 ± 11	38 ± 13[a]
Locus coeruleus	289 ± 19	98 ± 13[a]
Principal sensory trigeminal nucleus	75 ± 6	15 ± 4[a]
Parasolitary nucleus	220 ± 15	57 ± 14[a]
Inferior olive	1328 ± 61	181 ± 32[a]
Medial amygdaloid nucleus	159 ± 8	94 ± 9[a]
Medial geniculate nucleus	338 ± 24	71 ± 6[a]
Group G		
Anterior pretectal nucleus	53 ± 8	ND
Nucleus of the optic tract	101 ± 13	ND
Ventral tegmental area	101 ± 11	ND
Posterodorsal tegmental nucleus	110 ± 21	ND
Hypoglossal nucleus	141 ± 11	ND
Central medial and paracentral thalamic nuclei	202 ± 14	ND
Laterodorsal thalamic nucleus	110 ± 10	ND
Oculomotor nucleus	98 ± 13	ND

Note: Values are means ± SE in fmol/mg protein; n = 6 individually measured animals per group. The concentration of [^{125}I]Sar^1-AngII used was 3×10^{-9} *M*. ND, Below the limit of sensitivity of the method (results not significantly different from nonspecific binding).

[a] $p < 0.05$, 8-week-old vs. 2-week-old rats.

An intriguing possibility has recently emerged. Based on the selective localization of AT_2 receptors in cerebral arteries,[64] a series of preliminary investigations has revealed that AT_2 receptors contribute to modulate cerebral blood flow autoregulation.[80] If these results are confirmed, pharmacological manipulation of brain AT_2 receptors could have therapeutic implications.

REFERENCES

1. **Volicer, L. and Loew, C.G.,** Penetration of angiotensin II into the brain, *Neuropharmacology,* 10, 631, 1971.
2. **Harding, J.W., Sullivan, M.J., Hanesworth, J.M., Cushing, L.L., and Wright, J.W.,** Inability of [^{125}I]Sar^1,Ile^{8}angiotensin II to move between the blood and cerebrospinal fluid compartments, *J. Neurochem.,* 50, 554, 1988.
3. **Bickerton, R.K. and Buckley, J.P.,** Evidence for a central mechanism of angiotensin induced hypertension, *Proc. Soc. Exp. Biol. Med.,* 106, 834, 1961.

4. **Van Houten, M., Schiffrin, E.L., Mann, J.F.E., Posner, B.I., and Boucher, R.,** Radioautographic localization of specific binding sites for blood-borne angiotensin II in the rat brain, *Brain Res.*, 186, 480, 1980.
5. **Bennett, J.P. and Snyder, S.H.,** Angiotensin II binding to mammalian brain membranes, *J. Biol. Chem.*, 251, 7423, 1976.
6. **Sirett, N.E., McLean, A.S., Bray, J.J., and Hubbard, J.I.,** Distribution of angiotensin II receptors in rat brain, *Brain Res.*, 122, 299, 1977.
7. **Bennett, J.P. and Snyder, S.H.,** Receptor binding interactions of the angiotensin II antagonist, ^{125}I-[sarcosine^1, leucine^8] angiotensin II, with mammalian brain and peripheral tissues, *Eur. J. Pharmacol.*, 67, 11, 1980.
8. **Sirett, N.E., Thornton, S.N., and Hubbard, J.I.,** Angiotensin binding and pressor activity in the rat ventricular system and midbrain, *Brain Res.*, 166, 139, 1979.
9. **Mann, J.F.E., Schiffrin, E.L., Schiller, P.W., Rascher, W., Boucher, R., and Genest, J.,** Brain receptor binding and central actions of angiotensin analogs in rats, *Am. J. Physiol.*, 241, R124, 1981.
10. **Harding, J.W., Stone, L.P., and Wright, J.W.,** The distribution of angiotensin II binding sites in rodent brain, *Brain Res.*, 205, 265, 1981.
11. **Mendelsohn, F.A.O., Quirion, R., Saavedra, J.M., Aguilera, G., and Catt, K.J.,** Autoradiographic localization of angiotensin II receptors in rat brain, *Proc. Natl. Acad. Sci. U.S.A.*, 81, 1575, 1984.
12. **Saavedra, J.M.,** Brain and pituitary angiotensin, *Endocr. Rev.*, 13, 1, 1992.
13. **Mendelsohn, F.A.O., Allen, A.M., Chai, S.Y., McKinley, M.J., Oldfield, B.J., and Paxinos, G.,** The brain angiotensin system. Insights from mapping its components, *Trends Endocrinol. Metab.*, 1, 189, 1990.
14. **Gehlert, D.R., Speth, R.C., and Wamsley, J.K.,** Distribution of [^{125}I]angiotensin II binding sites in the rat brain: a quantitative autoradiographic study, *Neuroscience*, 18, 837, 1986.
15. **Speth, R.C., Wamsley, J.K., Gehlert, D.R., Chernicky, C.L., Barnes, K.L., and Ferrario, C.M.,** Angiotensin II receptor localization in the canine CNS, *Brain Res.*, 326, 137, 1985.
16. **Healy, D.P. and Printz, M.P.,** Localization of angiotensin II binding sites in rat septum by autoradiography, *Neurosci. Lett.*, 44, 167, 1984.
17. **Mendelsohn, F.A.O., Allen, A.M., Clevers, J., Denton, D.A., Tarjan, E., and McKinley, M.J.,** Localization of angiotensin II receptor binding in rabbit brain by in vitro autoradiography, *J. Comp. Neurol.*, 270, 372, 1988.
18. **Saavedra, J.M., Israel, A., Plunkett, L.M., Kurihara, M., Shigematsu, K., and Correa, F.M.A.,** Quantitative distribution of angiotensin II binding sites in rat brain by autoradiography, *Peptides*, 7, 679, 1986.
19. **Walters, D.E. and Speth, R.C.,** Neuronal localization of specific angiotensin II binding sites in the rat inferior olivary nucleus, *J. Neurochem.*, 50, 812, 1988.
20. **Diz, D.I., Barnes, K.L., and Ferrario, C.M.,** Contribution of the vagus nerve to angiotensin II binding sites in the canine medulla, *Brain Res. Bull.*, 17, 497, 1986.
21. **Lewis, S.J., Allen, A.M., Verberne, A.J.M., Figdor, R., Jarrott, B., and Mendelsohn, F.A.O.,** Angiotensin II receptor binding in the rat nucleus tractus solitarii is reduced after unilateral nodose ganglionectomy or vagotomy, *Eur. J. Pharmacol.*, 125, 305, 1986.
22. **Rowe, B.P., Kalivas, P.W., and Speth, R.C.,** Autoradiographic localization of angiotensin II receptor binding sites on noradrenergic neurons of the locus coeruleus of the rat, *J. Neurochem.*, 55, 533, 1990.
23. **Israel, A., Correa, F.M.A., Niwa, M., and Saavedra, J.M.,** Quantitative determination of angiotensin II binding sites in rat brain and pituitary gland by autoradiography, *Brain Res.*, 322, 341, 1984.

24. **Israel, A., Correa, F.M.A., Niwa, M., and Saavedra, J.M.,** Quantitative measurement of angiotensin II (AII) receptors in discrete regions of rat brain, pituitary and adrenal gland by autoradiography, *Clin. Exp. Hypertens.*, A6, 1761, 1984.
25. **Israel, A., Plunkett, L.M., and Saavedra, J.M.,** Quantitative autoradiographic characterization of receptors for angiotensin II and other neuropeptides in individual brain nuclei and peripheral tissues from single rats, *Cell. Mol. Neurobiol.*, 5, 211, 1985.
26. **Nazarali, A.J., Gutkind, J.S., and Saavedra, J.M.,** Calibration of ^{125}I-polymer standards with ^{125}I-brain paste standards for use in quantitative receptor autoradiography, *J. Neurosci. Methods,* 30, 247, 1989.
27. **Peach, M.J. and Dostal, D.E.,** The angiotensin II receptor and the actions of angiotensin II, *J. Cardiovasc. Pharmacol.*, 16 (Suppl. 4), S25, 1990.
28. **Pobiner, B.F., Hewlett, E.L., and Garrison, J.C.,** Role of Ni in coupling angiotensin receptors to inhibition of adenylate cyclase in hepatocytes, *J. Biol. Chem.*, 260, 16200, 1985.
29. **Gunther, S.,** Characterization of angiotensin II receptor subtypes in rat liver, *J. Biol. Chem.*, 259, 7622, 1984.
30. **Chiu, A.T., Duncia, J.V., McCall, D.E., Wong, P.C., Price, W.A., Thoolen, M.J.M.C., Carini, D.J., Johnson, A.L., and Timmermans, P.B.M.W.M.,** Nonpeptide angiotensin II receptor antagonists. III. Structure-function studies, *J. Pharmacol. Exp. Ther.*, 250, 867, 1989.
31. **Whitebread, S., Mele, M., Kamber, B., and De Gasparo, M.,** Preliminary biochemical characterization of two angiotensin II receptor subtypes, *Biochem. Biophys. Res. Commun.*, 163, 284, 1989.
32. **Dudley, D.T., Panek, R.L., Major, T.C., Lu, G.H., Bruns, R.F., Klinkfefus, B.A., Hodges, J.C., and Weishaar, R.E.,** Subclasses of angiotensin II binding sites and their functional significance, *Mol. Pharmacol.*, 38, 370, 1990.
33. **Timmermans, P.B.M.W.M., Carini, D.J., Chiu, A.T., Duncia, J.V., Price, W.A., Wells, G.J., Wong, P.C., Johnson, A.L., and Wexler, R.R.,** The discovery of a new class of highly specific nonpeptide angiotensin II receptor antagonists, *Am. J. Hypertens.*, 4, 275S, 1991.
34. **Timmermans, P.B.M.W.M., Carini, D.J., Chiu, A.T., Duncia, J.V., Price, W.A., Wells, G.J., Wong, P.C., Wexler, R.R., and Johnson, A.L.,** Nonpeptide angiotensin II receptor antagonists: a novel class of antihypertensive agents, *Blood Vessels,* 27, 295, 1990.
35. **Timmermans, P.B.M.W.M., Wong, P.C., Chiu, A.T., and Herblin, W.F.,** Nonpeptide angiotensin II receptor antagonists, *TIPS,* 12, 55, 1991.
36. **Bumpus, F.M., Catt, K.J., Chiu, A.T., De Gasparo, M., Goodfriend, T., Husain, A., Peach, M.J., Taylor, D.G., Jr., and Timmermans, P.B.M.W.M.,** Nomenclature for angiotensin receptors. A report of the nomenclature committee of the Council for High Blood Pressure Research, *Hypertension,* 17, 720, 1991.
37. **Tsutsumi, K. and Saavedra, J.M.,** Heterogeneity of angiotensin II AT2 receptors in the rat brain, *Mol. Pharmacol.*, 41, 290, 1992.
38. **Tsutsumi, K., Zorad, S., and Saavedra, J.M.,** The AT_2 subtype of the angiotensin II receptors has differential sensitivity to dithiothreitol in specific brain nuclei of young rats, *Eur. J. Pharmacol. — Mol. Pharmacol.*, 226, 169, 1992.
39. **Gibson, R.E., Thorpe, H.H., Cartwright, M.E., Frank, J.D., Schorn, T.W., Bunting, P.B., and Siegl, P.K.S.,** Angiotensin-II receptor subtypes in renal cortex of rats and Rhesus monkeys, *Am. J. Physiol.*, 261, F512, 1991.
40. **Balla, T., Baukal, A.J., Eng, S., and Catt, K.J.,** Angiotensin-II receptor subtypes and biological responses in the adrenal cortex and medulla, *Mol. Pharmacol.*, 40, 401, 1991.

41. **Chiu, A.T., McCall, D.E., Price, W.A., Wong, P.C., Carini, D.J., Duncia, J.V., Wexler, R.R., Yoo, S.E., Johnson, A.L., and Timmermans, P.B.M.W.M.,** Nonpeptide angiotensin II receptor antagonists. VII. Cellular and biochemical pharmacology of DuP 753, an orally active antihypertensive agent, *J. Pharmacol. Exp. Ther.*, 252, 711, 1990.
42. **Wong, P.C., Hart, S.D., Zaspel, A.M., Chiu, A.T., Ardecky, R.J., Smith, R.D., and Timmermans, P.B.M.W.M.,** Functional studies of nonpeptide angiotensin II receptor subtype-specific ligands: DuP 753 (All-1) and PD123177 (All-2), *J. Pharmacol. Exp. Ther.*, 255, 584, 1990.
43. **Wright, G.B., Alexander, R.W., Ekstein, L.S., and Gimbrone, M.A., Jr.,** Sodium, divalent cations, and guanine nucleotides regulate the affinity of the rat mesenteric artery angiotensin II receptor, *Circ. Res.*, 50, 462, 1982.
44. **Viswanathan, M., Tsutsumi, K., Correa, F.M.A., and Saavedra, J.M.,** Changes in expression of angiotensin receptor subtypes in the rat aorta during development, *Biochem. Biophys. Res. Commun.*, 179, 1361, 1991.
45. **Pucell, A.G., Hodges, J.C., Sen, I., Bumpus, F.M., and Husain, A.,** Biochemical properties of the avarian granulosa cell type-2 angiotensin II receptor, *Endocrinology,* 128, 1947, 1991.
46. **Chiu, A.T., Herblin, W.F., McCall, D.E., Ardecky, R.J., Carini, D.J., Duncia, J.V., Pease, L.J., Wong, P.C., Wexler, R.R., Johnson, A.L., and Timmermans, P.B.M.W.M.,** Identification of angiotensin receptor subtypes, *Biochem. Biophys. Res. Commun.*, 165, 196, 1989.
47. **Rogg, H., Schmid, A., and de Gasparo, M.,** Identification and characterization of angiotensin II receptor subtypes in rabbit ventricular myocardium, *Biochem. Biophys. Res. Commun.*, 173, 416, 1990.
48. **Tsutsumi, K., Strömberg, C., Viswanathan, M., and Saavedra, J.M.,** Angiotensin-II receptor subtypes in fetal tissues of the rat — autoradiography, guanine nucleotide sensitivity, and association with phosphoinositide hydrolysis, *Endocrinology,* 129, 1075, 1991.
49. **Rowe, B.P., Grove, K.L., Saylor, D.L., and Speth, R.C.,** Discrimination of angiotensin II receptor subtype distribution in the rat brain using non-peptidic receptor antagonists, *Regul. Peptides,* 33, 45, 1991.
50. **Song, K., Allen, A.M., Paxinos, G., and Mendelsohn, F.A.O.,** Angiotensin II receptor subtypes in rat brain, *Clin. Exp. Pharmacol. Physiol.*, 18, 93, 1991.
51. **Leung, K.H., Smith, R.D., Timmermans, P.B.M.W.M., and Chiu, A.T.,** Regional distribution of the two subtypes of angiotensin II receptor in rat brain using selective nonpeptide antagonists, *Neurosci. Lett.*, 123, 95, 1991.
52. **Gehlert, D.R., Gackenheimer, S.L., Reel, J.K., Lin, H.S., and Steinberg, M.I.,** Nonpeptide angiotensin II receptor antagonists discriminate subtypes of ^{125}I-angiotensin II binding sites in the rat brain, *Eur. J. Pharmacol.*, 187, 123, 1990.
53. **Wamsley, J.K., Herblin, W.F., Alburges, M.E., and Hunt, M.,** Evidence for the presence of angiotensin II-type 1 receptors in brain, *Brain. Res. Bull.*, 25, 397, 1990.
54. **Tsutsumi, K. and Saavedra, J.M.,** Characterization and development of angiotensin-II receptor subtypes (AT_1 and AT_2) in rat brain, *Am. J. Physiol.*, 261, R209, 1991.
55. **Tsutsumi, K., and Saavedra, J.M.,** Increased dithiothreitol-insensitive, type 2 angiotensin II receptors in selected brain areas of young rats, *Cell. Mol. Neurobiol.*, 11, 295, 1991.
56. **Tsutsumi, K. and Saavedra, J.M.,** Quantitative autoradiography reveals different angiotensin II receptor subtypes in selected rat brain nuclei, *J. Neurochem.*, 56, 348, 1991.
57. **Tsutsumi, K. and Saavedra, J.M.,** Differential development of angiotensin II receptor subtypes in the rat brain, *Endocrinology,* 128, 630, 1991.
58. **Plunkett, L.M., Shigematsu, K., Kurihara, M., and Saavedra, J.M.,** Localization of angiotensin II receptors along the anteroventral-third ventricle area of the rat brain, *Brain Res.*, 405, 205, 1987.

59. **Shigematsu, K., Saavedra, J.M., Plunkett, L.M., Kurihara, M., and Correa, F.M.A.,** Angiotensin II binding sites in the anteroventral-third ventricle (AV3V) area and related structures of the rat brain, *Neurosci. Lett.*, 67, 37, 1986.
60. **Castrén, E. and Saavedra, J.M.,** Repeated stress increases the density of angiotensin II binding sites in rat paraventricular nucleus and subfornical organ, *Endocrinology,* 122, 370, 1988.
61. **Castrén, E. and Saavedra, J.M.,** Angiotensin II receptors in paraventricular nucleus, subfornical organ, and pituitary gland of hypophysectomized, adrenalectomized, and vasopressin-deficient rats, *Proc. Natl. Acad. Sci. U.S.A.*, 86, 725, 1989.
62. **Castrén, E. and Saavedra, J.M.,** Effect of corticoids on brain angiotensin II receptors, in *Recent Advances in Pharmacology and Therapeutics,* Velasco, M., Israel, A., Romero, E., and Silva, H., Eds., Elsevier, Amsterdam, 1989, 211.
63. **Tsutsumi, K. and Saavedra, J.M.,** Angiotensin II receptor subtypes in median eminence and basal forebrain areas involved in regulation of pituitary function, *Endocrinology,* 129, 3001, 1991.
64. **Tsutsumi, K. and Saavedra, J.M.,** Characterization of AT_2 angiotensin-II receptors in rat anterior cerebral arteries, Am. *J. Physiol.*, 261, H667, 1991.
65. **Speth, R.C., Rowe, B.P., Grove, K.L., Carter, M.R., and Saylor, D.,** Sulfhydryl reducing agents distinguish two subtypes of angiotensin II receptors in the rat brain, *Brain Res.*, 548, 1, 1991.
66. **Gehlert, D.R., Gackenheimer, S.L., and Schober, D.A.,** Angiotensin II receptor subtypes in rat brain: dithiothreitol inhibits ligand binding to All-1 and enhances binding to All-2, *Brain Res.*, 546, 161, 1991.
67. **Strömberg, C., Tsutsumi, K., Viswanathan, M., and Saavedra, J.M.,** Angiotensin II AT_1 receptors in rat superior cervical ganglia: characterization and stimulation of phosphoinositide hydrolysis, *Eur. J. Pharmacol. — Mol. Pharmacol.*, 208, 331, 1991.
68. **Bottari, S.P., Taylor, V., King, I.N., Bogdal, Y., Whitebread, S., and De Gasparo, M.,** Angiotensin II AT_2 receptors do not interact with guanine nucleotide binding proteins, *Eur. J. Pharmacol. — Mol. Pharmacol.*, 207, 157, 1991.
69. **Sasaki, K., Ymano, Y., Bardhan, S., Iwai, N., Murray, J.J., Hasegawa, M., Matsuda, Y., and Inagami, T.,** Cloning and expression of a complementary DNA encoding a bovine adrenal angiotensin II type-1 receptor, *Nature,* 351, 230, 1991.
70. **Murphy, T.J., Alexander, R.W., Griending, K.K., Runge, M.S., and Bernstein, K.E.,** Isolation of a cDNA encoding the vascular type-1 angiotensin II receptor, *Nature,* 351, 233, 1991.
71. **Garcia-Sainz, J.A.,** Angiotensin II receptors: one type coupled to two signals or receptor subtypes?, *TIPS,* 8, 47, 1987.
72. **Tamura, C.S. and Speth, R.C.,** Effects of angiotensin II on phosphatidylinositide hydrolysis in rat brain, *Neurochem. Int.*, 17, 475, 1990.
73. **Sharma, R.K., Smith, J.R., and Moore, G.J.,** Inhibition of bovine brain calmodulin-dependent cGMP phosphodiesterase by peptide and non-peptide angiotensin receptor ligands, *Biochem. Biophys. Res. Commun.*, 179, 85, 1991.
74. **Sumners, C., Tang, W., Zelezna, B., and Raizada, M.K.,** Angiotensin II receptor subtypes and coupled with distinct signal transduction mechanisms in neurons and astrocytes from rat brain, *Proc. Natl. Acad. Sci. U.S.A.*, 88, 7567, 1991.
75. **Millan, M.A., Carvallo, P., Izumi, S., Zemel, S., Catt, K.J., and Aguilera, G.,** Novel sites of expression of functional angiotensin II receptors in the late gestation fetus, *Science,* 244, 1340, 1989.
76. **Grady, E.F., Sechi, L.A., Griffin, C.A., Schambelan, M., and Kalinyak, J.E.,** Expression of AT2-receptors in the developing rat fetus, *J. Clin. Invest.*, 88, 921, 1991.
77. **Raizada, M.K., Yang, J.W., and Fellows, R.E.,** Rat brain cells in primary culture: characterization of angiotensin binding sites, *Brain Res.*, 207, 343, 1981.

78. **Tsutsumi, K., Viswanathan, M., Strömberg, C., and Saavedra, J.M.,** Type-1 and type-2 angiotensin-II receptors in fetal rat brain, *Eur. J. Pharmacol.*, 198, 89, 1991.
79. **Baxter, C.R., Horvath, J.S., Duggin, G.G., and Tiller, D.J.,** Effect of age on specific angiotensin II-binding sites in rat brain, *Endocrinology,* 106, 995, 1980.
80. **Strömberg, C. et al.,** unpublished observations.

Chapter 15

ANGIOTENSIN II RECEPTOR SUBTYPES IN NEURONAL CELLS

C. Sumners and M.K. Raizada

TABLE OF CONTENTS

0-8493-4622-3/93/$0.00 + $.50

I. INTRODUCTION

Numerous studies have demonstrated that angiotensin II (AngII) acts at specific receptors within the central nervous system (CNS) and elicits profound physiological/pharmacological changes such as increased water and salt intake, increased blood pressure, and increased secretion of vasopressin.[1-4] In an attempt to understand the cellular and molecular properties of these CNS AngII receptors, many investigators have used either primary cultures from brain tissue or neural/glial cell lines as *in vitro* systems. The use of these approaches has enabled the investigation of the properties of CNS AngII receptors in a relatively defined environment, and has proven advantageous for many reasons. For example, the use of brain cell cultures has demonstrated the presence of specific AngII receptors on both neurons and astroglia in the CNS.[5] The purpose of this chapter is to review the studies that have been performed on the properties of AngII receptors (and more recently, AngII receptor subtypes) in neuronal cells, both primary cultures and neural cell lines.

II. ANGIOTENSIN II RECEPTOR SUBTYPES IN THE BRAIN

Since the discovery of specific receptors for AngII within the CNS,[6,7] numerous studies have demonstrated that there is a discrete localization of these receptor sites in the brain.[8-11] AngII receptors were recently classified into two major subtypes, based on their differential affinities for nonpeptide drugs (losartan [formerly DuP 753], PD123177, PD123319, WL-19) or peptides (CGP 42112A, p-NH_2-phe^6-AngII).[12-15] AngII type 1 (AT_1) receptors have a high affinity for losartan, while AngII type 2 (AT_2) receptors have a high affinity for PD123177, PD123319, WL-19, CGP 42112A, and p-NH_2-Phe^6-AngII.[13-15] Autoradiographic analyses have determined that there is a mostly exclusive localization of AT_1 and AT_2 receptors in the brain.[16-20]

This area is given extensive coverage by Saavedra et al.[21] in the current volume, and thus will not be detailed here. However, within the context of this chapter, two points concerning AngII receptor subtypes in the brain are worth noting. First, it is apparent that both fetal (18-day-old) and young (1- to 2-week-old) rats display increased expression of the AT_2 receptor sites compared with adult rats.[22-25] This is relevant because many of the studies described in this chapter use brain tissue from neonatal or one-day-old rats. As will be seen, the primary neuronal cultures prepared from one-day-old rat brain indeed contain high levels of AT_2 receptors. Second, it is important to note that the neonatal rat brain areas used in many of our primary brain cell culture studies (e.g., diencephalon, brainstem) do contain high levels of both AT_1 and AT_2 receptors.[22-25]

III. ANGIOTENSIN II RECEPTOR SUBTYPES IN NEURONAL CULTURES

The aim of this section is to provide an extensive review of the properties of AngII receptor subtypes in neuronal cell cultures derived from neonate rat brain. First, it is necessary to discuss the characteristics of these cultured brain cells, and the rationale for using this *in vitro* system.

A. PRIMARY NEURONAL CULTURES: AN *IN VITRO* APPROACH FOR STUDYING ANGIOTENSIN II RECEPTORS

Several years ago we began to use primary neuronal or astroglial cultures prepared from neonate rat brain as a novel approach for investigating the properties of CNS AngII receptors. Our rationale for switching to this *in vitro* approach was that the complexity of the *in vivo* situation may make the results of studies performed on brain AngII receptors difficult to interpret. For example, studies on the regulation of brain AngII receptors *in vivo* require treatment of rats with hormones or dietary regimes, followed by analysis of AngII receptors in brain membranes or by autoradiography. The drawback is that nonspecific factors such as stress, locomotion, anesthesia, or even tissue preparation may in some way influence (directly or indirectly) brain AngII receptor expression. Thus, in the *in vivo* situation, the effects of a specific factor on brain AngII receptors is analyzed in the presence of many potential nonspecific influences. The use of primary brain cell cultures provides an advantage of working in relatively defined conditions, i.e., the effects of a particular hormone on AngII receptors can be analyzed to the exclusion of other influences. Further, this system allows analysis of AngII receptors on different types of brain cells (neurons, astroglia), something which is not possible by traditional methods. While brain cell cultures have many advantages over *in vivo* approaches, there are, of course, drawbacks to this *in vitro* approach. For example, the system is purely *in vitro*. In addition, the neuronal cultures can be prepared from only neonate rats, and so it might be argued that data from these cells perhaps represent only the picture in neonate brain tissue. In fact, the increased expression of AT_2 receptors (relative to AT_1 sites) seen in 10-day-old (days *in vitro*) neuronal cultures, resembles the picture seen in the neonatal or developing rat brain.[22,23] However, as will be discussed in subsequent sections, we believe that the data from neuronal cultures is very comparable to that from adult brain. We will also present evidence that with longer culture times (e.g., 25 days *in vitro*), changes occur in the expression of neuronal AngII receptor subtypes which make this *in vitro* system even more comparable with the *in vivo* situation. Nonetheless, in the final summation, brain cell cultures must ultimately be used in combination with *in vivo* approaches to give a full understanding of the properties of central AngII receptors.

B. GENERAL CHARACTERISTICS OF NEURONAL CULTURES

In this section, the properties of the primary neuronal cultures used in the studies presented below will be discussed.

In all cases, primary neuronal cultures were prepared from the brains of one-day-old rat pups. Brains were obtained from either Sprague-Dawley (SD), Wistar Kyoto (WKY), or spontaneously hypertensive (SH) rats, depending upon the study. In a few cases, whole brain cultures were used, but in most cases cultures were prepared from the hypothalamus and brainstem. The "hypothalamic" or diencephalic block used contains the paraventricular (PVN), supraoptic, anterior, lateral, posterior, dorsomedial, and ventromedial nuclei, and also parts of the thalamus and septum. The brainstem regions used are the medulla oblongata and the pons. The rationale for choosing these regions and for preparing cocultures is as follows. First, these brain regions contain the major concentrations of AngII receptors,[8-10] and also AT_1 and AT_2 receptors.[16-20] Second, the hypothalamus is innervated by catecholaminergic neurons derived from the brainstem.[26,27] The presence of these neurons was (and is) important for our studies on the effects of catecholamines on AngII receptor regulation. Third, the lateral nucleus and PVN project AngII-immunoreactive neurons back towards the brainstem, terminating in catecholamine-rich regions.[28] Fourth, neurons grown in culture in the presence of their target cells exhibit increased maturation in terms of morphology and neurochemical properties.[29]

Details of the preparation of primary neuronal cultures are presented elsewhere.[30,31] In most cases, these cells are used between 10 and 15 d *in vitro,* but in a few experiments cultures of 25 d *in vitro* age were used. Regardless of *in vitro* age, the cultures consist of many phase bright cells which contain classical neuronal morphology, and these cells overlie a bed of astroglia. We have determined, with the use of specific antibodies against neurofilament proteins and also glial fibrillary acidic protein (specific astroglia marker), that these primary neuronal cultures contain ~90% neuronal cells.[5,30,31] Most of the remaining cells are astroglia, with small numbers of microglia also present. Figure 1 shows a fluorescence micrograph of SD hypothalamus/brainstem neuronal cultures stained with a synaptophysin antibody. Specifically, this is a monoclonal antibody against a 38-kDa synaptophysin protein, and it was prepared by immunizing mice with a rat retinal synaptosomal preparation. The discrete pattern of immunostaining seen in Figure 1 (neuronal cell bodies, neurites, terminals) is consistent with the distribution of this protein.

C. PROPERTIES OF ANGIOTENSIN II RECEPTORS IN NEURONAL CULTURES

In this section, the discussion will be limited to the properties of AngII receptors in neuronal cultures prepared from normotensive (i.e., SD, WKY) rat brain. In addition, the major emphasis will be on newer discoveries with AT_1 and AT_2 receptors.

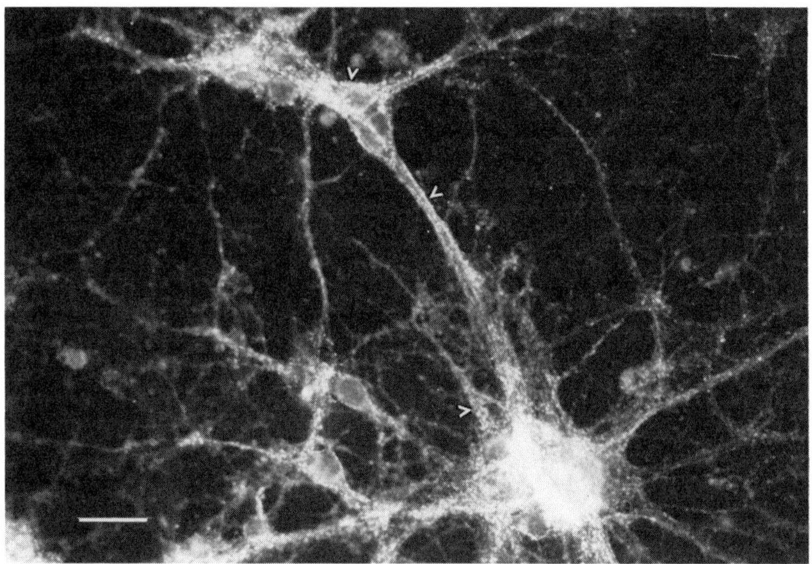

FIGURE 1. Immunofluorescent staining of neuronal cultures prepared from SD rat hypothalamus and brainstem using an antibody to synaptophysin. Arrows indicate specific staining. Bar, 10 μm.

1. Receptor Binding

Prior to the discovery of subtypes of AngII receptors, [^{125}I]AngII, [^{125}I]Sar^1,Ile^8-AngII, or [^{125}I]Sar^1-AngII were used to assess the levels and affinity of AngII binding sites in various tissues. Likewise, we utilized [^{125}I]AngII to characterize AngII receptor binding in primary neuronal cultures. These studies have been reviewed previously,[5] and so will not be detailed here. In brief, with the use of [^{125}I]AngII, we determined that neuronal cultures from SD or WKY rat whole brain contain a single population of binding sites (or heterogeneous binding sites with similar affinities) with K_d and B_{max} values of 1.0 n*M* and ~25 fmol/mg protein, respectively.[32,33] Hypothalamus/brainstem cocultures exhibited similar K_d values for [^{125}I]AngII binding (0.69 to 1.13 n*M*). However, the B_{max} values for [^{125}I]AngII binding in hypothalamus/brainstem neuronal cultures ranged from 94 to 163 fmol/mg protein, higher than in whole brain neuronal cultures.[34-37] This is not surprising, considering that the hypothalamus and brainstem contain the highest concentrations of AngII receptors in rat brain.[8-11] It should also be noted that these kinetic properties of AngII receptors in neuronal cultures, and also the pharmacological properties (i.e., competition/inhibition profiles) of these sites, are similar to the characteristics of brain AngII receptors.[6,7]

We have now determined that 10- to 15-d-old (*in vitro*) whole brain — and hypothalamus/brainstem neuronal cultures — contain both AT_1 and AT_2

receptors.[38] Thus, the more realistic conclusion from the above [^{125}I]AngII binding studies should have been that these cells contain a heterogeneous population of AngII receptors, with similar affinities. In whole brain neuronal cultures, competition/inhibition experiments using AT_1 and AT_2 receptor-selective ligands revealed the presence of predominantly AT_2 sites.[38] This was concluded from data which showed that the AT_2 receptor-selective ligands PD123177, CGP 42112A and *p*-NH_2-phe^6-AngII were potent displacers of [^{125}I]AngII-specific binding in these cells (IC_{50} values ranging from 0.92 for CGP 42112A to 14.9 n*M* for PD123177). In fact, PD123177 (1 μ*M*) displaced over 80% of the [^{125}I]AngII-specific binding. By contrast, the AT_1 receptor-selective drug losartan was a weak displacer of [^{125}I]AngII-specific binding, but was able to reduce the binding by ~20%. In the light of these results, we investigated the competition of [^{125}I]AngII binding in neuronal cultures by losartan in the presence of 1 μ*M* PD123177 to completely block AT_2 receptor binding. This concentration of PD123177 reduced [^{125}I]AngII-specific binding by 73 to 84%, and the residual binding was completely inhibited by losartan with IC_{50} values of 25 to 26 n*M*. These data led to the conclusion that whole brain neuronal cultures contain ~80% AT_2 receptors and ~20% AT_1 sites. In further studies, we have analyzed the levels of AT_1 and AT_2 receptors in 10- to 15-d-old (*in vitro*) hypothalamus and brainstem neuronal cultures by competition of [^{125}I]AngII binding with PD123319 (AT_2 selective) and losartan. First, we determined the competition of [^{125}I]AngII binding by losartan in the presence of 1 μ*M* PD123319 to eliminate [^{125}I]AngII binding to AT_2 receptors. PD123319 (1 μ*M*) reduced [^{125}I]AngII-specific binding by 66%. The remaining binding (34%) was completely inhibited in a concentration-dependent manner by losartan, with an IC_{50} value of ~20.8 n*M* (Figure 2A). In the reverse experiment, we tested the competition of [^{125}I]AngII binding by PD123319 in the presence of 1 μ*M* losartan, to completely antagonize binding to AT_1 receptors. In these experiments, 1 μ*M* losartan reduced [^{125}I]AngII-specific binding by 31%, and the remaining [^{125}I]AngII-specific binding (69%) was completely antagonized by PD123319 in a concentration-dependent manner, with an IC_{50} value of 17.2 n*M* (Figure 2B). These data indicate that the hypothalamus/brainstem neuronal cultures contain 31 to 34% AT_1 and 66 to 69% AT_2 receptors. Thus, the ratio of AT_1:AT_2 receptors is higher than found in the whole brain neuronal cultures, but his is not surprising considering the high levels of AT_1 receptors that are localized in the hypothalamus and brainstem regions.[20]

In additional experiments we have directly estimated the presence of AT_1 receptors in 10- to 15-d-old (*in vitro*) hypothalamus/brainstem neuronal cultures with the use of the AT_1 receptor-specific ligand [^{3}H]DuP 753.[39] Figure 3 shows saturation/Scatchard analyses of [^{3}H]DuP 753-specific binding to these cultures, revealing K_d and B_{max} values of 21.1 ± 1.7 n*M* and 11.7 ± 2.4 fmol/mg protein, respectively (n = 3 experiments).

The relative levels of AT_1 and AT_2 receptors in either whole brain or hypothalamus/brainstem neuronal cultures resemble the expression of these

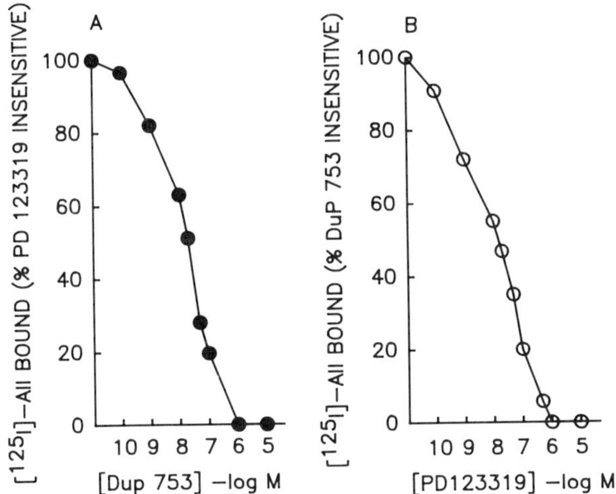

FIGURE 2. Competition of [^{125}I]AngII binding by nonpeptide AngII receptor blockers in hypothalamus/brainstem neuronal cultures. (A) Competition of [^{125}I]-AngII binding by unlabeled DuP 753 in the presence of 1 μM PD123319. $IC_{50} \simeq 20.8$ n*M* (n = 3 experiments). (B) Competition of [^{125}I]AngII binding by PD123319 in the presence of 1 μM DuP 753. $IC_{50} \simeq 17.2$ n*M* (n = 4 experiments).

receptor subtypes in the brains of fetal and neonatal rats,[22-25] and this is perhaps not surprising because the cultures are made from tissue derived from one-day-old rats. The cultures used in the above studies are 10 to 15 d old (*in vitro* age), so it appears that even after this time in culture the relative levels of neuronal AT_1/AT_2 receptors resemble those in the brains of young rats. However, the levels of AT_1 receptors in neurons maintained in culture for 25 to 30 days are significantly increased, as indicated by higher levels of [^{3}H]DuP 753-specific binding than are present in 10- to 15-d-old cultures.[133] Thus, it appears that even *in vitro* there are developmental changes occurring in AT_1 receptor expression, and these changes resemble those seen *in vivo,* i.e., increased levels of AT_1 receptors in the brains of juvenile and adult rats.[22,23]

These data clearly indicate the presence of AT_1 and AT_2 receptors in whole brain — and hypothalamus/brainstem neuronal cultures. Cultured astroglia also contain high densities of AT_1 receptors.[38] Thus, it might be argued that the AT_1 receptors found in neuronal cultures are associated with the ~10% astroglia which are present. However, we will present substantial evidence in subsequent sections which demonstrates that the neurons present in neuronal cultures do indeed contain a population of AT_1 receptors.

One other point to note is that the [^{125}I]AngII binding in neuronal cultures is not altered by the presence of guanine nucleotides [e.g., Gpp (NH)p].[5] This is interesting, since AT_1 receptors, both in other tissues[40,41] and in these cultures (Section III.C.4) are coupled to inositol phospholipid (IP) hydrolysis, a G-protein-coupled event.[40,41] Perhaps the insensitivity of [^{125}I]AngII binding

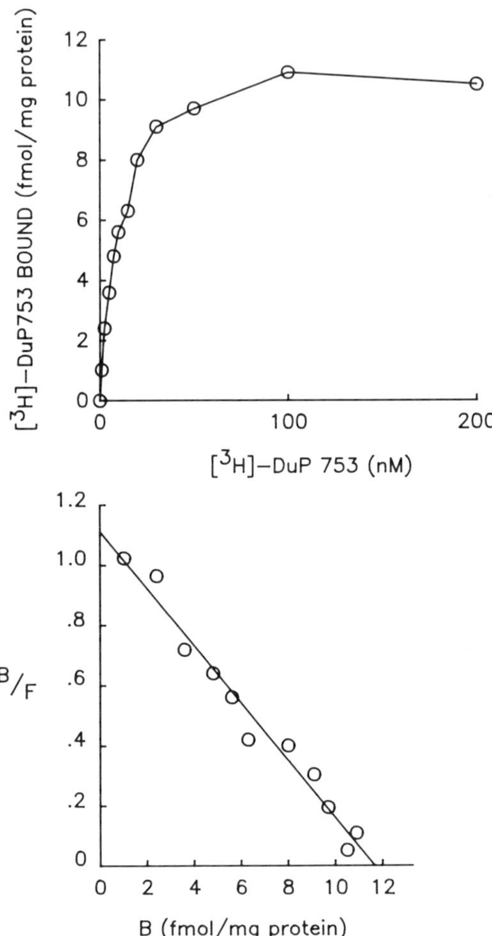

FIGURE 3. Top: Specific binding of [^{3}H]DuP 753 to hypothalamus/brainstem neuronal cultures as a function of radioligand concentration. Bottom: Scatchard analysis of the above saturation data.

in neuronal cultures to Gpp (NH)p is due to the fact that these cells contain predominantly AT_2 receptors,[38] and any effect of the GTP analog on AT_1 receptors may be too small to detect when analyzing total [^{125}I]AngII binding (i.e., both subtypes). This explanation is supported by the findings that at least one population of AT_2 receptors is not coupled to G-proteins.[42,43] Taken together, this data might indicate that the AT_2 receptors present in primary neuronal cultures are of the recently proposed AT_{2B} subtype, i.e., not G-protein coupled.[43] However, it should also be noted that in AT_1 receptor-rich astroglia,[38] Gpp (NH)p also has no effects on [^{125}I]AngII binding.[5] Therefore, sensitivity to Gpp (NH)p may not be a good index, in all tissues and cells, of whether a receptor is G-protein coupled.

2. Localization of AngII Receptors in Neuronal Cultures

The presence of AngII receptors on cultured neurons has also been confirmed by fluorescence and autoradiographic methods. One early study used fluorescein-labeled AngII to demonstrate the presence of AngII receptors on cell bodies and neurites of neurons in mixed cultures.[44] This localization was confirmed by studies which used light microscopic autoradiography along with [^{125}I]AngII or [^{3}H]AngII to label AngII receptors on cultured neurons.[33,45,46] In preliminary studies we have determined, using light microscopic autoradiography, that the AT_1 receptor selective ligand [^{3}H]DuP 753 labels sites on both the cell bodies and axons of cultured neurons.

3. Regulation

The regulation of AngII receptors in neuronal cultures has been reviewed in detail previously.[5] In this section, our review will be restricted to the effects of mineralocorticoids and catecholamines on AngII receptor regulation, including some new information on AT_1 and AT_2 receptors. Steroid hormones such as mineralocorticoids (e.g., aldosterone, deoxycorticosterone acetate [DOCA]) and β-estradiol have profound effects on brain AngII receptors and their responsiveness.[47-51] For example, DOCA hypertensive rats and also rats infused with DOCA (nonhypertensive) exhibit increased levels of brain AngII receptors.[47-49] In addition, infusion of DOCA into rats enhances the pressor and dipsogenic actions of centrally injected AngII.[47] Treatment of whole brain or hypothalamus/brainstem neuronal cultures with aldosterone or DOCA results in an upregulation of [^{125}I]AngII-specific binding. This mineralocorticoid action is dependent upon protein synthesis (maximal effect following a 20-h treatment; blocked by cycloheximide),[36,47] and so may involve synthesis of new receptors or synthesis of a protein which is necessary for the expression of AngII receptors. The exact cellular mechanisms involved in the upregulatory effects of mineralocorticoids on neuronal AngII receptors are unknown at present. However, it is clear that mineralocorticoids increase the levels of functional neuronal AngII receptors.[5] AngII stimulates neuronal reuptake of norepinephrine (NE) in these cultures,[52] and this effect is amplified by mineralocorticoid pretreatment of the cells.[5] Thus, the actions of mineralocorticoids on AngII receptors in rat brain and in primary neuronal cultures are identical — i.e., these hormones stimulate an increase in the level of functional AngII receptors.[47-49] These findings thus serve to validate the use of neuronal cultures as an *in vitro* model for studying AngII neuronal receptors, at least in terms of mineralocorticoid effects.

The question remains as to whether AT_1 or AT_2 receptors (or both subtypes) are increased by mineralocorticoid treatment. So far, there have been no reports of mineralocorticoid effects on AT_1 and AT_2 receptors in the brain. However, in preliminary studies we have clearly shown that treatment of hypothalamus/brainstem neuronal cultures with aldosterone increases the specific binding of [^{3}H]DuP 753 to these cells (Figure 4), suggesting an increase in AT_1 receptor levels. Mineralocorticoids also amplify AngII-stimulated neuronal NE uptake.[5] Since this AngII-stimulated effect in cultured neurons appears to involve AT_1 receptors (Section III.C.6), these findings suggest that mineralocorticoid hormones modulate at least this AngII receptor subtype. As yet we have no information on whether these hormones are involved in the regulation of AT_2 receptors. In summary, these data suggest that mineralocorticoid hormones upregulate neuronal AngII receptors (AT_1 subtype) and AT_1 receptor-mediated effects in these cells. This is, perhaps, logical, considering a number of different *in vivo* findings. First, the pressor and dipsogenic actions of AngII are mediated by AT_1 receptors,[53,54] and involve activation of brain catecholamine pathways.[55-57] In fact, a recent report has determined that AngII stimulates NE release from the PVN, mediated by AT_1 receptors.[58] Mineralocorticoids are important in the maintenance of extracellular fluid volume (ECFV), primarily via conservation of sodium. It is also possible that mineralocorticoid hormones can enhance the effects of AngII in conserving the ECFV by increasing the number of neuronal AT_1 receptors for this peptide. This is important because the pressor and dipsogenic effects elicited by centrally acting AngII help to maintain ECFV. Since the regulation of central AngII receptors by mineralocorticoids is a long-term action (e.g., 20 h in neurons), this regulation would be important in long-term decreases in ECFV.

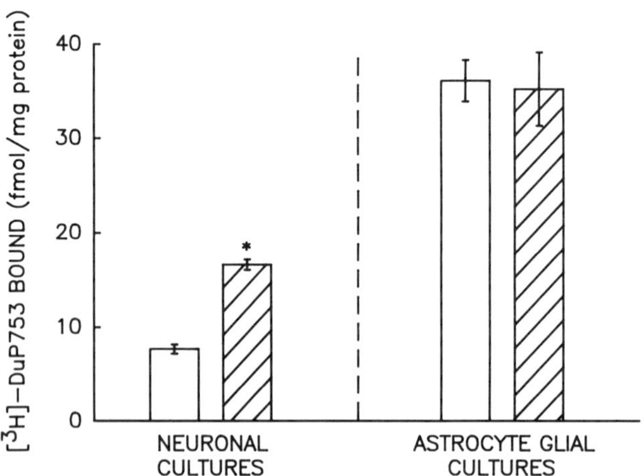

FIGURE 4. Effect of aldosterone on [^{3}H]DuP 753 binding to hypothalamus/brainstem neuronal and astroglial cultures. Cultures were incubated with control solvent (open bar) or 5 n*M* aldosterone (hatched bar) for 20 h at 37°C, followed by analysis of [^{3}H]DuP 753-specific binding. Data are means ± SEM, n = 3 experiments (*) $p < 0.01$, ANOVA1 and Newman-Keuls test.

From all of the above, it may be hypothesized that AngII acts at neuronal AT_1 receptors and, via activation of brain catecholamine pathways, helps to regulate ECFV and blood pressure. In addition, these AT_1 receptors/responses are regulated by mineralocorticoids. Further *in vitro* and *in vivo* studies are required to solidify these ideas.

Since neuronal cultures contain approximately 10% astroglia, which themselves contain AT_1 receptors, it could be argued that the increase in [^{3}H]DuP 753-specific binding elicited by aldosterone is due to effects on astroglial AT_1 sites. However, we have determined in other experiments that incubation of astroglial cultures with mineralocorticoids does not alter [^{125}I]AngII-specific binding.[36] Further, in preliminary studies we have shown that aldosterone (5 n*M*, 20 h) does not alter [^{3}H]DuP 753-specific binding in astroglial cultures (Figure 4). These data suggest that this stimulatory action of mineralocorticoids is on neuronal AT_1 receptors.

As stated previously, the pressor and dipsogenic actions of centrally administered AngII involve activation of brain catecholamine pathways,[55-57] including stimulatory effects on NE release.[58-60] These data suggest a neuromodulatory role of AngII on brain catecholamines. It is also apparent that AngII receptors and catecholaminergic cells/terminals are colocalized in certain brain regions.[27,28,61,62] Considering this, it is possible that brain catecholamines are able to regulate or modulate brain AngII systems, including AngII receptors, and there are two conflicting reports on this area in the literature.[5,63]

In primary neuronal cultures, we have identified two regulatory actions of catecholamines on neuronal AngII receptors, effects which are distinct from the actions of mineralocorticoids. Incubation of hypothalamus/brainstem neuronal cultures with NE or epinephrine (EPI) for short time periods (15 to 60 min) at concentrations which cause a large stimulation of IP hydrolysis, results in a rapid increase in the number of [^{125}I]AngII-specific binding sites, with no change in K_d.[37] This effect of NE and EPI is mediated via α_1-adrenergic receptors, stimulation of IP hydrolysis, and activation of protein kinase C (PKC).[35,37,64] We have now determined that this NE-induced upregulation of neuronal AngII receptors is mediated via α_{1B}-adrenergic receptors. This is based on the finding that NE-induced increases in [^{125}I]AngII-specific binding are inhibited by the α_{1B}-receptor-selective inhibitor chloroethylclonidine (CEC),[65] but not by the α_{1a}-receptor-selective drug, 5-methylurapidil[66] (see Figure 5A). It should also be noted that this action of NE does not involve protein synthesis, in contrast to the effects of mineralocorticoids.[5]

NE and EPI also have downregulatory actions on neuronal AngII receptors.[34] Incubation of hypothalamus/brainstem neuronal cultures with NE or EPI for longer time periods (2 to 4 h) at concentrations which do not alter IP hydrolysis, results in a significant decrease in the number of [^{125}I]AngII-specific binding sites, with no change in K_d.[34] This downregulatory action of NE is also mediated via α_1-adrenergic receptors, but does not involve IP hydrolysis or activation of PKC.[5] Rather, this NE effect is probably mediated

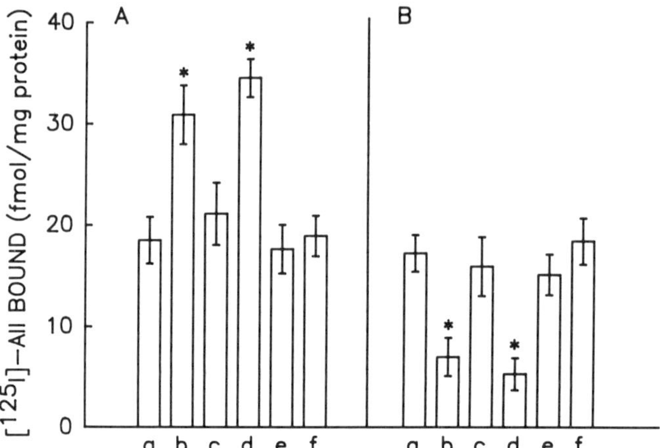

FIGURE 5. Effects of selective α_1-adrenergic receptor blockers on NE-induced changes in AngII receptors in neuronal cultures. (A) NE-induced increase in [^{125}I]AngII-specific binding. Hypothalamus/brainstem neuronal cultures were incubated for 60 min at 37°C with control solution (a), 10 μM NE (b), NE plus 10 μM CEC (c), NE plus 10 μM 5-methylurapidil (d), 10 μM CEC (e), or 10 μM 5-methylurapidil (f), followed by analysis of [^{125}I]AngII-specific binding. Data are means ± SEM, n = 3 experiments. (*) $p < 0.01$, ANOVA1 and Newman-Keuls test. (B) NE-induced decrease in [^{125}I]AngII-specific binding. Hypothalamus/brainstem neuronal cultures were incubated for 4 h at 37°C with control solution (a), 500 nM NE (b), NE plus 1 μM CEC (c), NE plus 1 μM 5-methylurapidil (d), 1 μM CEC (e), or 1 μM 5-methylurapidil (f), followed by analysis of [^{125}I]-AngII-specific binding. Data are means ± SEM, n = 3 experiments. (*) $p < 0.05$, ANOVA1 and Newman-Keuls test.

via calcium influx into the cells.[5,67] In addition, our recent studies suggest that the downregulation of [^{125}I]AngII binding by NE is mediated via α_{1A}-adrenergic receptors. This is based on the finding that NE-induced decreases in [^{125}I]AngII binding are inhibited by 5-methylurapidil, but not by CEC (Figure 5B).

At present, the exact mechanisms of NE-induced up- and downregulation of AngII receptors in neuronal cultures are unknown. Likewise, it is not known whether NE alters AT_1 or AT_2 receptors, or both subtypes in these effects. Since we have identified both AngII receptor subtypes in these neuronal cultures, along with intracellular processes, which are modulated by these receptors (see Sections III.C.4 and III.C.5), the question of whether AT_1 or AT_2 receptors are influenced by NE can be easily approached. Further, it will be possible to determine whether NE alters *functional* AT_1 or AT_2 receptors.

One other area that should be addressed is the physiological significance of regulation of AngII receptors by NE or EPI. We have determined that AngII has a neuromodulatory role on NE neurons in these cultures, i.e., AngII stimulates synthesis, release, reuptake, and ultimately, metabolism of NE. It is quite possible that the effects of NE on AngII receptors may act to regulate this neuromodulation of NE neurons by AngII. For example, if we propose that NE alters functional AngII receptors, the following scenario may exist. AngII causes release of NE which (in addition to other physiological effects) acts in the short term to upregulate AngII receptors, so amplifying the actions of AngII, i.e., a positive feedback. In the longer term, the released NE acts to reduce AngII receptors and decrease the neuromodulatory actions of AngII, i.e., a negative feedback. In addition, we now have evidence that the neuromodulatory actions of AngII are mediated by AT_1 receptors (see Section III.C.6). Thus, if the above scenario is true, then it is not unreasonable to propose that NE regulates at least neuronal AT_1 receptors. At this point, however, we cannot exclude the possibility that NE influences the expression of AT_2 receptors and nonfunctional AngII receptor subtypes, and this will be addressed in future studies.

The effects of catecholamines and mineralocorticoid hormones on AngII receptor regulation in neuronal cultures appear to be distinct events. This is because the upregulatory actions of aldosterone on neuronal AngII receptors are not altered by α_1-receptor blockers, nor are the regulatory actions of NE modulated by type 1 mineralocorticoid receptor blockers.[134]

In summary, we have discussed the regulatory actions of mineralocorticoids and catecholamines on AngII receptors in cultured neurons, and (where possible) have addressed the relationship of these data to *in vivo* findings and also the putative physiological relevance.

4. Intracellular Coupling

As discussed in Section III.C.1, hypothalamus/brainstem and whole brain neuronal cultures contain both AT_1 and AT_2 receptors. One major area of our investigations has been the intracellular coupling of these neuronal receptor subtypes, and our studies will be reviewed. First, it is pertinent to briefly review the signal transduction mechanism(s) coupled to each receptor subtype in brain and peripheral tissues, for the sake of comparison with data from cultured brain cells.

Starting with the AT_1 receptors, it appears that this subtype is responsible for mediating the well-documented intracellular effects of AngII in peripheral tissues. For example, depending upon the peripheral tissue or cell, AngII-stimulated IP hydrolysis,[40,68-70] AngII-induced calcium mobilization,[70,71] and AngII-induced reductions in cyclic AMP[70,72] are all mediated by AT_1 receptors. The picture in central tissues is less clear, mostly due to a lack of investigation. One study has shown that AngII stimulates IP hydrolysis in hypothalamic slices from estrogen-treated female rats.[135] It is likely that brain AT_1 receptors are coupled to a stimulation of IP hydrolysis, similar to certain peripheral counterparts, but this fact remains to be established.

In astroglia cultured from neonatal or adult rat brain, cells which contain high densities of AT_1 receptors,[38] AngII stimulates significant increases in IP hydrolysis.[5,38,73,74] These effects are blocked by losartan, but not by PD123177, and thus are AT_1 receptor mediated.[38,74] Therefore, astroglia AT_1 receptors are coupled to a stimulation of IP hydrolysis, similar to the AT_1 receptors in certain peripheral tissues.[40,68-70] In neuronal cultures prepared from rat hypothalamus and brainstem or whole brain, we previously reported that AngII had little or no significant stimulatory effect on IP hydrolysis.[5] However, these studies were performed prior to the discovery of two subtypes of AngII receptors. We now know that these primary neuronal cultures contain predominantly AT_2 receptors, and that AT_1 sites amount to only 20 to 30% of the total AngII receptors present. Thus, it is probably not surprising that we saw minor or no effect of AngII on IP hydrolysis in these cells. We have now used a much more sensitive IP hydrolysis assay, including separation of the different inositol phosphates produced, by which to assess the effects of AngII. In this way, we have determined that AngII elicits significant increases in IP, IP_2 (Ins[1,4]P_2), IP_3 (Ins[1,4,5]P_3), and IP_4 (Ins[1,3,4,5]P_4), effects which are blocked by losartan (Figure 6). These effects of AngII are not altered by the presence of PD123177, and so are clearly an AT_1 receptor-mediated event. Thus, the use of more sensitive IP hydrolysis assay conditions has forced us to revise our former conclusion that AngII showed little or no stimulation of IP hydrolysis in cultured neurons.[5] It is possible that the astroglia present in neuronal cultures are responsible for the AngII-stimulated

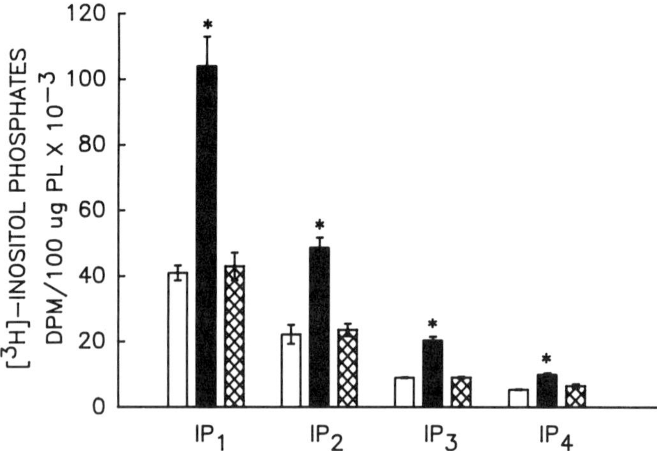

FIGURE 6. Effects of AngII on inositol phospholipid (IP) hydrolysis in SD hypothalamus/brainstem neuronal cultures. Cultures were incubated for 3 days with [^{3}H]myo-inositol (15 μCi/dish). After this, cells were incubated with control solution (open bar; KRB), 10 n*M* AngII (solid bar), or 10 n*M* AngII plus 1 μ*M* losartan (crosshatched bar) for 1 min. Following these incubations the levels of IP_1, IP_2, IP_3, and IP_4 were analyzed by Dowex chromatography. Data are means ± SEM from 3 experiments, and are expressed as DPM [^{3}H]inositol phosphate per 100 μg of phospholipid (PL). *, Significantly different from control, $p < 0.01$, ANOVA1 and Newman-Keuls test.

IP hydrolysis in these cells, because astroglia contain high densities of AT_1 receptors.[38] Evidence against this idea comes from the fact that treatment of neuronal cultures with aldosterone (5 n*M*, 20 h) significantly enhances AngII-stimulated IP hydrolysis in these cells, but similar treatments of pure astroglial cultures do not alter AngII-induced IP hydrolysis in those cells.[136] Along with the data presented in Section III.C.3 on the effects of aldosterone on [^{3}H]DuP 753 binding in neurons and astroglia, this evidence clearly suggests the presence of a population of *neuronal* AT_1 receptors which are coupled to IP hydrolysis. Future studies will include further characterization of AT_1 receptor-mediated effects in neuronal cultures, e.g., mobilization of intracellular calcium.

AT_2 receptors are clearly present in the brain[17-20] and in selected peripheral cells and tissues.[15,40,75,76] However, the physiological function(s) of these AT_2 sites and their intracellular coupling remain something of a mystery, and it is apparent that AT_2 receptors are not coupled to the same transduction pathways as AT_1 receptors (e.g., IP hydrolysis, calcium mobilization, reduction in cAMP).[75,76] In fact, there are no reports from brain or peripheral *tissues* of signal transduction processes which are modulated by AT_2 receptors. Studies from neural tumor cell lines have yielded some information on this subject, and will be discussed in detail in Section IV. Studies from primary brain cell cultures have yielded considerable information. In whole brain or hypothalamus/brainstem neuronal cultures, AngII elicits a concentration-dependent decrease in the basal cellular levels of cGMP.[38,77] This effect is blocked by PD123177, but not by losartan, indicating that it is mediated by AT_2 receptors. Preliminary studies suggest that this AT_2 receptor-mediated effect on cGMP in primary neuronal cultures may involve a calcium-dependent phosphodiesterase enzyme, though the exact mechanisms involved remain to be elucidated. In addition, it will be interesting to determine whether AngII and natriuretic peptides interact in modulating cellular cGMP levels in these cells. We have recently determined that of the guanylate cyclase coupled receptors present in cultured neurons, the ANP-B subtype predominates.[78] Studies are in progress to assess whether atrial natriuretic peptide (ANP) and c-type natriuretic peptide (CNP-22; ANP-B selective ligand) interact with AngII in regulating cellular cGMP levels in neuronal cultures. Interestingly, a recent report has suggested that AT_2 receptors mediate an inhibitory effect of AngII on basal and ANP-stimulated particulate guanylate cyclase activity in rat adrenal glomerulosa cells and in PC12W cells.[79] These authors suggest that AngII acts at AT_2 receptors to stimulate a phosphotyrosine phosphatase, which in turn inhibits particulate guanylate cyclase.[79] In essence, both of these studies demonstrate an inhibitory effect of AngII on cGMP, mediated by AT_2 receptors, though the cellular mechanism involved in each case must be identified.

In summary, while the signal transduction pathways that are coupled to AT_1 receptors are mostly established, the intracellular events modulated by AT_2 receptors are not. In fact, AT_2 receptors do not seem to be coupled intracellularly to the "traditional" AngII-induced signal transduction pathways.[75,76] Aside from the above-mentioned studies, which indicate cellular effects mediated by AT_2 receptors,[38,77,79] it appears that the AT_2 receptors present in many cells are functionally dormant.[75,76] The lack of measurable effects of AngII in many AT_2 receptor-rich tissues/cells might lead one to question whether this is simply a binding site. However, in the next section evidence from electrophysiological studies support the above intracellular messenger data in showing that the AT_2 site is indeed functional, at least in the brain.

5. Electrophysiological Effects

The binding of AngII to receptors on neurons in the hypothalamus and brainstem modulates the activity of neuronal pathways and ultimately leads to physiological changes such as increased blood pressure and water intake.[1-4] Neuronal activity and action potentials (and their constituent membrane ionic currents) are the basis of these physiologic effects. In spite of this, only a few electrophysiological studies have investigated the receptor-mediated effects of AngII on neuronal activity and membrane ionic currents. These studies can be divided into the effects of AngII on neurons *in situ,* in brain slices, and also on neurons in culture. Iontophoretic procedures have been used to apply AngII to neurons *in situ,* in the subfornical organ, the supraoptic nucleus, septum, or medulla. In each case, AngII produced an increase in neuronal firing rate which was blocked by AngII receptor antagonists that are not subtype selective.[80-84] In addition, iontophoretic application of AngII onto the organum vasculosum of the lamina terminalis (in hypothalamic brain slices) resulted in increases in cell firing rate which were blocked by Sar^1,Ile^8-AngII.[85] In contrast, AngII was shown to inhibit neuronal firing in neurons contained in hippocampal slices.[86] A recent study by Barnes et al. has determined that AngII evokes an excitatory effect in nucleus tractus solitarius neurons following pressure injection.[87] Finally, Legendre et al. have shown that AngII hyperpolarized cultured mouse spinal cord neurons, possibly due to activation of a transmembrane Cl^- current.[88]

The discovery of two distinct subtypes of AngII receptors complicates matters, however, because AT_1 and AT_2 receptors appear to be functionally distinct, at both the intracellular coupling and physiological levels. Thus, it is likely that the AT_1 and AT_2 receptor-mediated changes in neuronal activity and membrane ionic currents are exclusive effects. Recent iontophoretic experiments have clearly demonstrated both AT_1 and AT_2 receptor-mediated actions of AngII on neuronal activity. In the AT_1 receptor-rich PVN,[16-20] it has been shown that AngII produces excitatory effects on neuronal activity.[89] Losartan is much more potent than either PD123177 or CGP 42112A in blocking this AngII action, which is thus AT_1 receptor mediated.[89] In the AT_2 receptor-rich inferior olivary nucleus,[16-20] iontophoretic application of AngII also modulates neuronal activity, but this effect is AT_2 (rather than AT_1)

receptor mediated.[90] Since neuronal activity and action potentials are the basis of all brain-mediated physiological actions and behaviors, these studies provide excellent evidence that AT_2 receptors (at least in the brain) are functional.

Recent studies in hypothalamus/brainstem neuronal cultures have used whole cell patch-clamp methodologies to define the AT_1 and AT_2 receptor modulation of membrane ionic currents, which are the basis of action potentials and thus neuronal activity. Kang et al.[91] have recorded the effects of AngII on net outward ionic current (I_{no}) in cultured neurons. I_{no} is the sum of all inward and outward membrane ionic currents (minus sodium, which is blocked by tetrodotoxin), that occur during the repolarization phase of the action potential. In these cultured neurons, AngII elicits two different effects on I_{no}. There is an increase in I_{no}, an effect mediated by AT_2 receptors.[91] This appears to be a stimulatory effect on I_K (delayed rectifier potassium current) and I_A (transient potassium current), with no effects on calcium current.[92,93] In addition, this AT_2 receptor-mediated effect is enhanced by the presence of intracellular calcium, i.e., removal of intracellular calcium severely blunts (but does not extinguish) this response.[92] An increase in I_K and I_A will probably be inhibitory in terms of neuronal activity, due to hyperpolarizing effects. One question is whether this stimulatory action of AngII on I_{no} (I_K and I_A) is related to the fall in basal cGMP also elicited by AngII in these cells? These two events are perhaps associated because it has been demonstrated in other cells (enterocytes and cultured insect neurons) that cGMP inhibits potassium current.[94,95] Thus, if AngII induces a reduction in cGMP in cultured neurons, then this may result in an increase in potassium current. This is obviously hypothetical and further studies are needed to determine whether the AT_2 receptor-mediated changes in I_{no} and cGMP are part of the same event or are separate entities.

In another population of cultured neurons, AngII elicits a decrease in I_{no}, an effect mediated by AT_1 receptors. This appears to be an inhibitory effect on I_K, but whether or not I_A, $I_{Ca,K}$, and I_{Ca} contribute to this response has yet to be established.[92] A decrease in I_K will probably be excitatory in terms of neuronal activity, due to depolarizing effects. Increases in neuronal activity mediated by AT_1 receptors would be consistent with the AT_1 receptor-mediated excitatory actions of AngII in the PVN.[89] As shown in the previous section, AngII stimulates IP hydrolysis in neuronal cultures. It is quite possible that the AngII stimulation of IP hydrolysis and the AngII-induced inhibition of I_{no} (I_K) in neuronal cultures are part of the same event. This is suggested because in both central neurons and in peripheral sympathetic ganglia various studies have shown that muscarinic-induced inhibition of potassium currents is mediated by inositol phosphates (with subsequent calcium mobilization) and/or PKC activation.[96-98] Studies from the superior cervical ganglion also support this contention. In this AT_1 receptor-rich tissue, AngII stimulates IP hydrolysis,[41] and also inhibits potassium current.[99]

In summary, the studies on membrane ionic currents in cultured neurons and their modulation by AT_1 and AT_2 receptors will provide fundamental information on the basic cellular functions of these receptor subtypes. It will be interesting to determine which type of cell/neurotransmitter is mediated by AT_1 and AT_2 receptors. For the AT_1 receptors this will likely include NE (see Section III.C.6), but for the AT_2 receptors we can only speculate at this time. This electrophysiologic approach may help to uncover the function of AT_2 receptors.

6. Receptor-Mediated Neuromodulatory Effects

It is clear that AngII has a neuromodulatory effect on brain NE systems. For example, AngII stimulates release of NE from certain brain regions,[58-60,100] and in addition modulates neuronal re-uptake of NE.[101] AngII-stimulated release of NE from the PVN is mediated by AT_1 receptors.[60] These neuromodulatory actions of AngII on brain NE neurons are involved in the pressor effects which result from central injection of this peptide.[55-57] In neuronal cultures, AngII also has neuromodulatory effects on NE neurons. Incubation of hypothalamus/brainstem or whole brain neuronal cultures with AngII results in receptor-mediated increases in NE synthesis[102] and release.[103] The stimulation of NE synthesis appears to involve activation of dopamine β-hydroxylase (DβH).[102] Further, in the short term (1 to 5 min), there is an increase in neuronal re-uptake of NE, along with an induction of monoamine oxidase (MAO).[52,104] In the longer term (15 to 30 min), there is an inhibition of neuronal NE uptake by AngII.[52] This latter observation parallels the findings of Palaic and Khairallah, who determined that AngII inhibits brain uptake of NE over 30 min.[101] The effects of AngII on NE uptake appear to be a purely neuronal phenomenon, because this peptide does not alter the uptake of NE into astroglia.[5,52]

Considering all of the above, and the findings that the pressor effects of centrally injected AngII are mediated by AT_1 receptors,[54] it is likely that the neuromodulatory actions of AngII on NE in cultured neurons involve this receptor subtype. This appears to be the case. In preliminary experiments we have determined that the AngII-stimulated increase in neuronal [^{3}H]NE uptake is antagonized by losartan, but not by PD123177, suggesting the involvement of AngII receptors in this response (Figure 7). Studies will continue in order to establish the role of AngII receptor subtypes in the neuromodulatory effects of AngII on NE in cultured neurons.

D. INTEGRATED VIEW OF ANGII RECEPTOR REGULATION/CELLULAR FUNCTION IN CULTURED NEURONS

In the preceding section we discussed the properties of AT_1 and AT_2 receptors in neuronal cultures from neonate rat brain. In this section we will attempt, where possible, to fit this information into models for the regulation and cellular functions of AngII receptor subtypes in neurons. As will be seen, some aspects of these models are speculative, but this speculation is based upon the information available.

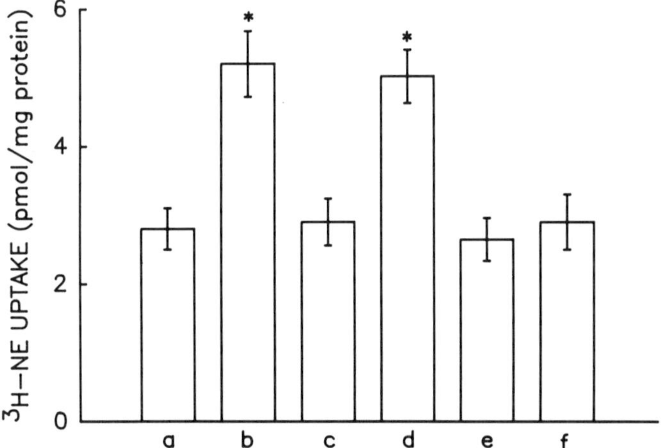

FIGURE 7. Effects of AngII on neuronal [^{3}H]NE uptake in hypothalamus/brainstem neuronal cultures in the presence of nonpeptide AngII receptor blockers. Neuronal cultures were incubated with control solution (a; phosphate-buffered saline, pH 7.4), AngII (b; 10 n*M*), AngII plus 1 μ*M* losartan (c), AngII plus 1 μ*M* PD123177 (d), 1 μ*M* losartan (e), or 1 μ*M* PD123177 (f) for 10 min at 37°C. Following these incubations, media and drugs were removed from all dishes, and neuronal [^{3}H]NE uptake was analyzed as detailed in Reference 52. Data are means ± SEM from 3 experiments. (*) Significantly different from controls ($p < 0.01$), ANOVA1 and Newman-Keuls test.

1. AT_1 Receptors

Neuronal cultures prepared from rat hypothalamus and brainstem contain approximately 30% AT_1 receptors (Figure 2), and a significant population of these sites is associated with neurons. Our data are consistent with the following hypothesis, which is depicted in diagram form in Figure 8. First, AngII binds to its AT_1 receptor on a neuronal cell body and elicits an increase in IP hydrolysis (Figure 5). This results in calcium mobilization from intracellular stores (via IP_3) and also activation of PKC (via diacylglycerol). Either PKC or increased cytosolic free calcium, or both, inhibit I_K and result in an excitatory effect (i.e., depolarization). The depolarization sets up an action potential which is propagated along the axon to the neuron terminal. Depolarization of the terminal causes calcium influx, which results in release of stored NE from synaptic vesicles. The released NE will modulate the activity of other neurons (e.g., via α_1-adrenergic receptors), but will also undergo reuptake back into the neuron (uptake 1) and be metabolized by MAO. All of these processes are induced by AngII in cultured neurons. In the above sequence, the AT_1 receptors are presumed to be present on the neuronal cell body. However, a similar sequence of events (minus action potential propagation) may result from the stimulation of AT_1 receptors located at the neuron terminals (Figure 9). A further AT_1 receptor-mediated action of AngII at the

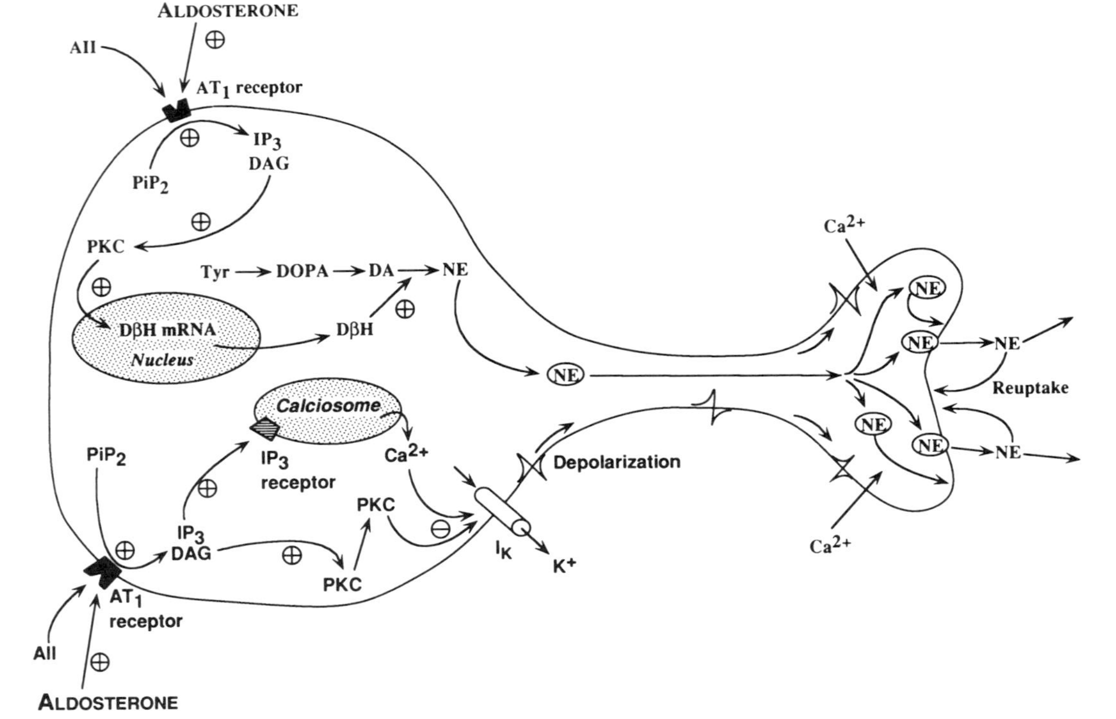

FIGURE 8. Schematic of putative AT_1 receptor-mediated effects in cultured neurons. P_iP_2, polyinositol biphosphate; IP_3, Ins(1,4,5)P_3; PKC, protein kinase C; DAG, diacylglycerol; DβH, dopamine β-hydroxylase; NE, norepinephrine; I_K, delayed rectifier K^+ current.

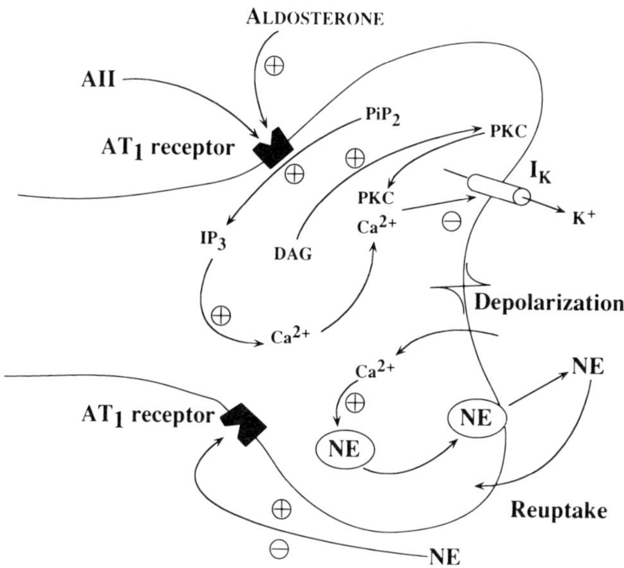

FIGURE 9. Schematic of putative AT_1 receptor-mediated effects in a cultured neuron terminal. Key as in legend to Figure 8.

cell body level could be the PKC-dependent induction of DβH mRNA. This would result in increased levels of DβH molecules, and increased synthesis of NE. The newly synthesized NE would be packaged into vesicles and transported to the neuronal terminals ready for release.

These sequences may be under regulatory control by both mineralocorticoids and NE itself. First, the released NE may act in the short term (via α_{1B}-adrenergic receptors) to upregulate AT_1 receptors at the neuron terminals, and thus enhance the actions of AngII (Figure 9). In the longer term, NE acts (via α_{1A}-adrenergic receptors) to downregulate AT_1 receptors, providing a negative feedback for this sequence of events (Figure 9). It is likely that NE exerts its regulatory effects by mediating the availability of AT_1 receptors at the cell surface. The regulatory effects of mineralocorticoids appear quite distinct from those of NE, at least in terms of mechanism. Aldosterone upregulates AT_1 receptors via a protein synthetic event (perhaps synthesis of new receptors) and also increases AT_1-mediated responses. This is perhaps one way in which the central AngII system can be modulated by a peripheral hormone which is of major importance in controlling ECFV.

As stated previously, these sequences are based on both experimental data and on speculation. Further studies are needed to substantiate (or modify) these ideas, but at least they provide a working model for AT_1 receptor-mediated neuromodulatory effects in cultured neurons.

2. AT_2 Receptors

Neuronal cultures prepared from rat hypothalamus and brainstem contain approximately 70% AT_2 receptors. The relatively high proportion of these receptors (compared with AT_1 sites) is not surprising. This is because the neuronal cultures are prepared from neonatal rat brain, which contains higher levels of AT_2 than AT_1 receptors.[22-25] Our data are consistent with the following hypothesis for the cellular actions mediated by AT_2 receptors in neurons, and these ideas are depicted in Figure 10. AngII acts at AT_2 receptors on cultured neurons and reduces the level of cellular cGMP. This probably involves activation of a calcium-dependent enzyme, perhaps a calcium-dependent phosphodiesterase (type I PDE). The reduction in cGMP results in an increase in I_K and I_A, which have been suppressed by higher levels of cGMP. Increases in I_K and I_A would reduce neuronal activity (i.e., be inhibitory), due to a hyperpolarizing action.

While the data on the effects of AngII on cGMP and on I_K and I_A are solid, the cellular mechanisms proposed above are speculative. In addition, the functional consequences of these cellular effects mediated by AT_2 receptors in neurons are unclear. We have no evidence that these effects are neuromodulatory, at least with respect to NE systems. In fact, the physiological function of AT_2 receptors in neurons (and indeed other tissues) remains unknown. Considering that AT_1 and AT_2 receptors mediate opposite effects of AngII on potassium currents, it may be possible that one of these receptors modulates or inhibits the activity of the other. For example, in some cultured neurons it has been determined that blockade of AT_1-induced decreases in I_{no} with losartan results in previously unseen AngII-stimulated increases in I_{no} (At_2 receptor mediated).[91] This argues for the presence of "interacting" AT_1

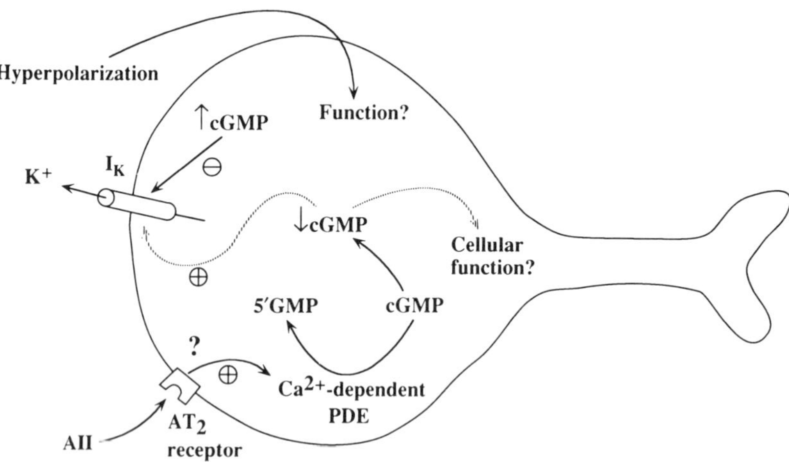

FIGURE 10. Schematic of putative AT_2 receptor-mediated effects in cultured neurons. cGMP, cyclic GMP; PDE, phosphodiesterase; I_K, delayed rectifier K^+ current.

and AT_2 receptors on the same neuron. However, the idea of mutual inhibitory effects between AT_1 and AT_2 sites depends upon the colocalization of these receptors in tissues, and in many cases (especially in brain) there appears to be distinct localization of each receptor subtype.[17-20] In conclusion, while the above may be one explanation for the functional role of AT_2 receptors, this area needs much clarification.

E. ANGIOTENSIN II RECEPTORS IN NEURONAL CULTURES FROM SPONTANEOUSLY HYPERTENSIVE RATS

Spontaneously hypertensive (SH) rats exhibit increased levels of AngII receptors in specific brain regions,[105-107] and also show enhanced pressor responses to centrally injected AngII,[108,109] in comparison with Wistar Kyoto (WKY) normotensive rats. These *in vivo* studies suggest that an overactivity of brain AngII receptors may contribute to the increased blood pressure in these animals.

Studies from our laboratories have used neuronal cultures prepared from SH and WKY rat brain to study these changes in brain AngII receptor densities and function at the cellular level. An advantage of this approach is that any observed changes in AngII receptors in SH neuronal cultures are likely to be genetically predetermined, rather than as a consequence of the sustained raised blood pressure in SH rats. This is because the cultures are prepared from one-day-old rat brain tissue, and in these animals hypertension is certainly not established. Neuronal cultures prepared from SH rat brain exhibit greater numbers (B_{max}) of [^{125}I]AngII-specific binding sites compared with WKY cells,[33] and this is in agreement with the above-mentioned *in vivo* studies.[105-109] Further, this increase in [^{125}I]AngII binding represents functional receptors, because AngII-stimulated [^{3}H]NE uptake is also increased in SH neuronal cultures compared with WKY cells.[5] The increase in the number of functional AngII receptors in SH neuronal cultures is perhaps related to a lack of downregulatory effects of NE on [^{125}I]AngII binding in these cells.[110] However, SH neuronal cultures also exhibit increased numbers of total α_1-adrenergic receptors,[111] and so the lack of downregulatory effect of NE is hard to explain at this level. Since the downregulatory effects of NE on AngII receptors appear to be mediated by α_{1A}-adrenergic receptors (Figure 5), it is possible that this receptor subtype is reduced or nonfunctional in SH neurons. Such a reduction would have been overlooked in previous studies which analyzed total α_1-adrenergic receptors.[111]

All of the above studies were performed prior to the discovery of AngII receptor subtypes. However, considering that AT_1 receptors mediate the pressor effects of centrally injected AngII,[54] it is likely that this receptor subtype will be increased in SH rat brain and in SH rat neuronal cultures. As yet, there are no reports of the relative levels of AT_1 and AT_2 receptors in SH and WKY rat brain. In preliminary experiments we have determined that AT_1 receptor gene expression is increased in SH rat hypothalamus/brainstem neuronal cultures as compared to WKY cells.[112] Further, these SH rat neuronal cultures exhibit a doubling of the number of [^{3}H]-DuP 753-specific binding sites present compared with WKY neuronal cultures, with no change in affinity.[112] These increases in AT_1 receptor mRNA and AT_1 receptor levels also appear to be associated with an increase in AngII-stimulated IP hydrolysis. It is worth noting that astroglial cultures prepared from one-day-old SH and WKY rats exhibit no differences in the level or affinity of [^{3}H]DuP 753 binding.[112] Thus, the increases in AT_1 receptors seen in neuronal cultures are of AT_1 receptors on neurons, and not on the astroglia which are present.

The above data demonstrate changes in neuronal AT_1 receptors which occur prior to the development of sustained hypertension, i.e., genetically induced changes. This suggests that changes in neuronal AT_1 receptors may be involved in the development of hypertension in SH rats. Even though the SH neuronal cultures are prepared from one-day-old rats and reflect prehypertensive changes in neuronal AT_1 receptors, the observed increases in AT_1 receptors in these cells correlate well with the increases seen in adult SH rats.[105-107] The use of neuronal cultures prepared from adult SH rats would confirm whether the observed changes in AT_1 receptors in SH neurons persist into adulthood, and perhaps contribute to the maintenance of hypertension. The finding that AT_1 receptor levels in SH and WKY astroglia from one-day-old rats are similar suggests that these receptors are not involved in the development of hypertension. However, the levels of AT_1 receptors in SH astroglia decrease as a function of the rat's age, compared with WKY astroglia. This suggests that AT_1 receptors on astroglia may be involved in the maintenance of hypertension, and this is reviewed in detail Chapter 18.[113]

From all of the above data we are able to speculate on the changes in AngII receptors and responses in SH rat brain. It appears that the neuromodulatory actions of AngII on NE systems are enhanced in SH rat brain due to an increase in the level of functional AT_1 receptors on neurons. Whether or not this increased AT_1 receptor responsiveness is manifested at multiple levels (e.g., IP hydrolysis, potassium currents, NE synthesis and release; see Figures 8 and 9), as well as the receptor itself remains to be seen. The reasons for this increase in neuronal AT_1 receptors probably include a lack of downregulatory actions by NE (long term). It will also be interesting to determine if the short-term upregulatory effects of NE on AngII receptors are enhanced in SH rat neuronal cultures.

The above discussion has focused mainly on AT_1 receptors, because this subtype is involved in mediating the pressor effects of AngII. Even though the functional role of AT_2 receptors is unknown, it will also be important to assess the expression of this receptor subtype in SH neuronal cultures.

IV. ANGIOTENSIN II RECEPTORS IN NEURAL CELL LINES

In the preceding section we discussed the use of primary brain cell cultures for studying the characteristics and properties of neuronal AngII receptors.

Specific AngII receptors are also present in neural cell lines, and in the current section we will discuss the studies which have been performed on AngII receptors in two types of CNS derived cells, the NG108-15 cells and the N1E-115 cells.

The NG108-15 cell line is a mouse neuroblastoma × glioma hybrid which has been used extensively as an experimental "neuronal" model system.[114] Undifferentiated NG108-15 cells contain specific high-affinity binding sites for [^{125}I]AngII, though at fairly low densities.[115] Incubation of these cells with AngII results in a stimulation of IP hydrolysis,[116,117] and a mobilization of calcium from intracellular stores,[117,118] effects mediated (not surprisingly) by AT_1 receptors.[117-119] These data indicate that undifferentiated NG108-15 cells contain mostly AT_1 receptors. NG108-15 cells can be induced to differentiate by treatment with dibutyl cyclic AMP or by incubation with culture medium containing low levels of fetal bovine serum (0.5%) and 1.5% dimethylsulfoxide. Differentiation results in cells that contain an extensive network of neurites. These differentiated NG108-15 cells also exhibit increased levels of either [^{125}I]AngII- or [^{125}I]Sar^1,Ile^8-AngII-specific binding[116-118] (i.e., increased levels of total AngII receptors). It has been clearly demonstrated that this increase in AngII receptors in differentiated NG108-15 cells is an increased expression of AT_2 receptors.[117,118] A similar expression of AT_2 receptors is induced by differentiation of murine neuroblastoma N1E-115 cells.[120] These data, along with the findings that primary neuronal cultures from one-day-old rats contain primarily AT_2 receptors,[38] may suggest a role of the AT_2 receptor subtype in neuronal differentiation. The expression of high levels of AT_2 receptors in differentiated NG108-15 cells may provide a rich source of this receptor subtype for cloning and biochemical investigations. As a cautionary note, these cells are a neuron/glia hybrid cell line, and their use as a system for studying neuronal AngII receptors is therefore questionable, especially since both neurons and astroglia contain AngII receptors.[5]

N1E-115 cells were originally derived from the mouse C1300 neuroblastoma, and have been used extensively by Fluharty and colleagues[121-125] as a model system for investigating biochemical properties of neuronal AngII receptors. With the use of [^{125}I]-Sar^1,Ile^8-AngII, it has been clearly demonstrated that undifferentiated N1E-115 cells contain a homogenous population of high-affinity AngII receptors.[121] These receptors are regulated by guanine nucleotides and are coupled to a stimulation of IP hydrolysis with subsequent mobilization of intracellular calcium.[122] Differentiation of these cells by exposure to 1.5% dimethylsulfoxide and 0.5% fetal bovine serum resulted in a large increase in the number of AngII receptors present.[123] These studies also determined that following differentiation N1E-115 cells contained at least two populations of high-affinity AngII receptors.[123] One population was identical to that found in the undifferentiated cells, while the other (much larger) population was not G-protein coupled.[123] From these studies it appeared that undifferentiated and differentiated N1E-115 cells contain a population of

AngII receptors with all the characteristics of the AT_1 receptor subtype. Differentiated N1E-115 cells also contained a much larger population of AngII receptors that were not characteristic of the AT_1 subtype and it is now known that these are AT_2 receptors.[120] Biochemical analysis of the AngII receptors present in differentiated N1E-115 cells revealed two distinct [^{125}I]AngII binding sites, with molecular masses of 111 and 68 kDA,[124] which are comparable to the masses of [^{125}I]AngII-binding sites from rat brain membranes.[125] Recent studies have cloned and sequenced the AT_1 receptor,[126,127] and the molecular mass predicted from the cDNA sequence suggests that this subtype is ~40 kDa. Taking into account glycosylation, this resembles the lower molecular weight [^{125}I]-AngII-binding site from N1E-115 cells and rat brain.[124,125] It is thus possible that the higher molecular mass [^{125}I]AngII-binding site represents the AT_2 receptor, and this area awaits further investigation. From a functional point of view, in differentiated N1E-115 cells AngII also stimulates an increase in cellular cGMP levels in addition to the stimulation of IP hydrolysis.[128,129] This effect of AngII on cGMP appears to be mediated by both AT_1 and AT_2 receptors,[129] and is also inhibited by a nitric oxide (NO) synthetase antagonist, suggesting an involvement of NO in this response.

As can be seen from the above discussion, N1E-115 cells have been used to investigate the biochemical properties of AT_1 and AT_2 receptors in neurons, and the overexpression of AT_2 receptors in differentiated N1E-115 cells may help in identifying this elusive subtype, by providing a rich source of receptor.

V. CONCLUSIONS

In this review we have discussed the properties of AngII receptor subtypes on neural cells, both primary neuronal cultures from rat brain and neural cell lines. The use of these cells has allowed for significant progress to be made in our understanding of AT_1 and AT_2 receptors in neurons. There is obviously a long way to go, however, before a complete picture of AngII receptor subtypes in neurons is obtained. In fact, this area will become even more complicated with the recent discovery of two subtypes of AT_1 receptors (AT_{1A} and AT_{1B}),[130,131] two putative subtypes of AT_2 receptors (AT_{2A} and AT_{2B}),[43] and a new AngII receptor subtype which appears to be neither AT_1 nor AT_2[132] (perhaps AT_3?).

ACKNOWLEDGMENTS

The authors are grateful to Dr. William Dunn for the immunocytochemistry, to Dr. Phil Posner for many useful discussions, and to Dr. Wei Tang and Mrs. Tammy Gault for their invaluable contributions. This work was supported by grants from the National Institutes of Health (NS-19441, HL-33610) and from DuPont-Merck Pharmaceutical Company.

REFERENCES

1. **Epstein, A.N., Fitzsimons, J.T., and Rolls, B.J.,** Drinking induced by injection of angiotensin into the brain of the rat, *J. Physiol. (London),* 238, 457, 1970.
2. **Keil, L.C., Summy-Long, J., and Severs, W.B.,** Release of vasopressin by angiotensin II, *Endocrinology,* 96, 1063, 1975.
3. **Severs, W.B., Daniels, A.E., and Buckley, J.P.,** On the central hypertensive effect of angiotensin II, *Int. J. Pharmacol.,* 6, 199, 1967.
4. **Phillips, M.I.,** Functions of angiotensin in the central nervous system, *Annu. Rev. Physiol.,* 49, 413, 1987.
5. **Sumners, C., Myers, L.M., Kalberg, C.J., and Raizada, M.K.,** Physiological and pharmacological comparisons of angiotensin II receptors in neuronal and astrocyte glial cultures, *Prog. Neurobiol.,* 34, 355, 1990.
6. **Bennett, J.P., Jr. and Snyder, S.H.,** Angiotensin II binding to mammalian brain membranes, *J. Biol. Chem.,* 251, 7423, 1976.
7. **Sirrett, N.E., McLean, A.S., Bray, J.J., and Hubbard, J.I.,** Distribution of angiotensin II receptors in rat brain, *Brain Res.,* 122, 299, 1977.
8. **Mendelsohn, F.A.O., Quirion, R., Saavedra, J.M., Aguilera, G., and Catt, K.J.,** Autoradiographic localization of angiotensin II receptors in rat brain, *Proc. Natl. Acad. Sci. U.S.A.,* 81, 1575, 1984.
9. **Gehlert, D., Speth, R.C., and Walmsley, J.K.,** Distribution of [^{125}I]-angiotensin II binding sites in the rat brain: a quantitative autoradiographic study, *Neuroscience,* 18, 837, 1986.
10. **Healy, D.P., Maciejewski, A.R., and Printz, M.P.,** Localization of central angiotensin II receptors with [^{125}I]-Sar^{1}Ile^8-angiotensin II: periventricular sites of the anterior third ventricle, *Neuroendocrinology,* 44, 15, 1986.
11. **Walters, D.E. and Speth, R.C.,** Neuronal localization of specific angiotensin II binding sites in the rat inferior olivary nucleus, *J. Neurochem.,* 50, 812, 1988.
12. **Chiu, A.T., Herblin, W.F., McCall, D.E., Ardecky, R.J., Carini, D.J., Duncia, J.V., Pease, L.J., Wong, P.C., Wexler, R.R., Johnson, A.L., and Timmermans, P.B.M.W.M.,** Identification of angiotensin II receptor subtypes, *Biochem. Biophys. Res. Commun.,* 165, 196, 1989.
13. **Chang, R.S.L., Lotti, V.J., Chen, T.B., and Faust, K.A.,** Two angiotensin II binding sites in rat brain revealed using [^{125}I]-Sar^{1}Ile^8-angiotensin II and selective non-peptide antagonists, *Biochem. Biophys. Res. Commun.,* 171, 813, 1990.
14. **Whitebread, S., Mele, M., Kamber, B., and de Gasparo, M.,** Preliminary biochemical characterization of two angiotensin II receptor subtypes, *Biochem. Biophys. Res. Commun.,* 163, 284, 1989.
15. **Speth, R.C. and Kim, K.H.,** Discrimination of two angiotensin II receptor subtypes with a selective agonist analogue of angiotensin II, p-aminophenylalanine^6 angiotensin II, *Biochem. Biophys. Res. Commun.,* 170, 997, 1990.
16. **Rowe, B.P., Grove, K.L., Saylor, D.L., and Speth, R.C.,** Angiotensin II receptor subtypes in the rat brain, *Eur. J. Pharmacol.,* 186, 339, 1990.
17. **Walmsley, J.K., Herblin, W.F., Alburges, M.E., and Hunt, M.,** Evidence for the presence of angiotensin II type-1 receptors in brain, *Brain Res. Bull.,* 25, 397, 1990.
18. **Gehlert, D.R., Gackenheimer, S.L., and Schober, D.T.,** Autoradiographic localization of subtypes of angiotensin II antagonist binding in rat brain, *Neuroscience,* 44, 501, 1991.
19. **Tsutsumi, K. and Saavedra, J.M.,** Quantitative autoradiography reveals different angiotensin II receptor subtypes in selected rat brain nuclei, *J. Neurochem.,* 56, 348, 1991.
20. **Song, K., Allen, A.M., Paxinos, G., and Mendelsohn, F.A.O.,** Mapping of angiotensin II receptor subtype heterogeneity in rat brain, *J. Comp. Neurol.,* 316, 467, 1992.

21. **Saavedra, J.M., Tsutsumi, K., Strömberg, C., Seltzer, A., Michels, K., Zorad, S., and Viswanathan, S.,** Localization, characterization, development and function of brain angiotensin II receptor subtypes, Chapter 13, this volume.
22. **Tsutsumi, K. and Saavedra, J.M.,** Characterization and development of angiotensin II receptor subtypes (AT_1 and AT_2) in rat brain, *Am. J. Physiol.*, 261, R209, 1991.
23. **Millan, M.A., Jacobowitz, D.M., Aguilera, G., and Catt, K.J.,** Differential distribution of AT_1 and AT_2 angiotensin II receptor subtypes in the rat brain during development, *Proc. Natl. Acad. Sci. U.S.A.*, 88, 11440, 1991.
24. **Tsutsumi, K., Vishwanathan, M., Stromberg, C., and Saavedra, J.M.,** Type-1 and type-2 angiotensin II receptors in fetal rat brain, *Eur. J. Pharmacol.*, 198, 89, 1991.
25. **Cook, V.I., Grove, K.L., McMenamin, K.M., Carter, M.R., Harding, J.W., and Speth, R.C.,** The AT_2 angiotensin receptor subtype predominates in the 18 day gestation fetal rat brain, *Brain Res.*, 560, 334, 1991.
26. **Moore, R.Y. and Bloom, F.E.,** Central catecholamine neuron systems: anatomy and physiology of the norepinephrine and epinephrine systems, *Annu. Rev. Neurosci.*, 2, 113, 1979.
27. **Swanson, L.W., Sawchenko, P.E., Bèrod, A., Hartman, B.K., Helle, K.B., and Vanorden, D.E.,** An immunohistochemical study of the organization of catecholaminergic cells and terminal fields in the paraventricular and supraoptic nuclei of the hypothalamus, *J. Comp. Neurol.*, 196, 271, 1981.
28. **Lind, R.W., Swanson, L.W., and Ganten, D.,** Organization of angiotensin II immunoreactive cells and fibers in the rat central nervous system, *Neuroendocrinology*, 40, 2, 1985.
29. **Prochiantz, A., DiPorzio, U., Kato, A., Berger, B., and Glowinski, J.,** In vitro maturation of mesencephalic dopaminergic neurons is enhanced in the presence of their striatal target cells, *Proc. Natl. Acad. Sci. U.S.A.*, 76, 5387, 1979.
30. **Richards, E.M., Sumners, C., Chou, Y.-C., Raizada, M.K., and Phillips, M.I.,** α_2-Adrenergic receptors in neuronal and glial cultures: characterization and comparison, *J. Neurochem.*, 53, 287, 1989.
31. **Chou, Y.-C., Luttge, W.G., and Sumners, C.,** Characterization of glucocorticoid type II receptors in neuronal and glial cultures from rat brain, *J. Neuroendocrinol.*, 2, 29, 1990.
32. **Sumners, C. and Raizada, M.K.,** Catecholamine-angiotensin II receptor interaction in primary cultures of rat brain, *Am. J. Physiol.*, 246 (Cell Physiol. 15), C502, 1984.
33. **Raizada, M.K., Muther, T.F., and Sumners, C.,** Increased angiotensin II receptors in neuronal cultures from hypertensive rat brain, *Am. J. Physiol.*, 247 (Cell Physiol. 16), C364, 1984.
34. **Sumners, C., Watkins, L.L., and Raizada, M.K.,** α_1-Adrenergic receptor-mediated downregulation of angiotensin II receptors in neuronal cultures, *J. Neurochem.*, 47, 1117, 1986.
35. **Sumners, C., Rueth, S.M., Crews, F.T., and Raizada, M.K.,** Protein kinase C agonists increase the expression of angiotensin II receptors in neuronal cultures, *J. Neurochem.*, 48, 1954, 1987.
36. **Sumners, C. and Fregly, M.J.,** Modulation of angiotensin II binding sites in neuronal cultures by mineralocorticoids, *Am. J. Physiol.*, 256 (Cell Physiol. 25), C121, 1989.
37. **Myers, L.M. and Sumners, C.,** Regulation of angiotensin II binding sites in neuronal cultures by catecholamines, *Am. J. Physiol.*, 257 (Cell Physiol. 26), C706, 1989.
38. **Sumners, C., Tang, W., Zelezna, B., and Raizada, M.K.,** Angiotensin II receptor subtypes are coupled with distinct signal transduction mechanisms in neurons and astrocytes from rat brain, *Proc. Natl. Acad. Sci. U.S.A.*, 88, 7567, 1991.
39. **Chiu, A.T., McCall, D.E., Aldrich, P.E., and Timmermans, P.B.M.W.M.,** [^{3}H]-DuP753, a highly potent and specific radioligand for the angiotensin II-1 receptor subtype, *Biochem. Biophys. Res. Commun.*, 172, 1195, 1990.

40. **Dudley, D.T., Panek, R.L., Major, T.C., Lu, G.H., Bruns, R.F., Klinkefus, B.A., Hodges, J.C., and Weishaar, R.E.,** Subclasses of angiotensin II binding sites and their functional significance, *Mol. Pharmacol.,* 38, 370, 1990.
41. **Strömberg, C., Tsutsumi, K., Viswanathan, M., and Saavedra, J.M.,** Angiotensin II AT_1 receptors in rat superior cervical ganglia: characterization and stimulation of inositol phospholipid hydrolysis, *Eur. J. Pharmacol.,* 208, 331, 1991.
42. **Bottari, S.P., Taylor, V., King, I.N., Bogdal, Y., Whitebread, S., and de Gasparo, M.,** Angiotensin II AT_2 receptors do not interact with guanine nucleotide binding proteins, *Eur. J. Pharmacol. Mol. Pharmacol.,* 207, 157, 1991.
43. **Tsutsumi, K. and Saavedra, J.M.,** Heterogeneity of angiotensin II AT_2 receptors in the brain, *Mol. Pharmacol.,* 41, 290, 1992.
44. **Stamler, J.F., Landas, S., Raizada, M.K., and Phillips, M.I.,** Neurons in brain cell culture specifically bind fluorescein-labelled angiotensin II, *Neuropeptides,* 1, 421, 1981.
45. **Simonnet, G., Legendre, P., Laribi, C., Allard, M., and Vincent, J.D.,** Location of angiotensin II binding sites on neuronal and glial cells of cultured mouse spinal cord: an autoradiographic study, *Brain Res.,* 443, 403, 1988.
46. **Hösli, E. and Hösli, L.,** Autoradiographic localization of binding sites for vasoactive intestinal peptide and angiotensin II on neurons and astrocytes of cultured central nervous system, *Neuroscience,* 31, 463, 1989.
47. **Wilson, K.M., Sumners, C., Hathaway, S., and Fregly, M.J.,** Mineralocorticoids modulate central angiotensin II receptors in rats, *Brain Res.,* 382, 87, 1986.
48. **Gutkind, J.S., Kurihara, M., and Saavedra, J.M.,** Increased angiotensin II receptors in brain nuclei of DOCA-salt hypertensive rats, *Am. J. Physiol.,* 255, H646, 1988.
49. **Wilson, S.K., Lynch, D.R., and Ladenson, P.W.,** Angiotensin II and atrial natriuretic factor binding sites in various tissues in hypertension: comparative receptor localization and changes in different hypertension models in the rat, *Endocrinology,* 124, 2799, 1989.
50. **Jonklaas, J. and Buggy, J.,** Angiotensin-estrogen interaction: localization and mechanisms, *Brain Res.,* 326, 239, 1985.
51. **Fregly, M.J., Rowland, N.E., Sumners, C., and Gordon, D.B.,** Reduced dipsogenic responsiveness to intracerebroventricularly administered angiotensin II in estrogen-treated rats, *Brain Res.,* 338, 115, 1985.
52. **Sumners, C. and Raizada, M.K.,** Angiotensin II stimulates norepinephrine uptake in hypothalamus-brainstem neuronal cultures, *Am. J. Physiol.,* 250 (Cell Physiol. 19), C236, 1986.
53. **Koepke, J.P., Bovy, P.R., McMahon, E.G., Olins, G.M., Reitz, D.B., Salles, K.S., Schuh, J.R., Trapani, A.J., and Blaine, E.D.,** Central and peripheral actions of a non-peptide angiotensin II receptor antagonist, *Hypertension,* 15, 841, 1990.
54. **Hogarty, D.C., Speakman, E.A., Puig, V., and Phillips, M.I.,** The role of AT_1 and AT_2 receptors in the pressor, drinking and vasopressin responses to central angiotensin II, *Brain Res.,* 586, 389, 1992.
55. **Fitzsimons, J.T. and Setler, P.E.,** The relative importance of catecholaminergic and cholinergic mechanisms in drinking in response to angiotensin and other thirst stimuli, *J. Physiol. (London),* 250, 613, 1975.
56. **Gordon, F.J., Brody, M.J., Fink, G.D., Buggy, J., and Johnson, A.K.,** Role of central catecholamines in the control of blood pressure and drinking behavior, *Brain Res.,* 178, 161, 1979.
57. **Sumners, C. and Phillips, M.I.,** Central injection of angiotensin II alters catecholamine activity in rat brain, *Am. J. Physiol.,* 244 (Reg. Int. Comp. Physiol. 13), R257, 1983.
58. **Chevillard, C., Duchene, N., Pasquier, R., and Alexandre, J.M.,** Relation of the centrally evoked pressor effect of angiotensin II to central noradrenaline in the rabbit, *Eur. J. Pharmacol.,* 58, 203, 1979.
59. **Meldrum, M.J., Xue, C.S., Budine, L., and Westfall, T.C.,** Angiotensin facilitation of noradrenergic neurotransmission in central tissues of rats: effects of sodium restriction, *J. Cardiovasc. Pharmacol.,* 6, 989, 1984.

60. **Stadler, T., Veltmar, A., Quadri, F., and Unger, T.,** Angiotensin II evokes noradrenaline release from the paraventricular nucleus in conscious rats, *Brain Res.*, 569, 117, 1992.
61. **Fuxe, K., Ganten, D., Anderson, K., Calza, L., Agnati, L.F., Lang, R.E., Poulsen, K., Hökfelt, T., and Bernard, P.,** Immunocytochemical demonstration of angiotensin II and renin-like immunoreactive nerve cells in the hypothalamus: angiotensin peptides as comodulators in vasopressin and oxytocin neurons and their regulation of various types of catecholamine nerve terminals, in *The Renin-Angiotensin System in the Brain,* Ganten, D., Printz, M.P., Phillips, M.I., and Scholkens, B.A., Eds., Springer-Verlag, New York, 1982, 208.
62. **Rowe, B.P., Kalivas, P.W., and Speth, R.C.,** Autoradiographic localization of angiotensin II receptor binding sites on noradrenergic neurons of the locus coeruleus of the rat, *J. Neurochem.*, 55, 533, 1990.
63. **Walters, D.E. and Speth, R.C.,** Monoamine depletion does not alter angiotensin II binding sites in the rat brain, *Brain Res. Bull.*, 22, 283, 1989.
64. **Kalberg, C.J. and Sumners, C.,** Regulation of angiotensin II binding sites in neuronal cultures by protein kinase C, *Am. J. Physiol.*, 258 (Cell Physiol. 27), C610, 1990.
65. **Minneman, K.P., Han, C., and Abel, P.W.,** Comparison of α_1-adrenergic receptor subtypes distinguished by chloroethylclonidine and WB101, *Mol. Pharmacol.*, 33, 509, 1988.
66. **Gross, G., Hanff, G., and Rugevics, C.,** 5-Methylurapidil discriminates between subtypes of the α_1-adrenoreceptor, *Eur. J. Pharmacol.*, 151, 333, 1988.
67. **Sumners, C., Rueth, S.M., Myers, L.M., Kalberg, C.J., Crews, F.T., and Raizada, M.K.,** Phorbol ester-induced upregulation of angiotensin II receptors in neuronal cultures is potentiated by a calcium ionophore, *J. Neurochem.*, 51, 153, 1988.
68. **Garcia-Sainz, J.A. and Macias-Silva, M.,** Angiotensin II stimulates phosphoinositide turnover and phosphorylase through AII-1 receptors in isolated rat hepatocytes, *Biochem. Biophys. Res. Commun.*, 172, 780, 1990.
69. **Pfeilschifter, J.,** Angiotensin II B-type receptor mediates phosphoinositide hydrolysis in mesangial cells, *Eur. J. Pharmacol.*, 184, 201, 1990.
70. **Bauer, P.H., Chiu, A.T., and Garrison, J.C.,** DuP753 can antagonize the effects of angiotensin II in rat liver, *Mol. Pharmacol.*, 39, 579, 1991.
71. **Chiu, A.T., McCall, D.E., Price, W.A., Wong, P.C., Carini, D.J., Duncia, J.V., Wexler, R.R., Yoo, S.E., Johnson, A.L., and Timmermans, P.B.M.W.M.,** Nonpeptide angiotensin II receptor antagonists. VII. Cellular and Biochemical Pharmacology of DuP753, an orally active antihypertensive agent, *J. Pharmacol. Exp. Ther.*, 252, 711, 1990.
72. **Balla, T., Baukal, A.J., Eng, S., and Catt, K.J.,** Angiotensin II receptor subtypes and biological responses in the adrenal cortex and medulla, *Mol. Pharmacol.*, 40, 401, 1991.
73. **Raizada, M.K., Phillips, M.I., Crews, F.T., and Sumners, C.,** Distinct angiotensin II receptor in primary cultures of glial cells from rat brain, *Proc. Natl. Acad. Sci. U.S.A.*, 84, 4655, 1987.
74. **Rydzewski, B., Zelezna, B., Tang, W., Sumners, C., and Raizada, M.K.,** Angiotensin II stimulation of plasminogen activator inhibitor-1 gene expression in astroglial cells from the brain, *Endocrinology,* 130, 1255, 1992.
75. **Dudley, D.T., Hubbell, S.E., and Summerfelt, R.M.,** Characterization of angiotensin II (AT_2) binding sites in R3T3 cells, *Mol. Pharmacol.*, 40, 360, 1991.
76. **Pucell, A.G., Hodges, J.C., Sen, I., Bumpus, F.M., and Husain, A.,** Biochemical properties of the ovarian granulosa cell type 2-angiotensin II receptor, *Endocrinology,* 128, 1947, 1991.
77. **Sumners, C. and Myers, L.M.,** Angiotensin II decreases cGMP levels in neuronal cultures from rat brain, *Am. J. Physiol.*, 260 (Cell Physiol. 29), C79, 1991.

78. **Sumners, C. and Tang, W.,** Atrial natriuretic peptide receptor subtypes in rat neuronal and astrocyte glial cultures, *Am. J. Physiol.*, 262 (Cell Physiol. 31), C1134, 1992.
79. **Bottari, S.P., King, I.N., Reichlin, S., Dahlstrom, I., Lydon, N., and de Gasparo, M.,** The angiotensin AT_2 receptor stimulates protein tyrosine phosphatase activity and mediates inhibition of particulate guanylate cyclase, *Biochem. Biophys. Res. Commun.*, 183, 206, 1992.
80. **Nicoll, R.A. and Barker, J.L.,** Excitation of supraoptic neurosecretory cells by angiotensin II, *Nature New Biol.*, 233, 172, 1971.
81. **Felix, D. and Schlegel, W.,** Angiotensin-receptive neurons in the subfornical organ: structure-activity relations, *Brain Res.*, 149, 107, 1978.
82. **Felix, D. and Schelling, P.,** Increased sensitivity of neurons to angiotensin II in SHR as compared to WKY rats, *Brain Res.*, 252, 63, 1982.
83. **Suga, T., Suzuki, M., and Suzuki, M.,** Effects of angiotensin II on the medullary neurons and their sensitivity to acetylcholine and catecholamines, *Jpn. J. Pharmacol.*, 29, 541, 1979.
84. **Harding, J.W. and Felix, D.,** Angiotensin-sensitive neurons in the rat paraventricular nucleus: relative potencies of angiotensin II and angiotensin III, *Brain Res.*, 410, 130, 1987.
85. **Knowles, W.D. and Phillips, M.I.,** Angiotensin II-responsive cells in the organum vasculosum lamina terminalis (OVLT) recorded in hypothalamic brain slices, *Brain Res.*, 197, 256, 1980.
86. **Palovcik, R.A. and Phillips, M.I.,** Saralasin increases activity of hippocampal neurons inhibited by angiotensin II, *Brain Res.*, 323, 345, 1984.
87. **Barnes, K.L., Knowles, W.D., and Ferrario, C.M.,** Angiotensin II and angiotensin (1-7) excite neurons in the canine medulla in vitro, *Brain Res. Bull.*, 24, 275, 1990.
88. **Legendre, P., Simmonet, G., and Vincent, J.-D.,** Electrophysiological effects of angiotensin II on cultured mouse spinal cord neurons, *Brain Res.*, 297, 287, 1984.
89. **Ambühl, P., Felix, D., Imboden, H., Khosla, M.C., and Ferrario, C.M.,** Effects of angiotensin analogues and angiotensin receptor antagonists on paraventricular neurons, *Reg. Peptides,* 38, 111, 1992.
90. **Ambühl, P., Felix, D., Imboden, H., Khosla, M.C., and Ferrario, C.M.,** Effects of angiotensin II and its selective analogues on inferior olivary neurons, *Regul. Peptides,* 41, 19, 1992.
91. **Kang, J., Sumners, C., and Posner, P.,** Modulation of net outward current in cultured neurons by angiotensin II: involvement of AT_1 and AT_2 receptors, *Brain Res.*, 580, 317, 1992.
92. **Kang, J., Posner, P., and Sumners, C.,** Angiotensin II type-1 (AT_1) and angiotensin type-2 (AT_2) receptor-mediated changes in potassium currents in cultured neurons: role of intracellular calcium, *FASEB J.*, 6 (Abstr. 443), 1992.
93. **Kang, J., Sumners, C., and Posner, P.,** Angiotensin II (AII) type-2 (AT_2) receptor-mediated effects on potassium currents in cultured neurons, *Soc. Neurosci. Abstr.*, 18 (Abstr. 488.1), 1992.
94. **Zufall, F., Stengl, M., Franke, C., Hildebrand, J., and Hatt, H.,** Ionic currents of cultured olfactory receptor neurons from antennae of male manduca sexta, *J. Neurosci.*, 11, 956, 1991.
95. **O'Grady, S.M., Cooper, K.E., and Rae, J.L.,** Cyclic GMP regulation of a voltage-activated K channel in dissociated enterocytes, *J. Membr. Biol.*, 124, 159, 1991.
96. **Dutar, P. and Nicoll, R.A.,** Stimulation of phosphatidylinositol (PI) turnover may mediate the muscarinic suppression of the M-current in hippocampal pyramidal cells, *Neurosci. Lett.*, 85, 89, 1988.
97. **Brown, D.A.,** G-proteins and potassium currents in neurons, *Annu. Rev. Physiol.*, 52, 215, 1990.

98. **Kirkwood, A., Simmons, M.A., Mather, R.J., and Lisman, J.,** Muscarinic suppression of the M-current is mediated by a rise in internal Ca^{2+} concentration, *Neuron,* 6, 1009, 1991.
99. **Constanti, A. and Brown, D.A.,** M-currents in voltage-clamped mammalian sympathetic neurones, *Neurosci. Lett.,* 24, 289, 1981.
100. **Qadri, F., Badoer, E., Stadler, T., and Unger, T.,** Angiotensin II-induced noradrenaline release from anterior hypothalamus in conscious rats: a brain microdialysis study, *Brain Res.,* 563, 137, 1991.
101. **Palaic, D. and Khairallah, P.,** Effect of angiotensin on uptake and release of norepinephrine by brain, *Biochem. Pharmacol.,* 5, 2291, 1967.
102. **MacLean, M.R., Raizada, M.K., and Sumners, C.,** The influence of angiotensin II on catecholamine synthesis in neuronal cultures from rat brain, *Biochem. Biophys. Res. Commun.,* 167, 492, 1990.
103. **Sumners, C., Phillips, M.I., and Raizada, M.K.,** Angiotensin II stimulates changes in the norepinephrine content of primary cultures of rat brain, *Neurosci. Lett.,* 36, 305, 1983.
104. **Sumners, C., Shalit, S.L., Kalberg, C.J., and Raizada, M.K.,** Norepinephrine metabolism in neuronal cultures is increased by angiotensin II, *Am. J. Physiol.,* 252 (Cell Physiol. 21), C650, 1987.
105. **Stamler, J.F., Raizada, M.K., Fellows, R.E., and Phillips, M.I.,** Increased specific binding of angiotensin II in the organum vasculosum of the lamina terminalis area of the spontaneously hypertensive rat brain, *Neurosci. Lett.,* 17, 173, 1980.
106. **Ashida, T., Ohuchi, Y., Saito, T., and Yazaki, Y.,** Effects of dietary sodium on brain angiotensin II receptors in spontaneously hypertensive rats, *Jpn. Circ. J.,* 46, 1328, 1982.
107. **Saavedra, J.M., Correa, F.M., Plunkett, L.M., Israel, A., Kurihara, M., and Shigematsu, K.,** Binding of angiotensin and atrial natriuretic peptide in the brain of hypertensive rats, *Nature (London),* 320, 758, 1986.
108. **Johnson, A.K., Simon, W., Schaz, K., Ganten, U., Ganten, D., and Mann, J.F.E.,** Increased blood pressure responses to central angiotensin II in spontaneously hypertensive rats, *Klin. Wochenschr.,* 56, 47, 1978.
109. **Hoffman, W.E., Phillips, M.I., and Schmid, P.,** Central angiotensin II-induced responses in spontaneously hypertensive rats, *Am. J. Physiol.,* 232 (Heart Circ. Physiol. 1), H426, 1977.
110. **Raizada, M.K. and Sumners, C.,** Lack of $alpha_1$-adrenergic receptor-mediated down-regulation of angiotensin II receptors in neuronal cultures from spontaneously hypertensive rats, *Mol. Cell. Biochem.,* 91, 111, 1990.
111. **Feldstein, J.B., Pacitti, A.J., Sumners, C., and Raizada, M.K.,** $Alpha_1$-adrenergic receptors in neuronal cultures from rat brain: increased expression in the spontaneously hypertensive rats, *J. Neurochem.,* 47, 1190, 1986.
112. **Tang, W., Lu, D., Raizada, M.K., and Sumners, C.,** Increased AT_1 receptor gene expression in neuronal cultures from spontaneously hypertensive (SH) rats, *Hypertension,* in press.
113. **Rydzewski, B., Wozniak, M., Sumners, C., and Raizada, M.K.,** Plasminogen activator system and its interaction with angiotensin II in the brain, Chapter 18, this volume.
114. **Nelson, P.G., Christian, C., and Nirenberg, M.,** Synapse formation between clonal neuroblastoma × glioma cells and striated muscle cells, *Proc. Natl. Acad. Sci. U.S.A.,* 73, 123, 1976.
115. **Weyhenmeyer, J.A. and Hwang, C.-J.,** Characterization of angiotensin II binding sites on neuroblastoma × glioma hybrid cells, *Brain Res. Bull.,* 14, 409, 1985.
116. **Carrithers, M.D., Raman, V.K., Masuda, S., and Weyhenmeyer, J.A.,** Effect of angiotensin I and II on inositol polyphosphate production in differentiated NG108-15 hybrid cells, *Biochem. Biophys. Res. Commun.,* 167, 1200, 1990.

117. **Tallant, E.A., Diz, D.I., Khosla, M.C., and Ferrario, C.M.,** Identification and regulation of angiotensin II receptor subtypes on NG108-15 cells, *Hypertension,* 17, 1135, 1991.
118. **Bryson, S.E., Warburton, P., Wintersgill, H.P., Drew, G.M., Michel, A.D., Ball, S.G., and Balmforth, A.J.,** Induction of the angiotensin AT_2 receptor subtype expression by differentiation of the neuroblastoma × glioma hybrid, NG108-15, *Eur. J. Pharmacol. Mol. Pharmacol.,* 225, 119, 1992.
119. **Ransom, J.T., Sharif, N.A., Dunne, J.F., Momiyama, M., and Melching, G.,** AT_1 angiotensin receptors mobilize intracellular calcium in a subclone of NG108-15 neuroblastoma cells, *J. Neurochem.,* 58, 1883, 1992.
120. **Ades, A.M., Slogoff, F., and Fluharty, S.J.,** Differential regulation of type-1 and type-2 angiotensin II receptors in murine neuroblastoma N1E-115 cells, *Soc. Neurosci. Abstr.,* 17 (Abstr. 321.8), 1991.
121. **Fluharty, S.J. and Reagan, L.P.,** Characterization of binding sites for the angiotensin II antagonist ^{125}I-[Sar^1,Ile^8]-angiotensin II on murine neuroblastoma N1E-115 cells, *J. Neurochem.,* 52, 1393, 1989.
122. **Monck, J.R., Williamson, R.E., Rogulja, I., Fluharty, S.J., and Williamson, J.R.,** Angiotensin II effects on cytosolic free Ca^{2+} concentration in N1E-115 neuroblastoma cells: kinetic properties of the Ca^{2+}-transient measured in single fura-2 loaded cells, *J. Neurochem.,* 54, 278, 1990.
123. **Reagan, L.P., Ye, X., Mir, R., DePalo, L.R., and Fluharty, S.J.,** Up-regulation of angiotensin II receptors by in vitro differentiation of murine N1E-115 neuroblastoma cells, *Mol. Pharmacol.,* 38, 878, 1991.
124. **Siemens, I.R., Adler, H.J., Addya, K., Mah, S.J., and Fluharty, S.J.,** Biochemical analysis of solubilized angiotensin II receptors from murine neuroblastoma N1E-115 cells by covalent crosslinking and affinity purification, *Mol. Pharmacol.,* 40, 717, 1991.
125. **Siemens, I.R., Swanson, G.N., Fluharty, S.J., and Harding J.W.,** Solubilization and partial characterization of angiotensin II receptors from rat brain, *J. Neurochem.,* 57, 690, 1991.
126. **Murphy, T.J., Alexander, R.W., Griendling, K.K., Runge, M.S., and Bernstein, K.E.,** Isolation of a cDNA encoding a vascular type-1 angiotensin II receptor, *Nature (London),* 351, 233, 1991.
127. **Sasaki, K., Yamano, Y., Bardham, S., Iwai, N., Murray, J.J., Hasegawa, M., Matsuda, Y., and Inagami, T.,** Cloning and expression of a complementary DNA encoding a bovine adrenal angiotensin II type-1 receptor, *Nature (London),* 351, 230, 1991.
128. **Gilbert, J.A., Pfenning, M.A., and Richelson, E.,** The effect of angiotensins I, II, and III on formation of cyclic GMP in murine neuroblastoma clone N1E-115, *Biochem. Pharmacol.,* 33, 2527, 1984.
129. **Zarahn, E.D., Ye, X., Ades, A.M., Reagan, L.P., and Fluharty, S.J.,** Angiotensin-induced cGMP production is mediated by multiple receptor subtypes and nitric oxide in N1E-115 neuroblastoma cells, *J. Neurochem.,* 58, 1960, 1992.
130. **Iwai, N. and Inagami, T.,** Identification of two subtypes of the rat type-1 angiotensin II receptor, *FEBS Lett.,* 298, 257, 1992.
131. **Kakar, S.S., Sellers, J.C., Devor, D.C., Musgrove, L.C., and Neill, J.D.,** Angiotensin II type-1 receptor subtype cDNAs: differential tissue expression and hormonal regulation, *Biochem. Biophys. Res. Commun.,* 183, 1090, 1992.
132. **Chaki, S. and Inagami, T.,** Identification and characterization of a new binding site for angiotensin II in mouse neuroblastoma neuro-2A cells, *Biochem. Biophys. Res. Commun.,* 182, 388, 1992.
133. **Sumners, C.,** unpublished data.
134. **Myers, L.M. and Sumners, C.,** unpublished data.
135. **Steele, M.K. et al.,** personal communication.
136. **Sumners, C. et al.,** unpublished data.

Chapter 16

DEVELOPMENTAL EXPRESSION OF ANGIOTENSIN II RECEPTORS: DISTRIBUTION, CHARACTERIZATION, AND COUPLING MECHANISMS

G. Aguilera, M.C. Johnson, P. Feuillan, and M. Millan

TABLE OF CONTENTS

0-8493-4622-3/93/$0.00 + $.50

I. INTRODUCTION

Evidence is accumulating to indicate that the role of the renin-angiotensin system extends beyond its effects in circulatory homeostasis. Components of the system have been found in a number of tissues unrelated to the control of blood pressure and mineralocorticoid secretion, including the liver, pituitary gland, reproductive system, and the placental-fetal unit.[1-8] A number of studies have suggested that angiotensin II (AngII) has a role in growth and development. The possibility that AngII may affect cell replication was first proposed by Kharairallah et al.,[9] with the demonstration of increased DNA content in atria, ventricles, and adrenal cortex and medulla of rats following an intravenous infusion of AngII. More recently, AngII has been shown to have stimulatory[10-12] and inhibitory effects in cell proliferation, depending upon the conditions under which the cells are exposed to the peptide.[13] AngII has been shown to stimulate the expression of platelet-derived growth factor and growth-related oncogenes, as well as to potentiate the mitogenic effect of epidermal growth factor in cultured tubular kidney cells.[14-21]

AngII exerts its physiological effects in the target tissues through interaction with plasma membrane receptors.[2] Two distinct AngII receptors (AT_1 and AT_2) have been identified according to their sensitivity to dithiothreitol (DTT) and their ability to bind to nonpeptide[22] and peptide[23] analogs. AT_1 receptors bind to DuP 753, are negatively affected by DTT, and mediate most of the traditional effects of AngII, whereas AT_2 receptors bind to the nonpeptide analog PD123177 and to the peptide analog CGP42112A; their binding is increased by DTT and their function is not well defined.[23-25] Renewed interest in the potential effects of AngII in development has arisen from the discovery that there is transient expression of abundant AngII receptors at novel sites in late-gestation rodent and primate fetus.[6,26-28] AngII receptors are also present at critical times during postnatal development of organs such as the brain.[29]

An important but elusive problem has been the exact physiological role of AngII during development. Since the receptor is the primary site of action of AngII, defining the properties and distribution of AngII binding sites during ontogeny will be essential for understanding the functions of the peptide in growth and maturation. This chapter will review the current knowledge on the timing of expression, topographic distribution, receptor subtypes, and mechanism of coupling of AngII receptors during fetal and postnatal development.

II. ANGIOTENSIN II RECEPTORS IN THE FETUS

A. TEMPORAL EXPRESSION AND TOPOGRAPHIC DISTRIBUTION

Studies in membranes from whole rat conceptuses have shown detectable AngII binding at day 10 of gestation. The binding progressively increases up to day 18 and remains stable until birth.[6,26] Autoradiographic studies of the binding of radioiodinated AngII to slide-mounted sections have made it possible to define the topographic distribution of AngII binding in the fetus.[6,27-31] A sagittal autoradiogram from a 19-d-old fetus is shown in Figure 1A. Organs are fully formed at this stage and AngII receptors are readily detected in known target tissues, including the adrenal glomerulosa and medulla, kidney, liver, and smooth muscle of the bronchi, blood vessels, and gastrointestinal tract. Particularly striking, however, is the intense binding in areas not normally expected to contain AngII receptors, such as the subepidermal layer of the skin, mesenchymal and connective tissues, and skeletal muscle, especially the tongue. Little or no binding is detectable in cartilage, bone, heart, and fat. In the sections analyzed, binding is also absent in the brain and spinal cord. The binding is specific, since all autoradiographic staining is abolished in the presence of 1 μM AngII (Figure 1B), but unchanged by excess amounts of corticotropin releasing hormone, arginine vasopressin, and adrenocorticotropic hormone (ACTH) (not shown).

Binding studies in the presence of saturating amounts of AngII receptor subtype-specific analogs have shown that the unique fetal AngII receptors are AT_2.[30-32] As shown in Figure 1C, blockade of AT_1 sites with DuP 753 abolished the binding in liver and lung, whereas there is no significant change in the intensity of the binding in the skin and skeletal muscle. In contrast, in the presence of the AT_2 antagonist PD123177 (Figure 1D), binding in skin, skeletal muscle, mesenchymal tissue, and adrenal medulla, but not in the liver, is reduced. In the adrenal capsule, vascular smooth muscle, and parts of the kidney, binding is partially displaced by both antagonists, indicating the presence of similar proportions of both receptor subtypes in these tissues in the fetus. This is in contrast to the adult, in which the majority of AngII receptors are AT_1.[22] In the presence of 10 μM of both DuP 753 and PD123177, binding is decreased to near-nonspecific levels in all areas (not shown). Addition of DTT during incubation with the radioligand decreases the intensity of the binding in the liver and lung and enhances binding in skin, muscle, and adrenal medulla.

Autoradiographic analysis of the binding in the fetus at earlier gestational ages indicates similar distribution of the binding from day 15 of gestation, the age at which organogenesis is completed (Figure 2). Consistent with studies in membranes, binding density in the autoradiography was similar from day 18 until birth (day 21). The presence of AngII receptors with binding properties and a distribution pattern identical to those described in the rat has also been shown in fetal mouse,[27] and in primates, including the human.[28]

A condition for a binding site to be of physiological importance is the availability of the ligand. All components of the renin-angiotensin system are present in the fetal-placental unit.[33,34] Studies in several species, including humans, have shown the presence of renin and of angiotensin-converting enzyme in the fetus from the second third of gestation.[8,27,35-37] Angiotensinogen is present in the rat fetus as early as day 9.[38,39] Another possible source

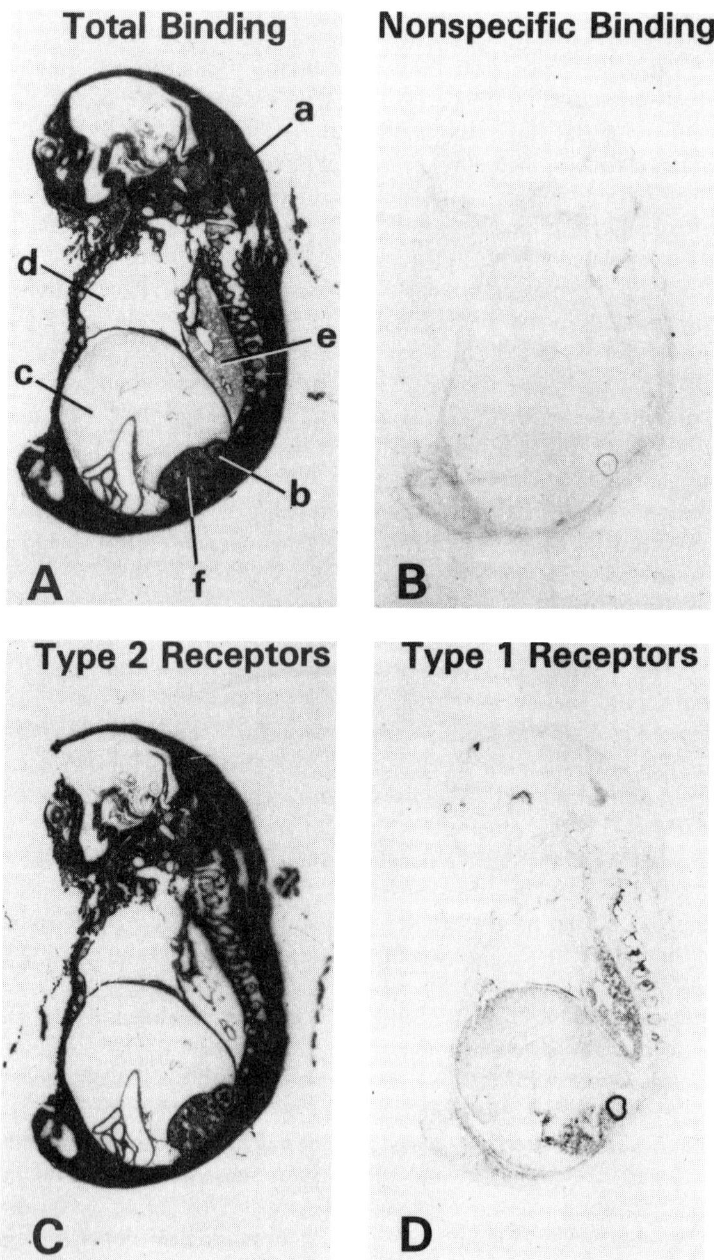

FIGURE 1. Autoradiographic analysis of ^{125}I[Sar^1,Ile^8]AngII binding to 20-μm frozen sections of an 18-d gestation rat fetus. (A) Total binding measured in the absence of AngII or antagonists: a, skin; b, adrenal; c, liver; d, heart; e, lung; f, kidney. (B) Nonspecific, measured in the presence of 1 μ*M* AngII. (C) AT_2 sites, measured in the presence of the AT_1 antagonist, DuP 753 (10 μ*M*). (D) AT_2 sites, measured in the presence of the AT_2 analog, PD123177 (10 μ*M*)

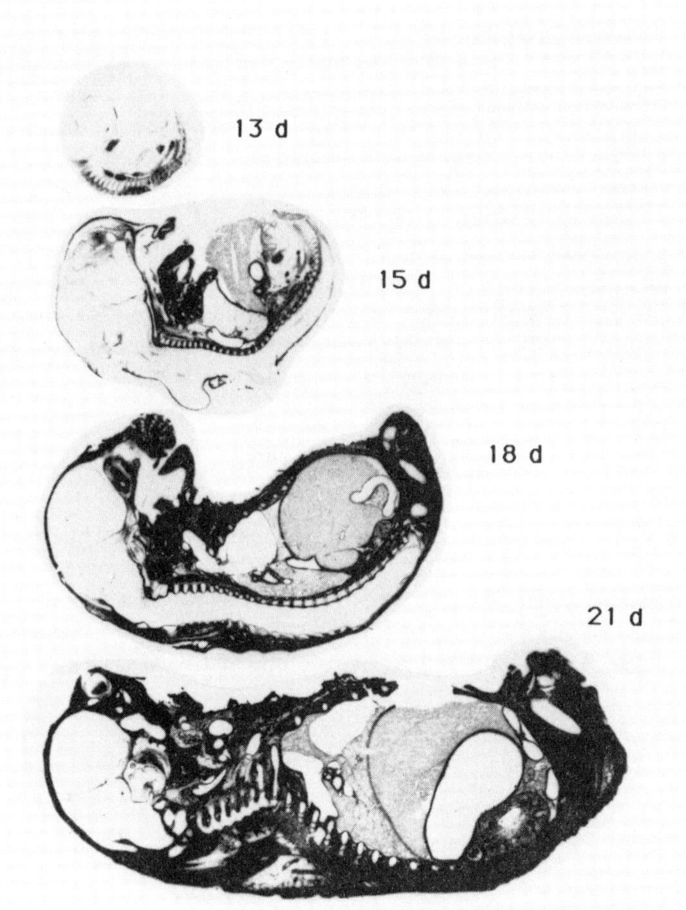

FIGURE 2. Autoradiographic analysis of AngII binding in 20-μm slide-mounted sections from rat fetuses of gestational age 13, 15, 18, and 21 d.

of AngII production for the fetus is the placenta, which is known to produce renin and other components of the renin-angiotensin system.[33] Consistent with the existence of an active renin-angiotensin system in the fetus is the presence of bioactive AngII in acid extracts of eviscerated rat and mouse fetuses.[26,27]

B. BINDING PROPERTIES OF THE FETAL ANGII RECEPTOR

The properties of the fetal AngII receptor have been studied in membrane-rich fractions from whole eviscerated fetuses,[26] and tissues shown by autoradiography to contain predominantly AT_1 or AT_2.[32] Regardless of the membrane preparation, tissue or species, binding is invariably associated with a single class of high-affinity sites, with an affinity in the nanomolar range.[6,26-28,32] The binding affinity, as well as the binding specificity, are similar to those described for traditional AngII target tissues in the adult[2] (Figures 3 and 4). Studies of AngII binding in fetal skin or liver membranes in the presence of saturating concentrations of the heterologous analog have also revealed similar binding affinity and specificity for both receptor subtypes.[32] Consistent with the autoradiographic studies, Scatchard analysis of the binding data in fetal skin shows predominantly AT_2, while fetal liver membranes contain mostly AT_1 (Table 1). Addition of the sulfydryl reducing agent DTT to skin membranes increases AngII binding, due to a significant increase in affinity of the AT_2 sites, without change in receptor number. In contrast, in liver membranes, DTT decreases binding due to a reduction in receptor concentration of the AT_1 sites (Table 1).

Early studies have shown that the unique AngII receptors in skeletal muscle and skin are markedly different from the adult type of receptor with respect to their sensitivity to guanyl nucleotides.[28] Thus, in contrast to the adult receptor in the adrenal, liver, or smooth muscle, in which binding affinity is markedly decreased by GTP or its analogs, AngII binding to fetal skin or muscle is unaffected. More recent studies have shown that the differences in guanyl nucleotide sensitivity correlate with the proportion of receptor subtypes in the tissue. As shown in Figure 5A, addition of 10 μM GTP-τ-S at binding

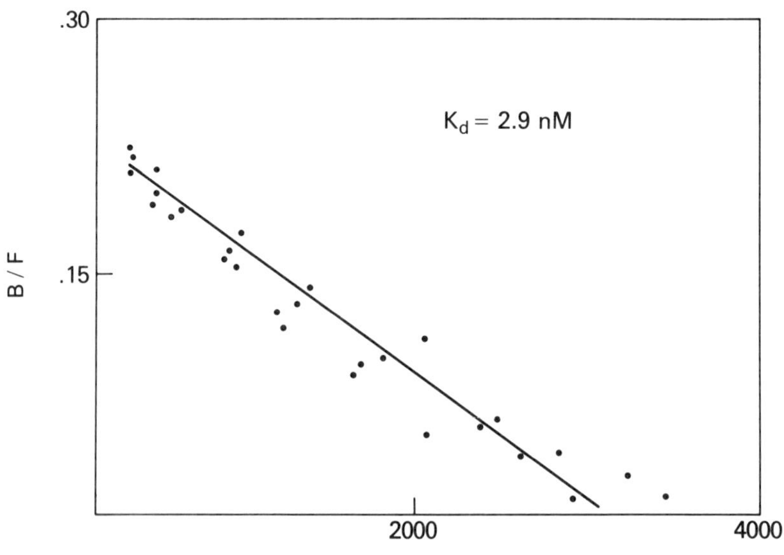

FIGURE 3. Scatchard analysis of the binding of $^{125}I[Sar^1,Ile^8]$AngII to membranes from eviscerated fetal mice at 16 to 18 d of gestation. B/F, bound-to-free ratio.

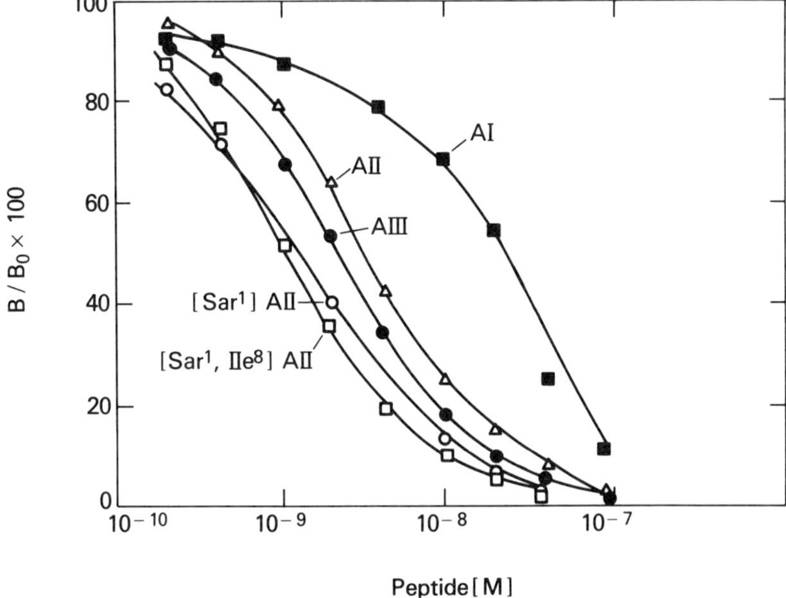

FIGURE 4. Ligand specificity of AngII binding sites in membranes from fetal mice at 16 to 18 d of gestation, as shown by inhibition of ^{125}I[Sar^1,Ile^8]AngII binding by AngII analogs. B/Bo, ratio bound in the presence of unlabeled peptide to bound in the absence of unlabeled peptide.

equilibrium has no effect on the binding of radioiodinated AngII to fetal skin membranes. Addition of AngII caused 35% dissociation after one hour, and this effect was not enhanced by the GTP analog. In the presence of complete blockade of AT_1, GTP-τ-S also failed to cause dissociation (Figure 5B). However, when AT_2 receptors were blocked with PD123177, GTP-τ-S caused marked dissociation of the binding, indicating that the residual AT_1 receptors are sensitive to guanyl nucleotides (Figure 5C). In contrast, in fetal liver membranes, where AT_1 receptors are predominant, addition of GTP-τ-S results in marked inhibition of binding and enhances the dissociation caused by AngII. Similar selective sensitivity of AT_1, but not AT_2, sites for guanyl nucleotides has been described in membranes from ovarian granulosa cells, uterus, adult liver, and PC-12 cells, and by autoradiography in fetal skin and muscle.[25,30,40,41]

C. RECEPTOR COUPLING TO SIGNAL TRANSDUCTION SYSTEMS

The mechanisms of coupling of the fetal AngII receptor to signaling transduction systems has been investigated in fetal skin slices and cultures.[6,30,42] Enzyme-dispersed fetal skin cultures on the first and second passage appear at light microscopy examination as a homogeneous cell population

TABLE 1
Number and Affinity (K_d) of Fetal Angiotensin II Binding

	Without DTT		With 2 m*M* DTT	
	Receptors (pmol/mg)	Affinity (K_d, n*M*)	Receptors (pmol/mg)	Affinity (K_d, n*M*)
Skin				
Total	8.20 ± 3.9	3.01 ± 1.9	4.79 ± 2.2	1.13 ± 0.34[a]
AT_2	7.9 ± 3.4	2.75 ± 1.4	4.64 ± 2.1	1.11 ± 0.24[a]
AT_1	0.86 ± 0.6[b]	6.21 ± 2.8	—	—
Liver				
Total	2.26 ± 0.6	7.0 ± 4.5	0.78 ± 0.43[c]	7.2 ± 4.4
Type 2	0.07 ± 0.1[b]	2.0 ± 1.4	0.04 ± 0.03	1.9 ± 1.3
Type 1	2.15 ± 0.6	7.7 ± 4.2	0.69 ± 0.28[c]	8.5 ± 4.8

Note: Binding sites determined by Scatchard analysis of binding inhibition curves. Type 2 (AT_2) receptors were studied in the presence of 10 μ*M* DuP 753, a type 1 receptor antagonist, and type 1 (AT_1) receptors were studied in the presence of 10 μ*M* PD123177, a type 2 receptor antagonist. Radioligand binding to type 1 sites in skin was too low for accurate estimation of receptor number of K_d. Values are the mean ± SD from 6 experiments.

[a] $p > 0.05$, compared to K_d without DTT.
[b] $p > 0.01$, compared to total number of receptors.
[c] $p > 0.01$, compared to receptor number without DTT.

with fibroblast-like characteristics. These cells contain AngII receptors with affinity and specificity similar to those in membrane preparations and a receptor concentration of about 500 fmol/10^6 cells, similar to that in adrenal glomerulosa cells.[6,42]

In other systems, the mechanism of action of AngII involves phospholipid breakdown and calcium mobilization.[43] Cultured fetal skin fibroblasts and slices have been used to investigate whether occupancy of fetal AngII receptors results in phospholipid breakdown and calcium mobilization.[6,30,42,44] Endogenous phosphoinositide pools were labeled by preincubation of the cells with [^{3}H]inositol prior to incubation with AngII. As shown in other systems, treatment of the cells with AngII results in rapid increases in inositol phosphates, as determined by high-performance liquid chromatography (HPLC) or after Dowex column separation. In the presence of the AT_1 antagonist, DuP 753, the stimulatory effect of AngII on inositol phosphate accumulation is almost abolished, whereas the AT_2 antagonist, PD123177, has no effect (Figure 6), indicating that this effect of AngII is mediated by AT_1. Studies in isolated hepatocytes have also shown that the stimulatory effect of AngII on inositol phosphate formation is mediated by AT_1[45]

Studies in cultured fetal skin fibroblasts preloaded with the fluorescent indicator Fura-2 show that nanomolar concentrations of AngII cause rapid increases in cytosolic calcium, an effect which is abolished by exposure of

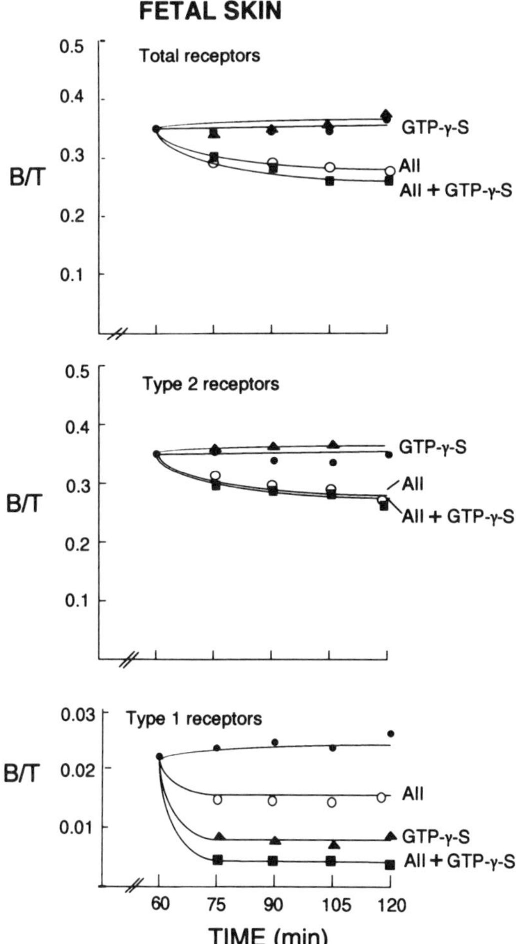

FIGURE 5. Dissociation of ^{125}I-AngII from membrane-rich fractions of fetal skin. AngII (1 μM), GTP-τ-S (10 μM), or their combination were added after 60 min preincubation and the reaction terminated at the time indicated. Total receptors were studied in the absence of analogs, AT_2 in the presence of 10 μM of the AT_1 antagonist DuP 753, and AT_1 in the presence of 10 μM of the AT_2 analog PD123177. B/T: bound cpm to total cpm ratio.

the cells to the AngII antagonist [Sar^1,Ile^8]AngII, 5 min prior to and during incubation with AngII[6] (Figure 7). There is no information on the effect of receptor subtype-specific analogs in this system. However, since the increase in cytosolic calcium is secondary to inositol phosphate formation, it is likely that the effect is also mediated by AT_1. This has been shown to be the case in other systems, such as vascular smooth muscle, where the AngII-induced increase in cytosolic calcium is blocked by the AT_1 antagonist, DuP 753.[46]

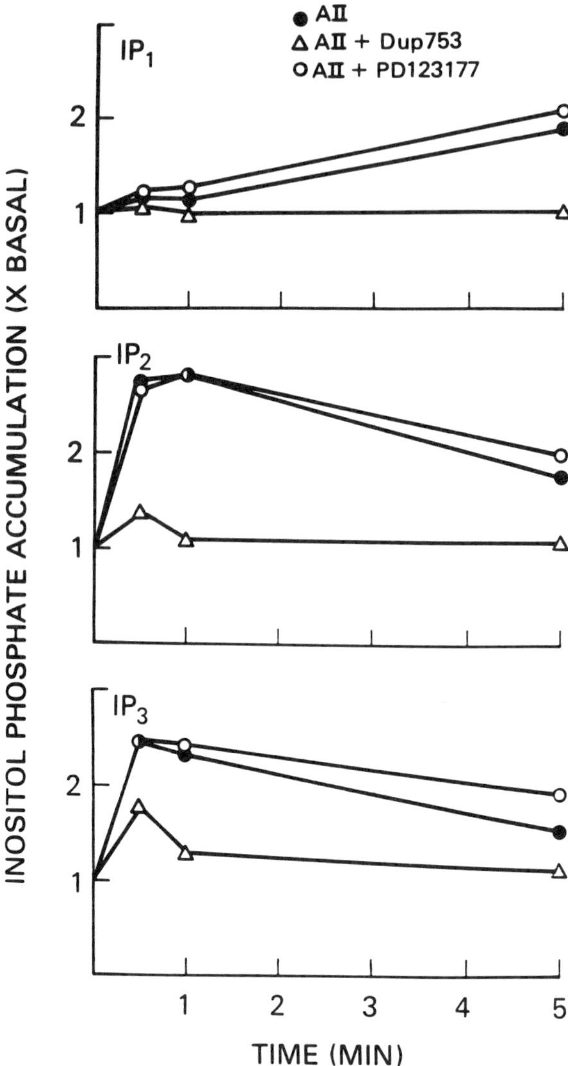

FIGURE 6. Effect of AngII receptor subtype analogs on AngII-stimulated inositol monophosphate (IP_1), Inositol bisphosphate (IP_2), and inositol *tris*phosphate (IP_3). Cells were incubated in the absence or in the presence of 10 μM antagonists for 5 min prior to and during exposure to AngII. Incubations are terminated by addition of 200 μl 0.1 *N* perchloric acid and inositol phosphates separated using Dowex columns.

In contrast to other systems, exposure of fetal fibroblasts to AngII results in small but consistent stimulation of basal and isoproterenol- and cholera toxin-stimulated cAMP accumulation.[42,44] This effect was observed in the presence of phosphodiesterase inhibitors, suggesting that it was due to stim-

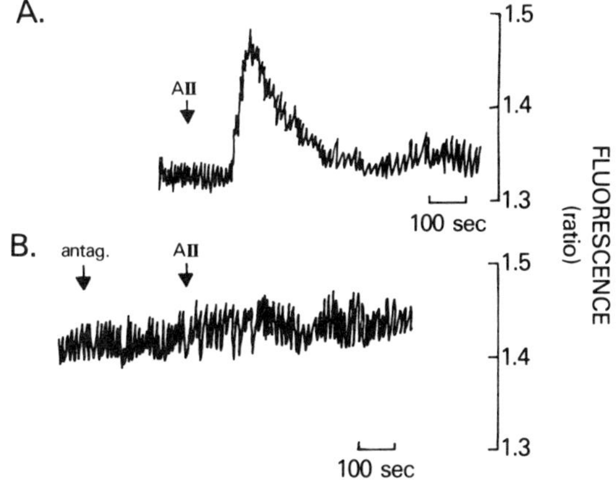

FIGURE 7. Changes in cytosolic calcium in fetal skin fibroblasts perfused with AngII in the absence (A) and in the presence (B) of 1 μM [Sar^1,Ala^8]AngII commenced 250 s before addition of AngII. Intracellular calcium was measured with the fluorescent indicator Fura-2.

ulation of adenylate cyclase rather than to inhibition of degradation of the cyclic nucleotide. This stimulatory effect of AngII on cAMP production differs from the effects of the peptide elsewhere. In most systems, including liver, anterior pituitary, proximal tubular cell of the kidney, myocardial sarcolema, and rat adrenal glomerulosa cells, AngII inhibits cAMP production by interacting with the inhibitory guanyl nucleotide binding protein, G_i.[47-50] Recent studies have show that AngII has no effect on adenylate cyclase activity, measured as conversion of [α-^{32}P]ATP to cAMP in cell homogenates.[44] This indicates that the effect of AngII on adenylate cyclase is indirect, probably through phosphorylation of a component of the system. In the pituitary corticotroph, there is evidence to suggest that vasopressin and phorbol esters potentiate the stimulatory effect of corticotropin releasing hormone on cAMP production by protein kinase C-dependent phosphorylation of G_i.[51] However, in the fetal fibroblast, pretreatment with pertussis toxin does not prevent the stimulatory effect of AngII on cAMP accumulation, indicating that inactivation of G_i is not likely to be involved.[44] In the distal tubule of the nephron, AngII stimulates cAMP production, an effect mediated through prostaglandin formation.[52] Such a mechanism does not appear to be involved in fetal fibroblasts, since the stimulatory effect of AngII on cAMP production is not blocked by the cyclooxygenase inhibitor, indomethacin.[44] The precise mechanism by which AngII stimulates adenylate cyclase in fetal fibroblasts is under current investigation. Similar to the effect on inositol phosphates, the stimulatory effect of AngII on cAMP production is mediated by AT_1, as shown by the ability of DuP 753 and not PD123177 to inhibit the effect.[42]

D. RECEPTOR SUBTYPES IN CULTURED FETAL FIBROBLASTS

The finding that the effects of AngII on inositol phosphate and cAMP production in fetal fibroblasts are mediated by AT_1 is not consistent with the demonstration by autoradiography or conventional binding techniques in membranes that AngII receptors in fetal skin are predominantly AT_2. A recent study shows that this apparent discrepancy is the result of dramatic changes in the AngII receptor subtype present in the cells during culture.[42] Thus, consistent with reports in membranes and autoradiography, in freshly isolated cells, the majority of the receptors are AT_2. However, culture and subculture of the cells result in an increasing proportion of AT_1. It is not clear whether the change in receptor subtype is due to selective growth of different cell types containing distinct receptors or if the receptor subtype changes in the same cell. During primary culture, the reversal in AngII receptor subtype is due to parallel increases in AT_1 and decreases in AT_2. Although changes in receptor type could be attributed to a decrease in epithelial cells and an increase in fibroblasts, autoradiographic studies have shown that the epidermis does not contain AngII receptors.[6] Consequently, it is likely that AngII receptors are only in the fibroblasts. Supporting this view is the finding that in secondary cultures, where the cell population is homogeneous,[53] the relative proportion of AT_1 also increases during culture.[42] Studies using probes based on the structure of the recently cloned AT_1[54,55] show no hybridization with RNA from tissues containing predominantly AT_2, suggesting that the two AngII receptor subtypes are the products of different genes.[71,72] Since, due to the effect of trypsin, binding is almost absent immediately following passage, the increase in number of AT_1 and AT_2 observed in secondary cultures indicates that both receptor subtypes are newly expressed during culture.[42] The increase in receptors during culture is inhibited by cycloheximide, indicating that the expression of both receptor subtypes requires protein synthesis.

The changes in AngII receptor subtype during culture are prevented by treatment of the cells with actinomycin D, but this effect is due to an increase in the number of AT_2 rather than a decrease in the synthesis of AT_1 (Table 2). The increase in the absolute number of AT_2 following complete blockade of transcription suggests that receptor synthesis is mainly controlled at the posttranscriptional level, probably through regulation of factors involved in the degradation of mRNA. Such a mechanism has been previously described for other proteins, including apolipoprotein II, tubulin, and the α-adrenergic and transferrin receptors.[56-59] Alternatively, actinomycin D may block the synthesis of a protein involved in regulating the translation of mRNA, as has been shown for ferritin.[60,61] An effect at the level of receptor degradation is less likely, because the marked decrease in binding caused by cycloheximide suggests that in the absence of protein synthesis the receptor protein has a short half-life. Actinomycin D did not decrease AT_1, indicating that, though not as prominent as for AT_2, posttranscriptional mechanisms are also involved in the regulation of this receptor subtype.

TABLE 2
Effect of Actinomycin D or Cycloheximide Treatment on the Changes in AngII Receptor Subtype During Culture in Fetal Skin Fibroblasts

Experimental Conditions	Angiotensin II receptors (fmol/10^5 cells)								
	Day 2			Day 4			Day 6		
	Total	Type 1	Type 2	Total	Type 1	Type 2	Total	Type 1	Type 2
Control	60 ± 7	13 ± 2	44 ± 6	43 ± 2	22 ± 5	30 ± 3	54 ± 2	31 ± 1	24 ± 1
Actinomycin D, 25 μg/ml	56 ± 5	12 ± 3	36 ± 5	89 ± 5*	35 ± 4	72 ± 7*	84 ± 9*	31 ± 2	57 ± 7*
Cycloheximide, 20 μg/ml	15 ± 4*	3 ± 1*	9 ± 3*	32 ± 1	15 ± 3	18 ± 2*	31 ± 3*	16 ± 1*	17 ± 1

Note: After removal of nonviable cells, cycloheximide treatment resulted in a decrease in the number of both receptor subtypes. Actinomycin D treatment had no significant effect on the concentration of either receptor subtype at day 2 of culture, but at days 4 and 6, when the major changes in receptors occur, the number of type 1 sites was not significantly affected and the absolute number of type 2 receptors was markedly increased.

III. POSTNATAL CHANGES IN ANGII RECEPTOR EXPRESSION

Binding studies in membranes or by autoradiography have shown dramatic reduction in AngII binding in the skin and skeletal muscle as early as one day after birth.[6] A similar temporal pattern of expression has been described for insulin-like growth factor (IGF) II, which is present during the last third of gestation and markedly declines immediately after birth.[62,63] Quantitative studies in skin and tongue membranes have shown that AngII binding decreases to 20, 10, and 5% after 1, 7, and 14 d after birth, respectively.[6] Binding is virtually undetectable by day 21. The dramatic decrease in receptor concentration in skin and skeletal muscle was in marked contrast to the persistent high receptor levels observed in smooth muscle, adrenal, kidney, and pituitary, known target tissues in the adult.[2] For example, using urinary bladder membranes as a model of smooth muscle, the number of binding sites is 2- to 4-fold that of adult levels, 7 d after birth. In contrast to the adult in which 95% of the binding sites are AT_1, AT_2 constitutes about 50% of the total at day 7, falling to 35, 20, and 15% by days 14, 21, and 31, respectively.[73]

Another site in which AngII receptor content and distribution changes dramatically during development is the brain.[29,64,65] In the adult brain, AngII receptors are located at discrete sites, mainly in the circumventricular organs, some hypothalamic nuclei, and areas related to autonomic activity.[66] However, studies in the rat have shown that AngII binding is markedly higher at early ages of postnatal development.[29,64,65] The developmental expression of AngII binding in the brain appears to follow three different patterns. One is illustrated by the circumventricular organs, paraventricular nucleus, and some components of the noradrenergic and parasympathetic system, in which binding is present at birth and remains at relatively constant levels throughout development. A second pattern is represented by a small number of structures, such as the striatum, central nucleus of the amygdala, some thalamic and cerebellar nuclei, olivary and locus coeruleus, in which prominent binding at birth becomes relatively low or absent after two weeks. A third pattern is shown by a number of structures, of which the most prominent is the ventral nucleus of the thalamus, in which intense binding is transiently expressed around two weeks of age and becomes low or undetectable at later ages.

The development of the central nervous system is not completed until the third week after birth, and it is noteworthy that in several areas the highest receptor concentration is observed at the time of maximum synaptogenic activity.[67] For example, it is remarkable that the concentration of AngII receptors in components of the noradrenergic pathway, such as the A1, A2, and A6 regions and the ventral bundle, is extremely high during the first two weeks of life, a critical time in the maturation of the system. In the hippocampus where in general neurogenesis occurs into adulthood, AngII receptors are present up to 4 weeks.[68] Also in the rat, after birth, ascending neurons from the inferior olive reach the cerebellum to establish one-to-one connections with Purkinje cells.[69,70] On the other hand, the formation of dendritic prolongations from Purkinje cells giving origin to the molecular layer and synapses to the cerebellar nuclei peaks by day 14 and is completed at day 21.[69] The parallelism between the development of the olivary cerebellar pathway and the expression of AngII receptors in the components of the pathway is striking.[29] With respect to AngII receptor subtype, as occurs with the fetal receptors, in general in the brain, AngII receptors in circumventricular organs and other areas in which receptors persist through adulthood are AT_1, while the receptors transiently expressed during development are AT_2.[64,65] The transient expression of AngII receptors at critical times in the maturation of certain pathways suggests that, in addition to the actions of AngII in circulatory homeostasis, the peptide is involved in neuronal development.

IV. CONCLUSIONS

The recent discovery of abundant specific binding sites for AngII at novel sites in the fetus and neonate, and the finding that all components of the renin-angiotensin system are present in the developing organism suggest that AngII not only plays a role in maintaining homeostasis, but may also be instrumental for somatic maturation.

The newly available nonpeptide, receptor subtype-specific AngII analogs have revealed that the predominant receptors which express transiently in fetal and neonatal tissues are AT_2. Consistent with findings in other systems, AT_2 in the fetus are not coupled to the traditional signaling transduction systems. In addition to AT_2, a small proportion of AT_1 are also present at the unique fetal sites and, as in adult tissues, they are linked to phospholipid turnover and calcium mobilization. It has been shown that AT_1 mediate AngII-induced growth responses in cultured vascular smooth muscle and mesangial cells, but the extent to which they have a similar role in the fetus remains to be elucidated. Although the mechanism of coupling and functional significance of AT_2 are still under investigation, their abundance and transient expression at critical times in maturation suggest that they have an important role in developmental processes.

REFERENCES

1. **Garrison, J.C., Borland, M.K., Florio, V.A., and Twible, D.A.,** The role of calcium ion as a mediator of the effects of angiotensin II, catecholamines and vasopressin on the phosphorylation and activity of enzymes in isolated hepatocytes, *J. Biol. Chem.*, 25, 7147, 1979.
2. **Catt, K.J., Mendelsohn, F.A.O., Millan, M.A., and Aguilera, G.,** The role of angiotensin II receptors in vascular regulation, *J. Cardiol. Pharmacol.*, 6, S575, 1984.

3. **Deschepper, C.F., Mellon, S.H., Cumin, F., Baxter, J.D., and Ganong, W.F.,** Analysis by immunocytochemistry and in situ hybridization of renin and its mRNA in kidney, testes, adrenal and pituitary of the rat, *Proc. Natl. Acad. Sci. U.S.A.*, 83, 7552, 1986.
4. **Husain, A., Bumpus, M.F., DeSilva, P., and Speth, R.C.,** Localization of angiotensin II receptors in ovarian follicles and the identification of angiotensin II in rat ovaries, *Proc. Natl. Acad. Sci. U.S.A.*, 84, 2489, 1987.
5. **Aguilera, G., Millan, M.A., and Harwood, J.P.,** Angiotensin II receptors in the gonads, *Am. J. Hypertens.*, 2, 395, 1989.
6. **Millan, M.A., Carvallo, P., Izumi, S.-I., Catt, K.J., and Aguilera, G.,** Novel sites of expression of functional angiotensin II receptors in the late gestation fetus, *Science*, 244, 1340, 1989.
7. **Lightman, A., Tarlatzis, B.C., Rzasa, P.J., Culler, M.D., Caride, V.J., Negro-Vilar, A.F., Lennard, D., DeCherney, A.H., and Naftolin, F.,** The ovarian renin-angiotensin system: renin like activity and angiotensin II/III immunoreactivity in gonadotropin stimulated human follicular fluid, *Am. J. Obstet. Gynecol.*, 156, 806, 1987.
8. **Taylor, G.M., Peart, S.W., Porter, K.A., Kondek, L.H., and Zondek, T.,** Concentration and molecular forms of active and inactive renin in human fetal kidney, amniotic fluid and adrenal gland: evidence for renin angiotensin system hyperactivity in 2nd trimester of pregnancy, *J. Hypertens.*, 4, 121, 1986.
9. **Khairallah, P.A., Robertson, A.L., and Davila, D.,** Effects of angiotensin II on DNA, RNA and protein synthesis, in *Hypertension*, Genest, J. and Koiw, E., Eds., Springer-Verlag, Berlin, 1972, 212.
10. **Geisterfer, A.A.T., Peach, M.J., and Owens, G.K.,** Angiotensin II induces hypertrophy, not hyperplasia, of cultured rat aortic smooth muscle cells, *Circ. Res.*, 62, 749, 1988.
11. **Berk, B.C., Vekshtein, V., Gordon, H.M., and Tsuda, T.,** Angiotensin II-stimulated protein synthesis in cultured smooth muscle cells, *Hypertension*, 13, 305, 1989.
12. **Powell, J.S., Clozel, J.-P., Muller, R.K.M., Kuhn, H., Hefti, F., Honsag, M., and Baumgartner, H.R.,** Inhibitors of converting enzyme prevent myointimal proliferation after vascular injury, *Science*, 245, 186, 1989.
13. **Dzau, V.J., Gibbons, G.H., and Pratt, R.E.,** Molecular mechanisms of vascular renin-angiotensin system in myointimal hyperplasia, *Hypertension*, 18, II-100, 1991.
14. **Wolf, G. and Neilson, E.G.,** Angiotensin II induces cellular hypertrophy in cultured murine proximal tubular cells, *Am. J. Physiol.*, 259, F768, 1990.
15. **Taubman, M.B., Berk, B.C., Izumo, S., Tsuda, T., Alexander, R.W., and Nadal-Ginard, R.,** Angiotensin II induces c-fos mRNA in aortic smooth muscle, *J. Biol. Chem.*, 264, 526, 1989.
16. **Bobik, A., Grinpukel, S., Grooms, A., and Jackman, G.,** Angiotensin II and noradrenaline increase PDGF-BB receptors and potentiate PDGF-BB stimulated DNA synthesis in vascular smooth muscle, *Biochem. Biophys. Res. Commun.*, 166, 580, 1990.
17. **Naftilan, A.J., Pratt, R.E., and Dzau, V.J.,** Induction of platelet derived growth factor A-chain and c-myc gene expression by angiotensin II in cultured rat smooth muscle cells, *J. Clin. Invest.*, 83, 1419, 1989.
18. **Kawahara, Y., Sunako, M., Tsuda, T., Fukuzaki, H., Fukomoto, Y., and Takai, Y.,** Angiotensin II induces expression of the c-fos gene through protein kinase C activation and calcium ion mobilization, *Biochem. Biophys. Res. Commun.*, 150, 52, 1988.
19. **Naftilan, A.J., Gilliland, G.K., Eldridge, C.S., and Kraft, A.S.,** Induction of the protooncogene c-jun by angiotensin II, *Mol. Cell. Biol.*, 10, 5536, 1990.
20. **Viard, I., Hall, S.H., Jaillard, C., Berthelon, M.C., and Saez, J.M.,** Regulation of c-fos, c-jun and jun-B messenger ribonucleic acids by angiotensin II and corticotropin in ovine and bovine adrenocortical cells, *Endocrinology*, 130, 1193, 1992.
21. **Norman, J., Badie-Dezfoly, B., Nord, E.P., Kurtz., I., Schlosser, A., Chaudhari, A., and Fine, L.G.,** EGF-induced mitogenesis in proximal tubular cells: potentiation by angiotensin II, *Am. J. Physiol.*, 253, F299, 1987.

22. **Chiu, A.T., McCall, D.E., Nguyen, T.T., Carini, D.J., Duncia, J.V., Herblin, W.F., Uyeda, R.T., Wong, P.C., Wexler, R.R., Johnson, A.L., and Timmermans, P.B.M.W.M.,** Discrimination of angiotensin II receptor subtypes by dithiothreitol, *Eur. J. Pharmacol.,* 170, 117, 1989.
23. **Whitebread, S., Mele, M., Kamber, B., and de Gasparo M.,** Preliminary biochemical characterization of two angiotensin II receptor subtypes, *Biochem. Biophys. Res. Commun.,* 163, 284, 1989.
24. **Chiu, A.T., Herblin, W.F., McCall, D.E., Ardecky, R.J., Carini, D.J., Duncia, J.V., Pease, L.J., Wong, P.C., Wexler, R.R., Johnson, A.L., and Timmermans, P.B.M.W.M.,** Identification of angiotensin II receptor subtypes, *Biochem. Biophys. Res. Commun.,* 165, 196, 1989.
25. **Dudley, D.T., Panek, R.L., Major, T.C., Lu, G.H., Bruns, R.F., Klinkenfus, B.A., Hodges, J.C., and Weishaar, R.E.,** Subclasses of angiotensin II binding sites and their functional significance, *Mol. Pharmacol.,* 38, 370, 1990.
26. **Jones, C., Millan, M.A., Naftolin, F., and Aguilera, G.,** Characterization of angiotensin II receptors in the rat fetus, *Peptides,* 10, 459, 1989.
27. **Zemel, S., Millan, M.A., and Aguilera, G.,** Distribution of angiotensin II receptors and renin in the mouse fetus, *Endocrinology,* 124, 1774, 1989.
28. **Zemel, S., Millan, M.A., Feuillan, P., and Aguilera, G.,** Characterization and distribution of angiotensin II receptors in the primate fetus, *J. Clin. Endocrinol. Metab.,* 71, 1003, 1990.
29. **Millan, M.A., Kiss, A., and Aguilera, G.,** Developmental changes in brain angiotensin II receptors in the rat, *Peptides,* 12, 723, 1991.
30. **Tsusumi, K., Stromberg, C., Viswanathan, M., and Saavedra, J.M.,** Angiotensin II receptor subtypes in fetal tissues of the rat: autoradiography, guanyl nucleotide sensitivity, and association with phosphoinositide hydrolysis, *Endocrinology,* 129, 1075, 1991.
31. **Grady, E.F., Sechi, L.A., Griffin C.A., Schambelan, M., and Kalinyak, J.E.,** Expression of AT_2 receptors in the developing rat fetus, *J. Clin. Invest.,* 88, 921, 1991.
32. **Feuillan, P., Millan, M.A., and Aguilera, G.,** Angiotensin II receptor subtypes in the rat fetus, *Fed. Proc.,* 5, 2915, 1991.
33. **Pipkin, F.B. and Symonds, E.M.,** Renin-angiotensin system in early life, in *Fetal Physiology and Medicine: The Basis of Perinatology,* Beard, R.W. and Nathanielz, P.W., Eds., Marcel Dekker, New York, 1984, 459.
34. **Wilkes, B.M., Krim, E., and Mento, P.E.,** Evidence for a functional renin angiotensin system in the full term fetoplacental unit, *Am. J. Physiol.,* 249, E366, 1985.
35. **Wallace, K.B., Bailie, M.D., and Hook, J.B.,** Development of angiotensin converting enzyme in fetal rat lungs, *Am. J. Physiol.,* 236, R57, 1979.
36. **Richoux, J.P., Amsaguine, P.S., Grignon, G., Bouhnik, J., Menard, J., and Corvol, P.,** Earliest renin containing cell differentiation during ontogenesis in the rat, *Histochemistry,* 88, 41, 1987.
37. **Wigger, H.J. and Stalcup, S.A.,** Distribution and development of angiotensin converting enzyme in the fetal and new born rabbit. An immunofluorescence study, *Lab. Invest.,* 38, 581, 1978.
38. **Gomez, R.A., Cassis, A.L., Lynch, K.R., Chevalier, R.L., Wilfong, N., Carey, R.M., and Peach, M.J.,** Fetal expression of the angiotensinogen gene, *Endocrinology,* 123, 2298, 1988.
39. **Lee, H.U., Campbell, D.J., and Habener, J.F.,** Developmental expression of angiotensinogen gene in rat embryos, *Endocrinology,* 121, 1335, 1987.
40. **Pucell, A.G., Hodges, J.C., Sen, I., Bumpus, M.F., and Husain, A.,** Biochemical properties of the ovarian granulosa cell type 2 angiotensin II receptor, *Endocrinology,* 128, 1947, 1991.

41. **Speth, R.C. and Kim, K.H.,** Discrimination of two angiotensin II receptor subtypes with a selective antagonist analogue of angiotensin II, *Biochem. Biophys. Res. Commun.*, 169, 997, 1990.
42. **Johnson, M.C. and Aguilera, G.,** Angiotensin II receptor subtypes and coupling to signaling systems in cultured fetal fibroblasts, *Endocrinology,* 129, 1266, 1991.
43. **Catt, K.J., Carson, M.C., Hausdorff, W.P., Leach-Harper, C.M., Baukal, A.J., Guillemette, G., Bala, T., and Aguilera, G.,** Angiotensin II receptors and mechanisms of action in adrenal glomerulosa cells, *J. Steroid Biochem.*, 77, 915, 1987.
44. **Johnson, M.C. and Aguilera, G.,** Interactions between angiotensin II and regulators of adenylate cyclase in rat fetal fibroblasts, The Endocrine Society 73rd Annual Meeting, Washington, D.C., 1991, 621.
45. **Garcia-Sainz, J.A. and Macias-Silva, M.,** Angiotensin II stimulates phosphoinositide turnover and phosphorylase through AII-1 receptors in isolated rat hepatocytes, *Biochem. Biophys. Res. Commun.*, 172, 780, 1990.
46. **Chiu, A.T., Carini, D.J., Duncia, J.V., Leung, K.H., McCall, D.E., Price, W.A., Wong, P.C., Smith, R.D., Wexler, R.R., and Timmermans, P.B.M.W.M.,** DuP532: a second generation of nonpeptide angiotensin II receptor antagonists, *Biochem. Biophys. Res. Commun.*, 177, 209, 1991.
47. **Marie, J. and Jard, S.,** Angiotensin II inhibits adenylate cyclase from adrenal cortex glomerulosa zone, *FEBS Lett.*, 159, 97, 1983.
48. **Marie, J., Gaillard, R.C., Schoenenberg, P., Jard, S., and Bockaert, J.,** Pharmacological characterization of angiotensin II receptor negatively coupled with adenylate cyclase in rat anterior pituitary gland, *Endocrinology,* 116, 1044, 1985.
49. **Hausdorff, W.P., Secura, R.D., Aguilera, G., and Catt, K.J.,** Control of aldosterone production by angiotensin II is mediated by two guanine nucleotide regulatory proteins, *Endocrinology,* 120, 1668, 1987.
50. **Anand-Srivastava, M.B.,** Angiotensin II receptors negatively coupled to adenylate cyclase in rat myocardial sarcolema, *Biochem. Pharmacol.*, 38, 489, 1989.
51. **Abou-Samra, A.-B., Harwood, J.P., Manganiello, V.C., Catt, K.J., and Aguilera, G.,** Phorbol 12-myristate 13 acetate and vasopressin potentiate the effect of corticotropin releasing factor on cyclic AMP production in rat anterior pituitary cells, *J. Biol. Chem.*, 262, 1129, 1987.
52. **Douglas, J.G.,** Angiotensin II receptor subtypes of the kidney cortex, *Am. J. Physiol.*, 253, F1, 1987.
53. **Tulkens, P., Beaufay, H., and Trouet, A.,** Analytical fractionation of homogenates from cultured rat embryo fibroblasts, *J. Cell Biol.*, 63, 383, 1974.
54. **Murphy, T.J., Alexander, R.W., Griendling, K.K., Runge, M.S., and Bernsteinn, K.E.,** Isolation of a cDNA encoding the vascular type-1 angiotensin II receptor, *Nature,* 351, 233, 1991.
55. **Sasaki, K., Yamano, Y., Bardhan, S., Iwai, N., Murray, J.J., Hasegawa, M., Matsuda, Y., and Inagami, T.,** Cloning and expression of a complementary DNA encoding a bovine adrenal angiotensin II type-1 receptor, *Nature,* 351, 230, 1991.
56. **Gordon, D.A., Shelness, G.S., Nicosia, M., and Williams, D.L.,** Estrogen induced destabilization of yolk precursor protein mRNAs in avian liver, *J. Biol. Chem.*, 263, 2625, 1988.
57. **Izzo, N.J., Seidman, C.E., Collins, S., and Colucci, W.S.,** Alpha-1-adrenergic receptor mRNA is regulated by norepinephrine in rabbit aortic smooth muscle cells, *Proc. Natl. Acad. Sci. U.S.A.*, 87, 6268, 1990.
58. **Cassey, J.L., Koeller, D.M., Ramin, V.C., Klausner, R.D., and Harford, J.B.,** Iron regulation of transferrin receptor mRNA levels requires iron responsive elements and rapid turnover determinant in the 3′ untranslated region of mRNA, *EMBO J.*, 8, 3693, 1989.

59. **Yen, T.J., Gay, D.A., Pachter, J.S., and Cleveland, D.W.,** Autoregulated changes in stability of polyribosome-bound β-tubulin mRNAs are specified by the first 13 translated nucleotides, *Mol. Cell. Biol.*, 8, 1224, 1988.
60. **Leibold, E.A. and Munro, H.N.,** Cytoplasmatic protein binds in vitro to highly conserved sequence in the 5′ untranslated region of ferritin heavy- and light-subunit mRNA, *Proc. Natl. Acad. Sci. U.S.A.*, 85, 2171, 1988.
61. **Rouault, T.A., Hentze, M.W., Caughman, S.W., Harford, J.B., and Klausner, R.D.,** Binding of a cytosolic protein to the iron responsive element of human ferritin messenger RNA, *Science*, 241, 1207, 1988.
62. **Brown, A.L., Graham, D.E., Nissley, S.P., Hill, D.J., Stain, A.J., and Rediles, M.M.,** Developmental regulation of insulin-like growth factor II mRNA in different rat tissues, *J. Biol. Chem.*, 161, 13144, 1988.
63. **Beck, F., Samani, N.J., Penschow, J.D., Thorley, B., Tregear, G.W., and Coghlan, J.P.,** Histochemical localization of IGF I and II mRNA in the developing rat embryo, *Development*, 101, 175, 1987.
64. **Millan, M.A., Jacobowitz, D.M., Aguilera, G., and Catt, K.J.,** Differential distribution of AT_1 and AT_2 angiotensin II receptor subtypes in the rat brain during development, *Proc. Natl. Acad. Sci. U.S.A.*, 88, 11440, 1991.
65. **Tsusumi, K. and Saavedra, J.M.,** Differential development of angiotensin II receptor subtypes in the rat brain, *Endocrinology*, 128, 630, 1991.
66. **Mendelsohn, F.A.O., Quirion, R., Saavedra, J.M., Aguilera, G., and Catt, K.J.,** Autoradiographic localization of angiotensin II receptors in rat brain, *Proc. Natl. Acad. Sci. U.S.A.*, 8, 1575, 1984.
67. **Slotkin, T.A., Cowdery, T.S., Orband, L., Pachman, S., and Whitmore, W.L.,** Effects of neonatal hypoxia on brain development in the rat: immediate and long term biochemical alterations in discrete regions, *Brain Res.*, 347, 63, 1986.
68. **Bayer, S.A. and Altman, J.,** Hippocampal development in the rat: cytogenesis and morphogenesis examined by autoradiography and low level X-irradiation, *J. Comp. Neurol.*, 158, 55, 1974.
69. **Altman, J.,** Postnatal development of the cerebellar cortex in the rat, *J. Comp. Neurol.*, 150, 399, 1972.
70. **Carpenter, M.B. and Sutin, J.,** *Human Neuroanatomy,* 8th ed., Williams & Wilkins, Baltimore, 1983.
71. **Inagami, T.,** personal communication.
72. **Clark, A.J., Johnson, M.C., and Aguilera, G.,** unpublished data.
73. **Feuillan, P. and Aguilera, G.,** unpublished data.

Chapter 17

INTERACTIONS OF ANGIOTENSIN II AND ATRIAL NATRIURETIC PEPTIDE

E.L. Schiffrin, R. Garcia, and E.M. Konrad

TABLE OF CONTENTS

0-8493-4622-3/93/$0.00 + $.50

I. INTRODUCTION

Atrial natriuretic peptides (ANP) are a family of different peptides originally identified in the heart.[1] Three distinct peptides may be distinguished at present: ANP, brain natriuretic peptide (BNP),[2] and C-type natriuretic peptide (CNP).[3] Originally identified in the atria because of the potent natriuretic action of ANP,[1] these peptides have been found in different tissues and exert actions on the kidney to promote salt and water excretion[1,4] and inhibit renin secretion;[5] they relax blood vessels independently of effects on endothelium,[6] act on the adrenal gland to inhibit aldosterone secretion,[7,8] on the brain to inhibit drinking stimulated by angiotensin II,[9] and dehydration and hemorrhage-induced vasopressin release,[10] among many other effects; and on the peripheral nervous system to block neurotransmission,[11] to mention some of the most important activities of the ANP system.

Since the discovery of ANP, it has become apparent that this family of peptides has actions which suggest that they may act as physiological antagonists of angiotensin II (AngII). Initial observations showed that ANP relaxed blood vessels contracted by AngII.[12] Shortly after, it was shown that ANP blunted the aldosterone secretory response to AngII.[7,8] Later, studies of the distribution of ANP binding sites gave further support to this possibility when it was shown that it was topographically similar to the distribution of AngII binding.[13] Indeed, not only did ANP exert actions on key organs involved in blood pressure regulation and water and sodium balance, but furthermore, these actions resulted in a physiological antagonism of AngII. Together with the overlap in distribution of binding sites for both peptides in the same tissues, these data have strongly suggested that indeed the blood pressure lowering, vasorelaxant, aldosterone-inhibiting, and natriuretic actions of ANP are a countervailing mechanism of the pressor, vasoconstrictor, aldosterone-stimulating, and antinatriuretic effects of the renin-angiotensin system. The basis for these interactions has become more complex as the biochemistry and physiology of the ANP receptor system has become better known, and evidence of direct interactions or ''cross-talk'' between AngII and ANP receptors has been provided by more recent studies. We will review the interactions between these two opposing systems involved in regulation of blood pressure and water and electrolyte balance at the organ, tissue, and cellular level. We will successively describe the overlapping distribution of ANP and AngII receptors, then the interactions of both peptides in different organs, such as blood vessels, adrenal gland, kidney, and brain, and finally, briefly describe current knowledge on molecular interactions of ANP and AngII receptors.

II. DISTRIBUTION OF ANP AND ANGIOTENSIN II RECEPTORS

As expected from the actions of ANP briefly described in preceding paragraphs, ANP receptors have been found by different techniques (*in vitro* binding to particulate fractions or by *in vitro* or *in vivo* autoradiography) in organs involved in key roles in regulation of blood pressure and sodium and water excretion, as well as other functions. These receptors are present in different organs, particularly the kidney,[14,15] blood vessels,[16,17] the adrenal,[17] brain,[18-20] hypophysis,[18-20] gonads,[21] intestine,[22] pancreas,[23] liver,[24] the ciliary body of the eye,[18] and platelets[25] in different species. Binding sites are located in specific structures in some of these organs. In blood vessels, receptors are located in smooth muscle cells[26,27] and in endothelial cells.[28] In the adrenal, ANP sites are found predominantly in the glomerulosa in the cortex, and in the medulla.[19,29,30] In the kidney, receptors for ANP may be found in glomeruli, on arcuate and interlobar arteries in the cortex, and in tubules and vasa recta in the papilla.[19,29,31] In the brain, receptors have been localized to the plexiform layer of the olfactory bulb, the pia-arachnoid, choroid plexus, subfornical organ, median preoptic nucleus, supraoptic nucleus, paraventricular nucleus, and area postrema.[18,20,32] A detailed distribution of ANP sites in other organs will not be discussed.

These ANP receptors have been shown to belong to different subtypes, namely guanylate-cyclase-containing and non-guanylate cyclase-containing receptors. The latter have been called R_2 receptors, C-receptors, or ANPR-C (for clearance receptors),[33] and have since been cloned.[34] These ANP receptors lack an intracellular catalytic domain which is replaced by a short tail. Cloning and pharmacological characteristics have also further allowed guanylate cyclase-coupled receptors to be distinguished as R_1-A and R_1-B or ANP-A and B-type receptors (ANPR-A and ANPR-B).[35,36] These receptors have an extracellular binding domain, a short transmembrane region, and an intracellular catalytic domain, which contains guanylate cyclase activity and a sequence with high homology to protein kinases. There is 43% homology for the extracellular domain and 78% for the intracellular domain between A- and B-type ANP receptors. Their extracellular domain, in turn, exhibits 33 and 29% homology, respectively, to the extracellular sequence of C-type ANP receptors. It appears at present that ANP is the preferred ligand of ANPR-A while CNP is the preferred ligand of ANPR-B.[36]

The different types of ANP receptors are widely distributed in many tissues in variable proportions. Some tissues such as the renal inner medulla and papilla appear to contain predominantly A receptors although B receptors are also present,[31,37-39] whereas in the brain, A and B receptors predominate, while C receptors have been localized only in choroid plexus and pia-arachnoid.[40-44] In blood vessels and in kidney glomeruli, C receptors outnumber A receptors.[37-39] Intrarenal arcuate and interlobar arteries appear to express A receptors more abundantly than the C type, in contrast to other vascular beds.[37] Distribution of subtypes of AngII receptors is reviewed in other chapters of this volume and will only be briefly discussed. However, it should be mentioned that the two main classes of AngII receptors currently recognized (AT_1 and AT_2)[45-47] have different topographical distribution. The AT_1 receptor is found in rat peripheral vessels and is the predominant subtype in the rat adrenal cortex.[45] In the rat and human kidney, AT_1 is predominant, specially in glomeruli and medullary vessels.[48] In the adult human kidney, the AngII receptor antagonist PD123177 displaces AngII in preglomerular vessels.[49] This predominance of AT_2 receptors in preglomerular vessels of the kidney is surprising, since the contractile effect of AngII in these vessels has been well demonstrated.[50] The apparent absence of AT_1 receptors in these vessels may be the result of a lack of sensitivity of the method employed (autoradiography) and not a true absence of AT_1 receptors. A brief description of distribution of AT_1 and AT_2 receptors in the brain is provided later in this chapter.

The relative distribution and overlap of ANP and AngII receptors in kidney, adrenal, and brain were reviewed by Mendelsohn et al.[13] In the adrenal, both AngII and ANP exhibit high density of binding to the zona glomerulosa,[13,19,29] while there are numerous AngII sites in the medulla[13] but fewer ANP sites.[30] In the kidney, ANP[19,29] and AngII binding[51,52] has been shown to have similar distribution, that is, high density of ANP and AngII sites in glomeruli (although relative number of binding sites on mesangial and epithelial cells may be different for both peptides[52]), moderately high binding in the inner medulla over vasa recta bundles for both ANP and AngII, and a particularly high density of ANP sites at the tip of the papilla, where in contrast there are no AngII binding sites. On the other hand, the outer medulla is richer in AngII binding than in ANP sites.[13,19,29] Thus, although distribution is similar, relative densities of receptors in each topographical area in the kidney may be different for both peptides. This suggests that physiological interactions in some of these different structures may be the result of exerting opposing effects which may counterbalance each other rather than direct intracellular interaction of signal transduction mechanisms, although the latter is not excluded.

In the brain, receptors for both peptides have been localized in the lateral olfactory tract, nucleus of the accessory olfactory tract, olfactory tubercule, cortical areas, cerebellum, corpus callosum, lateral septum, caudate/putamen, globus pallidus, nucleus accumbens, amygdala, hippocampus, median preoptic nucleus, suprachiasmatic nucleus, periventricular and paraventricular hypothalamic nuclei, subfornical organ, organum vasculosum of the lamina terminalis, median eminence, dorsomedial hypothalamic nucleus, medial habenular nucleus, mammillary nuclei, medical geniculate nucleus, ventral posterodorsal thalamic nucleus, superior colliculus, periaqueductal central gray, parabranchial nucleus, locus coeruleus, pontine nuclei, inferior olive, inferior colliculus, dorsal motor nucleus of the vagus, nucleus of the solitary tract, area postrema, choroid plexus, and anterior pituitary.[13,18,20,32,41-44,54] The localization of different AngII receptor subtypes (AT_1 and AT_2) in the brain has been demonstrated,[55-57] but is described in detail elsewhere in this book. While the AT_1 subtype is found in the organum vasculosum lamina terminalis, subfornical organ, median preoptic nucleus, area postrema, and nucleus of the solitary tract, AT_2 receptors are found in the locus coeruleus, ventral and dorsal parts of lateral septum, superior colliculus, subthalamic nucleus, some nuclei of the thalamus, and the nuclei of the inferior olive. The specific overlap of AT_1 and AT_2 receptors with ANP sites is later summarized in Table 1. ANP but no AngII receptors are found in the external plexiform layer of the olfactory bulb. Thus, as in the kidney and adrenal, in the brain there is overlap of binding sites, possibly allowing interactions of signal transduction intracellularly for those neurons bearing receptors for both peptides, and/or intercellular interactions if cells responding to ANP and AngII are in relative vicinity. Finally, some areas may have interactions simply as a result of antagonistic physiological effects where there is no overlap of receptors for both systems, when such interactions can be detected, as described below.

III. INTERACTIONS OF ANGIOTENSIN II AND ANP IN DIFFERENT TISSUES AND ORGANS

A. BLOOD VESSELS

One of the first reports of an extrarenal action of ANP demonstrated that ANP relaxed rat aortic strips contracted by AngII.[12] Similar results were shown with ANP on rabbit aorta contracted with AngII.[58] Thus, ANP may modulate

TABLE 1
Overlapping Distribution of Angiotensin II and ANP Receptors in Brain

ANP and AT_1 receptors	ANP and AT_2 receptors	ANP, AngII receptors, subtype not established
Subfornical organ	Accessory olfactory tract	Olfactory tract
Organum vasculosum lamina terminalis	Lateral septum	Olfactory tubercule
Median eminence	Amygdala	Cortical areas
Area postrema	Medial geniculate nucleus	Corpus callosum
Anterior pituitary	Ventral posterolateral thalamic nucleus	Hippocampus
Median preoptic nucleus	Superior colliculus	Caudate/putamen
Suprachiasmatic nucleus	Locus coeruleus	Globus pallidus
Periventricular, paraventricular, and dorsomedial hypothalamic nuclei	Inferior olive	Nucleus accumbens
Periaqueductal central gray	Inferior colliculus	Medial habenular nucleus
Parabrachial nucleus	Cerebellum	Mammillary nuclei
Pontine nuclei		Choroid plexus
Nucleus of the solitary tract		
Motor nucleus of the vagus		

Data from References 13, 18, 20, 32, 41, 42, 44, and 54 to 57.

AngII vascular effects. There is, however, little evidence of how important this countervailing influence of the vasorelaxant action of ANP on AngII-mediated effects is in different conditions, particularly when ANP concentrations in plasma are elevated, such as in many forms of experimental hypertension or heart failure. However, it is evident that ANP could play a buffering role on exaggerated contractile effects of AngII. When ANP is infused intravenously, it selectively antagonizes the contractile effect of exogenous AngII in the rat microcirculation,[59] and this effect, as well as antagonism of the vasoconstrictor action of endogenous AngII, is particularly evident in the renal circulation of the rat.[60] In sheep,[61] as in rats,[59] ANP was not effective *in vivo* in antagonizing the effect of arginine vasopressin (AVP), while it did attenuate responses to AngII and norepinephrine. In AngII-induced hypertension in rats, blood pressure decreased slightly after infusion of ANP, further supporting the hypothesis that ANP modulates the vasopressor effect of AngII.[62] The extent to which elevated AngII may affect ANP-induced vasorelaxation is unclear. In some experimental models of hypertension, such as the deoxycorticosterone acetate (DOCA)-salt[63] and the 1-kidney, 1-clip Goldblatt hypertensive rat,[64] the relaxation response of aorta to ANP is blunted, in agreement with depressed density of ANP binding sites, while in the 2-kidney, 1-clip Goldblatt hypertensive rat, relaxation of aorta by ANP is also decreased but ANP receptor density in blood vessels is similar to that of control rats.[64,65] In the 2-kidney, 1-clip Goldblatt hypertensive rat, plasma AngII is elevated, in contrast to the two former hypertensive models, and thus, AngII could be involved in the decrement in vascular relaxation found.

Mechanisms underlying this antagonism between ANP and AngII actions on blood vessels or vascular smooth muscle have been studied by various investigators, often providing conflicting results. There is some evidence that ANP may lower cytosolic calcium in cultured vascular smooth muscle cells.[66] Although the transient initial rise in calcium following AngII was unaffected, the sustained response of cytosolic calcium to AngII or the response to KCl were dose-dependently decreased by ANP. Other investigators examining rabbit aorta failed to find changes in the cytosolic calcium response to AngII, although its contractile effect was blunted.[67] There have been reports indicating that ANP may inhibit intracellular calcium release elicited by agents such as norepinephrine or histamine,[68-72] vasopressin,[68,72] or AngII[68-72] in smooth muscle cells, while other investigators have been unable to document an antagonistic effect of ANP on calcium kinetics in vascular smooth muscle.[73] Cyclic guanosine monophosphate (cGMP), the intracellular mediator of ANP, has been proposed as a regulator of cell calcium through a cGMP-dependent protein kinase which activates a Ca^{2+}-ATPase and removal of calcium from vascular smooth muscle cells.[74,75] Other mechanisms for inhibition by ANP of AngII effects include possibly inhibition of phosphatidylinositol hydrolysis, as shown on rat aorta contracted with norepinephrine,[76] although ANP has also been reported paradoxically to stimulate phospholipid breakdown,[77] and to inhibit the agonist-induced increase in extent of myosin light chain phosphorylation, demonstrated in rabbit aorta precontracted with AngII or histamine.[67]

Mechanisms whereby AngII may affect ANP-induced cGMP accumulation are varied. There is evidence that AngII may modulate ANP receptor density. The possibility of heterologous regulation of ANP receptors was raised initially by Hirata[78] who showed that phorbol esters, presumably via activation of protein kinase C, inhibited ANP binding and cGMP generation in cultured rat vascular smooth muscle cells. Similar results were shown by Nambi et al.[79] in A-10 cells. Simultaneously, AngII and AVP were shown to decrease cGMP generation by ANP in cultured aortic vascular smooth muscle cells.[80] Because the calcium ionophore A23187 mimicked these effects, it was concluded that they were mediated by the transient increase in cytosolic calcium, probably via a calcium-activated cGMP phosphodiesterase. Based on these data, we infused AngII, which will stimulate protein kinase C as part of its signal-transduction mechanism, at a pressor dose intravenously into rats and found that high plasma concentration of AngII, ANP sites in blood vessels were decreased.[81,82] However, plasma ANP was simultaneously increased and could account for the reduced ANP vascular receptor numbers found (Figure 1). The reduction in vascular ANP sites at pressor doses of AngII comprised only the R_2 (or C-type) sites, while R_1 (or A or B type) were spared (Table 2).[82] Interestingly, AngII at a nonpressor dose did not result in changes in plasma ANP or in ANP binding sites. However, relaxation responses of blood vessels to ANP were sensitized (Figure 2). This agrees with the exaggerated biological response to ANP (cGMP production) described in cultured vascular smooth muscle cells after exposure to angiotensin II.[83,84] Chabrier et al.[83] has reported that incubation of cultured aortic smooth vascular muscle cells with AngII for 18 h resulted in decreased ANP binding together with exaggerated cGMP production in response to ANP. In contrast to enhanced cGMP response to ANP after AngII, epinephrine, norepinephrine, histamine, serotonin, and AVP were ineffective. The authors concluded that this effect was specific for AngII and suggested that guanylate cyclase-coupled ANP receptors were upregulated while uncoupled sites were downregulated. In contrast, Hirata et al.[80] recently demonstrated that AngII and AVP both decreased ANP binding and cGMP accumulation in response to ANP, the latter in agreement with previous results.[85] Our data[84] agree in part with those of Chabrier et al.,[83] indicating that AngII produces a sensitization of the cGMP response to ANP (Figures 3 and 4). However, we found no evidence of downregulation of either subtype of ANP receptor in cultured vascular smooth muscle cells, in contrast to what other reports suggest,[83,85,86] with either AngII or AVP.[84] Furthermore, when another vasoconstrictor peptide, endothelin, which acts in part through a similar signal transduction system (phospholipase C-inositol phosphate-calcium-protein kinase C), was used, we could not find ANP receptor downregulation or effects on cGMP generation

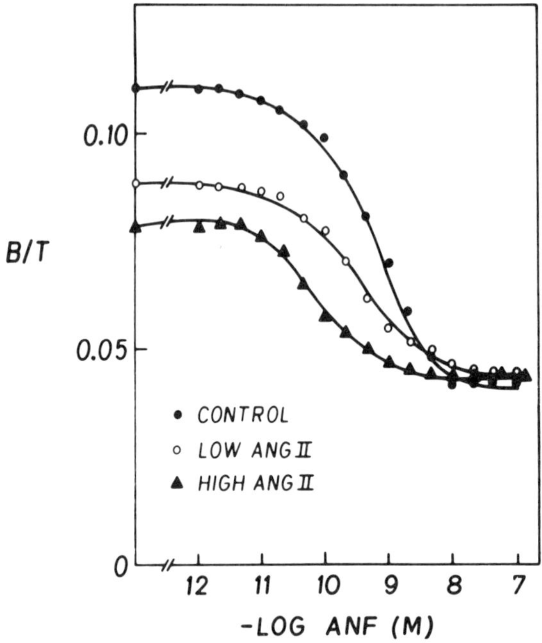

FIGURE 1. Representative curves of ^{125}I-atrial natriuretic peptide (ANP or ANF) binding to mesenteric artery membranes from rats infused with angiotensin II, in the presence of increasing concentrations of unlabeled ANP, expressed as bound/total (B/T) v. unlabeled ANP. ●, control; ○, 200 ng/kg/min angiotensin II; ▲, 800 ng/kg/min angiotensin II. (From Gauquelin, G. et al., *J. Hypertens.*, 9, 1151, 1991. With permission.)

in response to ANP, or on the ratio of ANP receptor subtypes.[84] The differences reported between all these studies may relate to different protocols, origin of cultured rat vascular smooth muscle cells (aortic vs. mesenteric arteries), or other unknown factors. The exaggerated responses to ANP in the presence of AngII may explain results observed in different physiological conditions. In adrenalectomized rats[87] and in a heart failure model, the cardiomyopathic hamster,[88] vascular ANP binding sites are decreased but relaxation of aorta is more sensitive to ANP (Figure 5). These two conditions are associated with increased circulating AngII. As mentioned earlier, similar results were observed in rats infused with subpressor doses of AngII, which also exhibit exaggerated sensitivity of aortic relaxation in response to ANP in spite of similar density of ANP receptors.[81] AngII may be acting to sensitize ANP receptors in these conditions. Alternatively, the relative proportions of guanylate cyclase-containing and ''clearance'' receptors may change *in vivo*,[82] which does not seem to be the case *in vitro*.[84]

A final recently reported expression of the antagonism of ANP and AngII on blood vessels is the finding that both ANP and BNP inhibit the stimulation

TABLE 2
Effects of AngII Infusion on the Density (fmol/mg Protein) of Glomerular and Vascular ANP Receptor Subtypes (R_1 and R_2)

Group	Glomerular ANP receptors Δ B_{max} (fmol/mg protein)		Mesenteric artery ANP receptors ΔΔ B_{max} (fmol/mg protein)	
	R_1	R_2	R_1	R_2
Control	146 ± 1	358 ± 3	113 ± 15	193 ± 25
800 ng AngII per kg/min	116 ± 1[a]	273 ± 3[a]	136 ± 29	43 ± 9[a]

Note: Values are means ± SEM. Δ, Each value represents the means of 4 experiments ± SEM; ΔΔ, each value represents the means of 5 experiments ± SEM. R_1 corresponds to A- and/or B-type ANP receptors, R_2 to C-type ANP receptors.

[a] $p < 0.001$ vs. controls.

From Gauquelin, G. et al., *J. Hypertens.*, 9, 1151, 1991. With permission.

of endothelin secretion by AngII acting upon porcine aorta.[89] The physiopathological significance of such a finding remains to be determined, but it could clearly contribute to blunting by ANP of AngII actions. The mechanism of this effect is probably similar to mechanisms described for the interference of ANP with AngII action in other circumstances.

B. ADRENAL GLAND

The first two reports demonstrating the inhibitory effect of ANP on aldosterone secretion showed that ANP blunted the stimulation of aldosterone biosynthesis induced by AngII *in vitro*[7,8,90,91] and *in vivo*.[7] Subsequent studies confirmed these findings *in vitro*[92] and *in vivo*.[93] The stimulation of aldosterone after activation of the endogenous renin-angiotensin system by salt depletion was also inhibited in rats by infusion of ANP.[94] Other studies in rats,[95] dogs,[96] and humans[97-100] have also demonstrated that ANP may inhibit AngII-induced aldosterone secretion. This interaction results in a clearcut decrease in the maximum response to AngII,[101,102] in contrast to the inhibition of adrenocorticotrophic hormone (ACTH)-induced aldosterone secretion, for which the dose response is displaced to the right both in *in vitro*[101-103] and *in vivo*.[103]

The nature of the interaction between ANP and AngII in the adrenal cortex has been the subject of a number of studies yielding controversial data and remains unexplained. There is little doubt that ANP increases cGMP in the adrenal[104] and that AngII stimulates phospholipase C, calcium release, and calcium entry into the adrenal cell.[105-108] What has generated controversy

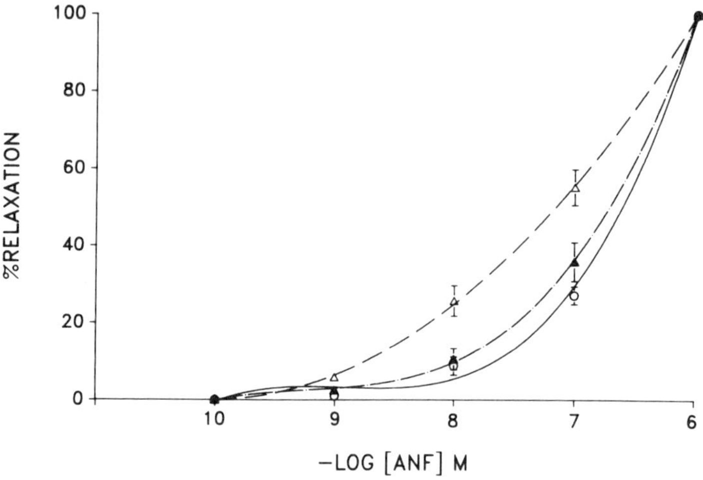

FIGURE 2. Atrial natriuretic peptide (ANP or ANF)-induced relaxation of norepinephrine-precontracted aorta strips in AngII-infused rats. ○, control; △, 200 ng/kg/min; ▲, 800 ng/kg/min. (From Cachofeiro, V. et al., *J. Hypertens.*, 8, 1077, 1990. With permission.)

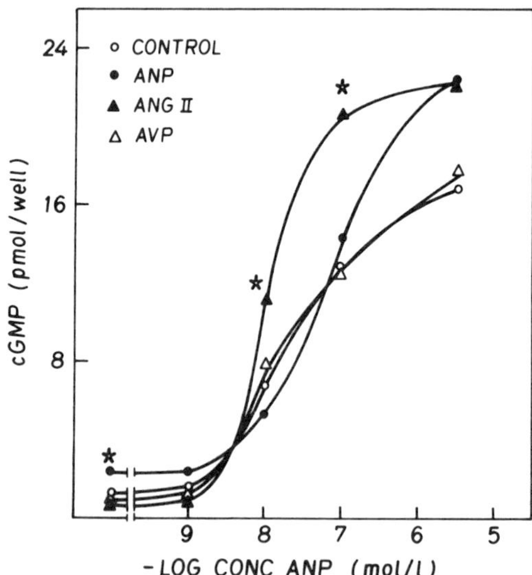

FIGURE 3. Dose response of vascular smooth muscle cells to increasing concentrations of ANP in cultured rat vascular smooth muscle cells, which have been exposed for 18 h to 100 nmol/l ANP, 100 nmol/l AngII, or 1 μmol/l arginine vasopressin (AVP). cGMP measured is intracellular cGMP accumulated during a 10-min stimulation in presence of 0.5 mmol/l 3-isobutyl-1-methylxanthine. Results are means of 6 experiments. $*p < 0.05$ vs. control. (From Schiffrin, E.L. et al., *Am. J. Physiol.*, 260, H58, 1991. With permission.)

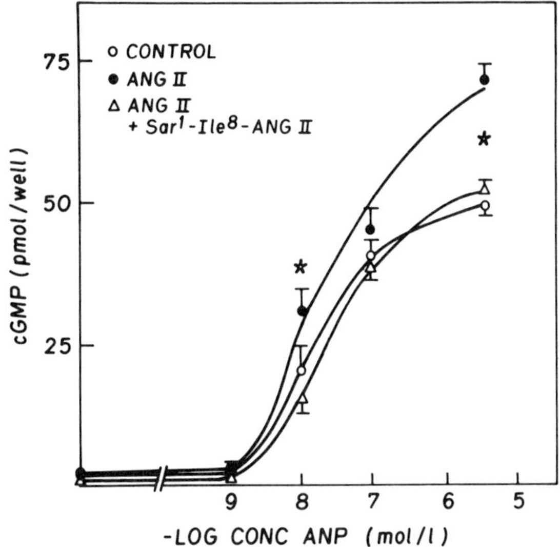

FIGURE 4. Dose-response of cGMP to increasing concentrations of ANP in cultured rat vascular smooth muscle cells that have been exposed for 18 h to 100 nmol/l AngII with or without 10 μmol/l Sar[1],Ile[8]-AngII, a specific AngII antagonist. Results are means ± SE of 4 to 6 experiments. $p < 0.05$, higher mean values vs. other values by a post hoc least significant difference test following analysis of variance. (From Schiffrin, E.L. et al., *Am. J. Physiol.*, 260, H58, 1991. With permission.)

is how these two intracellular second messenger systems interact to result in the physiological effects which have been described. Doubts have even been raised on the actual role of cGMP in the inhibitory effect of ANP on aldosterone secretion,[109] although it is accepted that indeed cGMP is generated by interaction of ANP with ANPR-A in the adrenal glomerulosa.

Because of the similarity of the effects of ANP and dihydropyridine calcium channel blockers on AngII, ACTH, and potassium-induced aldosterone biosynthesis, it was proposed that ANP could affect calcium influx in the adrenal cortex.[101] ^{45}Ca influx stimulated by AngII was indeed blocked by ANP,[101] although other investigators were unable to detect this,[110,111] or inhibition of cytosolic calcium transients or ^{45}Ca efflux stimulated by AngII.[111,112] Since ANP blocked calcium ionophore A23187-stimulated aldosterone secretion, whether the latter was combined or not with a phorbol ester, it was suggested that ANP selectively and noncompetitively inhibited an intracellular step necessary for calcium-dependent stimulation of aldosterone secretion,[111,113] as had been previously suggested by other investigators.[92] ANP did not affect the rise in diacylglycerol levels in bovine adrenal cells stimulated with AngII, further supporting the notion that phospholipase C activation by angiotensin was unaffected by ANP.[111] More recently, on the other hand,

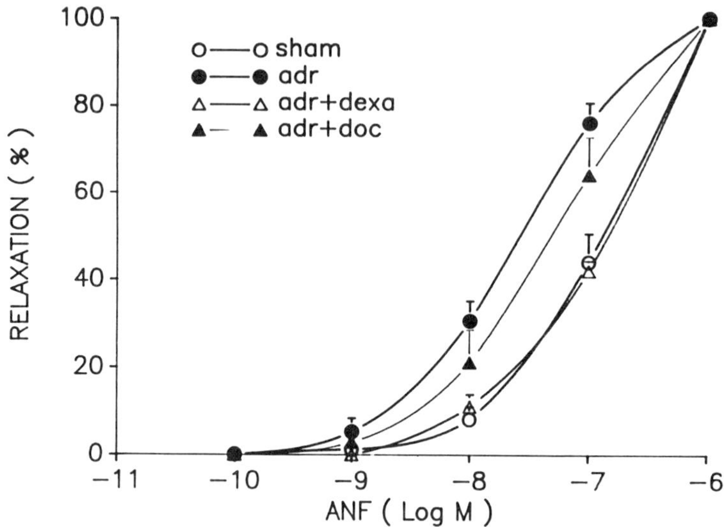

FIGURE 5. Relaxation of norepinephrine-precontracted rat aorta with atrial natriuretic peptide (ANP or ANF) in adrenalectomized rats (adr). Calculated ED_{50} in adrenalectomized rats was significantly lower (1.7×10^{-8} *M*) than in sham-operated (1.5×10^{-7} *M*) or adrenalectomized rats receiving dexamethasone (dexa) replacement treatment (2.1×10^{-7} *M*, $p < 0.001$); doc = deoxycorticosterone acetate. Eight aortic strips per group were examined. (From Cachofeiro, V. et al., *Am. J. Physiol.*, 256, R1250, 1989. With permission.)

using patch-clamp techniques, McCarthy et al.[114] and Barrett et al.[115] showed, in zona glomerulosa cells, that ANP inhibited (by shifting inactivation to more negative potentials) T-type calcium channels, which appear to be of critical importance for the physiological action of AngII. At the same time, these authors reported that L-currents in glomerulosa cells were stimulated.[114,115] Since the resting membrane potential of glomerulosa cells favors the involvement of T- rather than L-currents in response to stimulation, the inhibition of T-type channels could underlie previous findings of ^{45}Ca influx experiments.[101] Under other conditions, with cells under a greater degree of depolarization, L-type channels may acquire greater importance, obscuring the effects of ANP on ^{45}Ca influx,[110,111] which could explain the reported discrepancies and in part some of the interactions between ANP and AngII in the adrenal glomerulosa. It is likely that as a consequence of the action of AngII, the phosphorylation status of some intracellular proteins is changed as part of the chain of events leading to stimulation-response coupling. Phosphorylation of a 17,600-Da protein stimulated by AngII, which is reduced if cells are also exposed to ANP, has been described.[116] However, other investigators have failed to confirm this finding, suggesting that phosphorylation of this protein is not an obligatory step in the regulation of aldosterone production affected in opposite directions by ANP and AngII.[117]

ANP has been shown to inhibit adenylate cyclase in the adrenal.[118] However, since AngII also inhibits adenylate cyclase in the adrenal,[119] it is unlikely that these effects are involved in the ANP-mediated inhibition of AngII-stimulated steroidogenesis. Recent findings of possible interactions of AT_2 and ANP receptors in the rat adrenal are discussed in the section on molecular interactions between both types of receptors.

C. KIDNEY

The mechanism(s) whereby ANP increases water and sodium excretion by the kidney is not completely understood. Several mechanisms contributing to the increase in sodium excretion after ANP administration have been suggested. It has been reported that ANP-induced natriuresis is dependent on an elevation in glomerular filtration rate (GFR).[4,5,120] It has been suggested also that ANP may have a direct inhibitory effect on sodium transport in the proximal[121,122] or distal nephron segments, especially along the medullary collecting ducts.[123,124] Others have suggested that the renal effect of ANP could be associated with an increase in medullary blood flow.[125] The fact that no ANP binding sites has been described in the proximal tubule[19] has caused some doubts about a direct effect of ANP on sodium transport by epithelia of this tubular segment. On the other hand, AngII induces a decrease in GFR which seems to be associated with a fall in the glomerular capillary ultrafiltration coefficient,[126] probably secondary to a shrinkage of the glomerular tuft by contraction of mesangial cells.[127] AngII may stimulate sodium tubular transport by a direct epithelial effect.[128,129] The *in vivo* infusion of AngII induces an increase in renal vascular resistance, the afferent arteriole being responsible for about 50% of the rise in resistance, whereas pre-afferent resistance increase is secondary to interlobular and arcuate constriction,[130,131] suggesting an important contribution of preglomerular arterioles in the regulation of cortical blood flow by AngII.

Evidence of a modulatory role for AngII in natriuresis induced by ANP in the whole animal have been contradictory. An early report[132] suggested that the enhanced effect of exogenous AngII in ANP-induced natriuresis could be explained in part by an intrarenal interaction of both peptides. Other investigators have demonstrated an attenuated natriuretic response to ANP when AngII was infused intrarenally, associated with a blunted ANP-induced increase in GFR.[133] The role of the inhibition of the intrinsic renin-angiotensin system by ANP on the natriuresis induced by the latter has been evaluated also. On one hand, it has been suggested that the inhibition of angiotensin formation in the kidney by ANP may account for only a minor part of the natriuretic effect of ANP.[134] On the other hand, when the intrarenal renin-angiotensin system is inhibited and plasma AngII kept at fixed levels, the renal response to ANP at high renal perfusion pressure is abolished,[135] suggesting that renal hemodynamics and renal excretory response is modulated, at least in part, by the intrarenal renin-angiotensin system. This apparent discrepancy may be due to the pharmacological[134] or pathophysiological[135] doses of AngII infused.

These interactions between ANP and AngII can occur at multiple levels in the nephron. The glomerulus is a major target of ANP. It has been postulated that one of the mechanisms by which ANP and AngII may modify GFR is by their effect on the ultrafiltration coefficient, by modifying the glomerular capillary surface area. AngII reduces,[136] whereas ANP increases,[137] the ultrafiltration coefficient, thus affecting GFR in opposite directions. AngII contracts glomerular mesangial cells,[138] while ANP relaxes either AngII-precontracted mesangial cells[139] or whole glomeruli.[140] The *in vitro* relaxing effect of ANP on AngII-precontracted mesangial cells correlates well with the *in vitro* production of cGMP or *in vivo* changes in GFR.[139] Preincubation of cultured mesangial cells with ANP inhibits the AngII-induced production of inositol trisphosphate, AngII-stimulated increases in cytosolic free Ca^{2+}, and AngII-induced prostaglandin E_2 synthesis.[141] Since these effects are mimicked by sodium nitroprusside and membrane-permeable 8-bromo cGMP, it was suggested that the inhibitory effect of ANP on AngII-induced intracellular effects are modulated by ANP-induced elevation of cGMP. On the other hand, it has been reported that preincubation of cultured mesangial cells with AngII inhibits ANP-stimulated cGMP production in an initial phase by activation of Ca^{2+}-dependent, calmodulin-stimulated cyclic nucleotide phosphodiesterase, and in a later phase by a mechanism regulated via protein kinase C.[142] Despite the *in vitro* evidence of AngII-induced contraction of glomeruli and cultured mesangial cells, morphometric analysis has substantiated an *in vivo* contractile effect of AngII in afferent and efferent arterioles, but has not demonstrated glomerular contraction,[143] casting doubts on whether or not the decrease in GFR induced by AngII is due to a reduction of the capillary area in the glomerular tuft.

AngII not only modifies sodium excretion by the kidney through hemodynamic changes, but, as well, by a direct effect on Na^+ transport in the proximal nephron mediated by AT_1 receptors.[144] On the other hand, ANP has no direct effect on proximal tubular sodium reabsorption,[145] but inhibits that induced by AngII,[146,147] and this inhibition seems to be mediated by cGMP.[146] In addition to inhibiting AngII-induced sodium reabsorption, ANP abolishes proximal tubule reabsorption of bicarbonate induced by AngII,[148] participating, thus, in proximal tubule acidification mechanisms. In contrast, other investigators have failed to demonstrate an effect of ANP on proximal tubule transport, either in the presence or absence of AngII.[149] An effect of ANP on sodium reabsorption by the proximal tubule is not surprising, since microdissected proximal tubules are able to produce cGMP when stimulated with ANP.[150] Furthermore, guanyl cyclase-coupled ANP receptor mRNA has been localized along the proximal and distal nephron in the rat kidney.[151]

One major renal structure where interactions between ANP and AngII seem to be particularly important are intrarenal microvessels. Biologically active guanylate cyclase-coupled ANP receptors are predominant in rat renal microvessels.[31,152] AngII receptors are abundant in vasa recta bundles, although the presence of AT_1 receptors in preglomerular vessels has not been demonstrated by autoradiography.[48,49] Their presence, however, is suggested by the inhibition of AngII-induced contraction of both efferent and afferent arterioles by an AT_1-specific antagonist.[50] However, preglomerular vascular AngII receptors seem to be regulated differently from their peripheral counterparts.[153] As mentioned already, exogenous administration of AngII induces an increase in renal vascular resistance, which is in part a result of constriction of the afferent arteriole and in part a result of constriction of interlobar and arcuate arteries.[130,131] Although AngII causes greater reduction in afferent arteriolar diameter, the vascular resistance induced is higher in efferent vessels.[143] These effects may play an important role in GFR and intrarenal blood flow distribution, and suggest that intracellular signal transduction for angiotensin may not be identical for efferent and afferent arterioles.[154] AngII may also increase sodium reabsorption by decreasing papillary blood flow, without any change in GFR or renal plasma flow.[155]

There is little doubt that the effects of ANP on intrarenal microvessels may affect the overall magnitude of its action on the renal excretion of water and solutes. A specific vasorelaxant effect of ANP has been demonstrated in rings of rat renal arcuate arteries.[156] The effects of ANP on afferent and efferent arterioles are more controversial. A dose-dependent dilation of arcuate interlobular and afferent vessels and constriction of the efferent arterioles during ANP administration has been reported.[157-160] The constrictor effect of ANP in efferent arterioles have been confirmed by others.[158,159] Direct videometry in the normal rat kidney has demonstrated a vasodilatory effect of preglomerular vessels by ANP, without any change in efferent arterioles.[160] There is agreement about the preglomerular dilatory effect of ANP in the presence[159] or absence[157-161] of AngII. However, the fact that the constrictor effect of ANP on efferent arterioles is attained with pharmacological doses of ANP casts doubts on its physiological or pathophysiological relevance. ANP may have a preferential effect on medullary blood flow,[162] suggesting that by elevation of vasa recta hydraulic pressure it may produce an imbalance between papillary vascular and tubular structures, which may be an important mechanism in ANP-induced diuresis and natriuresis.[163]

One of the major interactions of ANP and AngII in the kidney is the acute[5,95] or chronic[164] *in vivo* inhibition of renin release, also demonstrated *in vitro*.[165-168] Inhibition of renin release by ANP has been shown to be a cGMP-mediated process,[165,166] probably through interaction with calcium-calmodulin-dependent steps affecting the mechanisms of regulation of renin secretion.[167,168] The physiological significance of inhibition of renin release by ANP has been the subject of many studies but its importance in physiological and pathophysiological situations remains to be established. In both animals[5,93-95] and humans,[169-172] basal and stimulated renin secretion is inhibited by ANP. It has been proposed that the inhibition of renin release by ANP in renin-dependent models of hypertension may relate to the efficacy of ANP infusion to lower blood pressure in these.[173,174] Other studies also attribute this efficacy to the state of ANP receptors in renin-dependent and non-renin-dependent hypertension.[65] The inhibition of stimulated renin by infusion of ANP is also found in other physiopathological conditions, such as in cardiac failure, in humans.[175,176]

Interactions of AngII and ANP may occur through receptor regulation in the kidney, as described earlier in this chapter in relation to blood vessels. Previous results from our and other laboratories have suggested that the density of ANP receptors may be reciprocally regulated by plasma ANP levels.[63-65,177-180] However, we have recently reported that the number of glomerular ANP receptors may change without parallel changes in plasma ANP levels,[180,181] suggesting that factors other than plasma ANP may regulate ANP receptors. Furthermore, in heart failure, high plasma ANP levels failed to downregulate glomerular and renal papillary ANP receptors, whereas at the same time peripheral vascular receptors were decreased in density, suggesting a tissue-specific regulation.[182] *In vivo,* chronic infusion of AngII induces downregulation of glomerular ANP receptors (Figure 6).[81,82] Both guanylate cyclase-coupled, high-molecular-weight (R_1 corresponding to A- and/or B-type) and guanylate cyclase-uncoupled, low-molecular-weight (R_2 or C) glomerular ANP receptor subtypes are decreased (Table 2).[82] However, since this downregulation was associated with an elevation of plasma ANP levels, a direct or indirect effect of AngII on ANP receptor regulation could not be substantiated. More recently, we have demonstrated that chronic infusion of AngII in the rat not only blunts glomerular ANP receptor upregulation induced by renal denervation, but downregulates glomerular ANP receptors in intact rats in the absence of any change in plasma ANP levels,[243] suggesting a direct AngII effect on glomerular ANP-glomerular receptor regulation as described in smooth muscle by some investigators.[83] Since phorbol esters, potent activators of protein kinase C, induce ANP receptor downregulation with attenuation of ANP-stimulated cGMP generation,[78,142] this suggests that protein kinase C may be involved in the mechanism of heterologous downregulation of ANP receptors and its effector system by AngII.

During experimental situations in which high plasma levels of AngII occur, ANP receptors in glomeruli and vascular tissue may not behave in a parallel fashion. High levels of either endogenous or exogenous AngII shift to the left the relaxation induced by ANP on AngII-precontracted vascular strips (Figures 2 and 5), in spite of downregulation of vascular ANP receptors.[81,82] On the other hand, the infusion of AngII induces downregulation of glomerular ANP receptors (Figure 6),[81,82] shifting to the right cGMP production by isolated glomeruli upon stimulation with ANP (Figure 7).[82] These results emphasize the tissue specificity which may characterize the intracellular interaction of ANP and AngII.

The overlapping distribution of ANP and AngII receptors in the kidney, as in other target organs, suggests a functional interaction which has already

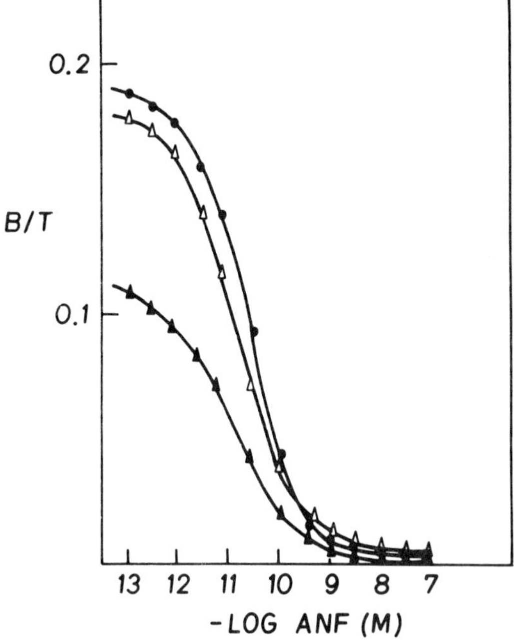

FIGURE 6. Representative curves of ^{125}I-atrial natriuretic peptide (ANP or ANF) binding to a glomerular preparation from rats infused with AngII, in the presence of increasing concentrations of unlabeled ANF, expressed as bound/total (B/T) vs. unlabeled, ANF. ●, Control; △, 200 ng/kg/min AngII; ▲, 800 ng/kg/min AngII. (From Gauquelin, G. et al., *J. Hypertens.*, 9, 1151, 1991. With permission.)

been referred to. High density for AngII receptors has been identified in glomeruli,[48,49,51,52,183-185] medulla,[184,185] brush border and basolateral membranes,[186,187] and in preglomerular vessels.[188] Most glomerular AngII receptors are localized in mesangial cells,[52,189] in contrast to those for ANP, which are predominantly distributed in the podocytes of the visceral epithelium and only partially on mesangial cells.[52] This differential localization of AngII and ANP glomerular receptors may explain some of their respective glomerular and hemodynamic effects as discussed earlier. It has been accepted that glomerular AngII receptors are regulated in a classical fashion. Maneuvers such as a high-sodium diet, which decreases endogenous plasma AngII levels, upregulate glomerular AngII receptors.[181,191,192] On the other hand, high circulating levels of AngII by either infusion or a low-salt diet, may induce downregulation of glomerular AngII receptors.[190,192] This glomerular angiotensin receptor regulation is reciprocal to that reported for glomerular ANP receptors under similar experimental situations.[65,177-180] This classical conception of glomerular AngII receptors has been somewhat shaken by the demonstration

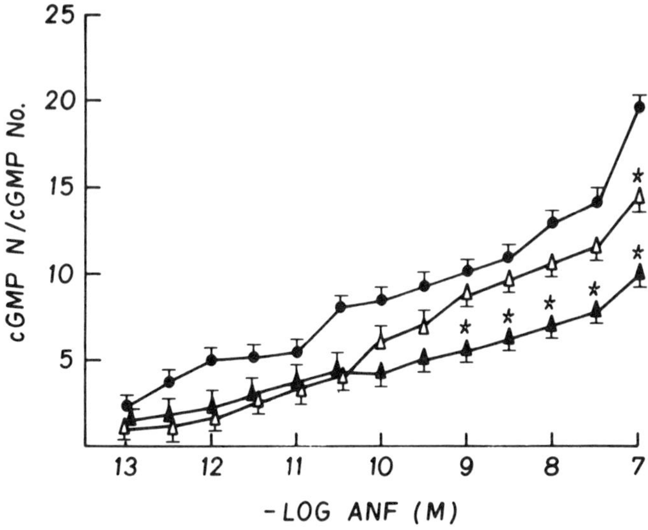

FIGURE 7. Atrial natriuretic peptide (ANP or ANF)-stimulated cGMP accumulation by rat glomeruli in AngII-infused rats. ●, Control; △, 200 ng/kg/min AngII; ▲, 800 ng/kg/min AngII. $*p < 0.05$, vs. control. (From Gauquelin, G. et al., *J. Hypertens.*, 9, 1151, 1991. With permission.)

that infusionof subpressor doses of AngII during several hours actually increased not only the density and affinity of glomerular AngII receptors, but AngII postreceptor effects as well, including an enhanced glomerular contraction when exposed to AngII.[192] It would seem, thus, that in glomeruli, AngII and ANP receptors may share some regulatory mechanisms: AngII infusion downregulates ANP receptors and at the same time upregulates its own receptors.[193] On the other hand, renal denervation increases the density of both ANP[181] and AngII glomerular receptors,[194] the latter probably mediated by the intrarenal renin-angiotensin system.[195]

D. BRAIN

The antagonistic relationship between AngII and ANP described in the peripheral tissues can be further extended to the central nervous system. Biological actions elicited by AngII at the level of central nervous system, such as effects on blood pressure, water and salt intake, and release of pituitary hormones, are opposite to those of ANP. This close AngII-ANP relationship is further strengthened by the overlapping localization of their receptors in a number of brain structures, which constitutes a morphological basis for functional antagonism between these two peptides. The mechanism underlying the AngII-ANP interaction in the brain is largely unresolved, but it is known that AngII and ANP do not interact with the same receptors. Hence, the biological interactions are probably not due to occupation of ANP receptors by AngII (or opposite, AngII receptors by ANP), but are due to postreceptor interactions. It seems that these antagonistic interactions may take place on single neurons (if they are bearing receptors for both peptides) and/or on distinct neurons (if they harbor receptors for one of the peptides). These two possibilities became apparent in experiments designed to measure the effects of AngII and ANP on activity of neurons in the subfornical organ in rat brain slice preparations.[196] In these experiments, some neurons responded to both AngII (with excitation) and ANP (with inhibition) whereas other neurons responded only to either one of these peptides or were unresponsive. ANP selectively depressed the AngII-induced excitation in almost all subfornical organ neurons tested, although it had only very weak inhibitory effects on the spontaneous activity. The mechanism whereby ANP exerts this action is unknown but may include presynaptic inhibition exerted by ANP acting as a neuromodulator. ANP could possibly inhibit release or stimulate uptake of potential neurotransmitters, e.g., certain monoamines. This probably is the case in the septum and hypothalamus. In these structures, ANP decreased the concentration of dopamine and its metabolites, in contrast to AngII, which increased their levels in these brain regions.[197] The simultaneous administration of AngII and ANP resulted in a marked reduction of the AngII-induced increases in dopamine and its metabolites, suggesting that the antagonistic AngII-ANP interaction is likely to be mediated, at least in part, by the dopaminergic system.[197] Changes in central norepinephrine stores may also account for the antagonistic relationship between these peptides. It has been demonstrated in brain slice preparations that AngII decreased while ANP increased neuronal norepinephrine uptake in the hypothalamus and medulla oblongata.[198] In these preparations ANP reversed the effects produced by AngII, suggesting that some AngII-ANP interactions could be mediated by the central noradrenergic system.

1. Effects on Blood Pressure

The mutually antagonistic relationship between AngII and ANP is apparent in the regulation of blood pressure. A number of laboratories have reported that the pressor effect of AngII injected intracerebroventricularly (icv) into the forebrain ventricles was attenuated by concurrent icv injection of ANP.[199-205] This effect has not been observed when AngII was given via the intravenous route at the same dose, suggesting that the attenuation of blood pressure elevation is centrally mediated and involves an interaction between brain AngII and ANP.[201] Brain AngII controls three major mechanisms of blood pressure elevation: activation of the sympathetic nervous system, inhibition of the baroreceptor reflex, and stimulation of AVP secretion.[206,207] Brain ANP has been shown to produce effects opposite to that of brain AngII on all these three variables.

Evidence for changes in sympathetic nervous activity in response to central AngII or ANP is based, among others, on pharmacological experiments, on direct measurement of the levels of catecholamines in the blood, or on recordings of electrical activity of sympathetic nerves. It has been shown that AngII facilitates noradrenergic neurotransmission by increasing the release of norepinephrine.[208] ANP, in contrast, inhibits norepinephrine release from peripheral sympathetic nerve terminals.[209-211] Moreover, it has been demonstrated that centrally administered AngII caused hypertension with elevation of plasma epinephrine level.[203] ANP given by the same route blunted the central AngII-induced pressor response in conscious, freely moving rats. The attenuation of the plasma epinephrine response by ANP may indicate inhibition of the sympathetic nervous system activity which causes the depressor effect.[203] Furthermore, it has been observed that in rats with sinoaortic denervation, AngII upon icv administration causes an increase in renal sympathetic nerve activity (RSNA), whereas ANP applied in the same way decreased this activity.[202] These results are in agreement with previous studies performed by Schultz et al.[212] which indicated that centrally administered ANP decreases sympathetic outflow. These investigators pointed at the importance of sinoaortic denervation when performing such studies, since in animals with intact baroreceptor mechanisms there would be a reflex change in sympathetic nerve activity secondary to the pressor or depressor responses of tested peptides. This may explain the contradictory reports on the effect of AngII[213] or ANP[200] on sympathetic nerve activity. It is worth mentioning that the reduction in RSNA induced by icv ANP may contribute not only to the lowering of blood pressure but also to natriuresis which is produced by central administration of ANP.[214]

As noted before, brain AngII and ANP may exert their antagonistic effects on regulation of blood pressure by a reciprocal action on the baroreceptor reflex. The first synapse of most vagal primary afferent fibers involved in baroreflex occurs in the nucleus of the solitary tract. AngII and ANP and their respective receptors are present in this structure.[32,54,215,216] Microinjection of AngII into the region of the nucleus of the solitary tract yielded a significant dose-dependent elevation in blood pressure, suggesting that neuronally released AngII may control blood pressure to some extent by dampening the relay synapse of the baroreflex located in the nucleus of the solitary tract.[217] In contrast, microinjections of ANP into this brain region produced a decline in arterial pressure.[218,219] Moreover, single neurons excited by the injection of ANP (which was followed by a decline in blood pressure) were also stimulated by the activation of arterial baroreceptors and inhibited by baroreceptor unloading,[219] which supports the hypothesis that ANP acts in the nucleus of the solitary tract to produce facilitated baroreceptor responses.

Brain AngII and ANP may also result in their antagonistic effects on blood pressure regulation by their influence on AVP release. It has been demonstrated that AngII introduced icv potentiated AVP release,[220] whereas icv infusion of ANP significantly inhibited AVP secretion in normally hydrated rats.[221] The increase in the concentration of AVP in plasma induced by AngII was dose-dependently inhibited by the icv injection of ANP.[220] Despite conflicting reports, the site of action of ANP influencing AVP secretion appears to be in the hypothalamus,[221,222] rather than the neural terminals of AVP-containing neurons which lie outside the blood-brain barrier in the posterior pituitary. AVP-producing cells, such as neurons in the paraventricular hypothalamic nucleus, contain receptors for both of the peptides[32,54] and are responsive to the stimulation of AngII[222] and ANP,[223] as revealed by electrophysiological studies. Indirect effects of these peptides may also be mediated through activation of neurons in the subfornical organ (also containing a high density of AngII and ANP receptors) projecting to the magnocellular neurons in the paraventricular hypothalamic nucleus via the median preoptic nucleus.[224] It is worth noting that AVP secretion produced by icv administration of ANP may be responsible also in part for increase in urinary volume which is produced by icv ANP.[204,205,214]

2. Effects on Water Intake

Antagonism between brain AngII and ANP is very striking in regulation of water intake. There are a number of reports indicating that drinking induced by AngII is antagonized by centrally applied ANP in rats.[9,225,226] Studies with antibodies neutralizing ANP[227] suggest that ANP may have a modulatory rather than a direct regulatory effect on water intake and may act only when drinking mechanisms are stimulated. Since AngII and ANP and their respective receptors are present in brain areas known to be involved in the central control of drinking,[32,54,215,216] the antagonistic regulation of water intake by brain AngII and ANP may be mediated by the presence of their receptors at common sites.

A proposed model for the neural circuitry involved in drinking behavior[228] suggests that the anteroventral third ventricular (AV3V) region of the brain, particularly the median preoptic nucleus, receives information from receptors present in the subfornical organ and organum vasculosum of the lamina terminalis (for circulating peptides) and from receptors present in the organum vasculosum of the lamina terminalis and median preoptic nucleus (for brain peptides), integrates information derived from them and from peripheral pressor/volume receptors, and sends the information to higher regions involved in drinking behavior.

3. Effects on Sodium Intake

An antagonistic relationship between brain AngII and ANP is also evident in the modulation of sodium appetite. In sodium-depleted rats (which have an exaggerated urge to consume salted water over tap water) sodium appetite can be significantly blunted by icv ANP.[229-231] Similarly, in spontaneously hypertensive rats, which also have exaggerated sodium appetite, icv infusion of ANP causes a preferential intake of water over hypertonic saline. No significant effects were observed in control Wistar Kyoto rats.[232] In contrast, icv administration of AngII has been demonstrated to induce sodium intake and alter drinking preference from water to saline.[233,234] A probable site for this action lies in the AV3V region of the brain, which is known to be critical for salt appetite and which was shown to contain a high density of receptors for both AngII and ANP.[32,54]

4. Effects on Pituitary Hormone Secretion

Brain AngII and ANP exert functionally antagonistic effects on secretion of pituitary hormones, e.g., AVP (reviewed earlier) or ACTH. Centrally administered AngII increased ACTH release while ANP administered by the same route attenuated AngII-induced ACTH secretion in freely moving rats.[203] With regard to the mechanism of the observed antagonism, it may involve at least two substances released from the hypothalamus: corticotropin-releasing factor (CRF, the most important and potent ACTH-releasing factor) as well as AVP, which also has ACTH-releasing activity and is able to potentiate the release of ACTH directly or via CRF. The central AngII-induced ACTH release is thought to be mediated by CRF secretion from the hypothalamus.[235] ANP has also been shown to directly inhibit CRF release from the hypothalamus *in vitro*.[236] Thus, AngII and ANP may interact at the hypothalamic level to inhibit ACTH secretion by regulating CRF and AVP release.[203]

IV. MOLECULAR INTERACTIONS OF ANGIOTENSIN II AND ANP RECEPTORS

The mechanisms of interaction of ANP and AngII have been discussed when reviewing interactions in different organs. We will now summarize some of these results and review some recent data on possibly important new molecular interactions between ANP and AngII receptor subtypes. Current information suggests that molecular interactions between both systems can occur through heterologous regulation of ANP receptors or through modulation of signal transduction through ''cross-talk'' between receptors. As mentioned previously, there is evidence that AngII may or may not down-regulate ANP receptors in blood vessels and glomeruli.

When vascular smooth muscle cells are exposed to AngII, they develop an enhanced sensitivity to ANP and respond to it with exaggerated generation of cGMP (Figures 3 and 4),[83,84] without change in receptor density or proportion of receptor subtypes (guanylate cyclase-containing or not) to explain these results.[84] Recent evidence showing that ATP regulates guanylate cyclase activity of ANP receptors provides a model for understanding possible regulation of cGMP generation.[237] A protein kinase-like domain functions as a regulatory element modulated by ATP, and the latter markedly stimulates guanylate cyclase activity. Deletion of this domain results in loss of the regulatory effects of ATP, although the cyclase remains active. Recently, Bottari et al.[238] have shown that angiotensin AT_2 receptors in rat adrenal glomerulosa and PC12W cells inhibit particulate guanylate cyclase activity, both basal and ANP stimulated. Orthovanadate, but not okadaic acid, inhibits this effect, suggesting that a phosphotyrosine phosphatase is involved. AngII induced a rapid orthovanadate-sensitive dephosphorylation of phosphotyrosine-containing proteins. This suggests that angiotensin AT_2 receptors stimulate protein tyrosine phosphatase activity to produce inhibition of guanylate cyclase via the phosphorylation state of a tyrosine residue in the protein kinase domain of the ANP receptor. This is a novel mechanism for heterologous modulation of ANP receptor activity. In mesangial cells AngII inhibits ANP-induced cGMP accumulation by an early, transient, calcium-dependent, calmodulin-stimulated cyclic nucleotide phosphodiesterase. AngII-induced inhibition of cGMP was blunted by W7 and by isobutylmethylxanthine, which restored the generation of cGMP.[142] This data is in agreement with findings of Smith and Lincoln[239] in vascular smooth muscle cells supporting the hypothesis that AngII decreases cGMP accumulation by stimulating its hydrolysis by a calcium-activated phosphodiesterase. These effects may be acute, in contrast to more chronic effects (over 18 h) showing sensitization by AngII of cGMP responsiveness to ANP in vascular smooth muscle cells (Figures 3 and 4).[83,84] In mesangial cells, there is also a relatively rapid effect (obtained by a 15-min treatment with AngII prior to ANP) which was protein kinase C dependent (inhibited by H7),[142] as previously mentioned. Inhibition of renal glomerular ANP receptor-coupled guanylate cyclase by partially purified protein kinase C has also been reported.[240]

The effects of ANP on AngII responses are induced by cGMP and may involve decreased generation of inositol phosphates,[241] interference with calcium fluxes,[67-71,101] particularly through T-type calcium channels,[114,115] intracellular calcium release,[68-72] calcium extrusion from the cell,[74,75] blockade of protein kinase C[242] or of protein phosphorylation.[67,116] It is difficult at present to conclude which is the most important site for molecular interaction of the signal transduction systems of ANP and AngII receptors.

V. CONCLUSION

The overlap of AngII and ANP functions and the opposing effects of both systems which attenuate each other's actions suggest indeed that AngII and ANP exert a physiological antagonism which contributes to homeostasis of blood pressure and body fluid balance. Although there is a dearth of information available, the exact cellular and molecular mechanisms involved in interactions between these peptides require further study.

ACKNOWLEDGMENTS

The authors thank Ms. Angie Poliseno for typing this manuscript. The work of the authors' laboratories was supported by grants MA 9112 and MT 10667 and a group grant to the Multidisciplinary Research Group on Hypertension, from the Medical Research Council of Canada, and grants from the Heart and Stroke Foundation of Quebec.

REFERENCES

1. **De Bold, A.J., Borenstein, H.B., Veress, A.T., and Sonnenberg, H.,** A rapid and potent natriuretic response to intravenous injection of atrial myocardial extract in rats, *Life Sci.,* 28, 89, 1981.
2. **Sudoh, T., Kangawa, K., Minamino, N., and Matsuo, H.,** A new natriuretic peptide in porcine brain, *Nature,* 332, 78, 1988.
3. **Sudoh, T., Minamino, N., Kangawa, K., and Matsuo, H.,** C-type natriuretic peptide (CNP) a new peptide of natriuretic peptide family identified in porcine brain, *Biochem. Biophys. Res. Commun.,* 168, 863, 1990.
4. **Camargo, M.J.F., Kleinert, H.D., Atlas, S.A., Sealey, J.E., Laragh, J.H., and Maack, T.,** Ca-dependent hemodynamic and natriuretic effects of atrial extraction in isolated kidney, *Am. J. Physiol.,* 246, F447, 1984.
5. **Burnett, J.C., Jr., Granger, J.P., and Opgenorth, T.G.,** Effects of synthetic atrial natriuretic factor on renal function and renin release, *Am. J. Physiol.,* 247, F863, 1984.
6. **Currie, M.G., Geller, D.M., Cole, B.R., Boylan, J.G., Scheng, W.Y., Holmberg, S.W., and Needleman, P.,** Bioactive cardiac substances: potent vaso-relaxant activity in mammalian atria, *Science,* 221, 71, 1983.
7. **Chartier, L., Schiffrin, E.L., Thibault, G., and Garcia, R.,** Atrial natriuretic factor inhibits the stimulation of aldosterone secretion by angiotensin II, ACTH and potassium in vitro and angiotensin II-induced steroidogenesis in vivo, *Endocrinology,* 115, 2026, 1984.
8. **Atarashi, K., Mulrow, P.J., Franco-Saenz, R., Snajdar, R., and Rapp, J.,** Inhibition of aldosterone production by an atrial extract, *Science,* 224, 992, 1984.
9. **Antunes-Rodrigues, J., McCann, S.M., Rogers, L.C., and Samson, W.K.,** Atrial natriuretic factor inhibits dehydration and angiotensin II-induced water intake in the conscious, unrestrained rat, *Proc. Natl. Acad. Sci. U.S.A.,* 82, 8720, 1985.
10. **Samson, W.K.,** Atrial natriuretic factor inhibits dehydration and hemorrhage-induced vasopressin release, *Neuroendocrinology,* 40, 277, 1985.
11. **Debinski, W., Kuchel, O., Garcia, R., Buu, N.T., Racz, K., Cantin, M., and Genest, J.,** Atrial natriuretic factor inhibits the sympathetic nervous activity in one-kidney, one clip hypertension in the rat, *Proc. Soc. Exp. Biol. Med.,* 181, 173, 1986.
12. **Garcia, R., Thibault, G., Cantin, M., and Genest, J.,** Effects of a purified atrial natriuretic factor on rat and rabbit vascular strips and vascular beds, *Am. J. Physiol.,* 247, R34, 1984.
13. **Mendelsohn, F.A.O., Allen A.M., Chai, S.Y., Sexton, P.M., and Figdor, R.,** Overlapping distributions of receptors for atrial natriuretic peptide and angiotensin II visualized by in vitro autoradiography: morphological basis of physiological antagonism, *Can. J. Physiol. Pharmacol.,* 65, 1517, 1987.

14. **Ballermann, B.J., Hoover, R.L., Karnovsky, M.J., and Brenner B.M.,** Physiologic regulation of atrial natriuretic factor peptide receptors in rat renal glomeruli, *J. Clin. Invest.,* 76, 2049, 1985.
15. **Gauquelin, G., Garcia, R., Carrier, T., Cantin, M., Gutkowska, J., Thibault, G., and Schiffrin, E.L.,** Glomerular ANF receptor regulation during changes in sodium and water metabolism, *Am. J. Physiol.,* 254 (Renal Fluid Electrolyte Physiol.), F51, 1988.
16. **Napier, M.A., Vandlen, R.L., Albers-Schonberg, G., Nutt, R.F., Brady, S., Lyle, T., Winquist, R., Faison, E.P., Heinel, L.A., and Blaine, E.H.,** Specific membrane receptors for atrial natriuretic factor in renal and vascular tissues, *Proc. Natl. Acad. Sci. U.S.A.,* 81, 5946, 1984.
17. **Schiffrin, E.L., Chartier, L., Thibault. G., St-Louis, J., Cantin, M., and Genest, J.,** Vascular and adrenal receptors for atrial natriuretic factor in the rat, *Circ. Res.,* 56, 801, 1985.
18. **Quirion, R., Dalpé, M., De Léan, A., Gutkowska, J., Cantin, M., and Genest, J.,** Atrial natriuretic factor (ANF) binding sites in brain and related structures, *Peptides,* 5, 1167, 1984.
19. **Bianchi, C., Gutkowska, J., Thibault, G., Garcia, R., Genest, J., and Cantin, M.,** Radioautographic localization of ^{125}I-atrial natriuretic factor (ANF) receptors in rat tissues, *Histochemistry,* 82, 441, 1985.
20. **Quirion, R., Dalpé, M., and Dam, T.V.,** Characterization and distribution of receptors for the atrial natriuretic peptides in mammalian brain, *Proc. Natl. Acad. Sci. U.S.A.,* 83, 174, 1986.
21. **Mukhopadhyay, A.K., Bohnet, H.G., and Leidenberger, F.A.,** Testosterone production by mouse Leydig cells is stimulated in vitro by atrial natriuretic factor, *FEBS Lett.,* 202, 111, 1986.
22. **Bianchi, C., Thibault, G., De Léan, A., Genest, J., and Cantin, M.,** Atrial natriuretic factor binding sites in the jejunum, *Am. J. Physiol.,* 256 (Gastrointestinal Liver Physiol.), G436, 1989.
23. **Chabot, J.G., Morel, G., Kopelman, H., Belles-Isles, M., and Heisler, S.,** Atrial natriuretic factor and exocrine pancreas, *Pancreas,* 2, 404, 1987.
24. **Nair, B.G., Steinke, L., Yu, Y.-M., Rashed, H.M., Seyer, J.M., and Patel, T.B.,** Increase in the number of atrial natriuretic hormone receptors in regenerating rat liver, *J. Biol. Chem.,* 266, 567, 1991.
25. **Schiffrin, E.L., Deslongchamps, M., and Thibault, G.,** Platelet binding sites for atrial natriuretic factor in humans. Characterization and effects of sodium intake, *Hypertension,* 8 (Suppl. II), 6, 1986.
26. **Hirata, Y., Tomita, M., Takada, S., and Yoshimi, H.,** Vascular receptor binding activities and cyclic GMP responses by synthetic human and rat atrial natriuretic peptide (ANP) and receptor down-regulation by ANP, *Biochem. Biophys. Res. Commun.,* 128, 538, 1985.
27. **Schiffrin, E.L., Poissant, L., Cantin, M., and Thibault, G.,** Receptors for atrial natriuretic factor in cultured vascular smooth muscle cells, *Life Sci.,* 38, 817, 1986.
28. **Leitman, D.C. and Murad, F.,** Comparison of binding and cyclic GMP accumulation by atrial natriuretic peptides in endothelial cells, *Biochim. Biophys. Acta,* 885, 74, 1986.
29. **Chai, S.Y., Sexton, P.M., Allen, A.M., Figdor, R., and Mendelsohn, F.A.O.,** In vitro autoradiographic localization of ANP receptor in rat kidney and adrenal gland, *Am. J. Physiol.,* 250, F753, 1986.
30. **Morel, G., Chabot, J.G., Garcia-Caballero, T., Gossard, F., Dihl, F., Belles-Isles, M., and Heisler, S.,** Synthesis, internalization and localization of atrial natriuretic peptide in rat adrenal medulla, *Endocrinology,* 123, 149, 1988.
31. **Rutherford, R.A.D., Wharton, J., Needleman, P., and Polak, J.M.,** Autoradiographic discrimination of brain and atrial natriuretic peptide-binding sites in the rat kidney, *J. Biol. Chem.,* 266, 5819, 1991.

32. **Gibson, T.R., Wildey, G.M., Manaker, S., and Glembostki, C.C.,** Autoradiographic localization and characterization of atrial natriuretic peptide binding sites in the rat central nervous system and adrenal gland, *J. Neurosci.*, 6, 2004, 1986.
33. **Maack, T., Suzuki, M., Almeida, F.A., Nussenzveig, D., Scarborough, R.M., McEnroe, G.A., and Lewicki, J.A.,** Physiological role of silent receptors of atrial natriuretic factor, *Science,* 228, 675, 1987.
34. **Fuller, G., Porter, J.G., Arfsten, A.E., Miller, J., Schilling, J.W., Scarborough, R.M., Lewicki, J.A., and Schenk, D.B.,** Atrial natriuretic peptide clearance receptor. Complete sequence and functional expression of cDNA clones, *J. Biol. Chem.*, 163, 9395, 1988.
35. **Chinkers, M., Garbers, D.L., Chang, M.-S., Lowe, D.G., Chiu, H., Goeddel, D.W., and Schulz, S.,** Molecular cloning of a new type of cell surface receptor: a membrane form of guanylate cyclase is an atrial natriuretic peptide receptor, *Nature (London),* 338, 78, 1989.
36. **Koller, K.J., Lowe, D.G., Bennett, G.L., Minamino, N., Kangawa, K., Matsuo, H., and Goeddel, D.V.,** Selective activation of the B natriuretic peptide receptor by C-type natriuretic peptide (CNP), *Science,* 252, 120, 1991.
37. **Brown, J., Salas, S.P., Singleton, A., Polak, J.M., and Dollery, C.T.,** Autoradiographic localization of atrial natriuretic peptide receptor sub-types in rat kidney, *Am. J. Physiol.*, 259, F26, 1990.
38. **Martin, E.R., Lewicki, J.A., Scarborough, R.M., and Ballerman, B.J.,** Expression and regulation of ANP receptor sub-types in rat renal glomeruli and papillae, *Am. J. Physiol.*, 257, F649, 1989.
39. **Nuglozeh, E., Gauquelin, G., Garcia, R., Tremblay, J., and Schiffrin, E.L.,** Atrial natriuretic peptide in renal papillae in DOCA-salt hypertensive rats, *Am. J. Physiol.*, 259, F130, 1990.
40. **Brown, J. and Czarnecki, A.,** Distribution of atrial natriuretic peptide receptor subtypes in rat brain, *Am. J. Physiol.*, 258, R1078, 1990.
41. **Konrad, E.M., Thibault, G., Pelletier, S., Genest, J., and Cantin, M.,** Brain natriuretic peptide binding sites in rats: in vitro autoradiographic study, *Am. J. Physiol.*, 259, E246, 1990.
42. **Konrad, E.M., Thibault, G., Schiffrin, E.L., and Cantin, M.,** Atrial natriuretic factor receptor subtypes in the rat central nervous system, *Hypertension,* 17, 1144, 1991.
43. **Konrad, E.M., Thibault, G., and Schiffrin, E.L.,** Autoradiographic visualization of the natriuretic peptide receptor-β in rat tissues, *Regul. Peptides,* 39, 177, 1992.
44. **Konrad, E.M., Thibault, G., and Schiffrin, E.L.,** Atrial natriuretic factor binding sites in rat area postrema: autoradiographic study, *Am. J. Physiol.*, 263, R747, 1992.
45. **Chiu, A.T., Hublin, W.F., McCall, D.F., Ardecky, R.J., Carini, J.V., Pease, L.J., Wong, P.C., Wexler, R.R., Johnson, A.L., and Timmermans, P.B.M.W.M.,** Identification of angiotensin II receptor subtypes, *Biochem. Biophys. Res. Commun.*, 165, 196, 1989.
46. **Chiu, A.T., McCall, D.E., Ardecky, R.J., Duncia, J.V., Nguyen, T.T., and Timmermans, P.M.B.W.M.,** Angiotensin II receptor sub-types and their selective ligands, *Receptor,* 1, 33, 1990.
47. **Bumpus, F.M., Catt, K.T., Chiu, A.T., De Gasparo, M., Goodfriend, T., Husain, A., Peach, M.J., Taylor, D.G., Jr., and Timmermans, P.B.M.W.M.,** Nomenclature for angiotensin receptors. A report of the nomenclature committee of the Council of High Blood Pressure Research, *Hypertension,* 17, 720, 1991.
48. **Sechi, L.A., Grady, E.F., Griffin, C.A., Kalinyak, J.E., and Schambelan, M.,** Distribution of angiotensin II receptor sub-types in rat and human kidney, *Am. J. Physiol.*, 262, F236, 1992.
49. **Gröne, H.-J., Simon, M., and Fuchs, E.,** Autoradiographic characterization of angiotensin receptor subtypes in fetal and adult human kidney, *Am. J. Physiol.*, 262, F326, 1992.

50. **Loutzenhiser, R., Epstein, M., Hayashi, K., Takenaka, T., and Forster, H.,** Characterization of the renal microvascular effects of angiotensin II antagonist, DuP 753: studies in isolated perfused hydronephrotic kidneys, *Am. J. Hypertens.*, 4, S309, 1991.
51. **Mendelsohn, F.A.O., Dunbar, M., Allen, A., Chou, S.-T., Millan, M.A., Aguilera, G., and Catt, K.J.,** Angiotensin II receptors in the kidney, *Fed. Proc. Fed. Am. Soc. Exp. Biol.*, 45, 1420, 1986.
52. **Bianchi, C., Gutkowska, J., Thibault, G., Garcia, R., Genest, J., and Cantin, M.,** Distinct localization of atrial natriuretic factor and angiotensin II binding sites in the glomerulus, *Am. J. Physiol.*, 251, F594, 1986.
53. **Bianchi, C., Ballak, M., Gutkowska, J., Charbonneau, C., Genest, J., and Cantin, M.,** Localization of ^{125}I-atrial natriuretic factor (ANF) binding sites in rat renal medulla. A light and electron microscopic radio-autographic study, *J. Histochem. Cytochem.*, 35, 149, 1987.
54. **Mendelsohn, F.A.O., Quirion, R., Saavedra, J.M., Aguilera, G., and Catt, K.J.,** Autoradiographic localization of angiotensin II receptors in rat brain, *Proc. Natl. Acad. Sci. U.S.A.*, 81, 1575, 1984.
55. **Gehlert, D.R., Gackenheimer, S.L., Reel, J.K., Lin, H.-S., and Steinberg, M.I.,** Non-peptide angiotensin II receptor antagonists discriminate subtypes of ^{125}I-angiotensin II binding sites in the rat brain, *Eur. J. Pharmacol.*, 187, 123, 1990.
56. **Song, K., Allen, A.M., Paxinos, G., and Meldelsohn, F.A.O.,** Angiotensin II receptor subtypes in rat brain, *Clin. Exp. Pharmacol. Physiol.*, 18, 93, 1991.
57. **Rowe, B.P., Grove, K.L., Saylor, D.L., and Speth, R.C.,** Discrimination of angiotensin II receptor subtype distribution in the brain using non-peptidic receptor antagonistics, *Regul. Peptides,* 33, 45, 1991.
58. **Cornwell, T.L. and Lincoln, T.M.,** Regulation of phosphorylase A formation and calcium content in aortic smooth muscle and smooth muscle cells: effects of atrial natriuretic peptide II, *J. Pharmacol. Exp. Ther.*, 247, 524, 1988.
59. **Proctor, K.G. and Bealer, S.L.,** Selective antagonism of hormone-induced vasoconstriction by synthetic atrial natriuretic factor in the rat microcirculation, *Circ. Res.*, 61, 42, 1987.
60. **Lappe, R.W., Todt, J.A., and Wendt, R.L.,** Effects of atrial natriuretic factor on the vasoconstrictor actions of the renin-angiotensin system in conscious rats, *Circ. Res.*, 61, 134, 1987.
61. **Parkes, D.G., Coghlan, J.P., McDougall, J.G., and Scoggins, B.A.,** Effects of atrial natriuretic factor on pressor responsiveness to angiotensin II, norepinephrine, and vasopressin in conscious sheep, *J. Cardiovasc. Pharmacol.*, 15, 16, 1990.
62. **Yasujima, M., Abe, K., Kohzuki, M., Tanno, M., Kasai, Y., Sato, M., Omata, K., Kudo, K., Takeuchi, K., Hiwatari, M., Kimura, T., Yoshinaga, K., and Inagami, T.,** Effect of atrial natriuretic factor on angiotensin II-induced hypertension in rats, *Hypertension,* 8, 748, 1986.
63. **Schiffrin, E.L. and St-Louis, J.,** Decreased density of vascular receptors for atrial natriuretic peptide in DOCA-salt hypertensive rats, *Hypertension,* 9, 504, 1987.
64. **Schiffrin, E.L., St-Louis, J., Garcia, R., Thibault, G., Cantin, M., and Genest, J.,** Vascular and adrenal binding sites for atrial natriuretic factor. Effects of sodium and hypertension, *Hypertension,* 8 (Suppl. I), 141, 1986.
65. **Garcia, R., Gauquelin, G., Cantin, M., and Schiffrin, E.L.,** Glomerular and vascular atrial natriuretic factor receptors in saralasin-sensitive and -resistant two-kidney, one clip hypertensive rats, *Circ. Res.*, 63, 503, 1988.
66. **Hassid, A.,** Atriopeptin II decreases cytosolic free Ca in cultured vascular smooth muscle cells, *Am. J. Physiol.*, 251, C681 – C686, 1986.
67. **Paglin, S., Takuwa, Y., Kamm, K.E., Stull, J.T., Gavras, H., and Rasmussen, H.,** Atrial natriuretic peptide inhibits the agonist-induced increase in extent of myosin light chain phosphorylation in aortic smooth muscle, *J. Biol. Chem.*, 263, 13120, 1988.

68. **Meisheri, K.D., Taylor, C.J., and Saneii, H.,** Synthetic atrial peptide inhibits intracellular calcium release in smooth muscle, *Am. J. Physiol.*, 250, C171, 1986.
69. **Hassid, A.,** Atriopeptins decrease resting and hormone-elevated cytosolic Ca in cultured mesangial cells, *Am. J. Physiol.*, 253, F1077, 1987.
70. **Appel, R.G., Dubyak, G.R., and Dunn, M.J.,** Effect of atrial natriuretic factor on cytosolic free calcium in rat glomerular mesangial cells, *FEBS Lett.*, 224, 396, 1987.
71. **Meyer-Lehnert, H., Caramelo, C., Tsai, P., and Schrier, R.W.,** Interaction of atriopeptin III and vasopressin on calcium kinetics and contraction of aortic smooth muscle cells, *J. Clin. Invest.*, 82, 1407, 1988.
72. **Cornwell, T.L. and Lincoln T.M.,** Regulation of phosphorylase A formation and calcium content in aortic smooth muscle cells: effects of atrial natriuretic peptide II, *J. Pharmacol. Exp. Ther.*, 247, 524, 1988.
73. **Capponi, A.M., Lew, P.D., Wüthrich, R., and Vallotton, M.B.,** Effects of atrial natriuretic peptide on the stimulation by angiotensin II of various target cells, *J. Hypertens.*, 4, S61, 1986.
74. **Rashatwar, S.S., Cornwell, T.L., and Lincoln, T.M.,** Effects of 8-bromo-cGMP on Ca^{2+} levels in vascular smooth muscle cells: possible regulation of Ca^{2+}-ATPase by cGMP-dependent protein kinase, *Proc. Natl. Acad. Sci. U.S.A.*, 84, 5685, 1987.
75. **Cornwell, T.L. and Lincoln, T.M.,** Regulation of intracellular Ca^{2+} levels in cultured vascular smooth muscle cells. Reduction of Ca^{2+} by atriopeptin and 8-bromo-cyclic GMP is mediated by cyclic GMP-dependent protein kinase, *J. Biol. Chem.*, 264, 1146, 1989.
76. **Rapoport, R.M.,** Cyclic guanosine monophosphate inhibition of contraction may be mediated through inhibition of phosphatidylinositol hydrolysis in rat aorta, *Circ. Res.*, 58, 407, 1986.
77. **Hirata, M., Chang, C.H., and Murad, F.,** Stimulatory effects of atrial natriuretic factor on phosphoinositide hydrolysis in cultured bovine smooth muscle cells, *Biochim. Biophys. Acta,* 1010, 346, 1989.
78. **Hirata, Y.,** Heterologous down-regulation of vascular atrial natriuretic peptide receptors by phorbol esters, *Biochem. Biophys. Res. Commun.*, 152, 1097, 1988.
79. **Nambi, P., Whitman, M., Aiyar, N., Stassen, F., and Crooke, S.T.,** An activator of protein kinase C (phorbol dibutyrate) attenuates atrial-natriuretic-factor-stimulated cyclic GMP accumulation in smooth-muscle cells, *Biochem. J.*, 244, 481, 1987.
80. **Hirata, Y., Emori, T., Ohta, K., Scichiri, M., and Marumo, F.,** Vasoconstrictor-induced heterologous down-regulation of vascular atrial natriuretic peptide receptor, *Eur. J. Pharmacol.*, 164, 603, 1989.
81. **Cachofeiro, V., Schiffrin, E.L., Bonhomme, M.-C., Cantin, M., and Garcia, R.,** In vivo heterologous regulation of rat glomerular and vascular atrial natriuretic factor receptors by angiotensin II, *J. Hypertens.*, 8, 1077, 1990.
82. **Gauquelin, G., Schiffrin, E.L., and Garcia, R.,** Downregulation of glomerular and vascular atrial natriuretic factor receptor subtypes by angiotensin II, *J. Hypertens.*, 9, 1151, 1991.
83. **Chabrier, P.E., Roubert, P., Lonchampt, M.-O., Plas, P., and Braquet, P.,** Regulation of atrial natriuretic factor receptors by angiotensin II in rat vascular smooth muscle cells, *J. Biol. Chem.*, 263, 13199, 1988.
84. **Schiffrin, E.L., Turgeon, A., Tremblay, J., and Deslongchamps, M.,** Effects of ANP, angiotensin, vasopressin, and endothelin on ANP receptors in cultured rat vascular smooth muscle cells, *Am. J. Physiol.*, 260, H58, 1991.
85. **Roubert, P., Lonchampt, M.O., Chabrier, P.E., Plas, P., Goulin, J., and Braquet, P.,** Down-regulation of atrial natriuretic factor receptors and correlation with cGMP stimulation in rat cultured vascular smooth muscle cells, *Biochem. Biophys. Res. Commun.*, 148, 61, 1987.
86. **Bellemann, P. and Neuser, D.,** Modulation of ANP receptor-mediated cGMP accumulation by atrial natriuretic peptides and vasopressin in A10 vascular smooth muscle cells, *J. Receptor Res.*, 8, 407, 1988.

87. **Cachofeiro, V., Schiffrin, E.L., Cantin, M., and Garcia, R.**, Glomerular and vascular atrial natriuretic factor receptors in adrenalectomized rats, *Am. J. Physiol.*, 256, R1250, 1989.
88. **Cachofeiro, V., Schiffrin, E.L., Cantin, M., and Garcia, R.**, Glomerular and vascular atrial natriuretic factor receptors in the cardiomyopathic hamster: correlation with the peptide biological effects, *Cardiovasc. Res.*, 24, 843, 1990.
89. **Kohno, M., Yokokawa, K., Horio, T., Yasunari, K., Murakawa, K.-I., and Takeda, T.**, Atrial and brain natriuretic peptides inhibit the endothelin-1 secretory response to angiotensin II in porcine aorta, *Circ. Res.*, 70, 241, 1992.
90. **De Léan, A., Racz, K., Gutkowska, J., Buu, N.T., Cantin, M., and Genest, J.**, Specific receptor-mediated inhibition by synthetic atrial natriuretic factor of hormone-stimulated steroidogenesis in cultured bovine adrenal cells, *Endocrinology,* 115, 1636, 1984.
91. **Goodfriend, T.L., Elliott, M.E., and Atlas, S.A.**, Actions of synthetic atrial natriuretic factor on bovine adrenal glomerulosa, *Life Sci.*, 35, 1675, 1984.
92. **Campbell, W.B., Currier, M.G., and Needleman, P.**, Inhibition of aldosterone biosynthesis by atriopeptins in rat adrenal cells, *Circ. Res.*, 57, 113, 1985.
93. **Hirata, Y., Ishii, M., Sugimoto, T., Matsuoka, H., Ishimitsu, T., Atarashi, K., Sugimoto, T., Miyata, A., Kangawa, K., and Matsuo, H.**, Relationship between the renin-aldosterone system and atrial natriuretic polypeptide in rats, *Clin. Sci.*, 72, 165, 1987.
94. **Chartier, L. and Schiffrin, E.L.**, Atrial natriuretic peptide inhibits the effect of endogenous angiotensin II on plasma aldosterone in conscious sodium-depleted rats, *Clin. Sci.*, 72, 31, 1987.
95. **Brands, M.W. and Freeman, R.H.**, Aldosterone and renin inhibition by physiological levels of atrial natriuretic factor, *Am. J. Physiol.*, 254, R1011, 1988.
96. **Metzler, C.H. and Ramsay, D.J.**, Physiological doses of atrial peptide inhibit angiotensin II-stimulated aldosterone secretion, *Am. J. Physiol.*, 256, R1155, 1989.
97. **Ohashi, M., Fujio, N., Kato, K., Nawata, H., Ibayashi, H., and Matsuo, H.**, Effect of human α-atrial natriuretic polypeptide on adrenocortical function in man, *J. Endocrinol.*, 110, 287, 1986.
98. **Vierhapper, H., Nowotny, P., and Waldhäusl, W.**, Prolonged administration of human atrial natriuretic peptide in healthy men. Reduced aldosteronotropic effect of angiotensin II, *Hypertension,* 8, 1040, 1986.
99. **Anderson, J.V., Struthers, A.D., Payne, N.N., Slater, J.D.H., and Bloom, S.R.**, Atrial natriuretic peptide inhibits the aldosterone response to angiotensin II in man, *Clin. Sci.*, 70, 507, 1986.
100. **Shenker, Y.**, Atrial natriuretic hormone and aldosterone regulation in salt-depleted state, *Am. J. Physiol.*, 257, E583, 1989.
101. **Chartier, L. and Schiffrin, E.L.**, Role of calcium in effects of atrial natriuretic peptide on aldosterone production in adrenal glomerulosa cells, *Am. J. Physiol.*, 252, E485, 1987.
102. **Aguilera, G.**, Differential effects of atrial natriuretic factor on angiotensin II- and adrenocorticotropin-stimulated aldosterone secretion, *Endocrinology,* 120, 299, 1987.
103. **Chartier, L. and Schiffrin, E.L.**, Atrial natriuretic peptide inhibits the stimulation of aldosterone secretion by ACTH in vitro and in vivo, *Proc. Soc. Exp. Biol. Med.*, 182, 132, 1986.
104. **Waldman, S.A., Rapoport, R.M., Fiscus, R.R., and Murad, F.**, Effects of atriopeptin on particulate guanylate cyclase from rat adrenal, *Biochim. Biophys. Acta,* 845, 298, 1985.
105. **Farese, R.V., Larson, R.E., and Davis, J.**, Rapid effects of angiotensin II on polyphosphoinositide metabolism in the rat adrenal glomerulosa, *Endocrinology,* 114, 302, 1984.

106. **Enyedi, P., Bucki, B., Mucsi, I., and Spät, A.,** Polyphosphoinositide metabolism in adrenal glomerulosa cells, *Mol. Cell Endocrinol.*, 41, 105, 1985.
107. **Capponi, A.M., Lew, P.D., Jornot, L., and Valloton, M.B.,** Correlation between cytosolic free Ca^{2+} and aldosterone production in bovine adrenal glomerulosa cells. Evidence for a difference in the mode of action of angiotensin II and potassium, *J. Biol. Chem.*, 259, 8863, 1984.
108. **Kojima, I., Kojima, K., and Rasmussen, H.,** Characteristics of angiotensin II-, K^+, and ACTH-induced calcium influx in adrenal glomerulosa cells. Evidence that angiotensin II, K^+, and ACTH may act upon a common calcium channel, *J. Biol. Chem.*, 260, 9171, 1985.
109. **Matsuoka, H., Ishii, M., Hirata, Y., Atarashi, K., Sugimoto, T., Kangawa, K., and Matsuo, H.,** Evidence for lack of a role of cGMP in effect of α-hANP on aldosterone inhibition, *Am. J. Physiol.*, 252, E643, 1987.
110. **Takagi, M., Takagi, M., Franco-Saenz, F., Shier, D., and Mulrow, P.J.,** Effect of atrial natriuretic factor on calcium fluxes in adrenal glomerulosa cells, *Hypertension*, 11, 433, 1988.
111. **Isales, C.M., Bollag, W.B., Kiernan, L.C., and Barrett, P.Q.,** Effect of ANP on sustained aldosterone secretion stimulated by angiotensin II, *Am. J. Physiol.*, 256, C89, 1989.
112. **Apfeldorf, W.J., Isales, C.M., and Barrett, P.Q.,** Atrial natriuretic peptide inhibits the stimulation of aldosterone secretion but not the transient increase in intracellular free calcium concentration induced by angiotensin II addition, *Endocrinology*, 122, 1460, 1988.
113. **Lotshaw, D.P., Franco-Saenz, R., and Mulrow, P.J.,** Atrial natriuretic peptide inhibition of calcium ionophore A23187-stimulated aldosterone secretion in rat adrenal glomerulosa cells, *Endocrinology*, 129, 2305, 1991.
114. **McCarthy, R.T., Isales, C.M., Bollag, W.B., Rasmussen, H., and Barrett, P.Q.,** Atrial natriuretic peptide differentially modulates T- and L-type calcium channels, *Am. J. Physiol.*, 258, F473, 1990.
115. **Barrett, P.Q., Isales, C.M., Bollag, W.B., and McCarthy, R.T.,** Ca^{2+} channels and aldosterone secretion: modulation by K^+ and atrial natriuretic peptide, *Am. J. Physiol.*, 261, F706, 1991.
116. **Elliott, M.E. and Goodfriend, T.L.,** Effects of atrial natriuretic peptide, angiotensin, cyclic AMP, and potassium on protein phosphorylation in adrenal glomerulosa cells, *Life Sci.*, 41, 2517, 1987.
117. **Takagi, M., Takagi, M., Franco-Saenz, R., Mulrow, P.J., and Reimann, E.M.,** Effects of dibutyryl adenosine 3′,5′-monophosphate angiotensin II, and atrial natriuretic factor on phosphorylation of a 17,600-Dalton protein in adrenal glomerulosa cells, *Endocrinology*, 123, 2419 – 2423, 1988.
118. **Anand-Srivastava, M.B., Franks, D.J., Cantin, M., and Genest, J.,** Atrial natriuretic factor inhibits adenylate cyclase activity, *Biochem. Biophys. Res. Commun.*, 121, 855, 1984.
119. **Woodcock, E.A. and Johnston, C.,** Inhibition of adenylate cyclase in rat adrenal glomerulosa cells by angiotensin II, *Endocrinology*, 155, 337, 1984.
120. **Cogan, M.C.,** Atrial natriuretic factor can increase renal solute excretion primarily by raising glomerular filtration, *Am. J. Physiol.*, 250, F710, 1986.
121. **Salazar, F.J., Fiksen-Olsen, M.J., Opgenorth, T.J., Granger, J.P., Burnett, J.C., Jr., and Romero, J.C.,** Renal effects of ANP without changes in glomerular filtration rate and blood pressure, *Am. J. Physiol.*, 251, F532, 1986.
122. **Hammond, T.G., Yusupi, A.H.K., Knox, F.G., and Dousa, T.P.,** Administration of atrial natriuretic factor inhibits sodium-coupled transport in proximal tubules, *J. Clin. Invest.*, 75, 1983, 1985.
123. **Briggs, J.P., Steipe, B., Schubert, G., and Schnerman, J.,** Micropuncture studies of the renal effects of atrial natriuretic substance, *Pfluegers Arch.*, 395, 271, 1982.

124. **Sonnenberg, H., Cupples, W.A., de Bold, A.J., and Veress, A.T.,** Intrarenal localization of the natriuretic effect of cardiac atrial extract, *Can. J. Physiol., Pharmacol.,* 60, 1149, 1982.
125. **Borenstein, H.B., Cupples, W.A., Sonnenberg, H., and Veress, A.T.,** The effect of atrial extract on renal haemodynamics and urinary secretion in anaesthetized rats, *J. Physiol. (London)* 334, 138, 1983.
126. **Ichikawa, I., Miele, J.F., and Brenner, B.M.,** Reversal of renal cortical actions of angiotensin II by verapamil and manganese, *Kidney Int.,* 16, 137, 1979.
127. **Hornych, H., Beaufils, M., and Richet, G.,** The effect of exogenous angiotensin on superficial and deep glomeruli in the rat kidney, *Kidney Int.,* 2, 336, 1972.
128. **Harris, P.J. and Young, J.A.,** Dose-dependent stimulation and inhibition of proximal sodium reabsorption, *Pfluegers Arch.,* 367, 295, 1977.
129. **Schuster, V.L., Kokko, J.P., and Jacobson, H.R.,** Angiotensin II directly stimulates sodium transport in rabbit proximal convoluted tubules, *J. Clin. Invest.,* 73, 507, 1984.
130. **Bokman, L., Ericson, A.-C., Åherg, B., and Vefendahl, H.R.,** Flow resistance of the interlobular artery in the rat kidney, *Acta Physiol. Scand.,* 111, 159, 1981.
131. **Källsbog, O., Linbonr, L.O., Ulfendahl, H.R., and Wolgast, M.,** Hydrostatic pressures within the vascular structures of the rat kidney, *Pfluegers Arch.,* 363, 205, 1956.
132. **Trippodo, N.C., MacPhee, A.A., and Cole, F.A.,** Natriuretic response to atrial natriuretic factor enhanced by angiotensin II and vasopressin, *J. Hypertens.,* 2 (Suppl. 3), 289, 1984.
133. **Showalter, C.J., Zimmerman, R.S., Schwab, T.R., Edwards, B.S., Opgenorth, T.J., and Burnett, J.C., Jr.,** Renal response to atrial natriuretic factor is modulated by intrarenal angiotensin II, *Am. J. Physiol.,* 254, R453, 1988.
134. **Salazar, F.J., Granger, J.P., Fiksen-Olsen, M.J., Bentley, M.D., and Romero, J.C.,** Possible modulatory role of angiotensin II on atrial peptide-induced natriuresis, *Am. J. Physiol.,* 253, F880, 1987.
135. **Mizelle, H.L., Hall, J.E., and Hildebrandt, D.A.,** Atrial natriuretic peptide and pressure natriuresis: interactions with the renin-angiotensin system, *Am. J. Physiol.,* 257, 1169, 1989.
136. **Blantz, R.C., Ronnen, K.S., and Tucker, B.J.,** Angiotensin II effects upon the glomerular microcirculation and ultrafiltration coefficient of the rat, *J. Clin. Invest.,* 56, 419, 1976.
137. **Fried, T.A., McCoy, R.N., Osgood, R.W., and Stein, J.H.,** Effect of atriopeptin II on determinants of glomerular filtration rate in the in vitro perfused dog glomerulus, *Am. J. Physiol.,* 250, F1119, 1986.
138. **Ausiello, D.A., Kreisberg, J.I., Roy, C., and Karnowsky, M.J.,** Contraction of cultured rat glomerular cells of apparent mesangial origin after stimulation with angiotensin II and arginine vasopressin, *J. Clin. Invest.,* 65, 754, 1980.
139. **Appel, R.G., Wang, J., Swenson, M.S., and Dunn, M.J.,** A mechanism by which atrial natriuretic factor mediates its glomerular actions, *Am. J. Physiol.,* 251, F1036, 1986.
140. **Barrio, V., De Arriba, G., Lopez-Novoa, J.M., and Rodriguez-Puyol, D.,** Atrial natriuretic factor inhibits glomerular contraction induced by angiotensin II and platelet activating factor, *Eur. J. Pharmacol.,* 135, 93, 1987.
141. **Burnett, R., Ortiz, P.A., Blaufox, S., Singer, S., Nord, E.P., and Ramsammy, L.,** Atrial natriuretic factor alters phospholipid metabolism in mesangial cells, *Am. J. Physiol.,* 258, C37, 1990.
142. **Haneda, M., Kikkawa, R., Maeda, S., Togawa, M., Koya, D., Horide, N., Kajiwara, N., and Shigeta, Y.,** Dual mechanism of angiotensin II inhibits ANP-induced mesangial cyclic GMP accumulation, *Kidney Int.,* 40, 188, 1991.
143. **Denton, K.M., Fennessy, P.A., Alcorn, D., and Anderson, W.P.,** Morphometric analysis of the actions of angiotensin II on renal arterioles and glomeruli, *Am. J. Physiol.,* 262, F367, 1992.

144. **Xie, M.-H., Liu, F.-Y., Wong, P.C., Timmermans, P.B.M.W.M., and Cogan, M.G.,** Proximal nephron and renal effects of DuP 753, a nonpeptide angiotensin II antagonist, *Kidney Int.*, 38, 473, 1990.
145. **Baum, M. and Toto, R.D.,** Lack of direct effect of atrial natriuretic factor in the rabbit proximal tubule, *Am. J. Physiol.*, 250, F66, 1986.
146. **Harris, P.J., Thomas, D., and Morgan, T.O.,** Atrial natriuretic peptide inhibits angiotensin-stimulated proximal tubular sodium and water reabsorption, *Nature (London)*, 326, 697, 1987.
147. **Garvin, J.L.,** Inhibition of Jv by ANP in rat proximal straight tubules requires angiotensin, *Am. J. Physiol.*, 257, F907, 1989.
148. **Gomes, G.N. and Aires, M.M.,** Interaction of atrial natriuretic factor and angiotensin II in proximal HCO_3 reabsorption, *Am. J. Physiol.*, 262, F303, 1992.
149. **Liu, F.-Y. and Cogan, M.G.,** Atrial natriuretic factor does not inhibit basal or angiotensin II-stimulated proximal transport, *Am. J. Physiol.*, 255, F434, 1988.
150. **Nonoguchi, H., Knepper, M.A., and Manganello, V.C.,** Effects of atrial natriuretic factor on cyclic guanosine monophosphate and cyclic adenosine monophosphate accumulation in microdissected nephron segments from rats, *J. Clin. Invest.*, 79, 500, 1987.
151. **Terada, Y., Moriyama, T., Martin, B.M., Knepper, M.A., and Garcia-Perez, A.,** RT-PCR microlocalization of mRNA for guanylyl cyclase-coupled ANP receptor in rat kidney, *Am. J. Physiol.*, 261, F1080, 1991.
152. **De Leon, H., Gauquelin, G., Thibault, G., and Garcia, R.,** Characterization of ANP receptors in rat renal microvessels, *Hypertension*, 20, 434A, 1992.
153. **Brown, G.P. and Venuto, R.C.,** Angiotensin II receptors in rabbit renal preglomerular vessels, *Am. J. Physiol.*, 255, E16, 1988.
154. **Carmines, P.K. and Fleming, J.T.,** Control of the renal microvasculature by vasoactive peptides, *FASEB J.*, 4, 3300, 1990.
155. **Faubert, P.F., Chou, S.-Y., and Porush, J.G.,** Regulation of papillary plasma flow by angiotensin II, *Kidney Int.*, 32, 472, 1987.
156. **Aalkjaer, C., Mulvany, M.J., and Nyborg, N.C.B.,** Atrial natriuretic factor causes specific relaxation of rat renal arcuate arteries, *Br. J. Pharmacol.*, 86, 447, 1985.
157. **Marin-Grez, M., Fleming, J.T., and Steinhausen, M.,** Atrial natriuretic peptide causes preglomerular vasodilation and post glomerular vasoconstriction in rat kidney, *Nature (London)*, 342, 473, 1986.
158. **Kimura, K., Hirata, Y., Nanba, S., Tojo, A., Matsuoka, H., and Sugimoto, T.,** Effects of atrial natriuretic peptide on renal arterioles: morphometric analysis using microvascular casts, *Am. J. Physiol.*, 259, F936, 1990.
159. **Lanese, D.M., Yuan, B.H., Falk, S.A., and Conger, J.D.,** Effects of atriopeptin III on isolated rat afferent and efferent arterioles, *Am. J. Physiol.*, 161, F1102, 1991.
160. **Veldkamp, P.J., Carmines, P.K., Inscho, E.W., and Navar, L.G.,** Direct evaluation of the microvascular actions of ANP in juxtamedullary nephrons, *Am. J. Physiol.*, 254, F440, 1988.
161. **Ohishi, K., Hishida, A., and Honda, N.,** Direct vasodilatory action of atrial natriuretic factor on canine glomerular afferent arterioles, *Am. J. Physiol.*, 255, F415, 1988.
162. **Takezawa, K., Cowley, A.W., Jr., Skelton, M., and Roman, R.J.,** Atriopeptin III alters renal medullary hemodynamics and the pressure-diuresis response in rats, *Am. J. Physiol.*, 252, F992, 1987.
163. **Mendez, R.E., Dunn, B.R., Troy, J.L., and Brenner, B.M.,** Atrial natriuretic factor and furosemide effects on hydraulic pressure in renal papillae, *Kidney Int.*, 34, 36, 1988.
164. **Garcia, R., Thibault, G., Gutkowska, J., Hamet, P., Cantin, M., and Genest, J.,** Effect of chronic infusion of synthetic atrial natriuretic factor (ANP 8-33) in conscious two-kidney, one-clip hypertensive rats, *Proc. Soc. Exp. Biol. Med.*, 178, 155, 1985.

165. **Obana, K., Naruse, M., Naruse, K., Sakurai, H., Demura, H., Inagami, T., and Shizume, K.**, Synthetic rat atrial natriuretic factor inhibits in vitro and in vivo renin secretion in rats, *Endocrinology,* 117, 1282, 1985.
166. **Kurtz, A., Bruna, R.D., Pfeilschifter, J., Taugner, R., and Bauer, C.**, Atrial natriuretic peptide inhibits renin release from juxtaglomerular cells by a cyclic GMP-mediated process, *Proc. Natl. Acad. Sci. U.S.A.*, 83, 4769, 1986.
167. **Antonipillai, I., Vogelsang, J., and Horton, R.**, Role of atrial natriuretic factor in renin release, *Endocrinology,* 119, 318, 1986.
168. **Henrich, W.L., McAlister, E.A., Smith, P.B., Lipton, J., and Campbell, W.B.**, Direct inhibitory effect of atriopeptin III on renin release in primate kidney, *Life Sci.*, 41, 259, 1987.
169. **Cuneo, R.C., Espiner, E.A., Nicholls, M.G., Yandle, T.G., Joyce, S.L., and Gilchrist, N.L.**, Renal, hemodynamic and hormonal responses to atrial natriuretic peptide infusions in normal man, and effect of sodium intake, *J. Clin. Endocrinol. Metab.*, 63, 946, 1986.
170. **Struthers, A.D., Anderson, J.V., Payne, N., Causon, R.C., Slater, J.D.H., and Bloom, S.R.**, The effect of atrial natriuretic peptide on plasma renin activity, plasma aldosterone and urinary dopamine in man, *Eur. J. Clin. Pharmacol.*, 31, 223, 1986.
171. **Anderson, J.V., Donckier, J., Payne, N.N., Beacham, J., Slater, J.D.H., and Bloom, S.R.**, Atrial natriuretic peptide: evidence of action as a natriuretic hormone at physiologic plasma concentrations in man, *Clin. Sci.*, 72, 305, 1987.
172. **McMurray, J.J. and Struthers, A.D.**, Atrial natriuretic factor inhibits isoproterenol- and furosemide-stimulated renin release in humans, *Hypertension,* 13, 9, 1989.
173. **Volpe, M., Odell, G., Kleinert, H.D., Camargo, M.J., Laragh, J.H., Lewicki, J.A., Maack, T., and Vaughan, E.D., Jr.**, Antihypertensive and aldosterone-lowering effect of synthetic atrial natriuretic factor in renin-dependent renovascular hypertension, *J. Hypertens.*, 2 (Suppl. III), 13, 1984.
174. **Garcia, R., Gutkowska, J., Cantin, M., and Thibault, G.**, Renin dependency of the effect of chronically administered atrial natriuretic factor in two-kidney, one-clip rats, *Hypertension,* 9, 88, 1987.
175. **Crozier, J.G., Nicholls, M.G., Ikram, H., Espiner, E.A., Gomez, H.J., and Warner, J.**, Haemodynamic effects of atrial peptide infusion in heart failure, *Lancet,* 2, 1242, 1986.
176. **Cody, R.J., Atlas, S.A., Laragh, J.H., Kubo, S.H., Couitt, A.B., Ryman, K.S., Shankovich, A., Pondolfino, K., Clark, M., Camargo, M.J.F., Scarborough, R.M., and Lewicki, J.A.**, Atrial natriuretic factor in normal subjects and heart failure patients: plasma levels and renal, hormonal and hemodynamic responses to peptide infusion, *J. Clin. Invest.*, 78, 1362, 1986.
177. **Ballerman, B.J., Bloch, K.D., Seidman, J.G., and Brenner, B.M.**, Atrial natriuretic peptide transcription, secretion and glomerular receptor activity during mineralocorticoid escape in the rat, *J. Clin. Invest.*, 78, 840, 1986.
178. **Gauquelin, G., Thibault, G., Cantin, M., Schiffrin, E.L., and Garcia, R.**, Glomerular atrial natriuretic factor receptors during rehydration: plasma NH_2- and COOH-terminal levels, *Am. J. Physiol.*, 255, F621, 1988.
179. **Gauquelin, G., Schiffrin, E.L., Cantin, M., and Garcia, R.**, Specific binding of atrial natriuretic factor to renal glomeruli in DOCA and DOCA-salt treated rats: correlation with atrial and plasma levels, *Biochem. Biophys. Res. Commun.*, 145, 522, 1987.
180. **Garcia, R., Gauquelin, G., Cantin, M., and Schiffrin, E.L.**, Renal glomerular atrial natriuretic factor receptors in one-kidney, one clip rats, *Hypertension,* 63, 56, 1988.
181. **De Leon, H. and Garcia, R.**, Regulation of atrial natriuretic factor receptor sub-types by renal sympathetic nerves, *Am. J. Physiol.*, 260, R1043, 1991.

182. **Garcia, R., Bonhomme, M.-C., and Schiffrin, E.L.,** Divergent regulation of atrial natriuretic factor receptors in high-output heart failure, *Am. J. Physiol.*, in press.
183. **Sraer, J.D., Sraer, J., Ardaillou, R., and Minoune, O.,** Evidence for renal glomerular receptors for angiotensin II, *Kidney Int.*, 6, 241, 1974.
184. **Mendelsohn, F.A.O., Dunbar, M., Allen, A., Chou, S.T., Millan, M.A., Aguilera, G., and Catt, K.J.,** Angiotensin II receptors in the kidney, *Fed. Proc.*, 45, 1420, 1986.
185. **Yamada, H., Sexton, P.M., Chai, S.Y., Adam, W.R., and Mendelsohn, F.O.A.,** Angiotensin II receptors in the kidney. Localization and physiological significance, *Am. J. Hypertens.*, 3, 250, 1990.
186. **Brown, G.P. and Douglas, J.D.,** Angiotensin II binding sites on isolated rat renal brush border membranes, *Endocrinology,* 111, 1830, 1982.
187. **Cox, H.M., Munday, K.A., and Poat, J.A.,** Location of [^{125}I]-angiotensin II receptors on rat kidney cortex epithelial cells, *Br. J. Pharmacol.*, 82, 891, 1984.
188. **Brown, G.P. and Venuto, R.C.,** Angiotensin II receptors in rabbit renal preglomerular vessels, *Am. J. Physiol.*, 255, E16, 1988.
189. **Osborne, M.J., Droz, B., Meyer, P., and Morel, F.,** Angiotensin II: renal localization in glomerular mesangial cells by autoradiography, *Kidney Int.*, 8, 245, 1975.
190. **Beaufils, M., Sraer, J., Lepreux, C., and Ardaillou, R.,** Angiotensin II binding to renal glomeruli from sodium-loaded and sodium depleted rats, *Am. J. Physiol.*, 230, 1187, 1976.
191. **Skorecki, K.L., Ballerman, B.J., Rennke, H.G., and Brenner, B.M.,** Angiotensin II receptor regulation in isolated renal glomeruli, *Fed. Proc.*, 42, 3064, 1983.
192. **Bellemi, A. and Wilkes, B.M.,** Mechanism of sodium modulation of glomerular angiotensin receptors in the rat, *J. Clin. Invest.*, 74, 1593, 1984.
193. **Douglas, G.J.,** Subpressor infusions of angiotensin II alter glomerular binding, prostaglandin E_2, and cyclic AMP production, *Hypertension,* 9 (Suppl. III), 49, 1987.
194. **Tucker, B.J., Mundy, C.A., Mariejewski, A.R., Printz, M.P., Ziegler, M.G., Pelayo, J.C., and Blantz, R.C.,** Changes in glomerular hemodynamic response to angiotensin II after subacute renal denervation, *J. Clin. Invest.*, 78, 680, 1986.
195. **Wilkes, B.M., Pion, I., Sollo, S., Michaels, S., and Hiesel, G.,** Intrarenal renin-angiotensin system modulates glomerular angiotensin receptors in the rat, *Am. J. Physiol.*, 254, F345, 1988.
196. **Hattori, Y., Kasai, M., Uesugi, S., Kawata, M., and Yamashita, H.,** Atrial natriuretic polypeptide depresses angiotensin II induced excitation of neurons in the rat subfornical organ in vitro, *Brain Res.*, 443, 355, 1988.
197. **Nakao, K., Katsuura, G., Morii, N., Itoh, H., Shiono, S., Yamada, T., Sugawara, A., Sakamoto, M., Saito, Y., Eigyo, M., Matsushita, A., and Imura, H.,** Mechanism of central action of atrial natriuretic polypeptide (ANP), in *Advances in Atrial Peptide Research,* Vol. 2, Brenner, B.M. and Laragh, J.H., Ed., Raven Press, New York, 1988, 327–331.
198. **Fernandez, B.E., Dominguez, A.E., Vatta, M.S., Mendez, M.A., Bianchiotti, L.G., and Martinez Seeber, A.,** Atrial natriuretic peptide increases norepinephrine uptake in the central nervous system, *Arch. Int. Physiol. Biochim.*, 98, 127, 1990.
199. **Shimizu, T., Katsuura, G., Nakamura, M., Nakao, K., Morii, N., Itoh, Y., Shiono, S., and Imura, H.,** Effect on intracerebroventricular atrial natriuretic polypeptide on blood pressure and urine production in rats, *Life Sci.*, 39, 1263, 1986.
200. **Kannan, H., Ueta, Y., Nakamura, T., Yamashita, H., and Hayashida, Y.,** Effects of centrally administered atrial natriuretic polypeptide on sympathetic nerve activity and blood flow to the kidney in conscious rats, *Neurosci. Lett.*, 116, 123, 1990.
201. **Casto, R., Hilbig, J., Schroeder, G., and Stock G.,** Atrial natriuretic factor inhibits central angiotensin II pressor responses, *Hypertension,* 9, 473, 1987.
202. **Stelle, M.K., Gardner, D.G., Xie, P., and Schultz, H.D.,** Interactions between ANP and ANG II regulating blood pressure and sympathetic outflow, *Am. J. Physiol.*, 260, R1145, 1991.

203. **Makino, S., Hashimoto K., and Ota, Z.,** Atrial natriuretic polypeptide attenuates central angiotensin II-induced catecholamine and ACTH secretion, *Brain Res.*, 501, 84, 1989.
204. **Yoshida, K., Kawano, Y., Kawamura, M., Kuramochi, M., and Omae, T.,** Effects of intracerebroventricular atrial natriuretic factor on angiotensin II- or sodium-induced blood pressure elevation and natriuresis, *J. Hypertens.*, 7, 639, 1989.
205. **Al-Barazanji, K.A. and Balment, R.J.,** The renal and vascular effects of central angiotensin II and atrial natriuretic factor in the anaesthetized rat, *J. Physiol. (London),* 423, 485, 1990.
206. **Phillips, M.I.,** Function of angiotensin in the central nervous system, *Annu. Rev. Physiol.*, 49, 413, 1987.
207. **Unger, T., Rascher, W., Schuster, C., Pavlovitch, R., Schomig, A., Dietz, R., and Ganten, D.,** Central blood pressure effect of substance P and angiotensin II: role of the sympathetic nervous system and vasopressin, *Eur. J. Pharmacol.*, 71, 33, 1982.
208. **Meldrum, M.J., Xue, C.S., Badino, L., and Westfall, T.C.,** Angiotensin facilitation of noradrenergic neural transmission in central tissue of the rat: effects of sodium restriction, *J. Cardiovasc. Pharmacol.*, 6, 989, 1984.
209. **Kuchel, O., Debinski, W., Pach, K., Buu, N.T., Garcia, R., Cusson, J.R., Larochelle, P., Cantin, M., and Genest, J.,** An emerging relationship between peripheral sympathetic nervous activity and atrial natriuretic factor, *Life Sci.*, 40, 1545, 1987.
210. **Kakamaru, M. and Inagami, T.,** Atrial natriuretic factor inhibits norepinephrine release evoked by sympathetic nerve stimulation in isolated perfused rat mesenteric arteries, *Eur. J. Pharmacol.*, 123, 459, 1986.
211. **Drewett, J.G., Trachte, G.J., and Marchand, G.R.,** Atrial natriuretic factor inhibits adrenergic and purinergic neurotransmission in the rabbit isolated vas deferens, *J. Pharmacol. Exp. Ther.*, 248, 135, 1988.
212. **Schultz, H.D., Stelle, M.K., and Gardner, D.G.,** Central administration of atrial peptide decreases sympathetic outflow in rats, *Am. J. Physiol.*, 258, R1250, 1990.
213. **Unger, T., Becker, H., Petty, M., Demmert, G., Schneider, B., Ganten, D., and Lang, R.E.,** Differential effects on central angiotensin II and substance P on sympathetic nerve activity in conscious rats, *Circ. Res.*, 56, 563, 1985.
214. **Shoji, M., Kimura, T., Matsui, K., Ota, K., Iitake, K., Inoue, M., Yasujima, M., Abe, K., and Yoshinaga, K.,** Effects of centrally administered atrial natriuretic peptide on renal functions, *Acta Endocrinol. (Copenhagen),* 115, 433, 1987.
215. **Skofisch, G., Jacobowitz, D.M., Eskay, R.L., and Zamir, N.,** Distribution of atrial natriuretic factor-like immunoreactive neurons in the rat brain, *Neuroscience,* 16, 917, 1985.
216. **Lind, R.W., Swanson, L.W., and Ganten, D.,** Organization of angiotensin II immunoreactive cells and fibers in the rat central nervous system. An immunological study, *Neuroendocrinology,* 40, 2, 1985.
217. **Casto, R. and Phillips, M.I.,** Mechanisms of pressor effects by angiotensin in the nucleus tractus solitarius of rats, *Am. J. Physiol.*, 247, R575, 1984.
218. **McKitrick, D.J. and Calaresu, F.R.,** Cardiovascular responses to microinjection of ANF into dorsal medulla of rats, *Am. J. Physiol.*, 255, R182, 1988.
219. **Ermirio, R., Ruggeri, P., Cogo, C.E., Molinari, C., and Calaresu, F.R.,** Neuronal and cardiovascular responses to ANF microinjected into the solitary nucleus, *Am. J. Physiol.*, 256, R577, 1986.
220. **Yamada, T., Nakao, K., Morii, N., Itoh, H., Shiono, S., Sakamoto, M., Sugawara, A., Saito, Y., Ohno, H., Kanai, A., Katsuura, G., Eigyo, M., Matsushita, A., and Imura, H.,** Central effect of atrial natriuretic polypeptide on angiotensin II stimulated vasopressin secretion in conscious rats, *Eur. J. Pharmacol.*, 125, 453, 1986.
221. **Samson, W.K., Aguila, M.C., Martinovic, J., Antunes-Rodrigues, J., and Norris, M.,** Hypothalamic action of atrial natriuretic factor to inhibit vasopressin secretion, *Peptides,* 8, 449, 1987.

222. **Harding, J.W. and Felix, D.,** Angiotensin-sensitive neurons in the rat paraventricular nucleus: relative potencies of angiotensin II and angiotensin III, *Brain Res.,* 410, 130, 1987.
223. **Standaert, D.G., Cechetto, D.F., Needleman, P., and Saper, C.B.,** Inhibition of the firing of vasopressin neurons by atriopeptin, *Nature,* 329, 151, 1987.
224. **Tanaka, J., Saito, H., and Kaba, H.,** Subfornical organ and hypothalamic paraventricular nucleus connections with median preoptic nucleus neurons: an electrophysiological study in the rat, *Exp. Brain Res.,* 68, 579, 1987.
225. **Nakamura, M., Takayanagi, R., and Inagami, T.,** Effect of atrial natriuretic factor on central angiotensin II-induced responses in rats, *Peptides,* 7, 373, 1986.
226. **Masotto, C. and Negro-Vilar, A.,** Inhibition of spontaneous or angiotensin II-stimulated water intake by atrial natriuretic factor, *Brain Res. Bull.,* 15, 523, 1985.
227. **Franci, C.R., Kozlowski, G.P., and McCann, S.M.,** Water intake in rats subjected to hypothalamic immunoneutralization of angiotensin II, atrial natriuretic peptide, vasopressin, or oxytocin, *Proc. Natl. Acad. Sci. U.S.A.,* 86, 2952, 1989.
228. **Johnson, A.K. and Cunningham, J.T.,** Brain mechanisms and drinking: the role of lamina terminalis-associated systems in extracellular thirst, *Kidney Int.,* 21, S35, 1987.
229. **Antunes-Rodrigues, J., McCann, S.M., and Samson, W.K.,** Central administration of atrial natriuretic factor inhibits saline preference in the rat, *Endocrinology,* 118, 1726, 1986.
230. **Fitts, D.A., Thunhorst, R.L., and Simpson, J.B.,** Diuresis and reduction of salt appetite by lateral ventricular infusions of atriopeptin II, *Brain Res.,* 348, 118, 1985.
231. **Tarjan, E., Denton, D.A., and Weisinger, R.S.,** Atrial natriuretic peptide inhibits water and sodium intake in rabbits, *Regul. Peptides,* 23, 63, 1988.
232. **Itoh, H., Nakao, K., Katsuura, G., Morii, N., Shiono, S., Sakamoto, M., Yamada, T., Saito, Y., Matsushita, A., and Imura, H.,** Centrally infused atrial natriuretic polypeptide attenuates exaggerated salt appetite in spontaneously hypertensive rats, *Circ. Res.,* 59, 342, 1986.
233. **Buggy, J. and Fisher, A.E.,** Evidence for a dual central role for angiotensin in water and sodium intake, *Nature,* 250, 733, 1974.
234. **Bryant, R.W., Epstein, A.N., Fitzsimons, J.T., and Fluharty, S.J.,** Arousal of a specific and persistent sodium appetite in the rat with continuous intracerebroventricular infusion of angiotensin II, *J. Physiol. (London),* 301, 365, 1980.
235. **Rivier, C. and Vale, W.,** Effect on angiotensin II on ACTH release in vivo: role of corticotropin-releasing factor, *Regul. Peptides,* 7, 253, 1983.
236. **Takao, T., Hashimoto, K., and Ota, Z.,** Effect of atrial natriuretic peptide on acetylcholine-induced release of corticotropin-releasing factor from rat hypothalamus in vitro, *Life Sci.,* 42, 1199, 1988.
237. **Chinkers, M. and Garbers, D.L.,** The protein kinase domain of the ANP receptor is required for signaling, *Science,* 245, 1392, 1989.
238. **Bottari, S.P., King, I.N., Reichlin, S., Dahlstroem, N.Y., and de Gasparo, M.,** The angiotensin AT_2 receptor stimulates protein tyrosine phosphatase activity and mediates inhibition of particulate guanylate cyclase, *Biochem. Biophys. Res. Commun.,* 183, 206, 1992.
239. **Smith, J.B. and Lincoln, T.M.,** Angiotensin decreases cyclic GMP accumulation produced by atrial natriuretic factor, *Am. J. Physiol.,* 253, C147, 1987.
240. **Ballermann, B.J., Marala, R.B., and Sharma, R.K.,** Characterization and regulation by protein kinase C of renal glomerular atrial natriuretic peptide receptor-coupled guanylate cyclase, *Biochem. Biophys. Res. Commun.,* 157, 755, 1988.
241. **Hirata, M., Kohse, K.P., Chang, C.-H., Ikebe, T., and Murad, F.,** Mechanism of cyclic GMP inhibition of inositol phosphate formation in rat aorta segments and cultured bovine aortic smooth muscle cells, *J. Biol. Chem.,* 265, 1268, 1990.
242. **Sauro, M.D. and Fitzpatrick, D.F.,** Atrial natriuretic peptides inhibit protein kinase C activation in rat aortic smooth muscle, *Peptide Res.,* 3, 138, 1990.

Section V

Angiotensin Regulation of Gene Expression

Chapter 18

MOLECULAR AND CELLULAR BIOLOGY OF ANGIOTENSIN-MEDIATED GROWTH OF THE CARDIOVASCULAR SYSTEM

R.E. Pratt and V.J. Dzau

TABLE OF CONTENTS

0-8493-4622-3/93/$0.00 + $.50

I. INTRODUCTION

Control of vascular smooth muscle cell (VSMC) and myocyte growth is key to the pathophysiology of many vascular diseases such as restenosis, atherosclerosis, and hypertension-induced hypertrophy. This growth control is highly complex and involves the combined action of neural, hormonal, and mechanical factors. This chapter will focus on the hormonal influences on vascular and cardiac structure. Numerous *in vivo* studies have implicated angiotensin II (AngII) in the chronic changes in vascular structure that occur during hypertension.[1,2] Similar studies have implicated AngII in changes of cardiac structure. Multiple lines of investigation suggest that the influence of AngII on growth is due to the induction by AngII of several autocrine growth factors. The molecular and cellular biology of these processes are reviewed in this chapter.

II. CONTROL OF VASCULAR SMOOTH MUSCLE CELL GROWTH

Cultured rat aortic smooth muscle cells have been used as a model system for the studies of the growth-promoting effects of AngII. These cells are isolated from the aortas of young rats by enzymatic dispersal[3] and are maintained in culture for 10 to 12 passages. They are highly differentiated, expressing many of the same muscle-specific proteins found *in vivo.*[3] These cells grow and divide when grown in 10% fetal calf serum but can be made quiescent by incubation in a defined serum-free media containing transferrin, ascorbic acid, and insulin.[4] When quiescent, these cells become more differentiated and are highly responsive to AngII.[5-7] Activation of the vascular smooth muscle AngII receptor results in an increase in phospholipase C activity which generates inositol triphosphate and diacyl glycerol.[8] These metabolites initiate a series of intracellular events such as the transient increase in intracellular calcium and activation of protein kinase C that mediate the immediate events such as contraction. Presumably, these events also mediate the long-term or delayed effects of AngII, such as regulation of gene expression, increases in protein, DNA, and RNA syntheses, and the resultant increase in cell growth and alteration of vascular structure.

Cellular hypertrophy is marked by an increase in the rate of RNA synthesis and subsequently by an increase in protein synthesis, protein content per cell, and cell size. Evidence that AngII is directly involved in the hypertrophic growth of smooth muscle cells has been provided by *in vitro* and *in vivo* studies. *In vitro* AngII treatment of VSMC in culture results in a rapid and sustained increase in the rates of RNA and protein syntheses,[5,9-12] leading to an overall increase in cellular size and protein content as measured by fluorescent-activated cell sorting. Consistent with the prior observations that AngII was a hypertrophic stimulus, little or no increase in the rate of DNA synthesis was observed.

Many of the events associated with cell growth have been examined in other systems. When quiescent cells are induced to proliferate by the addition of serum, a sequence of events is initiated.[13-15] Initially, there is a rapid and transient increase in the expression of the early response genes. The role of these genes in cell growth has been demonstrated by experiments using antisense oligomers or antisense expression vectors showing that inhibition of these genes results in the inhibition of proliferation.[16,17] Moreover, the overexpression of these genes can in some cases lead to a transformed phenotype. Furthermore, the unregulated expression of certain of these genes has been noted in many tumors and neoplasms, which led to their designation as protooncogenes.[13,14]

Accordingly, the effect of AngII on the expression of these proto-oncogenes has been examined. Initially, the expression of *c-fos* and *c-jun* were examined since these are among the earliest genes induced and because the function of these gene products in regulating the expression of other genes has been extensively studied.[18] Treatment of quiescent VSMC resulted, indeed, in the rapid and transient increase in c-fos and c-jun mRNAs.[7,9,19] The kinetics of this induction was similar to that noted after stimulation by serum or growth factors. This was of interest in that, as noted, AngII differs from these other factors in that it induces hypertrophy rather than hyperplasia or true proliferation.

The products of the fos and jun oncogenes form a heterodimeric transcription complex termed AP-1 that result in the regulation of multiple target genes.[32] This regulation occurs via the interaction of AP-1 with a cis sequence within these target genes which is termed phorbol ester (TPA) responsive element or TRE. To investigate if the induction of the fos and jun mRNAs was sufficient to lead to the formation of functional AP-1, a test expression vector containing a 3× tandom repeat of the consensus TRE linked to a reporter gene (choramphenicol acetyl transferase, CAT) was introduced into cells in culture. Subsequent treatment with AngII resulted in a significant increase in CAT production, indicating the presence of functional AP-1.[20] AngII had no effect on CAT expression of a control vector. This AngII-induced stimulation was abolished by a protein kinase C (PKC) inhibitor, staurosporine, and the AngII receptor antagonist, salarlasin. These data suggest that the TRE plays a crucial role in AngII-induced gene expression that is mediated by PKC and we concluded that TRE is one of the AngII-responsive elements.

To examine further the DNA:protein interaction mediating the action of AP-1, we studied renin gene expression as a model system, since putative TREs are present in the human renin gene and this gene is regulated by AngII.[21] Nuclear extracts were prepared from Hep G2 cells treated with phorbol ester (PMA) or AngII for 3 h. Gel mobility shift assays with radiolabeled synthetic human renin TREs and human collagenase TRE (used as a positive control) identified two DNA protein complexes. These complexes were induced by both AngII and PMA treatment. The complexes were specific since their formation was inhibited by synthetic TRE (human renin or collagenase gene), human renin gene fragments containing the TREs, but not by fragments that did not contain TRE, nor by a synthetic human renin cyclic AMP-responsive element. The function of the human renin TRE was examined in a transient transfection system. ptKCAT plasmids containing human renin gene fragments were transfected into Hep G2 cells and the cells treated with AngII. Twelve-hour treatment with AngII resulted in an increase in CAT expression relative to untreated cells, whereas 24- and 48-h treatment with AngII suppressed the expression of CAT. The expression of the plasmids containing fragments lacking a TRE were unaffected by AngII. Taken together, these data indicate that AngII can regulate renin gene expression via TREs presumably due to AP-1 binding.

We next examined the direct transcriptional consequences of fos and jun.[22] Hep G2 cells were transfected with the renin gene-CAT constructs together with a combination of c-fos, c-jun, and/or jun-B expression vectors. We observed a threefold increase in CAT expression of the construct containing TREs, when transfected with the c-fos plus c-jun expression vectors. This effect was abolished by the jun-B expression vector. There was no significant change in the expression of ptkCAT (control vector) or renin-CAT construct which lacks TREs. The above data indicate that AP-1 binds to TREs. Moreover, it suggests that gene transcription can be stimulated by some member of AP-1 (c-jun and c-fos), but negatively regulated by others (jun-B).

The role of these proto-oncogenes in the hypertrophic response such as that to AngII is controversial. On the one hand, Berk and colleagues have shown that down-regulation of PKC with phorbol esters prior to AngII treatment attenuates the AngII induction of c-fos,[11] but has little effect on the subsequent AngII-induced increase in protein synthesis.[21] On the other hand, Rainer et al.[23] and Naftilan et al.[24] have presented preliminary evidence that inhibition of c-fos production with antisense oligomers or expression vectors will attenuate the AngII-induced increase in protein synthesis. This discrepancy may be related to the degree of inhibition of fos production in each case. This important issue clearly requires further investigation.

To provide further information concerning AngII regulation of cellular hypertrophy, the potential regulation of autocrine growth factors was examined. In particular, the expression of platelet-derived growth factor (PDGF) A-chain is of interest, since it has been shown to be expressed by VSMC[25] and is a rather potent growth stimulus for these cells. Indeed, we observed that AngII-treated smooth muscle cells express 10- to 20-fold higher levels of PDGF A-chain mRNA compared to control cultures and that this increase in mRNA resulted in a parallel increase in the synthesis and secretion of the protein.[6] To examine if PDGF expression mediates the AngII-induced hypertrophy of VSMC, we examined the effect of antibodies against PDGF.[27] This antibody, which is specific for PDGF-AA, could inhibit completely the PDGF activity found in the smooth muscle cell conditioned media. Co-administration of the antibody and AngII resulted in a decrease in both protein and

RNA synthesis compared to AngII alone. However, despite a complete inhibition of the secreted PDGF, the decrease in RNA and protein synthesis was only 20%. Thus, these data provide evidence for the role of endogenous PDGF as an autocrine mediator of AngII-mediated vascular hypertrophy. However, other factors must also contribute to this hypertrophic response.

To further examine if the PDGF expression mediated the AngII-induced hypertrophy, the ability of antisense oligonucleotides complementary to PDGF mRNA to suppress the endogenous production of PDGF and to inhibit the AngII-induced hypertrophy of VSMC[27] was examined. Two different antisense oligonucleotides (15mers) directed at the translation initiation site of PDGF-A mRNA were introduced into VSMC by cationic liposome-mediated transfection, prior to AngII exposure. We demonstrated that the antisense oligomers blocked completely the production of PDGF stimulated by AngII. The antisense oligomers significantly attenuated the AngII-induced RNA and protein syntheses by 50 to 60%. This inhibitory effect of the antisense oligomers was greater than that observed with neutralizing anti-PDGF antibodies, suggesting that the action of the endogenously produced PDGF in the smooth muscle cells occurred by both an autocrine and intracrine pathways. Taken together, these data provide evidence that endogenous PDGF plays a role in AngII-mediated vascular hypertrophy.

It is interesting to note that AngII induces the expression of proto-oncogenes and the mitogen PDGF, yet fails to promote smooth muscle cell proliferation. This apparent paradox led us to hypothesize that the growth-promoting effects of AngII may be counteracted by the parallel activation of an antiproliferative pathway.[10] To examine this postulated bifunctional property of AngII, we studied its interaction with the mitogen, basic fibroblast growth factor (bFGF). We observed that the simultaneous incubation of AngII and bFGF resulted in a potentiated mitogen response. However, either pretreatment with AngII prior to bFGF administration, or sequential incubation with AngII after one-hour exposure to bFGF, inhibited bFGF-stimulated DNA synthesis. Similarly, prolonged sequential exposure to AngII increased the proportion of cells in G_0/G_1 arrest and markedly attenuated the increase in cell density induced by bFGF. The inhibitory effect of AngII appeared to be mediated by the protein kinase C (PKC) pathway in that it was mimicked by phorbol myristate acetate (PMA) but not by an inactive phorbol ester, and it was inhibited by PKC downregulation by prolonged PMA pretreatment. We conclude that AngII is a bifunctional modulator of VSMC growth. In addition to its growth-promoting effects, AngII also activates an antiproliferative pathway mediated by PKC.

Since transforming growth factor-beta 1 (TGF-β1) has bifunctional effects on smooth muscle cell growth similar to those of AngII and its synthesis is regulated by PKC activation, we hypothesized that AngII exerted its antiproliferative effect via the autocrine production of TGF-β1 (Figure 1). Indeed, our experiments demonstrated that AngII induced a significant increase in

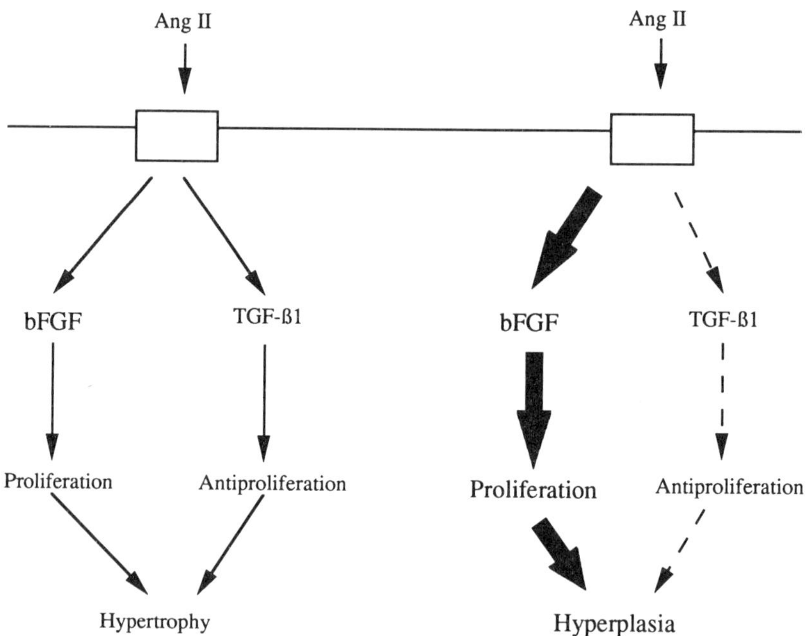

FIGURE 1. Effects of angiotensin II on vascular smooth muscle cells. Under most conditions, angiotensin II induces both a proliferative and an antiproliferative pathway, mediated by bFGF and TGF-β, respectively. This results in a hypertrophic response. However, in some conditions, an imbalance of the proliferative and antiproliferative signals results in cellular hyperplasia.

TGF-β1 mRNA levels in the smooth muscle cells.[11] This response was detectable within 4 h of incubation and was sustained for at least 20 h. The response was dose dependent, with a threshold of approximately 10^{-9} *M*. In order to demonstrate the role of TGF-β1 as a modulator of AngII-induced smooth muscle cell growth, we examined the effect of co-incubation with a neutralizing anti-TGF-β1 antibody.[11] In the presence of preimmune IgG, AngII induced a 100% increase in the rate of protein synthesis as measured by tritiated proline incorporation. Co-incubation with the anti-TGF antibody had no significant effect on the AngII-induced increase in protein synthesis. However, co-incubation with anti-TGF antibody markedly enhanced AngII-stimulated DNA synthesis, as measured by tritiated thymidine incorporation, when compared to the preimmune IgG control. Similar results were obtained when an antisense oligo, directed towards TGF-β mRNA, was employed.[28] Furthermore, while AngII had no effect on cell density when incubated with preimmune IgG, a 50% increase in cell number was observed in the presence of TGF antibody. These data suggest that AngII activates simultaneously a proliferative pathway as well as an antiproliferative pathway (Figure 1). The AngII-induced antiproliferative pathway is mediated by TGF-β1 production, since inhibition of the action of TGF-β1 using specific antibodies or inhibition of TGF-β1 synthesis with antisense oligomer uncovered the proliferative action of AngII.

Further evidence suggesting that the PKC-mediated induction of TGF-β mediates the antiproliferative pathway of AngII has recently been obtained.[29] Exposure of most AngII-responsive cells in culture to AngII results in an increase in particulate PKC levels. As stated above, this increase in PKC mediates the increase in TGF-β mRNA levels. Interestingly, a cell line was isolated which failed to increase PKC in response to AngII. Moreover, these cells respond hyperplastically to AngII, exhibiting a 4- to 5-fold increase in DNA synthesis and a 2- to 3-fold increase in cell number. Accordingly, we examined the induction of growth factors by AngII in these cells. AngII resulted in the expected increase in bFGF and PDGF. However, no active TGF-β was observed either basally or after AngII exposure. This suggested that the mitogenic response to AngII was due to the lack of active TGF-β induction. Consistent with this, exogenously added TGF-β blocked the mitogenic response to AngII in this cell line.

The growth factors which mediate the proliferative pathway induced by AngII were examined next. Based on the above experiments using PDGF, it was initially assumed that this factor mediated the proliferative pathway. However, when the action or synthesis of TGF-β was inhibited with antisera or antisense oligomers, respectively, the resultant increase in DNA synthesis could not be blocked with antisera or antisense oligos directed towards PDGF.[28] Thus, while the data suggest that autocrine PDGF may mediate the hypertrophic effects of AngII, the hyperplastic effects of AngII must be mediated by another growth factor. Further investigations showed that AngII does in fact induce other proliferative growth factors. Northern blot analysis revealed that bFGF is synthesized in the smooth muscle cells and that AngII induces this synthesis twofold.[28] Antisense oligomers directed towards this mRNA inhibited DNA synthesis both in quiescent and AngII-stimulated smooth muscle cells.[28] Moreover, the increase in DNA synthesis observed when the synthesis of TGF-β was inhibited could be blocked with the FGF antisense oligomers.[28] Thus, taken together, these results suggest that the proliferative pathway is mediated by bFGF and the antiproliferative pathway is mediated by TGF-β. An imbalance in these autocrine loops could possibly result in proliferation.

The above results, demonstrating that under certain conditions AngII can result in VSMC mitogenesis, are intriguing in light of the recent report by Powell et al.[30] These investigators demonstrated that the neointimal proliferative lesion produced by vascular injury was attenuated by treatment with angiotensin-converting enzyme (ACE) inhibitors. This antiproliferative activity of ACE inhibition was not due only to the reduction in blood pressure, since verapamil did not affect this lesion despite an equal lowering of pressure. Recent data suggest that this antiproliferative activity is in fact due to the blockade of AngII production and not to other effects of ACE inhibitors, since the specific AT-1 receptor antagonist Dup 753 produced a similar effect.[59] Taken together, the data suggest a direct role for AngII in the neointimal proliferation. We hypothesize that vascular injury results in an imbalance between the proliferative and antiproliferative pathways induced by AngII, thereby favoring cellular proliferation.

We have also hypothesized that the local generation of AngII may play a role in the development of the neointima. Moreover, we suggest that the components of the local renin-angiotensin system may be induced following injury. We examined this using *in situ* hybridization and immunohistochemistry of the rat abdominal aorta and carotid artery balloon injury model. Abdominal aorta of Sprague-Dawley rats were studied before or after injury with a balloon catheter. Neointimal hyperplasia developed, as documented by a progressive increase in the ratio of intimal to media thickness from essentially 0 in the control of 0.17 at 1 week and 1.17 at 6 weeks post-injury. Angiotensinogen mRNA was clearly detected by *in situ* hybridization in the adventitia and media of control and injured aorta.[31] Control sense probe or RNAse treatment resulted in very low background. However, at one week after injury, the media to adventitia angiotensinogen mRNA ratio was higher in the injured aorta, suggesting increased gene expression in the media compared to control. Of potential importance, is the observation that angiotensinogen mRNA was also detected in the neointima of the injured aorta, and this was also highest at one week after injury.

In parallel studies, we also detected ACE immunostaining and mRNA in the intima. Abdominal aorta of Sprague-Dawley rats were injured with a 2F balloon catheter. Morphometrical changes, ACE enzymatic activity, and localization of ACE by immunohistochemistry and *in situ* hybridization in injured and uninjured aorta were analyzed.[32] Vascular ACE activity in the injured aorta was significantly higher than in the uninjured aorta, while serum and lung ACE levels were not different between the two groups. The cellular distribution of ACE protein and mRNA in the neointima was similar to that of alpha smooth muscle actin but differed from those of endothelial (vWF) or monocytes/macrophages (ED-1) markers, demonstrating that ACE was expressed in neointimal smooth muscle cells. These data demonstrate that vascular injury results in the induction of vascular ACE and suggest that the inhibition of vascular ACE as opposed to serum ACE may be important in the prevention of restenosis after balloon injury. To examine this in more detail, we treated rats with increasing doses of the converting enzyme inhibitor, quinapril.[33] Direct analysis of blood pressure was performed via an indwelling catheter one day prior to euthanasia. After euthanasia, serum and carotid artery ACE levels and the degree of neointimal formation was measured. The results demonstrate a dose-dependent inhibition of both serum and tissue ACE activities and neointimal formation. However, the IC_{50} for blood pressure decrease and serum ACE inhibition was significantly lower than that for tissue ACE inhibition and inhibition of neointimal formation. The degree of neointimal formation showed a greater correlation with tissue ACE than with serum ACE or blood pressure. These results demonstrate a highly significant correlation between tissue ACE and the formation of the neointima following vascular injury. These results suggest that the local synthesis of AngII plays a major role in neointimal formation.

III. CONTROL OF CARDIAC HYPERTROPHY

Left ventricular hypertrophy, an adaptive process in chronic hypertension, is an independent risk factor for coronary heart disease and sudden death. The evidence suggests that both mechanical and humoral factors participate in the development of hypertrophy.[34] Unlike vascular smooth muscle cells, the cardiac myocyte is terminally differentiated. Thus, cardiac hypertrophy is caused by true cellular hypertrophy with no increase in cell number. This cellular hypertrophy is marked by an overall increase in cellular protein, especially the contractile proteins.[35] Interestingly, there is a switch in isoforms of the proteins such that the proteins usually expressed during embryogenesis are now re-expressed and accumulate in the hypertrophic heart.[36-40] In addition, other non-contractile proteins which are expressed during embryogenesis are also re-expressed during hypertrophy.[40-42]

As in the case of vascular hypertrophy, the role of AngII as a cardiotrophic agent was first shown pharmacologically. In the spontaneously hypertensive rat (SHR), the development of hypertension is accompanied by cardiac hypertrophy.[43-45] Interestingly, the normalization of blood pressure, once elevated, or the blockade of the elevation using vasodilators did not regress or inhibit this hypertrophy, suggesting that hemodynamic factors are not solely responsible for the hypertrophy. However, treatment of the hypertension with converting enzyme inhibitors did block or regress the hypertrophy.[43-45] Similar conclusions were recently drawn from experiments with the hypertrophy induced by aortic constriction.[46] Moreover, infusions of subpressor doses of AngII results in cardiac hypertrophy.[47] Thus, *in vivo* evidence suggests that AngII plays a role in cardiac hypertrophy. Evidence from *in vitro* experiments with cultured myocytes support this conclusion. Baker and co-workers have shown that in embryonic chick and neonatal rat myocytes, AngII increases protein synthesis and cell size.[34,48,49]

As discussed above in vascular hypertrophy, cardiac hypertrophy appears to involve the increased expression of proto-oncogenes. In pressure-overloaded hearts, induced by aortic constriction, there appears to be a rapid increase in the expression of *c-fos, c-myc,* and *c-Ha-ras.*[38,50,51] Similarly, *in vitro* studies have demonstrated the increase in expression of proto-oncogenes during hypertrophy in mechanically[52] and hormonally stimulated myocytes.[53-55] One of these, *Egr*-1, appears central to the hypertrophic response. For instance, both α- and β-adrenergic agents will increase the rates of protein synthesis and will increase the synthesis of fos and jun in myocardial cells. Moreover, both lead to an increase in some cardiac contractile proteins. However, other contractile proteins whose synthesis is increased during *in vivo* hypertrophy (MLC-2, cardiac and skeletal actin, and ANF) are induced to a greater extent by α- than β-adrenergic agents. Interestingly, Egr-1 is only induced by α-agonists.[54] Moreover, Egr-1 is induced in pressure-overloaded hearts *in vivo*.[56]

The mechanism by which AngII induces cardiac hypertrophy has not been examined in detail. However, it has been hypothesized that PKC plays an important role in cardiac hypertrophy, since PKC activation regulates the expression of many of the cardiac proteins which are induced during hypertrophy. In addition, growth factors such as TGF-β[57] and FGF,[58] which have been shown to be induced in vascular smooth muscle cells in response to AngII, have been implicated in cardiac hypertrophy. Therefore, while the evidence is not strong, one may speculate that the model developed for vascular hypertrophy in response to AngII (Figure 1) may also explain the role of AngII in cardiac hypertrophy.

REFERENCES

1. **Owens, G.K.,** Influence of blood pressure on development of aortic medial smooth muscle hypertrophy in spontaneously hypertensive rats, *Hypertension,* 9, 178, 1987.
2. **Limas, C., Westrum, B., and Limas, C.J.,** Comparative effects of hydralazine and captopril on the cardiovascular changes in spontaneously hypertensive rats, *Am. J. Pathol.,* 117, 360, 1984.
3. **Owens, G.K., Loeb, A., Gordon, D., and Thompson, M.M.,** Expression of smooth muscle-specific alpha-isoactin in cultured vascular smooth muscle cells: relationship between growth and cytodifferentiation, *J. Cell Biol.,* 102, 343, 1986.
4. **Libby, P. and O'Brien, R.V.,** Culture of quiescent arterial smooth muscle cells in a defined serum-free medium, *J. Cell. Physiol.,* 115, 217, 1983.
5. **Geisterfer, A.A.T., Peach, M.J., and Owens, G.K.,** Angiotensin II induces hypertrophy, not hyperplasia of cultured rat aortic smooth muscle cells, *Circ. Res.,* 62, 749, 1988.
6. **Naftilan, A.J., Pratt, R.E., and Dzau, V.J.,** Induction of platelet-derived growth factor A-chain and c-myc gene expressions by angiotensin II in cultured rat vascular smooth muscle cells, *J. Clin. Invest.,* 83, 1419, 1989.
7. **Naftilan, A.J., Pratt, R.E., Eldridge, C.S., Lin, H.L., and Dzau, V.J.,** Angiotensin II induces c-fos expression in smooth muscle cell via transcriptional control, *Hypertension,* 13, 706, 1989.
8. **Griendling, K.K., Berk, B.C., Ganz, P., Gimbrone, M.A., Jr., and Alexander, R.W.,** Angiotensin II stimulation of vascular smooth muscle phosphoinositide metabolism. State of the art lecture, *Hypertension,* 9 (Suppl. III), 181, 1987.
9. **Taubman, M.B., Berk, B.C., Izumo, S., Tsuda, T., Alexander, R.W., and Nadal-Ginard, B.,** Angiotensin II induces c-fos mRNA in aortic smooth muscle. Role of Ca^{2+} mobilization and protein kinase C activation, *J. Biol. Chem.,* 264, 526, 1989.

10. **Gibbons, G.H., Pratt, R.E., and Dzau, V.J.,** Angiotensin II is a bifunctional vascular smooth muscle cell growth factor, *Hypertension,* 14 (Abstr.), 358, 1989.
11. **Gibbons, G.H., Pratt, R.E., and Dzau, V.J.,** Transforming growth factor: beta expression modulates the bifunction of growth response of vascular smooth muscle cells to angiotensin II, *J. Clin. Invest.,* in press.
12. **Itoh, H., Pratt, R.E., and Dzau, V.J.,** Atrial natriuretic polypeptide inhibits hypertrophy of vascular smooth muscle cells, *J. Clin. Invest.,* 86, 1690, 1990.
13. **Marx, J.L.,** The fos gene as "master switch", *Science,* 237, 854, 1987.
14. **Verma, I.M. and Sassone-Corsi, P.,** Proto-oncogene fos: complex but versatile regulation, *Cell,* 51, 513, 1987.
15. **Greenberg, M.E. and Ziff, E.B.,** Stimulation of 3T3 cells induces transcription of the c-fos proto-oncogene, *Nature,* 311, 433, 1984.
16. **Holt, J.T., Gopal, T.V., Moulton, A.D., and Nienhuis, A.W.,** Inducible production of c-fos antisense RNA inhibits 3T3 cell proliferation, *Proc. Natl. Acad. Sci. U.S.A.,* 83, 4794, 1986.
17. **Nishikura, K. and Murray, J.M.,** Antisense RNA of protooncogene c-fos blocks renewed growth of quiescent 3T3 cells, *Mol. Cell Biol.,* 7, 639 – 649, 1987.
18. **Curran, T. and Franza, B.R., Jr.,** Fos and jun: the AP-1 connection, *Cell,* 55, 395, 1988.
19. **Berk, B.C., Vekhstein, V., Gordon, H.M., and Tsuda, T.,** Angiotensin II-stimulated protein synthesis in cultured vascular smooth muscle cells, *Hypertension,* 13, 305, 1989.
20. **Takeuchi, K., Nakamura, N., Cook, N.S., Pratt, R.E., and Dzau, V.J.,** Angiotensin II can regulate gene expression by the AP-1 binding sequence via a protein kinase C-dependent pathway, *Biochem. Biophys. Res. Commun.,* 72, 1189, 1990.
21. **Takeuchi, K., Pratt, R.E., and Dzau, V.J.,** Angiotensin II regulation of human renin gene expression via AP-1 binding site, *Clin. Res.,* 39, 349A, 1991.
22. **Chen, S., Takeuchi, K., Dzau, V.J., and Pratt, R.E.,** AP-1 DNA interaction in human renin gene, *Hypertension,* 1991.
23. **Rainer, R.S., Eldridge, C.S., Gilliland, G.K., and Naftilan, A.J.,** Antisense oligonucleotide to c-fos blocks the angiotensin II-induced stimulation of protein synthesis in rat aortic smooth muscle cells, *Hypertension,* 16, 326, 1990.
24. **Naftilan, A.J., Gilliland, G.K., and Eldridge, C.S.,** Induction of antisense c-fos in rat aortic smooth muscle cells blocks the angiotensin II induced increased in protein synthesis, *Circulation,* 84 (Suppl. II), 338, 1991.
25. **Sejersen, T., Betsholtz, C., Sjolund, M., Heldin, C.H., Westermark, B., and Thyberg, J.,** Rat skeletal myoblasts and arterial smooth muscle cells express the gene for the A chain but not the gene for the B chain (c-sis) of platelet-derived growth factor (PDGF) and produce a PDGF-like protein, *Proc. Natl. Acad. Sci. U.S.A.,* 83, 6844, 1986.
26. **Itoh, H., Pratt, R.E., Gibbons, G.H., and Dzau, V.J.,** Endogenous platelet-derived growth factor as a paracrine/autocrine mediator for angiotensin II-induced vascular remodeling, *Circulation,* submitted.
27. **Itoh, H., Pratt, R.E., and Dzau, V.J.,** Antisense oligonucleotides complementary to PDGF mRNA attenuate angiotensin II-induced vascular hypertrophy, *Hypertension,* 16, 325, 1990.
28. **Itoh, H., Pratt, R.E., Gibbons, G.H., and Dzau, V.J.,** Angiotensin II modulates proliferation of VSMC via dual autocrine loops of TGFb and bFGF, *Hypertension,* 18, 396, 1991.
29. **Koibuchi, Y., Gibbons, G.H., Lee, W.S., and Pratt, R.E.,** Role of protein kinase C in angiotensin II induced growth of vascular smooth muscle cells, *Clin. Res.,* 40, 220A, 1992.

30. **Powell, J.S., Clozel, J.P., Muller, R.K.M., Kuhn, H., Hefti, F., Hosang, M., and Baumgartner, H.R.,** Inhibitors of angiotensin-converting enzyme prevent myointimal proliferation after vascular injury, *Science,* 245, 186, 1989.
31. **Rakugi, H., Jacob, H.J., Krieger, J.E., Ingelfinger, J.R., and Pratt, R.E.,** Vascular injury induces angiotensinogen gene expression in the media and neointima, *J. Clin. Invest.,* in preparation.
32. **Rakugi, H., Krieger, J.E., Wang, D.S., McCook, O., Dzau, V.J., and Pratt, R.E.,** Induction of angiotensin converting enzyme in the neointima after vascular injury: possible role in restenosis, *J. Clin. Invest.,* submitted.
33. **Rakugi, H., Dzau, V.J., and Pratt, R.E.,** Importance of tissue angiotensin converting enzyme (ACE) in neointimal hyperplasia, *Circulation,* submitted.
34. **Morgan, H.E. and Baker, K.M.,** Cardiac hypertrophy: mechanical, neural and endocrine dependance, *Circulation,* 83, 13, 1991.
35. **Lee, H., Henderson, S., Reynolds, R., Dunnmon, P., Yuan, D., and Chien, K.R.,** α-1 Adrenergic stimulation of cardiac gene transcription in neonatal rat myocardial cells: effects on myosin light chain-2 gene expression, *J. Biol. Chem.,* 263, 7352, 1988.
36. **Izumo, S., Lompre, A.M., Matsuoka, R., Koren, G., and Schwartz, K.,** Myosin heavy chain messenger RNA and protein isoform transitions during cardiac hypertrophy: interaction between hemodynamic and thyroid hormone-induced signals, *J. Clin. Invest.,* 79, 970, 1987.
37. **Schwartz, K., De la Bastie, D., Bouveret, P., Oliviero, P., Alonso, S., and Buckingham, M.,** α-Skeletal muscle actin mRNAs accumulate in hypertrophied adult rat hearts, *Circ. Res.,* 59, 551, 1986.
38. **Izumo, S., Nadal-Ginard, B., and Mahdavi, V.,** Protooncogene induction and reprogramming of cardiac gene expression produced by pressure overload, *Proc. Natl. Acad. Sci. U.S.A.,* 85, 339, 1988.
39. **Long, C.S., Ordahl, C.P., and Simpson, P.C.,** Alpha I-adrenergic receptor stimulation of sarcomeric actin isogene transcription in hypertrophy of cultured rat heart muscle cells, *J. Clin. Invest.,* 83, 1078, 1989.
40. **Knowlton, K.U., Baracchini, E., Ross, R.S., Henderson, S.A., Evans, S.M., Glembotski, C.C., and Chien, K.R.,** Co-regulation of atrial natriuretic factor and cardiac myosin light chain-2 genes during α-adrenergic stimulation of neonatal rat ventricular cells, *J. Biol. Chem.,* 266, 7759, 1991.
41. **Arai, H., Nakao, K., Saito, Y., Morii, N., Sugawara, A., Yamada, T., Itoh, H., Shiono, S., Mykoyama, M., Ohkubo, H., Nakanishi, S., and Imura, H.,** Augmented expression of atrial natriuretic polypeptide gene in ventricles of spontaneously hypertensive rats (SHR) and SHR-stroke prone, *Circ. Res.,* 62, 926, 1988.
42. **Lee, T.R., Bloch, K.D., Pfeffer, J.M., Pfeffer, M.A., Neer, E.J., and Seidman, C.E.,** Atrial natriuretic factor gene expression in ventricles of rats with spontaneous ventricular hypertrophy, *J. Clin. Invest.,* 81, 431, 1988.
43. **Oparil, S., Rinoff, L., and Cuttilleta, A.,** Catecholamines, blood pressure, renin and myocardial function in the spontaneously hypertensive rat, *Clin. Sci. Mol. Med. Suppl.,* 3, 455S, 1976.
44. **Sen, S., Tarazi, R.C., Khairallah, P.A., and Bumpus, F.M.,** Cardiac hypertrophy in spontaneously hypertensive rats, *Circ. Res.,* 35, 775, 1974.
45. **Pfeffer, J.M., Pfeffer, M.A., Mirsky, I., and Braunwald, E.,** Regression of left ventricular hypertrophy and prevention of left ventricular dysfunction by captopril in the spontaneously hypertensive rat, *Proc. Natl. Acad. Sci. U.S.A.,* 79, 3310, 1982.
46. **Baker, K.M., Chernin, M.I., Wixson, S.K., and Aceto, J.F.,** Renin-angiotensin system involvement in pressure overload cardiac hypertrophy in rats, *Am. J. Physiol.,* 259, H324, 1990.

47. **Khairallah, P.A. and Kanabus, J.,** Angiotensin and myocardial protein synthesis, in *Perspectives in Cardiovascular Research,* Tarazi, R.C. and Dunbar, J.B., Eds., Raven Press, New York, 1983, 337.
48. **Aceto, J.F. and Baker, K.M.,** [Sar^1]angiotensin II receptor-mediated stimulation of protein synthesis in chick heart cells, *Am. J. Physiol.,* 258, H806, 1990.
49. **Baker, K.M. and Aceto, J.F.,** Angiotensin II stimulation of protein synthesis and cell growth in chick heart cells, *Am. J. Physiol.,* 259, H610, 1990.
50. **Komuro, I., Kurabayashi, M., Takaku, F., and Yazaki, Y.,** Expression of cellular oncogenes in the myocardium during the developmental stage and pressure-overloaded hypertrophy of the rat heart, *Circ. Res.,* 62, 1075, 1988.
51. **Mulvagh, S.L., Michael, L.H., Perryman, M.B., Roberts, R., and Schneider, M.D.,** A hemodynamic load in vivo induces cardiac expression of the cellular oncogene, c-myc, *Biochem. Biophys. Res. Commun.,* 147, 627, 1987.
52. **Komuro, I., Kaida, T., Shibazaki, Y., Kurabayashi, M., Katoh, Y., Hoh, E., Takaku, F., and Yasaki, Y.,** Stretching cardiac myocytes stimulates protooncogene expression, *J. Biol. Chem.,* 265, 3595, 1990.
53. **Starksen, N.F., Simpson, P.C., Bishopric, N., Coughlin, S.R., Lee, W.M.F., Escobedo, J.A., and Williams, L.T.,** Cardiac myocyte hypertrophy is associated with c-myc protooncogene expression, *Proc. Natl. Acad. Sci. U.S.A.,* 83, 8348, 1986.
54. **Iwaki, K., Sukhatme, V.P., Shubeita, H.E., and Chien, K.R.,** α and β adrenergic stimulation induce distinct patterns of immediate early gene expression in neonatal rat myocardial cells: fos/jun expression is associated with sarcomere assembly; Egr-1 induction is primarily an a1-mediated response, *J. Biol. Chem.,* 265, 13809, 1990.
55. **Neyses, L., Nouskas, J., and Vetter, H.,** Endothelin-1 (Endo) and angiotensin II (AII)-induced myocardial protein synthesis involves the early growth response gene-1 (Egr-1), *Circulation,* 84 (Suppl. II), 3951, 1991.
56. **Ross, R.S., Rockman, H.A., Harris, A.N., Knowlton, K.U., Field, L., Ross, J., and Chien, K.R.,** In vivo hypertrophy in transgenic mice segregates inducible from tissue specific expression of the ANF gene, *Clin. Res.,* 39, 296A, 1990.
57. **Parker, T.G., Packer, S.E., and Schneider, M.D.,** Peptide growth factors can provoke "fetal" contractile protein gene expression in rat cardiac myocytes, *J. Clin. Invest.,* 85, 507, 1990.
58. **Schneider, M.P. and Packer, T.G.,** Cardiac myocytes as targets for the action of peptide growth factors, *Circulation,* 81, 1443, 1990.
59. **Wang, D.S., Rakugi, H., Dzau, V.J., and Pratt, R.E.,** unpublished observations.

Chapter 19

THE PLASMINOGEN ACTIVATOR SYSTEM AND ITS INTERACTIONS WITH ANGIOTENSIN II IN THE BRAIN

B. Rydzewski, M. Wozniak, C. Sumners, and M.K. Raizada

TABLE OF CONTENTS

0-8493-4622-3/93/$0.00 + $.50

I. INTRODUCTION

The structural and functional complexity of the brain requires an intricate mechanism of cellular communication involving various chemical messengers. Abnormalities in any of the hormonal signaling systems usually result in aberrant development and/or functioning of the brain. In spite of extensive studies on the hormonal and enzymatic systems that are involved in the development of the brain, the cellular and molecular bases of maladies such as essential hypertension or neurodegenerative disorders are poorly understood. In recent years angiotensin II (AngII) and the plasminogen activator system (PAS) of the brain have been implicated in the control of central nervous system (CNS) development and function. Elucidation of the interactions between these two systems may aid our understanding of brain function.

AngII has been shown to play a major role in cell-to-cell interactions of the brain. During the last decade the brain renin-angiotensin system has been implicated in the central regulation of blood pressure, achieved through regulation of fluid intake[1,2] and the release of vasopressin,[1,3,4] as well as through its centrally mediated effects on blood vessel tone.[1] The presence of abnormalities in the brain AngII system is thought to be an important factor in the etiology of hypertension.[1,2,5,6] In addition, developmental changes in the expression of the brain AngII system suggest its significance in the control of CNS maturation.[7,8]

The maturation of the nervous system is also dependent on the expression of certain proteases that are responsible for the regulation of tissue growth and remodeling.[9-13] During the last decade, evidence has accumulated that plasminogen activators (PAs) and plasminogen activator inhibitors, the components of the PAS, are synthesized in the mammalian brain. Originally, their functions in the CNS were thought to be limited to the control of hemostasis and the regulation of their expression was poorly understood. The recent molecular cloning of the components of the PAS[14-24] and the application of brain-derived cell cultures to study their expression has greatly advanced our knowledge of this area.

Studies on brain-derived cell cultures have demonstrated that the AngII system may play a major role in the regulation of the PAS.[25,26] In addition, AngII expression is dependent on multiple proteolytic steps involving plasmin[27-30] and plasminogen activators.[31] Thus, AngII and the PAS of the brain seem to be interactive. In this chapter, studies that have addressed the expression and function of the PAS in the brain will be reviewed and its connections with the renin-angiotensin system will be discussed.

II. THE PLASMINOGEN ACTIVATOR SYSTEM (PAS) IN PERIPHERAL TISSUES

A. PLASMIN FORMATION CASCADE

Plasmin is a broad-spectrum protease predominantly present in the extracellular space. Figure 1 shows an enzymatic cascade responsible for the formation of this enzyme. Plasmin is generated from its inactive precursor, plasminogen, by the proteolytic action of urokinase (uPA) and tissue plasminogen activator (tPA).[32,33] Both of these enzymes are rapidly inactivated by plasminogen activator inhibitor 1 (PAI-1)[34-40] and a relatively weaker inhibitor, PAI-2.[40] Changes in the expression of PAI-1 and PAI-2 reciprocally control the extent of plasmin-dependent proteolysis[41,42] and thus constitute an important regulatory mechanism.

B. FUNCTIONS OF THE PAS IN PERIPHERAL TISSUES

The PAS is primarily responsible for degradation of fibrin and thrombin deposits, thus controlling the extent of blood clot elimination in peripheral tissues.[43] In addition, it is also responsible for the degradation of proteins of the extracellular matrix (ECM) and basement membranes.[44,45] This activity is of major significance in cell migration, cancerous growth, and metastasis.[46-52] The PAS also participates in the control of other physiologic processes such

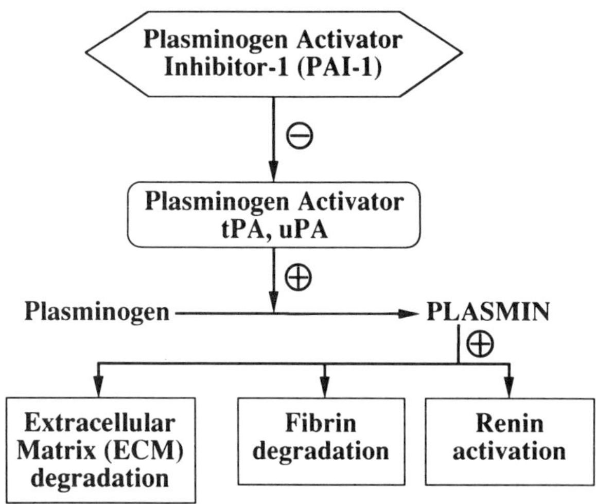

FIGURE 1. Plasmin activation cascade.

as spermatogenesis,[53,54] ovulation,[55] and the invasive stage of embryo implantation during the early stages of pregnancy.[56-58] The role that plasminogen activators play in the invasion of tissue has been confirmed by a study relating decreased PA activity to abnormalities in blastocyst implantation in nude mice.[59] Increased levels of tPA, uPA, PAI-1, and PAI-2 are observed later in pregnancy, suggesting their significance in the control of gestational processes.[60] uPA and tPA have also been implicated in neurite outgrowth of peripheral nerves, a process crucial for the development of the peripheral nervous system.[13,61-65]

C. THE PAS IN PATHOLOGICAL CONDITIONS

Abnormal levels of the components of the PAS have been associated with various pathological conditions. For example, PA and PAI-1 levels are changed in patients with malignant tumors, where they may contribute to invasive and metastatic properties of neoplasm.[46-52] Increased plasma levels of PAI-1 resulting in hypofibrinolysis have been observed in patients with cardiovascular or thromboembolic disease.[42,66-70] In addition, elevated PAI-1 levels constitute a risk factor for the development of cardiovascular disease in diabetic patients.[71-73] PAS has also been implicated in the pathogenesis of neuroimmunologic disease.[74] Finally, patients show high plasma levels of PAI-1 during the postoperative phase, after trauma, and also after extracorporeal circulation.[75] In summary, the PAS participates in the control of many physiological processes in peripheral tissues, and abnormalities in its expression may have serious pathological consequences.

III. EXPRESSION OF THE PAS IN THE BRAIN

The presence of the PAS in the human brain has been demonstrated as early as the 1960s. In 1966 Okada et al. demonstrated the presence of PA activity in cerebrospinal fluid.[76] Early reports implicated endothelial cells as the source of PA activity in the brain. Takashima et al.[77] showed PA activity in endothelial cells of the brain vessels, the meninges, and the chorioid plexus. In addition, these authors also demonstrated that the PA activity was dependent on the stage of development of human infant brain. Menon et al.[78] proposed that the brain contributes to the fibrinolytic properties of the blood through the brain's blood flow-dependent synthesis and release of PA. Since the blood-brain barrier prevents protein compounds synthesized by neurons and glia from entering the blood stream, the observed effects were due rather to the activity of the endothelial cells of the brain vessels. Finally, studies performed on human brain tumors of astroglial and neuronal origin have shown that these neoplastic cells synthesize uPA, tPA, and PAI-1.[79-81]

Animal studies also strongly support the presence of PA activity in brain tissue. In 1970, Glas and Astrup[82] reported high levels of PA activity in homogenates from various regions of rabbit brain. In 1981 the expression of

PAs was demonstrated in brain tissue using a fibrin degradation assay.[83] Later *in vivo* studies have confirmed the presence of PAs in the brains of mice and rats.[11,83-85] In fact, brain contains a relatively active plasmin-activating system when compared with other tissues.[84] According to Soreq and Miskin,[83] the highest PA activity was observed in the thalamic area of adult mouse brain, whereas cerebral and cerebellar cortices and the brainstem demonstrated only approximately one third to one half as much activity. In contrast to adult mouse brain, brains of neonate animals expressed relatively high levels of PA in the cerebellum and in the brainstem. This suggests a developmental regulation of PA activity. However, later studies showed high PA activity in the murine diencephalon/brainstem and the cerebral cortex.[84] The cerebellum had approximately 1/14 as much activity as the diencephalon/brainstem region. These observations clearly demonstrate that PAs are expressed in the brain tissue; however, the level of PA activity in discrete anatomical areas of the brain must undergo further analysis in order to resolve the discrepancies between existing studies.

During the last decade, molecular cloning of tPA and uPA,[15,16,19] as well as PAI-1[14,17,21,23,24] has provided tools with which the expression of these proteins at the transcriptional level can be analyzed. We have used Northern blot analysis to determine whether tPA and uPA genes are expressed in various anatomical areas of the rat brain. Detectable levels of tPA and uPA mRNA could be observed throughout the brain, with especially high expression of these mRNA species in the hypothalamus and the brainstem of the adult rat brain (Figures 2 and 3).

Sawdey and Loskutoff have demonstrated PAI-1 mRNA expression in rodent brain *in vivo*.[86] The level of brain PAI-1 mRNA was markedly lower than that observed in heart, lung, or adipose tissue. To our knowledge, no reports have addressed the issue of differential PAI-1 mRNA expression in various areas of the brain. Using PAI-1-specific cDNA probes, we demonstrated the presence of PAI-1 mRNA throughout the brain. Interestingly, the brainstem and the hypothalamus, which express high levels of uPA and tPA mRNAs (Figures 2 and 3), also contain higher levels of the mRNAs coding for PAI-1, compared with other brain areas (Figure 4). These observations indicate that mRNA levels for PA and PAI-1 are differentially expressed in anatomically discrete brain areas. Since PAI-1 inhibits both tPA and uPA in equimolar concentrations, and the mRNA data do not necessarily reflect activity of the nascent protein, the question arises as to how increased transcriptional expression of both PAs and their inhibitors affects the PA/PAI ratio and resultant PA activity in these brain areas. Direct measurement of PA activity in the brain is needed to answer this question.

Cell culture techniques have proven to be useful tools in studying the expression of the PAS in the brain. Numerous studies have shown that neuronal cells synthesize both tPA and uPA.[64,87-92] There are, however, conflicting reports describing the presence of PAI-1 in neuronal cells. PAI-1 synthesis

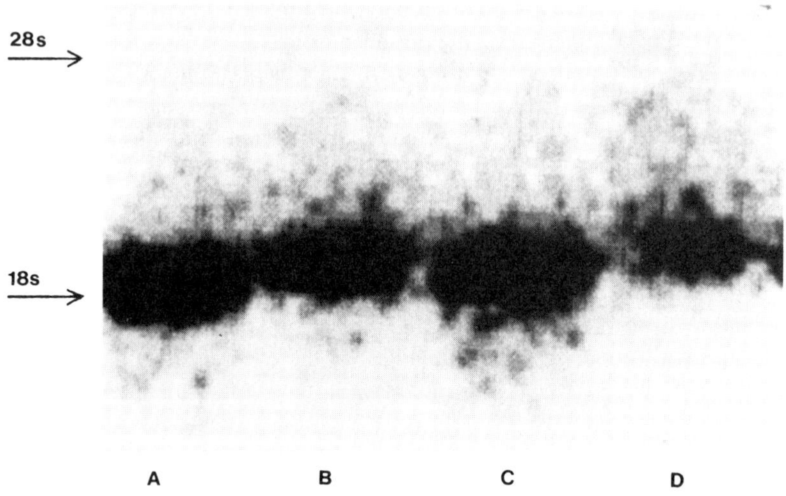

FIGURE 2. Steady-state levels of uPA mRNA in the rat brain. Total RNA was extracted from the brainstem (A), cerebellum (B), hypothalamic area (C), and cerebral cortex (D) of WKY male rats using phenol/chloroform extraction as described previously.[26] Northern blot analysis with a specific uPA cDNA probe was used to quantify 2.4-kb uPA transcript levels in these areas.

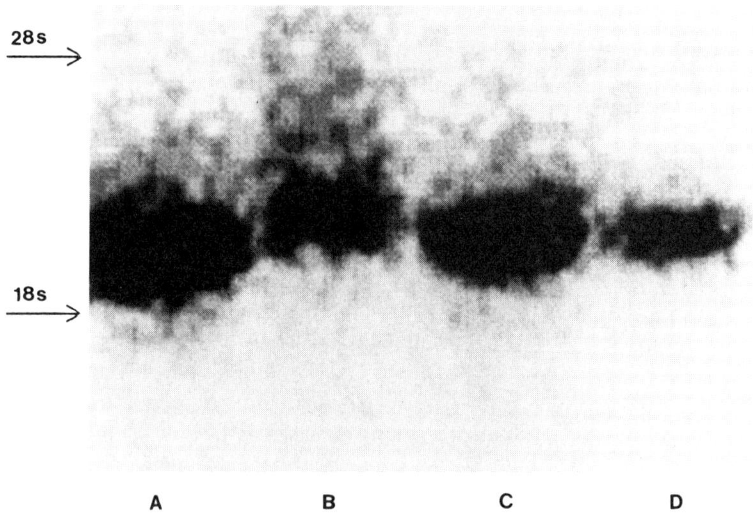

FIGURE 3. Steady-state levels of tPA mRNA in the rat brain. Total RNA was extracted from the brainstem (A), cerebellum (B), hypothalamic area (C), and cerebral cortex (D) of WKY male rats. Northern blot analysis with a specific tPA cDNA probe was used to quantify a 2.7-kb tPA transcript, essentially as described previously.[26]

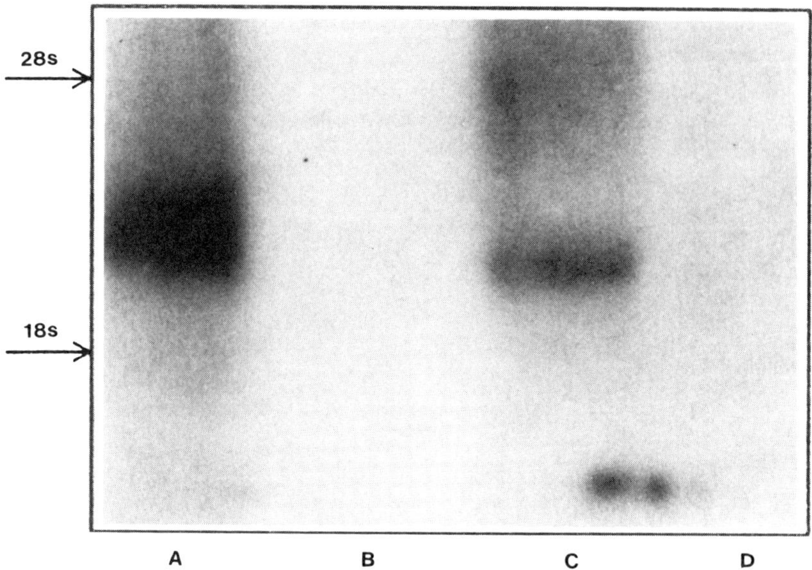

FIGURE 4. Steady-state levels of PAI-1 mRNA in the brain. Total RNA was extracted from the brainstem (A), cerebellum (B), hypothalamic area (C), and cerebral cortex (D) of WKY male rats. Northern blot analysis with a specific PAI-1 cDNA probe was used to quantify 3.1-kb PAI-1 mRNA levels in these areas, as described in Figure 2.

has been demonstrated in cell cultures derived from human fetal brain.[92] Other reports, however, demonstrated only negligible amounts of PAI-1 mRNA in neuronal cultures derived from the rat brain.[26,93] This discrepancy can perhaps be explained by the different origins of the neuronal cells, their developmental age, or, alternatively, by the different culturing techniques used in these studies.

In contrast to neurons, there is a consensus opinion concerning the role of astroglial cells in the production of both PAs and PAI-1 in the brain. Cultures derived from noncancerous tissue, as well as from brain tumors, have clearly determined that astroglial cells are an excellent source of these proteins.[12,25,26,94,95] In addition, primary astroglial, and to an even greater extent glioma, cell cultures synthesize PAI-1 and other protease inhibitors.[25,26,96-100]

The purity and thus the results obtained from astroglial culture preparations are sometimes questioned, due to the possibility of contamination of these cultures with other cells, such as fibroblasts, microglia, or endothelial cells. In order to verify that PAI-1 release is an astroglia-specific phenomenon, double-staining immunocytochemical experiments using specific antibodies against glial fibrillary acidic protein (GFAP, an astroglia-specific marker) and PAI-1 were performed. The results of these experiments show 95 to 98% of the cells in our glial cultures are GFAP-positive, and are thus of astroglial origin. In addition, all the GFAP-positive cells are also stained with anti-PAI-1 antibody (Figure 5).

In summary, astroglial cells of the brain are believed to be critical in the regulation of the PAS in the CNS, and astroglia-derived PAI-1 is thought to modulate uPA and tPA in the brain.

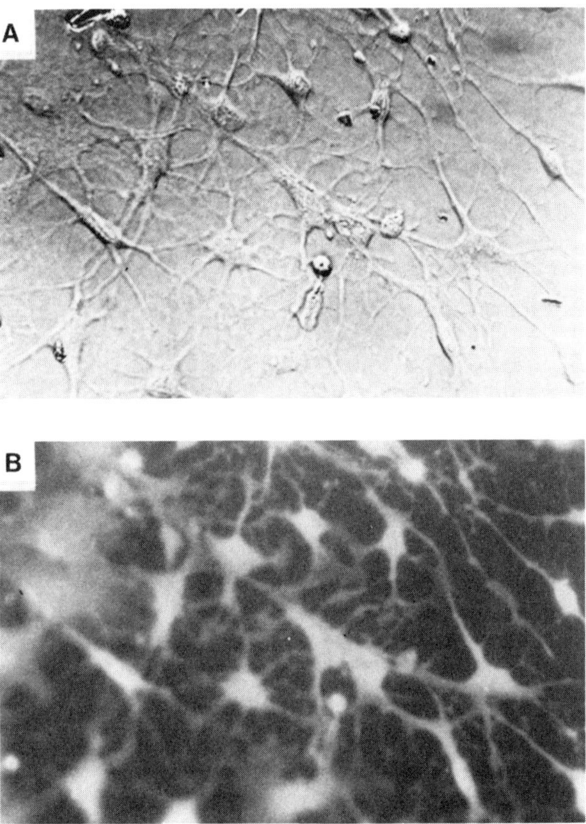

FIGURE 5. Colocalization of PAI-1 and GFAP immunoreactivity in astroglial cultures from rat brain. Astroglial cell cultures derived from the brains of 21-d-old rats were treated with 1 μM AngII for 24 h. Medium was aspirated, cells were washed twice with phosphate-buffered saline (PBS) and fixed with ice-cold 100% methanol for 7 min. A mixture of rabbit anti PAI-1 IgG (1:200), and mouse anti-GFAP IgG (1:100) was applied overnight at 4°C. The next day, cells were washed extensively with PBS and treated with secondary antibodies at 4°C for 3 h. Rhodamine-treated anti-mouse IgG (produced in donkey) and fluorescein-tagged anti-rabbit IgG (produced in sheep) were used. After extensive washes with PBS, cells were photographed under fluorescent light. Panel A shows a phase-contrast micrograph of a representative group of astroglial cells. Panel B represents a fluorescence micrograph of the same group of cells showing positive GFAP staining. Panel C is another fluorescence micrograph which shows this group to be PAI-1-positive as well. It is evident that all GFAP-positive cells seen here are also PAI-1-positive.

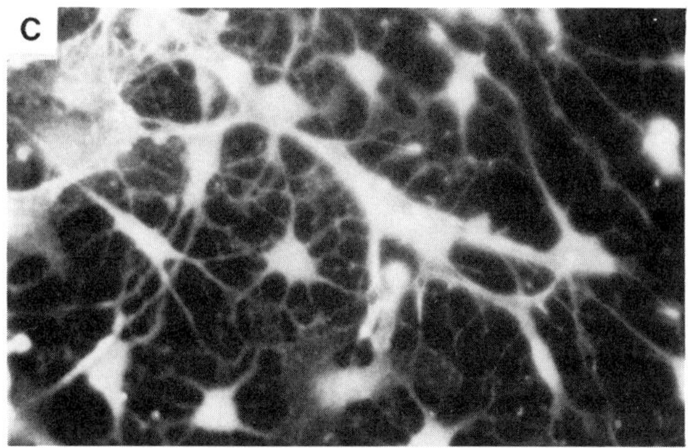

FIGURE 5. (*continued*)

IV. CONTROL OF PAS EXPRESSION IN THE BRAIN

The expression of uPA, tPA, and PAI-1 in the periphery has been studied extensively and has been shown to be controlled by multiple factors. Substances such as glucocorticoids,[101] endotoxins,[102] cytokines,[103,104] cAMP activators,[105,106] and various growth factors[107-112] have all been found to control PAI-1 release from different cell types. Thus, control of the peripheral PAS seems to be complex and relatively well understood. In contrast, studies on the regulation of PAS activity in the brain are still in their infancy. This area of research attracts substantial attention, due to a potential role of the PAS in the control of cell migration and growth.

A. CONTROL OF tPA AND uPA EXPRESSION

Both neuronal and astroglial cells appear to be highly regulatable sources of PAS components. One of the major pathways leading to the induction of tPA and uPA genes in both neuronal and astroglial cells is the activation of protein kinase C (PKC). Since the effect of PKC activators on tPA mRNA steady-state levels are blocked by cycloheximide (CHX), some authors have suggested that c-fos and c-jun, two immediate early gene products, might participate in this regulation.[113] Thus, hormones whose receptors are coupled with the inositol phospholipid hydrolysis (IP)-PKC activation-immediate early gene transduction pathway, such as AngII and norepinephrine, may play an important role in the regulation of the PAS. PA activity in astroglia is also regulated by glucocorticoids which decrease the proteolytic ability of the cultures.[95]

B. REGULATION OF PAI-1 EXPRESSION

PAI-1 gene expression in glia is also regulated by the IP-PKC transduction pathway;[26,92,110] however, the involvement of a protein synthesis step in this regulation has not been established. CHX itself causes a marked elevation of PAI-1 mRNA steady-state levels.[107,112,114] The mechanism of this effect of CHX on PAI-1 mRNA expression is unclear. It may be related to an inhibition of the synthesis of a selective RNAse or repression of PAI-1 gene transcription. Thus, the apparent inability of CHX to block the PKC-related stimulation of PAI-1 mRNA levels is difficult to interpret at the present.

Rogister et al.[115] report that the release of PAI but not PA from cultured astroglial cells is stimulated by interleukin-1 (IL-1), a polypeptide mediating the acute phase of immunological response at certain brain centers.[103,116-118] Glioma cells in culture are also susceptible to this regulation. In contrast, PAI-1 secretion from hippocampal neurons in culture is not affected by IL-1. IL-1 is a potent growth factor for endothelial cells and fibroblasts.[119] Since both of these cell types are PAI-1 producers, these actions of IL-1 may indirectly result in an even greater elevation of PAI-1 in the CNS. The mitogenic actions of IL-1 on astroglial cells are uncertain at present.

C. ROLE OF VITRONECTIN IN THE CONTROL OF PAI-1 ACTIVITY

In addition to control at the level of protein synthesis, PAI-1 activity can be regulated in the extracellular space by the ECM protein, vitronectin. Vitronectin binds and stabilizes secreted PAI-1, thus preventing its inactivation.[120-126] Binding with vitronectin extends the half-life of PAI-1 from 3 h, as observed in conditioned medium, to 24 h after association with the protein (reviewed in Reference 86). Vitronectin, and possibly other components of the ECM, may therefore be able to profoundly modulate the activity of the PAS system. It is pertinent to note, however, that vitronectin is rather scarce in the brains of healthy subjects.[127,128] In contrast, the brains of Alzheimer's disease patients have been shown to contain abnormally high levels of vitronectin.[128] This increase in vitronectin levels may result in a relatively higher PAI-1 activity in the brains of Alzheimer's patients. The importance of this phenomenon is not clear at the present time.

D. ANGIOTENSIN II AND THE PAS OF THE BRAIN

As discussed previously, the components of the peripheral PAS are significantly altered in certain diseases, including cardiovascular disorders. This observation, combined with the role of AngII in the control of cardiovascular function, stimulated the study of possible interactions between AngII and the PAS.

The presence of AngII in the CNS has been documented by numerous studies. In fact, all of the essential components of the AngII system, such as angiotensinogen and its mRNA,[129-132] immunoreactive AngI, AngII, renin,[133-135] and also the specific synthetic and degradation pathways for AngII,[136] have been demonstrated in the brain. Central AngII has been implicated in the control of the cardiovascular system, achieved through regulation of fluid intake and regulation of vasopressin release, as well as through its centrally mediated effects on blood vessel tone.[1,137] Despite extensive studies, the cellular and molecular mechanisms of AngII actions in the brain remain poorly understood. Studies with cell culture systems have helped to define the role of brain cell types in AngII actions. For example, specific receptors on neuronal cells have been implicated in an AngII-induced neuromodulatory function. This subject has been extensively covered in Chapter 14.

Raizada et al.[138] were the first to demonstrate that astroglial cells prepared from the hypothalamus and brainstem of neonatal rat brain possess specific, high-affinity AngII receptors. Activation of these receptors results in an increase of IP formation, suggesting that they are coupled to the phospholipase C-IP cascade second messenger system.[138] Further studies demonstrated that specific AngII receptors present in whole-brain cultures are pharmacologically and physiologically identical to those from hypothalamus-brainstem cultures. The availability of nonpeptide AngII receptor antagonists enabled us to determine the AngII receptor subtypes in astroglial cells in culture. Astroglia from whole brain as well as from the hypothalamus-brainstem area of the rat brain were shown to express AngII receptors that were 90% type 1 (AT_1), the remainder perhaps being type 2 (AT_2). AngII-induced hydrolysis of IP is completely blocked by DuP 753, a specific AT_1 antagonist, suggesting that AT_1 receptors are coupled to the IP hydrolysis second messenger system.[139] The finding that astroglial cells express predominantly AT_1 receptors was confirmed by studies on human astroglia.[140] AT_1 receptors in these cells mediate both release of prostaglandins and mobilization of intracellular calcium, whereas AT_2 receptors are only coupled to the release of prostaglandins. These observations, taken together with our studies, conclusively demonstrate that astroglial cells express physiologically active AngII receptors. In contrast to astroglia, neuronal cells express AngII receptors that are 70 to 80% AT_2 receptors and only 20 to 30% AT_1.[139]

The demonstration of AngII receptors in astroglia raises the question of their physiological significance in the central actions of AngII. Studies summarized below provide evidence that astroglial AT_1 receptors play a role in the regulation of the PAS in the CNS.

The initial indication that AngII and the PAS are interactive came from the studies by Olson et al.,[25] who showed that AngII stimulates the release of PAI-1 from astroglial cells derived from the rat brain. Rydzewski et al. corroborated this observation by demonstrating the effect of AngII on PAI-1 gene expression.[26] AngII-induced stimulation of PAI-1 gene expression is mediated through AT_1 receptors, since it is completely blocked by DuP 753, an AT_1 receptor-specific antagonist, but not by PD123177, an AT_2 receptor antagonist.

In contrast to astroglia, AngII fails to stimulate PAI-1 gene expression in neuronal cells. This lack of response could be attributed to a relatively low number of AT_1 receptors in neurons.[139] If this notion were correct then direct stimulation of PKC (bypassing AT_1 receptors) should result in a stimulation of PAI-1 gene expression. However, direct stimulation of PKC by phorbol esters in neuronal cultures derived from the rat brain did not result in an elevation of PAI-1 mRNA levels.[26] Thus, it would appear that the low number of AT_1 receptors on rat-derived neurons is probably not the only factor determining their insensitivity to AngII and PKC stimulators. In contrast, another report by Presta et al.[92] showed that a direct stimulation of PKC in cells derived from human fetal brains yielded a stimulation of PAI-1 expression. This suggests that steps distal to the AT_1 receptor and its coupling to PKC may be important in the regulation of this response.

The question also arises whether stimulation of other astroglial receptors that are coupled with PKC activation stimulates PAI-1 expression. In order to answer this question we treated astroglial cultures with norepinephrine (NE), an α_1-adrenergic receptor agonist, and carbachol, a synthetic muscarinic cholinergic receptor (M_1) agonist. Both of these compounds increased PAI-1 mRNA steady-state levels, thus verifying our conclusion that activation of PKC is a primary regulatory step in the expression of the PAI-1 gene (Figure 6). This takes on added significance, considering that AngII stimulates the release of NE from neuronal cells and that α_1-adrenergic receptors are present in astroglia (see Chapter 14). Based on these facts, one could suggest that the mechanism by which AngII controls PAI-1 release from astroglial cells is twofold: (1) the direct interaction of AngII with astroglial AT_1 receptors,

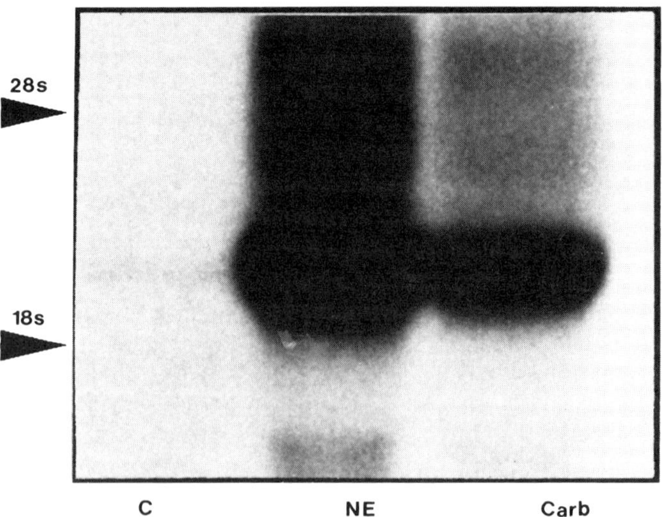

FIGURE 6. Effect of norepinephrine and carbachol on the steady-state levels of PAI-1 mRNA in astroglial cultures. Astroglial cultures prepared from neonate WKY rats were not treated (A), treated with 1 m*M* norepinephrine (B), or 500 μ*M* carbachol (C) for 2 h. Total RNA was extracted and analyzed using specific PAI-1 cDNA probe.[26]

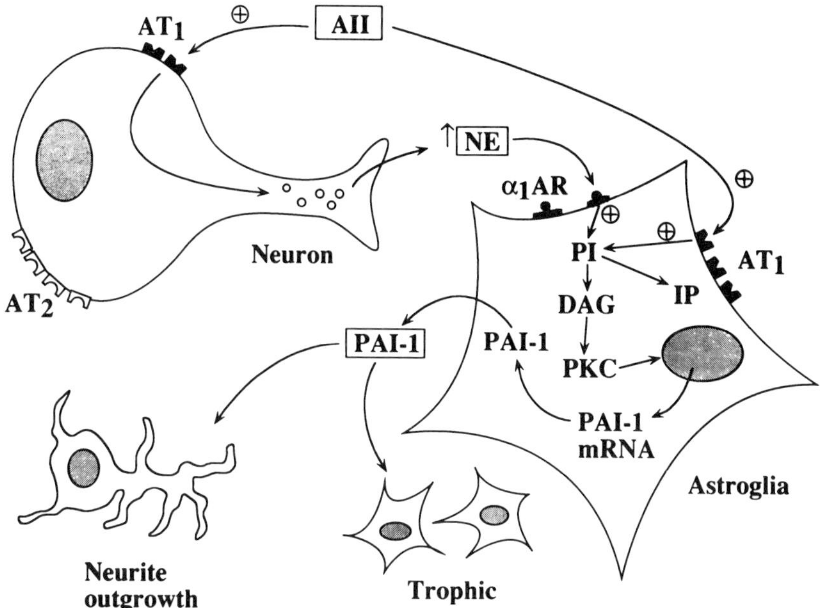

FIGURE 7. Proposed mechanism of control of PAI-1 release by AngII and norepinephrine.

as described above, and (2) the release of neuronal NE that in turn interacts with α_1-adrenergic receptors on the surface of astroglial cells (Figure 7).

AngII may also influence the activity of the PAS through the regulation of tPA and uPA expression. AngII directly elevates the uPA and tPA mRNA levels in rat astroglial cells. Treatment with this hormone for 4 h results in approximately four- and twelvefold increases in the levels of uPA and tPA mRNA, respectively (Figures 8 and 9). Interestingly, a long-term (24 h) effect of AngII seems to be a suppression of both uPA and tPA mRNA levels, which reach levels below those observed in control cells (unpublished observation).

From the studies presented above, one could conclude that AngII may regulate the activity of the PAS in the brain by the control of the expression of tPA, uPA, and PAI-1 genes. In addition to these direct effects, AngII can also influence PAS activity indirectly, through the stimulation of vasopressin release. The release of vasopressin from hypothalamic neurons to the blood stream is controlled by AngII derived from neurons in the CNS.[3,4] Vasopressin in turn stimulates uPA and tPA expression in peripheral tissues, thus changing the fibrinolytic properties of plasma.[141,142] It is therefore possible that centrally acting AngII plays a role in the control of peripheral uPA and tPA release.

Since AngII stimulates the expression of both plasminogen activators and their inhibitor, its net effect on the PAS depends on the resultant change in the ratio of PA to PAI activity. Our data indicate that the effect of AngII on PAI-1 expression is more pronounced (18- to 20-fold) than that on uPA or

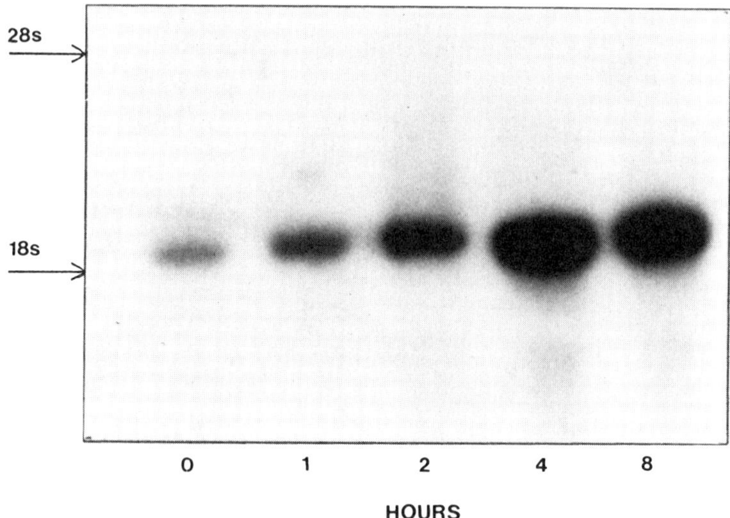

FIGURE 8. Effect of AngII on steady-state levels of uPA mRNA in rat astroglial cultures. Astroglial cultures prepared from the brains of 21-d-old WKY rats were treated with 100 n*M* AngII for the specified time periods. Total RNA was extracted and analyzed using a specific uPA cDNA probe, as described in Figure 3.

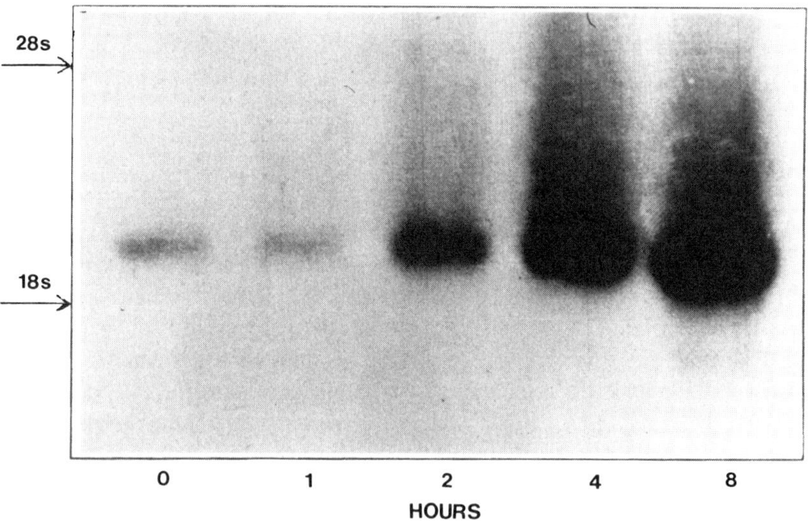

FIGURE 9. Effect of AngII on steady-state levels of tPA mRNA in astroglial cultures. Astroglial cultures prepared from the brains of 21-d-old WKY rats were treated with 100 n*M* AngII for the specified time periods. Total RNA was extracted and analyzed using a specific tPA cDNA probe, as described in Figure 2.

tPA (4- to 12-fold) (see Reference 26). In addition, since uPA and tPA mRNA levels are decreased by a prolonged treatment with AngII, the long-term effect would seem to be a reduction of the PA/PAI ratio. The relative instability of uPA and tPA, whose half-lives in plasma are measured in minutes (as compared with hours for PAI-1), will decrease the ratio even further. Thus, AngII in the brain would ultimately result in a decrease of PA activity.

V. FUNCTIONS OF THE PAS IN THE BRAIN

The first reports showing the presence of PA activity in the CNS suggested that the PAS is responsible for the control of hemostasis through its regulation of the fibrinolytic activity of plasmin.[76] Later studies suggested that meningeal and ependymal cells present in the CNS provide the plasminogen activators to the cerebrospinal fluid and plasma, where they are needed for thrombolysis, whereas PAs secreted by neuronal and glial cells have different functions.[83] Some potential functions of the brain PAS are presented below.

A. ROLE OF THE PAS IN THE DEVELOPMENT OF THE NERVOUS SYSTEM

Studies by Krystosek and Seeds[13] indicated that plasminogen activators play a role in neurite outgrowth and neuronal development. PAs were present at the growth cones of cells bearing neurites, suggesting that their release from the tip of a neurite might be essential for neurite growth and neuronal migration. These results were later confirmed by other studies which implicated the PAS in the development of the central as well as peripheral nervous systems.[11,63,88,96,143,144] Studies performed on sympathetic neurons[91] have indicated that neurite outgrowth can be diminished by anti-uPA antibodies which inhibit the activity of the enzyme. One would expect to find a similar effect using specific inhibitors of PA instead of anti-uPA antibodies. In fact, PA-mediated promotion of neuronal migration and neurite outgrowth could be inhibited by the addition of protease inhibitors.[144] In this study, synthetic inactivators of plasminogen as well as artificial serine protease inhibitors and antiplasmins all inhibited PA-mediated stimulation of neurite outgrowth. This observation suggests that the release of PAI-1 *in vivo* may have an important role in controlling the growth-promoting function of plasminogen activators.

However, several cell culture studies did not support these observations. On the contrary, some inhibitors of PAs seem to promote the neurite outgrowth.[97] This unexpected effect was originally attributed to the distinct extracellular environment present in neuronal cultures, compared with brain tissue *in vivo*.[97] More recent studies, however, showed that the neurite-promoting activity of some glia-derived protease inhibitors, such as protease nexin-1, is probably unrelated to their inhibition of uPA and depends on their ability to inhibit thrombin and therefore to prevent or reverse the thrombin-related retraction of neurites.[145,146] Ehrlich et al.[147] report that PAI-1, after binding with vitronectin, is able (similar to protease nexin-1) to inhibit thrombin activity.

Our own experiments indicate that PAI-1 acts as a potent trophic factor in neuronal cultures (Figure 10). Treatment of neurons with PAI-1 for 4 d resulted in increased development of the neurite network. An increase in the number of neurites and an extension of the life-span of neuronal cells were also observed. In addition, PAI-1 also acted as a potent mitogen in astroglial cultures, where it caused a marked increase in the cell number (Figure 11). *In vivo* studies indicate that the expression of PAI-1 in the brain may be developmentally regulated.[11,12] *In vitro* experiments suggest that AngII-stimulated PAI-1 gene expression in astroglia is elevated for days and weeks postnatally (unpublished observation). This coincides with the end of neuronal migration and an extensive glial proliferation.[148] Thus, increased release of PAI-1 by AngII during this time may play a major role in the control of timing of these developmental events. Interestingly, changes in AngII levels in the brain have been observed during development.[7,8] In addition, this hormone has been shown to possess some growth-promoting activity in the embryonic

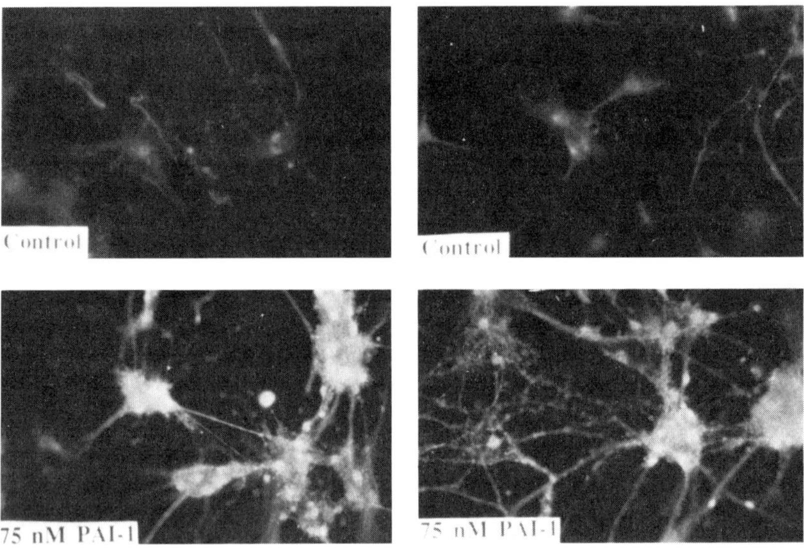

FIGURE 10. Effect of recombinant rat PAI-1 on neuronal cultures. Neuronal cultures were prepared from 1-d-old rats, essentially as described previously.[136] On day 5, after the removal of cytosine arabinoside, cultures were incubated with Dulbecco's modified Eagle's medium (DMEM) containing 10% plasma-derived horse serum (PDHS) or DMEM + PDHS containing 75 n*M* recombinant PAI-1 for 96 h at 37°C. Cultures were fixed and subjected to immunofluorescence with a neurofilament-specific antibody.[154] It is clearly evident that PAI-1 treatment of neuronal cultures stimulates development of neurite growth and neuronal processes (B) compared with cultures not treated with PAI-1 (A).

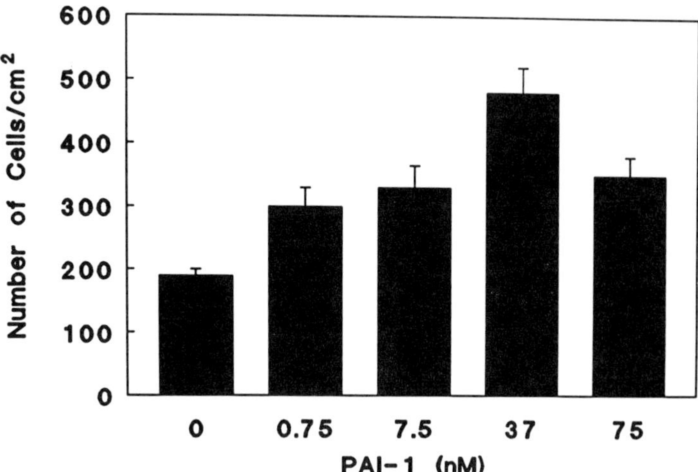

FIGURE 11. Effect of recombinant rat PAI-1 on astroglial cultures. Astroglial cells were plated in 35-mm tissue culture dishes in DMEM containing 10% fetal bovine serum (FBS). Cells were allowed to attach themselves to dishes for 24 h. Culture medium was removed, cells washed free of FBS by DMEM and incubated with DMEM without serum or DMEM containing the indicated concentrations of recombinant PAI-1 for 96 h at 37°C. Cells were dissociated by trypsin treatment and counted in a hemocytometer.

CNS.[149] These observations taken together suggest that the growth-promoting effects of AngII may be related to its interaction with PAI-1.

Analysis of the changes in PAI-1 levels in various pathophysiological conditions may be helpful in elucidating the functions of this protein. In fact, dramatic alterations in plasma PAI-1 levels have been observed in numerous disease states. Kruithof et al.[75] have proposed that PAI-1 might be involved in the acute phase of immunological response. This notion has recently been supported by studies indicating that IL-1 increases PAI-1 synthesis by cultured astrocytes.[115,150] The physiological relevance of this finding needs to be investigated.

B. PROPOSED ROLE OF THE PAS IN REGULATION OF ANGIOTENSIN II SYNTHESIS

AngII stimulates the synthesis and release of PAI-1 from astroglial cultures.[25,26] This release is likely to decrease PA activity in the brain, as discussed above. The physiological significance of this regulation is, however, unclear. We propose that PAI-1 may play a role in a negative feedback inhibition of AngII synthesis. This notion is based on the fact that many enzymes involved in AngII formation (renin, angiotensin-converting enzyme, plasmin) are proteases. Plasmin and PAs, whose activities are controlled by PAI-1, are the factors responsible for the conversion of inactive prorenin to its active form, renin.[27-31] Renin, in turn, cleaves angiotensinogen to angiotensin I, a crucial step in AngII formation. Increased PAI-1 levels induced by AngII can inhibit plasmin formation and thus decrease synthesis of AngII (Figure 12). This enzymatic interplay would enable other hormones that stimulate PAI-1 synthesis, such as NE or acetylcholine, to actively influence AngII levels.

C. ABNORMAL REGULATION OF PAI-1 EXPRESSION BY ANGIOTENSIN II IN THE BRAINS OF SPONTANEOUSLY HYPERTENSIVE RATS AND ITS SIGNIFICANCE

Brains from spontaneously hypertensive (SH) rats contain higher levels of AngII compared with those from normotensive animals.[7,8,151,152] In addition, we have shown that both basal and AngII-stimulated levels of PAI-1 mRNA in astroglia of SH rat brain are lower than those observed in cultures from normotensive Wistar Kyoto (WKY) animals (unpublished observations). These changes in PAI-1 expression are not secondary to hypertension because: (1) cell cultures were prepared from 21-d-old rats (where hypertension is not established), and (2) high blood pressure does not alter the expression of PAI.[153] Should the notion of PAI-1 acting in a negative feedback loop to inhibit AngII synthesis be true, a reduced sensitivity of the PAI-1 gene to AngII stimulation could be an underlying reason for abnormally high AngII concentrations in the brains of SH rats. Consequently, the abnormal levels of AngII observed in certain brain regions of SH rats may be a result of an abnormal negative feedback inhibition due to low PAI-1 levels. Changes in PAI-1 may influence the remodeling of the ECM. In addition, it is tempting to speculate, based on the potential trophic actions of PAI-1, that alterations in AngII-induced PAI-1 gene expression may lead to an altered neurotrophic

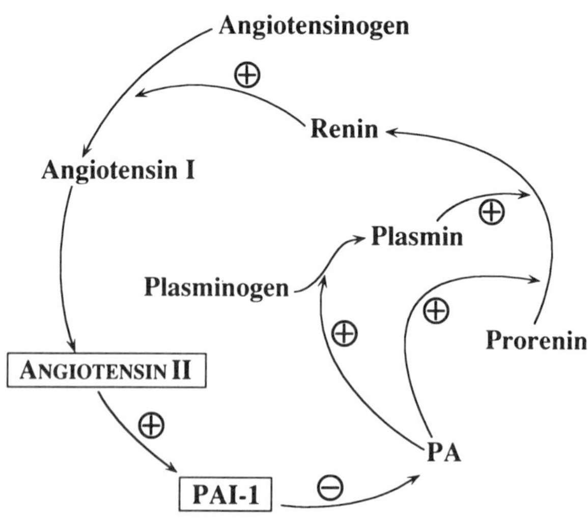

FIGURE 12. Proposed regulation of AngII synthesis by PAS.

activity during development of the SH rat CNS. These are attractive speculations that need further experimental verification.

VI. CONCLUSIONS AND FUTURE DIRECTIONS

In this review we have discussed the evidence in favor of the existence of the PAS and its potential functions in the development and differentiation of the brain. We have demonstrated that AngII exerts a profound stimulation of PAI-1 gene expression. This stimulation is exclusive to astroglia and is mediated by AT_1 receptors. Attempts have been made to develop a concept that the influence of AngII on PAI-1 may be an important determinant in the regulation of the ECM in the brain and thus may regulate neuronal-glial interactions. In addition, centrally mediated actions of AngII have been proposed to regulate both the brain and peripheral PAS. Finally, we have hypothesized that altered effects of AngII on PAI-1 gene expression in astroglia from SH rat brains may lead to profound changes in neuronal-glial interactions and thus are pertinent to the development and establishment of the hypertensive state in SH rats. This, obviously, is an attractive hypothesis and is primarily based on *in vitro* and some *in vivo* experiments. The future directions of our research, summarized below, will attempt to answer key questions to support or refute this hypothesis.

In Vitro Studies

- Does the PAS regulate AngII levels?
- Does AngII modulate the expression of ECM proteins such as vitronectin and thrombospondin?
- What is the cellular mechanism of PAI-1-induced trophic actions on neurons?
- Does PAI-1 induced the expression of GAP43, a protein associated with neurite outgrowth?
- Are some or all effects of AngII mediated via modulation of the PAS?

In Vivo Studies

- Are components of the PAS (i.e., PAI-1, uPA, tPA) uniquely distributed in the CNS? Are there differences between WKY and SH rat brains?
- Does AngII regulate the PAS in the brain?

ACKNOWLEDGMENTS

We are grateful to Dr. Thomas D. Gelehrter (University of Michigan, Ann Arbor, MI) for providing a cDNA probe for rat PAI-1 and to Dr. Jay L. Degen (Children's Hospital Research Foundation, Cincinnati, OH) for providing rat cDNA probes for uPA and tPA. We thank Dr. Thomas M. Reilly

(DuPont Company, Wilmington, DE) for providing us with the recombinant rat PAI-1. We also appreciate the expert assistance of Ms. Elizabeth Brown in the preparation of brain cell cultures. This work was supported by grants from the National Institutes of Health (HL-33610 and NS-19441) and the American Heart Association, Florida Affiliate.

REFERENCES

1. **Buckley, J.P.,** The central effects of the renin-angiotensin system, *Clin. Exp. Hypertens. A,* 10(1), 1, 1988.
2. **Phillips, M.I.,** Functions of angiotensin in the central nervous system, *Annu. Rev. Physiol.,* 49, 413, 1987.
3. **Brooks, V.L., Keil, L.C., and Reid, I.A.,** Role of the renin-angiotensin system in the control of vasopressin secretion in conscious dogs, *Circ. Res.,* 58(6), 829, 1986.
4. **Shoji, M., Kimura, T., Matsui, K., Ota, K., Iitake, K., Inoue, M., and Yoshinaga, K.,** Role of intracerebral angiotensin receptors in the regulation of vasopressin release and the cardiovascular system, *Neuroendocrinology,* 43(2), 239, 1986.
5. **Frohlich, E.D., Iwata, T., and Sasaki, O.,** Clinical and physiologic significance of local tissue renin-angiotensin systems, *Am. J. Med.,* 87(6B), 19S, 1989.
6. **Unger, T., Badoer, E., Ganten, D., Lang, R.E., and Rettig, R.,** Brain angiotensin, pathways and pharmacology, *Circulation,* 77(6 Pt. 2), I40, 1988.
7. **Phillips, M.I. and Kimura, B.,** Brain angiotensin in the developing spontaneously hypertensive rat, *J. Hypertens.,* 6(8), 607, 1988.
8. **Meyer, J.M., Felten, D.L., and Weyhenmeyer, J.A.,** Measurement of immunoreactive angiotensin II levels in microdissected brain nuclei from developing spontaneously hypertensive and Wistar Kyoto rats, *Exp. Neurol.,* 107(2), 164, 1990.
9. **Soreq, H. and Miskin, R.,** Screening of the protease plasminogen activator in the developing mouse brain, in *Drug Receptors in the Central Nervous System,* Littauer, P. et al., Eds., John Wiley & Sons, London, 1981, 559.
10. **Baron Van Evercooren, A., Leprince, P., Rogister, B., Lefebvre, P.P., Delree, P., Selak, I., and Moonen, G.,** Plasminogen activators in developing peripheral nervous system, cellular origin and mitogenic effect, *Brain Res.,* 433(1), 101, 1987.
11. **Kalderon, N. and Williams, C.A.,** Extracellular proteolysis, developmentally regulated activity during chick spinal cord histogenesis, *Brain Res.,* 390(1), 1, 1986.
12. **Kalderon, N., Ahonen, K., and Fedoroff, S.,** Developmental transition in plasticity properties of differentiating astrocytes, age-related biochemical profile of plasminogen activators in astroglial cultures, *GLIA,* 3(5), 413, 1990.
13. **Krystosek, A. and Seeds, N.W.,** Plasminogen activator release at the neuronal growth cone, *Science,* 213(4515), 1532, 1981.
14. **Antalis, T.M., Clark, M.A., Barnes, T., Lehrbach, P.R., Devine, P.L., Schevzov, G., Goss, N.H., Stephens, R.W., and Tolstoshev, P.,** Cloning and expression of a cDNA coding for a human monocyte-derived plasminogen activator inhibitor, *Proc. Natl. Acad. Sci. U.S.A.,* 85(4), 985, 1988.
15. **Belin, D., Vassalli, J.D., Combepine, C., Godeau, F., Nagamine, Y., Reich, E., Kocher, H.P., and Duvoisin, R.M.,** Cloning, nucleotide sequencing and expression of cDNAs encoding mouse urokinase-type plasminogen activator, *Eur. J. Biochem.,* 148(2), 225, 1985.

16. **Bell, L.D., Smith, J.C., Derbyshire, R., Finlay, M., Johnson, I., Gilbert, R., Slocombe, P., Cook, E., Richards, H., Clissold, P. et al.,** Chemical synthesis, cloning and expression in mammalian cells of a gene coding for human tissue-type plasminogen activator, *Gene,* 63(2), 155, 1988.
17. **Ginsburg, D., Zeheb, R., Yang, A.Y., Rafferty, U.M., Andreasen, P.A., Nielsen, L., Dano, K., Lebo, R.V., and Gelehrter, T.D.,** cDNA cloning of human plasminogen activator-inhibitor from endothelial cells, *J. Clin. Invest.,* 78(6), 1673, 1986.
18. **Harris, T.J., Patel, T., Marston, F.A., Little, S., Emtage, J.S., Opdenakker, G., Volckaert, G., Rombauts, W., Billiau, A., and De Somer, P.,** Cloning of cDNA coding for human tissue-type plasminogen activator and its expression in Escherichia coli, *Mol. Biol. Med.,* 3(3), 279, 1986.
19. **Hung, P.P.,** The cloning, isolation and characterization of a biologically active human enzyme, urokinase, in E. coli, *Adv. Exp. Med. Biol.,* 172, 281, 1984.
20. **Ny, T., Leonardsson, G., and Hsueh, A.J.,** Cloning and characterization of a cDNA for rat tissue-type plasminogen activator, *DNA,* 7(10), 671, 1988.
21. **Ny, T., Sawdey, M., Lawrence, D., Millan, J.L., and Loskutoff, D.J.,** Cloning and sequence of a cDNA coding for the human beta-migrating endothelial-cell-type plasminogen activator inhibitor, *Proc. Natl. Acad. Sci. U.S.A.,* 83(18), 6776, 1986.
22. **Pennica, D., Holmes, W.E., Kohr, W.J., Harkins, R.N., Vehar, G.A., Ward, C.A., Bennett, W.F., Yelverton, E., Seeburg, P.H., Heyneker, H.L., Goeddel, D.V., and Collen, D.,** Cloning and expression of human tissue-type plasminogen activator cDNA in E. coli, *Nature,* 301(5897), 214, 1983.
23. **Wun, T.C. and Kretzmer, K.K.,** cDNA cloning and expression in E. coli of a plasminogen activator inhibitor (PAI) related to a PAI produced by Hep G2 hepatoma cell, *FEBS Lett.,* 210(1), 11, 1987.
24. **Zaheb, R. and Gelehrter, T.D.,** Cloning and sequencing of cDNA for the rat plasminogen activator inhibitor-1, *Gene,* 73(2), 459, 1988.
25. **Olson, J.A., Jr., Shiverick, K.T., Ogilvie, S., Buhi, W.C., and Raizada, M.K.,** Angiotensin II induces secretion of plasminogen activator inhibitor 1 and a tissue metalloprotease inhibitor-related protein from rat brain astrocytes, *Proc. Natl. Acad. Sci. U.S.A.,* 88(5), 1928, 1991.
26. **Rydzewski, B., Zelezna, B., Tang, W., Sumners, C., and Raizada, M.K.,** AngiotensinII stimulation of plasminogen activator inhibitor-1 gene expression in astroglial cells from the brain, *Endocrinology,* 130, 1255, 1992.
27. **Leckie, E.J.,** The reversible activation of inactive renin in human plasma, role of acid and of plasma kallikrein and plasmin, *Clin. Exp. Hypertens. A.,* 4(11 – 12), 2133, 1982.
28. **Schalekamp, M.A. and Derkx, F.H.,** Plasma kallikrein and plasmin as activators of prorenin, links between the renin-angiotensin system and other proteolytic systems in plasma, *Clin. Sci.,* 61(1), 15, 1981.
29. **Sealey, J.E., Atlas, S.A., Laragh, J.H., Silverberg, M., and Kaplan, A.P.,** Plasmin can activate plasma prorenin but is not required for the alkaline phase of acid activation, *Clin. Sci.,* 57 (Suppl. 5), 97s, 1979.
30. **Hiwada, K., Imamura, Y., and Kokubu, T.,** Activation and molecular weight changes of human plasma inactive renin by proteolytic enzymes, *Jpn. Circ. J.,* 47(10), 1193, 1983.
31. **Derkx, F.H., Brommer, E.J., Boomsma, F., and Schalekamp, M.A.,** Activation of plasma prorenin by plasminogen activators in vitro and increase in plasma renin after stimulation of fibrinolytic activity in vivo, *Clin. Exp. Hypertens. A,* 4(11 – 12), 2247, 1982.
32. **Camiolo, S.M., Thorsen, S., and Astrup, T.,** Fibrinogenolysis and fibrinolysis with tissue plasminogen activator, urokinase, streptokinase-activated human globulin, and plasmin, *Proc. Soc. Exp. Biol. Med.,* 138(1), 277, 1971.

33. **Summaria, L., Ling, C.M., Groskopf, W.R., and Robbins, K.C.,** The active site of bovine plasminogen activator. Interaction of streptokinase with human plasminogen and plasmin, *J. Biol. Chem.*, 243(1), 144, 1968.
34. **Travis, J. and Salvesen, G.S.,** Human plasma proteinase inhibitors, *Annu. Rev. Biochem.*, 52, 655, 1983.
35. **Kruithof, E.K., Tran-Thang, C., Ransijn, A., and Bachmann, F.,** Demonstration of a fast-acting inhibitor of plasminogen activators in human plasma, *Blood*, 64(4), 907, 1984.
36. **Wiman, B. and Chmielewska, J.,** A novel fast inhibitor to tissue plasminogen activator in plasma, which may be of great pathophysiological significance, *Scand. J. Clin. Lab. Invest. Suppl.*, 177, 43, 1985.
37. **Hekman, C.M. and Loskutoff, D.J.,** Kinetic analysis of the interactions between plasminogen activator inhibitor 1 and both urokinase and tissue plasminogen activator, *Arch. Biochem. Biophys.*, 262(1), 199, 1988.
38. **Pannekoek, H., Veerman, H., Lambers, H., Diergaarde, P., Verweij, C.L., van Zonneveld, A.J., and van Mourik, J.A.,** Endothelial plasminogen activator inhibitor (PAI), a new member of the Serpin gene family, *EMBO J.*, 5(10), 2539, 1986.
39. **Thorsen, S. and Philips, M.,** Isolation of tissue-type plasminogen activator-inhibitor complexes with from human plasma. Evidence for a rapid plasminogen activator inhibitor, *Biochim. Biophys. Acta*, 801(1), 802(1), 111, 1984.
40. **Thorsen, S., Philips, M., Selmer, J., Lecander, I., and Astedt, B.,** Kinetics of inhibition of tissue-type and urokinase-type plasminogen activator by plasminogen-activator inhibitor type 1 and type 2, *Eur. J. Biochem.*, 175(1), 33, 1988.
41. **Travis, J. and Salvesen, G.,** Control of coagulation and fibrinolysis by plasma proteinase inhibitors, *Behring Inst. Mitt.*, (73), 56, 1983.
42. **Hamsten, A., de Faire, U., Walldius, G., Dahlen, G., Szamosi, A., Landou, C., Blomback, M., and Wiman, B.,** Plasminogen activator inhibitor in plasma, risk factor for recurrent myocardial infarction, *Lancet*, 2(8549), 3, 1987.
43. **Collen, D.,** On the regulation and control of fibrinolysis. Edward Kowalski Memorial Lecture, *Thromb. Haemost.*, 43(2), 77, 1980.
44. **Alitalo, K. and Vaheri, A.,** Pericellular matrix in malignant transformation, *Adv. Cancer Res.*, 37, 111, 1982.
45. **Dano, K., Andreasen, P.A., Grondahl-Hansen, J., Kristensen, P., Nielsen, L.S., and Skriver, L.,** Plasminogen activators, tissue degradation, and cancer, *Adv. Cancer Res.*, 44, 139, 1985.
46. **Sumiyoshi, K., Baba, S., Sakaguchi, S., Urano, T., Takada, Y., and Takada, A.,** Increase in levels of plasminogen activator and type-1 plasminogen activator inhibitor in human breast cancer, possible roles in tumor progression and metastasis, *Thromb. Res.*, 63(1), 59, 1991.
47. **Sawaya, R., Ramo, O.J., Shi, M.L., and Mandybur, G.,** Biological significance of tissue plasminogen activator content in brain tumors, *J. Neurosurg.*, 74(3), 480, 1991.
48. **Quax, P.H., van Muijen, G.N., Weening-Verhoeff, E.J., Lund, L.R., Dano, K., Ruiter, D.J., and Verheijen, J.H.,** Metastatic behavior of human melanoma cell lines in nude mice correlates with urokinase-type plasminogen activator, its type-1 inhibitor, and urokinase-mediated matrix degradation, *J. Cell Biol.*, 115(1), 191, 1991.
49. **Maeda, Y., Souma, M., and Kasakura, S.,** [Plasma PAI-1 levels in patients with various tumors with or without metastasis], *Rinsho Ketsueki*, 32(9), 927, 1991.
50. **Foucre, D., Bouchet, C., Hacene, K., Pourreau-Schneider, N., Gentile, A., Martin, P.M., Desplaces, A., and Oglobine, J.,** Relationship between cathepsin D, urokinase, and plasminogen activator inhibitors in malignant vs benign breast tumours, *Br. J. Cancer*, 64(5), 926, 1991.
51. **Frame, M.C., Freshney, R.I., Vaughan, P.F., Graham, D.I., and Shaw, R.,** Interrelationship between differentiation and malignancy-associated properties in glioma, *Br. J. Cancer*, 49(3), 269, 1984.

52. **Sier, C.F., Verspaget, H.W., Griffioen, G., Verheijen, J.H., Quax, P.H., Dooijewaard, G., De Bruin, P.A., and Lamers, C.B.,** Imbalance of plasminogen activators and their inhibitors in human colorectal neoplasia. Implications of urokinase in colorectal carcinogenesis, *Gastroenterology,* 101(6), 1522, 1991.
53. **Lacroix, M., Smith, F.E., and Fritz, I.B.,** Secretion of plasminogen activator by Sertoli cell enriched cultures, *Mol. Cell Endocrinol.,* 9(2), 227, 1977.
54. **Vihko, K.K., Kristensen, P., Dano, K., and Parvinen, M.,** Immunohistochemical localization of urokinase-type plasminogen activator in Sertoli cells and tissue-type plasminogen activator in spermatogenic cells in the rat seminiferous epithelium, *Dev. Biol.,* 126(1), 150, 1988.
55. **Beers, W.H., Strickland, S., and Reich, E.,** Ovarian plasminogen activator, relationship to ovulation and hormonal regulation, *Cell,* 6(3), 387, 1975.
56. **Strickland, S., Reich, E., and Sherman, M.I.,** Plasminogen activator in early embryogenesis, enzyme production by trophoblast and parietal endoderm, *Cell,* 9(2), 231, 1976.
57. **Cajander, S.B.,** Periovulatory changes in the ovary. Morphology and expression of tissue-type plasminogen activator, *Prog. Clin. Biol. Res.,* 296, 91, 1989.
58. **Lala, P.K. and Graham, C.H.,** Mechanisms of trophoblast invasiveness and their control, the role of proteases and protease inhibitors, *Cancer Metastasis Rev.,* 9(4), 369, 1990.
59. **Axelrod, H.R.,** Altered trophoblast functions in implantation-defective mouse embryos, *Dev. Biol.,* 108(1), 185, 1985.
60. **Kruithof, E.K., Tran-Thang, C., Gudinchet, A., Hauert, J., Nicoloso, G., Genton, C., Welti, H., and Bachmann, F.,** Fibrinolysis in pregnancy, a study of plasminogen activator inhibitors, *Blood,* 69(2), 460, 1987.
61. **Kalderon, N.,** Migration of Schwann cells and wrapping of neurites in vitro, a function of protease activity (plasmin) in the growth medium, *Proc. Natl. Acad. Sci. U.S.A.,* 76(11), 5992, 1979.
62. **Krystosek, A. and Seeds, N.W.,** Peripheral neurons and Schwann cells secrete plasminogen activator, *J. Cell Biol.,* 98(2), 773, 1984.
63. **Patterson, P.H.,** On the role of proteases, their inhibitors and the extracellular matrix in promoting neurite outgrowth, *J. Physiol. Paris,* 80(4), 207, 1985.
64. **Krystosek, A. and Seeds, N.W.,** Normal and malignant cells, including neurons, deposit plasminogen activator on the growth substrata, *Exp. Cell Res.,* 166(1), 31, 1986.
65. **McGuire, P.G. and Seeds, N.W.,** Degradation of underlying extracellular matrix by sensory neurons during neurite outgrowth, *Neuron,* 4(4), 633, 1990.
66. **Ettenger, M.M., MacCarthy, E.P., Glas-Greenwalt, P., Clyne, D.H., and Pollak, V.E.,** Abnormalities of fibrinolysis in essential hypertension, *J. Hypertens. Suppl.,* 2(3), S175, 1984.
67. **Landin, K., Tengborn, L., and Smith, U.,** Elevated fibrinogen and plasminogen activator inhibitor (PAI-1) in hypertension are related to metabolic risk factors for cardiovascular disease, *J. Intern. Med.,* 227(4), 273, 1990.
68. **van Wersch, J.W., Rompelberg-Lahaye, J., and Lustermans, F.A.,** Plasma concentration of coagulation and fibrinolysis factors and platelet function in hypertension, *Eur. J. Clin. Chem. Clin. Biochem.,* 29(6), 375, 1991.
69. **Jansson, J.H., Johansson, B., Boman, K., and Nilsson, T.K.,** Hypo-fibrinolysis in patients with hypertension and elevated cholesterol, *J. Intern. Med.,* 229(4), 309, 1991.
70. **Jansson, J.H., Nilsson, T.K., and Olofsson, B.O.,** Tissue plasminogen activator and other risk factors as predictors of cardiovascular events in patients with severe angina pectoris, *Eur. Heart J.,* 12(2), 157, 1991.
71. **Auwerx, J., Bouillon, R., Collen, D., and Geboers, J.,** Tissue-type plasminogen activator antigen and plasminogen activator inhibitor in diabetes mellitus, *Arteriosclerosis,* 8(1), 68, 1988.

72. **Juhan-Vague, I., Alessi, M.C., and Vague, P.,** Increased plasma plasminogen activator inhibitor 1 levels. A possible link between insulin resistance and atherothrombosis. *Diabetologia,* 34(7), 457, 1991.
73. **Schneider, D.J. and Sobel, B.E.,** Augmentation of synthesis of plasminogen activator inhibitor type 1 by insulin and insulin-like growth factor type I, implications for vascular disease in hyperinsulinemic states, *Proc. Natl. Acad. Sci. U.S.A.,* 88(22), 9959, 1991.
74. **Paterson, P.Y., Koh, C.S., and Kwaan, H.C.,** Role of the clotting system in the pathogenesis of neuroimmunologic disease, *Fed. Proc.,* 46(1), 91, 1987.
75. **Kruithof, E.K., Gudinchet, A., and Bachmann, F.,** Plasminogen activator inhibitor 1 and plasminogen activator inhibitor 2 in various disease states, *Thromb. Haemost.,* 59(1), 7, 1988.
76. **Okada, Y., Tsuchiya, T., Tada, T., Mishima, T., and Suzuki, T.,** [Experimental study on plasminogen activator and trypsin inhibitor system in the cerebrospinal fluid and brain in cases of trauma and shock], *No To Shinkei,* 18(8), 784, 1966.
77. **Takashima, S., Koga, M., and Tanaka, K.,** Fibrinolytic activity of human brain and cerebrospinal fluid, *Br. J. Exp. Pathol.,* 50, 533, 1969.
78. **Menon, I.S., Muscat-Baron, J., Weightman, D., and Dewar, H.A.,** The brain as contributor of plasminogen activator to the blood, *Clin. Sci.,* 38(1), 85, 1970.
79. **Franks, A.J. and Ellis, E.,** Immunohistochemical localisation of tissue plasminogen activator in human brain tumours, *Br. J. Cancer,* 59(3), 462, 1989.
80. **Sawaya, R. and Highsmith, R.,** Plasminogen activator activity and molecular weight patterns in human brain tumors, *J. Neurosurg.,* 68(1), 73, 1988.
81. **Tucker, W.S., Kirsch, W.M., Martinez-Hernandez, A., and Fink, L.M.,** In vitro plasminogen activator activity in human brain tumors, *Cancer Res.,* 38(2), 297, 1978.
82. **Glas, P. and Astrup, T.,** Thromboplastin and plasminogen activator in tissues of the rabbit, *Am. J. Physiol.,* 219(4), 1140, 1970.
83. **Soreq, H. and Miskin, R.,** Plasminogen activator in the rodent brain, *Brain Res.,* 216(2), 361, 1981.
84. **Danglot, G., Vinson, D., and Chapeville, F.,** Qualitative and quantitative distribution of plasminogen activators in organs from healthy adult mice, *FEBS Lett.,* 194(1), 96, 1986.
85. **Bruesch, M.R., Johnson, G.L., Palackdharry, C.S., Weber, M.J., and Carl, P.L.,** Plasminogen activator in normal and tumor-bearing mice, *Int. J. Cancer,* 32(1), 121, 1983.
86. **Sawdey, M.S. and Loskutoff, D.J.,** Regulation of murine type 1 plasminogen activator inhibitor gene expression in vivo. Tissue specificity and induction by lipopolysaccharide tumor necrosis factor-α and transforming growth factor-β, *J. Clin. Invest.,* 88(4), 1364, 1991.
87. **Wachsman, J.T. and Biedler, J.L.,** Fibrinolytic activity associated with human neuroblastoma cells, *Exp. Cell Res.,* 86, 264, 1974.
88. **Krystosek, A. and Seeds, N.W.,** Plasminogen activator secretion by granule neurons in cultures of developing cerebellum, *Proc. Natl. Acad. Sci. U.S.A.,* 78(12), 7810, 1981.
89. **Soreq, H., Miskin, R., Zutra, A., and Littauer, U.Z.,** Modulation in the levels and localization of plasminogen activator in differentiating neuroblastoma cells, *Brain Res.,* 283(2–3), 257, 1983.
90. **Soreq, H. and Miskin, R.,** Plasminogen activator in the developing rat cerebellum, biosynthesis and localization in granular neurons, *Brain Res.,* 313(2), 149, 1983.
91. **Pittman, R.N., Ivins, J.K., and Buettner, H.M.,** Neuronal plasminogen activators, cell surface binding sites and involvement in neurite outgrowth, *J. Neurosci.,* 9(12), 4269, 1989.
92. **Presta, M., Ennas, M.G., Torelli, S., Ragnotti, G., and Gremo, F.,** Synthesis of urokinase-type plasminogen activator and of type-1 plasminogen activator inhibitor in neuronal cultures of human fetal brain, stimulation by phorbol ester, *J. Neurochem.,* 55(5), 1647, 1990.

93. **Wagner, S.L., Lau, A.L., Nguyen, A., Mimuro, J., Loskutoff, D.J., Isackson, P.J., and Cunningham, D.D.**, Inhibitors of urokinase and thrombin in cultured neural cells, *J. Neurochem.*, 56(1), 234, 1991.
94. **Toshniwal, P.K., Firestone, S.L., Barlow, G.H., and Tiku, M.L.**, Characterization of astrocyte plasminogen activator, *J. Neurol. Sci.*, 80(2 – 3), 277, 1987.
95. **Toshniwal, P.K., Tiku, M.L., Tiku, K., and Skosey, J.L.**, Secretion of plasminogen activator by cerebral astrocytes and its modulation, *J. Neurol. Sci.*, 80(2 – 3), 307, 1987.
96. **Gross, J.L., Behrens, D.L., Mullins, D.E., Kornblith, P.L., and Dexter, D.L.**, Plasminogen activator and inhibitor activity in human glioma cells and modulation by sodium butyrate, *Cancer Res.*, 48(2), 291, 1988.
97. **Guenther, J., Nick, H., and Monard, D.**, A glia-derived neurite-promoting factor with protease inhibitory activity, *EMBO J.*, 4(8), 1963, 1985.
98. **Rehemtulla, A., Murphy, P., Dobson, M., and Hart, D.A.**, Purification and partial characterization of a plasminogen activator inhibitor from the human glioblastoma, U138, *Biochem. Cell Biol.*, 66(12), 1270, 1988.
99. **Sawaya, R., Ramo, O.J., Glas-Greenwalt, P., and Wu, S.Z.**, Plasma fibrinolytic profile in patients with brain tumors, *Thromb. Haemost.*, 65(1), 15, 1991.
100. **Keohane, M.E., Hall, S.W., VandenBerg, S.R., and Gonias, S.L.**, Secretion of alpha 2-macroglobulin, alpha 2-antiplasmin, and plasminogen activator inhibitor-1 by glioblastoma multiforme in primary organ culture, *J. Neurosurg.*, 73(2), 234, 1990.
101. **Andreasen, P.A., Christensen, T.H., Huang, J.Y., Nielsen, L.S., Wilson, E.L., and Dano, K.**, Hormonal regulation of extracellular plasminogen activators and Mr approximately 54,000 plasminogen activator inhibitor in human neoplastic cell lines, studied with monoclonal antibodies, *Mol. Cell Endocrinol.*, 45(2 – 3), 137, 1986.
102. **Crutchley, D.J. and Conanan, L.B.**, Endotoxin induction of an inhibitor of plasminogen activator in bovine pulmonary artery endothelial cells, *J. Biol. Chem.*, 261(1), 154, 1986.
103. **Dinarello, C.A.**, The biology of interleukin 1 and comparison to tumor necrosis factor, *Immunol. Lett.*, 16(3 – 4), 227, 1987.
104. **Schleef, R.R., Bevilacqua, M.P., Sawdey, M., Gimbrone, M.A., Jr., and Loskutoff, D.J.**, Cytokine activation of vascular endothelium. Effects on tissue-type plasminogen activator and type 1 plasminogen activator inhibitor, *J. Biol. Chem.*, 263(12), 5797, 1988.
105. **Ny, T., Bjersing, L., Hsueh, A.J., and Loskutoff, D.J.**, Cultured granulosa cells produce two plasminogen activators and an antiactivator, each regulated differently by gonadotropins, *Endocrinology*, 116(4), 1666, 1985.
106. **Santell, L. and Levin, E.G.**, Cyclic AMP potentiates phorbol ester stimulation of tissue plasminogen activator release and inhibits secretion of plasminogen activator inhibitor-1 from human endothelial cells, *J. Biol. Chem.*, 263(32), 16802, 1988.
107. **Lund, L.R., Riccio, A., Andreasen, P.A., Nielsen, L.S., Kristensen, P., Laiho, M., Saksela, O., Blasi, F., and Dano, K.**, Transforming growth factor-beta is a strong and fast acting positive regulator of the level of type-1 plasminogen activator inhibitor mRNA in WI-38 human lung fibroblasts, *EMBO J.*, 6(5), 1281, 1987.
108. **Laiho, M., Saksela, O., and Keski-Oja, J.**, Transforming growth factor beta alters plasminogen activator activity in human skin fibroblasts, *Exp. Cell Res.*, 164(2), 399, 1986.
109. **Laiho, M., Saksela, O., Andreasen, P.A., and Keski-Oja, J.**, Enhanced production and extracellular deposition of the endothelial-type plasminogen activator inhibitor in cultured human lung fibroblasts by transforming growth factor-beta, *J. Cell Biol.*, 103(6 Pt. 1), 2403, 1986.
110. **Thalacker, F.W. and Nilsen-Hamilton, M.**, Specific induction of secreted proteins by transforming growth factor-beta and 12-O-tetradecanoylphorbol-13-acetate. Relationship with an inhibitor of plasminogen activator, *J. Biol. Chem.*, 262(5), 2283, 1987.

111. **Konkle, B.A. and Ginsburg, D.,** The addition of endothelial cell growth factor and heparin to human umbilical vein endothelial cell cultures decreases plasminogen activator inhibitor-1 expression, *J. Clin. Invest.*, 82(2), 579, 1988.
112. **Lucore, C.L., Fujii, S., Wun, T.C., Sobel, B.E., and Billadello, J.J.,** Regulation of the expression of type 1 plasminogen activator inhibitor in Hep G2 cells by epidermal growth factor, *J. Biol. Chem.*, 263(31), 15845, 1988.
113. **Kooistra, T., Bosma, P.J., Toet, K., Cohen, L.H., Griffioen, M., van den Berg, E., le Clercq, L., and van Hinsbergh, V.W.,** Role of protein kinase C and cyclic adenosine monophosphate in the regulation of tissue-type plasminogen activator, plasminogen activator inhibitor-1, and platelet-derived growth factor mRNA levels in human endothelial cells. Possible involvement of proto-oncogenes c-jun and c-fos, *Arterioscler. Thromb.*, 11(4), 1042, 1991.
114. **van den Berg, E.A., Sprengers, E.D., Jaye, M., Burgess, W., Maciag, T., and van Hinsbergh, V.W.,** Regulation of plasminogen activator inhibitor-1 mRNA in human endothelial cells, *Thromb. Haemost.*, 60(1), 63, 1988.
115. **Rogister, B., Leprince, P., Delree, P., Van Damme, J., Billiau, A., and Moonen, G.,** Enhanced release of plasminogen activator inhibitor(s) but not of plasminogen activators by cultured rat glial cells treated with interleukin-1, *GLIA*, 3(4), 252, 1990.
116. **Sipe, J.D.,** The molecular biology of interleukin 1 and the acute phase response, *Adv. Intern. Med.*, 34, 1, 1989.
117. **Dinarello, C.A.,** Interleukin-1 and the pathogenesis of the acute-phase response, *N. Engl. J. Med.*, 311(22), 1413, 1984.
118. **Krueger, J.M., Walter, J., Dinarello, C.A., Wolff, S.M., and Chedid, L.,** Sleep-promoting effects of endogenous pyrogen (interleukin-1), *Am. J. Physiol.*, 246(6 Pt. 2), R994, 1984.
119. **Mizel, S.G.,** The interleukins, *FASEB J.*, 3, 2379, 1989.
120. **Declerck, P.J., De Mol, M., Alessi, M.C., Baudner, S., Paques, E.P., Preissner, K.T., Muller-Berghaus, G., and Collen, D.,** Purification and characterization of a plasminogen activator inhibitor 1 binding protein from human plasma. Identification as a multimeric form of S protein (vitronectin), *J. Biol. Chem.*, 263(30), 15454, 1988.
121. **Wiman, B., Almquist, A., Sigurdardottir, O., and Lindahl, T.,** Plasminogen activator inhibitor 1 (PAI) is bound to vitronectin in plasma, *FEBS Lett.*, 242(1), 125, 1988.
122. **Mimuro, J. and Loskutoff, D.J.,** Purification of a protein from bovine plasma that binds to type 1 plasminogen activator inhibitor and prevents its interaction with extracellular matrix. Evidence that the protein is vitronectin, *J. Biol. Chem.*, 264(2), 936, 1989.
123. **Wun, T.C., Palmier, M.O., Siegel, N.R., and Smith, C.E.,** Affinity purification of active plasminogen activator inhibitor-1 (PAI-1) using immobilized anhydrourokinase. Demonstration of the binding, stabilization, and activation of PAI-1 by vitronectin, *J. Biol. Chem.*, 264(14), 7862, 1989.
124. **Preissner, K.T.,** Specific binding of plasminogen to vitronectin. Evidence for a modulatory role of vitronectin on fibrin(ogen)-induced plasmin formation by tissue plasminogen activator, *Biochem. Biophys. Res. Commun.*, 168(3), 966, 1990.
125. **Hekman, C.M., and Loskutoff, D.J.,** Bovine plasminogen activator inhibitor 1, specificity determinations and comparison of the active, latent, and guanidine-activated forms, *Biochemistry*, 27(8), 2911, 1988.
126. **Keijer, J., Ehrlich, H.J., Linders, M., Preissner, K.T., and Pannekoek, H.,** Vitronectin governs the interaction between plasminogen activator inhibitor 1 and tissue-type plasminogen activator, *J. Biol. Chem.*, 266(16), 10700, 1991.
127. **Seiffert, D., Keeton, M., Eguchi, Y., Sawdey, M., and Loskutoff, D.J.,** Detection of vitronectin mRNA in tissues and cells of the mouse, *Proc. Natl. Acad. Sci. U.S.A.*, 88(21), 9402, 1991.

128. **Akiyama, H., Kawamata, T., Dedhar, S., and McGeer, P.L.,** Immunohistochemical localization of vitronectin, its receptor and beta-3 integrin in Alzheimer brain tissue, *J. Neuroimmunol.*, 32(1), 19, 1991.
129. **Kumar, A., Rassoli, A., and Raizada, M.K.,** Angiotensinogen gene expression in neuronal and glial cells in primary cultures of rat brain, *J. Neurosci. Res.*, 19, 287, 1988.
130. **Stornetta, K.L., Hawelu-Johnson, C.L., Guyenet, P.G., and Lynch, K.R.,** Astrocytes synthesize angiotensinogen in brain, *Science,* 242, 1446, 1988.
131. **Milsted, A., Barna, B.P., Ransohoff, R.M., Brosnihan, K.B., and Ferrario, C.M.,** Astrocyte cultures derived from human brain tissue express angiotensinogen mRNA, *Proc. Natl. Acad. Sci. U.S.A.*, 87, 5720, 1990.
132. **Intebi, A.D., Flaxman, M.S., Ganong, W.F., and Deschepper, C.F.,** Angiotensinogen production by rat astroglial cells in vitro and in vivo, *Neuroscience,* 32, 545, 1990.
133. **Hermann, K., Raizada, M.K., Sumners, C., and Phillips, M.I.,** Immunocytochemical and biochemical characterization of angiotensin I and II in cultured neuronal and glial cells from rat brain, *Neuroendocrinology,* 47(2), 125, 1988.
134. **Hermann, K., Phillips, M.I., Hilgenfeldt, U., and Raizada, M.K.,** Biosynthesis of angiotensinogen and angiotensins by brain cells in primary culture, *J. Neurochem.*, 51(2), 398, 1988.
135. **Changaris, D.G., Demers, L.M., Keil, L.C., and Severs, W.B.,** Immunopharmacology of angiotensin I in brain, in *Central Actions of Angiotensin and Related Hormones,* Buckley, J. and Ferrario, C., Eds., Pergamon Press, Oxford, 1977, 233.
136. **Hermann, K., Raizada, M.K., Sumners, C., and Phillips, M.I.,** Presence of renin in primary neuronal and glial cells from rat brain, *Brain Res.*, 437, 205, 1987.
137. **Unger, T., Gohlke, P., Kotrba, M., Rettig, R., and Rohmeiss, P.,** Angiotensin II and atrial natriuretic peptide in the brain, effects on volume and Na+ balance, *Resuscitation,* 18(2 – 3), 309, 1989.
138. **Raizada, M.K., Phillips, M.I., Crews, F.T., and Sumners, C.,** Distinct angiotensin II receptor in primary cultures of glial cells from rat brain, *Proc. Natl. Acad. Sci. U.S.A.*, 84, 4655, 1987.
139. **Sumners, C., Tang, W., Zelezna, B., and Raizada, M.K.,** Angiotensin II receptor subtypes are coupled with distinct signal transduction mechanism in neurons and astroglia from rat brain, *Proc. Natl. Acad. Sci. U.S.A.*, 88, 7567, 1991.
140. **Tallant, E.A., Jaiswal, N., Diz, D.I., and Ferrario, C.M.,** Human astrocytes contain two distinct angiotensin receptor subtypes, *Hypertension,* 18, 32, 1991.
141. **Cash, J.D., Gader, A.M., and da Costa, J.,** Proceedings, The release of plasminogen activator and factor VIII to lysine vasopressin, arginine vasopressin, I-desamino-8-d-arginine vasopressin, angiotensin and oxytocin in man, *Br. J. Haematol.*, 27(2), 363, 1974.
142. **Hariman, H., Grant, P.J., Hughes, J.R., Booth, N.A., Davies, J.A., and Prentice, C.R.,** Effect of physiological concentrations of vasopressin on components of the fibrinolytic system, *Thromb. Haemost.*, 61(2), 298, 1989.
143. **Kalderon, N.,** Schwann cell proliferation and localized proteolysis, expression of plasminogen-activator activity predominates in the proliferating cell populations, *Proc. Natl. Acad. Sci. U.S.A.*, 81(22), 7216, 1984.
144. **Moonen, G., Grau-Wagemans, M.P., and Selak, I.,** Plasminogen activator-plasmin system and neuronal migration, *Nature,* 298(5876), 753, 1982.
145. **Gurwitz, D. and Cunningham, D.D.,** Neurite outgrowth activity of protease nexin-1 on neuroblastoma cells requires thrombin inhibition, *J. Cell. Physiol.*, 142(1), 155, 1990.
146. **Cavanaugh, K.P., Gurwitz, D., Cunningham, D.D., and Bradshaw, R.A.,** Reciprocal modulation of astrocyte stellation by thrombin and protease nexin-1, *J. Neurochem.*, 54(5), 1735, 1990.

147. **Ehrlich, H.J., Gebbink, R.K., Keijer, J., Linders, M., Preissner, K.T., and Pannekoek, H.,** Alteration of serpin specificity by a protein cofactor. Vitronectin endows plasminogen activator inhibitor 1 with thrombin inhibitory properties, *J. Biol. Chem.*, 265(22), 13029, 1990.
148. **Schuerch-Rathgeb, Y. and Mongard, D.,** Brain development influences the appearance of glial factor-like activity in rat brain primary cultures, *Nature*, 273(5660), 308, 1978.
149. **Iwasaki, Y., Kinoshita, M., Ikeda, K., Shiojima, T., Kurihara, T., and Appel, S.H.,** Trophic effect of angiotensin II, vasopressin and other peptides on the cultured ventral spinal cord of rat embryo, *J. Neurol. Sci.*, 103(2), 151, 1991.
150. **Murphy, P. and Hart, D.A.,** Modulation of plasminogen activator and plasminogen activator inhibitor expression in the human U373 glioblastoma/astrocytoma cell line by inflammatory mediators, *Exp. Cell Res.*, 198(1), 93, 1992.
151. **Robberecht, W. and Denef, C.,** Enhanced Ang II activity in anterior pituitary cell aggregates from hypertensive rats, *Am. J. Physiol.*, 255(3 Pt. 2), R407, 1988.
152. **Gutkind, J.S., Kurihara, M., Castren, E., and Saavedra, J.M.,** Increased concentration of angiotensin II binding sites in selected brain areas of spontaneously hypertensive rats, *J. Hypertens.*, 6(1), 79, 1988.
153. **Kokolis, N., Ploumis, T., and Smokovitis, A.,** The effect of experimental chronic hypertension on tissue plasminogen activator activity, plasminogen activator inhibition and plasmin inhibition, *Thromb. Res.*, 56(4), 523, 1989.
154. **Richards, E.M., Sumners, C., Chou, Y.-C., Raizada, M.K., and Phillips, M.I.,** α_2-Adrenergic receptors in neuronal and glial cultures: characterization and comparison, *J. Neurochem.*, 53, 287, 1989.

Section VI

Renin-Angiotensin and Human Hypertension

Chapter 20

DERANGEMENTS IN RENIN-ANGIOTENSIN REGULATION IN THE PATHOGENESIS OF HYPERTENSION

G.H. Williams and N.K. Hollenberg

TABLE OF CONTENTS

0-8493-4622-3/93/$0.00 + $.50

I. INTRODUCTION

Blood pressure sensitivity to sodium intake is present in 50 to 60% of patients with hypertension.[1,2] Epidemiologic studies suggest that there is a variety of pathophysiologic mechanisms leading to sodium-sensitive hypertension. For example, when sodium sensitivity of blood pressure has been evaluated in large numbers of hypertensive patients, a unimodal distribution is observed.[3] Second, on theoretical grounds, a number of differing pathophysiologic mechanisms have been described, including low renin essential hypertension, bilateral renal parenchymal damage, and nonmodulating essential hypertension.[4-6] Even in these broad subgroups, heterogeneity of mechanisms likely exists, i.e., low renin levels can occur in older subjects, probably reflecting vascular damage within the kidney, or in patients with diabetes, secondary to the metabolic derangements. Contrarily, some subjects with low renin hypertension demonstrate an enhanced adrenal response to angiotensin II.[7-9] In these subjects, the sodium sensitivity of the blood pressure is produced by mechanisms presumably analogous to those present in patients with primary aldosteronism.

The most recently described group of sodium-sensitive hypertensives are nonmodulators.[6,10-12] This group differs from other forms of sodium-sensitive hypertensives in that in a sodium-restricted state, their plasma renin activities are normal or high rather than low. Furthermore, it has been proposed that these subjects have a functional derangement in angiotensin II (AngII) target tissues which may be mediated by changes in the regulation of the local renin-angiotensin system.

II. THE RENIN-ANGIOTENSIN-ALDOSTERONE SYSTEM AND SODIUM HOMEOSTASIS

Several lines of evidence suggest that the renin-angiotensin system plays a prominent role in the maintenance of normal sodium homeostasis. This evidence includes the preferential sensitivity of the renal vasculature over the peripheral vasculature for AngII,[13] the role of AngII as a major secretagogue for aldosterone secretion,[14] and the likely role of AngII as a direct regulator of renal tubular function.[15] Contrariwise, sodium intake is a major factor regulating renin and aldosterone secretion.[16,17] A third dimension of this relationship is the effect dietary sodium intake has on the responsiveness of target tissues to AngII. In the late 1960s to early 1970s, several groups documented this effect both in the vascular system and on aldosterone secretion. Kaplan and Silah were among the first to report that sodium restriction reduces the blood pressure response to AngII.[18] Strewler et al. reported similar observations when they studied blood flow in rabbit leg vessels.[19] The renal vasculature responses are particularly unique: its sensitivity is much greater than the peripheral vasculature and sodium intake effect is more pronounced.[20]

Ganong and Boryczka made the initial observation that sodium restriction enhanced the adrenal response to AngII using the dog as their model.[21] Davis and colleagues demonstrated the opposite effect, i.e., a decreased sensitivity of the adrenal gland to AngII, when dogs were placed on a high salt intake.[22] Although initially controversial, it is now the consensus of most investigators that sodium intake has a similar effect in humans.[23-25] Furthermore, the sensitivity of the human adrenal to AngII is similar to the renal vasculature and, therefore, considerably more sensitive than the peripheral vasculature.[23]

Not only does sodium intake influence the responsive of these two AngII target tissues, but also this change is in the direction which increases its effectiveness in maintaining sodium homeostasis. With sodium restriction, the response of the adrenal to AngII is enhanced while the response of the renal vasculature is reduced, thereby promoting sodium retention. With sodium loading, the opposite occurs — the adrenal response is reduced and the renal vascular response is increased. Thus, sodium intake induces a reciprocal change in the responsiveness of the vasculature and the adrenal to AngII (Figure 1).

The mechanisms responsible for these changes have been extensively evaluated. For the vascular system, this change can be explained by a change in AngII receptor number. Studies *in vivo* have documented that as the circulating AngII levels rise, the responsiveness of the vasculature to infused AngII is reduced.[26] This occurs regardless of the mechanism used to change the AngII level, i.e., estrogen therapy, changes in dietary potassium intake, administration of converting enzyme inhibitors, or changes in dietary sodium intake.[27] Studies *in vitro* provide additional support to the hypothesis that AngII downregulates its own receptor,[28] and this is the mechanism mediating the change in vascular responses with changes in dietary sodium intake.

Several lines of evidence suggest that a similar mechanism is not likely to underlie the sodium intake-induced change in aldosterone responsiveness.

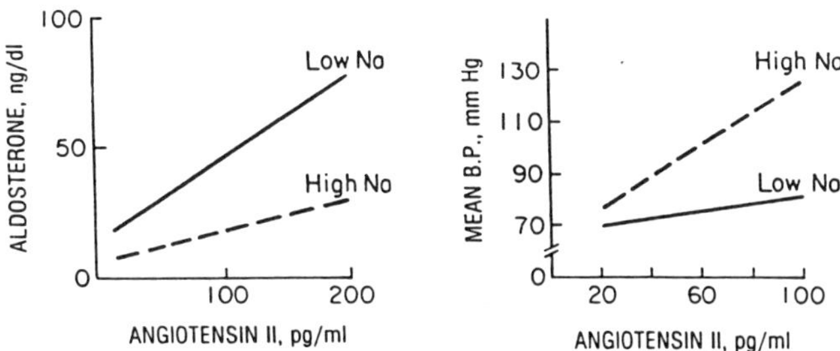

FIGURE 1. The reciprocal influence of sodium intake on adrenal and vascular responses to AngII. Vascular responses apply to both blood pressure and renal blood flow. With sodium loading, vascular responses are enhanced and adrenal responses are reduced. (From Hollenberg, N.K., Chenitz, W.R., Adams, D.F., and Williams, G.H., *J. Clin. Invest.*, 54, 34, 1974. By copyright permission of the Society for Clinical Investigation.)

First, the impact of sodium restriction or AngII infusion on the number of adrenal AngII receptors depends on the species studied. In rats, these manipulations increase the number of AngII-binding sites.[29] However, in primates, sodium restriction reduces them, similar to the effect of sodium restriction on vascular AngII receptors.[20] Second, in normal subjects on a low-sodium intake, administration of a converting enzyme inhibitor to acutely reduced circulating AngII levels does not change the responsiveness of the adrenal to AngII, even though it completely reverses the reduced responsiveness of the renal vasculature.[31] Third, if an increase in the number of adrenal AngII receptors is the mechanism mediating the enhanced adrenal response with sodium restriction, then one would anticipate that with sodium restriction, the efficacy of a competitive antagonist of AngII would also increase. Using glomerulosa cells obtained from rats maintained on either a high- or a low-sodium intake, Williams et al.[32] clearly demonstrated that the number of functional AngII receptors on the glomerulosa cell decreased with sodium restriction — despite the previously described increased number of AngII-binding sites with sodium restriction in this species. Furthermore, this study strongly supported the hypothesis that an event distal to binding of AngII to its receptor was the major mediator of the sodium restriction-induced enhancement in aldosterone responsiveness to AngII (Figure 2).

A possible mediator of this effect is the adrenal renin-angiotensin system. The presence of all of the components of this system in the adrenal, including the converting enzyme, has been documented.[33-36] Indeed, it has been reported that AngII is secreted from isolated perfused glomerulosa cells.[37] Sodium intake modifies the adrenal renin-angiotensin system in a predictable fashion, i.e., sodium restriction increases its activity[38] and even appears to increase the activity of the converting enzyme (Figure 3). Both AngII and potassium increase adrenal renin activity in contrast to their inhibitory effects on juxtaglomerular renin secretion.[36,39]

Another possible site of regulation of aldosterone secretion by sodium intake is at one or more transduction steps. Several lines of evidence suggest that the major transduction mechanism utilized by AngII in increasing aldosterone secretion is the hydrolysis of phosphoinositol by phospholipase C with the formation of inositol trisphosphate (IP_3) and diacylglycerol (DAG).[40,41] Activation of phospholipase C is rapid, as evidenced by the reduction in [^{3}H]inositol-labeled PIP_2 and the appearance of radioactively labeled IP_3 within 5 s of AngII application.[42] Most studies suggest that IP_3 levels remain elevated following AngII stimulation; however, only a few have measured the actual mass of this compound. One such study reported that the increase in the level of IP_3 mass is extremely short (less than 12 s).[43] Of interest is a study that has reported a change in the IP_3 response to AngII in sodium-restricted vs. sodium-loaded animals: sodium restriction increases the IP_3 response.[44] There is little evidence to support or refute a role for changes in DAG levels or cytosolic calcium, as additional factors regulated by modification of dietary sodium intake.

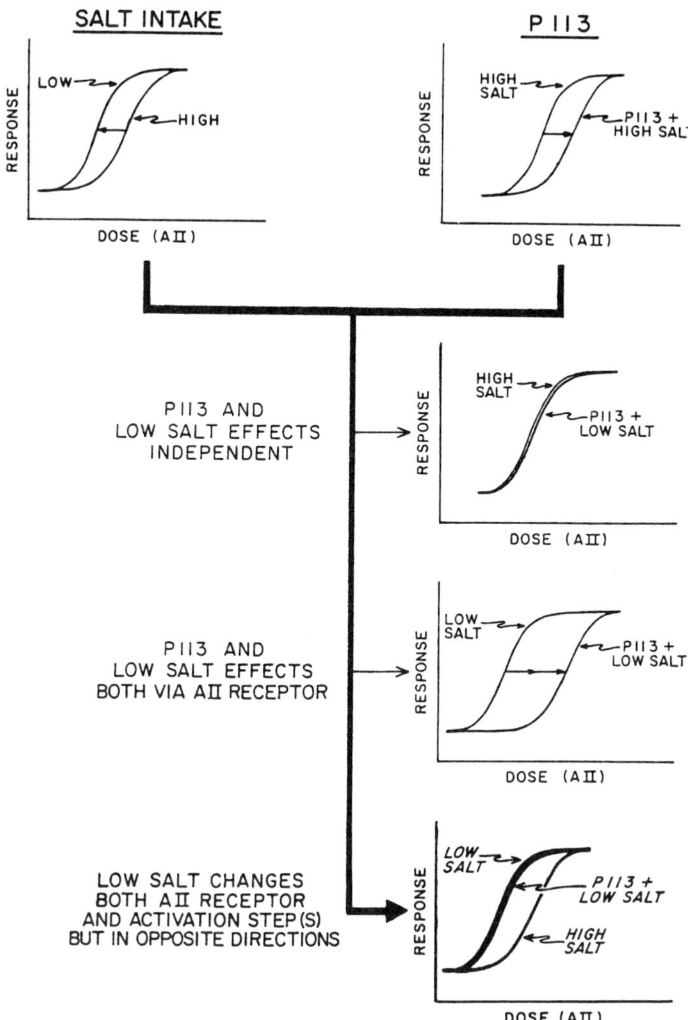

FIGURE 2. Theoretical interactions of sodium intake, AngII, and an AngII competitive antagonist (P113). *Top panel:* Restriction of sodium intake increases the glomerulosa cell sensitivity to angiotensin II; P113, the competitive antagonist, has the opposite effect. *Lower panels:* There are three possible outcomes if P113 is given simultaneously with AngII to glomerulosa cells obtained from animals on a low salt diet. (1) If they act independently, their effects should cancel each other and the angiotensin dose-response curve should be the same on the low- as on the high-salt diet (upper panel). (2) If sodium restriction acts via a change in receptor, then with sodium restriction P113 should be more efficacious (middle panel). (3) If sodium restriction has two opposite effects (one reducing binding of AngII to its receptor and a second activating a post-receptor step) then with sodium restriction the adrenal will be more sensitive to AngII but P113 would be less efficacious (bottom panel). The experimental data supported the third hypothesis. (From Williams, G.H., Hollenberg, N.K., and Braley, L.M., *Endocrinology,* 98, 1343, 1976. With permission from The Endocrine Society.)

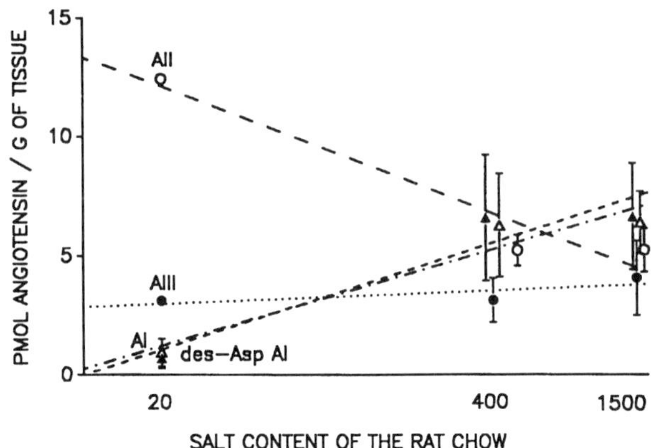

FIGURE 3. Peptide content in glomerulosa cells after five days of a fixed sodium intake. Correlations between the logarithm of the dietary salt content and the tissue levels of the angiotensins were all highly significant ($p < 0.005$) except for AIII. (AI, angiotensin I; AII, angiotensin II; AIII, Des-Asp angiotensin II; and Des-Asp AI, Des-Asp angiotensin I.) (From Kifor, I. et al., *Endocrinology,* 128, 1277, 1991. With permission of the Endocrine Society.)

A third potential site where sodium intake could regulate aldosterone secretion is at the level of an aldosterone biosynthetic enzyme. Aldosterone biosynthesis is regulated at two steps in its pathway: (1) early (conversion of cholesterol to pregnenolone), and (2) late (conversion of corticosterone to aldosterone).[45] All acute stimuli alter the early pathway of aldosterone biosynthesis which is mediated by the P450 side chain cleavage enzyme (P450scc).[45] By increasing corticosterone, which via an allosteric effect increases aldosterone production, acute stimuli may also indirectly modify the late pathway.[45,46] However, it is uncertain whether acute stimuli directly influence aldosterone biosynthesis except at the early pathway. In contrast, chronic stimuli such as sodium restriction appear to preferentially increase the late pathway activity. Indeed, this activity is increased by four- to tenfold with sodium restriction.[47] A similar effect of dietary sodium intake on the early pathway activity has not been described. However, small changes may have been missed in earlier studies addressing this issue.

The enzyme responsible for catalyzing the late pathway is P450 aldosterone synthetase (P450aldo). This enzyme is encoded by a gene on human chromosome 8q in close proximity to another steroidogenic enzyme — P450 11β-hydroxylase (P450c11).[48] Even though these enzymes are closely associated, it is only the aldosterone synthetase activity that is modified by the level of dietary sodium intake.

P450scc and aldosterone synthetase are mitochondrial enzymes which use adrenodoxin and adrenodoxin reductase to receive electrons from NADPH.[49]

Thus, the chronic adaptive response of the glomerulosa cell to sodium restriction could be secondary either to an increase in the number of aldosterone synthetase enzyme units and/or to an increase in the functional capacity of these units by increasing the flow of electrons to them through an increase in the adrenodoxin-adrenodoxin reductase system.

III. NONMODULATING HYPERTENSION

In 1982, Shoback and colleagues[10] reported that in a subgroup of patients with essential hypertension sodium intake did not modify the adrenal and renal vascular responses to AngII. They, therefore, termed these patients ''nonmodulators,'' since dietary sodium did not modulate their target tissue responses to AngII. Evidence published before their report gave clues to support their hypothesis. As early as 1972 some patients with essential hypertension have been reported to not increase aldosterone secretion when placed on a low-salt diet and/or given a diuretic challenge.[50,51] In some of these patients it is likely that the failure to modify aldosterone secretion was secondary to the failure to activate the renin-angiotensin system. However, in others, the renin-angiotensin system, if anything, appeared to display an exuberant response.[51] In other studies it was reported that some patients with essential hypertension have a reduced renal blood flow level on a high-salt diet compared to normotensive individuals, even when the values were adjusted for age and gender.[52] Third, patients who have abnormal renal blood flows seem also to have abnormalities in the regulation of aldosterone secretion.[53]

Of interest, in Shoback's original study, patients with low renin hypertension had been excluded from analysis, thereby eliminating the possibility of the blunted aldosterone response being secondary to low renin levels. In subsequent studies over the past decade, the fraction of essential hypertensive population that are nonmodulators has been most precisely determined by an evaluation of the adrenal response to either endogenous or exogenous AngII (Figures 4 and 5). Using this technique, the distribution of responses in the normal and high renin essential hypertensive population is bimodal, with the portion of patients being nonmodulators varying between 25 and 50%.[54] The correlation between the adrenal and renal vascular responses to AngII is excellent. In nearly 60 subjects, in which both were assessed, agreement in defining the patient as a modulator or nonmodulator occurred in 85% of the subjects (Figure 6).[54]

A. DEFINITION OF NONMODULATION

Nonmodulators are defined as those individuals in whom sodium intake does not modify the adrenal or renal vascular responses to AngII. The patients may be identified by assessing the adrenal response to AngII in subjects on

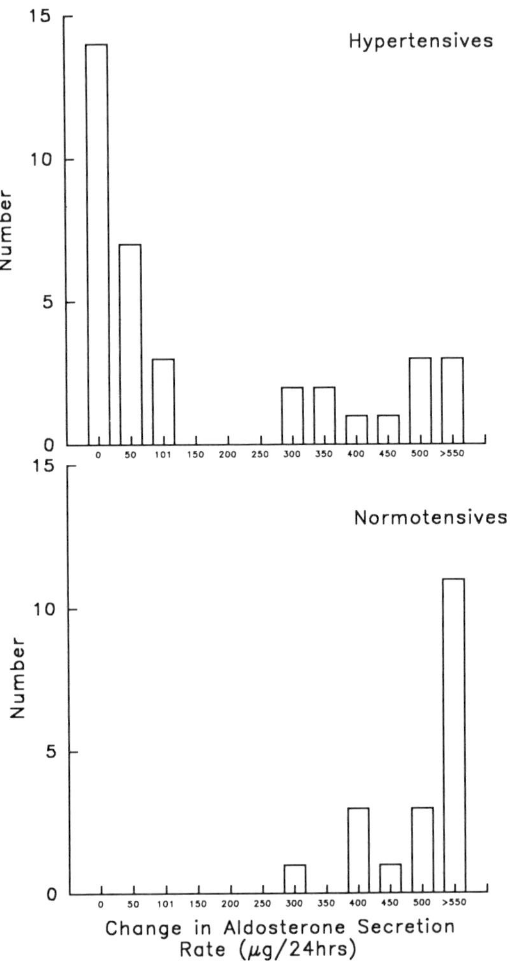

FIGURE 4. Aldosterone secretory response to furosemide. Increments in aldosterone secretion in normotensive and hypertensive subjects following 24 h of volume depletion induced by furosemide. All subjects were in balance on a 10 mEq sodium intake. Furosemide was administered in divided doses sufficient to produce a 2% reduction in body weight (18 normotensives and 26 hypertensive patients were studied). (From Williams, G.H. et al., *Hypertension,* in press. With permission of the American Heart Association.)

a 10 mEq sodium diet (nonmodulators having an increment of less than 15 ng/dl).[54] Alternatively, renal vascular responses to AngII as assessed by PAH clearance, can be determined on a high (200 mEq) sodium intake. A nonmodulator has been defined as an individual who has a decrement in PAH clearance under these circumstances of less than 100 ml/min/1.73 m^2 when adjusted for age.[54] While assessing the adrenal response to AngII is more

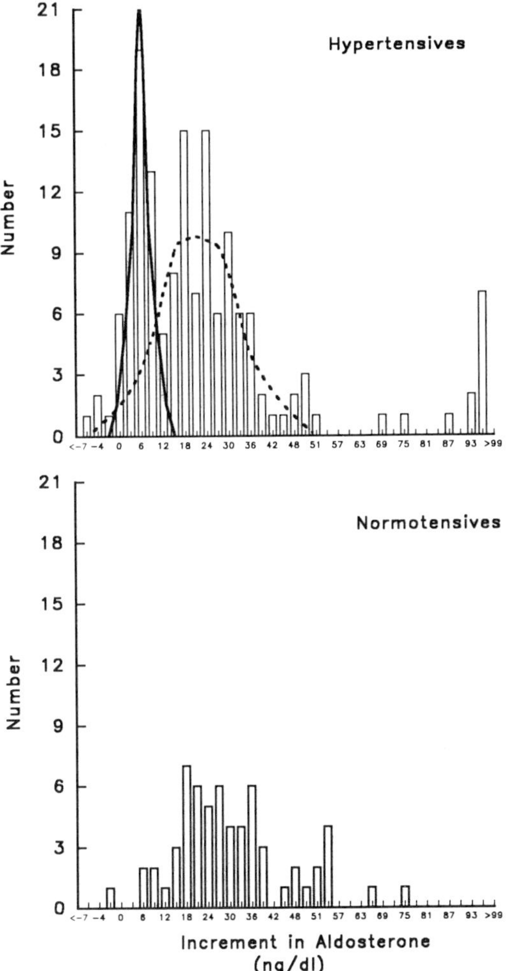

FIGURE 5. The response of 150 normal or high renin essential hypertensive patients (top panel) and 61 normotensive subjects (bottom panel) to a 3-ng AngII/kg/min infusion. Plasma aldosterone levels were measured before and after the infusion. All subjects were studied when in balance on a 10 mEq sodium intake. In the normotensive subjects, there was a highly significant bimodal distribution (p <0.00009; maximum likelihood analysis). (From Williams, G.H. et al., *Hypertension,* in press. With permission of the American Heart Association.)

precise, the evaluation of the renal vascular response to AngII is more convenient and is nearly as accurate. Of importance is the precision of sodium intake. For example, in normal subjects, changing the sodium intake from 10 to 30 mEq/d produces a 30 to 50% change in the aldosterone response to AngII. Likewise, changing sodium intake from 100 to 200 mEq/d produces a 20 to 30% change in the renal vascular response to AngII.[55] Thus, in

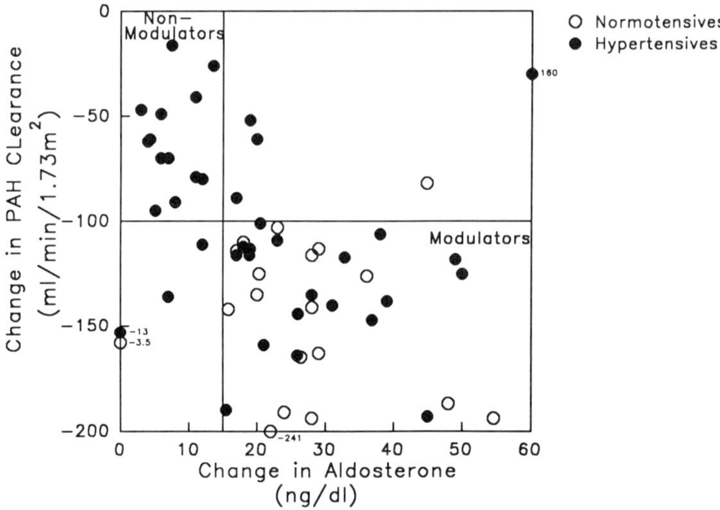

FIGURE 6. Correlation between increments in aldosterone in subjects on a 10 mEq sodium intake and decrements in PAH clearance as an index of renal plasma flow in the same subjects on a 200 mEq sodium intake when infused with AngII (3 ng AngII/kg/min). Fifty-nine subjects are depicted. In 50 of the 59 subjects (85%) the two indices were in agreement. (From Williams G.H. et al., *Hypertension*, in press. With permission of the American Heart Association.)

performing these studies, it is of critical importance to accurately fix and assess sodium intake to avoid misclassification.

Nonmodulators do not differ from other essential hypertensive patients in age, duration of hypertension, sodium and potassium balance, renal function, cardiac output, and/or plasma volume at the time of assessment.[54] There appears to be no clinical evidence that these patients are more severely hypertensive based on blood pressure, physical examination, or electrocardiograms.[6,53,54] In addition to the two major features associated with nonmodulation, other features have been reported (Table 1). One of the most important was the early documentation that sodium intake not only fails to influence target tissue responsiveness to AngII, but also fails to increase basal renal plasma flow (Figure 7).[6,56] Second, with sodium restriction and an upright position, the plasma renin activity and the AngII levels are higher and the plasma aldosterone levels lower in nonmodulators.[54,57] Acutely, neither AngII nor sodium chloride suppress plasma renin activity in nonmodulators on a low-sodium diet.[58,59] Fourth, sodium-lithium countertransport in red blood cells is significantly elevated in nonmodulators.[60] Fifth, norepinephrine and dopamine levels in the plasma are increased in nonmodulating patients in response to both sodium restriction and upright posture, while urine dopamine excretion is fixed and does not vary with the level of sodium intake.[61,62]

TABLE 1
Characteristics Associated with Nonmodulation

1. With sodium loading, decreased renal plasma flow response to AngII
2. With sodium restriction, decreased aldosterone response to AngII
3. With salt loading, decreased renal plasma flow response
4. Decreased sodium excretion with chronic and acute salt loading
5. With chronic salt loading, increase in blood pressure
6. Reduced plasma renin activity response to acute sodium loading
7. Reduced plasma renin activity response to AngII infusion
8. Correction of 1. to 5. above with converting enzyme inhibitors
9. Increased Na/Li countertransport activity in red cells
10. Increased plasma and urine dopamine levels

Some, but not all, of these defects are reversed by the acute administration of a converting enzyme inhibitor (see Table 1).[56,58,59,63]

B. PATHOPHYSIOLOGY OF NONMODULATING HYPERTENSION

A plausible model providing a mechanistic link between the adrenal and renal abnormalities has been proposed. Given the dominant effect of the renin-angiotensin system on the response to sodium intake in both organs, it has been suggested that primary alterations in the local renin-angiotensin systems could account for the observed abnormalities. For example, constitutive local AngII production that is not suppressed by high sodium intake could result in renal vasoconstriction that persists on a high-sodium diet. Such AngII-mediated vasoconstriction also could explain the abnormal response of the renal vasculature to exogenously administered AngII. In support of this hypothesis, the presence of the local renin-angiotensin system has been established in both systems, as noted above, and the role of AngII in regulating the level of its own receptor in vascular smooth muscle has been documented.[28]

If the renal and adrenal defects are both mediated by abnormal production of local AngII, then one would predict that they should be reversed by converting enzyme inhibition (CEI). This hypothesis has been adjusted and proved correct for both the renal abnormalities described in nonmodulators.[56] Other groups have obtained similar results, documenting greater increases in blood flow after CEI in renal,[64] brachial,[65-67] and carotid[68] arteries in hypertensive than normotensive subjects. Similar results regarding partial correction of the adrenal defect have been obtained, with increased adrenal sensitivity to AngII after 72 h of CEI in subjects who initially had blunted adrenal responses to AngII on a low-salt diet.[63] A recent study by Dluhy and colleagues has suggested a complete correction of the adrenal defect after six weeks of CEI therapy.[69] Thus, these data suggest that there is a subgroup of patients with essential hypertension with specific abnormalities in adrenal and renal vascular responses to AngII. Furthermore, the renal and adrenal phenotypes are tightly coupled.

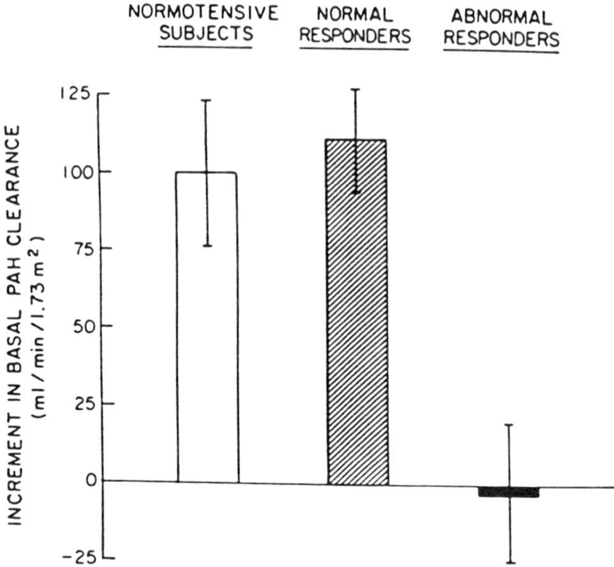

FIGURE 7. Changes in PAH clearance as an index of renal plasma flow (ml/min/1.73 m^2) were determined during a 10 mEq and a 200 mEq sodium intake. In normotensives and modulators (normal responders) renal plasma flow increased significantly with sodium loading ($p < 0.01$, Fisher exact test). No change was observed in non-modulators (abnormal responders). (From Shoback, D.M. et al., *J. Clin. Invest.*, 72, 2115, 1983. By copyright permission of the Society for Clinical Investigation.)

How do these abnormalities produce a rise in blood pressure? Two possible mechanisms exist — each active to varying degrees in each patient. First, there is abnormal sodium handling as a result of the renal blood flow defect when the individuals are on a high-salt diet; and second, there are abnormal increases in AngII levels when sodium restricted, secondary to the adrenal defect. Several studies provide support for these hypotheses. First, the non-modulators have lower basal aldosterone and higher basal renin and/or AngII levels when sodium restricted, particularly when the patients are studied upright.[54,57] Second, saralasin, a competitive antagonist of AngII, lowers blood pressure in sodium-restricted nonmodulators to a greater degree than in other essential hypertensive subjects who start with similar renin levels.[70] These results suggest that the nonmodulator's blood pressure in the sodium-restricted state is more dependent on AngII. Third, the time necessary to achieve low sodium balance is prolonged in nonmodulators.[71] The half time of disappearance of sodium from the urine when salt is removed from the diet is approximately 24 h in normotensive subjects but nearly 36 h in non-modulators. Fourth, nonmodulators require a longer period of time to achieve high salt balance and do so with a higher level of total accumulated sodium balance (Figure 8).[71] Furthermore, only nonmodulating normal and high renin

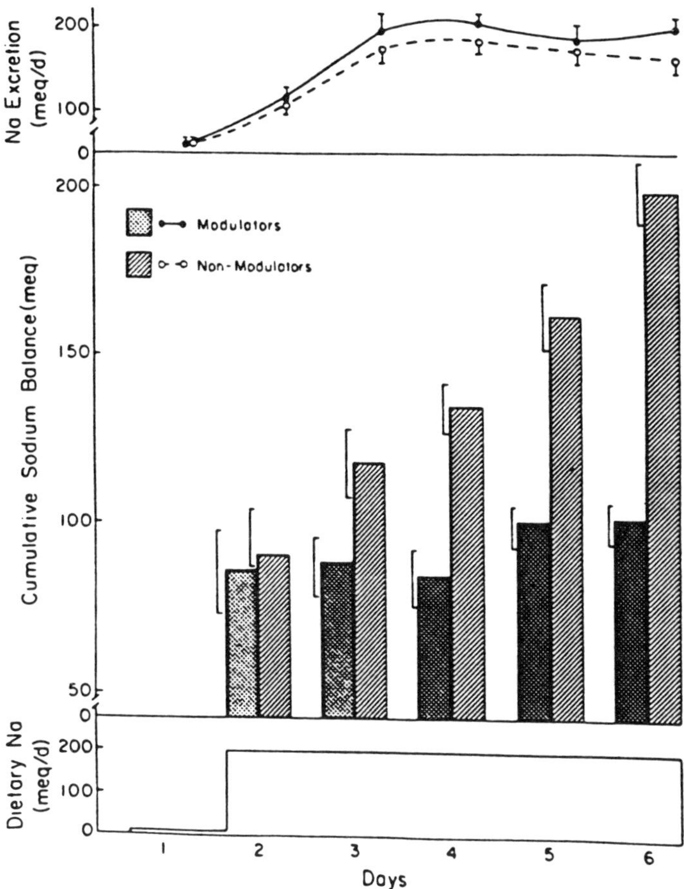

FIGURE 8. Response of hypertensive subjects to chronic sodium loading. The patients were characterized as modulators and nonmodulators according to the adrenal response to AngII. They were then placed in balance on a 10 mEq sodium intake, and the sodium intake was increased to 200 mEq/d (bottom panel). Sodium excretion in the two subgroups is depicted in the top panel. The middle panel presents the mean value for the cumulative sodium balance in the two groups. Nonmodulators had not reachieved sodium balance even after five days on the high sodium intake and had accumulated approximately 200 mEq of sodium. In contrast, the modulators required only two days to reestablish sodium balance with a net accumulation of 100 mEq. (From Hollenberg, N.K. and Williams, G.H., *Am. J. Med.*, 81, 412, 1986. By copyright permission of the York Publishing Group.)

essential hypertensives increase their blood pressure significantly when placed on a high sodium intake for five days.[56,71] Fifth, the sodium excretory response to an acute saline load is approximately half in nonmodulators when compared to modulating and normotensive individuals.[72] Finally, CEIs not only correct the renal blood flow response to AngII, they also correct the renal blood flow response to a high-sodium diet and the sodium excretory response to an acute sodium load.[56,72] Thus, these studies provide strong support for the conclusion that the blood pressure in nonmodulating hypertensive patients is secondary to abnormalities in sodium homeostasis, i.e., sodium-sensitive hypertension, if the individual is studied on a high-sodium intake, or to an increased activation of the circulating renin-angiotensin system, i.e., angiotensin-dependent hypertension, if the subject is studied on a low-sodium intake. Under most circumstances, except perhaps with the administration of a diuretic, the effect of a high-sodium intake will prevail. Thus, nonmodulation will usually express itself as a sodium-sensitive form of hypertension.

IV. NONMODULATION IN THE SPONTANEOUSLY HYPERTENSIVE RAT (SHR)

Many defects have been reported in the regulation of vascular and hormonal function in the SHR. Over the past several years, it has become increasingly apparent that some of these defects may simply reflect the heterogeneity of this strain of rats and/or the heterogeneity of the normal animals to which they have been compared. The most common comparison is with the Wistar Kyoto (WKY) rat, developed in Japan in the same facility in which the SHR was derived. Unfortunately, the WKY strain was not developed at the same time as the SHR. Thus, it is uncertain when a difference in a particular feature between SHR and WKY is observed whether such actually contributes to the hypertension in the SHR.[73]

There have been only a few studies that have examined adrenal steroidogenesis in the SHR.[74-79] Most reported no significant changes in basal aldosterone levels, although a few have reported that adrenal steroidogenesis may be decreased.[74,75] In nearly all such studies, young animals (less than 12 weeks of age) have been studied and the animals have been obtained from Charles River Laboratories. If these young animals are sodium restricted there is a reduced plasma aldosterone level, a decreased aldosterone output from isolated glomerulosa cells in response to AngII, and an increase in the plasma renin activity and/or AngII levels when responses are compared to either Sprague-Dawley or WKY animals[74,79] (Figure 9). While the abnormalities in the adrenal response to AngII when sodium intake is changed are not as dramatic in the SHR as in nonmodulating hypertensives, the data are highly suggestive that a common mechanism may exist.

The mechanisms responsible for the alteration in steroidogenesis in the SHR may be at one or more levels. First, AngII receptor regulation appears to be abnormal.[80] In contrast to the vascular smooth muscle, the number of receptors on the SHR adrenal are reduced, and sodium restriction does not produce the increase in their number as observed in normotensive animals.[81] Contrariwise, tissue renin levels in the SHR are increased but again do not appear to be modulated by the level of sodium intake. The actual production of AngII from SHR adrenals is decreased particularly when the animals are

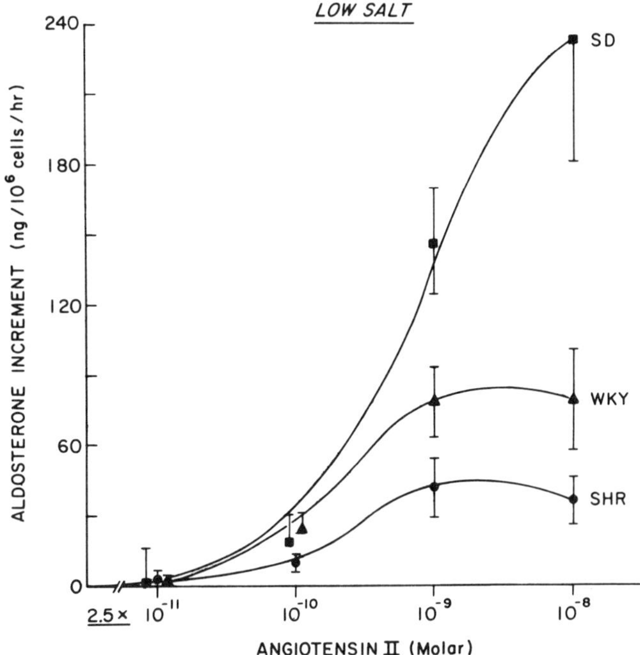

FIGURE 9. Response of isolated glomerulosa cells to AngII administration. Three groups of animals were studied: two normotensive strains (Sprague-Dawley [SD] and WKY) and one hypertensive strain (spontaneously hypertensive rats [SHR]). Cells were obtained from animals which had been maintained on a low-sodium intake for ten days before death (mean ± SEM of 12 experiments). Note the blunted response of the SHR glomerulosa cells to AngII compared to the two normotensive strains. (From Williams, G.H., Braley, L.M., and Menachery, A., *J. Clin. Invest.*, 69, 31, 1982. With permission from the American Society for Clinical Investigation.)

studied on a sodium-restricted intake, suggested an abnormality in the converting enzyme activity of the SHR adrenal. Second, there also appears to be abnormalities in the regulation of the mRNA for the steroidogenic enzymes in the SHR. In normotensive animals, the mRNA for aldosterone synthetase is increased during sodium restriction. This does not occur in the SHR.[82] In both normotensive and hypertensive animals, sodium intake appears to have minimal, if any, effect on the mRNA for the early pathway of aldosterone biosynthesis, i.e., P450scc.[82]

Thus, these data support the hypothesis that the abnormal regulation of aldosterone secretion in the SHR is secondary to a failure of sodium restriction to increase mRNA aldosterone synthetase levels in the adrenal of the SHR. Whether this failure is secondary to the reduced AngII production by the glomerulosa cell or some other unidentified defect is uncertain.

Several renal vascular abnormalities have also been described in the SHR. However, assessment of their relevance to the hypertensive process is clouded by the known effect of an increase in blood pressure per se in damaging the kidney and its vasculature. Therefore, the presence of primary defects needs to be assessed in the young SHR. In one such study, it has been reported that the renal blood flow level, as assessed by a flow probe technique, is lower in the SHR than in the WKY when the animals are comparably sodium loaded.[83] Furthermore, during a high-sodium state, the renal blood flow response to AngII infusion was smaller than in the SHR, and administration of a converting enzyme inhibitor substantially increases renal blood flow only in the SHR.[83] Thus, both of these features suggest that there are abnormalities in the renal vasculature in the SHR which are similar to those observed in nonmodulating hypertensive patients.

V. GENETICS OF NONMODULATING HYPERTENSION

Several lines of evidence suggest that nonmodulators are a discrete subgroup and that this defect may be a heritable trait. As noted earlier, the adrenal responses to AngII in the normal and high renin essential hypertensive populations distribute in a bimodal pattern.[54] The most precise technique to separate patients into modulators and nonmodulators is to determine their aldosterone secretory response to acute volume depletion after achieving balance on a sodium-restricted intake. The responses are clearly bimodal, both visually and statistically, using maximum likelihood analysis ($p < 0.000023$) (Figure 4).[54] A similar analysis was performed on the aldosterone increment in response to a 3-ng/kg/min AngII infusion in 150 sodium-restricted hypertensive subjects and 61 normotensive controls. Again, a highly significant ($p < 0.0009$) bimodal distribution of the response in the hypertensive but not normotensive patients was observed (Figure 5).[54] Renal blood flow responses to sodium loading have been evaluated by a similar technique. The distribution again was bimodal ($p < 0.01$) in the essential hypertensive patients. Thus, several features of the nonmodulating trait do distribute bimodally. This is in contrast to most other characteristics of the hypertensive population, in which the distribution has usually been continuous.

Additional lines of evidence suggest that this trait is genetically determined. First, there is a strong positive family history for hypertension in nonmodulators.[71] Eighty-five percent of nonmodulators had a positive family history for hypertension in contrast to 25 to 30% in the rest of the normal and high renin essential hypertensive patients. Second, the nonmodulating phenotype can be observed in normotensive subjects, particularly if there is a positive family history for hypertension. For example, in normotensive kidney donors, those individuals with a positive family history for hypertension have significantly lower plasma aldosterone levels when sodium restricted than those individuals with a negative family history for hypertension.[84] Second, Baretta-Piccoli et al.[85] have documented that aldosterone responses to AngII infusion in normotensive subjects with a positive family history for hypertension are significantly less than normotensive subjects with a negative family history for hypertension, particularly if the patient is studied in a sodium-restricted state. Third, Van Hooft et al.[86] reported that basal renal plasma flows on a high-sodium diet in normotensive Dutch subjects with a positive family history for hypertension are lower than in those individuals with a negative family history.[86] Fourth, Uneda et al.[87] have documented that the renal plasma flow response to CEI on a high-salt diet is increased in normotensive Japanese with a positive family history for hypertension compared to subjects with a negative family history.[87] Finally, in families in which sibling pairs with hypertension were tested for the nonmodulating trait, a high degree of concordance was observed ($p < 0.004$).[88] Thus, there is highly suggestive evidence from a number of sources that nonmodulators form a discrete subset of the essential hypertensive population and that this trait is likely to be inherited.

VI. CONCLUSIONS

Many studies have documented that blood pressure is distributed continuously in the general population. Furthermore, even in certain subsets of the hypertensive population, i.e., sodium sensitivity of blood pressure, the feature is continuously distributed. Because of these features, it has long been assumed that essential hypertension was a single disease entity. However, evidence accumulated over the past two decades suggests this is likely not to be the case. Rather, the continuous distribution of the several features associated with hypertension are the result of a limited number of common homeostatic responses to an increase in blood pressure or to the mechanisms underlying the hypertension. The establishment of a noncontinuous distribution of certain features associated with hypertension has profound implications both for treatment and for understanding the pathophysiology of this entity.

In this chapter we have summarized the features of one such entity — nonmodulators — who comprise a substantial fraction of the essential hypertensive population. They are defined as individuals in whom sodium intake fails to modulate the vascular and adrenal responses to AngII. As a result of this defect, there is a failure of sodium loading to increase basal renal blood flow levels, resulting in sodium retention and sodium-sensitive hypertension. Thus, nonmodulators share with other subsets of the sodium-sensitive hypertensive subgroup a hypertensive response to sodium loading. Yet, in contrast to most of the other members of this subgroup, nonmodulators do not have low plasma renin activities in response to sodium restriction and upright posture. Thus, unique to this subset of the salt-sensitive hypertensive population is the presence of normal and even high plasma renin activity if measured when the individual is sodium restricted. This difference results in a unique blood pressure response to sodium intake in nonmodulators: when sodium loaded, blood pressure rises. However, when sodium restricted, blood pressure may not fall to the same extent as observed in low renin hypertensive patients because of the limitation imposed on the fall in blood pressure by the abnormal activation of the renin-angiotensin system, secondary to the adrenal defect. Understanding the mechanisms underlying the defects observed in the nonmodulator has been enhanced since the recognition that the spontaneously hypertensive rat shares several features with nonmodulators. Thus, the defect in the nonmodulators appears to be related to an abnormality in the regulation of the local renin-angiotensin systems in the adrenal and the kidney, as administration of converting enzyme inhibitors corrects many of the abnormalities. This observation also forms the basis for specific therapy for this subset of the hypertensive population.

Recent studies have also documented that this trait is likely to be inherited, providing entree for the identification of the abnormal gene and the development of specific preventive measures. Finally, it is important to stress that nonmodulation is not the only pathogenic mechanism operating in patients with essential hypertension or even in the sodium-sensitive subset of the hypertensive population.

REFERENCES

1. **Chapman, B.,** Some effects of the fruit-rice diet in patients with essential hypertension, in *Hypertension, A Symposium,* Bell, E.T., Ed., University of Minnesota, Minneapolis, MN, 1951, 504.
2. **Kawasaki, T., Delca, C.S., Bartter, F.C., and Smith, H.,** The effect of high sodium and low sodium intakes on blood pressure and other related variables in human subjects with idiopathic hypertension, *Am. J. Med.,* 64, 193, 1978.
3. **Weinberger, M.H., Miller, J.Z., Luft, F.C., Grim, C.E., and Fineberg, M.S.,** Definitions and characteristics of sodium sensitivity and blood pressure resistance, *Hypertension,* 8, II127, 1986.
4. **Laragh, J.H.,** Vasoconstriction-volume analysis for understanding and treating hypertension: the use of renin and aldosterone profiles, *Am. J. Med.,* 55, 261, 1973.
5. **Fujita, T., Henry, W.L., Bartter, F.C., Lake, C.R., and Delea, C.S.,** Factors influencing blood pressure in salt-sensitive patients with hypertension, *Am. J. Med.,* 69, 334, 1980.
6. **Williams, G.H. and Hollenberg, N.K.,** Pathophysiology of essential hypertension, in *Cardiology,* Parmley, W.W. and Chatterjee, K., Eds., Lippincott, Philadelphia, 1990, 1.
7. **Kisch, E.S., Dluhy, R.G., and Williams, G.H.,** Enhanced aldosterone response to angiotensin II in human hypertension, *Circ. Res.,* 38, 502, 1979.
8. **Wisergof, M. and Brown, R.D.,** Increased adrenal sensitivity to angiotensin II in patients with low renin essential hypertension, *J. Clin. Invest.,* 63, 1456, 1979.
9. **Marks, A.D., Marks, D.B., Kanefsky, T.M., Adlin, V.E., and Channick, B.J.,** Enhanced adrenal responsiveness to angiotensin II in patients with low renin essential hypertension, *J. Clin. Endocrinol. Metab.,* 48, 266, 1980.
10. **Shoback, D.M., Moore, T.J., Dluhy, R.G., Hollenberg, N.K., and Williams, G.H.,** Defect in the sodium-modulated tissue responsiveness to angiotensin II in essential hypertension, *J. Clin. Invest.,* 72, 2115, 1983.

11. **Williams, G.H. and Hollenberg, N.K.,** Are non-modulating patients with essential hypertension a distinct subgroup? Implications for therapy, *Am. J. Med.*, 79, 3, 1985.
12. **Redgrave, J.E., Canessa, M., Gleason, R., Hollenberg, N.K., and Williams, G.H.,** Red blood cell lithium-sodium countertransport in non-modulating essential hypertension, *Hypertension*, 13, 721, 1989.
13. **Sokabe, H.,** Physiology of the renal effects of angiotensin, *Kidney Int.*, 6, 263, 1974.
14. **Laragh, J.H., Angers, M., Kelly, W.G., and Lieberman, S.,** Hypotensive agents and pressor substances: the effect of epinephrine, norepinephrine, angiotensin II, and others on the secretory rate of aldosterone in man, *JAMA*, 174, 234, 1960.
15. **Kibrough, H.M., Vaughn, E.D., Carey, R.M., and Ayers, C.R.,** Effect of intrarenal angiotensin II blockade on renal function in conscious dogs, *Circ. Res.*, 40, 174, 1977.
16. **Williams, G.H., Hollenberg, N.K., Brown, C., and Mersey, J.H.,** Adrenal responses to pharmacological interruption of the renin-angiotensin system in sodium restricted normal man, *J. Clin. Endocrinol. Metab.*, 47, 725, 1978.
17. **Hollenberg, N.K., Meggs, L.G., Williams, G.H., Katz, J., Garnic, J.D., and Harrington, D.P.,** Sodium intake and renal responses to captopril in normal man and essential hypertension, *Kidney Int.*, 20, 240, 1981.
18. **Kaplan, N.M. and Silah, J.G.,** Effect of angiotensin II on the blood pressure in humans with hypertensive disease, *J. Clin. Invest.*, 43, 659, 1964.
19. **Strewler, G.J., Hinrichs, K.J., Guiod, L.R., and Hollenberg, N.K.,** Sodium intake and vascular smooth muscle responsiveness to norepinephrine and angiotensin in the rabbit, *Circ. Res.*, 31, 758, 1972.
20. **Barraclough, M.A., Jones, N.F., Marsden, C.D., and Bradford, B.C.,** Renal and pressor actions of angiotensin in salt loaded and depleted rabbits, *Experimentia (Basel)*, 23, 553, 1967.
21. **Ganong, W.F. and Boryczka, A.T.,** Effect of a low sodium diet on aldosterone-stimulating activity of angiotensin II in dogs, *Proc. Soc. Exp. Biol. Med.*, 124, 1230, 1967.
22. **Davis, W.W., Burwell, L.R., and Bartter, F.C.,** Inhibition of the effects of angiotensin II on adrenal steroid production by dietary sodium, *Proc. Natl. Acad. Sci. U.S.A.*, 63, 718, 1969.
23. **Hollenberg, N.K., Chenitz, W.R., Adams, D.F., and Williams, G.H.,** Reciprocal influence of salt intake on adrenal glomerulosa and renal vascular responses to angiotensin II in normal man, *J. Clin. Invest.*, 54, 34, 1974.
24. **Oelkers, W., Brown, J.J., Fraser, R., Lever, A.F., Morton, J.J., and Robertson, J.I.,** Sensitization of the adrenal cortex to angiotensin II in sodium-deplete man, *Circ. Res.*, 34, 69, 1974.
25. **Steele, J.M., Jr., Neusy, A.J. and Lowenstein, G.,** The effects of des-Asp-angiotensin II on blood pressure, plasma aldosterone concentration, and plasma renin activity in the rabbit, *Circ. Res.*, 38, 113, 1976.
26. **Hollenberg, N.K., Solomon, H.S., Adams, D.F., Abrams, H.L., and Merrill, J.P.,** Renal vascular responses to angiotensin and norepinephrine in normal man, *Circ. Res.*, 31, 750, 1972.
27. **Hollenberg, N.K., Williams, G.H., Burger, B., Chenitz, W., Hooshmand, I., and Adams, D.F.,** Renal blood flow and its response to AII: an interaction between oral contraceptive agents, sodium intake, and the renin-angiotensin system in healthy young women, *Circ. Res.*, 38, 35, 1976.
28. **Gunther, S., Gimbrone, M.A., Jr., and Alexander, R.W.,** Regulation by angiotensin II of its receptors in resistance blood vessels, *Nature*, 287, 230, 1980.
29. **Douglas, J. and Catt, K.J.,** Regulation of angiotensin II receptors in the rat adrenal cortex by dietary electrolytes, *J. Clin. Invest.*, 58, 834, 1976.
30. **Platia, M.P., Catt, K.J., Hodgen, G.D., and Aguilera, G.,** Angiotensin II receptor regulation during altered sodium intake in primates, *Hypertension*, 8, 1121, 1986.

31. **Shoback, D.M., Williams, G.H., Hollenberg, N.K., Davies, R.O., Moore, T.J., and Dluhy, R.G.,** Endogenous angiotensin II as a determinant of sodium modulated change in tissue responsiveness to angiotensin II in normal man, *J. Clin. Endocrinol. Metab.,* 51, 764, 1983.
32. **Williams, G.H., Hollenberg, N.K., and Braley, L.M.,** Influence of sodium intake on vascular and adrenal angiotensin II receptors, *Endocrinology,* 98, 1343, 1976.
33. **Ganten, D., Ganten, U., Kubo, S., Granger, P., Nowaczyski, W., Boucher, R., and Genest, J.,** Influence of sodium, potassium, and pituitary hormones on iso-renin in rat adrenal glands, *Am. J. Physiol.,* 227, 224, 1974.
34. **Brecher, A.S., Shier, D.N., Dene, H., Wang, S.-M., Rapp, J.P., Franco-Saenz, R., and Mulrow, P.J.,** Regulation of adrenal renin messenger ribonucleic acid by dietary sodium chloride, *Endocrinology,* 124, 2907, 1989.
35. **Mizuno, K., Hoffman, L.H., McKenzie, J.C., and Inagami, T.,** Presence of renin secretory granules in rat adrenal gland and stimulation of renin secretion by angiotensin II but not by adrenocorticotropin, *J. Clin. Invest.,* 82, 1007, 1988.
36. **Doi, K., Atarashi, K., Franco-Saenz, R., and Mulrow, P.J.,** Effect of changes in sodium and potassium balance and nephrectomy on adrenal renin and aldosterone concentrations, *Hypertension,* 6, I29, 1984.
37. **Kifor, I., Moore, T.J., Fallo, F., Sperling, E., Chiou, C.-Y., Menachery, A., and Williams, G.H.,** Potassium-stimulated angiotensin release from superfused adrenal capsules and enzymatically dispersed cells of the zona glomerulosa, *Endocrinology,* 129, 823, 1991.
38. **Kifor, I., Moore, T.J., Fallo, F., Sperling, E., Chiou, C.-Y., Menachery, A., and Williams, G.H.,** The effect of sodium intake on angiotensin content of the rat adrenal gland, *Endocrinology,* 128, 1277, 1991.
39. **Nakamura, M., Misono, K.S., Naruse, M., Workman, R.J., and Inagami, T.,** A role for the adrenal renin-angiotensin system in the regulation of potassium-stimulated aldosterone production, *Endocrinology,* 117, 1772, 1985.
40. **Enyedi, P., Buki, B., Muscsi, I., and Spat, A.,** Polyphosphoinositide metabolism in adrenal glomerulosa cells, *Mol. Cell. Endocrinol.,* 41, 105, 1985.
41. **Farese, R.V.,** Phospholipids as intermediates in hormone action, *Mol. Cell. Endocrinol.,* 35, 1, 1984.
42. **Farese, R.V., Larson, R.E., and Davis, J.S.,** Rapid effects of angiotensin II on polyphosphoinositide metabolism in the rat adrenal glomerulosa, *Endocrinology,* 114, 302, 1984.
43. **Underwood, R.H., Greeley, R., Glennon, E.T., Menachery, A.I., Braley, L.M., and Williams, G.H.,** Mass determination of polyphosphoinositides and inositol triphosphate in rat adrenal glomerulosa cells with a microspectrophotometric method, *Endocrinology,* 123, 211, 1988.
44. **Underwood, R.H., Menachery, A.I., and Williams, G.H.,** Effect of sodium intake on phosphoinositides and inositol trisphosphate response to angiotensin II, K^+ and ACTH in rat glomerulosa cells, *J. Endocrinol.,* 122, 371, 1989.
45. **Quinn, S.J. and Williams, G.H.,** Regulation of aldosterone secretion, *Annu. Rev. Physiol.,* 50, 409, 1988.
46. **Williams, G.H., McDonnell, L.M., Tait, S.A.S., and Tait, J.F.,** The effect of medium composition and in vitro stimuli on the conversion of corticosterone to aldosterone in rat glomerulosa tissue, *Endocrinology,* 91, 948, 1972.
47. **Haning, R., Tait, S.A.S., and Tait, J.F.,** In vitro effects of ACTH, angiotensins, serotonin and potassium on steroid output and corticosterone to aldosterone by isolated adrenal cells, *Endocrinology,* 87, 1147, 1970.
48. **Mornet, E., DuPont, J., Vitek, A., and White, P.C.,** Characterization of two genes encoding human steroid 11 beta-hydroxylase (P-450) 11-beta, *J. Biol. Chem.,* 264, 20961, 1989.

49. **Tremblay, A. and Lehoux, J.G.,** Effects of dietary sodium restriction and potassium intake on cholesterol side chain cleavage cytochrome P-450 and adrenodoxin mRNA levels, *J. Steroid Biochem.*, 34, 385, 1989.
50. **Christlieb, A.R., Hickler, R.B., Lauler, D.P., and Williams, G.H.,** Hypertension with inappropriate aldosterone stimulation: a syndrome, *N. Engl. J. Med.*, 281, 128, 1969.
51. **Williams, G.H., Rose, L.I., Dluhy, R.G., McCaughn, D., Jagger, P.I., Hickler, R.B., and Lauler, D.P.,** Abnormal responsiveness of the renin aldosterone system to acute stimulation in certain patients with "essential hypertension", *Ann. Intern. Med.*, 72, 317, 1970.
52. **Hollenberg, N.K. and Merrill, J.P.,** Intrarenal perfusion in the young "essential" hypertensive: a subpopulation resistant to sodium restriction, *Trans. Assoc. Am. Physicians,* 83, 93, 1970.
53. **Williams, G.H., Tuck, M.L., Sullivan, J.M., Dluhy, R.G., and Hollenberg, N.K.,** Parallel adrenal and renal abnormalities in the young patients with essential hypertension, *Am. J. Med.*, 72, 907, 1982.
54. **Williams, G.H., Dluhy, R.G., Lifton, R.P., Moore, T.J., Gleason, R., Williams, R., Hunt, S.C., Hopkins, P.N., and Hollenberg, N.K.,** Non-modulation as an intermediate phenotype in hypertension, *Hypertension,* in press.
55. **Adler, G.K., Moore, T.J., Hollenberg, N.K., and Williams, G.H.,** Changes in adrenal responsiveness and potassium balance with shifts in sodium intake, *Endocr. Res.*, 13, 439, 1987.
56. **Redgrave, J.E., Rabinowe, S.L., Hollenberg, N.K., and Williams, G.H.,** Correction of renal blood flow response to angiotensin II by converting enzyme inhibition in essential hypertensives, *J. Clin. Invest.*, 75, 1285, 1985.
57. **Moore, T.J., Williams, G.H., Dluhy, R.G., Bavli, S.Z., Himathongkam, T., and Greenfield, M.,** Altered renin-angiotensin-aldosterone relationships in normal renin essential hypertension, *Circ. Res.*, 41, 167, 1977.
58. **Seely, E.W., Moore, T.J., Rogacz, S., Gordon, M.S., Gleason, R., Hollenberg, N.K., and Williams, G.H.,** Angiotensin-mediated renin suppression is altered in non-modulating hypertension, *Hypertension,* 13, 31, 1989.
59. **Rabinowe, S.L., Redgrave, J.E., Shoback, D.M., Podolsky, S., Hollenberg, N.K., and Williams, G.H.,** Renin suppression by saline is blunted in non-modulating essential hypertension, *Hypertension,* 10, 404, 1987.
60. **Redgrave, J.E., Canessa, M., Gleason, R., Hollenberg, N.K., and Williams, G.H.,** Red blood cell lithium-sodium countertransport in non-modulating essential hypertension, *Hypertension,* 13, 721, 1989.
61. **Schnurr, E., Lahme, W., and Kuppers, H.,** Measurement of renal clearance on inulin and PAH in the steady state without urine collection, *Clin. Nephrol.*, 13, 26, 1980.
62. **Conlin, P.R., Braley, L.M., Menachery, A.I., Hollenberg, N.K., and Williams, G.H.,** Abnormal norepinephrine and aldosterone responses to upright posture in non-modulating hypertension, *J. Clin. Endocrinol. Metab.*, in press.
63. **Taylor, T.T., Moore, T.J., Hollenberg, N.K., and Williams, G.H.,** Converting enzyme inhibition corrects the altered adrenal response to angiotensin II in essential hypertension, *Hypertension,* 6, 92, 1984.
69. **Dluhy, R.G., Smith, K., Taylor, T., Hollenberg, N.K., and Williams, G.H.,** Prolonged converting enzyme inhibition in non-modulating hypertension, *Hypertension,* 13, 371, 1989.
70. **Dluhy, R.G., Bavli, S.Z., Leung, F.K., Solomon, H.S., Moore, T.J., Hollenberg, N.K., and Williams, G.H.,** Abnormal adrenal responsiveness and angiotensin II dependency in high renin essential hypertension, *J. Clin. Invest.*, 64, 1270, 1979.

71. **Hollenberg, N.K., Moore, T.J., Shoback, D.M., Redgrave, J.E., Rabinowe, S., and Williams, G.H.,** Abnormal renal sodium handling in essential hypertension; relation to failure of renal and adrenal modulation of responses to angiotensin II, *Am. J. Med.*, 81, 412, 1986.
72. **Rystedt, L., Williams, G.H., and Hollenberg, N.K.,** The renal and endocrine response to saline infusion in essential hypertension, *Hypertension,* 8, 217, 1986.
73. **Rapp, J.P.,** Genetics of experimental and human hypertension, in *Hypertension, Physiopathology and Treatment,* 2nd ed., Genest, J., Kuchel, O., Hamet, P., and Cantin, M., Eds., McGraw-Hill, Montreal, 1983, chap. 10.
74. **Williams, G.H., Braley, L.M., and Menachery, A.,** Decreased adrenal responsiveness to angiotensin II: a defect present in spontaneously hypertensive rats, *J. Clin. Invest.*, 69, 31, 1982.
75. **Braley, L.M., Menachery, A.I., and Williams, G.H.,** Specificity of the alteration in aldosterone biosynthesis in the spontaneously hypertensive rat, *Endocrinology,* 112, 562, 1983.
76. **Moll, D., Dale, S.L., and Melby, J.C.,** Adrenal steroidogenesis in the spontaneously hypertensive rat (SHR), *Endocrinology,* 96, 416, 1975.
77. **Freeman, R.H., Davis, J.O., Aharon, N.V., Ulick, S., and Weinberger, M.H.,** Control of aldosterone secretion in the spontaneously hypertensive rat, *Circ. Res.*, 37, 66, 1975.
78. **Iams, S.G., McMurthy, J.P., and Wexler, B.C.,** Aldosterone, deoxycorticosterone, corticosterone, and prolactin changes during the lifespan of chronically and spontaneously hypertensive rats, *Endocrinology,* 104, 1357, 1979.
79. **Sowers, J.R., Sollars, E.G., Tuck, M.L., and Asp, N.D.,** Dopaminergic modulation of renin activity and aldosterone and prolactin secretion in the spontaneously hypertensive rat, *Proc. Soc. Exp. Biol. Med.*, 164, 598, 1980.
80. **Bradshaw, B. and Moore, T.J.,** Abnormal regulation of adrenal angiotensin II receptors in spontaneously hypertensive rats, *Hypertension,* 11, 49, 1988.
81. **Naruse, S. and Inagami, T.,** Markedly elevated specific renin levels in the adrenal in genetically hypertensive rats, *Proc. Natl., Acad. Sci. U.S.A.*, 79, 3295, 1982.
82. **Adler, G.K., Chen, R., Menachery, A.I., Braley, L.M., and Williams, G.H.,** Sodium restriction increases aldosterone biosynthesis and P-450 11 beta-hydroxylase mRNA levels in glomerulosa cells of normotensive but not hypertensive rats, *Endocrinology,* (Abstract), 1991.
83. **Guidi, E. and Hollenberg, N.K.,** Different reactivity to angiotensin II of peripheral and renal arteries in spontaneously hypertensive rats: effect of acute and chronic angiotensin converting enzyme inhibition, *J. Hypertens.*, 4, S480, 1986.
84. **Blackshear, J.L., Garnic, D., Williams, G.H., Harrington, D.P., and Hollenberg, N.K.,** Exaggerated renal vasodilator response to calcium entry blockade in first-degree relatives of essential hypertensive subjects, *Hypertension,* 9, 384, 1987.
85. **Baretta-Piccolo, C., Pusterla, C., Stadler, P., and Weidmann, P.,** Blunted aldosterone responsiveness to angiotensin II in normotensive subjects with familial predisposition to essential hypertension, *J. Hypertens.*, 6, 57, 1988.
86. **Van Hooft, I.M.S., Grobbee, D.E., Derkx, F.H.M., DeLeeuw, P.W., Schalekamp, M.A., and Hofman, A.,** Renal hemodynamics and the renin-angiotensin-aldosterone system in normotensive subjects with hypertensive and normotensive parents, *N. Engl. J. Med.*, 324, 1305, 1991.
87. **Uneda, S., Fujishima, S., Fujiki, Y., Tochikubo, O., Oda, H., Asahina, S., and Kaneko, Y.,** Renal hemodynamics and the renin-angiotensin system in adolescents genetically predisposed to essential hypertension, *J. Hypertens.*, 2, S437, 1984.
88. **Lifton, R.P., Hopkins, P.N., Williams, R.R., Hollenberg, N.K., Williams, G.H., and Dluhy, R.G.,** Evidence for heritability of non-modulating essential hypertension, *Hypertension,* 13, 884, 1989.

Chapter 21

GENES OF THE RENIN-ANGIOTENSIN SYSTEM AND THE GENETICS OF HUMAN HYPERTENSION

X. Jeunemaitre and R.P. Lifton

TABLE OF CONTENTS

0-8493-4622-3/93/$0.00 + $.50

I. GENETICS OF HUMAN HYPERTENSION

Despite intensive investigation of the physiology of blood pressure regulation, the primary abnormalities contributing to the pathogenesis of this common disorder remain unknown in the overwhelming majority of patients. This is due in large part to the nature of blood pressure as a quantitative trait which can only be measured *in vivo*. As a consequence of the complex interplay of the many systems which combine to regulate blood pressure, alteration in one of these systems invariably imparts secondary effects on others. As a result, while a wide variety of physiologic abnormalities have been described in hypertensive subjects, determining which, if any, of these are primary causes contributing to the pathogenesis of hypertension and distinguishing these from mere secondary consequences of underlying primary abnormalities has proved exceedingly difficult from physiologic analysis alone.

An alternative approach to pure physiologic analysis would be to identify mutations contributing to the pathogenesis of hypertension. Since such genetic abnormalities cannot be simply a consequence of hypertension, their identification should provide a useful starting point from which to elucidate pathophysiology. Furthermore, knowledge of the relevant mutations may permit simple genetic screening tests to identify individuals with particular disease alleles, permitting both early diagnosis and, potentially, intervention with specific therapy tailored to the underlying defect(s).

This approach presupposes that genetic factors contribute to the pathogenesis of human hypertension. Complementary evidence from a variety of study designs supports this contention, and has been reviewed by Ward.[1] Twin studies demonstrate greater concordance of blood pressures of monozygotic than dizygotic twins. Blood pressure levels show strong familial aggregation, beginning within a few months after birth and extending to late in life. Large population-based studies have demonstrated that this familial aggregation is due in large part to genetic factors rather than simply the influence of shared familial environment. These observations are corroborated and strengthened by adoption studies which demonstrate stronger correlations of blood pressures of natural sibling pairs living within the same household than adoptive sibling pairs or natural-adoptive pairs. Population-based studies estimate that 20 to 40% of the variance in blood pressure is explained by genetic factors.

While these studies demonstrate the importance of genetic factors in determining human blood pressure, they also demonstrate one major pitfall in the inheritance of blood pressure, namely, the inheritance is not monogenic. Thus, in contrast to simple Mendelian traits such as cystic fibrosis or neurofibromatosis, in which inheritance at a single genetic locus specifies the presence or absence of a trait, variation in blood pressure does not generally segregate in a fashion consistent with determination by a single gene, even within hypertensive pedigrees. Instead, the inheritance of blood pressure appears to be influenced by many genetic loci.[1] In human populations the true number of genes affecting blood pressure cannot be meaningfully estimated at present.

This non-Mendelian inheritance of blood pressure has several adverse consequences for genetic approaches to hypertension. Standard linkage analysis in pedigrees using blood pressure as the trait can be problematic, as even within individual pedigrees individuals may have hypertension due to inheritance of different genes. Take, for example, a pedigree in which alleles at three genetic loci contribute to the pathogenesis of hypertension, and inheritance of any two of these predisposing alleles will suffice to produce hypertension. In this case, comparing inheritance at any one of these loci to the segregation of hypertension will reveal some individuals who are hypertensive who have not inherited the predisposing allele at this locus as well as some individuals who have inherited the predisposing allele who are normotensive. Simply extending the pedigree to larger size may be of no help, since at each mating new alleles influencing blood pressure may be introduced.

II. APPROACHES TO IDENTIFY MUTATIONS CONTRIBUTING TO THE PATHOGENESIS OF HUMAN HYPERTENSION

A number of alternative approaches can be employed to attempt to identify genes and mutations contributing to the pathogenesis of human hypertension. The strengths and weaknesses of these will be discussed briefly herein.

A. ASSOCIATION STUDIES

Association studies look for a difference in the distribution of alleles at a marker locus in cases and controls. In order to find significant positive results, a large fraction of mutant alleles in the population studied must derive from the identical ancestral mutation or the marker must directly detect the mutation of interest. Because these criteria are not commonly met, many positive results of association studies are false positives which derive from imprecise matching of cases and controls in which subtle or overt differences in racial or ethnic origins account for observed differences in allele frequencies. While these experiments are relatively easy to perform, caution must consequently be observed in the interpretation of positive results.

In addition, because of the unusual requirement that a large fraction of mutations must derive from the same ancestral chromosome or that the marker directly detects the relevant mutation, association studies have zero power to exclude even common variants contributing to the pathogenesis of a trait. For example, familial hypercholesterolemia due to low density lipoprotein (LDL) receptor deficiency is caused by a wide variety of independent mutations in the LDL receptor. As a result, association studies of alleles at this locus comparing subjects with FH to controls would be negative, despite this trait being caused by mutation at a single genetic locus in all cases.

B. LINKAGE ANALYSIS

The principle of linkage tests is to compare the inheritance of a genetic marker to inheritance of a trait of interest, in order to test the hypothesis that the gene causing the trait lies near the marker being tested. In complex disorders such as hypertension, there are a number of crucial considerations in the design of linkage studies. These include the phenotypic trait used, the sampling design (i.e., pedigrees vs. affected sib pairs), and the markers to be employed (i.e., candidate genes vs. anonymous markers).

1. Phenotypic Classification

As noted above, blood pressure is a continuously distributed quantitative trait, providing no simple biological basis for dichotomous classification of subjects into hypertensive and normotensive populations. Power of linkage analysis could potentially be maximized by use of blood pressure as a quantitative trait; however, in most populations the most interesting subjects for linkage studies, i.e., those with significant hypertension, are on medication to reduce blood pressure. Withdrawal of such subjects from medication is possible but time consuming and costly, and potentially poses risk for severely hypertensive subjects. Classification according to diagnosis of hypertension or blood pressure greater than some percentile of the population norm adjusted for age and sex is an alternative approach to classification, but potentially loses some power compared with quantitative values, depending on the true and unknown model of inheritance. Classification is further confounded by delayed onset of hypertension and the possibility of incomplete penetrance, such that individuals inheriting a predisposing allele may be phenotypically normal at the time of study.

As indicated above, the true (and at present unknown) degree of etiologic heterogeneity of hypertension is likely the largest obstacle to successful linkage analysis. As a consequence, for pedigrees not displaying patent Mendelian segregation of blood pressure levels, linkage analysis by likelihood methods has much reduced power to detect linkage, and still less power to exclude linkage to a marker locus.

Consequently, it is apparent that efforts to increase the homogeneity of the test population for underlying genetic factors may be critical to the success of linkage studies. Heterogeneity can potentially be reduced by selecting for study pedigrees with increased likelihood of genetic contribution, such as early onset disease, or increased severity. In addition, a potentially important adjunct is the use of *intermediate phenotypes* to identify subgroups of the hypertensive population with more homogeneous causation.

Use of intermediate phenotypes can reduce etiologic heterogeneity and can potentially resurrect the power of linkage with traits showing patent Mendelian segregation. Ideally, such intermediate phenotypes would be traits which reflect the action of single genes that contribute to the pathogenesis of hypertension. In this case linkage analysis could be conducted using the intermediate phenotype as a surrogate for blood pressure. Even in the absence of a known Mendelian model of inheritance, intermediate phenotypes can still be used to stratify the hypertensive population into more etiologically homogeneous subsets, potentially increasing power to detect linkage.

Unfortunately, few experimentally useful intermediate phenotypes applicable to human essential hypertension have been defined, fewer still with direct relevance to the renin-angiotensin system.[2] Among potential intermediate phenotypes, erythrocyte sodium-lithium countertransport shows strong evidence for bimodality and evidence for major gene effect, though there may be several different genes with major effect. Urinary kallikrein, similarly, shows bimodality and evidence for major gene effect. Other traits, such as non-modulation of renal blood flow and adrenal response to angiotensin II, and plasma renin activity show association with subsets of the hypertensive population but major gene effects have not yet been documented. Still others, such as various measures of sodium sensitivity show continuous distribution in all populations studied, dampening enthusiasm that major gene effects will be found. A number of other traits, including markers of adrenergic receptors and ion transport have been suggested and are in varying stages of evaluation.

2. Linkage Strategies

a. Pedigrees vs. Affected Relative Pairs

For pedigrees bearing traits showing major gene effects or patent Mendelian segregation, classical linkage in pedigrees remains the strategy of choice. In the setting of common traits displaying high degrees of etiologic heterogeneity, however, linkage with affected sibs or relatives presents a number of advantages.[3-5] Most importantly, these methods can accommodate high degrees of etiologic heterogeneity as well as incomplete penetrance. This approach has the further appeal that it is model-free, requiring no *a priori* formulation of an explicit model of inheritance. This approach is thus well suited to many of the problems that confront linkage analysis in hypertension. The major disadvantage of this approach is the reduced power to detect linkage compared with linkage in pedigrees. Power can be maximized by using large numbers of affected pairs, using maximally informative markers with identity by descent inferred where possible from parental/other relative genotypes, and minimizing the recombination distance between the trait and marker loci. Experimental control of these three parameters can be achieved to varying extents.

b. Candidate Genes vs. Anonymous Markers

When nothing is known about the physiologic determinants of a trait, general linkage approaches with anonymous markers may be the most direct route to identify the responsible gene. Alternatively, when potential physiologic determinants of a trait have been identified, linkage with candidate genes may quickly lead to identification of the gene responsible for the trait. In the case of hypertension, the extensive study of the physiology of blood pressure regulation suggests a number of candidate genes for study. Given the central role of the renin-angiotensin system in the regulation of blood pressure in humans, these genes are excellent candidates in human hypertension. The short-term advantage of linkage with a candidate gene strategy is (1) the possibility of directly identifying the genes causing the trait of interest; (2) the increased power to detect linkage compared with anonymous markers (due to limiting degrees of freedom as well as reducing recombination fraction between trait and marker loci to zero); (3) the ability to test the specific hypothesis that inheritance at the candidate locus influences expression of the trait. In conjunction with a sib pair approach, a candidate gene strategy has the advantage that, in the absence of evidence for linkage, upper limits can be placed on the effect of inheritance at the candidate locus on the trait.

Success of the candidate gene strategy requires that the correct candidates have already been identified and merely await genetic analysis. Thus, while the candidate gene approach has some likelihood of success, the likelihood cannot be stated with accuracy. If candidate strategies fail, general linkage studies with relevant phenotypes may be contemplated. The rapidly improving human genetic map will doubtless improve the power of such endeavors; however, such approaches will require very large collections of well-characterized hypertensive pedigrees to have any reasonable power to detect linkage.

c. *Power Considerations*

The determinants of power of such studies are (1) the degree of homogeneity of locus specifying the phenotype and the magnitude of the effect exerted by the locus on the trait; (2) sample size; (3) the true model of inheritance of the trait; (4) distance (recombination fraction) between trait and marker loci; (5) the informativeness of the marker(s) used.

Consequently, it is readily apparent that increasing the homogeneity of the trait, studying large numbers of relevant subjects, and employing highly informative polymorphic markers whenever possible will maximize the power of such studies. Recombination between trait and marker loci can be reduced to zero by the use of candidate genes.

We estimate that if mutation at a single locus contributes to hypertension or an intermediate phenotype in 5 to 10% of a defined test population, available methods have some realistic chance to detect linkage. However, if no locus contributes this commonly to the trait, success of linkage analysis is doubtful unless large pedigrees segregating a limited number of contributing loci are found.

3. Marker Development

As indicated above, development of highly informative markers in candidate genes will increase the power of linkage studies. Until recently, however, most polymorphic markers available for linkage studies were simple site polymorphisms which score the presence or absence of a restriction endonuclease cleavage site. Such polymorphisms result in only two possible alleles, a maximum heterozygosity of 50%, and consequently waste vast amounts of information in genetic studies; this is particularly true in sib pair linkage studies using candidate genes. This situation may not be much improved by haplotyping a number of such two-allele systems at a given locus, since these are commonly in linkage disequilibrium such that little or no information is added; even when not in disequilibrium, complete construction of haplotypes of these systems may not be possible in the absence of considerable linkage data from each family studied.

An alternative to such two-allele systems is to develop highly informative markers for analysis wherever possible. We have taken two basic approaches to this problem. One is to identify highly informative markers in candidate genes based on simple repeat motifs exemplified by repeats of structure $(GT)_n$.[6,7] Such repeats are common in the human genome, and the precise number of repeats may be highly variable, providing multiallelic series with high heterozygosity. Cloning large segments of genomic DNA spanning candidate loci provides a ready substrate for development of such markers.

If no highly informative markers can be identified at a candidate locus, the increasing informativeness of the human genetic map may permit identification of surrogate markers, i.e., highly informative markers which are sufficiently close to the candidate locus to be either substituted for the candidate locus in linkage studies or used in conjunction with less informative markers at candidate loci. In this approach, a polymorphism of low informativeness in a candidate gene is located on the human genetic map by linkage, and highly informative closely linked markers are sought. The increasing density of highly informative markers on the human genetic map is improving the likelihood of success with this latter approach.

C. DIRECT SEARCH FOR MUTATIONS IN CANDIDATE GENES

The major competing strategy to linkage analysis is direct search for mutation in candidate genes in hypertensive subjects. There are two major advantages of this approach: first, it has power to detect uncommon or rare mutations which could be difficult or impossible to detect by linkage analysis; second, no family collections are required at the outset — one only must have a large number of unrelated subjects bearing the trait of interest. One disadvantage of this approach is that, in contrast to linkage analysis, negative studies, i.e., those in which no significant mutations are found, have no power to exclude the presence of relevant mutations in unscreened parts of the same locus. In this sense, linkage and direct search strategies provide complementary information.

Among potential means of detecting mutations, single-strand conformational polymorphism,[8] denaturing gradient gel electrophoresis,[9] heteroduplex mobility shifts,[10] and direct sequencing of genomic DNA show considerable promise. The major methodologic consideration is balancing the sensitivity to detect mutation with ability to screen large numbers of samples. Given that these techniques are relatively new, it is anticipated that advances in these areas will be forthcoming.

Once mutations are identified and characterized, their impact on the relevant phenotype must be assessed. This can be approached in three principal fashions: linkage in pedigrees segregating the mutation; association studies comparing the prevalence of the mutation in affected and unaffected populations; *in vitro* expression of the mutant gene with assessment of whether the function of the gene product has been altered.

III. GENETIC STUDIES OF GENES OF THE RENIN-ANGIOTENSIN SYSTEM IN HUMAN HYPERTENSION

The key role of the gene products of the renin-angiotensin system (RAS) in the regulation of blood pressure in humans strongly motivates the study of the genes of this system in human hypertension. These genes include those encoding renin, angiotensinogen, angiotensin-converting enzyme, and the angiotensin II receptor. In addition, genes involved in the responses to angiotensin II can be broadly considered to be part of the RAS system.

A. RENIN

The human renin gene (REN) is an attractive candidate in the etiology of essential hypertension. In addition to the role of the RAS in salt-water homeostasis and vascular tone, several arguments reinforce the potential involvement of this gene in blood pressure (BP) regulation: (1) renin is the limiting step in the RAS enzymatic cascade; (2) an increase in renin production can generate a major increase in BP as illustrated by renin-secreting tumors and renal artery stenosis; (3) genetic studies have shown that renin is associated with the development of hypertension in several rat strains;[11,12] and (4) transgenic animals bearing either a foreign renin gene alone[13] or in combination with the angiotensinogen gene[14] develop precocious and severe hypertension.

1. Polymorphisms

REN has been diversely assigned to chromosome 1 q25, q32, and q42 regions by *in situ* hybridization.[15-17] Several diallelic restriction fragment length polymorphisms (RFLPs) have been characterized at this locus, with heterozygosity ranging from 20 to 40%.[18-20] These RFLPs have been used to test for linkage disequilibrium between hypertensives and control groups.

More recently, we have identified an informative dinucleotide repeat flanking the 3′ end of the human renin gene, composed of an uninterrupted (GT)17 repeat.[53] DNA from 144 unrelated individuals of the CEPH reference families were genotyped, revealing nine alleles and 77% heterozygosity. The Mendelian codominant inheritance of alleles of this marker was established in 15 3-generation CEPH reference pedigrees, which demonstrated complete

TABLE 1
Human Case-Controls Studies Involving the Renin Gene

Association study	Renin RFLP	Allele Frequencies Hypertensives	Controls	Significance
Sydney[a]	HindIII	.55/.45	.60/.40	ns
Paris[b]	HindIII	.69/.31	.68/.32	ns
	TaqI	.89/.11	.91/.09	ns
	HinfI	.81/.19	.78/.22	ns

[a] Data from Reference 21.
[b] Data from Reference 22.

linkage of this marker to a HindIII site polymorphism at the REN locus[18] (lod score of 8.42 for complete linkage).

2. Association Studies

Morris and Griffiths[21] compared the frequencies of the HindIII RFLP of 29 adult subjects with confirmed hypertension to 202 controls. The two groups had similar allele frequencies (Table 1) and no relation was found between the genotypes and plasma renin measurements, although most of them were obtained on medication.

In a more systematic approach, Soubrier et al.[22] compared the frequencies of three diallelic RFLPs in 102 hypertensive subjects with familial history of hypertension and 120 normotensive controls with similar age and sex distributions. Indistinguishable frequencies of the TaqI, HindIII and HinfI polymorphisms were found in both groups (Table 2). The combination of these

TABLE 2
Linkage Analysis of Essential Hypertension and hGH-A1819: Analysis of Utah Hypertensive Pairs

Hypertensive groups	Pairs n	Alleles shared observed/expected	Excess of alleles n	95% CI (%)	t	*p*
Entire group	237	255/254.8	+0.2	[−6.9, +6.9]	0.01	ns
Age dx ≤40 years[a]	52	49/55.9	−6.9	[−26.9, +2.1]	<0	—
Rx ≥2 drugs[b]	31	34/33.3	+0.7	[−16.6, +20.6]	0.18	ns

Note: The excess of alleles shared by the corresponding hypertensive sib pairs is shown as absolute value (n) with the 95% confidence interval limits of the percentage of excess.

[a] Diagnosis of hypertension prior to age 40.
[b] Sib pairs in whom two or more different antihypertensive medications were required for BP control. For these two subgroups, only pairs in whom both sibs demonstrated levels in the upper half of the distribution were included.

three RFLPs allowed the generation of eight haplotypes; the frequencies of these haplotypes were also not significantly different between the two groups.

The limitations of such association studies have been addressed above, indicating the inability of such studies to exclude even common mutations at the renin locus in hypertensives.

3. Linkage Studies

A first attempt to test linkage between the renin gene and essential hypertension was performed by Naftilan et al.[23] They used four RFLPs at the REN locus and genotyped 68 persons from a large Utah pedigree with a high incidence of hypertension. Among the 9 related pedigree members with treated hypertension, the renin genotypes showed no evidence of cosegregation. Also, no relation was found between plasma renin measurements obtained in 59 untreated members of this pedigree and the renin genotypes. As discussed above, negative results from such a linkage study in a large pedigree do not exclude the renin locus as influencing blood pressure, even in the studied pedigree, since in such extended pedigrees homogeneity of causation of hypertension cannot be presumed.

For such reasons, applications of the affected sib-pair method — which can accommodate heterogeneity and incomplete penetrance — is becoming commonplace in complex inherited diseases like hypertension.

Jeunemaitre et al.[24] have applied this methodology to test linkage between the renin gene and essential hypertension by the analysis of the concordance of renin haplotypes in 98 hypertensive French sib pairs coming from 57 independent sibships. The hypertensive sibs were ascertained for a strong family history of hypertension (at least one parent and one sibling), an early onset of the disease (mean 40.7 years), and established essential hypertension. Three different RFLPs located throughout the renin gene (TaqI, HindIII, HinfI) were used as genetic markers. The combination of these three RFLPs permitted definition of eight haplotypes of which six were observed in this study. Taking into account the incomplete heterozygosity of this renin marker (70%) and the absence of parental information in 40 of the 57 sibships, the alleles shared in common by the affected sibs were considered as identical by state (i.b.s.) and the appropriate statistical test was used.[5] We did not find any statistically significant difference between the observed frequencies of total, half, or null allelic concordances and those expected under the hypothesis of no linkage between the renin gene and hypertension. The mean number of marker alleles shared by the 98 hypertensive sib pairs showed only a nonsignificant 5.8% excess; the 95% confidence limits of this excess were, respectively, −3.0% and 14.6%. Given the level of heterozygosity of our haplotypes (PIC = 0.65), this study of 98 sib pairs had 50% power to detect a 15% excess in renin gene alleles identical by descent.

Powerful testing of linkage using the sib pair method requires highly polymorphic markers — that allow, at best, inference of identity by descent in the allele sharing — as well as a very large number of affected sib pairs.

Using the (GT)17 repeat at the renin locus, we have conducted another study[54] of 258 hypertensive sibling pairs selected from 141 Utah families ascertained through population-based sampling.[25] All sibling pairs but three were Caucasian and all affected individuals were on antihypertensive therapy for control of blood pressure. The preliminary results demonstrate a strict concordance between the observed (328) and expected (330) number of alleles shared by the sibling pairs. From these results obtained in a large number of hypertensive pairs, the 95% confidence limits exclude more than 10% excess allele sharing at the renin locus by the whole group of Utah hypertensives. It will be of interest to analyze different subsets of this hypertensive population separately, according to intermediate phenotypes such as indices of severity of the disease and biological measurements of the renin angiotensin system.

It is important to note that only Caucasian hypertensive populations have been studied thus far. Given the high prevalence of low-renin hypertension in African Americans, it will be of particular interest to test this locus in black populations as well.

B. ANGIOTENSINOGEN

The contribution to blood pressure regulation of the renin substrate, mainly synthesized in the liver, is still unclear. Indeed, this glycoprotein has a key position in the RAS but, except in certain pathologic conditions,[26] this substrate is in excess in the plasma and does not appear as a limiting step in the renin angiotensin enzymatic cascade. Plasma angiotensinogen levels are generally stable in individuals but an increase is observed during pregnancy or with exogenous source of estrogens.[27] No segregation analysis has yet been performed to test whether this plasma level is genetically determined. However, a correlation between plasma level and diastolic blood pressure has been found in a large study involving 574 subjects[28] and angiotensinogen levels have been reported to be higher in hypertensives and offspring of hypertensive parents compared to normotensives.[29] In addition, Gardes et al.[30] have demonstrated in rats on a normal sodium diet a sharp BP decrease induced by the administration of a rat angiotensinogen-specific antiserum (ref). Expression of the angiotensinogen (ANG) gene in tissues directly involved in BP regulation[31] and hypertension observed in transgenic mice bearing a rat ANG gene[32] are further arguments to support testing the contribution of this component of the RAS in the genetic determination of blood pressure.

To date, the only known diallelic polymorphism of the ANG gene is located in the exon 4, with an amino acid change (Leu to Met) which eliminates a site for restriction enzyme PstI.[33] More recently, Kotelevtsev et al.[34] have described a dinucleotide GT repeat located in the 3′ region of the ANG gene. This latter microsatellite is highly polymorphic (80% heterozygosity), displaying 10 alleles.[34]

Interestingly, the human ANG gene lies close to the REN gene. Indeed, ANG has been localized to chromosome 1q42-3 by *in situ* hybridization[35,36] while REN has been assigned and mapped to the chromosome 1q32 region.

This proximity raises the possibility of linkage between these two genes, a possibility that we have tested by linkage analysis with several markers in this region. We genotyped 168 individuals of 15 CEPH pedigrees with the Ren1-GT marker described above and the angiotensinogen GT repeat. The lod score for linkage of these loci was significant, with a maximum lod score for linkage of 4.89 at a recombination fraction of 0.26 from pairwise linkage (95% confidence interval for recombination fraction = 0.17 to 0.36). The location of the ANG locus on the genetic map of chromosome 1q was determined by multipoint analysis.[37] ANG showed strong evidence for linkage to marker D1S74 (lod score 21) and is located between D1S48 (recombination fraction = 0.03) and D1S74 (recombination fraction = 0.10). This analysis confirmed linkage of the REN locus with markers of the chromosome 1q32 region, D1S58 and the cluster CR1/DAF (Figure 1). While there is no synteny of the ANG and REN loci in rodents,[38,39] knowledge of such relation on human chromosome 1 and the recombination fraction between these two loci should prove useful in linkage studies of the RAS in human hypertension.

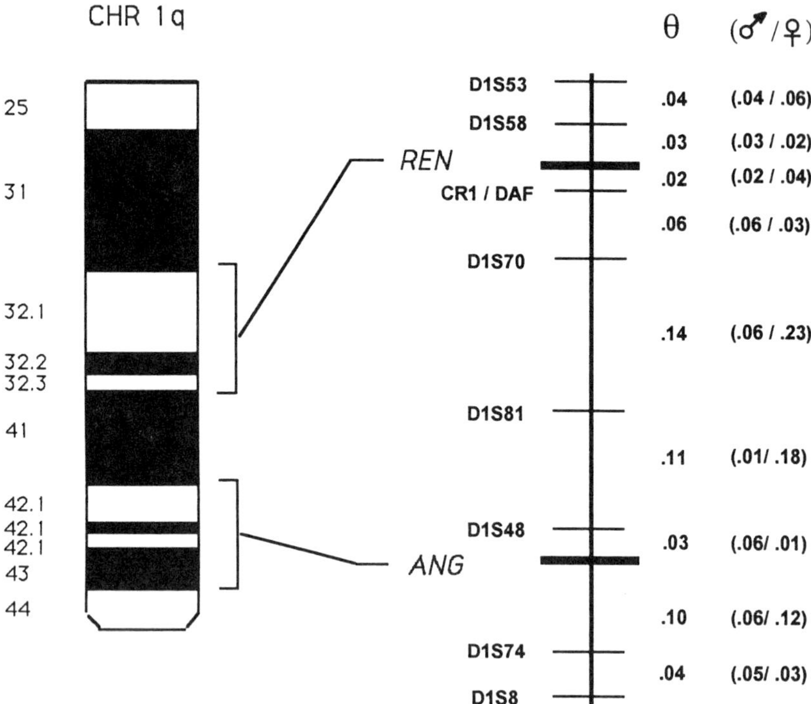

FIGURE 1. Location of the renin (*REN*) and angiotensinogen (*ANG*) genes on the genetic map of human chromosome 1. The sex-averaged genetic map of selected markers of chromosome 1, extracted from the CEPH consortium map of chromosome 1 is shown; the estimated recombination fractions for each sex is shown in parentheses. The odds ratios supporting the location of ANG in the interval shown were 35:1 and 10^{10}:1, respectively.

C. ANGIOTENSIN-CONVERTING ENZYME

The angiotensin-converting enzyme (ACE) is the second key enzyme of the RAS, generating the vasoactive peptide angiotensin II from the decapeptide angiotensin I. It also inactivates the vasodilator bradykinin which further increases its role on vascular tone, as demonstrated by the efficiency of ACE inhibitors in experimental and human hypertension.

1. Polymorphisms

A polymorphism consisting of the presence or absence of a 250-bp DNA fragment inside the ACE gene has been detected by Rigat et al.[40] These authors also demonstrated that this marker could explain about 40% of the variance of the ACE serum levels in 80 healthy Caucasian subjects.

Screening cloned cosmids spanning the ACE locus, we have not as yet identified an informative simple sequence repeat (unpublished results). However, we have demonstrated that a very polymorphic marker at the human growth hormone (hGH) locus can be used as a surrogate for the ACE locus.[41] ACE and hGH both have both been localized by *in situ* hybridization to chromosome 17q23.[42,43] The presence of a large number of Alu repetitive sequences in the hGH locus[44] enabled us to develop an extremely polymorphic marker based on AAAG and AG repeats lying between the 18th and 19th ALU sequences. The HGH-A1819 marker displayed 24 alleles and heterozygosity of 94.6% in 132 unrelated subjects and demonstrated complete linkage to the ACE locus in 109 meioses (lod score = 11.68) with a 95% confidence interval for recombination between these loci of ± 0.02.[41]

2. Association Studies

We have compared the allele frequencies at ACE and hGH loci in 132 controls (Utah grandparents belonging to the CEPH reference families) and 149 hypertensives (index patients of the Utah hypertensive pedigrees). The frequencies of the two ACE alleles were similar in the two groups (frequencies of the larger allele: 0.455 and 0.448, respectively) as were the frequencies of the 24 alleles at the hGH locus, indicating no linkage disequilibrium between the marker loci and hypertension.[41]

3. Linkage Studies

Two groups have recently demonstrated linkage between a chromosomal region spanning the ACE locus and elevated BP in a rat model of hypertension.[45,46] This finding has recently emphasized ACE as a candidate gene in human hypertension.

We have tested this hypothesis using a large number of sib pairs and the highly polymorphic hGH marker which shows no recombination with ACE. Linkage was tested by the affected sib pair method using the hGH marker as a surrogate for the ACE locus. The expected number of alleles shared by the 237 sib pairs — under the null hypothesis of no linkage of the marker locus and genes predisposing to hypertension — was 254.8 (1.075/sib pair); the observed number of alleles shared, 255, coincided with this expectation. The extremely high polymorphism of the hGH marker and the large number of sib pairs studied gave to this study 80% power to detect a 12.02 or 13.06% excess of alleles identical by descent (i.b.d) under a recessive or a dominant model, respectively. Results were also negative on pairs selected to have more severe hypertension or early onset of the disease (Table 2).[41]

In sum, these results demonstrate absence of linkage and of linkage disequilibrium between the ACE/hGH loci and human hypertension in the Utah hypertensive population and exclude the hypothesis that common variants at this locus could exert important effects on blood pressure. However, these results do not rule out the possibility that rare mutations of the ACE gene could have a major effect on the trait but account only for a small percentage of affected individuals in the population.

D. ALDOSTERONE SYNTHASE AND GLUCOCORTICOID-REMEDIABLE ALDOSTERONISM

Aldosterone synthase is normally expressed only in the adrenal glomerulosa, where it catalyzes the final two steps in the biosynthesis of aldosterone from corticosterone. The gene encoding aldosterone synthase is closely related to another gene involved in steroid metabolism, steroid 11β-hydroxylase. These two genes have identical intron-exon structure, are 95% identical in nucleic acid sequence, and are both on human chromosome 8.[47]

One of the rare human Mendelian hypertensive syndromes, glucocorticoid-remediable hypertension (GRA), displays unusual features which could result from mutation in the aldosterone synthase gene. Patients with this syndrome typically are diagnosed early with hypertension mediated via the aldosterone receptor. Aldosterone secretion is variably elevated, and plasma renin levels are suppressed. These patients invariably secrete large quantities of abnormal adrenal steroids 18-hydroxycortisol and 18-oxocortisol. These steroids have structures suggesting they result from aldosterone synthase acting on cortisol as a substrate. A diagnostic hallmark of GRA is the suppressibility of aldosterone and these abnormal steroids by exogenous glucocorticoids. GRA is transmitted as an autosomal dominant trait.[48-51]

Lifton et al.[52] have recently demonstrated that in a large GRA kindred, GRA segregates with a polymorphism identified by hybridization with an 11-OHase probe. The lod score was 5.23, indicating odds of 170,000:1 in favor of linkage, with the likelihood ratio being maximal for complete linkage of the marker with GRA. Further investigation has revealed a striking mutation: a gene duplication arising via unequal crossing over between 11-OHase and aldosterone synthase genes, fusing the 5′-regulatory region of 11-OHase to coding sequences of aldosterone synthase (Figure 2).

Lifton et al.[52] have proposed that this mutation can explain the known physiologic and genetic features of GRA. The 11-OHase gene is normally expressed in both adrenal glomerulosa and adrenal fasciculata. In this latter tissue, expression of this gene is increased by adrenocorticotrophic hormone

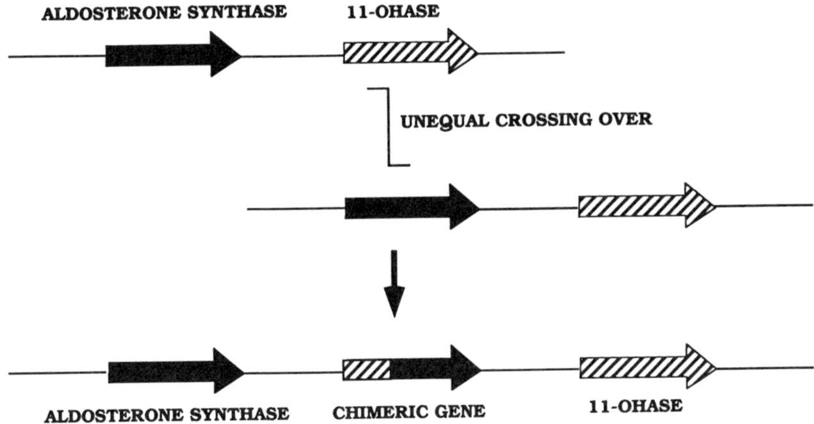

FIGURE 2. Unequal crossing over between aldosterone synthase and 11-OHase genes results in a chimeric gene duplication which causes GRA and hypertension. Two normal chromosomes, each bearing a normal aldosterone synthase and 11-OHase gene are shown undergoing unequal crossing over. One of the products of this recombination bears a chimeric gene duplication in which the 5′ end of 11-OHase is fused to coding sequences of aldosterone synthase.[41]

(ACTH). In GRA patients, the chimeric gene will be expressed in adrenal fasciculata by virtue of these 11-OHase regulatory sequences, and the gene product will have aldosterone synthase activity by virtue of the aldosterone synthase coding sequences. This ectopic expression of aldosterone synthase activity would result in activity of this enzyme on cortisol, leading to the abnormal steroids characteristic of GRA. Aldosterone would also be synthesized in the fasciculata from the large pool of corticosterone in this tissue. These hormones would all be under positive control of ACTH and suppressible by exogenous glucocorticoids. Finally, this neomorphic mutation would be expected to display autosomal dominant inheritance.[52] These are the characteristic features of GRA.

Further investigation will be required to determine whether other GRA pedigrees display similar chimeric gene duplications, and to determine the prevalence of such mutations in the hypertensive population. Nonetheless, this mutation represents the first mutation causing hypertension in otherwise phenotypically normal animals or humans, demonstrates the feasibility of identifying mutations causing hypertension, and, furthermore, highlights the utility of employing intermediate phenotypes for increasing homogeneity of causation within pedigrees.

IV. CONCLUSIONS AND FUTURE DIRECTIONS

Genetic approaches to hypertension remain in their infancy. Only a few loci have been carefully studied in hypertensive subjects; even within the renin-angiotensin system, linkage studies with the angiotensin and angiotensin II receptor genes have not as yet been performed. While the results cited above exclude common variants in renin and angiotensin-converting enzyme in the pathogenesis of hypertension in the populations studied, a number of important caveats to these studies must be noted: first, the interpretation of results is confined to the populations studied. No inference regarding populations of different ethnic and racial backgrounds can be made. Importantly, no black populations have been studied to date; collections of patients from diverse racial and ethnic backgrounds are desperately needed. Second, the 95% confidence interval from these linkage studies permits exclusion of more than about 10% excess allele sharing for the loci studied to date. Consequently, mutations at these loci could still contribute to hypertension in, for example, 5% of the population studied and not be detected by these methods. With 58 million hypertensives in the U.S. alone, such members would still be of profound significance from a public health standpoint.

As noted above, the major barrier to linkage studies in humans is the heterogeneity of the trait. Power can be vastly improved by decreasing the heterogeneity of the hypertensive population being studied. Consequently, development and implementation of new intermediate phenotypes in human hypertension is badly needed; these could permit identification of linkage in significant subsets of the hypertensive population.

In addition to linkage studies, application of a direct search for mutation in candidate genes can potentially identify even very uncommon mutations which could never be detected by large population linkage studies. The impact of variants detected by these methods can be subsequently assessed by linkage studies. This is an extremely promising approach which will doubtless be of importance in the near future.

The ability to map and subsequently identify genes contributing to the pathogenesis of hypertension in inbred animal models will undoubtedly provide hypotheses to be tested in human hypertension. Identification of new loci will motivate human linkage studies; in addition, identification of specific classes of mutations in animal models will motivate searches for comparable mutations in humans. The synergy between animal and human studies should be great in the ensuing years.

The promise of genetic approaches to identify mutations contributing to the pathogenesis of human hypertension is beginning to be fulfilled. Studies of the genes of the renin-angiotensin system as well as other physiologic systems known to be involved in the regulation of blood pressure will be of keen interest over the next several years.

ACKNOWLEDGMENTS

We would like to thank Jean-Marc Lalouel and Roger Williams, enthusiastic and generous collaborators, for support and encouragement. RPL is a Clinician-Scientist of the American Heart Association.

REFERENCES

1. **Ward, R.,** Familial aggregation and genetic epidemiology of blood pressure, in *Hypertension: Pathophysiology, Diagnosis and Management,* Laragh, J.H. and Brenner, B.M., Eds., Raven Press, New York, 1990, 81–100.
2. **Williams, R.R., Hunt, S.C., Hasstedt S. et al.,** Mutigenic human hypertension: evidence for subtypes and hope for haplotypes, *J. Hypertens.,* 8 (Suppl. 7), S39–S46, 1990.
3. **Blackwelder, W.C. and Elston, R.C.,** A comparison of sib-pair linkage tests for disease susceptibility loci, *Genet. Epidemiol.,* 2, 85–97, 1985.
4. **Lange, K.,** The affected sib-pair method using identity by state relations, *Am. J. Hum. Genet.,* 39, 148–150, 1986.
5. **Suarez, B.K. and Eerdewegh, P.V.,** A comparison of three affected sib-pair scoring methods to detect HLA-linked disease susceptibility genes, *Am. J. Hum. Genet.,* 18, 135–146, 1984.
6. **Weber, J.L. and May, P.E.,** Abundant class of human DNA polymorphisms which can be typed using the polymerase chain reaction, *Am. J. Hum. Genet.,* 44, 388–396, 1989.
7. **Litt, M. and Luty, J.A.,** A hypervariable microsatellite revealed by in vitro amplification of a dinucleotide repeat within the cardiac muscle actin gene, *Am. J. Hum. Genet.,* 44, 397–401, 1989.
8. **Orita, M., Suzuki, Y., Sekiya, T., and Hayashi, K.,** Rapid and sensitive detection of point mutations and DNA polymorphisms using the polymerase chain reaction, *Genomics,* 5, 874–880, 1989.
9. **Sheffield, V., Cox, D.R., Lerman, L.S., and Myers, R.M.,** *Proc. Natl. Acad. Sci. U.S.A.,* 86, 232–236, 1989.
10. **White, M.B., Carvalho, M., Derse, D. et al.,** Detecting single base substitutions as heteroduplex polymorphisms, *Genomics,* 12, 301–306, 1992.
11. **Rapp, J.P., Wang, S.M., and Dene, H.,** A genetic polymorphism in the renin gene of Dahl rats cosegregates with blood pressure, *Nature,* 243, 542–544, 1989.
12. **Kurtz, T.W., Simonet, L., Kabra, P.M., Wolfe, S., Chan, L., and Hjelle, B.L.,** Cosegregation of the renin allele of the spontaneously hypertensive rat with an increase in blood pressure, *J. Clin. Invest.,* 85, 1328–1332, 1990.
13. **Mullins, J.J., Peters, J., and Ganten, D.,** Fulminant hypertension in transgenic rats harboring the mouse ren-2 gene, *Nature,* 344, 541–544, 1990.
14. **Ohkubo, H., Kawakami, H., Kakehu, Y., Takumi, T., Arai, H., Yokota, Y., Iwai, M., Tanabe, Y., Masu, M., Hata, J., Iwao, H., Okamoto, H., Yokoyama, M., Nomura, T., Katsuki, M., and Nakanishi, S.,** Generation of transgenic mice with elevated blood pressure by introduction of the rat renin and angiotensinogen genes, *Proc. Natl. Acad. Sci. U.S.A.,* 87, 5153–5157, 1990.
15. **McGill, J.R., Chirgwin, J.M., Moore, C.M., and McCombs, J.L.,** Chromosome localization of the human renin gene (REN) by in situ hybridization, *Cytogenet. Cell. Genet.,* 45, 55–57, 1987.
16. **Cohen-Haguenauer, O., Soubrier, F., Van Cong, N., Serero, S., Turleau, C., Jegou, C., Gross, M.S., Corvol, P., and Frezal, J.,** Regional mapping of the human renin gene to 1q32 by in situ hybridization, *Ann. Genet.,* 32, 16–20, 1989.
17. **Nakai, H., Inoue, S., Miyazaki, H., Murakami, K., and Tada, K.,** Human renin gene assigned to chromosome band 1q42 by in situ hybridization, *Cytogenet. Cell. Genet.,* 47, 90–91, 1988.
18. **Frossard, P.M., Gonzalez, P.A., Fritz, L.C., Ponte, P.A., Fiddes, J.C., and Atlas, S.M.,** Two RFLPs at the human renin gene locus, *Nucleic Acids Res.,* 14, 4380, 1986.
19. **Frossard, P.M., Gonzalez, P.A., Dillan, N.A., Coleman, R.T., and Atlas, S.M.,** Human renin (Ren) gene locus: Bgl II, Rsa I and Taq I RFLPs, *Nucleic Acids Res.,* 14, 6778, 1986.

20. **Masharani, U.,** Hinf I RFLP at the human renin gene locus, *Nucleic Acids Res.*, 17, 467, 1989.
21. **Morris, B.J. and Griffiths, L.R.,** Frequency in hypertensives of alleles for a RFLP associated with the renin gene, *Biochem. Biophys. Res. Commun.*, 150, 219–224, 1988.
22. **Soubrier, F., Jeunemaitre, X., Rigat, B. et al.,** Similar frequencies of renin gene RFLPs in hypertensives and normotensives, *Hypertension,* 16, 712–717, 1990.
23. **Naftilan, A.J., Williams, R., Burt, D. et al.,** A lack of genetic linkage of renin gene restriction length polymorphisms with human hypertension, *Hypertension,* 14, 614–618, 1989.
24. **Jeunemaitre, X., Charru, A., Rigat, B. et al.,** Sib pair linkage analysis of renin gene haplotypes in human essential hypertension, *Hum. Genet.*, 88, 301–306, 1992.
25. **Williams, R.R., Hunt, S.C., Hopkins, P.N. et al.,** Familial dyslipidemia hypertension evidence from 58 Utah families for a syndrome present in approximately 12% of patients with essential hypertension, *JAMA,* 259, 3579–3586, 1988.
26. **Arnal, J.F., Cudek, P., Plouin, P.-F. et al.,** Low angiotensinogen levels are related to the severity and liver dysfunction of congestive heart failure: implications for renin measurements, *Am. J. Med.*, 90, 17–22, 1991.
27. **Newton, M.A., Sealey, J.E., Ledingham, J.G.G., and Laragh, J.H.,** High blood pressure and oral contraceptives. Changes in plasma renin, renin substrate and aldosterone excretion, *Am. J. Obstet. Gynecol.*, 101, 1037–1045, 1968.
28. **Walker, W.G., Whelton, P., Saito, H., Russel, R.P., and Hermann, J.,** Relation between blood pressure and renin, renin substrate, angiotensin II, aldosterone and urinary sodium and potassium in 574 ambulatory subjects, *Hypertension,* 1, 287–291, 1979.
29. **Fasola, A.F., Martz, B.L., and Helmer, O.M.,** Plasma renin activity during supine exercise in offspring of hypertensive parents, *J. Appl. Physiol.*, 25, 410–415, 1968.
30. **Gardes, J., Bouhnik, J., Clauser, E., Corvol, P., and Menard, J.,** Role of angiotensinogen in blood pressure homeostasis, *Hypertension,* 4, 185–189, 1982.
31. **Campbell, D.J.,** Tissue renin-angiotensin system: sites of angiotensin formation, *J. Cardiovasc. Pharmacol.*, 10 (Suppl. 7), S1–S8, 1987.
32. **Kimura, S., Mullins, J.J., and Bunneman, B. et al.,** High blood pressure in transgenic mice carrying the rat angiotensinogen gene, *EMBO J.*, 11, 821–827, 1992.
33. **Kunapuli, P. and Kumar, A.,** Difference in the nucleotide sequence of human angiotensinogen cDNA, *Nucleic Acids Res.*, 14, 7509, 1986.
34. **Kotelevtsev, Y.V., Clauser, E., Corvol, P., and Soubrier, F.,** Dinucleotide repeat polymorphism in the human angiotensinogen gene, *Nucleic Acids Res.*, 19, 6978, 1991.
35. **Gaillard-Sanchez, I., Mattei, M.G., Clauser, E., and Corvol, P.,** Assignment by in situ hybridization of the angiotensinogen gene to chromosome band 1q4, the same region as the human renin gene, *Genomics,* 8, 598–600, 1990.
36. **Isa, M.N., Boyd, E., Morrison, N. et al.,** Assignment of the human angiotensinogen gene to chromosome 1q42-q43 by nonisotopic in situ hybridization, *Genomics,* 8, 598–600, 1990.
37. **Lathrop, G.M., Lalouel, J.-M., Julier, C., and Ott, J.,** Multilocus linkage analysis in humans: detection of linkage and estimation of recombination, *Am. J. Hum. Genet.*, 37, 482–498, 1985.
38. **Mori, M., Ishizaki, K., Yamada, T. et al.,** Restriction fragment length polymorphisms of the angiotensinogen gene in inbred rat strains and mapping of the gene on chromosome 19q, *Cytogenet. Cell Genet.*, 50, 42–45, 1989.
39. **Pravenec, M., Simonet, L., Kren, V. et al.,** The rat renin gene: assignment to chromosome 13 and linkage to the regulation of blood pressure, *Genomics,* 9, 466–472, 1991.
40. **Rigat, B., Hubert, C., Alhenc-Gelas, F. et al.,** An insertion/deletion polymorphism in the angiotensin I-converting enzyme gene accounting for half the variance of serum enzyme levels, *J. Clin. Invest.*, 86, 1343–1346, 1990.

41. **Jeunemaitre, X., Lifton, R.L., Hunt, S.C., Williams, R.R., and Lalouel, J.-M.,** Absence of linkage between the angiotensin converting enzyme locus and human essential hypertension, *Nature Genet.*, 1, 72–75, 1992.
42. **Mattei, M.G., Hubert, C., Alhenc-Gelas, F. et al.,** Angiotensin I converting enzyme is on chromosome 17, *Cytogenet. Cell Genet.*, 51, 1041, 1989.
43. **Harper, M.E., Barrera-Saldana, H.A., and Saunders, G.F.,** Chromosomal localization of the human placental lactogen-growth hormone gene cluster to 17q22-24, *Am. J. Hum. Genet.*, 34, 227–234, 1982.
44. **Chen, E.Y., Liao, Y.-U., Smith, D.H. et al.,** The human growth hormone locus: nucleotide sequence, biology and evolution, *Genomics*, 4, 479–487, 1989.
45. **Hilbert, P., Lindpaintner, K., Beckmann, J.S. et al.,** Chromosomal mapping of two genetic loci associated with blood pressure regulation in hereditary hypertensive rats, *Nature*, 353, 521–529, 1991.
46. **Jacob, H.J., Lindpaintner, K., Lincoln, S.E. et al.,** Genetic mapping of a gene causing hypertension in the stroke-prone spontaneously hypertensive rat, *Cell*, 67, 213–224, 1991.
47. **Mornet, E., Dupont, J., Vitek, A., and White, P.C.,** Characterization of two genes encoding human steroid 11β-hydroxylase (P-$450_{11\beta}$), *J. Biol. Chem.*, 264, 20961–20967, 1989.
48. **Sutherland, D.J.A., Ruse, J.L., and Laidlaw, J.C.,** Hypertension, increased aldosterone secretion and low plasma renin activity relieved by dexamethasone, *Can. Med. Assoc. J.*, 95, 1109–1119, 1966.
49. **New, M.I. and Peterson, R.E.,** A new form of congenital adrenal hyperplasia, *J. Clin. Enocrinol. Metab.*, 27, 300–305, 1967.
50. **Ulick, S., Chu, M.D., and Land, M.,** Biosynthesis of 18-oxocortisol by aldosterone-producing adrenal tissue, *J. Biol. Chem.*, 258, 5498–5502, 1983.
51. **Ulick, S., Chan, C.K., Gill, J.R. et al.,** Defective fasciculata zone function as the mechanism of glucocorticoid-remediable aldosteronism, *J. Clin. Endocrinol. Metab.*, 71, 1151–1157, 1990.
52. **Lifton, R.P., Dluhy, R.G., Powers, M. et al.,** A chimaeric 11β-hydroxylase/aldosterone synthase gene causes glucocorticoid-remediable aldosteronism and human hypertension, *Nature*, 355, 262–265, 1992.
53. **Jeunemaitre, X., Lalouel, J.-M., and Lifton, R.P.,** submitted.
54. **Jeunemaitre, X., Lifton, R.P., and Lalouel, J.-M.,** in preparation.

Index

INDEX

A

B

D

H

I

M

S

T

U

V

W–Y

Z